PLANT ANATOMY

Fourth Edition

Titles of Related Interest

COOMBS *et al.*
Techniques in Bioproductivity and Photosynthesis, 2nd Edition

GOODWIN & MERCER
Introduction to Plant Productivity, 3rd Edition

HADER & TEVINI
General Photobiology

MAYER & POLJAKOFF-MAYBER
The Germination of Seeds, 4th Edition

WAREING & PHILIPS
Growth & Differentiation in Plants, 3rd Edition

Journals of Related Interest

Biochemical Systematics and Ecology

Current Advances in Ecological Sciences

Current Advances in Plant Science

Environmental and Experimental Botany

Photochemistry and Photobiology

Phytochemistry

Soil Biology and Biochemistry

(sample copy gladly sent on request)

PLANT ANATOMY

Fourth Edition

A. FAHN

Professor of Botany
The Hebrew University of Jerusalem,
Jerusalem, Israel

PERGAMON PRESS

Member of Maxwell Macmillan Pergamon Publishing Corporation

OXFORD · NEW YORK · BEIJING · FRANKFURT
SAO PAULO · SYDNEY · TOKYO · TORONTO

U.K.	Pergamon Press plc, Headington Hill Hall, Oxford OX3 0BW, England
U.S.A.	Pergamon Press, Inc., Maxwell House, Fairview Park, Elmsford, New York 10523, U.S.A.
PEOPLE'S REPUBLIC OF CHINA	Pergamon Press, Room 4037, Qianmen Hotel, Beijing, People's Republic of China
FEDERAL REPUBLIC OF GERMANY	Pergamon Press GmbH, Hammerweg 6, D-6242 Kronberg, Federal Republic of Germany
BRAZIL	Pergamon Editora Ltda, Rua Eça de Queiros, 346, CEP 04011, Paraiso, São Paulo, Brazil
AUSTRALIA	Pergamon Press Australia Pty Ltd., P.O. Box 544, Potts Point, N.S.W. 2011, Australia
JAPAN	Pergamon Press, 5th Floor, Matsuoka Central Building, 1-7-1 Nishishinjuku, Shinjuku-ku, Tokyo 160, Japan
CANADA	Pergamon Press Canada Ltd., Suite No. 271, 253 College Street, Toronto, Ontario, Canada M5T 1R5

Copyright © 1990 A. Fahn

First Edition 1967

Second Revised Edition 1974

Reprinted 1975, 1977

Third Revised Edition 1982

Reprinted with corrections 1987

Fourth Revised Edition 1990

Original Hebrew Edition published by
Hakkibutz Hameuhad Publishing House Ltd.

Library of Congress Cataloging-in-Publication Data
Fahn, A., 1916–
Plant anatomy/A. Fahn.—4th ed.
p.   cm.
First ed. translated from the Hebrew by Sybil Broido-Altman,
published under title: Anatomyah shel ha-tsemah.
Includes bibliographical references.
1. Botany—Anatomy.   I. Title.
QK641.F313   1990   582'.04—dc20   89-22889

British Library Cataloguing in Publication Data
Fahn, A. (Abraham)
Plant anatomy—4th ed.
1. Plants. Anatomy
I. Title
581.4

ISBN 0-08-037490-5 (Hardcover)
ISBN 0-08-037491-3 (Flexicover)

PREFACE TO THE FOURTH EDITION

THE knowledge of the structural characteristics of inanimate materials or living organisms is of primary importance in scientific research. The level at which the structure is studied varies with the object under consideration. The study may start with observations made with the naked eye, and extend to examination with the electron microscope and, for further details, the use of indirect methods.

Plant anatomy, dealing with the structure and development of tissues and cells, and their contents, is of primary importance for all lines of research in plant sciences: morphogenesis, physiology, ecology, taxonomy, evolution, genetics, reproduction, etc.

The continuous development of new and more precise optical and electronic instruments and other direct or indirect methods has made it possible to reinvestigate previous information on the various tissues and cells, and to carry out research on structures not studied before.

In order to keep the readers of this book informed of the new developments in plant anatomy three updated editions (including this edition) have been published since the appearance of the first English edition in 1967. (The book appears also in several other languages.)

Only part of the results of the new research has been included in the present edition. Many publications, although not less important, are not cited because of the size limitation of the book.

Plant anatomy is a science in its own merit. However, as mentioned above, plant anatomy is also of basic importance for all lines of research in plant science. As in previous editions, the functional interpretation of anatomical features is emphasized.

In order to retain the main purpose of this book and keep its size within reasonable limits the relation between structure and function is discussed only to such an extent as to stress the relationship between the two and to raise the curiosity of the reader to the problems. For further, deeper knowledge of these aspects it is of course advisable to turn to the specific literature.

In the present edition several topics, e.g. cell structure, ecological anatomy, cambial activity and some other subjects have been extended. In all chapters new literature is cited, a number of new illustrations have been added and changes in some previous ones have been made. The Glossary of Terms has been extended and about 550 new references have been added.

PREFACE TO THE FIRST EDITION

PLANT ANATOMY is a basic science and as such is of great importance to students of all the plant science. Without a thorough knowledge of this field the physiological processes carried out within the plant and the phylogenetic relationships between the various plant groups cannot be fully understood. The detailed study of the elements and tissues of which the plant is constructed enables a better understanding of adaptation to special functions as well as of the adaptation of entire plants to different environmental conditions. Without a thorough knowledge of the anatomical and histological structure of plants the results of physiological and ecological experiments, for instance, may be incorrectly interpreted. Also, today no conclusive opinions on evolutionary trends or taxonomic relationships can be suggested on the traditional basis of the study of external morphological characteristics alone; it is now necessary to support such work by the use of the many and varied anatomical and histological characters, which can be observed only from microscopic, and even submicroscopic, investigation.

Anatomy, which draws the attention of the student to the form, variability and structure of the tissues comprising the plant body, can be said to develop an aesthetic sense. In addition to this, the awareness of the regularity, and repetition, at different levels, of the structural patterns, as well as of the amazing correlation of structure and function, serves to make anatomy a rewarding field of research.

A large section of this book deals with the vegetative plant body. The first introductory chapter briefly presents the general structure of the higher plant. This is followed by the descriptions of the different types of cells and tissues that are present in the Tracheophyta. Later chapters describe how the vegetative plant body, both primary and secondary, is constructed of these various tissues. The last section deals with the structure of the flower, fruit and seed. In the chapter on the flower I have covered pollination, fertilization and embryo development. In my opinion this is necessary in order that a full and balanced picture of the development and structure of plants and their tissues can be obtained.

An effort has been made, when dealing with the structure of the elements, tissues and organs of the plant body, to employ the following approaches—ontogenetic, phylogenetic, physiological and ecological. Attention has also been paid, wherever possible, to such characteristics that are of importance to agriculture and industry.

In view of modern research in plant anatomy and biology as a whole, which has brought to light so many transitional forms, the need for flexibility in the definitions of the various elements and tissues is stressed throughout the book.

In many cases I have endeavoured to point out problems which as yet constitute serious gaps in our knowledge, and which await further research.

The inclusion of a large number of illustrations has enabled the text to be written in a concise form. The great majority of the micrographs are original and have been made from slides in the collection of the laboratory of Plant Anatomy at the Hebrew University of Jerusalem. The drawings are in part original or have been taken from previous publications of the author, and the rest have been redrawn and adapted from various books and articles. In the case of the latter the original author is cited in the legend and the reference is given at the end of the relevant chapter or Chapter 1. In a book of this size and scope it is impossible to deal with all the relevant facts in detail and therefore many references are given in the text. It is hoped that readers will refer to these and other articles, books, etc., in order to broaden their knowledge. For convenience, details of these references are given at the end of each chapter.

This book was originally written in Hebrew for the use of students studying in Israel. Thus many of the examples cited are of plants growing in this and neighbouring regions.

I express my thanks to all those who helped me in the preparation of the original Hebrew book. I am indebted to my friend S. Stoler for his critical reading of the manuscript and for his valuable suggestions; to Mrs. Ella Werker for her great help in the preparation of the manuscript and for seeing the book through press; to Mrs. Batya Amir for the careful and accurate execution of most of the drawings; to Y. Shchori for his aid in the preparation of most of the photographs appearing in this book; and to Mrs. Irena Fertig for her assistance in reading the proofs. My thanks are also extended to all those who have put at my disposal photographs and drawings, as well as to my colleagues at the Hebrew University who gave me valuable advice at times. I especially thank my students, throughout the years, who have encouraged me to write this book.

In connection with this revised English edition I am indebted to Dr. C. R. Metcalfe and Sir George Taylor who suggested and encouraged me to have my book translated. I greatly appreciate the criticism and advice that Dr. Metcalfe extended after having read the English manuscript. I thank Dr. F. A. L. Clowes who undertook to edit the English. I gratefully acknowledge the permission so generously granted to me by the Hakkibutz Hameuhad Publishing House Limited to translate the original text. I also thank Mrs. Sybil Broido-Altman for undertaking the translation. Once again I thank Mrs. Ella Werker, who assisted in the collation.

April, 1965

PREFACE TO THE SECOND EDITION

THIS revised edition is larger than the first, although I have tried to keep the book as concise as possible. Reconsidering the aim of this book and taking into account suggestions made by some fellow teachers of plant anatomy, I have enlarged some of the previous topics and have added short accounts of several new ones. All topics were revised and recent literature, including developmental and functional aspects of plant anatomy, as well as results of electron microscopic investigations, were taken into consideration.

I have enlarged Chapter 1, in order to enable students who have not previously taken a course in general botany to get acquainted with the general structure and organization of the tissues in the plant organs before reading the further chapters discussing the various topics in detail. Chapter 9 has been revised and now includes a paragraph on secretory ducts, in addition to laticifers. Other secretory structures are discussed, in other chapters, in more detail than previously. Additional topics in relatively brief form were added in some chapters, e.g. haustoria of parasites and mycorrhiza in the chapter on the root, adaptation to aquatic habitats in the chapter on the stem, and dendrochronology in the chapter on secondary xylem.

A fair number of figures have been added and many previous half-tone illustrations have been improved.

A.F.

PREFACE TO THE THIRD EDITION

IN RECENT years a relatively large amount of research was carried out in the field of plant anatomy. The results of this research lead to a better understanding of developmental processes and of the relation betwen structure and function. New methods of preparation of sections for light microscopy and the increasing use of transmission and scanning electron microscopy for plant-anatomical studies, enabled the discovery of new structural characteristics and the re-evaluation of previous observations and views.

As a result of the new discoveries, plant physiologists, plant ecologists, and phytopathologists became more aware of the importance of plant anatomy to their own research. It was, thus, felt that some of the information recently acquired should be brought to the attention of students of botany, and a new revised edition of this book was undertaken. The literature reviewed for this edition is up to date until the beginning of 1981. As the book is intended to be used mainly as a textbook, only a limited number of references are cited in it. A number of figures which appeared in the second edition were replaced by new ones. With the addition of new material to the text, the total number of figures has been increased. Among the new figures many are scanning electron micrographs.

ACKNOWLEDGEMENTS

I AM grateful to all who have kindly provided me with original illustrations and those who have given me permission to redraw figures from their publications. They are acknowledged in the captions of the specific figures.

I am also indebted to: Prof. A. M. Mayer, for critical reading of the manuscript of the revised paragraphs; my colleagues of the Plant Anatomy Group of the Department of Botany of the Hebrew University of Jerusalem, with whom I discussed from time to time the problems dealt with in the book; Mr P. Grossmann, for line drawings; Mr J. Gamburg, for printings of micrographs; my wife Nehama, for helping in the preparation of the manuscript of the revised paragraphs and reading the proofs; Dr. David F. Cutler for proof reading; Dr. Judy Cohen for preparing the Subject Index and Mrs Anat Neville for preparing the Author Index.

CONTENTS

CHAPTER 1

GENERAL STRUCTURE OF HIGHER PLANTS

THE vascular plants, or tracheophytes, which possess a specialized conducting system include four phyla of the plant kingdom: 1, *Psilopsida* (chiefly fossils); 2, *Lycopsida* (clubmosses); 3, *Sphenopsida* (horsetails); and 4, *Pteropsida* (ferns, gymnosperms or cone-bearing seed plants, and angiosperms or flower-bearing seed plants). The angiosperms, which represent the most recently evolved group of plants, form the main part of the natural and cultivated vegetation on the earth.

The general structures of a flower-bearing seed plant, starting with the seed, are outlined below.

The *seed* contains an *embryo plant*, enveloped and protected by a *seed coat*, and is supplied with a source of stored food.

The plant embryo contains a minute axis with two poles—the root growing point and the shoot growing point. On the minute axis occur laterally the *cotyledons* or seed leaves. The food required for the germinating plantlet may be stored in the cotyledons (Fig. 1, no. 1) or in a special tissue, the *endosperm* (Fig. 1, nos. 6, 7). Under suitable growing conditions the seed germinates and a young plant or *seedling* emerges. The seedling grows, extends its roots into the soil and its shoot (stems and leaves) into the atmosphere (Fig. 1, nos. 2–4). The growth of the shoots and roots is due to the formation of new cells by *meristematic* (embryonic) tissues of the *growing points*, followed by growth and differentiation of these cells. When the plant attains adult size, flowers are formed. After *pollination* (transfer of pollen grains from the stamens to the style) and *fertilization, fruit*, containing seeds, develops, thus completing the life cycle. Some plants die shortly after seed set (annual plants), others (perennials) continue to grow for many years and become shrubs or trees.

The plant organs, as the organs of animals, are composed of tissues (groups of cells which carry on specific activities). The cells of plant tissues are small compartments possessing living material, the *protoplasm*, enclosed by a cell wall. All metabolic processes take place in the cells.

THE PLANT ORGANS

The *roots* anchor the plant in the soil, take up water and mineral salts from it, and in many cases store food. The *shoot* consists of stem and leaves. The leaves produce food by photosynthesis and give off water vapour by transpiration. The stem supports the leaves and has the role of conducting water and mineral salts from the roots to the leaves and synthesized organic substances from the leaves to regions of growth or storage.

At the tips of the shoots and roots the *apical meristems* are situated. The cells that make up these meristems divide, grow, differentiate, and thus cause the extension growth of the plant.

The shoot apices, together with the young developing leaves, form the buds. The latter usually also include modified scale-leaves which protect the apical meristems.

The root

In grasses and many other monocotyledons the main roots form a cluster and are approximately of the same size. The main roots give off side roots which may branch again. Such a root system is termed a *fibrous root system* (Fig. 2, no. 1).

Carrot, radish, and most other dicotyledons possess a main root which grows downwards, perpendicular to the soil surface. From this root branches arise. A root system of this type is called a *tap-root system* (Fig. 2, no. 3).

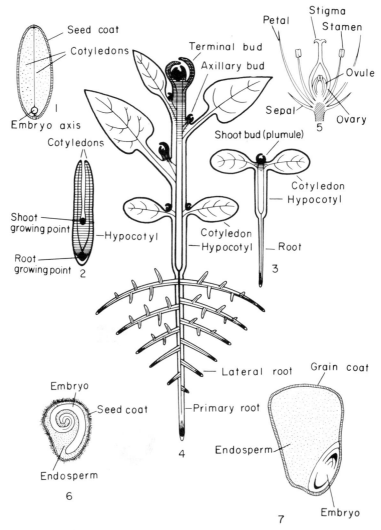

FIG. 1. Schematic drawings of longitudinal sections of various plant organs. 1, Bean seed. 2, 3, and 4, A dicotyledonous plant at three stages of development (embryo, seedling, mature plant). 5, Flower. 6, Tomato seed. 7, Corn grain.

Closer observations of young roots or young portions of roots reveal a region of hairs, *root hairs*, close to the root tip (Fig. 2, no. 2). These can best be seen in seedlings that have been grown on filter paper or in sand in a moist atmosphere. The root hairs are cell elongations of the root epidermis and are present in great numbers. They are relatively short-lived, and new ones are constantly formed close to the root tip as the latter continues to grow and push its way through the soil. The root hairs greatly increase the surface area of the root portions which are mainly involved in water and mineral salt uptake.

Internal structure of the root (Fig. 2, nos. 4–6)

The *epidermis* is a protective tissue and consists of a single layer of densely packed cells. Below the epidermis occurs a relatively thick region, the *cortex*. The cortex is composed chiefly of structurally unspecialized cells, *parenchyma* cells, with large intercellular spaces. The innermost layer of the cortex is a single row of cells, the *endodermis*. In the primary state the walls of all endodermal cells are thin except for a band-like thickening on the radial and transverse sides of the cell. These thick-

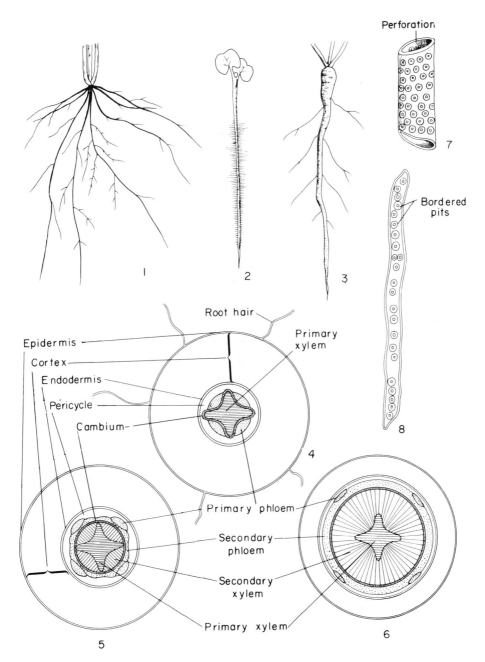

Fig. 2. 1, A fibrous root system. 2, A young seedling whose root is clothed in root hairs. 3, A tap-root system. 4–6, Diagrams of cross-sections of roots of dicotyledons at various distances from the root tip, showing successive stages in secondary thickening. 7, A vessel member. 8, A tracheid.

enings are known as the *Casparian strips*. Some investigators suggest that the water-impervious Casparian strips prevent diffusion of water along the wall and force the movement of solutions through the protoplasm.

The central region of the root is termed the *vascular cylinder*. It consists of the water-conducting tissue, the *xylem*, and the food-conducting tissue, the *phloem*. Between the vascular tissues (xylem and phloem) and the endodermis occurs a layer of unspecialized parenchyma cells, the *pericycle*, which originate from the same group of meristematic cells as the xylem and phloem (Fig. 2, nos. 4, 5). The pericycle which retains meristematic properties gives rise to lateral roots. The xylem consists of conducting cells—*tracheids* and *vessel members*—as well as of *fibres* and *parenchyma*. A mature tracheid is a single elongated cell (Fig. 2, no. 8). In the thickened walls there are thin places (*pits*) through which water can easily pass. Vessel members are also single cells with walls similar to those of the tracheids but completely perforated at their ends (Fig. 2, no. 7). They are usually shorter than tracheids and are arranged in vertical rows. A row of vessel members is termed a *vessel* or *trachea*. The fibres are elongated, tapering thick-walled cells functioning mainly in strengthening of tissues. The parenchyma represents a kind of filling tissue and functions in the storage of food.

The phloem consists of *sieve-tube members*, *companion cells*, *fibres*, and *parenchyma*. The sieve-tube members are living cells, arranged in vertical rows, which are known as *sieve tubes*, and function in the translocation primarily of organic substances. The companion cells are sister cells of the sieve-tube members and remain in close contact with them.

In gymnosperms and dicotyledons a meristematic tissue termed the *vascular cambium* surrounds the xylem (Fig. 2, nos. 4–6). While the apical meristem is responsible for the extension growth of the plant organs, the vascular cambium is responsible for their radial growth which brings about the thickening of the organs. The cells of the vascular cambium divide and produce xylem elements centripetally and phloem elements centrifugally. The vascular tissues originating from the cambium are regarded as secondary. They are termed the *secondary xylem* and the *secondary phloem*.

The stem

External structure

Many types of stems can be distinguished according to their external and internal characters, e.g. stems of grasses (wheat, corn), herbaceous dicotyledons (peas, sunflower, tomato), and woody stems of shrubs and trees. There are also unusual types of stem which underwent metamorphosis to special forms correlated to environment and function. To such types belong *bulbs* (onions) and *tubers* (potatoes) that store food, *succulent* stems of plants like cacti which store water and carry on photosynthesis, a process occurring in other plants almost exclusively in the leaves. Stems may also become modified to tendrils and thorns.

When studying the external structure of a shoot or twig (Fig. 3, no. 1) it can be seen that the leaves are distributed on the stem in a specific pattern. That part of the stem to which the leaf is attached is termed the *node* and that part of the stem between two nodes, the *internode*. Buds, which are actually minute shoots, also occur on the stem. The bud terminating the shoot is called the *terminal bud*. In the axils of the leaves occur *lateral buds* or *axillary buds* which are usually smaller than the terminal one. The buds may remain dormant for determined periods (e.g. during the winter in deciduous trees) or for many years (most of the axillary buds).

As long as the cells of the apical meristem of the terminal bud retain the ability to divide, most of the axillary buds remain dormant and the shoots which develop from axillary buds are limited in growth. The extent of growth of the side branches and the dormancy of the axillary buds is controlled by the terminal bud. In plants in which the terminal bud of the main axis remains active throughout the life of the plant, the branching of the stem is *monopodial* (Fig. 3, no. 2). In many plants the apical meristem of the terminal bud, after continuing for some time to contribute to the elongation of the stem and to producing leaves, may abort or become reproductive and produce a flower. In both these cases further growth is carried out by lateral buds, and the side branches take over the function of the main primary axis. Such branching is termed *sympodial* (Fig. 3, no. 3).

Internal structure of young stems

Below the apical meristem differentiation and

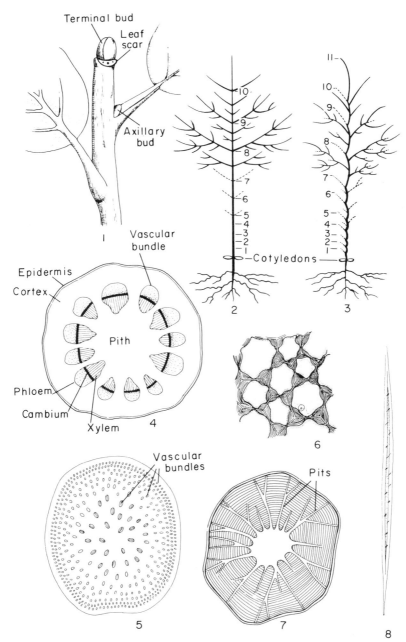

Fig. 3. 1, Portion of a stem. 2 and 3, Diagrams of monopodial and sympodial branching respectively (the numbers indicate the position of the tip at the end of each annual longitudinal increment). 4, Diagram of a cross-section of a young dicotyledonous stem. 5, Diagram of a cross-section of a monocotyledonous stem. 6, Cross-section of collenchyma. 7, An isodiametral sclereid. 8, A fibre. (Nos. 2 and 3 adapted from Troll, 1948.)

maturation of tissues occur. When examining young stem portions with the aid of a microscope the following arrangement of tissues from the outside towards the centre of the stem can be distinguished. The *epidermis* is coated by a very re- sistant substance of lipoid nature called *cutin*. This superficial layer of cutin is coined *cuticle*. It is, especially when thick, impermeable to water and gases. Gas exchange takes place mainly through *stomata*, which are described later in this chapter. Epidermal appendages such as hairs are often present. The zone between the epidermis and the vascular tissue is termed the *cortex*. In the cortex three types of tissues may occur: *parenchyma*, *collenchyma*, and *sclerenchyma*. The parenchyma may store food and in green stems also produce food by photosynthesis. The collenchyma and sclerenchyma serve as strengthening tissues. The collenchyma cells have unlignified walls which are thickened usually in the corners of the cell (Fig. 3, no. 6). The sclerenchyma cells have very thick, mostly lignified walls. Two types of sclerenchyma cells can be distinguished: *fibres*, which are long cells, tapering at their ends (Fig. 3, no. 8), and *sclereids*, which are more or less isodiametric (Fig. 3, no. 7) or branched.

Inside the cortex is the vascular cylinder, which contains the vascular tissues. The vascular tissues are generally organized—in the young stem portions—in strands called *vascular bundles*. Each bundle contains xylem on the inside facing the centre of the stem and phloem on the outside. In some cases, e.g. in pumpkin, phloem appears on both sides of the xylem. In the gymnosperms and dicotyledons the vascular bundles are arranged in a circle around a *pith*, which is usually parenchymatous (Fig. 3, no. 4), whereas in monocotyledons they are generally scattered throughout the stem cross-section (Fig. 3, no. 5). Another difference between the monocotyledons and the dicotyledons and gymnosperms is the appearance of a vascular cambium between the xylem and phloem only in the vascular bundles of the latter two.

Secondary thickening

The vascular cambium, which appears distant from the apices, produces additional xylem centripetally in continuation to the xylem of the vascular bundles and additional phloem centrifugally, thus pushing to the outside the phloem of the bundles. The tissues of the vascular bundles are produced directly by the apical meristems and are called *primary xylem* and *primary phloem*. Those produced by the vascular cambium are termed *secondary xylem* and *secondary phloem*. Cambium also develops between the primary vascular bundles, thus forming a whole ring of cambium,

as seen in cross-section (Fig. 4). From this stage the cambium starts to produce a solid mass of secondary xylem or *wood* which in some trees may reach several metres in diameter (Fig. 5). The secondary phloem is usually produced in smaller

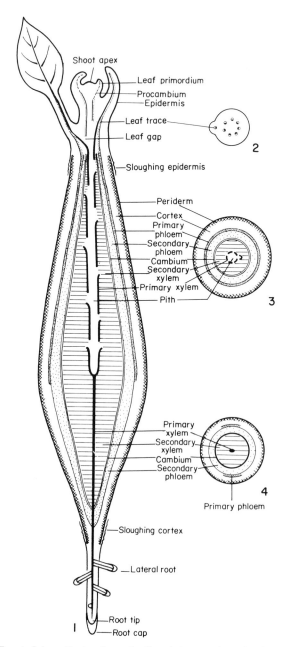

FIG. 4. Schematic drawings of a dicotyledonous plant showing the arrangement of the principal tissues. 1, Longitudinal section. 2, 3, and 4, Cross-sections at various levels. (Adapted from Esau, 1953.)

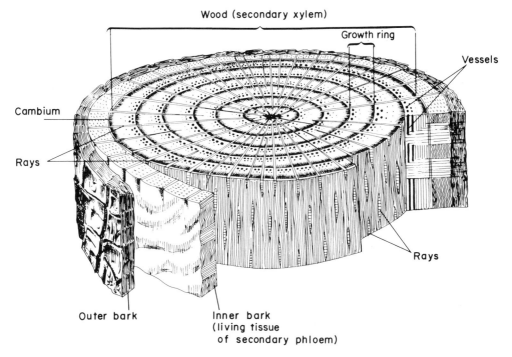

Wood (secondary xylem)

Growth ring

Vessels

Cambium

Rays

Rays

Outer bark

Inner bark
(living tissue
of secondary phloem)

FIG. 5. Diagram of a portion of a woody stem.

quantities and, together with remnants of the epidermis, cortex and additional tissues, which will be discussed later, it forms the *bark*.

The secondary xylem or wood consists basically of the same elements as the primary xylem but is usually richer in fibres and is penetrated radially by parenchymatous rays called vascular rays. In trees with seasonal growth cycles the wood exhibits rings when cross-sectioned. This is due to the denser structure of the xylem produced towards the end of the growth season.

While the cambium causes increase in width of the stem and root by adding more and more wood increments, extension growth of the young axes continues by the activity of the apical meristems.

The cork

With the increase in thickness of the wood, pressure is imposed on the epidermis and cortex which eventually rupture. A secondary protective tissue—the *cork layer* or *phellem*—then takes over the primary function of the epidermis. The cork is composed of dead flattened cells with no intercellular spaces. Its cells are lined by a lamella of a fatty substance, called *suberin*. This tissue is produced by the *cork cambium* or *phellogen* which, in the stems of most plants, arises in the outer cortical cells, and in the roots generally in the pericycle. In most long living plants, with the continuous increase in diameter of the axis and bursting of the first formed cork, successive cork cambia develop progressively further inwards, i.e. in the phloem. Since cork is an impermeable barrier, when formed in the secondary phloem it causes the death of that part of the phloem which is situated on its outside. All the tissues, from the vascular cambium outwards, form the bark of the axis. The cork divides the bark into the living *inner bark* and the dead *outer bark* (Fig. 5).

Oxygen supply to the living tissues which are covered by cork is effected by groups of cells with intercellular spaces called *lenticels*. These interrupt the continuity of the cork.

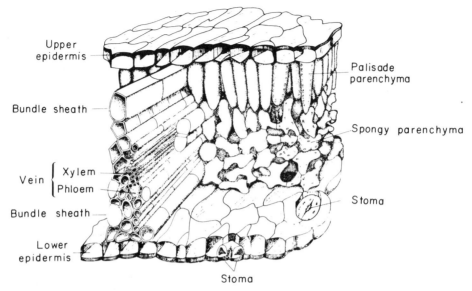

FIG. 6. A three-dimensional drawing of a leaf portion.

The leaf

The main function of the leaves is the synthesis of organic compounds using light as a source of the necessary energy, a process known as *photosynthesis*. This energy-converting process takes place within special cell organelles called *chloroplasts*, in which the pigment *chlorophyll* is present.

The external and internal structure of the leaf is correlated to its role in photosynthesis and *transpiration* (the loss of water as vapour). The leaf is flat and thin, thus enabling the solar rays to penetrate into all its cells. The high surface to volume ratio also enables successful gas exchange. The *veins* seen in the *leaf blade* (the expanded part of the leaf) contain vascular tissues. In the parenchymatous tissue (the *mesophyll*) occurring between the upper and lower epidermis of the leaf two zones can be distinguished: the upper—*palisade parenchyma*—consisting of elongated cells, and the lower—*spongy parenchyma*—consisting of irregularly shaped cells with large intercellular spaces (Fig. 6). The palisade parenchyma is more densely packed with chloroplasts. In the epidermis occur stomata (Fig. 6), which serve for gas exchange between the leaf tissues and the atmosphere. Each *stoma* consists of two *guard cells* surrounding an aperture. The stomata may open or close the aperture, and thus regulate the entrance and exit of gases to and from the leaf.

The flower

Flowers are the reproductive organs of the angiosperms. The flower is formed by a specialized apical meristem which has developed from a vegetative shoot apex after being induced to do so by internal or external factors.

The flower consists of a group of specialized leaves: *sepals*, *petals*, *stamens*, and *carpels* (Fig. 1, no. 5). All the sepals, which are usually green, constitute the *calyx*; all the petals, which are usually coloured and attractive, together constitute the *corolla*. Each stamen consists of a stalk—the *filament*—bearing an *anther* in which *pollen grains* develop. The pollen grains contain *male gametes* or *sperm cells*. The carpels occur singly or in groups, and they comprise the *pistil* in which three parts can be distinguished: the basal part—the *ovary*; the slender middle part—the *style*; and the top part—the *stigma*. The ovary bears in its locule or locules the *ovules* in which the female gametes—the *egg cells*—develop.

The pollen grains are transferred by wind or insects from the ripe anther to the stigma of the pistil. This process is called *pollination*. The pollen grain germinates on the stigma to form a *pollen tube*, which contains two sperm cells. When the pollen tube penetrates the ovule, one of the sperm cells *fertilizes* the egg cell and a *zygote* is formed. At this stage the carpels start to grow and form the *fruit* and the ovules form the seeds. The *embryo* develops from the zygote.

GENERAL READING

BARLOW, P. W. and CARR, D. J. (1984) *Positional Control in Plant Development.* Cambridge University Press, Cambridge.

BOUREAU, E. (1954, 1956, 1957) *Anatomie Végétale*, Vols. I-III. Presses Universitaires de France, Paris.

BURGESS, J. (1985) *An Introduction to plant Cell Development.* Cambridge University Press, Cambridge.

CARLQUIST, S. (1988) *Comparative Wood Anatomy.* Springer-Verlag, Berlin.

CUTLER, D. F. (1969) *Anatomy of the Monocotyledons.* IV. *Juncales.* Clarendon Press, Oxford.

CUTLER, D. F. (1978) *Applied Plant Anatomy.* Longman, London.

CUTLER, E. G. (1971) *Plant Anatomy: Experiment and Interpretation*, Part 2, *Organs.* Edward Arnold, London.

CUTLER, E. G. (1978) *Plant Anatomy*, Part 1, *Cells and Tissues*, 2nd edn. Edward Arnold, London.

DE BARY, A. (1877) *Vergleichende Anatomie der Vegetationsorgane.* W. Engelmann, Leipzig.

EAMES, A. J. (1961) *Morphology of the Angiosperms.* McGraw-Hill, New York.

EAMES, A. J. and MACDANIELS, L. H. (1974) *An Introduction to Plant Anatomy*, 2nd edn. McGraw-Hill, New York and London.

ESAU, K. (1965) *Plant Anatomy,* 2nd edn. Wiley, New York, London and Sydney.

GOEBEL, K. (1928–33) *Organographie der Pflanzen*, Vols. I–III. G. Fischer, Jena.

GUNNING, B. E. S. and STEER, M. W. (1975) *Ultrastructure and the Biology of Plant Cells*, Edward Arnold, London.

HABERLANDT, G. (1918) *Physiologische Pflanzenanatomie*, 5th edn. W. Engelmann, Leipzig.

KLEKOWSKI, E. J. Jr. (1988) *Mutation, Developmental Selection and Plant Evolution.* Columbia University Press, New York.

KOZLOWSKI, T. T. (1971) *Growth and Development of Trees,* Vols. I and II. Academic Press, New York.

MAUSETH, J. D. (1988) *Plant Anatomy.* The Benjamin/Cummings Publishing Company, Inc., Menlo Park, California.

MCLEAN, R. C. and IVIMEY-COOK, W. R. (1951, 1956) *Textbook of Theoretical Botany*, Vols. I and II. Longman, London.

METCALFE, C. R. (1960) *Anatomy of the Monocotyledons.* I. *Gramineae.* Clarendon Press, Oxford.

METCALFE, C. R. (1971) *Anatomy of the Monocotyledons.* V. *Cyperaceae.* Clarendon Press, Oxford.

METCALFE, C. R. (1987) *Anatomy of the Dicotyledons.* 2nd ed. Vol. III. Clarendon Press, Oxford.

METCALFE, C. R. and CHALK, L. (1950) *Anatomy of the Dicotyledons.* Vols. I and II. Clarendon Press, Oxford.

METCALFE, C. R. and CHALK, L. (1979 and 1983) *Anatomy of the Dicotyledons.* 2nd ed. Vols. I and II. Clarendon Press, Oxford.

SINNOT, E. W. (1960) *Plant Morphogenesis.* McGraw-Hill, New York.

STEEVES, T. A. and SUSSEX, I. M. (1972) *Patterns in Plant Development.* Prentice Hall, New Jersey.

TOMLINSON, P. B. (1961) *Anatomy of the Monocotyledons.* II. *Palmae.* Clarendon Press, Oxford.

TOMLINSON, P. B. (1969) *Anatomy of the Monocotyledons.* III. *Commelinales-Zingiberales.* Clarendon Press, Oxford.

TROLL, W. (1935, 1937, 1938, 1939) *Vergleichende Morphologie der höheren Pflanzen.* Gebr. Borntraeger, Berlin.

TROLL, W. (1954–7) *Praktische Einführung in die Pflanzenmorphologie*, Vols. I and II. G. Fischer, Jena.

WARDLAW, C. W. (1968) *Morphogenesis in Plants.* Methuen, London.

ZIMMERMANN, M. H. (1983) *Xylem Structure and the Ascent of Sap.* Springer-Verlag, Berlin.

ZIMMERMANN, M. H. and BROWN, C. L. (1971) *Trees—Structure and Function.* Springer-Verlag, Berlin.

REFERENCES FOR SOME OF THE DRAWINGS IN THIS AND OTHER CHAPTERS

PALLADIN, W. I. (1914) *Pflanzenanatomie.* B. G. Teubner, Leipzig and Berlin.

STRASBURGER, E. (1923) *Das botanische Praktikum*, 7th edn. G. Fischer, Jena.

TROLL, W. (1948) *Allgemeine Botanik.* F. Enke, Stuttgart.

CHAPTER 2

THE CELL

THE basic units of which organisms are constructed are the cells. The term *cellula* was first used by Robert Hooke in 1665. Hooke gave this term to the small cavities surrounded by walls that he saw in cork; later he observed cells in other plant tissues and noticed that they contained "juice" (Matzke, 1943).

Still later the *protoplasm*—the substance within the cell—was discovered. In 1880 Hanstein coined the term *protoplast* to indicate the unit of protoplasm found in a single cell. He also suggested that the term protoplast should be used instead of the term cell, but his suggestion is not generally accepted and cell is the accepted term. In plants the term cell includes the protoplast together with the wall.

The *cell wall* was, for a long time, regarded as a non-living excretion of the living cell matter, but recently more and more evidence has been found that organic unity exists between the protoplast and the wall, especially in young cells, and that the two together form a single biological unit.

In 1831 Robert Brown discovered the nucleus in an epidermal cell of an orchid plant. In 1846 Hugo von Mohl distinguished between the protoplasm and the cell sap, and in 1862 Kölliker introduced the term *cytoplasm*. From the end of the nineteenth century and during the twentieth century research on the cell has developed so rapidly and with such enormous strides that cytology has become a science of its own.

On the basis of internal organization *prokaryotic* and *eukaryotic* cells are distinguished. The eukaryotic cells, characteristic of all animals and plants, excluding blue–green algae and bacteria, have a membrane-bound nucleus and other organelles. Prokaryotic cells lack these compartments.

With the discovery of the electron microscope and the great progress in biochemical studies, a lot of knowledge has been accumulated on the structure and biology of the cell. There is a number of books dealing specifically with the ultrastructure and function of cells (e.g. Frey-Wyssling and Mühlethaler, 1965; Clowes and Juniper, 1968; Buvat, 1969; Ledbetter and Porter, 1970; Gunning and Steer, 1975; Lloyd, 1982; Alberts *et al.*, 1983; Thorpe, 1984).

The protoplast consists of membranous and non-membranous components. When suitably fixed the cross-sectioned membranes are seen under the electron microscope as two dark lines, each about 2.5 nm thick, separated by a lightly stained line, about 3.5 nm thick. This type of membrane is termed a *unit membrane*. It was generally considered that the unit membrane consists of a bimolecular lipid layer covered on each side with a layer of protein. Today the fluid mosaic model of functional membrane is widely accepted (Fig. 7).

The main components of the plant cell are the *cell wall*, *cytoplasm* and *nucleus*. The cytoplasm includes the *endoplasmic reticulum*, *Golgi apparatus*, *mitochondria*, *plastids*, *microbodies*, *ribosomes*, *spherosomes*, *microtubules*, *vacuoles* and *ergastic substances* (Fig. 8).

The cell usually contains a single nucleus, but in some cells, such as the sieve elements of the phloem which are adapted for translocation, the nucleus is generally absent from the mature cell. However, there are also cells which have numerous nuclei, e.g. in some laticifers. A multi-nucleated cell can comprise an entire organism as in some fungi and algae, or multi-nucleated cells may be a transitory stage in the development of a tissue, e.g. in the endosperm of many plants and sometimes in fibres. The accepted view in many cases is that each nucleus together with the pro-

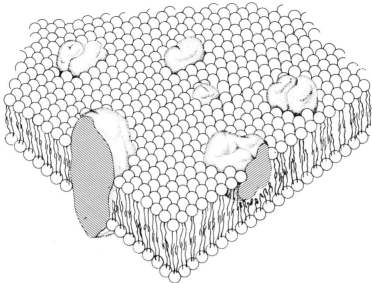

FIG. 7. A schematic three-dimensional drawing of the fluid mosaic model of a cell membrane (as presented by Singer and Nicolson, 1972). The phospholipid bilayer consists of: (a) ionic and polar head groups (the small circular structures in the drawing) of the phospholipid molecules which make contact with water, and (b) fatty acid chains (wavy lines). The large bodies with stippled cut surfaces represent the globular protein.

toplasm surrounding it forms a wall-less cell so that the entire multi-nucleate body comprises a group of protoplasmic units. Such a structure is called a *coenocyte*.

The coenocyte aroused much interest in phylogenetic and ontogenetic studies. Two theories exist which deal with the relation of the entire organism and the single cell. According to the *cell theory*, which was developed about the middle of the nineteenth century, the organism consists, both phylogenetically and ontogenetically, of a complex of an enormous number of cells each of which plays a role in determining the nature of the organism.

The theory contradicting the above is the *organismal theory*. This theory gives less importance to the individual cells and mainly stresses the unity of the protoplasmic mass of the entire organism. According to this theory the organism as a unit, to a large extent, determines the nature of the cells.

These two theories are important and, for the following reasons, attention was paid to both of them in histological and cytological research of plants. Many aspects of ontogeny, such as the processes of cell division, the origin of vessels and articulated laticifers, the development of idioblasts, etc., were investigated in the light of the cell

theory. However, the specialization of the different cells and tissues in the plant and the sites of appearance of the various types of cells and tissues can be explained only on the basis of the organismal theory which regards the organism as a unit.

THE CYTOPLASM

The cytoplasm comprises part of the protoplast. Physically it is a viscous substance which is more or less transparent in visible light. Chemically the structure of the cytoplasm is very complex even though the major component (85–90%) is water. Cytoplasmic streaming can frequently be seen with the aid of the light microscope in living cells. Various minute structures were discovered within the cytoplasm, first with the aid of the light microscope and later with the aid of the electron microscope. These discoveries complicated the exact use of the term *cytoplasm*.

It is now conventional to distinguish between the term *cytoplasm* in its wide sense, i.e. the whole protoplasmic substance surrounding the nucleus, including the membrane-bounded structures, and the terms *groundplasm, ground cytoplasm, cytoplasmic matrix, hyaloplasm* or *cytosol* for the compartment of the cytoplasm which includes

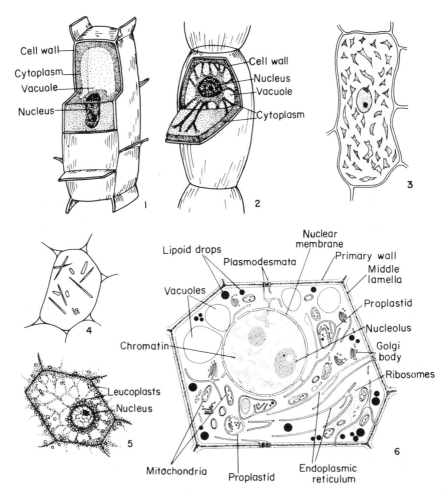

FIG. 8. 1, Three-dimensional diagram of a plant cell from which a portion has been removed to reveal a large central vacuole and the cytoplasm, which includes the nucleus, lining the cell wall. 2, As above, but of a cell in which the nucleus is located more or less centrally and in which the cytoplasm surrounding the nucleus is connected to the peripheral cytoplasm by cytoplasmic strands. 3, An adaxial epidermal cell from the calyx of *Tropaeolum majus* containing chromoplasts. 4, Chromoplasts in a carrot root cell. 5, Leucoplasts in a young endosperm cell of *Zea*. 6, Diagram of a meristematic plant cell. (No. 3 adapted from Strasburger, 1923; nos 4 and 5 adapted from Eames and MacDaniels, 1947; no. 6 adapted from Sitte, 1961.)

everything other than the membrane-bounded organelles.

In the cytosol there is the *cytoskeleton*, which gives the cell its shape, its ability to arrange its organelles and to move them. The cytoskeleton is composed of a network of protein filaments, the two, most important, of which are the actin *microfilaments* and the *microtubules*.

The cytoplasm is delimited from the cell wall by the plasma membrane, a unit membrane termed *plasmalemma* (Fig. 9, no. 2) and from the vacuole by another unit membrane, the *tonoplast*. The plasmalemma frequently exhibits regions which are folded to form an elaborate system of evaginations in tubular form. Such evaginations, together with portions of cytoplasm which they contain, may bud off from the surface of the protoplast. In electron microscopical sections these regions are often seen as pockets of vesicles and tubules between the plasmalemma and cell wall, and are often referred to as *multivesicular structures, plasmalemmasomes,* or *paramural bodies*. It was suggested that these structures may be involved in secretory processes (Fahn, 1979) including deposi-

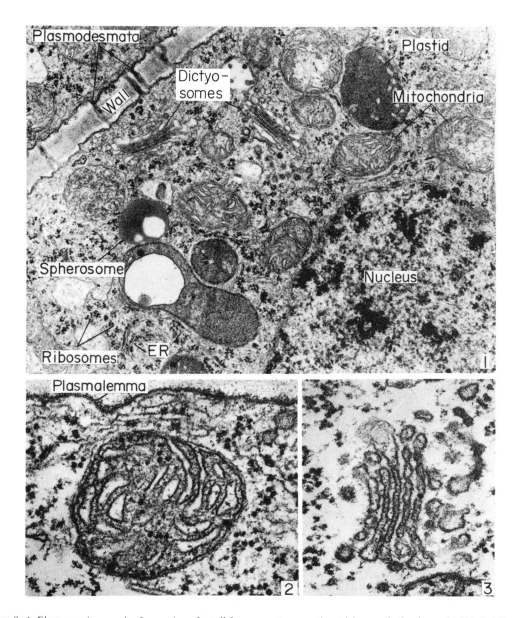

FIG. 9. 1, Electron micrograph of a portion of a cell from a nectar-secreting trichome of *Abutilon*. × 28,000. 2, Mitochondrion. × 77,000. 3, Dictyosome. × 77,000.

tion of wall matrix polysaccharides (Cox and Juniper, 1973).

ENDOPLASMIC RETICULUM

One of the membranous structures occurring in the cytoplasm is the endoplasmic reticulum (ER) (Fig. 9, no. 1; Fig. 269, no. 2). This is a complex system which consists of two unit-membranes enclosing a narrow space between them. The ER appears in the form of extended cisternae, tubules, or as fenestrated sheets. It may be connected with the nuclear envelope. In a special form the ER occurs in the cytoplasmic strands, *plasmodesmata* (see p. 39) traversing the walls of neighbouring cells (Fig. 8, no. 6; Fig. 9, no. 1; Fig 11, no. 3). The

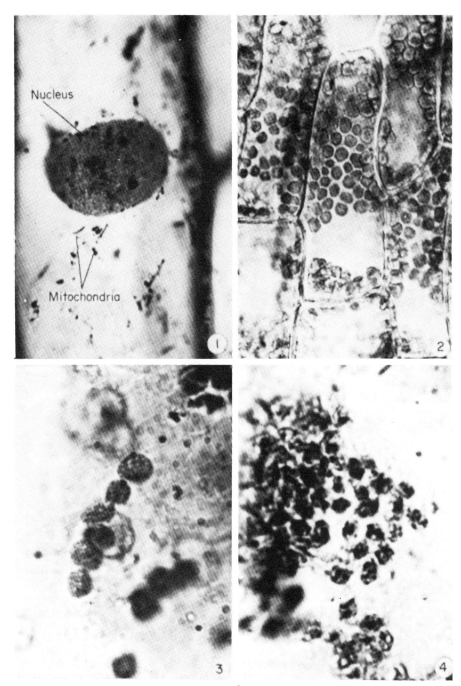

FIG. 10. 1, Portion of an epidermal cell of a bulb scale of *Allium cepa* showing the nucleus and mitochondria. × 1000. 2, Cells of *Elodea canadensis* showing chloroplasts. × 750. 3, Chloroplasts in a subepidermal cell of the green fruit of *Lycopersicon esculentum*; grana can be distinguished as darker areas in the chloroplasts. × 800. 4, As above, but in a mature fruit where the chloroplasts have become changed into chromoplasts. × 800. (Nos. 3 and 4 courtesy of Y. Ben-Shaul.)

portion of the ER situated in the centre of the plasmodesma, in tubular form, is called *desmotubule* (Fig. 11, no. 3; Fig. 18, no. 11).

When ribosomes adhere to the surface of the ER it is termed *granular* or *rough* (RER). When ribosomes are absent the ER is termed *agranular* or *smooth* (SER). The association with ribosomes is interpreted as evidence that the RER is involved in protein synthesis. In secretory tissues SER has been suggested to take part in the production of lipophilic substances. The ER is also supposed to serve in intracellular transport of secreted materials (Fahn, 1979, 1988). The ER cisternae may become dilated and accumulate protein and other products. There are suggestions that by dilation ER cisternae or ER vesicles form vacuoles. There are also views that the ER gives rise to membranes of dictyosomes and microbodies (Gunning and Steer, 1975; Chrispeels, 1980; Mollenhauer and Morré, 1980).

GOLGI APPARATUS

The Golgi apparatus consists of a system of stacks of flat circular cisternae each bound by a smooth unit membrane. Each stack is termed *Golgi body* or *dictyosome* (Fig. 9, nos. 1, 3) (Mollenhauer and Morré, 1980).

The margins of the cisternae may be simple in shape or fenestrated. Near the dilated edges of the cisternae there is usually a number of vesicles positioned in such a way as to suggest that they have budded off from the cisternal margins. In some dictyosomes rod-shaped elements in parallel arrangement have been seen in the very narrow spaces (about 10 nm) between the individual cisternea. It has been suggested that these elements might link the cisternae (Gunning and Steer, 1975). The dictyosomes are concerned with secretion processes and have a polar structure. In active bodies, in addition to the production of many vesicles, distal cisternae (on the *maturing* or *secreting face*) may break up into vesicles. New cisternae are formed on the opposite face (*proximal face* or *forming face*) of the dictyosome. The new cisternae appear to be produced from membrane components derived from the ER.

The Golgi bodies are mainly involved in the secretion of sugar (in nectar secretion), poly-saccharides (cell-wall material, mucilage), and polysaccharide protein complexes (certain mucilages). In certain cases the dictyosomes are also involved in the secretion of lipophilic materials (Fahn, 1979, 1988). It has been reported, that in plants the *coated vesicles* originate from the dictyosomes. The lattice-like coats of these vesicles are composed predominantly of a particular type of polypeptide which has been termed *clathrin* (Newcomb, 1980).

MITOCHONDRIA

Mitochondria are organelles which can be seen with the light microscope when living cells are stained with Janus Green B (Fig. 10, no. 1). In electron-microscopical sections they appear in various forms: spherical, elongated and sometimes lobed (Fig. 9, nos. 1, 2). They are about 0.5–1.0 μm in diameter and about 3 μm in length and are bound by an envelope consisting of two unit-membranes. The inner membrane forms invaginations (*cristae*) into the matrix. The matrix is mainly made of proteins. The mitochondria contain ribosomes which are smaller than those of the cytoplasm and DNA fibrils, but their genetic capability is limited. The mitochondria provide ATP as principal energy source of the cell, and to varying degrees participate in intermediary metabolism (Hanson and Day, 1980). Glandular cells, which are highly active, contain numerous mitochondria. In some cases mitochondria were suggested to take part in the process of secretion (Werker and Fahn, 1981; Fahn, 1979, 1988).

The formation of new mitochondria is most likely to occur by division.

PLASTIDS

Plastids are organelles characteristic of plant cells and have no homologues in the animal cell. They vary in form, size, and pigmentation (Kirk and Tinley-Basset, 1978). The principal types of plastids are *chloroplasts, chromoplasts,* and *leucoplasts* (Fig. 10, nos. 2–4; Fig. 15, no. 9). Chloroplasts are green as a result of the pigment chlorophyll which predominates in them. Chromoplasts are usually yellow, orange, or red because of the

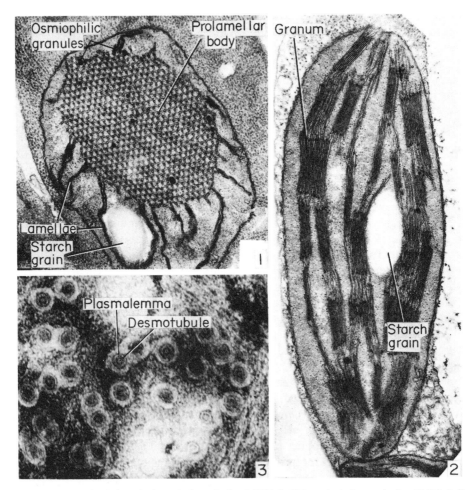

FIG. 11. Electron micrographs. 1, An etioplast of a bean leaf. × 36,000. 2, Chloroplast of a pea leaf. × 25,000. 3, Face view of a portion of a wall between two cells of a salt secreting gland of *Tamarix aphylla*, showing cross-sections of plasmodesmata with their membranes. × 90,000. (Nos. 1 and 2, courtesy of S. Klein.)

carotenoid pigments. They occur in petals, in some ripe fruits, and in some roots (e.g. carrot roots). Leucoplasts are non-pigmented plastids, usually located in tissues not exposed to light, and they store plant products such as starch (*amyloplasts*), proteins (*proteinoplasts*), and fats (*elaioplasts*). The *statoliths* are a special type of amyloplast, which occur in the root cap and in the nodes of some shoots, and are involved in the perception of gravity.

Lipids in the form of globules (*plastoglobuli*) and *phytoferritin* (an iron-protein complex) (Cresti *et al.*, 1978) may be present in various plastids, including chloroplasts. Among other inclusions reported to occur in plastids are membrane-bounded amorphous protein bodies (Bosabalidis, 1987a). Plastids take also part in the process of secretion, especially in the secretion of lipophilic materials (Fahn, 1979, 1988).

Leucoplasts of tissues which become exposed to light may develop into chloroplasts (e.g. in the potato tuber).

All types of plastids are derived from minute bodies, *proplastids* which are present in egg cells and meristematic cells. Proplastids and plastids in young cells multiply by division. The plastids are bound by an envelope consisting of two unit-membranes. Internally the plastids contain a membrane system embedded in a proteinaceous matrix or *stroma*. The stroma contains short circular

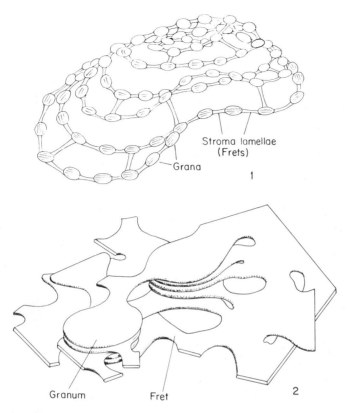

Stroma lamellae
(Frets)

Grana

1

Granum   Fret   2

FIG. 12. Schematic drawings of the arrangement of grana and frets (stroma lamellae) in chloroplasts. 1, A drawing of a sketch presented by Wildman *et al.* (1980) based on light microscopic analysis and suggests a spiral arrangement of frets and grana. 2, Portion of the inner lamellar system of a chloroplast based on EM observations; redrawn from Wehrmeyer (1964).

fibrils of DNA which is not complexed with histones and ribosomes, which are smaller than those of the cytoplasm. The membrane system is in the form of flattened sacs called *thylakoids*. The thylakoid system is developed to various degrees in the different types of plastids. It is least developed in proplastids where only a few or no thylakoids are present. With the differentiation of plastids flattened vesicles bud off from the inner membrane of the envelope, proliferate, and form thylakoids which in the chloroplasts align themselves into characteristic systems. The number of ribosomes increases during the differentiation of the chloroplastids, whereas in the leucoplasts the production of complete ribosomes ceases and their number may even become reduced, already in the early stages of their development (Whatley, 1979; Cheniclet and Carde, 1988).

In *chloroplasts* the thylakoid system, consists of *grana* and *frets* (Fig. 11, no. 2; Fig. 12) (stroma thylakoids). Each granum is composed of a series of disc-like thylakoids stacked one upon the other like a pile of coins. The grana are interconnected by the frets that traverse the stroma. It is now accepted that these two types of thylakoids are interconnected in such a way that the spaces within them are continuous. In addition to the light harvesting system the chloroplasts contain the enzymes responsible for the fixation of carbon dioxide into sugar.

In plants grown in the dark the developing leaves and stems lack the green colour. They become *etiolated*. In such a case the vesicles derived from the inner membrane of the plastid envelope develop into a paracrystalline lattice called *prolamellar body*. Plastids of this kind are called *etioplasts* (Fig. 11, no. 1). When a dark-grown plant is placed in light the prolamellar body gives rise to the thylakoid system characteristic to chloroplasts (Fig. 11, no. 2) (Hoober, 1984).

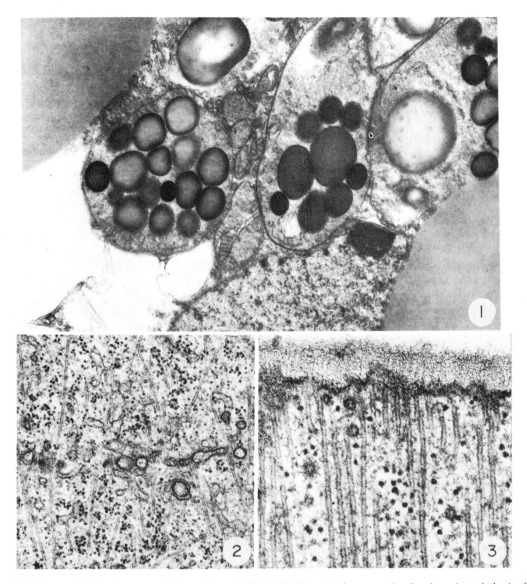

FIG. 13. 1, Chromoplastids of an orange juice vesicle. × 36,000. 2, Cytokinesis in bean root tip, showing microtubules in the spindle zone and aggregating vesicles. The cell plate is formed by fusion of these vesicles. × 36,000. 3, Microtubules running parallel to one another beneath a primary wall of a cell of a bean root tip. Vesicles can be seen among the microtubules. × 62,000. (No. 1 courtesy of I. Shomer; nos. 2 and 3, courtesy of E. H. Newcomb—no. 2 micrograph by B. A. Palevitz.)

Chromoplasts vary greatly in shape and size and are often derived from chloroplasts (e.g. in many fruits), but may also develop directly from proplastids (Fig. 8, nos. 3, 4; Fig. 10, no. 4) (Whatley and Whatley, 1987). The differentiation of chromoplasts involves synthesis of carotenoid pigments. These pigments may be stored in the lipid globules (Fig. 13, no. 1) (e.g. in petals of *Ranunculus repens* and yellow fruits of *Capsicum* and *Citrus*), or in protein fibrils (e.g. in red fruits of *Capsicum*). In the carrot root the caroten is in a crystalloid form surrounded by lipo-protein envelope (Ben-Shaul *et al.*, 1968). In tomato fruits two kinds of pigment bodies were found—long, narrow crystaloids and round globules (Mohr, 1979). Chromoplasts in many plant organs are capable of reverse differentiation into chloroplast (e.g. in carrot roots).

As mentioned above, both plastids and mitochondria contain DNA and ribosomes. Thus the two organelles are potentially autonomous. The indication of autonomy has led some researchers to the hypothesis that plastids and mitochondria originated, in the course of evolution, from primitive prokaryotes (e.g. some blue-green algae), which were enclosed in primitive eukaryotic cells and became there established as permanent symbiotic bodies (Schiff, 1980; Hoober, 1984).

MICROBODIES

Microbodies are small bodies, 0.5–1.5 μm in diameter, which occur in the cytoplasm of a variety of tissues. They are bound by a single membrane and their matrix appears granular or fibrillar. They contain enzymes that vary in accordance to the type of cell or tissue in which they are present (Frederick et al., 1975; Tolbert, 1980). In leaves of higher plants, closely associated with chloroplasts, microbodies called *peroxisomes* are found. These microbodies are sites for oxidation of glycollic acids, a product of carbon dioxide fixation. The oxidation of this compound results in a release of carbon dioxide and oxygen. Microbodies which occur during the germination of those seeds which store fats as a reserve, and contain the enzymes necessary for the breakdown of fatty acids to acetyl-CoA and the synthesis of succinate from acetyl CoA, are called *glyoxysomes*. Microbodies may contain a solitary crystalline inclusion, e.g. in tobacco leaves and in the potato tuber (Yoo et al., 1979). The crystalline inclusions possess catalase activity (Colman, 1977).

RIBOSOMES

Ribosomes (Fig. 9, no. 1) are small particles (17–20 nm in diameter) that occur free in the cytoplasm, on the outside of the membranes of the endoplasmic reticulum, and in the nucleus, chloroplasts, and mitochondria. They consist of RNA and protein, mainly histone. Ribosomes united into clusters (*polyribosomes* or *polysomes*) are involved in protein synthesis, i.e. in the assembly of polypeptide chains from amino acids (Alberts et al., 1983). The information specifying the amino-acid sequence of the polypeptide chain is brought from the nucleus by the messenger RNA molecules. When the messenger RNA molecules associate with the ribosomes, the information it contains is translated by transfer RNA molecules located in the cytoplasm.

SPHEROSOMES

Spherosomes are spherical lipid bodies which become electron opaque after fixation with osmium tetroxide (Fig. 9, no. 1) (Frey-Wyssling and Mühlethaler, 1965). There is no unanimous interpretation of the nature and structure of these bodies. In the view of some investigators the spherosomes are surrounded by a membrane, whereas in the view of others a surface skin consisting of an outer layer of oriented lipid molecules is formed in response to the aqueous cytoplasm surrounding them. It has been suggested that spherosomes originate as oil-containing vesicles detached from the ER.

MICROTUBULES

Microtubules, components of the cytoskeleton, are straight, elongated, hollow structures composed of globular subunits (Fig. 13, nos. 2, 3). Their average diameter is 23–27 nm (Lloyd, 1982; Alberts et al., 1983). Microtubules occur in the peripheral cytoplasm close to cell walls still growing in area and thickness, in the mitotic and meiotic spindles, and in the phragmoplast that arises between the daughter nuclei at the telophase. Similarity in the alignment of microtubules in the cytoplasm close to the plasmalemma and the cellulose microfibrils in the cell wall, led to the view that the microtubules might direct the orientation of the developing microfibrils.

MICROFILAMENTS

As mentioned before, the cytoskeleton consists of microtubules and microfilaments. The latter are in the form of long filaments 5–7 nm in diameter. They are composed of cortical proteins, such as actin (Fig. 14) which is similar to that of muscles. Myosin and other proteins have been implicated as components of the system in some cells. It has been suggested that microfilaments play a role in the cytoplasmic streaming (Lloyd, 1982; Parthasarathy *et al.*, 1985; Witztum and Parthasarathy, 1985).

NUCLEUS

The nucleus is usually more or less spherical, though nuclei with other shapes have also been observed. The nucleus is surrounded by an enve- lope (*nuclear envelope*) and contains the nuclear matrix (*nucleoplasm, karyolymph, nuclear sap*) and one or more *nucleoli* (Fig. 9, no. 1; Fig. 269, no. 1). In the nucleoplasm chromosomes consisting of deoxyribonucleic acid (DNA) and proteins are present. The complex of DNA and protein in the chromosomes which has an affinity to basic dyes is called *chromatin*. In the interphase (between nuclear divisions) the chromosomes are uncoiled and cannot be distinguished in the light micro- scope. At this phase condensed, deeply staining chromatin masses are often seen in the nucleo- plasm. This chromatin is called *heterochromatin*. The chromatin which stains less deeply than heterochromatin is called *euchromatin*. During the interphase DNA synthesis takes place in prepar- ation for the process of replication of chromo- somes. The nuclear envelope consists of two unit- membranes. The narrow space between these membranes is called *perinuclear space*. Connections exist between the nuclear envelope and the ER. The nuclear envelope has pores which are usually distributed in a more or less regular pattern.

The nucleoli are very dense, granular, and fibril- lar in structure, and are not bound by a membrane. They are often seen associated with some chrom- atin. They contain RNA, DNA, and proteins. Light-stained regions, which are often seen in the nucleoli, are commonly referred to as "vacuoles".

The nucleus carries the information for the cell protein in its DNA. The special kind of RNA, the messenger RNA, is transcribed on the DNA. The latter is transported to the cytoplasmic ribosomes where the synthesis of proteins, mainly enzymes, occurs. According to the genes present in the chromosomes the particular kinds of messenger RNA are synthesized. Differentiation can occur without differences in the DNA, as different cell types undergo different courses of development, according to which genes are active or depressed (Alberts *et al.*, 1983).

VACUOLES

Vacuoles occupy more than 90% of the volume of most mature plant cells (Fig. 8, nos. 1, 2, 6; Fig. 66). A vacuole is a watery cell compartment sur- rounded by a membrane, the *tonoplast*. It contains a variety of organic and inorganic substances, such as sugars, proteins, organic acids, phosphatides, tannins, flavonoid pigments, and calcium oxalate. Some substances in the vacoule may occur in solid form (e.g. tannins, protein bodies) and may even be crystalline.

Meristematic cells possess many minute vacu- oles. With growth and differentiation of a cell the vacuoles enlarge and fuse. In mature parenchy- matous cells usually a large central vacuole is present which is surrounded by a thin layer of cytoplasm. As a result of wounding, cells in the vicinity of the wound become mitotically active, the vacuoles subdivide and become reduced in volume (Schulz, 1988). In response to infestation by some insects, parenchyma cells become hyper- trophic, the amount of cytoplasm increases and the volume of vacuoles decreases (Chessen and Fahn, 1988).

If more than one vacuole are present in a cell, they are often termed collectively *vacuome*. Various views exist as to the initiation of the vacuoles: (1) from pre-existing vacuoles which multiply by fission, and after cell division each daughter cell receives a number of vacuoles; (2) by a *de novo* process, by attraction of water to a certain local- ized region in the cytoplasm and the formation of membrane around it; (3) from Golgi vesicles; (4) by dilation of ER cisternae or vesicles derived from the ER. In any case it seems that there is more than one way by which vacuoles originate (Buvat and Robert, 1979).

The vacuoles function in regulation of the water and solute content of the cell, i.e. in osmo- regulation, in storage, and in digestion. There is evidence that the vacuoles are active participants in cellular metabolism (Marty *et al.*, 1980). Vacuoles contain digestive enzymes capable of breaking down cytoplasmic components and metabolites. The hydrolitic activity of vacuoles resembles that of *lyosomes* of animal cells. The origin of the digestive enzymes of the vacuoles may be in the ER and the Golgi apparatus from where they are transported to the vacuole by membrane-bound vesicles. The amount of enzymes can change during the life of the cell and may be produced to different extents in different cells. It is even possible that some vacuoles completely lack digestive enzymes. The initiation of digestion may in some cases start in the ER.

ERGASTIC SUBSTANCES

Reserve materials and substances, which are produced and stored in plant cells, but do not reenter the metabolism of the plant are called *ergastic substances*.

Starch

Starch is a carbohydrate composed of long-chain molecules. It appears in the form of grains, which commonly stain bluish-black with a solution of iodine in potassium iodide. Starch grains are first formed in chloroplasts. Later the starch is broken down and moves as sugar to storage tissues where it is resynthesized in amyloplasts. Starch grains commonly show layering around a point termed *hilum* (Fig. 15, nos. 1–11). The hilum may be centrally situated (e.g. in *Triticum durum*) or it may be eccentric (e.g. in the potato tuber). The layering in the starch grains is seen as a result of more densely packed molecules at the beginning of growth of each layer and gradual thinning of them in the outer part of the layer, which becomes more hydrated. In the starch grains of cereals the number of layers corresponds to the number of days during which the grain grows. In potato starch-grains the periodicity of formation of layers depends on endogenous factors (Frey-Wyssling and Mühlethaler, 1965). In the grain the starch molecules are arranged radially in such a manner as to exhibit crystalline properties. When viewing a starch grain in a microscope with crossed polarizers it appears luminous except for a dark cross, the arms of which meet at the hilum of the grain. The position of the

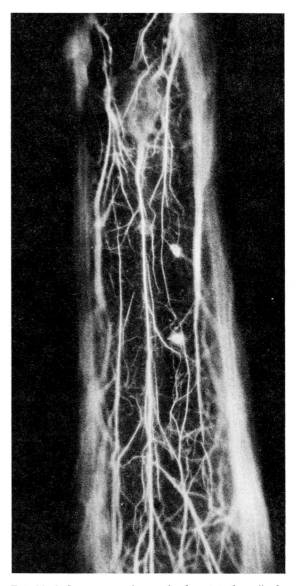

FIG. 14. A fluorescence micrograph of a part of a cell of a tomato stem-hair, stained with Rhodamine-phalloidin, showing distribution of bundles of actin microfilaments, × 700. (Courtesy of M.V. Parthasarathy, figure in Parthasarathy *et al.*, 1985.)

hilum, the shape and size of the grains and their appearance, solitary or in aggregates (compound starch grains), make it possible to identify the plant species from which the starch was obtained. Compound starch grains, for instance, occur in *Oryza sativa, Ipomoea batatas, Fagopyrum esculentum,* and *Avena sativa*.

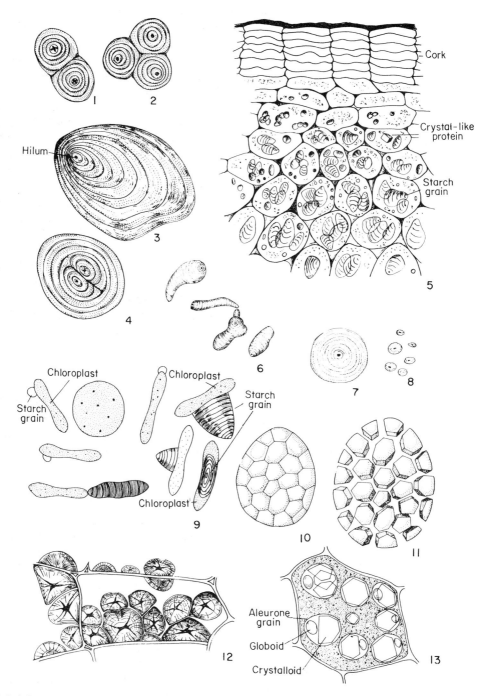

FIG. 15. 1–4, Potato starch grains. 1 and 2, Compound starch grains. 3, Simple starch grain. 4, Half-compound starch grain. 5, Cross-section of the outer portion of a potato tuber. 6, Banana starch grains. 7 and 8, Starch grains of *Triticum durum*. 9, Stages in development of starch grains in chloroplasts of *Phaius maculata*. 10, Compound starch grains of *Avena*. 11, As in no. 10, but disintegrating. 12, Sphaerocrystals of inulin in cells of a *Dahlia* tuber, precipitated when alcohol was added. 13, Aleurone grains in an endosperm cell of *Ricinus communis* from a section of material embedded in dilute glycerine. (Most of the figures adapted from Strasburger, Palladin, and Troll.)

Commercial starches are obtained from various plant parts, e.g. from endosperm of seeds of wheat, maize, and rice; from tubers of potato; from fleshy roots of *Manihot esculenta* (tapioca starch); from the stem of *Metroxylon sagu* (sago starch); from tuberous rhizomes of some scitaminous plants including *Canna edulis, Maranta arundinacea,* and *Curcuma angustifolia* (arrowroot starch).

Inulin is a polysaccharide found as a storage material in some Compositae, Campanulaceae and monocotyledons. In contrast to starch, it appears only in solution (Fig. 15, no. 12).

Proteins

Amorphous protein is found in the outermost endosperm layer, the aleuron layer, of the caryopsis of cereals (Fig. 303, no. 1). Protein, in the form of cuboidal crystalloids is found in the cells of the peripheral parenchyma of the potato tuber and in the fruit parenchyma of *Capsicum*. Crystalline and amorphous protein are found together in *protein bodies (aleurone grains)* in the endosperms and embryos of many seeds (Fig. 15, no. 13) (Lott, 1980).

The development of aleurone grains in the seed of *Ricinus* has been described in detail by Frey-Wyssling (1948). These grains are formed from readily soluble proteins with globular molecules and relatively low molecular weight which accumulate in vacuoles of the storage cells where they crystallize. From these liquid vacuoles water is lost by dehydration. This causes the various vacuolar components to precipitate according to their solubility. In *Ricinus* the first substance to be precipitated is the almost insoluble phytin (magnesium-potassium salt of inositol phosphoric acid); this substance forms the globoid. Next, the corpuscularly dispersed reserve proteins precipitate out in a lattice to fill the remaining space of the vacuole and so form the crystalloid part of the aleurone grain. Finally, the remaining liquid, which now contains soluble albumin, solidifies to form a homogeneous substance which surrounds both the globoid and crystalloid. On mobilization, these reserve substances are digested in the reverse order.

Oil, fats and waxes (Lipids)

Oils and fats are important reserve materials in plants. They are most commonly present in seeds and fruits. Both fats and oils are glycerides of fatty acids. The distinction between them is commonly based on physical properties; fats being solid and oils liquid at normal temperatures.

The fats and oils may be produced by elaioplasts or spherosomes. Waxes consist mainly of esters of long-chain fatty acids and long-chain monohydric alcohols. Waxes are widespread in plants where they usually form protective coatings on the epidermis of stems, leaves, and fruits. Very rarely waxes occur within cells. A liquid wax, similar in quality to the oil of the sperm whale, is produced in large quantities in the cotyledons of *Simmondsia chinensis* (jojoba) seeds (Rost *et al.,* 1977; Rost and Schmid, 1977).

Lipid compounds others than fats, oils, and waxes, e.g. terpenes and essential oils, are usually produced by specialized secretory tissues (see Chapter 9). Lipids stain a reddish colour when treated with Sudan III or IV.

Crystals and silica bodies

Many plants are known to deposit in their cells inorganic materials consisting mostly of calcium salts and silicon dioxide. The first occur as crystals and the second as silica bodies.

Crystals

Crystals of various forms are often present in plant cells. The most common crystals are those of calcium oxalate, which appear in different forms (Frey-Wyssling, 1981; Lersten, 1983; Metcalfe and Chalk, 1983) (Fig. 16, no. 1–3; Fig. 17; Fig. 116, no. 2; Fig 135, no. 1).

Solitary, rhomboidal or *prismatic* crystals, are formed in leaves of *Citrus, Begonia, Hyoscyamus niger, Vicia sativa* and *Pistacia palaestina* (in the bundle-sheath extensions).

Druses (Fig. 16, no. 1), spheroidal aggregates of prysmatic crystals are found in leaves of *Datura stramonium* and *Ruta graveolens*; in stems of *Opuntia ficus-indica*, in the fleshy cortex of *Anabasis articulata*, in the rhizome of *Rheum rhaponticum* and *Colocasia esculenta* (Sunell and Healey, 1979); in the roots of *Ipomoea batatas*.

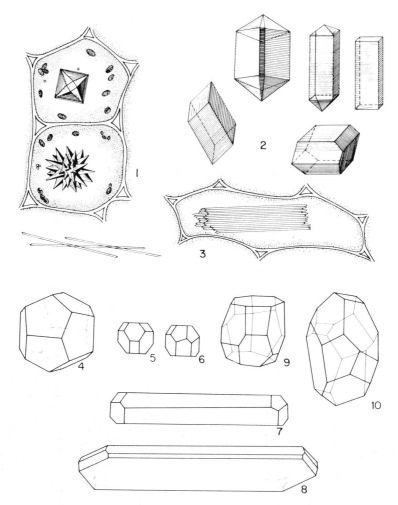

Fig. 16. 1, Two parenchyma cells from the petiole of *Begonia*; in the upper cell a solitary prismatic crystal and in the lower cell a druse. 2, Variously shaped prismatic crystals. 3, Individual raphides and a bundle of raphides. 4, A cell with pentagonal faces. 5–8, Various types of cell shapes with 14 faces. 9 and 10, Polyhedral parenchyma cells. (Nos. 1–3 adapted from Palladin, 1914; nos. 4–8 adapted from Frey-Wyssling, 1959; nos. 9 and 10 adapted from Marvin, 1944.)

Crystal sand, very small prysmatic crystals usually occurring in masses, are found in stems of *Sambucus nigra*, and *Aucuba japonica*, in leaves of *Atropa belladonna*.

Raphides and styloids. Raphides are thin elongated crystals which taper off at their both ends into a tip point. They are usually aggregated in bundles and are found in leaves of *Arum* and *Agave*, in leaves and stems of *Zebrina* (Fig. 17, nos. 1, 2) *Tradescantia*, and *Impatiens*, and in bulb scales of *Urginea maritima*. Styloids or pseudo-

raphides, as they are sometimes called, are long prismatic crystals tapered off at both ends into a blade. They occur in the cell solitary or in a small number. Styloids are found in Iridaceae, Agavaceae, and some species of Liliaceae, Aizoaceae, Rosaceae, Rutaceae and others (Metcalfe and Chalk, 1983). Styloids may be very large. Styloids occurring in the secondary xylem of *Cosmocalyx spectabilis* (Rubiaceae), for instance, were reported to be up to 50 μm wide and up to 200 μm long (Richter and Schmidt, 1987). The development of

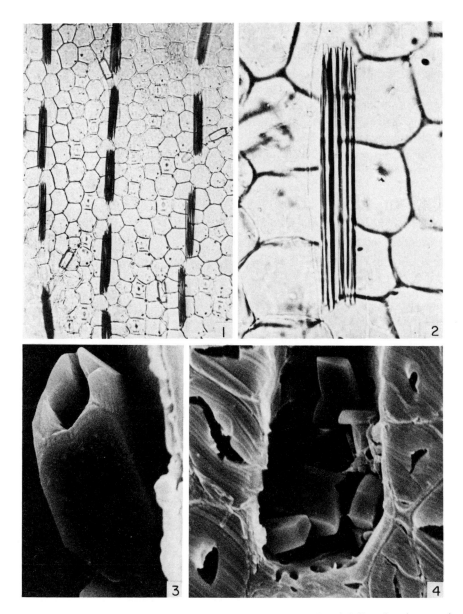

FIG. 17. Bundles of raphides in a leaf of *Zebrina pendula*. 1, × 125; 2, × 5330. 3 and 4, Scanning electron micrographs of crystals in ray cells of wood of some Oleaceae species. 3, A hollow crystal of *Schrebera arborea*, × 7000. 4. More or less rhomboidal crystals of *Nestegis sandwicensis*, × 2000. (Nos. 3 and 4 courtesy of P. Baas; Figs. occurring in Baas *et al.*, 1988.)

raphides and raphide-forming cells has been studied recently in young banana plants (Bruni *et al.*, 1982).

Crystals can be found in cells resembling those neighbouring them but lacking crystals, or they may be confined to special crystal-containing cells, i.e. *idioblasts* (Bosabalidis, 1987b). In some plants, e.g. in *Zebrina*, the idioblasts containing raphides

form long files (Wheeler, 1979).

Crystals are formed within vacuoles and are surrounded by an envelope. It has been suggested that in mature idioblasts the envelope is suberinic in nature (Arnott and Pantard, 1970; Schötz *et al.*, 1970; Wattendorff, 1976a, b, 1978). Calcium carbonate crystals are rare in higher plants. This type of crystals is associated in some plants with ingrowths of the cell wall known as *cystoliths* (Fig. 77, no. 1; Fig. 128, no. 2) (see Chapter 12). The appearance, location, and type of crystals is often used in taxonomic classification and in wood identification (Metcalfe, 1987; Metcalfe and Chalk, 1950, 1983; Fahn *et al.*, 1986).

Beside the crystals that occur in vacuoles there are others that are embedded in the cell walls or even on their outside (Fig. 132, nos. 2, 3), e.g. in the astrosclereids of *Nymphaea*.

Silica bodies

Silica bodies, deposits of silicon dioxide (Fig. 82, no. 1) occur in the epidermis of plants such as Gramineae, Cyperaceae (Metcalfe, 1960, 1971; Ollendorf *et al.*, 1987), Palmae (Tomlinson, 1961), Rapataceae (Carlquist, 1962), in the epidermis and hypodermis of some Cactaceae (Gibson and Horak, 1978) and in ray and axial parenchyma cells of the wood of some dicotyledons, e.g. species of the Lauraceae (Richter, 1980) and Dilleniaceae (Dickinson, 1984).

The silica bodies are mostly opaline, amorphous, even in cases when they superficially appear crystalline (Metcalfe and Chalk, 1983). Silica is also deposited in the cell walls of many plants (Hodson, 1986; Hodson and Sangster, 1988; Parry *et al.*, 1986).

Tannins

The tannins are a heterogenous group of phenol derivatives (Bate-Smith and Metcalfe, 1957; Goodwin and Mercer, 1983). In microscopical sections they usually appear as granular masses or bodies coloured yellow, red, or brown. Tannins can be found in the different parts of the plant, especially in leaves, in periderm, in vascular tissue, in unripe fruits, in seed coats, and in tissues of pathogenic growths. Tannin-containing cells may be interconnected or tannins may be found in isolated specialized cells (idioblasts), some of which may be coenocytes (Zobel, 1986). Within the cells the tannins may be found in the vacuole or in the form of droplets in the cytoplasm, and sometimes they penetrate into the cell wall, as, for instance, in cork tissue. Tannins are thought to protect the plant against dehydration, rotting, and damage by animals. Tannins are used commercially, especially in the industry of the tanning of animal skins to obtain leather.

Pigmentation

The plant pigments are usually found in the plastids and in the vacuole. The green colour is due to *chlorophyll* which is found in the chloroplasts. In the same plastids *carotenoids*, the yellow to red pigments, are also found but they are masked by the chlorophyll. The carotenoids become noticeable when there is little or no chlorophyll as is the case in chromoplasts. As mentioned before, chromoplasts develop also directly from proplastids.

Another group of pigments is the *flavonoids* (anthocyanins and flavones or flavonols) which are generally present in vacuoles (Goodwin and Mercer, 1972). These pigments are water soluble and give plant parts, especially many flowers and fruits, various colours, *Anthocyanins* give, red, pink, lilac, and blue colours. Because of the ionic character of the anthocyanins, their intensity and colour are dependent on the pH. In acid solution the colour varies from orange-red to lilac. As the solution approaches pH 7.0, colourless pseudo-bases are formed. In basic solutions blue-coloured anthocyanins are formed. *Flavones* or *flavonols* absorb strongly in the ultraviolet region of the spectrum and can be seen by insects. They confer a cream or ivory translucent appearance to petals.

Sometimes the visible colour of petals is the result of a few pigments occurring in the same cells. For instance, chromoplasts can be found together with anthocyanins and flavonoids (Whatley and Whatley, 1987).

White petals may be devoid of pigments, and the colour seen results from the reflection of light from the petals which are opaque due to the presence of numerous large intercellular spaces that are filled with air.

The colouring of autumn leaves is the result of various processes as well as the combination of different pigments. With the gradual death of the leaf the chlorophyll breaks down into colourless substances and the carotenoids become visible, making the leaf appear yellow. The red and purple colours are from oxidation products of flavonoids. These colours are most brilliant when formed in the presence of sugars in leaves exposed to strong light. Autumn colours, which result from the combination of small amounts of chlorophyll and carotenoids and greater amounts of anthocyanins together with tannins and various uncommon pigments and the browning of the cell wall, are best developed in the cold temperate zones.

THE CELL AS A TISSUE COMPONENT

Mature cells vary in size and shape. Cells may be ellipsoidal, ovate, cylindrical, flattened, prism-like, star-shaped, fibre-like, and lobed. Parenchyma cells are usually from 10 to 100 μm in diameter, but in fleshy fruits and the pith of stems larger cells can be found. Fibres are usually about 1–8 mm long but fibres 55 cm long are also known, e.g. in *Boehmeria*.

Basic shape and arrangement of cells

As the cell volume increases in the meristematic region the primary elastic cell wall tends to assume the smallest possible surface area, i.e. the form of a sphere. After mitosis, forces act that tend to impart to the cell a spherical form, but as there are no intercellular spaces in the meristem, the cells, which are densely arranged, become polyhedral in shape (Fig. 16, no. 4). The basic shape of cells is a 14-faced polyhedron (Matzke, 1946). However, in plant tissues, cells with 12, 13, 15, 16, or more faces are found. Most of the faces of the cell wall are, according to Matzke, pentagonal (Fig. 16, no. 4), but tetragonal and hexagonal faces can also be found (Fig. 16, nos. 5–8). Similar structure was found in bubbles of soap foam, and also in experiments where lead shot was subjected to sufficient pressure to cause the elimination of the air spaces (Marvin, 1939).

Accordingly, the basic shape of the cells of the apical meristems is that of a 14-faced polyhedron (Fig. 16, no. 9). In the apical meristem of *Anacharis densa*, Matzke (1956) found that during the interphase the average number of faces of the polyhedron increased from 13.85 to 16.84, and after

division the daughter cells had an average of 12.61 faces.

As a result of the continued increase of cell volume during growth, the number of wall faces increases above 14. This makes it impossible for all the sides to remain in contact with all the sides of the neighbouring cells and so *intercellular spaces* develop. In some tissues the intercellular spaces reach relatively large dimensions and then they are referred to as *air spaces, ducts*, etc. Such spaces can develop in two ways: (a) by the separation of neighbouring cell walls, as in the development of the resin ducts in *Pinus*; this type of development is known as *schizogenous* development (Fig. 45, nos. 1–4); (b) by the disintegration of the cells in the place where the space develops, as in the essential oil cavities in the peel of citrus fruits; this type of development is know as *lysigenous* development (Fig. 46, nos. 1–7). In some cases spaces are formed by these two methods together and then the development is known as *schizo-lysigenous* development. The intercellular spaces in the protoxylem are sometimes formed in this way.

The intercellular spaces can be irregular and variable in shape or they may form a distinct and permanent system as in many water plants, in the banana leaf, and other plants (Fig. 123, no. 3; Fig. 127, no. 1; Fig. 133, no. 2).

Interrelationship of cells during growth

All cells develop from existing cells by cell division. The young cells in the growing regions are all relatively small; as they mature their size and shape alter in accordance with their physiological function. As the cell wall is already present at the earliest stages of development of the cell it takes part, together with the protoplast, in the processes of cell growth.

When a group of cells grows together uniformly, the cells of the group take up different positions and shapes, but the relationships between the neighbouring cell walls do not change and no new areas of contact are formed between the cells; this type of growth is known as *symplastic growth*. In many cases, pit-free tips may grow individually along walls of neighbouring cells and form new areas of contact. In this type of growth, termed *intrusive growth*, a temporary loosening of the middle lamella between the cells in the growing region must occur. Intrusive growth takes place in many elongating fibres, in branches of some sclereids, in non-articulated laticifiers, and in the daughter cells of a fusiform initial of non-storied cambium which has divided anticlinally.

THE CELL WALL

The presence of a wall in plant cells distinguishes them from animal cells. The cell wall was discovered in the seventeenth century before the presence of the protoplast was recognized, and since then many researchers have investigated it. Much research has been done on cell walls because of their biological and economic importance. Various methods—chemical, biochemical, physical and morphological—have been used. In the study of the submicroscopical structure the investigators were assisted by advances in organic chemistry, X-rays, the use of the light, polarizing and electron microscopes. Recently much attention is being paid to the biosynthesis of the cell wall (Darvill *et al.*, 1980; Delmer and Stone, 1988).

Nature of the cell wall

The cell wall grows when in contact with the protoplast but outside of it. In pollen grains only it has been suggested that the outermost part of the wall, the exine, may develop by contributions from both sides of the young pollen wall, i.e. from the protoplast within the pollen grain and from the tapetum which surrounds the developing pollen grains.

The cell walls are composed of cellulose, hemi-cellulose, pectic substances, lignin and proteins (structural and enzymatic).

Gross structure of the wall

On the basis of the development and structure of plant tissues it is possible to distinguish the following principal three layers in the cell wall: (a) the *middle lamella* or the *intercellular layer* or *substance*; (b) the *primary wall*; and (c) the *secondary wall* (Fig. 18, nos. 1, 2).

The *middle lamella* is the cement that holds the individual cells together to form the tissues and, accordingly, it is found between the primary cell walls of neighbouring cells. It is an amorphous substance, inactive in term of polarized light (isotropic). In supporting tissues it may also fill the intercellular spaces. The middle lamella consists mainly of pectic substances. The enzyme pectinase and chemical reagents which solubilize pectins disintegrate tissues into their individual cells. This procedure is called *maceration*.

The *primary wall* (P-wall) is the first true cell wall which develops on the new cell. In many cells it is the only cell wall, as the middle lamella is regarded as intercellular substance and not a wall proper. The primary wall is that part of the cell wall that develops in cells or portions of them which are still growing. This wall is optically active (anisotropic).

The *secondary wall* (S-wall) is formed on the inner surface of the primary wall. It begins to develop in cells, or parts of them, that have ceased to grow. The secondary wall is very strongly anisotropic, and layering can be observed in it. The reason for the layering is discussed later in this chapter. In the majority of tracheids and fibres three layers—the *outer layer* (S_1), the *central layer* (S_2), and the *inner layer* (S_3)'can be discerned in the secondary wall. Of these layers the central layer is usually the thickest. In some cells, however, the number of layers may be more than three (Fig. 19). Some authors (e.g. Meier, 1957) use the term *tertiary wall* for the inner layer of the secondary wall. According to Frey-Wyssling (1976) an innermost lamella (*tertiary lamella*) with properties differing from those of the S-wall may be present. He suggests that this lamella may be differentiated into two strata—a membranogenoic stratum and a warty stratum.

It should be mentioned that some investigators use the term compound *middle lamella* when dealing with wood tissue. This term is used to refer to the complexes of lignified layers which appear more or less homogeneous when examined, without pretreatment, under the light microscope. The compound middle lamella may be three-layered when it refers to the middle lamella proper and the adjoining primary walls, or five-layered when it refers to the middle lamella proper, the primary

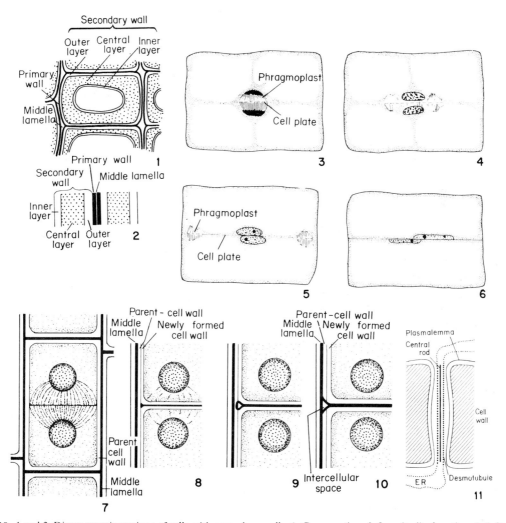

FIG. 18. 1 and 2, Diagrammatic sections of cells with secondary walls. 1, Cross-section. 2, Longitudinal section. 3–6, Stages in cell division showing the development and growth of the cell plate. 7–10, Schematic drawings showing the connection of the newly formed wall to the wall of the cell that underwent division. 11, Diagram of a plasmodesma. (Nos. 1 and 2 adapted from Kerr and Bailey, 1934; nos. 3–6 adapted from Sinnott and Bloch, 1941; nos. 7–10 adapted from Martens, 1937.)

walls, and the outer layer of the secondary walls of the adjoining cells (cf. Kerr and Bailey, 1934).

Formation of the wall

During mitosis, at the telophase, the *phragmoplast* widens and becomes barrel-shaped. At the same time, on the equatorial plane the *cell plate* i.e. the first-evident partition between the new protoplasts, begins to form inside the phragmoplast. In the area where the cell plate forms, the micro-tubules of the phragmoplast disappear but are successively regenerated at the circumference of the cell plate (Fig. 18, nos. 3–5). With the enlargement of the cell plate the microtubules of the phragmo-plast approach the wall of the dividing cell. In very long cells, such as the fusiform cells of the cambium, the cell plate soon reaches the side walls of the dividing mother cell, but contact with the end walls of the cells is delayed and thus it is possible to see the microtubules of the phragmoplast arranged in two zones perpendicular to the longi-tudinal axis of the cell (Fig. 175, nos. 3, 4). In such

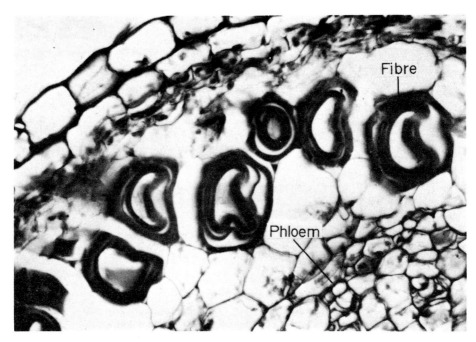

FIG. 19. Outer portion of a cross-section of a young stem of *Linum usitatissimum* showing maturing fibres in which the various layers of the secondary wall have separated from each other during sectioning. × 700.

cells the young nuclei almost reach the resting state, with an envelope and nucleoli, while the cell plate has not yet reached the end walls of the dividing cell. When the cell plate reaches all parts of the existing wall of the dividing cells, the phragmoplast disappears completely. At this stage the viscosity of the cell plate becomes higher. The cell plate gradually undergoes changes to form the intercellular substance referred to as the middle lamella.

With the aid of the electron microscope it has been shown that cell-plate formation is initiated by the concentration and fusion of a large number of vesicles (Fig. 13, no. 2) which are commonly accepted to be derived from dictyosomes (Frey-Wyssling et al., 1964; Esau and Gill, 1965; Hepler and Newcomb, 1967; O'Brien, 1972), but ER vesicles may also be involved in this process. The microtubules of the phragmoplast seem to be involved in directing the vesicles toward the equatorial region.

Individual plasmatic linkages remaining in the gaps of the cell plate form the plasmodesmata (Frey-Wyssling and Mühlethaler, 1965).

On both sides of the middle lamella thin lamellae

are laid down by the daughter protoplasts. Formation of these lamellae is the initial stage in the development of the new walls of the daughter cells. The walls consist of cellulose microfibrils (as described later) and of non-cellulosic matrix. The matrix of the wall consists mainly of pectic substances and hemicelluloses. It has been shown that the wall matrix is also secreted by Golgi vesicles (Northcote and Pickett-Heaps, 1966, and others). Some workers suggest that the ER may also, at least in some cases, play a role in the production of the matrix.

The microtubules of the peripheral cytoplasm are usually orientated parallel to the cellulose fibrils in contact with the plasmalemma, and are considered responsible for the alignment of these fibrils (Green, 1965; Pickett-Heaps, 1967; Itoh and Shimaji, 1976). Recently it has been shown that actin filaments co-operate with the microtubules in controlling the position and orientation of the wall thickenings of differentiating tracheary elements (Kobayashi et al., 1988).

The synthesis of the cellulose microfibrils is carried out by enzymes situated in the plasmalemma (Preston, 1974), which have been suggested to

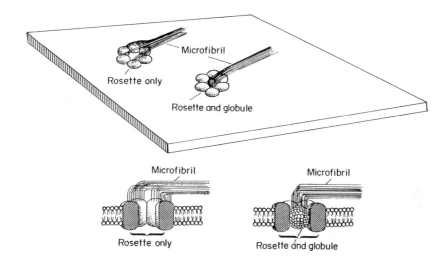

FIG. 20. Diagram presenting a putative cellulose synthesizing complex embedded in the plasma membrane. Example of two possible structures for rosette-type or rosette-globule-type complexes seen in freeze-fracture of some algae and a number of higher plants. (Adapted from Delmer and Stone, 1988.)

occur in the form of rosettes (Fig. 20) (Giddings *et al.*, 1980; Herth, 1984, 1985; Herth and Weber, 1984; Rudolph and Schnepf, 1988).

Along the line of contact of the new wall and the wall of the mother cell, the new and old middle lamella are separated by the primary wall of the mother cell (Fig. 18, no. 8). According to Martens (1937, 1938) the connection between these middle lamellae is effected in the following manner. In the primary wall of the mother cell a cavity, which is triangular in cross-section, develops all along the line of contact of the new and old walls. This cavity continues to enlarge till it reaches the middle lamella of the mother cell and so connection is made between the new and old middle lamellae. If the cavity continues to grow and the intercellular substance does not fill it, an intercellular space lined with intercellular substances is formed (Fig. 18, nos. 7–10). According to Priestley and Scott (1939) the middle lamellae are brought into contact after the stretching wall of the mother cell tears opposite the new wall.

The secondary wall develops on the inner surface of the primary wall. It is also composed of cellulose microfibrils, the matrix consisting of other polysaccharides including hemicellulose. In addition, deposits of lignin, suberin, cutin, waxes, tannins, inorganic salts such as calcium carbonate and calcium oxalate, silica, and other substances may occur in the secondary walls. Generally, the lignin first appears in the intercellular substance and primary wall from where it spreads centripetally through the secondary wall as this wall develops. However, in primary phloem fibres of *Phoradendron flavescens* the primary wall appears to remain largely non-lignified although the secondary wall bcomes lignified (Calvin, 1967). It has been suggested that the endoplasmic reticulum plays a role in lignin biosynthesis (Perez-Rodriguez and Catesson, 1982).

Fine structure of the cell wall

The fine structure of the cell wall, particularly that of the secondary wall, has been intensively studied in the last century. This research was stimulated because of its importance to the fibre, paper, and other industries. The researchers worked in two directions, i.e. from the morphological and physico-chemical approaches. By combining the results of these two fields of research a rather clear

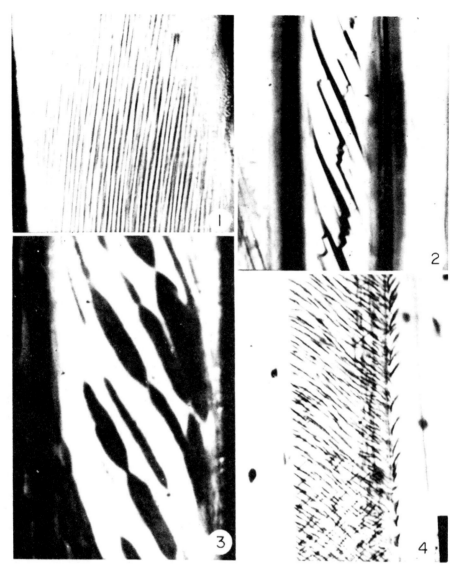

FIG. 21. 1, Striations seen on the surface of a tangential section of the secondary wall of a fibre-tracheid of *Siparuna bifida*. × 1650. 2, A longitudinal section through the secondary wall of a tracheid of *Pinus* showing the plane of mechanical cleavage. × 700. 3, A longitudinal section in the secondary xylem of *Pinus* showing the spiral arrangement of the cavities produced by enzymatic action of fungi on the secondary wall. × 350. 4, Longitudinal section of tracheids of *Larix* showing the orientation of iodine crystallized in the spaces between the microfibrils. × 720. (Courtesy of I. W. Bailey.)

picture of the fine structure of the cell wall has been derived.

Results of the morphological line of research

When fibres and tracheids are examined without special treatment under the ordinary light micro-

scope, layers are visible in cross-section. With special treatment and with the high-power magnification of the light microscope, finer lamellation can be seen. The coarser layers represent groups of lamellae. The lamellae can be concentric, radial, or have a complicated arrangement. When the cell wall is allowed to swell under the influence of suitable reagents, fine structures can be observed within the lamellae themselves (Fig. 21, no. 1). By such methods Bailey and others (Bailey and Kerr, 1935; Bailey and Vestal, 1937a, b; Bailey, 1957) found that the cell wall is built of a system of microscopic threads—the *fibrils*. It has been revealed that the wall consists of two continuous interpenetrating systems, one of which is the cellulose fibrils and the other the continuous system of microcapillary spaces. These spaces may be filled with lignin, cutin, suberin, hemicelluloses, and other organic substances, and even mineral crytals and in fresh tissue aqueous solutions. The material between the fibrils forms the non-cellulosic matrix.

In tracheary elements and sclerenchyma cells the layers seen with the ordinary light microscope are usually the result of the different quantities of lignin, pectic substances, hemicellulose, or other organic substances deposited in the interfibrillar spaces of the cellulose, or they may be due to the presence, in certain cells, of layers poor in cellulose, or of the different orientation of the microfibrils in the various wall layers.

The lamellation seen in the secondary wall is often the result of the different density of the fibrils. In the denser, darker areas, the fibrils are more numerous per unit area and they are more tightly packed. In the less dense, lighter areas, the fibrils are looser and the capillary spaces between the fibrils are larger.

Most fibres have lignified walls. Electron microscope studies of such fibres have confirmed previous assumptions that each lamella is composed of two parts, one consisting mainly of cellulose and one predominantly of lignin. It has also been suggested that each lamella is produced during a period of 24h (Anderson and Kerr, 1938; Anderson and Moore, 1937; Casperson, 1961a, b). Bobák and Nečesaný (1967) concluded that the two main components of the secondary wall are deposited at different periods of the day—cellulose in the afternoon, lignin after midnight. The lignin permeates the cellulosic part formed previously, impregnates it, and cements it to the preceding lamella.

In heavily lignified walls it is possible to dissolve the cellulose and retain the lignin only, or the lignin alone can be dissolved and the cellulose retained. In this way the component retained gives, as it were, a negative image of the component which has been dissolved. This phenomenon not only proves that the lignin is found in the elongated inter-fibrillar spaces of the cellulose but also that these capillary spaces are continuous.

In order to demonstrate the presence of the two parallel, three-dimensional systems, the cellulosic fibrillar network and the network of interfibrillar microcapillary spaces. Bailey also used methods other than those described above. For example, he succeeded in crystallizing iodine in the elongated, microcapillary spaces (Fig. 21, no. 4), thus demonstrating their presence and the orientation of the fibrils in the different layers of the wall.

Further clarification of the fine structure of the cell wall is based on the use of the electron microscope. The photographs, which were made with the electron microscope (Fig. 22, nos. 1, 2), revealed the fine *microfibrils* which cannot be seen by means of the ordinary light microscope. The results of research with the electron microscope have in general confirmed the theories of Bailey on the structure of the wall.

The morphological structure of the cellulose in the cell wall, as is known today, can be summaried in the following way. Within the cell wall differently built lamellae are recognized, each of which consists of fibrils. By using certain techniques it is possible to distinguish macrofibrils with the aid of the light microscope. These form a three-dimensional network. The network is interwoven with a parallel network of microcapillary spaces occupied by non-cellulose substance. The width of the macrofibrils (bundles of microfibrils) can be as much as 0.4 or 0.5 μm (Wilder, 1970); Frey-Wyssling, 1976), and that of the microfibrils 20–30 nm. The microfibrils have recently been found to be fasciations of *elementary fibrils* which are 3–5 nm thick.

Results of the physico-chemical line of research

The cellulose molecule consists of long chains of linked glucose residues. The chain molecules are arranged in bundles which were termed *micellae*. The hypothesis of the presence of micellae was proposed by Nägeli in the last century. According to him the micellae are the individual units arranged in a permanent order within an intermicellar matrix. With the aid of the polarizing

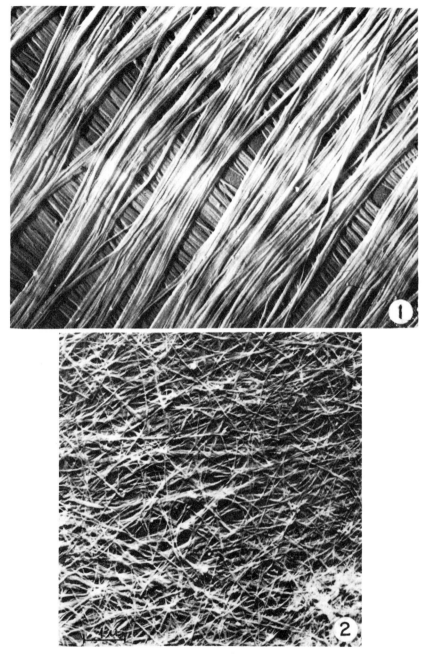

FIG. 22. 1. Electron micrograph showing the structure of the secondary wall of *Valonia*. × 10,500. 2, As above, but of the primary wall. × 12,000. (From Steward and Mühlethaler, 1953.)

microscope the crystal-like nature of the micellae was proven. From the results of various investigations, especially those made with X-rays, investigators came to the conclusion that the micellae consist of parallel chains of glucose residues which have characteristic and permanent distances between them. As a result of extensive research carried out by botanists, chemists, and physicists, several theories were suggested which attempted to explain the organization of the cellulose molecules in the cell wall (Roelofsen, 1959; Albersheim *et al.*, 1973; Albersheim, 1975; Frey-Wyssling, 1976). According to Frey-Wyssling and Mühlethaler (1965) the chain-like cellulose molecules are regularly arranged in bundles. Each such bundle, which forms an elementary fibril, consists of about 40 cellulose molecules and is about 3.5 nm wide and 3 nm thick. The elementary fibril is in greatest part crystalline. Only very small parts of it, which are presumably arranged at random, may be paracrystalline. The number of glucose residues in cellulose molecules of fibre cells was found to vary from

500 to 10,000 and the length of these molecules varies from 0.25 to 5 μm (Fig. 23). Most of the above is based on the results of research made on the secondary cell wall, but recently much attention had been paid to the structure of the primary wall. The primary wall is similar in structure to the secondary wall in that it consists of anisotropic (crystalline) cellulose microfibrils and a non-cellulosic matrix. Sometimes, as in the Phyco-mycetes, the microfibrils consist of chitin (Frey-Wyssling and Mühlethaler, 1950; Roelofsen, 1951) or of other substances. The interfibrillar matrix

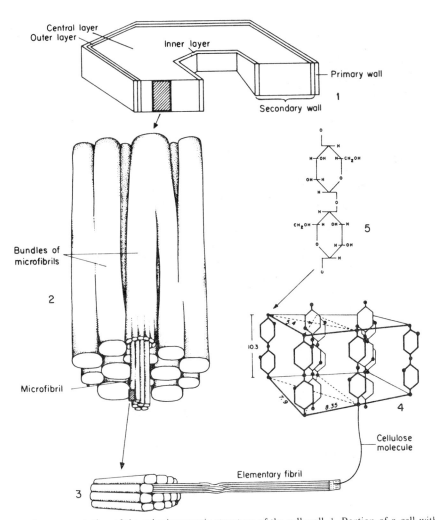

FIG. 23. Diagrammatic representation of the submicroscopic structure of the cell wall. 1, Portion of a cell with secondary wall layers. 2, Bundles of microfibrils which when swollen may be seen in the light microscope. 3, Portion of a microfibril composed of elementary fibrils. 4, Two unit cells of cellulose, as suggested by Meyer and Mark (Preston, 1952). 5, Two glucose residues.

usually contains pectic compounds and hemi-cellulose.

The outstanding mechanical property of the cellulose is tensile strength, while under compressive stress cellulose fibrils bend. This effect is prevented in supporting cells by replacement of the plastic interfibrillar matrix in the cell wall by solid substances which harden it. This introduction into the cell wall of additional substances in place of the matrix in which the cellulose framework is embedded is termed *incrustation*. The most important incrusting process in higher plants is *lignification*, but in many cells, substances such as suberin, cutin, waxes, quinones, tannins, and other organic and material substances may incrust the walls.

The primary wall of many cells has a lamellate structure. The angle at which the microfibrils cross in the primary wall differs in different cells as well as in the various lamellae and in different parts of the wall of a single cell (Fig. 22, no. 2).

The classical debate as to whether the growth of the cell wall is accomplished by *intussusception* or by *apposition* is still continued. According to the first opinion the material of the new wall is laid down between particles of the existing substance of the expanding wall. According to the second opinion the growth is due to the centripetal addition of new layers one upon the other.

New theories, based on electron microscopic research, have been developed as to the manner in which the primary wall grows. Frey-Wyssling and Stecher (1951), for instance, suggested that the primary cell wall grows in a way that has been termed *mosaic growth*. According to this view the fibrillar texture in certain wall areas becomes loosened as a result of turgor pressure and afterwards mended by deposition of new microfibrils in the gaps caused by the strain. The loosening of the fibrillar network requires the plasticizing of the wall matrix in the expanding area. Growth hormones available to the cell and proteins enzymes occurring within the cell wall are involved in regulating the progress of extension growth.

Another concept of growth is the theory of *multinet growth* (Roelofsen and Houwink, 1951; Houwink and Roelofsen, 1954). According to this theory the thickening and increase in surface area of the primary wall is brought about, in many cases, by the separation of the crossed microfibrils and alteration in their orientation, in the earliest formed lamellae, from being almost transverse to almost longitudinal. New lamellae with denser, crossed, and almost transversely orientated microfibrils are added centripetally (Fig. 24). This theory was strengthened by results of a recent study carried out with the aid of new methods for determining microfibril orientation (Sassen and Wolters-Arts, 1986).

Regarding the cellulose microfibrillar system it can be assumed that according to the multinet growth theory the fibrils are added to the growing cell wall by apposition, whereas in view of the mosaic growth concept intussusception also takes place. It has, of course, to be kept in mind that the wall matrix (hemicelluloses and pectins) is continously secreted not only into the lamellae adjacent to the cytoplasm but also into the outer lamellae.

Orientation of microfibrils, micellae, and cellulose chains

In order to discover the orientation of the microfibrils, the micellae, and the cellulose chains in the

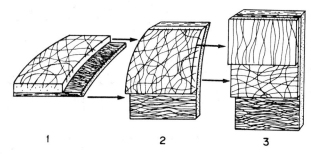

1   2   3

FIG. 24. Diagram of multinet growth showing passive changes in fibril orientation in the outer wall layers. Arrows indicate the corresponding layers with their textural changes. (Adapted from Frey-Wyssling and Mühlethaler, 1965.)

different layers of the cell wall, and particularly in the secondary wall, many investigations have been made (Bailey and Kerr, 1935, 1937; Bailey and Vestal, 1937a, b; Bailey and Berkley, 1942; Wardrop and Preston, 1947; Frey-Wyssling *et al.*, 1948; Frey-Wyssling, 1948, 1950; Bailey, 1957; Wardrop, 1954, 1958; and others). These investigations were based on all of the methods, both direct and indirect, that have been described here. In general, the results obtained by different methods of investigation on the same object have proved similar. The orientation of the microfibrils and of the bundles of microfibrils in two layers can be determined simultaneously in specially prepared sections (Fig. 22, no. 1). the orientation of iodine crystals in the interfibrillar spaces can also be seen easily with high magnification of the ordinary light microscope. Electron micrographs are also of value where the axes of the cells can be marked on the small portion of the wall seen in them.

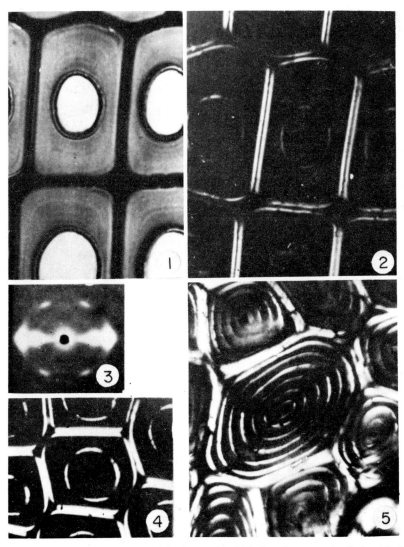

FIG. 25. 1, Micrograph of a cross-section of tracheids of *Pinus* in which three layers can be distinguished in the secondary wall. 2, As above, but as seen under the polarizing microscope. 3, X-ray diffraction pattern of delignified wood of *Pinus longifolia*. 4, As in no. 3, but as seen in a polarizing microscope. 5, Micrograph of a cross-section of fibres of *Pandanus* as seen in the polarizing microscope showing the numerous concentric layers in the secondary cell wall. (Courtesy of I. W. Bailey.)

When a thin cross-section through tracheids is examined with a polarizing microscope while the two polarizers are crossed, certain layers of the wall appear bright and others dark (Fig. 25, nos. 2, 4, 5). In the brightest layers, those with the strongest birefringence, the longitudinal axes of the crystals of cellulose are parallel to the surface of the section, i.e. perpendicular to the longitudinal axis of the tracheid. In the layers in which the cellulose crystals are perpendicular to the surface of the section the passage of light is not affected and the layers remain dark between crossed prisms. In these layers the crystals are parallel to the longitudinal axes of the tracheids. The bright layers, however, are not continuous on the circumference but are interrupted in four places. The brightest sections of such a layer are those that lie at an angle of 45° to the axis of the analyser and the polarizer of the microscope, and the darkest areas of the same layer are approximately parallel to these axes. The birefringence is only apparent when the longitudinal axes of the crystals are at an angle of 45° to the axes of the crossed analyser and polarizer. By the study of oblique sections, cut at different angles or on the basis of accurate calculations of the degree of the birefringence, it is possible to use this method to determine the accurate orientation of the cellulose crytals in the various layers of the wall (Preston, 1952).

From the results obtained from investigations using X-rays, the orientation of the cellulose crystals in the different layers cannot be determined; only conclusions as to the average orientation of the different wall layers of a large number of cells can be drawn. In order to make X-ray photographs, sections of tissue about 1 mm thick are used. The angle between the longitudinal axis of the cellulose crystals and that of the cell is calculated according to size of the preferred orientation of X-ray diffraction spots (Fig. 25, no. 3).

From the results of the above investigations, it is seen that the orientation of the microfibrils and the micellae in the secondary walls differs in different plants, in the cells of the different plant organs, in the different layers of the same cell wall, and sometimes even in the different lamellae of the same layer. In the walls of many vessel members, tracheids, and fibres that have been studied and that have three layers to the secondary wall, the

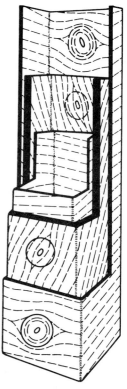

FIG. 26. Three-dimensional schematic drawing of a tracheid to show various secondary wall layers and the orientation of the microfibrils in them and around the bordered pits. The helicoidal orientation is indicated by broken lines. (Adapted from Brown *et al.*, 1949.)

following arrangement of the microfibrils and micellae has been found: the orientation in the outer and inner layers is almost horizontal or the microfibrils are orientated in a very low spiral; and the orientation in the central layer is almost parallel to the longitudinal axis of the cell, the microfibrils being arranged in a steep spiral (Fig. 26).

Bailey and Vestal (1937a) found that in the outer wall layer of the tracheids of early wood of conifers the orientation of the cellulose microfibrils around the large bordered pits is circular, while the microfibrils in the central layer are only slightly deflected around the pits.

In cotton fibres the largest portion of the secondary wall consists of helicoidally arranged microfibrils that are orientated at an angle of 45° or less to the longitudinal axis of the cell. In flax fibres the orientation of the microfibrils is helicoidal, but the direction of the helix in each of the numerous overlying layers opposes that in the adjacent layers.

Results of recent research led some authors to the conclusion that the morphogenesis of helicoidal walls is both very defined and flexible. Therefore it is adapted to varied programmes of differentiation and to different environmental conditions (Roland *et al.*, 1987).

Properties of the wall related to its structure

As has been mentioned previously, the wall in a cross-section appears to be built of layers, because of the different composition and structure of the lamellae which are added continuously as the wall grows. The difference in the structure is brought about, as explained above, by the differences in density and orientation of the cellulose microfibrils, the presence of different quantities of lignin, etc. These features result in differences in the refraction of light so that the layers of the wall are emphasized. Among the physical properties of the cell wall are tensility, strength, resistance to compression, swelling, and permeability properties (see Preston, 1974; Frey-Wyssling, 1976).

Because of the large amount of cellulose in the wall, the properties of the wall are mainly determined by those of cellulose; the other substances add some characteristics, or alter slightly, those of the cellulose. One of the important properties of cellulose is its ability to withstand stretching because of its elasticity. The lignin increases the resistance of the wall to pressure and so prevents the folding of the cellulose microfibrils. The orientation of the microfibrils in the different lamellae of the wall no doubt is an important factor in determining the strength of the wall.

Special structures of cell walls

Primary pit fields

The primary walls of young cells stretch and increase in surface area and thickness as the cell grows. However, certain areas of the wall generally remain thin; these portions are termed *primary pit fields* occasionally also called *primary pits* or *primordial pits* (Fig. 27, nos. 1, 2). Sometimes, when the primary pit fields are numerous and deeply sunken, the wall in which they occur appears beaded in cross-section.

A characteristic feature of the primary pits of living cells is the presence of concentrations of very thin protoplasmic strands, i.e. the *plasmodesmata*. The plasmodesmata connect the protoplasts of neighbouring cells (Fig. 9, no. 1; Fig. 11, no. 3). They are present almost in all living cells of higher plants, the living protoplats of which are, therefore, united to form a single unit. As a result of this the plant body can be divided into two major compartments: The interconnected protoplasts, called *symplast*, and the compartment external to the protoplasts, the *apoplast*. The latter consists of the cell walls, intercellular spaces, and the lumen of non-living cells, such as tracheary elements and most fibres. The diameter of the plasmodesmata is at the limit of resolution of the light microscope, and they can be observed only by suitable staining or by impregnation with heavy metals such as silver or mercury. Plasmodesmata are relatively easily seen in the endosperm of seeds, such as those of *Phoenix, Aesculus,* and *Diospyros* and in the cotyledons of some plants. On plasmolysis the protoplast withdraws usually from the wall in all places except where plasmodesmata are present.

Plasmodesmata usually occur in groups, but sometimes they may be more or less evenly distributed over the entire wall. When in groups the plasmodesmata usually occur in the primary pit fields as has been demonstrated in cambial cells (Kerr and Bailey, 1934; Livingston and Bailey, 1946). In mature living cells with secondary walls large groups of plasmodesmata traverse the pit membranes.

On the basis of electron microscope studies (Fig. 11, no. 3; Fig. 18, no. 11) it has recently been suggested that in the intercellular canals housing the plasmodesmata the plasmalemma of the two neighbouring cells is continuous and that through the centre passes a tubule, called *desmotubule*, which is considered to be a derivative of the ER and which was observed to be continuous with it. The desmotubules may contain an axial *central rod*. The space between the desmotubule and the inner face of the plasmalemma is called *cytoplasmic annulus*. The diameter of the lumen of plasmodesmata is about 30–60 nm. The external diameter

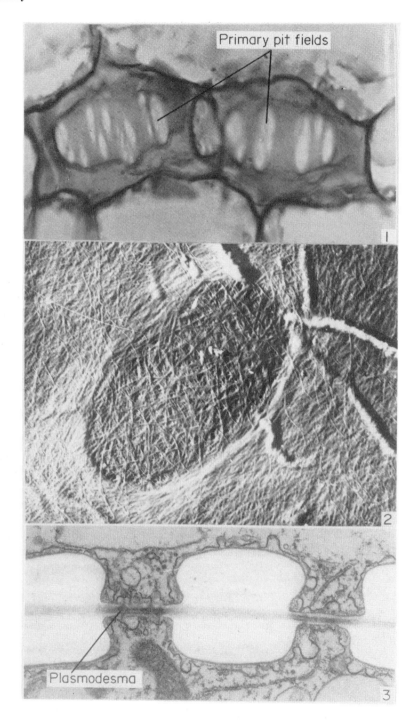

FIG. 27. 1, Micrograph of parenchyma cells from the pith of *Nicotiana tabacum* showing primary pit fields. × 950. 2, Electron micrograph of a primary pit field of *Zea mays.* × 30,000. 3, Simple pit-pairs of xylem ray cells of *Citrus*. Plasmodesmata crossing the pit membrane are seen. × 20,000. (No. 2 from Muhlethaler, 1950.)

of the desmotubule is 16–20 nm (Gunning and Robards, 1976; Gamalei, 1985).

The following explanation of the origin and development of the plasmodesmata has been given by Frey-Wyssling (1959, 1976). The nature of the cell plate is unknown but without doubt it is partly protoplasmic. It is thought that young growing walls are penetrated by cytoplasm. With the accumulation of the cellulose microfibrils (Fig. 27, no. 2) and pectic substances in the wall, the cytoplasmic connections gradually become narrower until they constitute thin threads, i.e. the plasmodesmata.

There is a controversy in the literature as to the possibility that new plasmodesmata can be produced between cells or portions of cell after completion of the cell plate, e.g. during intrusive growth, in graft junctions, between tyloses, between parasitic and host cells, etc.

In a variety of cells branched plasmodesmata have been observed to occur, e.g. between companion cells and sieve elements, in pit membranes of living fibres, e.g. of *Tamarix* and at other sites. According to Krull (1960) the branched plasmodesmata of the longitudinal walls of the cortical cells in *Viscum album* develop during wall extension, the branches being secondarily divided plasmodesmata. The branches depart from a median cavity or nodule to the cells on both sides. There is also a possibility that such branches are primary plasmodesmata which became joined by dissolution of a median portion of the common wall of neighbouring cells.

Plasmodesmata play an important role in the transport of materials and the relay of stimuli. It is also thought that viruses can pass from one cell to another via the plasmodesmata (Esau, 1961).

(For further reading on plasmodesmata, see Gunning and Robards, 1976; Gamalei, 1985.)

Pits

Certain portions of the cell wall remain thin even as the secondary wall is formed and they, therefore, consist only of primary wall material. These areas, which are of variable shape, are called *pits* (Fig. 28, nos. 1–6). Some authors use the term pit to refer only to the pit cavity together with the primary wall, which closes the pit. Others use the term pit to refer to the above structures together with that part of the secondary wall that surrounds the pit cavity. Pits can develop over the primary pit fields and then one or more pits may develop in the pit field, or pits develop on those parts of the primary wall devoid of pit fields. On the other hand, the primary pit fields can become completely covered by the secondary wall.

The pits are areas through which substances may pass from cell to cell. The concentration of plasmodesmata in living cells in the region of the pit membranes is an additional proof of the pit being a channel of exchange. Generally each pit has a complementary pit exactly opposite it in the wall of the neighbouring cell. Such pits form a morphological and functional unit called the *pit-pair* (Fig. 28, no. 2). The cavity formed by the break in the secondary wall is called the *pit cavity*. The membrane, built of the primary cell walls and middle lamella, that separates the two pit cavities of the pit-pair is called the *pit membrane* or *closing membrane*. The opening of the pit on the inner side of the cell wall, i.e. on that side facing the lumen of the cell, is called the *pit aperture* (Fig. 28, no. 2).

Two principal types of pits are recognized— *simple pits* and *bordered pits* (Fig. 28, nos. 1–7). The main characteristic of bordered pits is that the secondary wall develops over the pit cavity to form an overarching roof with a narrow pore in its centre. In a simple pit no such development of the secondary wall is present.

If the two pits of a pair are simple, a *simple pit-pair* is formed (Fig. 27, no. 3; Fig. 28, no. 2); if the two pits are bordered, a *bordered pit-pair* (Fig. 28, no. 3); if one of the pits is simple and the other bordered, a *half-bordered pit-pair* (Fig. 28, no. 4). If the pit has no complementary pit in the adjacent cell or if it is opposite an intercellular space it is termed a *blind pit* (Fig. 28, no. 1). Sometimes two or more pits are found opposite one large pit— such an arrangement is called *unilateral compound pitting*.

The pit cavity of a simple pit may have the same diameter over its entire depth, or it may widen or narrow towards the pit aperture. In places where the secondary wall is very thick, the pit cavity has the form of a canal. Sometimes this canal is branched towards the outer layers of the cell wall

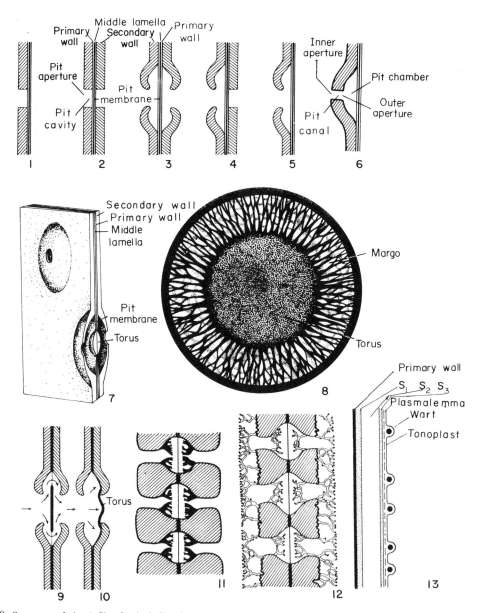

FIG. 28. Structure of pits. 1, Simple pit. 2, Simple pit-pair. 3, Bordered pit-pair. 4, Half-bordered pit-pair. 5 and 6, Bordered pits. 7, Three-dimensional diagram of a portion of the adjacent walls of two tracheids showing the structure of bordered pit-pairs. 8, Diagram of pit membrane and torus of *Pinus* showing the perforations in the membrane. 9 and 10, Longitudinal sections of bordered pit-pairs of a tracheid. Arrows indicate direction of water flow. 9, Torus and membrane in median position. 10, Torus closing one of the pit apertures. 11 and 12, Longitudinal section of the wall of adjacent vessels with vestured pits. 13, Diagram of a wall section showing a warty layer. (Nos. 8–12 after Bailey, 1913; no. 13 adapted from Liese, 1965b.)

and then the pit is called a *branched simple pit*. Such pits arise from the fusion of several pits during the centripetal addition of layers to the secondary wall (Fig. 51, no. 5).

Simple pits are usually found in parenchyma cells with thickened walls, in libriform fibres, and in sclereids. Bordered pits are found in the tracheary elements and in fibre-tracheids.

The bordered pit is more complicated than the simple in its structure and is variously shaped. In the bordered pit that part of the pit cavity that is formed by overarching of the secondary wall is called the *pit chamber* and the opening in the secondary wall that faces the cell lumen is called the *pit aperture*. If the secondary wall is very thick a canal—the *pit canal*—is formed between the cell lumen and the pit chamber. In the pit canal two openings are distinguished—that facing the cell lumen is termed the *inner aperture* and that nearest the pit chamber, the *outer aperture* (Fig. 28, no. 6).

In some plants there are bordered pit-pairs in which the pit membrane is thickened in its central portion; this thickening, which is of a primary nature, is disc-shaped and is termed the *torus* (Fig. 28, nos. 7–10). The diameter of the torus is wider than that of the pit aperture (see Liese, 1965a).

Bannan (1941) describes the occurrence of thickenings, other than the torus, on the pit membrane. These thickenings may be radial or tangential in relation to the torus. The torus found in the bordered pits of *Cedrus* is fringed or scalloped on its circumference (Fig. 185, no. 2). This feature is a characteristic of *Cedrus*, and as such it aids in the identification of the wood of this genus.

In tracheids of many conifers the pit membrane around the torus—the *margo*—is porous (Fig. 28, no. 8; Fig. 186). The presence of these pores was discovered in 1913 by Bailey in experiments that demonstrated the passage of a suspension of finely divided particles of carbon from one tracheid to another. This has been confirmed by electron micrographs (Liese and Fahnenbrock, 1952; Liese, 1954, 1965a; Frey-Wyssling et al., 1956; Côté and Krahmer, 1962). During differentiation of the margo the cell wall matrix is dissolved so that only the cellulosic microfibrils remain. The pit membrane is usually flexible and, under certain conditions, the torus can be pushed against one of the pit apertures (Fig. 28, nos. 9, 10). When the torus is in the median position, i.e. in the middle of a pit-pair, water can easily pass from one tracheid to another. In a pit-pair where the torus is in a lateral position, i.e. pressed against one of the pit apertures, the passage of water is prevented. Most of the tori in late wood and all of them in the heartwood are always in a lateral position and the flexibility of the pit membrane is lost. In this condition the pit is called *aspirate* .

During differentiation of the bordered pit the cellulose microfibrils of the margo become organized into a system of coarse radially oriented bundles. At the same time circular oriented fibrils form the origin of the torus (Frey-Wyssling et al., 1956). The microfibrillar system of the margo and the torus of differentiating tracheids is embedded in amorphous matrix, consisting of hemicelluloses and pectins. During the process of maturation the amorphous embedding substances disappear apparently by enzymatic action (Imamura et al., 1974) and the margo becomes perforated. Removal of non-cellulosic polysaccharides from the pit membranes of differentiating tracheary elements of dicotyledons has also been reported (O'Brien, 1970; Benayoun, 1983).

The presence of a torus is especially characteristic of the bordered pits of the Gnetales, of *Ginkgo*, and most of the Coniferales. Tori occur only rarely in the Ophioglossales (Bierhorst, 1960) and in the angiosperms.

In some dicotyledons thin, simple, or branched sculpturings are present on the secondary wall that forms the pit chamber or around the pit aperture. Such pits are called *vestured pits* (Fig. 28, nos. 11, 12) and the sculpturings may have various shapes. Because of the special properties of light refraction and of staining these pits appear as if porous or net-like in surface view and, thus, were once termed sieve-pits. Vestured pits are found in the tracheary elements of the secondary wood of certain dicotyledonous genera and species, such as some of the Leguminosae, Cruciferae, Myrtaceae, and Caprifoliaceae. Vestured pits are found in phylogenetically more developed xylem and therefore they are considered to be an advanced form of pit.

The shape of the pit aperture can be the same as that of the pit chamber or different from it. The pit aperture may be round, elliptic, or linear. As the walls continue to thicken, the pit chamber becomes smaller and the pit canal between the inner and outer apertures becomes longer. In such pits the inner aperture often becomes long and narrow, as seen in surface view, and, in very thick walls, its longitudinal axis may be longer than the diameter of the pit chamber. When the inner aperture is large and linear, narrow, or elliptic, and when the outer aperture is small and circular, the pit canal has the shape of a flattened funnel. The elongated inner apertures of such a bordered pit-pair may be

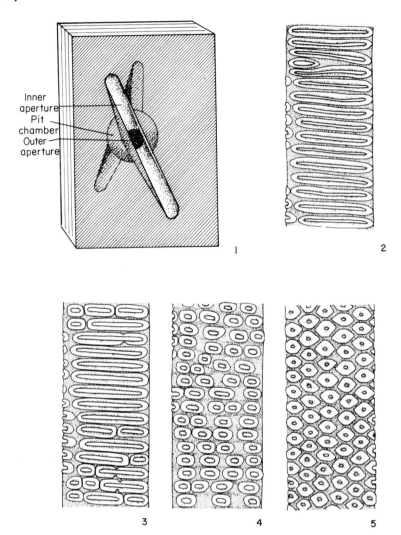

FIG. 29. 1, Portion of the common wall between two fibre-tracheids showing the type of bordered pit characteristic of these elements. 2–5, Types of pitting. 2, Scalariform pitting. 3, Transition from elongated pits in scalariform arrangement to shorter circular pits in opposite arrangement. 4, Opposite pitting. 5, Alternate pitting.

parallel or crossed. This type of pit occurs mainly in fibre-tracheids (Fig. 29, no. 1).

Bordered pits found in tracheary elements vary in shape and arrangement. When the pits are distinctly elongated or linear and arranged in ladder-like tiers the arrangement is termed *scalariform pitting*. When the pits are circular or only slightly elongated and the outlines are elliptic there are two possible ways in which they may be arranged on the wall: in horizontal lines, i.e. *opposite pitting*, or in diagonal lines, i.e. *alternate pitting*. When the pits are crowded the outline of the opposite pits becomes rectangular or square, and

that of the alternate pits hexagonal (Fig. 29, nos. 2–5).

Tracheary elements have especially well-developed bordered pits in those regions where they are adjacent to other tracheary elements. In the regions of contact with parenchymatous cells reduced bordered pits are sometimes found. Usually those parts of the wall adjacent to fibres are devoid of pits.

Other sculpturings on the cell wall

In addition to the pits many other sculpturings

exist on the cell walls. These include, for example, the perforations in the end walls of the vessel members, various thickenings on the inner surface of cell walls, such as wall thickenings in the protoxylem elements, spiral thickenings on the inner surfaces of pitted secondary walls, Casparian strips of endodermal cells, thickenings in the walls of the endothecial cells of pollen sacs, and external projections formed partly by the wall itself and partly by deposits, e.g. of cuticle on epidermal cells and of external layers on spores and pollen grains. The above features are discussed in later chapters. Here only three structures will be discussed—*crassulae, trabeculae,* and *wart structures.*

Crassulae are linear or crescent-shaped thickenings of the primary wall and middle lamella which occur between bordered pits or small groups of these pits. The crassulae may sometimes surround the pits. They represent the borders of the primary pit fields of the young cell from which the element developed. Crassulae are well developed in the tracheids of certain gymnosperms (Fig. 183, no. 4).

Trabeculae are rod-shaped thickenings of the wall which traverse the cell lumen radially. They usually appear in radial rows in the wood elements.

Wart structures are structures that have been observed on the inner surface of the secondary wall of conifer tracheids and of fibres and vessels of many dicotyledons (Wardrop *et al.,* 1959; Liese and Ledbetter, 1963; Liese, 1965b; Czaninski, 1967). The diameter of these structures varies between 0.1 μm and 0.5 μm. They develop after or towards the completion of differentiation and lignification of the secondary wall, and according to Wardrop and Davies (1962) they consist of remnants of the protoplast (Fig. 28, no. 13). Frey-Wyssling (1976) considers the wart content as remnants of lignin precursors brought to the inner wall surface towards the end of tracheid differentiation. Not being used, these remnants are deposited and polymerized in the warts.

Cystoliths

In some dicotyledonous families, such as the Moraceae and Urticaceae, stalked outgrowths of the wall that project into the cell lumen are present. These outgrowths are called *cystoliths.* They consist of cellulose and are impregnated with calcium carbonate. Cystoliths are irregular in shape and sometimes they almost completely fill the cell. Cystoliths may appear in parenchymatous cells in various parts of the plant including even the xylem and phloem rays, but they are usually found in the epidermis, in hairs or special large cells which are termed *lithocysts* (Fig. 77, no. 1; Fig. 128, no. 2).

REFERENCES

ALBERSHEIM, P. (1975) The walls of growing plant cells. *Scient. Am.* **232,** April: 80–95.

ALBERSHEIM, P., BAUER, W. D., KEEGSTRA, K., and TALMADGE, K. W. (1973) The structure of the wall of suspension cultured sycamore cells. In: *Biogenesis of Plant Cell Wall Polysaccharides* (ed. F. Loewus). Academic Press, New York, pp. 117–47.

ALBERTS, B., BRAY, D., LEWIS, J., RAFF, M., ROBERTS, K. and WATSON, J. D. (1983) *Molecular Biology of the Cell.* Garland, New York and London.

ANDERSON, D. B. and KERR, T. (1938) Growth and structure of cotton fibre. *Ind. Eng. Chem.* **30:** 49–54.

ANDERSON, D. B. and MOORE, J. H. (1937) The influence of constant light and temperature upon the structure of the walls of cotton fibers and colenchymatous cells. *Am. J. Bot.* **24:** 503–7.

ARNOTT, H. J. and PANTARD, F. C. F. (1970) Calcification in plants. In: *Biological Calcification: Cellular and Molecular Aspects* (ed. H. Schraer). Appleton–Century–Crofts, New York, pp. 375–446.

BAAS, P., ESSER, P. M., VAN DER WESTERN, M. E. T. and ZANDEE, M. (1988) Wood anatomy of the Oleaceae. *IAWA Bull.* **9:** 103–182.

BAILEY, I. W. (1913) The preservative treatment of wood: II, The structure of the pit membranes in the tracheids of conifers and their relation to the penetration of gases, liquids and finely divided solids into green and seasonal wood. *For Quart.* **11:** 12–20.

BAILEY, I. W. (1957) Aggregation of microfibrils and their orientations in the secondary wall of coniferous tracheids. *Am. J. Bot.* **44:** 415–18.

BAILEY, I. W. and BERKLEY, E. E. (1942) The significance of X-rays in studying the orientation of cellulose in the secondary wall of tracheids. *Am. J. Bot.* **29:** 231–41.

BAILEY, I. W. and KERR, T. (1935) The visible structure of the secondary wall and its significance in physical and chemical investigations of tracheary cells and fibres. *J. Arnold. Arb.* **16:** 273–300.

BAILEY, I. W. and KERR, T. (1937) The structural variability of the secondary wall as revealed by "lignin" residues. *J. Arnold Arb.* **18:** 261–72.

BAILEY, I. W. and VESTAL, M. R. (1937a) The orientation of cellulose in the secondary wall of tracheary cells. *J. Arnold Arb.* **18:** 185–95.

BAILEY, I. W. and VESTAL, M. R. (1937b) The significance of certain wood-destroying fungi in the study of the enzymatic hydrolysis of cellulose. *J. Arnold Arb.* **18:** 196–205.

BANNAN, M. W. (1941) Variability in wood structure in roots of native Ontario conifers. *Bull. Torrey Bot. Club* **68:** 173–94.

BATE-SMITH, E. C. and METCALFE, C. R. (1957) Leucoanthocyanins. 3. The nature and systematic distribution of tannins in dicotyledonous plants. *J. Linn. Soc. Bot.* **55:** 669–705.

BENAYOUN, J. (1983) A cytological study of cell wall hydrolysis in the secondary xylem of poplar (*Populus italica* Moench). *Ann. Bot.* **52:** 189–200.

BEN-SHAUL, Y., TREFFRY, T., and KLEIN, S. (1968) Fine structure studies of carotene body development. *J. Microscopie* **7:** 265–74.

BIERHORST, D. W. (1960) Observations on tracheary elements. *Phytomorphology* **10:** 249–305.

BOBÁK, M. and NEČESANÝ, V. (1967) Changes in the formation of lignified cell wall within a twentyfour hour period. *Biol. Plant.* **9:** 195–201.

BOSABALIDIS, A. M. (1987a) Origin, differentiation and cytochemistry of membrane-limited inclusion bodies in leucoplasts of leaf epidermal cells of *Origanum dictamnus L. Cytobios* **50:** 77–88.

BOSABALIDIS, A. M. (1987b) Origin, Ultrastructural estimation of the possible manners of growth and morphometric evaluation of calcium oxalate crystals in non-idioblastic parenchyma cells of *Tamarix aphylla* L. *J. Submicroscop. Cytol.* **19:** 423–432.

BROWN, H. P., PANSHIN, A. J., and FORSAITH, C. C. (1949) *Textbook of Wood Technology.* McGraw-Hill, New York.

BRUNI, A., DALL'OLIO, G. and TOSI, B. (1982) A study of the development of raphide-forming cells in *Musa paradisiaca* using fluorescence microscopy. *New Phytol.* **91:** 581–587.

BUVAT, R. (1969) *Plant Cells—An Introduction to Plant Protoplasm.* World University Library, London.

BUVAT, R. and ROBERT, G. (1979) Vacuole formation in the actively growing root meristem of barley (*Hordeum sativum*). *Am. J. Bot.* **66:** 1219–1237.

CALVIN, C. L. (1967) The vascular tissues and development of sclerenchyma in the stem of the mistletoe *Phoradendron flavescens. Bot. Gaz.* **128:** 35–59.

CARLQUIST, S. (1962) *Comparative Plant Anatomy.* Holt, Reinhart & Winston, New York.

CASPERSON, G. (1961a) Über die Bildung von Zellwänden bei Laubhölzern. II. Der zeitliche Ablauf der Sekundärwandbildung. *Z. Bot.* **49:** 289–309.

CASPERSON, G. (1961b) Licht- und elektronmikroskopische Untersuchungen über den zeitlichen Ablauf der Zellwandbildung bei Laubhölzern. *Ber. dt. bot. Ges.* **74:** 271–3.

CHENICLET, C. and CARDE, J. P. (1988) Differentiation of leucoplasts—Comparative transition of proplastids to chloroplasts or leucoplasts in trichomes of Stachys lanata leave. *Protoplasma* **143:** 74–83.

CHESSEN, G. and FAHN, A. (1988) Cell hypertrophy in stems of *Pinus halepenis* infested by *Matsucoccus josephi. Protoplasma* **143:** 111–117.

CHRISPEELS, M. J. (1980) Endoplasmic reticulum. In: The *Biochemistry of Plants—A Comprehensive Treatise*, Vol. 1 (ed. N. E. Tolbert). Academic Press, New York, pp. 389–412.

CLOWES, F. A. L. and JUNIPER, B. E. (1968) *Plant Cells.* Blackwell, Oxford.

COLMAN, B. (1977) Microbodies. In: *The Molecular Biology of Plant Cell.* Botanical Monographs. Blackwell Scientific Publications, Oxford, **14:** 136–59.

CÔTÉ, W. A. and KRAHMER, R. L. (1962) The permeability of coniferous pits demonstrated by electron microscopy. *Tappi* **45:** 119–22.

COX, G. C. and JUNIPER, B. E. (1973) Autoradiographic evidence for paramural-body function. *Nature New Biol.* **243:** 116–117.

CRESTI, M., CIAMPOLINI, F., PACINI, E., and SARFATTI, G. (1978) Phytoferritin in plastids of the style of *Olea europaea* L. *Acta bot. neerl.* **27:** 417–23.

CZANINSKI, Y. (1967) Observations infrastructurales sur les fibres libriformes du xylème du *Robinia pseudo-acacia. CR Acad. Sci. Paris,* Ser. D, **264:** 2754–6.

DARVILL, A., MCNEIL, M., ALBERSHEIM, P. and DELMER, D. P. (1980) The primary cell wall of flowering plants. In: *The Biochemistry of Plants—A Comprehensive Treatise*, Vol. 1 (ed. N. E. Tolbert). Academic Press, New York, pp. 91–162.

DAVEY, M. R. and MATHIAS, R. J. (1979) Close-packing of plasma membrane particles during wall regeneration by isolated higher plant protoplasts—fact or artefact? *Protoplasma* **100:** 85–99.

DELMER, D. P. and STONE, B. A. (1988) Biosynthesis of plant cell walls. In: *The Biochemistry of Plants*, Vol. 14 (ed. J. Preiss) Academic Press, New York, pp. 373–416.

DICKISON, W. C. (1984) On the occurrence of silicon grains in woods of *Hibbertia* (Dilleniaceae). *IAWA Bull.* **5:** 341–343.

EAMES, A. J. and MACDANIELS, L. H. (1947) *An Introduction to Plant Anatomy,* 2nd edn. McGraw-Hill, New York and London.

ESAU, K. (1961) *Plant Viruses and Insects.* Harvard Univ. Press, Cambridge, Mass.

ESAU, K. and GILL, R. H. (1965) Observations on cytokinesis. *Planta* **67:** 168–81.

FAHN, A. (1979) *Secretory Tissues in Plants.* Academic Press, London, New York, San Francisco.

FAHN, A. (1988) Secretory tissues in vascular plants. *New Phytol.* **108:** 229–257.

FAHN, A., WERKER, E. and BAAS, P. (1986) *Wood Anatomy and Identification of Trees and Shrubs from Israel and Adjacent Regions.* The Israel Academy of Sciences and Humanities, Jerusalem.

FREDERICK, S. E., GRUBER, P. J. and NEWCOMB, E. H. (1975) Plant microbodies. *Protoplasma* **84:** 1–29.

FREY-WYSSLING, A. (1948) *Submicroscopic Morphology of Protoplasm and its Derivatives.* Elsevier, New York.

FREY-WYSSLING, A. (1950) Physiology of cell wall growth. *A. Rev. Pl. Physiol.* **1:** 169–82.

FREY-WYSSLING, A. (1959) *Die Pflanzliche Zellwand.* Springer-Verlag, Berlin.

FREY-WYSSLING, A. (1969) The ultrastructure and biogenesis of native cellulose. In: *Progress in the Chemistry of Organic Natural Products* (ed. L. Zechmeister). Springer-Verlag, Wien, pp. 1–30.

FREY-WYSSLING, A. (1976) *The Plant Cell Wall.* Borntraeger, Berlin and Stuttgart.

FREY-WYSSLING, A. (1981) Crystalography of the two hydrates of crystalline Calcium oxalate in plants. *Am. J. Bot.* **68:** 130–141.

FREY-WYSSLING, A. and MÜHLETHALER, K. (1950) Der submikroskopische Feinbau von Chitinzell-wänden. *Vierteljahrsschr. Naturforsch. Ges. Zürich* **95:** 45–52.

FREY-WYSSLING, A. and MÜHLETHALER, K. (1965) *Ultrastructural Plant Cytology.* Elsevier, Amsterdam.

FREY-WYSSLING, A. and STECHER, H. (1951) Das Flächenwachstum der pflanzlichen Zellwände. *Experientia* **7:** 420–1.

FREY-WYSSLING, A., MÜHLETHALER, K., and WYCKOFF, R. W. G. (1948) Mikrofibrillenbau der pflanzlichen Zellwände. *Experientia* **4:** 475–6.

FREY-WYSSLING, A., BOSSHARD, H. H., and MÜHLETHALER, K. (1956) Die submikroskopische Entwicklung der Hoftüpfel.

Planta **47**: 115–26.

FREY-WYSSLING, A., LÓPEZ-SÁEZ, J. F. and MÜHLETHALER, K. (1964) Formation and development of the cell plate. *J. Ultrastruct. Res.* **10**: 422–31.

GAMALEI, Yu. V. (1985) Plasodesmata—Intercellular communication in plants. *Sov. Plant Physiol.* **32**: 134–148.

GIBSON, A. C. and HORAK, K. E. (1978) Systematic anatomy and phylogey of Mexican columnar cacti. *Ann. Missouri Bot. Gar.* **65**: 999–1057.

GIDDINGS, T. H. Jr., BROWER, D. L. and STACHELIN, L. A. (1980) Vizualisation of particle complexes in the plasma membrane of *Micrasterias denticulata* associated with the formation of cellulose fibrils in primary and secondary cell walls. *J. Cell Biology* **84**: 327–339.

GOODWIN, T. W. and MERCER, E. I. (1972) *Introduction to Plant Biochemistry*. Pergamon Press, Oxford.

GOODWIN, T. W. and MERCER, E. I. (1983) *Introduction to Plant Biochemistry*. 2nd ed. Pergamon Press, Oxford.

GREEN, P. B. (1965) Fibrous elements in plant morphogenesis. *Abstracts 11th Int. Congr. Cell Biol. Providence, Rhode Island, 1964, Excerpta Med. Intern. Congr. Ser.* **77**: 21.

GUNNING, B. E. S. and ROBARDS, A. W. (eds.) (1976) *Intercellular Communication in Plants: Studies on Plasmodesmata*. Springer-Verlag, Berlin.

GUNNING, B. E. S. and STEER, M. W. (1975) *Ultrastructure and the Biology of Plant Cells*. Edward Arnold, London.

HANSON, J. B. and DAY, D. A. (1980) Plant Mitochondria. In: *The Biochemistry of Plants: A Comprehensive Treatise*, Vol. 1 (ed. N. E. Tolbert). Academic Press, New York, pp. 355–358.

HEPLER, P. K. and NEWCOMB, E. H. (1967) The fine structure of cell plate formation in the apical meristem of *Phaseolus* roots. *J. Ultrastruct. Res.* **19**: 498–513.

HERTH, W. (1984) Oriented "rosette" alignment during cellulose formation in mung bean hypocotyl. *Naturwissenschaften* **71**: 216–217.

HERTH, W. (1985) Plasma-membrane rosettes involved in localized wall thickening during xylem vessel formation of *Lepidium sativum* L. *Planta* **164**: 12–21.

HERTH, W. and WEBER, G. (1984) Occurrence of the putative cellulose-synthesizing "rosettes" in the plasma membrane of *Glycine max* suspension culture cells. *Naturwissenschaften* **71**: 153–154.

HODSON, M. J. (1986) Silicon deposition in roots, culm and leaf of *Phalaris canariensis* L. *Ann. Bot.* **58**: 167–177.

HODSON, M. J. and SANGSTER, A. G. (1988) Silica deposition in the inflorescence bracts of wheat (*Triticum aestivum*). I. Scanning electron microscopy and light microscopy. *Can. J. Bot.* **66**: 829–838.

HOOBER, J. K. (1984) *Chloroplasts*. Plenum Press, New York.

HOUWINK, A. L. and ROELOFSEN, P. A. (1954) Fibrillar architecture of growing plant cell walls. *Acta bot. neerl.* **3**: 385–95.

IMAMURA, Y., HARADA, H., and SAIKI, H. (1974) Embedding substances of pit membranes in softwood tracheids and their degradation by enzymes. *Wood Sci. Tech.* **8**: 243–54.

ITOH, T. and SHIMAJI, K. (1976) Orientation of microfibrils and microtubules in cortical parenchyma cells of poplar during elongation growth. *Bot. Mag. Tokyo* **89**: 291–308.

JUNIPER, B. E. (1973) Autoradiographic evidence for paramural-body function. *Nature New Biol.* **243**: 116–17.

KERR, T. and BAILEY, I. W. (1934) The cambium and its derivative tissues: X. Structure, optical properties and chemical composition of the so-called middle lamella. *J. Arnold Arb.* **15**: 327–49.

KIRK, J. T. O. and TINLEY-BASSETT, R. A. E. (1978) *The Plastids—Their Chemistry, Structure, Growth and Inheritance*. Elsevier, North-Holland, Amsterdam.

KOBAYASHI, H., FUKADA, H. and SHIBAOKA, H. (1988) Interrelation between the spatial disposition of actin filaments and microtubules during the differentiation of tracheary elements in cultured *Zinnia* cells. *Protoplasma* **143**: 29–37.

KRULL, R. (1960) Untersuchungen über den Bau und die Entwicklung der Plasmodesmen im Rindenparenchym von *Viscum album*. *Planta* **55**: 598–629.

LEDBETTER, M. C. and PORTER, K. R. (1963) A "microtubule" in plant cell fine structure. *J. Cell Biol.* **19**: 239–50.

LEDBETTER, M. C. and PORTER, K. R. (1970) *Introduction to the Fine Structure of Plant Cells*. Springer-Verlag, Berlin, Heidelberg, and New York.

LERSTEN, N. R. (1983) Crystals of calcium compunds in Gramineae. *New Phytol.* **93**: 633–637.

LIESE, W. (1954) Der Feinbau der Hoftüpfel im Holz der Koniferen. *Proc. 3rd Int. Conf. Electron Microscopy, London 1954*, pp. 550–5.

LIESE, W. (1965a) The fine structure of bordered pits in softwoods. In: *Cellular Ultrastructure of Woody Plants* (ed. W. A. Côté Jr.). Syracuse University Press, Syracuse, NY, pp. 271–90.

LIESE, W. (1965b) The warty layer. In: *Cellular Ultrastructure of Woody Plants* (ed. W. A. Côté Jr.). Syracuse University Press, Syracuse, NY, pp. 251–69.

LIESE, W. and FAHNENBROCK, M. (1952) Elektronenmikroskopische Untersuchungen über den Bau der Hoftüpfel. *Holz*, Berlin, **10**: 197.

LIESE, W. and LEDBETTER, M. C. (1963) Occurrence of a warty layer in vascular cells of plants. *Nature* **197**: 201–2.

LIVINGSTON, L. G. and BAILEY, I. W. (1946) The demonstration of unaltered plasmodesmata in the cambium of *Pinus strobus* and in ray cells of *Sequoia sempervirens*. Abst. Am. J. Bot. **33**: 824.

LLOYD, C. W. (1982) *The Cytoskeleton in Plant Growth and Development*. Academic Press, London.

LÓPEZ-SÁEZ, J. F., GIMÉNEZ-MARTIN, G., and RISUEÑO, M. C. (1966) Fine structure of the plasmodesm. *Protoplasma* **61**: 81–84.

LOTT, J. N. A. (1980) Protein Bodies. In: *The Biochemistry of Plants—A Comprehensive Treatise*, Vol. 1 (ed. N. E. Tolbert). Academic Press, New York, pp. 589–623.

MARTENS, P. (1937) L'origine des espaces intercellulaires. *Cellule* **46**: 357–88.

MARTENS, P. (1938) Nouvelles recherches sur l'origine des espaces intercellulaires. *Beih. Bot. Zbl.* **58**: Abt. I: 349–64.

MARTY, F., BRANTON, D. and LEIGH, R. A. (1980) Plant vacuoles. In: *The Biochemistry of Plants—A Comprehensive Treatise*, Vol. 1 (ed. N. E. Tolbert). Academic Press, New York, pp. 625–658.

MARVIN, J. W. (1939) The shape of compressed lead shot and its relation to cell shape. *Am. J. Bot.* **26**: 280–8.

MARVIN, J. W. (1944) Cell shape and cell volume relations in the pith of *Eupatorium perfoliatum* L. *Am. J. Bot.* **31**: 208–18.

MATZKE, E. B. (1943) The concept of cells held by Hooke and Grew. *Science* **98**: 13–14.

MATZKE, E. B. (1946) The three-dimensional shape of bubbles of foam—an analysis of the role of surface forces in three-dimensional cell shape determination. *Am. J. Bot.* **33**: 58–80.

MATZKE, E. B. (1956) Progressive configurational changes, during cell division, of cells within the apical meristem. *Proc. Natn. Acad. Sci.* **42**: 26–33.

MEIER, H. (1957) Discussion of the cell wall organization of tracheids and fibres. *Holzforschung* **11**: 41–46.

METCALFE, C. R. (1960) *Anatomy of the Monocotyledons. Vol. I. Gramineae.* Clarendon Press, Oxford.

METCALFE, C. R. (1971) *Anatomy of the Monocotyledons. Vol. V. Cyperaceae.* Clarendon Press, Oxford.

METCALFE, C. R. (1987) *Anatomy of the Dicotyledons.* 2nd ed. Vol. III. Clarendon Press, Oxford.

METCALFE, C. R. and CHALK, L. (1950) *Anatomy of the Dicotyledons.* 2 Vols. Clarendon Press, Oxford.

METCALFE, C. R. and CHALK, L. (1983) *Anatomy of the Dicotyledons.* 2nd ed. Vol. II. Clarendon Press, Oxford.

MOHR, W. P. (1979) Pigment bodies in fruits of crimson and high pigment lines of tomatoes. *Ann. Bot.* **44**: 427–34.

MOLLENHAUER, H. H. and MORRE, D. J. (1980) The Golgi apparatus. In: *The Biochemistry of Plants—A Comprehensive Treatise*, Vol. 1 (ed. N. E. Tolbert). Academic Press, New York, pp. 437–488.

MÜHLETHALER, K. (1950) Electron microscopy of developing plant cell walls. *Biochim. biophys. Acta* **5**: 1.

MÜHLETHALER, K. (1960) Die Struktur der Grana- und Stromalamellen in Chloroplasten. *Z. Wiss. Mikroscop.* **64**: 444.

NEWCOMB, E. H. (1980) The General cell. In: *The Biochemistry of Plants—A Comprehensive Treatise*, Vol. 1 (ed. N. E. Tolbert). Academic Press, New York, pp. 1–54.

NORTHCOTE, D. H. and PICKETT-HEAPS, J. D. (1966) A function of the Golgi apparatus in polysaccharide synthesis and transport in root-cap cells of wheat. *Biochem. J.* **98**: 159–67.

O'BRIEN, T. P. (1970) Further observation on hydrolysis of the cell wall in the xylem. *Protoplasma* **69**: 1–14.

O'BRIEN, T. P. (1972) The cytology of cell-wall formation in some eukaryotic cells. *Bot. Rev.* **38**: 87–118.

OLLENDORF, A. L., MULHOLLAND, S. C. and RAPP, G. Jr. (1987) Phytoliths from Israeli sedges. *Isr. J. Bot.* **36**: 125–132.

PARRY, D. W., O'NEILL, C. H. and HODSON, M. J. (1986) Opaline silica deposition in the leaves of *Bideus pilosa* L. and their possible significance in cancer. *Ann. Bot.* **58**: 641–647.

PARTHASARATHY, M. V., PERDUE, T. D., WITZUM, A. and ALVERNAZ, J. (1985) Actin network as a normal component of the cytoskeleton in many vascular plant cells. *Am. J. Bot.* **72**: 1318–1323.

PEREZ-RODRIGUEZ, D. and CATESSON, A. M. (1982) La lignification du parenchyme vasculaire de Radis en survie: Evolution ultrastructurale et activities peroxydasiques parietales. *Ann. Sci. Nat. Bot. Paris 13 Serie*, **4**: 169–188.

PICKETT-HEAPS, J. D. (1966) Incorporation of radioactivity into wheat xylem walls. *Planta* **71**: 1–14.

PICKETT-HEAPS, J. D. (1967) The effect of colchicine on the ultrastructure of dividing plant cells, xylem, wall differentiation and distribution of cytoplasmic microtubules. *Dev. Biol.* **15**: 206–236.

PRESTON, R. D. (1952) *The Molecular Architecture of Plant Cell Walls.* Chapman & Hall, London.

PRESTON, R. D. (1974) *The Physical Biology of Plant Cell Walls.* Chapman and Hall, London.

PRIESTLEY, J. H. and SCOTT, L. I. (1939) The formation of a new cell wall at cell division. *Proc. Leeds Phil. Lit. Soc.* **3**:

532–45.

RAVEN, P. H. (1970) A multiple origin of plastids and mitochondria. *Science* **169**: 641–6.

RICHTER, H. G. (1980) Occurrence, morphology and taxonomic implications of crystalline and siliceous inclusions in the secondary xylem of the Lauraceae and related families. *Wood Sci. Tech.* **14**: 35–44.

RICHTER, H. G. and SCHMIDT, U. (1987) Unusual crystal formation in the secondary xylem of *Cosmocalyx spectabilis* Standl. (Rubiaceae). *IAWA Bull.* **8**: 323–329.

ROELOFSEN, P. A. (1951) Cell-wall structure in the growth-zone of *Phycomyces* sporangiophores: II, Double refraction and electron microscopy. *Biochim. biophys. Acta* **6**: 357–73.

ROELOFSEN, P. A. (1959) The plant cell-wall. In: K. Linsbauer, *Handbuch der Pflanzenanatomie*, Bd 3, T. 4. Gebr. Borntraeger, Berlin.

ROELOFSEN, P. A. and HOUWINK, A. L. (1951) Cell wall structure of staminal hairs of *Tradescantia virginica* and its relation with growth. *Protoplasma* **40**: 1–22.

ROLAND, J. C., REIS, D., VIAN, B., SATIAT-JEUNEMAITRE, B. and MOSINIAK, M. (1987) Morphogenesis of plant cell walls at the supramolecular level: Internal geometry and versatility of helecoidal expression. *Protoplasma* **140**: 75–91.

ROST, T. L. and SCHMID, R. (1977) Floral, fruit and seed anatomy of jojoba, and use of liquid wax during germination. In: *Jojoba Oil and Derivatives* (ed. J. Wisniak). Prog. Chem. Fats Other Lipids **15**: 167–218.

ROST, T. L., SIMPER, A. D., SCHELL, P., and ALLEN, S. (1977) Anatomy of jojoba (*Simmondsia chinensis*) seed and the utilization of liquid wax during germination. *Economic Bot.* **31**: 140–7.

RUDOLPH, U. and SCHNEPF, E. (1988) Investigation of the turnover of the putative cellulose-synthesizing particle "rosettes" within the plasma membrane of *Funaria hygrometrica* protonema. *Protoplasma* **143**: 63–73.

SASSEN, M. M. A. and WOLTERS-ARTS, A. M. C. (1986) Cell wall texture and cortical microtubules in growing staminal hairs of *Tradescantia virginiana. Acta Bot. Neeri.* **35**: 351–360.

SCHIFF, J. A. (1980) Development, inheritance and evolution of plastids and mitochondria. In: *The Biochemistry of Plants—A Comprehensive Treatise*, Vol. 1 (ed. N. E. Tolbert). Academic Press, New York, pp. 209–272.

SCHÖTZ, F., DIERS, L., and BATHELT, H. (1970) Zur Feinstruktur der Raphidenzellen: I, Die Entwicklung der Vakuolen und der Raphiden. *Z. Pflanzenphysiol.* **63**: 91–113.

SCHULZ, A. (1988) Vascular differentiation in root cortex of peas: Premitotic stages of cytoplasmic reactivation. *Protoplasma* **143**: 176–187.

SINGER, S. J. and NICOLSON, G. L. (1972) The fluid mosaic model of cell membranes. *Science* **175**: 720–31.

SINNOTT, E. W. and BLOCH, R. (1941) Division in vacuolate plant cells. *Am. J. Bot.* **28**: 225–32.

SITTE, P. (1961) Die submikroskopische Organisation der Pflanzenzelle. *Ber. dt. bot. Ges.* **74**: 177–206.

STEWARD, F. C. and MÜHLETHALER, K. (1953) The structure and development of the cell wall in Valoniaceae. *Ann. Bot.* **17**: 295–316.

SUNELL, L. A. and HEALEY, P. L. (1979) Distribution of calcium oxalate crystal idioblasts in corms of taro (*Colocasia esculenta*). *Am. J. Bot.* **66**: 1029–32.

THORPE, N. O. (1984) *Cell Biology.* John Wiley and Sons, New York.

TOLBERT, N. E. (1980) Microbodies—peroxisomes and glyoxysomes. In: *The Biochemistry of Plants—A Comprehensive Treatise*, Vol. 1 (ed. N. E. Tolbert). Academic Press, New York, pp. 359–388.

TOMLINSON, P. B. (1961) *Anatomy of the Monocotyledons. Vol. II. Palmae*. Clarendon Press, Oxford.

WARDROP, A. B. (1954) Observations on crossed lamellar structures in the cell walls of higher plants. *Aust. J. Bot.* **2**: 154–64.

WARDROP, A. B. (1958) The organization of the primary wall in differentiating conifer tracheids. *Aust. J. Bot.* **5**: 299–305.

WARDROP, A. B. and DAVIES, G. W. (1962) Wart structure of gymnosperm tracheids. *Nature*, **194**: 497–8.

WARDROP, A. B. and PRESTON, R. D. (1947) Organization of the cell walls of tracheids and wood fibers. *Nature* **160**: 911–13.

WARDROP, A. B., LIESE, W., and DAVIES, G. W. (1959) The nature of the wart structure in conifer tracheids. *Holzforschung* **13**: 115–20.

WATTENDORFF, J. (1976a) Ultrastructure of the suberized styloid crystal cells in *Agave* leaves. *Planta* **128**: 163–5.

WATTENDORFF, J. (1976b) A third type of raphide crystal in the plant kingdom: Six-sided raphides with laminated sheaths in *Agave americana* L. *Planta* **130**: 303–11.

WATTENDORFF, J. (1978) Feinbau und Entwicklung der Calciumoxalat-Kristallzellen mit suberinähnlichen Kristallscheiden in der Rinde und im sekundären Holz von *Acacia senegal* Willd. *Protoplasma* **95**: 193–206.

WEHRMEYER, W. (1964) Über Membranbildungsprozesse im Chloroplasten. II. Zur Entstehung der Grana durch Membranüberschiebung. *Planta* **63**: 13–30.

WERKER, E. and FAHN, A. (1981) Secretory hairs of *Inula viscosa* (L.) Ait.: development, ultrastructure and secretion. *Bot. Gaz.* **142**: 461–476.

WHATLEY, J. M. (1979) Plastid development in the primary leaf of *Phaseolus vulgaris*—Variation between diferent types of cell. *New Phytol.* **82**: 1–10.

WHATLEY, J. M. and WHATLEY, F. R. (1987) When is a chromoplast? *New Phytol.* **106**: 667–678.

WHEELER, G. E. (1979) Raphide files in vegetative organs of *Zebrina*. *Bot. Gaz.* **140**: 189–98.

WILDER, G. J. (1970) Structure of tracheids in three species of *Lycopodium*. *Am. J. Bot.* **57**: 1093–107.

WILDMAN, S. G., JOPE, C. A., and ATCHISON, B. A. (1980) Light microscopic analysis of the three-dimensional structure of higher plant chloroplasts. Position of starch grains and probable spiral arrangement of stroma lemellae and grana. *Bot. Gaz.* **141**: 24–36.

WITZTUM, A. and PARTHASARATHY, M. V. (1985) Role of actin in chloroplast clustering and banding in leaves of *Egeria, Elodea* and *Hydrilla*. *Eur. J. Cell. Biol.* **39**: 21–26.

YOO, B. Y., LAWRENCE, C. H. and CLARK, M. C. (1979) Ultrastructure of potato tuber microbodies. *Ann. Bot.* **44**: 373–5.

ZOBEL, A. M. (1986) Ontogenesis of tannin-containing coenocytes in *Sambucus racemosa* L. III. The mature coenocyte. *Ann. Bot.* **58**: 849–858.

CHAPTER 3

MERISTEMS

IN THE early stages of the development of the embryo all the cells undergo division, but with further growth and development cell division and multiplication become restricted to special parts of the plant which exhibit very little differentiation and in which the tissues remain embryonic in character and the cells retain the ability to divide. These embryonic tissues in the mature plant body are called *meristems*. Cell division can also occur in tissues other than meristems, for instance in the cortex of the stem and in young, developing vascular tissues. However, in these tissues the number of divisions is limited. On the other hand, the cells of the meristems continue to divide indefinitely and as a result new cells are continually added to the plant body. Meristems may also be found in a temporary resting phase, for instance in perennial plants that are dormant in certain seasons and in axillary buds that may be dormant even during the active phase of the plant.

The process of the growth and morpho-physiological specialization of the cells produced by the meristems is called *differentiation*. Theoretically, it was believed that the tissues that undergo differentiation gradually lose the embryonic characteristics of the meristem and acquire the mature state. Such tissues are called *mature* or *permanent*. However, it has been shown that the term permanent tissues can only be used in relation to certain cells which have undergone irreversible differentiation, as, for instance, sieve elements which have no nucleus and dead cells, such as tracheids, vessel elements and cork cells. All cells which contain nuclei possess, to a certain degree, the ability to grow and divide and redifferentiate if the appropriate stimulus is present (Bloch, 1941; Buvat, 1944, 1945; Gautheret, 1945, 1957; White, 1946;

Wetmore, 1954, 1956; Steward, 1970; Sachs, 1981; Vasil, 1984; Schulz, 1986).

CLASSIFICATION OF MERISTEMS

The classification of meristems is made on the basis of various criteria—their position in the plant body, their origin and the tissues which they produce, their structure, their stage of development, and their function.

According to the position of the meristems in the plant body they are divided into the following types: (a) *apical meristems*, which are found in the apices of the main and lateral shoots and roots; (b) *intercalary meristems*, which are found between mature tissues, as, for example, in the bases of the internodes of grasses; (c) *lateral meristems*, which are situated parallel to the circumference of the organ in which they are found, as, for instance, the vascular cambium and the phellogen.

It is customary to distinguish between *primary* and *secondary meristems*—a classification based on the origin of the meristems. Accordingly, primary meristems are those whose cells develop directly from the embryonic cells and so constitute a direct continuation of the embryo, while secondary meristems are those that develop from mature tissues which have already undergone differentiation.

The above definitions of primary and secondary meristems, however, are not always accurate. For example, the apical meristems of truly adventitious organs develop secondarily within relatively mature tissues as well as within secondary meristematic tissues, although according to their structure and function are primary meristems. On the other hand, a large part, or sometimes even the whole, of the vascular cambium, which is generally accepted to be a secondary meristem, develops, at a late stage, from the apical meristem, i.e. from a part of the procambium (see also Chapter 14).

Examples of secondary meristems, which can be determined as such without doubt, according to origin, are the phellogen which develops from parenchyma or collenchyma cells which have already undergone differentiation and callous tissue which develops in tissue cultures made from mature tissues.

From the above it can be seen that it is more correct to use the terms primary and secondary meristems to refer to the stage of development at which the meristems appear, and to the types of tissue that develop from them, and not to their origin. From the primary meristems the fundamental parts of the plant, such as epidermis, the cortical tissues of the stem and root, the mesophyll of the leaf, and the primary vascular tissues, develop, and from the secondary meristems the secondary vascular and protective tissues.

In certain monocotyledons, such as some palms, banana, *Veratrum,* and others, the thickening of the stem takes place near the apices and therefore is regarded as being of primary nature. The meristem responsible for this type of increase in thickness is termed *primary thickening meristem* (see secondary growth in monocotyledons, Chapter 18).

STAGES OF DEVELOPMENT OF PRIMARY MERISTEMS

Secondary meristems, e.g. the cambium and the phellogen, are homogeneous tissues in which different stages cannot be distinguished morphologically. In the primary meristems, however, different regions in various stages of differentiation can be distinguished. In primary apical meristems we distinguish a *promeristem* and a meristematic zone below it in which groups of cells have undergone a certain degree of differentiation. The promeristem consists of the *apical initials* together with the cells derived from them and which are still close to the initials. The promeristem is the least differentiated part of the apical meristem. Surgical experiments revealed that when portions of this part are cut away the cells regenerate and the apex continues to

grow (Rost and Jones, 1988). The partly differentiated meristematic zone consists of the following three meristems: the *protoderm* from which the epidermal system of the plant develops, the *procambium* from which the primary vascular tissues develop, and the *ground meristem* from which the ground tissues of the plant, as, for instance, the parenchyma and sclerenchyma of the cortex and pith and the collenchyma of the cortex, develop. The term initials in meristems refers to cells which always remain within the meristem. When an initial divides one of the daughter cells continues to fulfil the original function of an initial, whereas the other daughter cell, after several divisions, undergoes differentiation and maturation. Newman (1961, 1965) refers to an initial in the meristem as a *continuing meristematic residue.*

The position in which a cell is situated in the meristem determines it to be an initial. When an initial becomes eliminated naturally or artificially, a cell adjacent to it substitutes for it.

CYTOLOGICAL CHARACTERISTICS OF MERISTEMS

Meristematic cells are usually thin-walled, more isodiametric in shape than the cells of mature tissues, and relatively richer in protoplasm. However, it is not possible to find a general morphological criterion by which meristematic cells can be distinguished from unspecialized mature cells. Usually the protoplasts of meristematic cells are devoid of reserve materials and crystals, and the plastids are in the proplastid stage. However, the protoplasts of the phellogen, a secondary meristem, may contain these bodies. In most cells of apical meristems of a large number of plants, and especially among the angiosperms, the vacuoles are very small, not obvious, and are scattered throughout the protoplast. However, in pteridophytes and many spermatophytes at least some of the cells of the apical meristem contain conspicuous vacuoles. Also the cells of the vascular cambium are highly vacuolated (Bailey, 1930). In general, it is possible to state that the larger the meristematic cell, even if it is an initial cell or one close to the initials, the greater is the degree of vacuolization. The size of the meristematic cells varies. Also the ratio between the size of the cell and that of the nucleus varies very greatly in different meristematic cells. The wall of meristematic cells is usually thin, but certain cells in the apical meristems have thick walls, and cells of the vascular cambium have very thick radial walls at certain periods.

From the above it can be seen that morphological analysis alone is not sufficient to determine the meristematic nature of cells and the use of experimental methods is often necessary.

APICAL MERISTEMS

In the nineteenth century research workers mainly dealt with the problem of the number of the initials in the apices and the determination of the tissues that were derived from them. Thus the *histogen theory* of Hanstein (1868) and the *apical cell theory* of Nägeli (1878) were developed. Modern research on spermatophytes deals, in addition to the problem of initials, also with the cyto-histological division of the apex into zones and the activities of the cells of the various zones. Experimental research on apices has contributed to the clarification of these problems (Ball, 1947, 1960; Clowes, 1953, 1954; Wetmore, 1954, 1956; Wardlaw, 1957; Gifford and Tepper, 1962a,b). A considerable number of attempts have also been made to grow apices *in vitro* (Ball, 1946; Morel, 1965; Smith and Murashige, 1970; Romberger *et al.,* 1970; Jensen, 1971). The idea behind this work is that isolation will make possible an experimental analysis of the reactions of the apices.

Initials can be recognized by microscopical investigations and by the use of assumptions based on the orientation of cell divisions. Experiments have been made to determine the location and number of the initials by the application of colchicine. Using this substance it has been possible to increase the number of chromosomes in a few cells. As the derivatives of such cells possess the increased number of chromosomes it is possible to identify all the cells that are derived from the colchicine-affected cells by their enlarged nuclei. If the affected cell is an initial, entire regions of tissues, the cells of which have the increased chromosome number, result; thus polyploid chimeras are artificially formed. This phenomenon

makes it possible to identify the initials (Dermen, 1945, 1947, 1948, 1951; Satina *et al.,* 1940; Satina and Blakeslee, 1941, 1943; Satina, 1959; Klekowski and Kazarinova-Fukshansky, 1984). Although the initials are usually permanent, opinions exist that they may sometimes be replaced by new initials. In addition to the above-mentioned studies, the observations on variegated chimeras are of great importance in the study of apical meristems (Thielke, 1948, 1954, 1955, 1957, 1964; Bartels, 1960; Moh, 1961).

In this book the apical meristem will be divided, as already mentioned, into two main regions—the promeristem, which comprises the apical initials and neighbouring cells, and the meristematic zone behind it in which the three basic meristems (the protoderm, procambium, and ground meristem) of the tissue systems can be distinguished.

The following discussion deals mainly with the arrangement and function of the cells in the promeristem (see Clowes, 1961; Cutter, 1965; Nougarède, 1965, 1967; Schüepp, 1966; Gifford and Corson, 1971; Nougarède and Rembur, 1985).

Vegetative shoot apex

In 1759 Wolff discovered that the new leaves and tissues of the stem arise in the very apex of the stem. He termed this region the "punctum vegetationis". Today the term *shoot apex* is generally used (Fig. 30, no. 1) as it is the region of initiation of the primary organization of the shoot in which the processes of growth take place and which cannot be limited to a point. The shoot apex proper is considered as that terminal part of the shoot immediately above the uppermost leaf primordium. There are great differences in the shape and size of the shoot apices among the spermatophytes. In a median longitudinal section the apex generally appears more or less convex. In *Anacharis* and *Myriophyllum* and some grasses the shape of the apex is a narrow cone with a rounded tip (Fig. 31, nos. 1, 3; Fig. 32, no. 1), while in a few plants, e.g. *Drimys* and *Hibiscus syriacus,* it is slightly concave (Gifford, 1950; Tolbert, 1961).

Before the initiation of each leaf the apical meristem widens considerably and after the

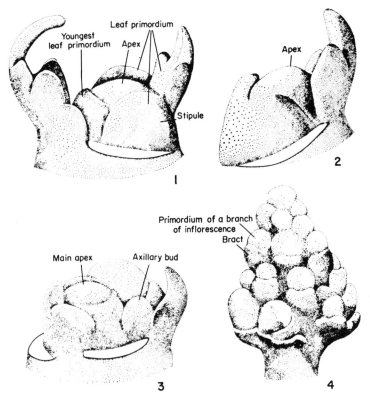

FIG. 30. Shoot apex of *Vitis vinifera*. 1, Vegetative shoot apex. 2, Vegetative shoot apex in which the tip of the apex is seen to be asymmetrical; this is apparently connected with the initiation of an axillary bud. 3, Apex in which the main apex and an axillary bud can be distinguished; the axillary bud is apparently reproductive. 4, Primordial inflorescence. (Drawings adapted from Z. Bernstein.)

appearance of the leaf primordium it again becomes narrow. This phenomenon is rhythmic, i.e. it recurs with the initiation of each leaf or pair of leaves. Schmidt (1924) introduced the terms minimal and maximal areas of the apex. For the period between the successive initiations of two leaves or two pairs of leaves he suggested the use of the term *plastochron* (Fig. 38, nos. 1–6) which had been used previously but with a much wider meaning. The shoot apices of dicotyledons with opposite leaves (such as *Lonicera, Coleus, Vinca, Ligustrum, Syringa,* and others) are particularly suitable for the study of plastochronic changes. Successive plastochrons, especially of genetically uniform plants grown in a controlled environment, may, at least during part of their vegetative growth, be of equal duration (Stein and Stein, 1960).

In the Angiospermae the shoot apices are usually small. The measurement of the diameter is taken as the width of the apex immediately above the youngest primordium. The diameter usually varies between 90 μm (in certain grasses) and 130–200 μm (in many dicotyledons). In the banana plant, however, the width of the apex reaches 280 μm, in certain Palmae and in *Nymphaea*, 500 μm in *Trichocereus* it is between 700–800 μm and in *Xanthorrhoea media* the maximum diameter reported was 1283 μm (Ball, 1941; Boke, 1941; Cutter, 1957; Fahn *et al.*, 1963; Staff, 1968). The differences in the diameter of the apices of gymnosperms are much greater (Kemp, 1943). The apices of the conifers are cone-shaped and fairly narrow and the dimensions of their diameter are similar to those typical for angiosperms. On the other hand, the apices in *Ginkgo* and *Cycas* are three to eight times as wide as they are high (Johnson, 1944). In *Cycas revoluta* the diameter of the maximal-area of the apex is 3.5 mm (Foster, 1940).

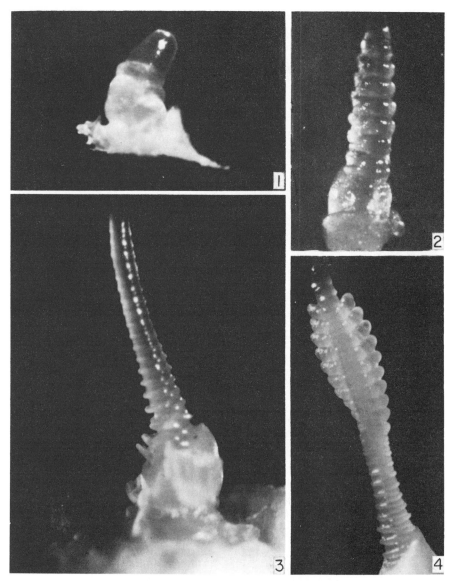

FIG. 31. Photographs of shoot apices. 1, Vegetative shoot apex of *Hordeum bulbosum*. 2, Early stage in development of inflorescence of *Hordeum bulbosum*. 3, Shoot apex of *Secale* at the time of floral induction. 4, Early stage of floral development in *Secale*. (Photographs courtesy of D. Koller.)

Shoot apex of pteridophytes

In Pteridophyta there are one or more initials which can usually be easily distinguished from the neighbouring cells. The latter, which derive directly from the initials, divide more rapidly than the initials. If only one initial is present it is termed the *apical cell* (Fig. 33, no. 2) and if more than one cell are present they are termed *apical initials* (Fig. 33,

no. 3). The single apical cell usually is tetrahedral in shape and its base is directed towards the surface of the apex. A single apical cell is found in the Psilotales, in *Equisetum* and in some ferns. The single apical cell divides in such a manner that the new cells are formed on all its sides with the exception of that on the surface of the apex. The apical cells of pteridophytes are usually four-sided,

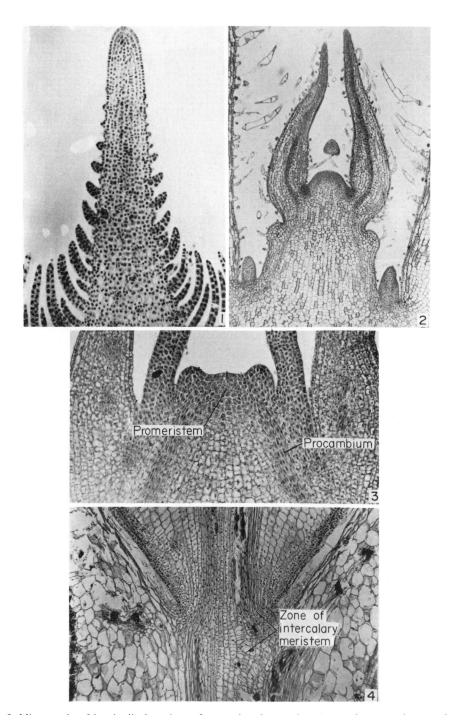

FIG. 32. 1–3, Micrographs of longitudinal sections of vegetative shoot apices. 1, *Anacharis canadensis.* × 35. 2, *Coleus blumei.* × 70. 3, *Vinca major.* × 90. 4, Portion of a median longitudinal section of the shoot of *Anabasis articulata* showing the intercalary meristem at the base of the internode. × 50.

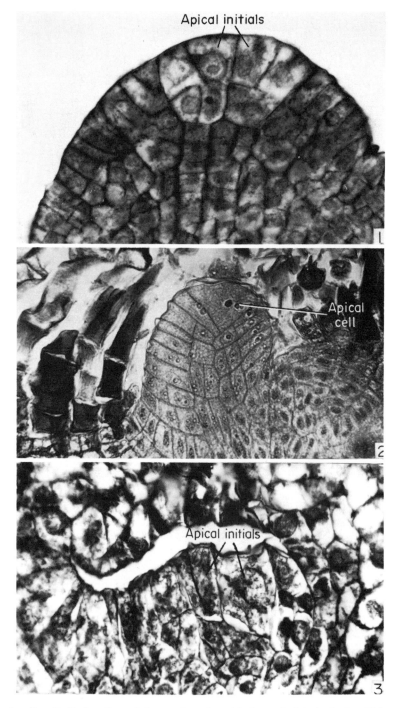

FIG. 33. Micrographs of longitudinal sections of shoot apices of pteridophytes. 1, *Selaginella*, in which two apical initials can be distinguished. × 1350. 2, *Marsilea* with a single apical initial. × 500. 3, *Ophioglossum lusitanicum* with several apical initials. × 450.

but in some water ferns, e.g. *Salvinia* and *Azolla*, and sometimes in *Selaginella*, they are only three-sided. In the former, new cells are produced on three sides while in the latter only on two sides (Fig. 34, nos. 1, 2).

It is thought that in the ferns (Filicinae) the genera with a single apical cell are evolutionarily more advanced than those with several apical initials.

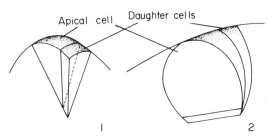

FIG. 34. Diagrams of apical cells to show the manner of division and addition of cells to the plant body. 1, Tetrahedral apical cell with base directed towards the surface of the apex and in which the planes of division are parallel to the other three faces. 2, Apical cell in which the planes of division are parallel to two faces only. (Adapted from Schüepp, 1926.)

Shoot apex of gymnosperms

It was believed that the tissue of the shoot apex of spermatophytes was a primordial meristem (promeristem) consisting of undifferentiated cells which are morphologically equal. Recent cytohistological research on the shoot apices of spermatophytes has disproved this theory and has shown that it is possible to distinguish, in these meristems, a complicated arrangement of groups of cells which are characterized by the following features—the size of the cell and the nucleus, differential staining, the relative thickness of the cell walls, and the frequency and orientation of cell division. The plane of these divisions may be anticlinal, i.e. at right-angles to the surface of the apex, or periclinal, i.e., parallel to the surface of the apex, or diagonal.

Since 1937 much research has been made on the structure of the shoot apex of gymnosperms (Korody, 1937; Foster, 1938, 1939a,b, 1940, 1943; Cross, 1939, 1942, 1943; Johnson, 1939, 1944; Gifford, 1943; Kemp, 1943; Majumdar, 1945; Sterling, 1945, 1946; Allen, 1947; Gifford and Wetmore, 1957; Parke, 1959; Guttenberg, 1961; Tribot, 1961; Pillai, 1963; Tepper, 1963; Vanden Born, 1963; Fosket and Miksche, 1966; Hanawa, 1966).

It is characteristic of all the gymnosperms that the direction of the cell divisions in the surface of the apex is both anticlinal and periclinal, and so the top layer represents the initiation zone of the entire apex and has been termed the *surface meristem*. In the cells of the summit of this meristem the periclinal divisions are more frequent. These cells are the *apical initials*. The striking feature in the structure of most gymnosperm apices is the occurrence of a distinct zone of *central mother cells*, which occurs in a median position below the surface layer. The central mother cells are relatively large, polyhedral, irregularly arranged, and their cell walls are thick, particularly in the angles of the cells (Fig. 35, nos. 3–5; Fig. 36).

The apical initials and the central mother cells contain numerous vacuoles and large light-staining nuclei. Cecich (1977) who has studied the shoot apical meristem of *Pinus banksiana* with the aid of the electron microscope has observed that the apical initials and the central mother cells each contain numerous lipid bodies and their nuclei contain very little, if any, heterochromatin. This author suggests that the vacuoles seen in the apical initials and central mother cells in light microscopical sections, prepared with standard histological techniques, represent sites of dissoluted lipid bodies. Thus the lack of heterochromatin in the nuclei and the dissolution of lipids in the cytoplasm appear to contribute most to the appearance of the cytohistological zonation of the shoot apex. Along the sides and the base of the central mother cell zone the other apical regions develop as a result of the diagonal and horizontal divisions of the central mother cells. In this way the *peripheral* or *flank meristem* is developed laterally and the *rib meristem* zone from the base. The peripheral meristem forms a kind of a short cylinder surrounding the rib meristem. The term rib meristem was introduced by Schüepp (1926) to describe that type of meristematic tissue that consists of vertical series of transversely dividing cells. The cytoplasm of the peripheral zone is dense. According to Cecich (1977) it contains numerous ribosomes. The cytoplasm of the rib meristem is less dense and contains numerous small vacuoles. According to Popham

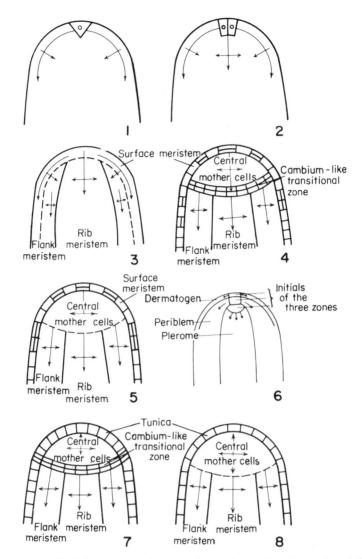

FIG. 35. Diagrams of cyto-histological zonation in the promeristem of vegetative shoot apices. 1, Pteridophyte type with single apical cell. 2, *Selaginella* type with 2–5 apical initials. 3, *Cycas* type. 4, *Ginkgo* type. 5, *Cryptomeria–Abies* type. 6, Schematic representation of the histogen theory of angiosperms. 7, *Opuntia* type. 8, Usual angiosperm type. (Flank meristem = Peripheral meristem.) (No. 1 adapted from Esau, 1953; nos. 2–5, 7 and 8 adapted from Popham, 1952.)

(1952), three principal types of gymnosperms can be distinguished on the basis of the structure of the shoot apex (Fig. 35, nos. 3–5; Fig. 36, nos. 1, 2).

1. The *Cycas* type (Fig. 35, no. 3). This type lacks the zone of central mother cells. Here three meristematic zones can be distinguished. (a) The *surface meristem* in which the cells divide anticlinally, periclinally, and diagonally. The cells of this zone are not uniform in appearance, and apical

initials have been distinguished in the centre of this zone in the seedlings of *Cycas revoluta,* but not in mature plants. The cells of this zone give rise to the epidermis and the other apical meristematic zones. (b) The *rib meristem* which is situated in the central region of the apex below the surface layer. In the upper region of this zone vertical rows of cells are obvious. The cells at the base of these rows divide periclinally, anticlinally, and diagonally, and they

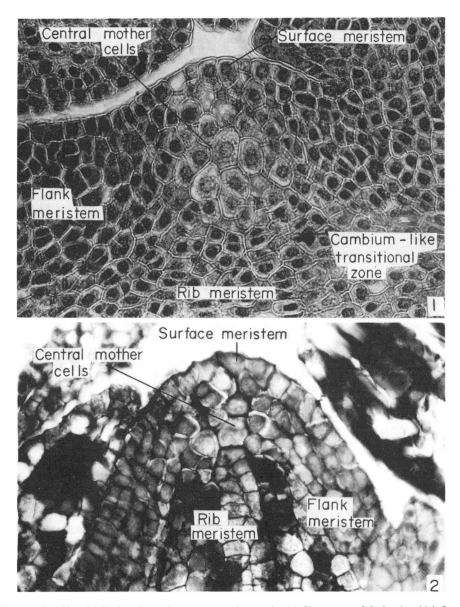

FIG. 36. Micrographs of longitudinal sections of gymnosperm shoot apices. 1, Shoot apex of *Ginkgo* in which five zones can be distinguished. × 350. 2, Shoot apex of *Pinus halepensis* in which four zones can be distinguished. × 350. (Flank meristem = Peripheral meristem.) (Dark-staining areas are tannin-filled cells.)

are usually large and contain large vacuoles. In *C. revoluta* the pith develops from this tissue. (c) The *peripheral meristem* which enlarges by cell division within the zone itself and by the addition of cells from the surface layer and from the periphery of the rib meristem. The cells of this zone are smaller than those of the rib meristem and they are generally elongated. In *C. revoluta* the cortex, the pro-

cambium, and the leaf primordia develop from this zone.

2. The *Ginkgo* type (Fig. 36, no. 1) in which five meristematic zones can be distinguished in the apex. (a) The *surface meristem*; (b) the zone of *central mother cells*; (c) the *rib meristem* from which the pith of the stem develops; (d) the *peripheral meristem,* and in addition, there is a cup-shaped

zone termed *cambium-like transitional zone*. This zone forms a transitional zone between the central mother cells and the rib and flank meristems, and is characterized by frequent cell divisions. Most of the divisions are periclinal in relation to the central mother cells and so cells are added to the zones below, i.e. to the flank and rib meristems. The number of cells in the flank meristem increases by division in the meristem itself as well as by the addition of cells from the surface meristem and the cambium-like transitional zone. The cortex, leaf primordia, procambium and, in certain plants (*Microcycas* and *Zamia*), the outer region of the pith develop from this zone. The following are some examples of plants with the *Ginkgo* type of apex: *Ginkgo biloba, Zamia* spp., *Sequoia sempervirens, Microcycas calocoma* (side branches), and *Pseudotsuga taxifolia*.

3. The *Cryptomeria–Abies* type (Fig. 36, no. 2). In this type four meristematic zones can be distinguished. The cambium-like transitional zone is absent and the remaining zones are as in the *Ginkgo* type. Of the plants with this type of apical meristem the following species should be mentioned: *Pinus montana, Sequoia gigantea, Metasequoia glyptostroboides, Abies concolor, Taxus baccata, Ephedra altissima*, and *Cryptomeria japonica*.

Shoot apex of angiosperms

At the beginning of the cyto-histological research on the apices of plants the *histogen theory* of Hanstein (1868) was put forward. According to Hanstein the following three zones (Fig. 35, no. 6) can be distinguished in the shoot apex of angiosperms: an outermost zone, the *dermatogen*; a central zone, *plerome,* which consists of irregularly arranged cells; and a hollow cylindrical zone of several layers of cells between the dermatogen and the plerome, the *periblem*. Hanstein stated that the dermatogen, periblem and plerome develop from independent groups of initials, which act as direct histogens. According to this theory, therefore, the meristems are destined from the beginning to produce certain tissues, i.e. the epidermis develops from the dermatogen, the cortex, and internal tissues of the leaf from the periblem, and the central cylinder from the plerome. The histogen

theory of Hanstein was accepted for a long time, but recent knowledge of the apical meristems disproves it. It is now accepted: 1. that all cells have basically equal potential of differentiation, and 2. that one zone of the apical meristem may contribute cells to another one. In 1924 the theory of Schmidt, which divides the apex into two regions, the *tunica* and the *corpus,* was postulated. According to this theory no constant relationship can be traced between the particular initials of the promeristem and the inner tissues of the shoot. The two regions recognized by this theory are usually distinguished by the planes of cell divisions in them. The tunica consists of the outermost layer or layers of cells which surround the inner cell mass— the corpus. The plane of cell division in the tunica is principally anticlinal. In the corpus the planes of cell division are in all directions. Recent electron microscope studies on shoot apices revealed that the ultrastructural differences between tunica and corpus cells are mainly quantitative (Lyndon and Robertson, 1976; Mauseth, 1980). The tunica enlarges in surface area and the corpus in volume. The *tunica-corpus theory* is a very adaptable one and today it is generally accepted in literature.

As already mentioned, the tunica consists of a single or a few layers of cells which surround the inner meristem. The number of layers is not always constant in a given genus or family, and not even in a species, and cases are known where the number of layers varies in a single plant during different stages of the development of the vegetative apex. The number of layers usually varies between one and nine. In *Xanthorrhoea media* Staff (1968) distinguished as many as 10 or sometimes even 18 layers. Sometimes, especially among the monocotyledons, a few periclinal divisions occur in the tunica, which is a contradiction of the original definition of the tunica and therefore Popham and Chan (1950) introduced the term *mantle* for all the outer layers of apex which can be distinguished histologically from the inner cell mass, the *core,* without taking into account the planes or division in these layers. They retained the term tunica for those layers in which only anticlinal cell divisions take place. In this book the term tunica is used in the broad sense, and is equivalent to the mantle of Popham and Chan.

Although the plane of cell division is generally

the same throughout the entire tunica, cytologically two zones can be distinguished in it. One zone is the *central apical zone* consisting of one or few *initials* which are usually larger and have larger nuclei and vacuoles than the other tunica cells and which, therefore, are also more lightly staining. The second zone is that region on the sides of the apex between the initials and the leaf primordia. It consists of smaller, more darkly staining cells which divide more frequently and among which periclinal division may occur close to the primordia.

The corpus is far less homogeneous than the tunica. Among the angiosperms two main types of corpus are recognized on the basis of internal arrangement (Popham, 1952).

1. The *usual angiosperm type* (Fig. 35, no. 8) in which three main zones can be distinguished in the corpus: (a) the zone of *central mother cells,* which represent the corpus initials, located below the apical portion of the tunica, i.e. below the tunica initials; (b) the *rib meristem*; and (c) the *peripheral meristem.* The latter two zones appear as continuations of the central mother cells.

2. The *Opuntia type* (Fig. 35, no. 7) in which, in addition to the above zones, a *cambium-like transitional zone* can be distinguished. This zone, which is cup-shaped, is found between the central mother cells and the rib and peripheral meristems (Fig. 37, nos. 1, 2). The cambium-like transitional zone differs from the other zones of the apical meristem in that its height and diameter vary considerably during the plastochron, reaching a maximum development close to a developing primordium (Fahn *et al.,* 1963). According to Philipson (1954) this zone is only a temporary feature in many of the plants in which it occurs, as it disappears towards the end of the plastochron.

The following are some examples of plants with an *Opuntia* type apex: *Phoenix dactylifera,* the Dwarf Cavendish banana, *Chrysanthemum morifolium, Opuntia cylindrica, Bellis perennis, Xanthium pennsylvanicum, Liriodendron tulipfera* and *Bougainvillea spectabilis.*

In some of the above-mentioned species the cambium-like zone can be seen in certain phases of the plastochron only, whereas in others it appears to occur consistently (Stevenson, 1978). It seems therefore preferable not to use this zone as a

criterion for classifying shoot apices of the angiosperms into two different types.

The cells of the peripheral meristem are usually derived from the initials of the corpus, but in some plants they arise also from the tunica.

The rib meristem usually consists of cells arranged in rows which narrow towards the apex but they may sometimes be irregularly arranged. The majority of divisions in this zone are horizontal, but diagonal divisions also occur. Series of cells that increase in size as they become further distant from the central mother cells can be distinguished. With the increased cell size the vacuoles also enlarge.

Johnson and Tolbert (1960) introduced the term *metrameristem* to include the initials of the tunica and corpus. In rare cases the number of corpus initials may sometimes be reduced to a great extent (Werker and Fahn, 1966).

Morphometric ultrastructural studies of shoot apical meristems have shown, that each zone has its own ultrastructure, but the differences between them are small (Mauseth, 1981, 1982; Berggren, 1984). The pattern of the ultrastructural organization differs between species (Mauseth, 1982; Berggren, 1984, 1985). Differences in ultrastructure of the apical meristem were observed between growing and dormant buds. The volume density of the vacuoles was lower, and the relative volumes of mitochondria, plastids, lipid bodies and starch grains was higher in the dormant than in the growing buds (Berggren, 1985).

From what is known today it can be concluded that the epidermis and its derivatives originate from the outermost layer of the tunica. The peripheral meristem contributes to the development of the leaf primordia, the cortex, all or part of the procambium, and sometimes also the outer region of the pith.

Regions of activity in the apical meristem

From the point of view of activity (cell division) two zones, which are parallel to the cyto-histological zones, are generally distinguished: (a) a central apical zone which includes the initials of the tunica and of the corpus and in which division is considered to occur rarely; and (b) a peripheral zone to which much mitotic activity is ascribed (see

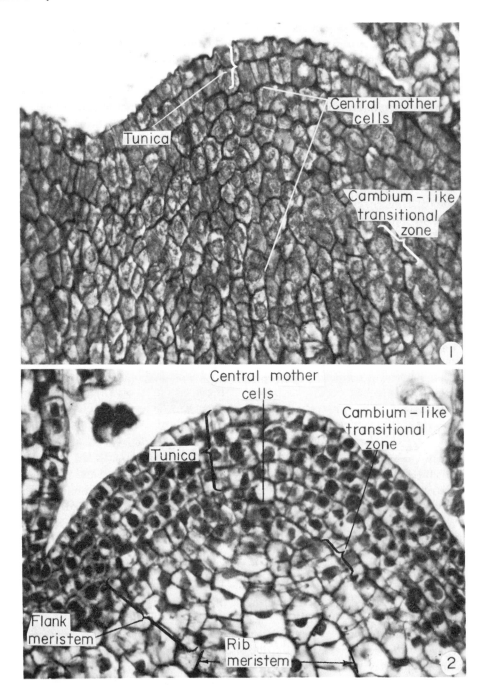

FIG. 37. 1, Micrograph of a longitudinal section of the upper portion of the vegetative shoot apex of the Dwarf Cavendish banana showing a two-layered tunica. × 600. 2, Micrograph of a longitudinal section of the shoot apex of *Coleus blumei* showing a four-layered tunica. × 580. (Flank meristem = Peripheral meristem.)

Gifford, 1954; Gifford and Corson, 1971; Nougarède, 1965; Cutter, 1965; Marc and Palmer, 1982).

Extreme views as to the activity of these two zones are held by Buvat (1952a, b) and other French workers who, contrary to most investigators, do not regard the central apical zone as having the role of cell-producing cells during the vegetative development of the plant (Fig. 38, no. 7), but as being active only when the vegetative apex becomes reproductive. This view is based on Plantefol's (1947) theory on phyllotaxis which attributes the organogenic and histogenic role only to a lateral, subterminal zone, the *anneau initial* (initiating ring) at which the leaf parastichies (helices) end. The mitotically inactive central apical zone was termed *méristème d'attente* (waiting meristem). This view was strongly supported by Loiseau (1962) and Nougarède (1965). However, the findings of most other investigators contradict this theory of quiescence of the central apical zone.

The results of the study of natural and colchicine-induced polyploid chimeras (Satina and Blakeslee, 1941, 1943; Dermen, 1945, 1947, 1948, 1951; Dermen and Bain, 1944; Satina, 1959; Thompson and Olmo, 1963) favoured the concept that divisions do occur in this zone. The investigations on variegated chimeras (Thielke, 1954, 1955; Stewart and Dermen, 1970) lead to a similar conclusion. From histological investigations made on apical meristems of many species (Millington and Fisk, 1956; Popham, 1958; Fahn *et al.,* 1963; Hara, 1962), cell divisions in the zone of the central mother cells were demonstrated. With the use of microphotographs taken with a cine camera of the surfaces of culture-grown shoot apices of several plants, Ball (1960) came to the conclusion that all the cells in the surface layer, including those at the summit of the apex, divide and do so quite frequently. Also the application of radio-actively labelled precursors of DNA synthesis and other methods (Partanen and Gifford, 1958; Gifford and Tepper, 1962a, b; Clowes, 1959; Ball, 1979; Davis *et al.,* 1979; Clowes and MacDonald, 1987) did not reveal the existence of an inactive zone in the shoot apex.

In a later account the supporters of the *Méristème d'attente* theory agreed that a few mitoses may occur in the central apical zone but claim that their role is minimal in relation to that played by the *anneau initial* (Buvat, 1955; Nougaräde, 1965).

West and Gunckel (1968a, b), who made cyto-histochemical studies of the shoot tips of *Brachychiton* species, found no great difference in RNA synthesis between the central apical cells and the cells of the flanking zone. Denne (1966), who

studied the duration of the mitotic cycle in the vegetative shoot apex of *Trifolium,* reports of relatively small ($\pm$ 3:4) differences between the above zones.

In young embryo shoots of *Zea mays,* Clowes (1978a) found recently that the rates of division of cells at the summit of the apex do not differ greatly from those in the rest of the apex.

Lyndon (1976) summarizing data on relative cell division rates in the various zones of the shoot apex concluded that in all or most plants there exists a gradient in the rate of division and growth from a minimum at the summit of the apex to a maximum in the region of leaf initiation.

The view expressed by some authors that the apical cell of the pteridophytes serves as a unicellular quiescent centre, without histogenic potential, has also been contradicted by Polito (1979) and Gifford and Polito (1981) who worked on *Ceratopteris* and *Azolla.* These authors have used the colchicine-induced metaphase accumulation technique to show that this cell can be active, even to the extent of its immediate derivatives.

Branching

The branching of the shoot originates at the shoot apex. The branch buds arise exogenously from superficial cell layers in the axils of leaf primordia. The bud meristem protrudes above the surface and the apical meristem of the bud is organized. In many species, an arcuate zone of cambiform cells, referred to as *shell zone* (Fig. 38, no. 8), delimits the incipient bud from the shoot (Shah and Patel, 1972). In many pteridophytes and in a few angiosperms (e.g. *Asclepias syriaca, Hyphaene thebaica, Flagellaria indica* and some Cactaceae) branching is brought about by the equal division into two of the single apical cell or group of apical cells as a result of which two apices are formed. This type of branching is termed *dichotomous branching* (Nolan, 1969; Tomlinson, 1970; Tomlinson and Posluszny, 1977; Boke, 1976; Boke and Ross, 1978).

Accessory buds (extra buds in an axil) originate from sectors of the peripheral meristem of the axillary buds (Shah and Unikrishnan, 1970). Some branches may develop extra-axillary, e.g. the tendrils of the Vitaceae (Shah and Dave, 1970).

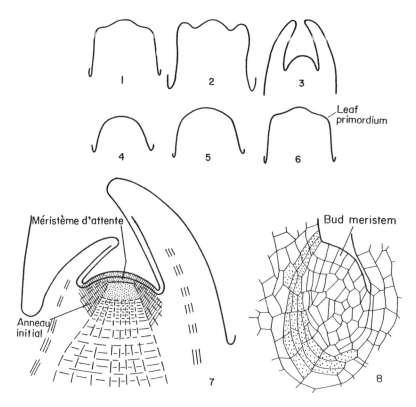

FIG. 38. 1–6, Diagrams of longitudinal sections of a shoot apex with opposite leaves showing the changes in shape and size of the apex during a plastochron. In nos. 1 and 6 the area of the apex is minimal, while in no. 5 it is maximal. 7, Diagram of the vegetative shoot apex after Buvat, showing the central inactive region (*méristème d'attente*) and a peripheral zone of high mitotic activity (*anneau initial*). 8, Longitudinal section of a developing axillary bud of *Amaranthus tricolor*, showing a shell zone (dotted). (Adapted from Shah and Patel, 1972.)

When axillary buds begin extension growth immediately after their formation, and there is no period of rest or dormancy between initiation of the axillary meristem and the development of a branch, the developmental process is termed *syllepsis*. The branches that develop in this manner are referred to as *sylleptic branches*. If there is a period of rest between axillary bud formation and the outgrowth of a branch, the developmental process is called *prolepsis* and the branch *proleptic branch* (Halle; *et al.*, 1978; Wheat, 1980).

Most conifers possess numerous blank leaf axils that lack bud-forming structures. However, in many *Araucaria* and *Agathis* species persistent, more or less deeply buried, meristems were reported to appear (Burrows, 1987). These meristems possess neither a bud-like organization nor vascular connections with the vascular cylinder.

Adventitious branches produced without relation to axillary buds develop from buds arising on roots (e.g. *Salix, Desmeria, Tripodanthus*) (Kuijt, 1981, 1985), on leaves (e.g. *Kalanchoë*), from wound callus, and from deep seated tissues in the stem bark (e.g. *Tilia platyphyllos, Acer pseudoplatanus, Fraxinus excelsior*). Inflorescences of *cauliflor trees*, which bear their flowers on mature trunks (e.g. *Artocarpus integrifolia, Swartzia schomburgkii, Couroupita guianensis*) also develop from buds formed in the bark (Fink, 1983). Many of the apparently adventitious branches arise from dormant axillary buds buried in the bark. These buds may be distinguishable on the outside of the trunk, as are for instance the clusters of flower buds occurring in some cauliflorous trees (e.g. *Theobroma cacao*) (Lent, 1966).

Reproductive apex

The reproductive apex, which produces a flower

or an inflorescence (Fig. 30, no. 4; Fig. 31, nos. 1–4), replaces the vegetative apex. According to the classical concept, traced to the writings of Goethe, the flowering shoot is homologous to the vegetative shoot. Grégoire (1938), on the basis of histo-genetical studies, opposed the theory that the flower is a metamorphosed shoot. According to Grégoire the reproductive meristem possesses a structure fundamentally different from the vege-tative meristem. A meristematic mantle is present in the apices of reproductive axes, but it is not formed from a transformation of the tunica-corpus layers present in the vegetative apex. The two superficial layers of the mantle give rise to the floral parts and contribute cells to a parenchymatous core. Grégoire's concept did not receive much support. Most investigators criticized it and con-cluded that at the time of flowering the vegetative shoot apex, terminal or lateral, undergoes various physiological and histological changes and becomes directly transformed into a reproductive apex. This may develop either a flower or an inflorescence.

There are inflorescences and even single flowers which, similarly to vegetative shoots, also exhibit plastochronic size fluctuations in the floral apex during successive stages of development of bracts, perianth, stamens, and carpels (Tucker, 1960). Some reproductive apices may retain the cyto-histological zonation of the vegetative apex, although their form may vary (Vaughan, 1955).

Plantefol (1947) considered that the reproductive apex originates by progressive metamorphosis of the vegetative apex. According to Plantefol the *anneau initial* functions during vegetative growth in production of foliage leaves. It forms also the sepals and even the petals. Stamens and carpels derive from the *méristème d'attente* (Bersillon, 1951, 1955; Buvat, 1952a, b; Lance, 1957). As mentioned earlier, however, most authors do not support the view of the existence of a "waiting meristem" which becomes active only when the apex becomes reproductive; they consider the apex as a whole to enter a new phase of development.

Philipson (1947, 1949) stated that the basic func-tion of the vegetative apex is to promote longi-tudinal growth of the axis, while that of the repro-ductive apex is to produce a meristematic envelope with a large surface area from which the parts of a flower or flowers develop. Inside this meristematic envelope there is a rib meristem consisting of relatively large, vacuolated cells. Many investi-gators (Boke, 1947; Popham and Chan, 1952; Wetmore *et al.*, 1959; Fahn *et al.*, 1963, and others) have shown that the transition from vegetative to reproductive apex is gradual.

The first noticeable change is the increase of mitotic activity on the boundary between the cen-tral mother cell zone and the rib meristem zone. Gradually this activity spreads into the central mother cell zone where the cells then become smaller and richer in protoplasm. In this way all the cells above the rib meristem are added to the tunica, the cells of which are more or less iso-diametric and are relatively small. Following these changes, mitotic activity and growth ceases, or almost so, in the cells of the rib meristem and of the pith below it. Thus in the apex a parenchymatous pith surrounded by meristematic cells develops (Fig. 39, no. 1). Depending on the species, only the flower parts or the bracts, the axillary branches of the inflorescence, and the flowers themselves deve-lop from these meristematic cells.

Many cyto-histological studies on the transition of the vegetative apex to reproductive state have been made (Gifford and Tepper, 1961; Gifford, 1963, 1969; Jacobs and Raghavan, 1962; Thomas, 1963; Corson and Gifford, 1969; Murty and Kumar, 1972). During transition to flowering a general increase in concentration of cytoplasmic basic proteins (histone), RNA, and total protein was found to occur in all zones of the apex.

Following is a summary of events occurring in the apical meristem during flower induction in *Sinapis alba* by transfer from short to long days, as described by Bernier *et al.* (1967): (1) rise in the mitotic index (percentage of nuclei observed to divide) culminating about 30 h after beginning of the long day; (2) a stimulation of DNA synthesis reaching a maximum at the 38th hour; (3) increase in a nucleolar diameter rising to a maximum at about the 54th hour; (4) increase in cell volume culminating at the 62 hour; (5) second rise in the mitotic index rising to a maximum at 62 h. Floral induction was suggested to occur during the first rise in mitotic activity. Flower buds were initiated in co-ordination with the second rise in mitotic activity. Ultrastructural changes were observed

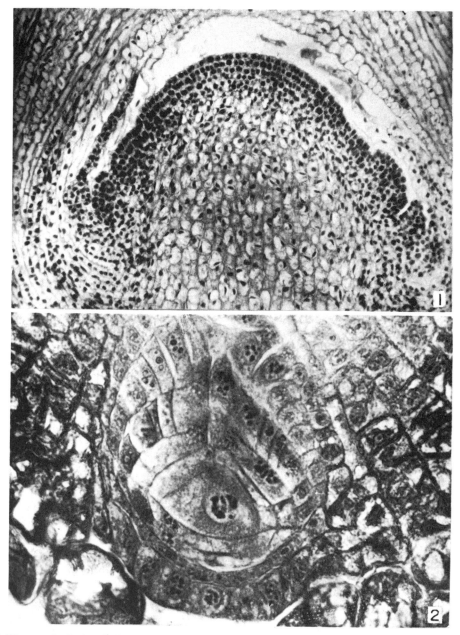

FIG. 39. 1, Micrograph of a longitudinal section of a reproductive shoot apex of *Chrysanthemum anethifolium* at an early stage of development. × 200. 2, Micrograph of a longitudinal section of a developing root of *Marsilea* in which it is possible to distinguish a single apical cell which contributes cells both towards the body of the root (upper side of micrograph) and towards the root cap (lower side of micrograph). × 750.

during floral induction (Healy, 1964; Gifford and Stewart, 1965; Havelange, 1980; Orr, 1981). These were: an increase in cytoplasmic volume, cytoplasmic matrix, ribosome density, number of mitochondria and dictyosomes as well as an increase in size of the nucleoli and changes in their structure. The cytohistological changes in the apex during induction of flowering recently described by many authors are no doubt preceded and accompanied by physiological and biochemical changes. One of these changes may be demonstrated by the fact that the dominance of the main apex, which suppresses the development of the lateral buds, is lost with the production of the inflorescence.

An apex in this stage of development ceases to elongate in plants with capitula or single flowers, and in other plants the rate of elongation is reduced. In certain plants, such as banana and pineapple, however, marked elongation takes place. In such apices the rib meristem is active in the process of elongation (Fahn *et al.*, 1963; Gifford, 1969).

The root apex

In the embryo, within the ripe seed, only the promeristem of the root or, sometimes, an embryonic radicle may be seen at the base of the hypocotyl. Usually the specific apical zonation can already be discerned in this stage. The promeristem of the lateral and adventitious roots have a similar structure to that of the primary root. The structure of the root promeristem has been intensively studied in order to discover the origin of the various tissues.

In some of the Pteridophyta, as, for instance, the Polypodiaceae, Ophioglossaceae and *Equisetum*, the entire root develops from a single apical cell (Fig. 39, no. 2; Fig. 40, no. 1), while in others, e.g. the Marattiaceae, a few initials are present. When there is only a single apical cell it is tetrahedral, and it divides in such a way as to add new cells to the body of the root from its upper three sides and to the root cap from its base.

In some pteridophytes, e.g. *Azolla,* after formation of the apical cell a division parallel to the distal face produces the precursor cell of the root cap. All subsequent divisions of the apical cell occur at the three proximal faces. Active division of the apical cell is confined to the early stage of root development. At later stages of development the apical cell becomes progressively more vacuolate and ceases to divide. Root growth in *Azolla* is determined and the apical cell divides about 55 times (Gunning *et al.*, 1978; Hardham, 1979).

According to various investigators (Guttenberg, 1940, 1947, 1960; Schade and Guttenberg, 1951; Guttenberg *et al.,* 1955) it appears that there is a single central initial or but a few initials in the root apex of the Spermatophyta (Fig. 41, no. 3; Fig. 43, no. 1). Other investigators, however, such as Clowes (1950, 1953, 1954, 1961), believe that there is a larger group of initials in the median region of the root apex.

Allen (1947), working on *Pseudotsuga* (Fig. 41, nos. 1, 2), distinguished a central group of permanent initials with three groups of temporary initials (mother cells) on its periphery. He observed that the meristem of the vascular cylinder developed from the first group of temporary initials, the meristem of the cortex from the second, and the *columella* from the third. The columella is a group of cells that forms the longitudinal axis of the root cap. In it the cells are arranged in longitudinal rows. Cells are added to the root cap from the columella by periclinal cell division on its periphery. The protoderm was seen to develop from the young cortex. According to Wilcox's work on *Abies procera* (1954) there appear to be two groups of temporary initials, one of which gives rise to the central cylinder and the other to the columella, from which the root cap and cortex develop.

Although research on the development of the histogens in the shoot apex has proved that they do not exist, many authors still use the terms *dermatogen*—meristem of the epidermis, *periblem*—meristem of the cortex, and *plerome*—meristem of the central cylinder, in connection with roots. However, the terms, as used today, have a meaning somewhat different from those as used by Hanstein. The mother cells of the various tissue systems of the root are replaced, at relatively long intervals, by new cells which are derived from the common permanent initials. In many cases more than one tissue develops from a group of temporary initials and so it is desirable to use, wherever possible, instead of histogens, the terms *protoderm, meristem of the cortex,* and *meristem of the vascular cylinder* for the meristems that are derived from the *promeristem,* i.e. from the zone of permanent and temporary initials of the root apex.

Adapting Guttenberg's view the meristems of the different tissue systems can be traced, in the root apex, at various distances from the central cells (i.e. the permanent initials). In some species the initials (temporary) of the various tissue systems are already discrete immediately adjacent to the central

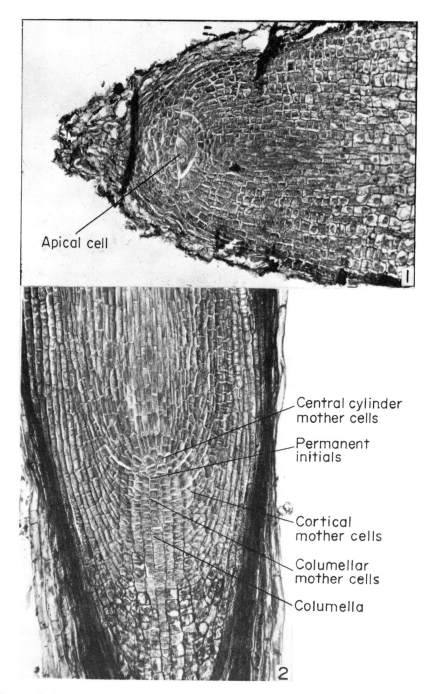

FIG. 40. 1, Micrograph of a longitudinal section of the root tip of *Ophioglossum lusitanicum* in which a single apical cell can be distinguished. × 140. 2, Portion of a longitudinal section of the root tip of *Pinus pinea*. × 155.

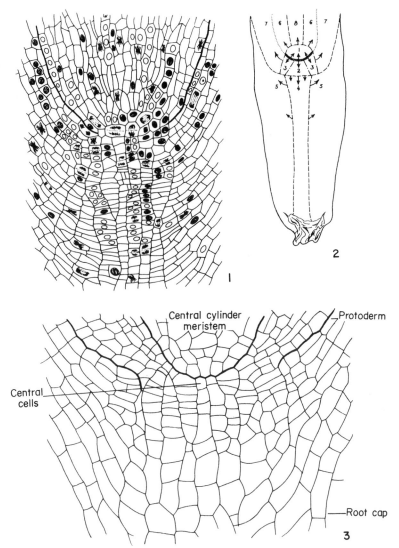

Fig. 41. 1 and 2, Median longitudinal sections of a root of a *Pseudotsuga* seedling. 1, Camera lucida drawing in which it is possible to trace the divisions in the permanent initials. × 140. 2, Diagram of entire root tip in which the zone of permanent initials is indicated by a thick line; (1) temporary initials from which the central cylinder, (6) and (8), develops; (2) temporary initials which give rise to the columella (4) from which the root cap (5) develops as a result of lateral divisions; (3) temporary initials of the cortex (7) from which the protoderm develops. 3, Longitudinal section of a young root tip of *Helianthus annuus*. (Nos. 1 and 2 adapted from Allen, 1947; no. 3 adapted from Guttenberg *et al.*, 1955.)

cells, i.e. *closed type*. These initials either represent those of the vascular cylinder, the cortex and the common initials of the protoderm and root cap, e.g. as in *Brassica,* or they represent common initials of the cortex and protoderm and separate initials of the vascular cylinder and of the root cap, e.g. as in *Zea* and *Triticum* (Fig. 42; Fig. 43, nos. 1, 3). The special initials of the root cap were termed *calyptrogen* by Janczewski (1874). In other species

the meristems of the different tissue systems finally become distinct only some distance away from the central cells, i.e. *open type*. In this type common initials for the cortex meristem, root cap, and protoderm (e.g. *Helianthus,* Fig. 41, no. 3), or for the meristems of all the tissue systems (e.g. *Allium*), appear on the periphery of the central cells. The importance of the above types is queried as both types may occur in the same species but in roots of

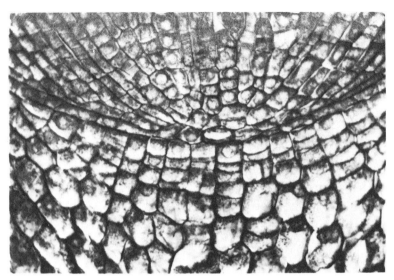

FIG. 42. Micrograph of a longitudinal section of the root tip of *Zea mays.* × 500.

different developmental stages (Byrne and Heimsch, 1970a).

Morphometric analysis of the ultrastructure of the calyptrogen, columella and peripheral cells of the root cap, showed that the cells of each zone are characterized by a distinctive ultrastructure (Moore and McClelen, 1983; Moore and Pasieniuk, 1984).

Investigations (Jensen, 1957; Clowes, 1958a, b, 1961, 1971, 1975, 1976; Jensen and Kavaljian, 1958; Pillai and Pillai, 1961; Byrne and Heimsch, 1970b) of the root apex have shown that a group of cells at the pole of the stele and cortex have a very low mitotic activity. This group of cells, which appears in the shape of a hemisphere or a disc, was termed the *quiescent centre* (Fig. 43, no. 2). It was found that the apex of a primary root of *Zea mays,* 1 mm in diameter, contains 110,000 cells, which are in one of the phases of mitotic circles, and 600 cells in the quiescent centre. The mitotically active cells laying over the surface of the quiescent centre (about 800 in primary roots of *Zea*) are regarded as temporary initials (Clowes, 1976).

The existence of the quiescent centre was revealed by various physiological and biochemical techniques, including irradiation and feeding of roots with labelled compounds involved in DNA synthesis (Torrey and Feldman, 1977; Clowes, 1978, 1982).

Experiments of *in vitro* culture of isolated quiescent centre showed that it retains the capacity to regenerate a new root, independent of influences derived from other portions of the root (Feldman and Torrey, 1976).

Various views have been expressed on possible causes of the appearance of a quiescent centre in roots. One view suggests control by hormones which may be synthesized in the quiescent centre. A substance may stimulate division at low concentration and inhibit at high concentration. Another view is that there is a competition between cells for supplies of nutrients or hormones. It was also suggested that the root cap controls quiescence in the particular portion of the meristem. One of the regulators of growth in plants is pressure. Pressure exerted by the rapidly dividing neighbouring cells was thus suggested to cause the inactivity in the quiescent centre (Webster and Langenauer, 1973). The way the files of cells in the root tip run towards the quiescent centre also suggests that the pressure extended by the expansion of the temporary initial (the cells around the quiescent centre) and their derivatives prevents the quiescent centre from expanding and thus its cells from dividing (Clowes, 1976).

It is also necessary to mention here the *Körper–Kappe theory* (Fig. 43, no. 3) which was put forward by Schüepp (1917). This theory, similar to the

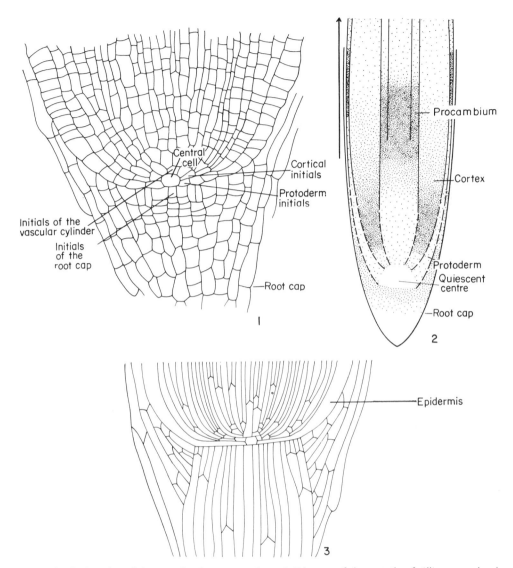

FIG. 43. 1, Longitudinal section of the root tip of *Triticum vulgare*. 2, Diagram of the root tip of *Allium cepa* showing, by means of shading, the gradation of mitotic activity in different zones; the most active zone is the most darkly shaded. 3, Diagram illustrating the Körper–Kappe pattern of the root apex of *Zea mays*. (No. 1, adapted from Schade and Guttenberg, 1951; no. 2, adapted from Jensen and Kavaljian, 1958; no. 3, adapted from Clowes, 1961.)

tunica-corpus theory of the shoot apex, is based on differences in the planes of cell division. According to the Körper–Kappe theory the cells divide in a pattern which was termed T-divisions. In the outer regions of the root apex the Kappe consists of cells in which, after the first horizontal division, the lower daughter cell divides longitudinally, i.e. at right angles to the plane of the first division. Thus the planes of the two divisions form a T in a

median longitudinal section of the root. In the Körper—the inner region of the apex—the T is inverted, i.e. the second division takes place in the upper daughter cell. In some families, such as the Gramineae, the position of the boundary between these two regions was found to be constant in relation to the meristems of the various tissue systems, but in other plants, e.g. *Fagus*, the boundary may occur in different positions in the root tip, such as in the middle of the cortex, between the cortex and the epidermis, etc. (Clowes, 1961). Körper-type T-divisions can also be seen in the central column of the root cap in *Fagus*.

INTERCALARY MERISTEMS

Extension growth is brought about by the primary meristems. Elongation of the shoot axis may be caused by uniform meristematic activity in the internodes or it may occur as a wave of growth extending progressively from the base of the inter-node in an acropetal direction, as, for instance, in *Helianthus annuus* (Garrison, 1973). Fisher and French (1976, 1978), who have studied the manner of extension growth of a large number of species belonging to various families of the mono-cotyledons, distinguished between two types of meristems in developing internodes—*uninterrupted meristems* (Fig. 44, no. 2) and *intercalary meristems* (Fig. 44, nos. 1, 3, 4). The uninterrupted meristem is progressively confined to the upper region of the internode and is continuous with the subapical meristematic region of the stem. Cell maturation within the internode is acropetal. The intercalary meristem represents isolated meristematic regions that are disjunct from the subapical meristematic region and the internodes mature basipetally. The nodes are the first to mature (Fig. 44, no. 1). In most plants with intercalary meristems the region with the cells showing the least degree of differen-tiation is at the base of the internode, but this region may sometimes be found at higher levels of the internode. In more mature stages the inter-calary meristems are separated from each other by fully matured tissues. Finally, they undergo com-plete differentiation and so disappear.

In the intercalary meristematic region vascular elements differentiate, but they become destroyed by stretching caused by the elongation of the ground tissue. The function of the destroyed tra-cheary elements is taken over by newly differen-tiating elements, as has been observed in *Eleo-charis acuta* (Evans, 1965). Destroyed sieve ele-ments are also replaced by newly differentiated ones. In the intercalary meristem of leaves of *Lolium perenne* and of shoots of *Triticum aestivum* total destruction of sieve elements without im-mediate replacement by new ones was reported.

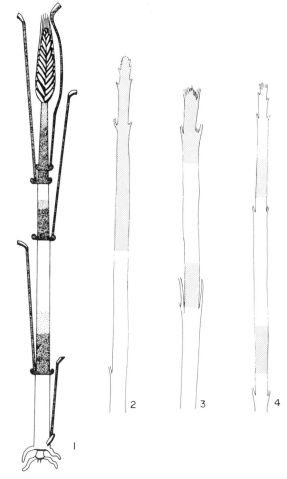

FIG. 44. Schematic drawings of shoot axes with intercalary (nos. 1, 3, 4) and uninterrupted (no. 2) meristems. Growing regions are shaded. (No. 1 adapted from Esau, 1953; nos. 2–4 adapted from Fisher and French, 1976.)

This produces a bottleneck in the transport, which in turn may enhance unloading of the su-crose into the meristematic sink (Forde, 1965; Patrick, 1972).

The best-known examples of intercalary meris-tems are those in the stems of grasses, some other monocotyledons, some species of the Caryophyl-laceae, Polygonaceae, and articulated species of the Chenopodiaceae and *Equisetum*. Intercalary meristems also occur in the peduncles of the inflorescences of certain plants, in the leaves of many monocotyledons (e.g. in the Gramineae and *Iris*), Pinaceae, and others. The gynophore of *Ara-chis* (groundnut) also elongates as a result of the activity of an intercalary meristem (Jacobs, 1947).

Internodal elongation in many grasses is brought about by an intercalary meristem, the cells of which divide to form parallel series of cells and which is, therefore, termed rib meristem (not to be confused with the rib meristem of apices). The enlargement of the derivatives of this meristem also contributes to the elongation of the internode (Miltényi, 1931; Kaufman, 1959). Miltényi states that the intercalary meristems of grass internodes have no fixed position, but that their position is altered as the internode elongates. At first the intercalary meristematic activity occurs throughout the internode, as has been mentioned above; but after the development of the lacunae, that are present in most grasses, this activity becomes restricted to the peripheral ground tissue in the proximity of and above the *nodal plates*, i.e. in the joint regions. The meristematic activity of the joints can be reactivated even in mature stems. Horizontally placed grass shoots show a rapid, negative geotropic reaction. This response is based on cell elongation on the lower side of the leaf-sheath base, which is called the *pulvinus*. The pulvinus of grasses has a radial symmetry and is made up of epidermal, vascular, parenchymatous, and collenchymatous tissues (Dayanandan *et al.*, 1976).

In the region of the intercalary meristem of the single elongated internode of *Cyperus alternifolius* (Fisher, 1970a, b) cell divisions in the epidermis were found to occur further from the base than in the ground parenchyma. In this species, as well as in *Avena sativa*, there is a gradual decrease in the stomata formed towards the base of the internode (Fisher, 1970a; Kaufman *et al.*, 1965).

It has been shown that, in those parts of plants that have already undergone a certain degree of differentiation, such as in flowers, fruits, leaves, and stems without special intercalary meristems, the cells continue to divide for a long time after they have been derived from the apical meristem. This type of growth can also be considered as intercalary, but the growth regions are less well defined.

In the last half century many investigators have been attracted to research on meristems, and especially on apical meristems. This interest is due to the great importance of the apex in which the pattern of initiation of leaves, buds, flowers, and the various tissues is determined. The continuity of the tissue systems of the shoot and root results from the activity of the apices. Because the systems of essential elements, such as the tracheary elements, cannot undergo further changes once they are fully differentiated, and because they must be stable and continuous in order to function, the structure of the plant must be determined in the apical meristem. In long-lived plants with secondary thickening, the continuity of the conducting systems between the young parts of the root and shoot is brought about by secondary vascular tissues which are produced by the vascular cambium, a lateral meristem.

The problem of the existence of initials has been widely discussed. The view of the author of this book is that initials are present in the meristems. They may divide at a similar rate to that of the other meristem cells or less frequently. The main factor determining which of the cells of the meristem will be a permanent initial is the location of the cell. When a permanent initial in the shoot apical meristem divides, one of the daughter cells, which is located in the proper place, will inherit its function, and the other daughter cell will continue to divide. The descendants of the latter cell will continue to divide until they are in a maturing region, where these cells will undergo differentiation and maturation. When a permanent initial becomes aborted or injured, another cell, which is in the proper position as related to the other cells in the tissue, will acquire the function of a permanent initial.

In the root apex the cells on the periphery of the quiescent centre are regarded as temporary initials. The fact that cells of the quiescent centre do not divide frequently may not exclude the view that they act as initials. They may be regarded as permanent initials which contribute cells at relatively long intervals to the periphery of the quiescent centre where they function as temporary initials.

The quiescent centre may serve as a reservoir of cells relatively invulnerable to accidental injury (Clowes, 1976) and in maintenance of the geometry of the apical meristem. Recent findings by Clowes (1984) that (except for the Gramineae and Cyperaceae) the volume and the number of constituent cells of the quiescent centre vary with the width of the root, may support this view. In the non-grass type of apex the number of cells in the quiescent centre approaches zero as the width of the root becomes smaller and nearer that of pteridophyte roots with a tetrahedral apical cell.

In conclusion it should be mentioned that although we have now relatively much information about the cytohistological zonation in the apical meristems, we do not yet know how the various zones relate to the morphogenetic events occurring just behind them.

REFERENCES

ALLEN, G. S. (1947) Embryogeny and the development of the apical meristem of *Pseudotsuga*: III, Development of the apical meristem. *Am. J. Bot.* **34**: 204–11.

BAILEY, I. W. (1930) The cambium and its derivative tissues: V, A reconnaissance of the vacuome in living cells. *Z. Zellforsch. mikrosk. Anat.* **10**: 651–82.

BALL, E. (1941) The development of the shoot apex and of the primary thickening meristem in *Phoenix canariensis* Chaub. with comparisons to *Washingtonia filifera* Wats. and *Trachycarpus excelsa* Wendl. *Am. J. Bot.* **28**: 820–32.

BALL, E. (1946) Development in sterile culture of stem tips and subjacent regions of *Tropaeolum majus* L. and of *Lupinus albus* L. *Am. J. Bot.* **33**: 301–18.

BALL, E. (1947) Isolation of the shoot apex of *Lupinus*. *Am. J. Bot.* **34**: Supp. 1a–2a.

BALL, E. (1960) Cell divisions in living shoot apices. *Phytomorphology* **10**: 377–96.

BALL, E. A. (1979) Direct absorption of C^{14} compounds by the center of the shoot apex. *La Cellule* **73**: 101–114.

BARTELS, F. (1960) Zur Entwicklung der Keimpflanze von *Epilobium hirsutum*: II, Die im Vegetationspunkt während eines Plastochrons ablaufenden Zellteilungen. *Flora* **149**: 206–24.

BERGGREN, B. (1984) Ultrastructure of the histological zones in growing vegetative buds of *Salix* spp. *Nord. J. Bot.* **4**: 771–789.

BERGGREN, B. (1985) Ultrastructure of dormant buds of *Salix* sp. in early winter. *Nord. J. Bot.* **5**: 475–488.

BERNIER, G., KINET, J. M., and BRONCHART, R. (1967) Cellular events at the meristem during floral induction in *Sinapis alba* L. *Physiol. Veg.* **5**: 311–24.

BERNIER, G., KINET, J. M., BODSON, Y., ROUMA, Y. and JACQMARD, A. (1974) Experimental studies on mitotic activity of the shoot apical meristem and its relation to floral evocation and morphogenesis in *Sinapis alba*. *Bot. Gaz.* **135**: 345–352.

BERSILLON, G. (1951) Sur le point végétatif de *Papaver somniferum* L.: structure et fonctionnement. *CR Acad. Sci.* (*Paris*) **232**: 2470–2.

BERSILLON, G. (1955) Sur la formation du bouton chez le *Papaver somniferum* L. *CR Acad. Sci.* (*Paris*) **240**: 903–5.

BLOCH, R. (1941) Wound healing in higher plants. *Bot. Rev.* **7**: 110–46.

BOKE, N. H. (1941) Zonation in the shoot apices of *Trichocereus spachianus* and *Opuntia cylindrica*. *Am. J. Bot.* **28**: 656–64.

BOKE, N. H. (1947) Development of the adult shoot apex and floral initiation in *Vinca rosea* L. *Am. J. Bot.* **34**: 433–9.

BOKE, N. H. (1976) Dichotomous branching in *Mammillaria* (Cactaceae). *Am. J. Bot.* **63**: 1380–1384.

BOKE, N. H. and ROSS, R. G. (1978) Fasciation and dichotomous branching in *Echinocereus* (Cactaceae). *Amer. J. Bot.* **65**: 522–530.

BURROWS, G. E. (1987) Leaf axil anatomy in the Araucariaceae. *Aust. J. Bot.* **35**: 631–640.

BUVAT, R. (1944) Recherches sur la dédifférentiation des cellules végétales: I, Plantes entières et boutures. *Annls Sci. nat., Bot.* ser. 11, **5**: 1–130.

BUVAT, R. (1945) Recherches sur la dédifférentiation des cellules végétales: II, Cultures de tissus et tumeurs. *Annls Sci. nat., Bot.* ser. 11, **6**: 1–119.

BUVAT, R. (1952a) Structure, évolution et fonctionnement du méristème apical de quelques dicotylédones. *Annls Sci. nat., Bot.* ser. 11, **13**: 199–300.

BUVAT, R. (1952b) L'organisation des méristèmes apicaux chez les végétaux vasculaires. *Bull. Un. Nat.* **40**: 54–66.

BUVAT, R. (1955) Le méristème apical de la tige. *Ann. Biol.* **31**: 596–656.

BYRNE, J. M. and HEIMSCH, C. (1970a) The root apex of *Malva sylvestris*: I, Structural development. *Am. J. Bot.* **57**: 1170–8.

BYRNE, J. M. and HEIMSCH, C. (1970b) The root apex of *Malva sylvestris*: II, The quiescent center. *Am. J. Bot.* **57**: 1179–84.

CECICH, R. A. (1977) An electron microscopic evaluation of cytohistological zonation in the shoot apical meristem of *Pinus banksiana*. *Am. J. Bot.* **64**: 1263–71.

CLOWES, F. A. L. (1950) Root apical meristems of *Fagus sylvatica*. *New Phytol.* **49**: 248–68.

CLOWES, F. A. L. (1953) The cytogenerative centres in roots with broad columellas. *New Phytol.* **52**: 48–57.

CLOWES, F. A. L. (1954) The promeristem and the minimal constructional centre in grass root apices. *New Phytol.* **53**: 108–16.

CLOWES, F. A. L. (1958a) Development of quiescent centres in root meristems. *New Phytol.* **57**: 85–88.

CLOWES, F. A. L. (1958b) Protein synthesis in root meristems. *J. Exp. Bot.* **9**: 229–38.

CLOWES, F. A. L. (1959) Adenine incorporation and cell division in shoot apices. *New Phytol.* **58**: 16–19.

CLOWES, F. A. L. (1961) *Apical Meristems*. Blackwell, Oxford.

CLOWES, F. A. L. (1971) The proportion of cells that divide in root meristems of *Zea mays* L. *Ann. Bot.* **35**: 249–61.

CLOWES, F. A. L. (1975) The quiescent centre. In: *The Development and Function of Roots*. (eds. J. G. Torrey and D. T. Clarkson). Academic Press, London, pp. 3–19.

CLOWES, F. A. L. (1976) The root apex. In: *Cell Division in Higher Plants* (ed. M. M. Yeoman). Academic Press, London, pp. 253–84.

CLOWES, F. A. L. (1978a) Development of the shoot apex in *Zea mays*. *New Phytol.* **81**: 663–9.

CLOWES, F. A. L. (1978b) Origin of the quiescent centre in *Zea mays*. *New Phytol.* **80**: 409–419.

CLOWES, F. A. L. (1982) The growth fraction of the quiescent centre. *New Phytol.* **91:** 129–135.

CLOWES, F. A. L. (1984) Size and activity of quiescent centres of roots. *New Phytol.* **96:** 13–21.

CLOWES, F. A. L. and MACDONALD, M. M. (1987) Cell cycling and fate of potato buds. *Ann. Bot.* **59:** 141–148.

CORSON, G. E., JR., and GIFFORD, E. M., JR. (1969) Histochemical studies of the shoot apex of *Datura stramonium* during transition to flowering. *Phytomorphology* **19:** 189–96.

CROSS, G. L. (1939) The structure and development of the apical meristem in the shoots of *Taxodium distichum. Bull. Torrey Bot. Club* **66:** 431–52.

CROSS, G. L. (1942) Structure of the apical meristem and development of the foliage leaves of *Cunninghamia lanceolata. Am. J. Bot.* **29:** 288–301.

CROSS, G. L. (1943) A comparison of the shoot apices of the sequoias. *Am. J. Bot.* **30:** 130–42.

CUTTER, E. G. (1957) Studies of morphogenesis in the Nymphaeaceae: I, Introduction: some aspects of the morphology of *Nuphar lutea* (L.) SM. and *Nymphaea alba* L. *Phytomorphology* **7:** 45–56.

CUTTER, E. G. (1965) Recent experimental studies of the shoot apex and shoot morphogenesis. *Bot. Rev.* **31:** 7–113.

DAVIS, E. L., RENNIE, P., and STEEVES, T. A. (1979) Further analytical and experimental studies on the shoot apex of *Helianthus annuus*; variable activity in the central zone. *Can. J. Bot.* **57:** 971–80.

DAYANANDAN, P., HEBARD, F. V. and KAUFMAN, P. B. (1976) Cell elongation in the grass pulvinus in response to geotropic stimulation and auxin application. *Planta* **131:** 245–52.

DENNE, M. P. (1966) Morphological changes in the shoot apex of *Trifolium repens* L.: I, Changes in the vegetative apex during the plastochron. *NZ J. Bot.* **4:** 300–14.

DERMEN, H. (1945) The mechanism of colchicine-induced cytohistological changes in cranberry. *Am. J. Bot.* **32:** 387–94.

DERMEN, H. (1947) Periclinal cytochimeras and histogenesis in cranberry. *Am. J. Bot.* **34:** 32–43.

DERMEN, H. (1948) Chimeral apple sports and their propagation through adventitious buds. *J. Hered.* **39:** 235–42.

DERMEN, H. (1951) Ontogeny of tissue in stem and leaf of cytochimeral apples. *Am. J. Bot.* **38:** 753–60.

DERMEN, H. and BAIN, H. F. (1944) A general cytological study of colchicine polyploidy in cranberry. *Am. J. Bot.* **31:** 451–63.

ESAU, K. (1953 and 1965) *Plant Anatomy,* 1st and 2nd eds. Wiley, New York, London, and Sydney.

EVANS, P. S. (1965) Intercalary growth in the aerial shoot of *Eleocharis acuta* R. Br. Prodr. I. Structure of the growing zone. *Ann. Bot.* **29:** 205–17.

FAHN, A., STOLER, S., and FIRST, T. (1963) Vegetative shoot apex in banana and zonal changes as it becomes reproductive. *Bot. Gaz.* **124:** 246–50.

FELDMAN, L. J. and TORREY, J. G. (1976) The isolation and culture *in vitro* of the quiescent center of *Zea mays. Am. J. Bot.* **63:** 345–55.

FINK, S. (1983) The occurrence of adventitious and preventitious buds within the bark of some temperate and tropical trees. *Am. J. Bot.* **70:** 532–542.

FISHER, J. B. (1970a) Development of the intercalary meristem of *Cyperus alternifolius. Am. J. Bot.* **57:** 691–703.

FISHER, J. B. (1970b) Control of the internodal intercalary meristem of *Cyperus alternifolius. Am. J. Bot.* **57:** 1017–26.

FISHER, J. B. and FRENCH, J. C. (1976) The occurrence of intercalary and uninterrupted meristems in the internodes of tropical monocotyledons. *Am. J. Bot.* **63:** 510–25.

FISHER, J. B. and FRENCH, J. C. (1978) Internodal meristems of monocotyledons: further studies and general taxonomic summary. *Ann. Bot.* **42:** 41–50.

FORDE, B. J. (1965) Differentiation and continuity of the phloem in the leaf intercalary meristem of *Lolium perenne. Am. J. Bot.* **52:** 953–61.

FOSKET, D. E. and MIKSCHE, J. P. (1966) A histochemical study of the seedling shoot apical meristem of *Pinus lambertiana. Am. J. Bot.* **53:** 694–702.

FOSTER, A. S. (1938) Structure and growth of the shoot apex in *Ginkgo biloba. Bull. Torrey Bot. Club* **65:** 531–56.

FOSTER, A. S. (1939a) Problems of structure, growth and evolution in the shoot apex of seed plants. *Bot. Rev.* **5:** 454–70.

FOSTER, A. S. (1939b) Structure and growth of the shoot apex of *Cycas revoluta. Am. J. Bot.* **26:** 372–85.

FOSTER, A. S. (1940) Further studies on zonal structure and growth of the shoot apex of *Cycas revoluta. Am. J. Bot.* **27:** 487–501.

FOSTER, A. S. (1943) Zonal structure and growth of the shoot apex of *Microcycas calocoma* (Mig.) A.DC. *Am. J. Bot.* **30:** 56–73.

GARRISON, R. (1973) The growth of internodes in *Helianthus. Bot. Gaz.* **134:** 246–55.

GAUTHERET, R. J. (1945) La culture des tissus. *L'Avenir de la Science* 21. Gallimard, Paris.

GAUTHERET, R. J. (1957) Histogenesis in plant tissue cultures. *J. Nat. Cancer Inst.* **19:** 555–90.

GIFFORD, E. M., JR. (1943) The structure and development of the shoot apex of *Ephedra altissima* Desf. *Bull. Torrey Bot. Club* **70:** 15–25.

GIFFORD, E. M., JR (1950) The structure and development of the shoot apex in certain woody Ranales. *Am. J. Bot.* **37:** 595–611.

GIFFORD, E. M., JR. (1954) The shoot apex in angiosperms. *Bot. Rev.* **20:** 477–529.

GIFFORD, E. M., JR. (1963) Developmental studies of vegetative and floral meristems. In: *Meristems and Differentiation, Brookhaven Symp. Biol.* **16:** 126–37.

GIFFORD, E. M., JR. (1969) Initiation and early development of the inflorescence in pineapple (*Ananas comosus*, "Smooth Cayenne") treated with acetylene. *Am. J. Bot.* **56:** 892–7.

GIFFORD, E. M., JR., and CORSON, G. E., JR. (1971) The shoot apex in seed plants. *Bot. Rev.* **37:** 143–229.

GIFFORD, E. M. Jr. and POLITO, V. S. (1981) Mitotic activity at the shoot apex of *Azolla filiculoides. Am. J. Bot.* 68: 1050–1055.

GIFFORD, E. M., JR., and STEWART, K. D. (1965) Ultrastructure of vegetative and reproductive apices of *Chenopodium album. Science* **149:** 75–77.

GIFFORD, E. M., JR., and TEPPER, H. B. (1961) Ontogeny of the inflorescence in *Chenopodium album. Am. J. Bot.* **48:** 657–67.

GIFFORD, E. M., JR., and TEPPER, H. B. (1962a) Histochemical and autoradiographic studies of floral induction in *Chenopodium album. Am. J. Bot.* **49:** 706–14.

GIFFORD, E., M., JR., and TEPPER, H. B. (1962b) Ontogenetic and histochemical changes in the vegetative shoot tip of *Chenopodium album. Am. J. Bot.* **49:** 902–11.

GIFFORD, E. M., JR., and WETMORE, R. H. (1957) Apical meristems of vegetative shoots and strobili in certain gymnosperms. *Proc. Natn. Acad. Sci.* **43:** 571–6.

GRÉGOIRE, V. (1938) La morphogénèse et l'autonomie morphologique de l'appareil floral: I, Le carpelle. *La Cellule* **47:**

287–452.

GUNNING, B. E. S., HUGHES, J. E., and HARDHAM, A. R. (1978) Formative and proliferative cell divisions, cell differentiation, and developmental changes in the meristem of *Azolla* roots. *Planta* **143**: 121–44.

GUTTENBERG, H. v. (1940) Der primäre Bau der Angiospermenwurzel. In: K. Linsbauer, *Handbuch der Pflanzenanatomie*, Bd. 8, Lief. 39. Gebr. Borntraeger, Berlin.

GUTTENBERG, H. v. (1947) Studien über die Entwicklung des Wurzelvegetationspunktes der Dikotyledonen. *Planta* **35**: 360–96.

GUTTENBERG, H. v. (1960) Grundzüge der Histogenese höheren Pflanzen: I, Die Angiospermen. In: K. Linsbauer, *Handbuch der Pflanzenanatomie*, Spez. Teil, Bd. 8, T. 3. Gebr. Borntraeger, Berlin.

GUTTENBERG, H. v. (1961) Grundzüge der Histogenese höheren Pflanzen: II, Die Gymnospermen. In: K. Linsbauer, *Handbuch der Pflanzenanatomie* Spez. Teil, Bd. 8, T. 4. Gebr. Borntraeger, Berlin.

GUTTENBERG, H. v., BURMEISTER, J., and BROSELL, H. J. (1955) Studien über die Entwicklung des Wurzelvegetationspunktes der Dikotyledonen: II. *Planta* **46**: 179–222.

HALLE, F., OLDEMAN, R. A. A. and TOMLINSON, P. B. (1978) *Tropical Trees and Forests: An Architectural Analysis.* Springer-Verlag, Heidelberg.

HANAWA, J. (1966) Growth and development in the shoot apex of *Pinus densiflora*: I, Growth periodicity and structure of the terminal vegetative shoot apex. *Bot. Mag. Tokyo* **79**: 736–46.

HANSTEIN, J. (1868) Die Scheitelzellgruppe im Vegetationspunkt der Phanerogamen. *Fetschr. Niederrhein. Ges. Natur- und Heilkunde* **1868**: 109–34.

HARA, N. (1962) Structure and seasonal activity of the vegetative shoot apex of *Daphne pseudomezereum*. *Bot. Gaz.* **124**: 30–42.

HARDHAM, A. R. (1979) Meristem, apical. In: *McGraw-Hill Yearbook of Science and Technology*, pp. 252–4.

HAVELANGE, A. (1980) The quantitative ultrastructure of the meristematic cells of *Xanthium strumarium* during the transition to flowering. *Am. J. Bot.* **67**: 1171–1178.

HEALY, P. L. (1964) Histochemistry and ultrastructure in the shoot of *Pharbitis* before and after induction. Ph.D. thesis, Univ. of Calif., Berkeley.

JACOBS, W. P. (1947) The development of the gynophore of the peanut plant, *Arachis hypogaea*: I, The distribution of mitoses, the region of greatest elongation and the maintenance of vascular continuity in the intercalary meristem. *Am. J. Bot.* **34**: 361–70.

JACOBS, W. P. and RAGHAVAN, V. (1962) Studies on the floral histogenesis and physiology of *Perilla*: I, Quantitative analysis of flowering in *P. frutescens* (L.) Britt. *Phytomorphology* **12**: 144–67.

JANCZEWSKI, E. DE (1874) Das Spitzenwachstum der Phanerogamenwurzel. *Bot. Ztg.* **32**: 113–27.

JENSEN, L. C. W. (1971) Experimental bisection of *Aquilegia* floral buds cultured *in vitro*: I, The effect on growth primordia initiation, and apical regeneration *Can. J. Bot.* **49**: 487–93.

JENSEN, W. A. (1957) The incorporation of C^{14} adenine and C^{14} phenylalanine by developing root-tip cells. *Proc. Natn. Acad. Sci.* **43**: 1039–46.

JENSEN, W. A. and KAVALJIAN, L. G. (1958) An analysis of cell morphology and the periodicity of division in root tip of *Allium cepa*. *Am. J. Bot.* **45**: 365–72.

JOHNSON, M. A. (1939) Structure of the shoot apex in *Zamia*.

Bot. Gaz. **101**: 189–203.

JOHNSON, M. A. (1944) On the shoot apex of the cycads. *Torreya* **44**: 52–58.

JOHNSON, M. A. and TOLBERT, R. J. (1960) The shoot apex of *Bombax*. *Bull. Torrey Bot. Club* **87**: 173–86.

KAUFMAN, P. B. (1959) Development of the shoot of *Oryza sativa* L.: III, Early stages in histogenesis of the stem and ontogeny of the adventitious root. *Phytomorphology* **9**: 382–404.

KAUFMAN, P. B., CASSELL, S. J., and ADAMS, P. A. (1965) On the nature of intercalary growth and cellular differentiation in internodes of *Avena sativa*. *Bot. Gaz.* **126**: 1–13.

KEMP, M. (1943) Morphological and ontogenetic studies on *Torreya californica* Torr.: I, The vegetative apex of the megasporangiate tree. *Am. J. Bot.* **30**: 504–17.

KLEKOWSKI, E. J. Jr. and KAZARINOVA-FUKSHANSKY, N. (1984) Shoot apical meristems and mutation: fixation of selectively neutral cell genotypes. *Am. J. Bot.* **71**: 22–27.

KORODY, E. (1937) Studien am Spross-Vegetationspunkt von *Abies concolor*, *Picea excelsa* und *Pinus montana*. *Beitr. Biol. Pfl.* **25**: 23–59.

KUIJT, J. (1981) Epicortical roots and vegetative reproduction in Loranthaceae (s.s.) of the New World. *Beitr. Biol. Pflanzen* **56**: 307–316.

KUIJT, J. (1985) Morphology, biology and systematic relationships of *Desmaria* (Loranthaceae). *Pl. Syst. Evol.* **151**: 121–130.

LANCE, A. (1957) Recherches cytologiques sur l'évolution de quelques méristèmes apicaux et sur ses variations provoquées par des traitements photopériodiques. *Annls Sci. nat., Bot. ser.* 11, **18**: 91–422.

LENT, R. (1966) The origin of the cauliflorous inflorescence of *Theobroma cacao*. *Turrialba* **16**: 352–358.

LOISEAU, J. E. (1962) Activité mitotique des cellules superficielles du sommet végétatif caulinaire. *Mém., Soc. Bot. France*, pp. 14–23.

LYNDON, R. F. (1976) The shoot apex. In: *Cell Division in Higher Plants* (ed. M. M. Yeoman). Academic Press, London, pp. 285–314.

LYNDON, R. F. and ROBERTSON, E. S. (1976) The quantitative ultrastructure of the pea shoot apex in relation to leaf initiation. *Protoplasma* **87**: 387–402.

MAJUMDAR, G. P. (1945) Some aspects of anatomy in modern research. Presid. Address to Sect. of Bot. 32nd Indian Sci. Congr., Nagpur.

MARC, J. and PALMER, J. H. (1982) Changes in mitotic activity and cell size in the apical meristem of *Helianthus annuus* L. during the transition to flowering. *Am. J. Bot.* **69**: 768–775.

MAUSETH, J. D. (1980) A morphometric study of ultrastructure of *Echinocereus engelmannii* (Cactaceae). I. Shoot apical meristems at germination. *Am. J. Bot.* **67**: 173–81.

MAUSETH, J. D. (1981) A morphometric study of the ultrastructure of *Echinocereus engelmannii* (Cactaceae). II. The mature, zonate shoot apical meristem. *Am. J. Bot.* **68**: 96–100.

MAUSETH, J. D. (1982) A morphometric study of the ultrastructure of *Echinocereus engelmannii* (Cactaceae). V. Comparison with the shoot apical meristems of *Trichocereus pachanoi (Cactaceae)*. *Am. J. Bot.* **69**: 551–555.

MILLINGTON, W. F. and FISK, E. L. (1956) Shoot development in *Xanthium pennsylvanicum*: I, The vegetative plant. *Am. J. Bot.* **43**: 655–6.

MILTÉNYI, L. (1931) Histologisch-entwicklungsgeschichtliche Untersuchungen an Getreidearten. *Bot. Közl.* **28**: 1–51.

MOH, C. C. (1961) Does a coffee plant develop from one initial cell in the shoot apex of an embryo? *Radiat., Bot.* **1**: 97–99.

MOORE, R. and MCCLELEN, C. E. (1983) A morphometric analysis of cellular differentiation in the root cap of *Zea mays. Am. J. Bot.* **70**: 611–617.

MOORE, R. and PASIENINK, J. (1984) Structure of columella cells in primary and lateral roots of *Ricinus communis* (Euphorbiaceae). *Ann. Bot.* **53**: 715–726.

MOREL, G. (1965) La culture du méristème caulinaire. *Bull. Soc. fr. Physiol. vég.* **11**: 213–24.

MURTY, Y. S. and KUMAR, V. (1972) Vegetative to reproductive state of apex in monocotyledons. *Adv. Pl. Morph.*: 291–303.

NÄGELI, C. W. (1878) Über das Scheitelwachstum der Phanerogamen. *Bot. Ztg.* **36**: 124–6.

NEWMAN, I. V. (1961) Pattern in the meristems of vascular plants: II, A review of shoot apical meristems of gymnosperms, with comments on apical biology and taxonomy, and a statement of some fundamental concepts. *Proc. Linn. Soc. NSW* **86**: 9–59.

NEWMAN, I. V. (1965) Pattern in the meristems of vascular plants: III, Pursuing the patterns in the apical meristem where no cell is a permanent cell. *J. Linn. Soc., Bot.* **59**: 185–214.

NOLAN, J. R. (1969) Bifurcation of the stem apex in *Asclepias syriaca. Am. J. Bot.* **56**: 603–9.

NOUGARÈDE, A. (1965) Organisation et fonctionnement du méristème apical des végétaux vasculaires. In: *Travaux dédiés au Lucien Plantefol.* Masson et Cie, Paris.

NOUGARÈDE, A. (1967) Experimental cytology of the shoot apical cells during vegetative growth and flowering. *Int. Rev. Cytol.* **21**: 203–351.

NOUGARÈDE, A. and REMBUR, J. (1985) Le point vegetatif en tant que modele pour l'etude du cycle cellulaire et des ses points de controle. *Bull. Soc. bot. Fr.* **132**: 9–34.

ORR, A. R. (1981) A quantitative study of cellular events in the shoot apical meristem of *Brassica campestris* (Cruciferae) during transition from vegetative to reproductive conditions. *Am. J. Bot.* **68**: 17–23.

PARKE, R. V. (1959) Growth periodicity and the shoot tip of *Abies concolor. Am. J. Bot.* **46**: 110–18.

PARTANEN, C. R. and GIFFORD, E. M., JR. (1958) Application of autoradiographic techniques to studies of shoot apices. *Nature* **182**: 1747–8.

PATRICK, J. W. (1972) Vascular system of the stem of the wheat plant. II. Development. *Aust. J. Bot.* **20**: 65–78.

PHILIPSON, W. R. (1947) Apical meristems of leafy and flowering shoots. *J. Linn. Soc., Bot.* **53**: 187–93.

PHILIPSON, W. R. (1949) The ontogeny of the shoot apex in dicotyledons. *Biol. Rev.* **24**: 21–50.

PHILIPSON, W. R. (1954) Organization of the shoot apex in dicotyledons. *Phytomorphology* **4**: 70–75.

PILLAI, S. K. (1963) Structure and seasonal study of the shoot apex of some *Cupressus* species. *New Phytol.* **62**: 335–41.

PILLAI, S. K. and PILLAI, A. (1961) Root apical organization in monocotyledons—Musaceae. *Indian Bot. Soc. J.* **40**: 444–55.

PLANTEFOL, L. (1947) Hélices foliaires, point végétatif et stèle chez les Dicotylédones. La notion d'anneau initial. *Rev. gén. Bot.* **54**: 49–80.

POLITO, V. S. (1979) Cell division kinetics in the shoot apical meristem of *Ceratopteris thalictroides* Brong. with special reference to the apical cell. *Am. J. Bot.* **66**: 485–93.

POPHAM, R. A. (1952) *Developmental Plant Anatomy.* Long, Columbus, Ohio.

POPHAM, R. A. (1958) Cytogenesis and zonation in the shoot apex of *Chrysanthemum morifolium. Am. J. Bot.* **45**: 198–206.

POPHAM, R. A. and CHAN, A. P. (1950) Zonation in the vegetative stem of *Chrysanthemum morifolium* Bailey. *Am. J. Bot.* **37**: 476–84.

POPHAM, R. A. and CHAN, A. P. (1952) Origin and development of the receptacle of *Chrysanthemum morifolium. Am. J. Bot.* **39**: 329–39.

ROMBERGER, J. A., VARNELL, R. J., and TABOR, C. A. (1970) *Culture of Apical Meristems and Embryonic Shoots of Picea abies—Approach and Techniques.* Technical Bulletin 1409, USDA Forest Service, 1–30.

ROST, T. L. and JONES, T. J. (1988) Pea root regeneration after tip excisions at different levels: Polarity of new growth. *Ann. Bot.* **61**: 513–523.

SACHS, T. (1981) The control of the patterned differentiation of vascular tissues. In: *Advances in Botanical Research.* Vol. 9 (ed. H. W. Woolhouse). Academic Press, New York, pp. 151–262.

SATINA, S. (1959) Chimeras. In: A. F. Blakeslee, *The Genus Datura* (eds. A. G. Avery, S. Satina, and J. Rietsema). Ronald Press, New York. pp. 132–51.

SATINA, S. and BLAKESLEE, A. F. (1941) Periclinal chimeras in *Datura stramonium* in relation to development of leaf and flower. *Am. J. Bot.* **28**: 862–71.

SATINA, S. and BLAKESLEE, A. F. (1943) Periclinal chimeras in *Datura* in relation to the development of the carpel. *Am. J. Bot.* **30**: 453–62.

SATINA, S., BLAKESLEE, A. F. and AVERY, A. G. (1940) Demonstration of the three germ layers in the shoot apex of *Datura* by means of induced polyploidy in periclinal chimeras. *Am. J. Bot.* **27**: 895–905.

SCHADE, C. and GUTTENBERG, H. v. (1951) Über die Entwicklung des Wurzelvegetationspunktes der Monokotyledonen. *Planta* **40**: 170–98.

SCHMIDT, A. (1924) Histologische Studien an phanerogamen Vegetationspunkten. *Bot. Arch.* **8**: 345–404.

SCHÜEPP, O. (1917) Untersuchungen über Wachstum und Formwechsel von Vegetationspunkten. *Jb. wiss. Bot.* **57**: 17–79.

SCHÜEPP, O. (1926) *Meristeme.* Gebr. Borntraeger, Berlin.

SCHÜEPP, O. (1966) Meristeme. *Experientia*, suppl. 11. Berkhaüser-Verlag, Basel.

SCHULZ, A. (1986) Wound phloem in transition to bundle phloem in primary roots of *Pisum sativum* L. I. Development of bundle-leaving wound-sieve tubes. *Protoplasma* **130**: 12–26.

SHAH, J. J. and DAVE, Y. S. (1970) Morpho-histogenic studies on tendrils of Vitaceae. *Am. J. Bot.* **57**: 363–73.

SHAH, J. J. and PATEL, J. D. (1972) The shell zone: its differentiation and probable function in some dicotyledons. *Am. J. Bot.* **59**: 683–690.

SHAH, J. J. and UNNIKRISHNAN, K. (1970) Ontogeny of axillary and accessory bud in *Mentha spicata* L. *Beitr. Biol. Pfl.* **46**: 355–69.

SMITH, R. H. and MURASHIGE, T. (1970) *In vitro* development of the isolated shoot apical meristem of angiosperms. *Am. J. Bot.* **57**: 562–8.

STAFF, I. A. (1968) A study of the apex and growth pattern in the shoot of *Xanthorrhoea media* R.Br. *Phytomorphology* **18**: 153–66.

STEIN, D. B. and STEIN, O. L. (1960) The growth of the stem tip of Kalanchoë cv. "Brilliant Star". *Am. J. Bot.* **47**: 132–40.

STERLING, C. (1945) Growth and vascular development in the

shoot apex of *Sequoia sempervirens* (Lamb.) Endl.: I, Structure and growth of the shoot apex. *Am. J. Bot.* **32**: 118–26.

STERLING, C. (1946) Organization of the shoot of *Pseudotsuga taxifolia* (Lamb.) Britt.: I, Structure of the shoot apex. *Am. J. Bot.* **33**: 742–60.

STEVENSON, D. WM. (1978) The shoot apex of *Bougainvillea spectabilis*. *Am. J. Bot.* **65**: 792–4.

STEWARD, F. C. (1970) From cultured cells to whole plants: the induction and control of their growth and morphogenesis. *Proc. R. Soc.*, ser. B, **175**: 1–30.

STEWART, R. N. and DERMEN, H. (1970) Determination of number and mitotic activity of shoot apical initial cells by analysis of mericlinal chimeras. *Am. J. Bot.* **57**: 816–26.

TEPPER, H. B. (1963) Dimensional and zonational variation in dormant shoot apices of *Pinus ponderosa*. *Am. J. Bot.* **50**: 589–96.

THIELKE, C. (1948) Beiträge zur Entwicklungsgeschichte und zur Physiologie panaschierter Blätter. *Planta* **36**: 2–33.

THIELKE, C. (1954) Die histologische Struktur des Sprossvegetationskegels einiger Commelinaceen unter Berücksichtigung panaschierter Formen. *Planta* **44**: 18–74.

THIELKE, C. (1955) Die Struktur des Vegetationskegels einer sektorial panaschierten *Hemerocallis fulva*. *Ber. dt. bot. Ges.* **68**: 233–8.

THIELKE, C. (1957) Chimären mit periklinal spaltender Oberhaut am Scheitel. *Acta Soc. Bot. Pol.* **26**: 247–53.

THIELKE, C. (1964) Histologische Untersuchungen am Sprossscheitel von *Saccharum*: IV, Der Sprossscheitel von *Saccharum officinarum* L. *Planta* **62**: 332–49.

THOMAS, R. G. (1963) Floral induction and the stimulation of cell division in *Xanthium*. *Science* **140**: 54–56.

THOMPSON, M. M. and OLMO, H. P. (1963) Cytohistological studies of cytochimeric and tetraploid grapes. *Am. J. Bot.* **50**: 901–6.

TOLBERT, R. J. (1961) A seasonal study of the vegetative shoot apex and the pattern of pith development in *Hibiscus syriacus*. *Am. J. Bot.* **48**: 249–55.

TOMLINSON, P. B. (1970) Dichotomous branching in *Flagellaria indica* (monocotyledones). In: *New Research in Plant Anatomy* (ed. N. K. B. Robson, D. F. Cutler, and M. Gregory). Academic Press, London. *J. Linn. Soc., Bot.* **63**, suppl. 1:1–14.

TOMLINSON, P. B. and POSLUSZNY, U. (1977) Features of dichotomizing apices in *Flagellaria indica* (monocotyledones). *Am. J. Bot.* **64**: 1057–65.

TORREY, J. G. and FELDMAN, L. J. (1977) The organization and function of the root apex. *Am. Sci.* **65**: 334–344.

TRIBOT, R. (1961) Le point végétatif du *Cryptomeria japonica* Don. Phyllotaxie et vascularisation des axes épicotylés.

Rev. Cytol. Biol. Vég. **23**: 1–48.

TUCKER, S. C. (1960) Ontogeny of the floral apex of *Michelia fuscata*. *Am. J. Bot.* **47**: 266–77.

VANDEN BORN, W. H. (1963) Histochemical studies of enzyme distribution in shoot tips of white spruce (*Picea glauca* (Moench.) Voss). *Can. J. Bot.* **41**: 1509–27.

VASIL, I. K. (1984) *Cell Culture and Somatic Cell Genetics of Plants*. Vol. 1. *Laboratory Procedures and Their Applications*. Academic Press, New York.

VAUGHAN, J. G. (1955) The morphology and growth of the vegetative and reproductive apices of *Arabidopsis thaliana* (L.) Heynh., *Capsella bursa-pastoris* (L.) Medic. and *Anagallis arvensis* L. *J. Linn. Soc., Bot.* **55**: 279–301.

WARDLAW, C. W. (1957) On the organization and reactivity of the shoot apex in vascular plants. *Am. J. Bot.* **44**: 176–85.

WEBSTER, P. L. and LANGENAUER, H. D. (1973) Experimental control of the activity of the quiescent center in excised root tips of *Zea mays*. *Planta* **112**: 91–100.

WERKER, E. and FAHN, A. (1966) Vegetative shoot apex and the development of leaves in articulated Chenopodiaceae. *Phytomorphology* **16**: 393–401.

WEST, W. C. and GUNCKEL, J. E. (1968a) Histochemical studies of the shoot of *Brachychiton*: I, Cellular growth and insoluble carbohydrates. *Phytomorphology* **18**: 269–82.

WEST, W. C. and GUNCKEL, J. E. (1968b) Histochemical studies of the shoot of *Brachychiton*: II, RNA and protein. *Phytomorphology* **18**: 283–93.

WETMORE, R. H. (1954) The use of *in vitro* cultures in the investigation of growth and differentiation in vascular plants. *Abnormal and Pathological Plant Growth, Brookhaven Symp. Biol.* **6**: 22–40.

WETMORE, R. H. (1956) Growth and development in the shoot system of plants. In: *Cellular Mechanism in Differentiation and Growth, Symp. Dev. Growth* **14**: 173–190.

WETMORE, R. H., GIFFORD, E. M., JR., and GREEN, M. C. (1959) Development of vegetative and floral buds. In: *Photoperiodism and Related Phenomena in Plants and Animals. Proc. Conf. Photoperiodism, Oct. 29–Nov. 2, 1957. AAAS, Washington*, pp. 255–73.

WHEAT, D. (1980) Sylleptic branching in *Myrsine floridana* (Myrsinaceae). *Am. J. Bot.* **67**: 490–499.

WHITE, P. R. (1946) Plant tissue cultures: II. *Bot. Rev.* **12**: 521–9.

WILCOX, H. (1954) Primary organization of active and dormant roots of noble fir, *Abies procera*. *Amer. J. Bot.* **41**: 812–21.

WOLFF, C. F. (1759) *Theoria Generationis*. W. Engelmann, Leipzig.

MATURE TISSUES

It is the usual practice to divide the plant body into different tissues, but with the accumulation of knowledge of tissue structure it becomes more and more difficult to give an exact definition of a tissue. The accepted definition of a tissue is a group of cells with common origin, structure, and function. However, this definition is not suitable for all cases. When dealing with the tissues of higher plants a more flexible definition is necessary. If we could base our descriptions on elements, i.e. the individual types of cells, it would be easier to define these units. However, difficulties would also arise from such a classification because of the transitional forms present. We also know now that experimental treatments can cause one type of cell to change into another. Parenchyma cells, for example, can be stimulated to redifferentiate to tracheary elements. For the sake of convenience, in this book the anatomical and histological structure of plants will be discussed on the basis of tissue classification. Today a complex of cells of common origin is generally understood by the term tissue. A tissue may consist of cells of different form and even different function, but in tissues consisting of different cell types the cell composition is always the same.

The tissues in the plant body are classified on the following bases: according to their position in the plant; the cell types of which they consist; their function; the manner and place of their origin; and their stage of development. Tissues are also divided into *simple* and *complex tissues* according to the number of cell types that they comprise. A simple tissue is homogeneous and consists of only one type of cell, while a complex tissue is heterogeneous and comprises two or more cell types. *Parenchyma, collenchyma,* and *sclerenchyma* are simple tissues, while *xylem, phloem,* and *epidermis* are complex tissues. Elements of parenchyma, sclerenchyma, and other types may be included in complex tissues. The classical classification of the main plant tissues (Chapters 4–8, 10) is based primarily on the segregation of cell complexes below the promeristem, though the structure and the function of the cells when they are mature is also taken into account. Some secretory structures which form complex systems and are usually associated with the conductive tissues will be discussed separately (Chapter 9). Other secretory structures will be dealt with in connection with the tissues and organs in which they occur.

CHAPTER 4

PARENCHYMA

THE parenchyma of the primary plant body develops from the ground meristem, and that connected with the vascular elements from the procambium or cambium. The phellogen in many plants also produces parenchyma (the phelloderm). Parenchyma consists of living cells of differing shape and with differing physiological functions.

By the term *parenchyma*, which was first introduced by Nehemiah Grew in 1682 (Metcalfe, 1979), we generally refer to tissues which exhibit relatively little specialization and which may be concerned with various physiological functions of the plant.

Parenchyma cells retain the ability to divide even when mature. They also play an important role in wound recovery and regeneration. Mature parenchyma cells may resume meristematic activity when their environment is artificially changed. It has been shown recently that a small group of parenchyma cells, or even one such cell, when grown in proper culture media, may produce whole plants which will flower and produce viable seeds (Steward, 1963; Steward *et al.*, 1958, 1964, 1970). Phylogenetically the parenchyma of the primary body is considered to be a primitive tissue as the lowest multicellular plants consist of parenchyma only. Ontogenetically parenchyma may also be considered primitive as its cells are morphologically similar to those of meristems.

Large portions of the plant, such as the pith, all or most of the cortex of the root and shoot, the pericycle, the mesophyll of the leaf, and the fleshy part of fruits, consist of parenchyma. Parenchyma cells also occur in the xylem and phloem.

SHAPE AND ARRANGEMENT OF PARENCHYMA CELLS

Many parenchyma cells are polyhedral and their diameter in the different planes is more or less equal (Fig. 16, nos. 9, 10) but many other shapes are common. Elongated parenchyma cells are found in the palisade tissue of the leaf, in the vascular rays, etc.; lobed cells are found in spongy mesophyll and in the palisade parenchyma of *Lilium* (Fig. 124, no. 1); and in the mesophyll of the Xanthorrhoeaceae the parenchyma cells have folds or projections (Fahn, 1954) Stellate parenchyma cells (Fig. 121, no. 1) are found in the stems of plants with well-developed air spaces, such as *Scirpus* and *Juncus*, for example. In *Juncus*, according to Geesteranus (1941), the stellate pith cells differentiate ontogenetically from a mass of cubo-octahedral cells which are arranged in vertical rows on their hexagonal faces. The mechanical stretching of the pith, which is mainly in a radial direction as a result of the growth of the surrounding tissues, as well as the special arrangement of the intercellular spaces, causes the development of the characteristic arms of the cells.

Parenchymal cells with inner wall protuberance (Fig. 135, no. 4) occur in various anatomical structures concerned with short distance transfer of solutes, such as in nectaries, salt glands and vascular parenchyma (Schnepf, 1965; Shimony and Fahn, 1968; Fahn and Rachmilevitz, 1970; Gunning and Pate, 1969, 1974). These cells were termed *transfer cells* by Gunning *et al.* (1968).

The medium-sized polyhedral parenchyma cells usually have 14 faces (Higinbotham, 1942; Lewis, 1944; Matzke, 1946; Hulbary, 1948); the number of sides is less in the smaller cells and greater in the larger cells (Marvin, 1944). The number and size of the intercellular spaces also affects the number of faces of the polyhedron as the presence of intercellular spaces reduces the planes of contact between the cells. The polyhedral shape of the parenchyma cells is the result of numerous factors among which are pressure and surface tension (see Chapter 2).

Mature parenchymatous tissue may be tightly packed and without intercellular spaces or it may have a well-developed system of intercellular spaces. For example, the parenchyma of the endosperm of most seeds is devoid, or almost so, of intercellular spaces, while in the stems and leaves of hydrophytes the intercellular spaces reach maximal development. The tissue with the prominent intercellular spaces occurring in plants, growing in waterlogged soils and aquatic habitats, is important in aerating the plant, and is called *aerenchyma* (see Chapters 11, 12, 13).

The development of intercellular spaces is either *schizogenous* or *lysigenous*. The schizogenous development of an intercellular space takes place as follows: at the time when the primary wall is formed between two new cells the middle lamella between the two new cell walls (Fig. 18, nos. 7–10) comes into contact only with the primary wall of the other cell and not with the middle lamella between it and the neighbouring cells. A small space develops where the new middle lamella comes into contact with the mother cell wall. That portion of the mother cell wall opposite this small

space disintegrates and so forms the intercellular space which can be enlarged by the formation of a similar space in the neighbouring cell. The intercellular space is lined with the substance of the middle lamella. These intercellular spaces may be further enlarged by divisions of the surrounding cells in a plane perpendicular to the circumference of the space. The resin ducts of the Coniferae, the secretory ducts of the Compositae, Umbelliferae, *Hedera helix* and the secretory cavities of *Eucalyptus* and *Lysimachia* (Fig. 45, nos. 1–4) are formed schizogenously (Carr and Carr, 1970; Fahn, 1979; Lersten, 1986). Lysigenous intercellular spaces are formed by the disintegration of entire cells. Examples of lysigenous intercellular spaces are the large spaces in water plants, in the roots of some monocotyledons and in the primary resin ducts of *Mangifera indica* (Fig. 46). Various views exist as to the manner of formation of the oil cavities in the genus *Citrus* (Fig. 290, no. 1) (see Fahn, 1979).

STRUCTURE AND CONTENT OF PARENCHYMA CELLS

Most parenchyma cells, e.g. those that contain chloroplasts and those that act as storage cells,

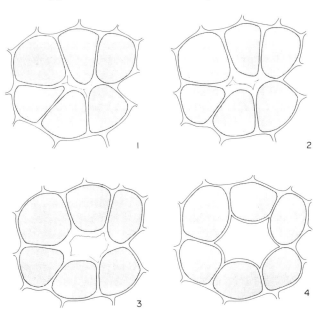

FIG. 45. Schematic drawings of successive stages (nos. 1–4) of schizogenous development of a resin duct of *Pinus halepensis*.

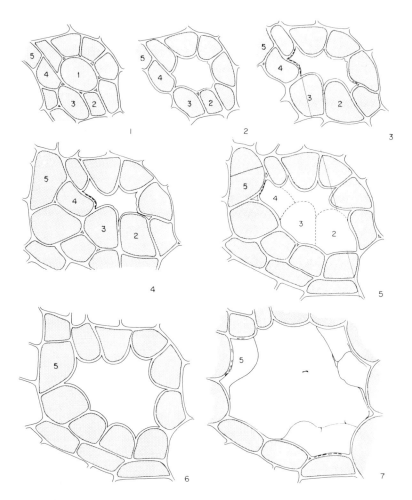

Fig. 46. A schematic drawing showing successive stages (nos. 1–7) of lysigenous development of a resin duct in the primary shoot of *Mangifera indica*. The disintegrating cells are numbered. (Adapted from Joel and Fahn, 1980.)

usually have thin primary walls, but parenchyma cells with thick primary walls also exist. Certain parenchyma storage cells, as, for example, the endosperm of *Phoenix* (Fig. 305, no. 3), *Diospyros*, *Coffea*, and *Asparagus*, have very thick walls in which hemicellulose, which serves as the reserve substance, accumulates. The walls of these cells gradually become thinner during germination. Parenchyma cells with relatively thick and lignified secondary walls are common, especially in the secondary xylem.

Lignification of parenchyma cell-walls occurs often in tissues infected by micro-organisms or in response to mechanical damage. It has been suggested that induced lignification may play a role in disease resistance (Vance *et al.*, 1980).

The internal structure of the parenchyma cell varies according to its function. Parenchyma cells which take part in photosynthesis contain chloroplasts and then the tissue they form is termed *chlorenchyma*. In the photosynthetic parenchyma there are usually single or numerous vacuoles. Certain parenchyma cells contain leucoplasts. Parenchyma cells may serve to store different reserve materials which may be found in solution in the vacuoles, or in the form of solid particles or liquid in the cytoplasm. Sugars or other soluble carbohydrates and nitrogenous substances may be found in the cell sap. Amides, proteins, and sugars are found dissolved in the cell sap, as, for example, in the roots of the sugar-beet and in the bulb scales of *Allium cepa*. Starch, protein, oils, and fats occur in the cytoplasm in the form of small particles. Protein and starch grains are found in the cytoplasm of the cells of the cotyledons of many species of the Leguminosae, and protein and oils are found in the endosperm of *Ricinus communis* and the cotyledons of *Glycine max*. In the parenchyma cells of the potato tuber, for example, amides and proteins are found in the vocuole and starch in the cytoplasm. Starch is the most common reserve material in plants and it is found in the endosperm, cotyledons, tubers, fruits, xylem and phloem parenchyma, the cortex, etc.

In succulent plants, parenchyma cells that store water are present. Such cells are usually large, thin-walled, and have only a thin layer of cytoplasm, and they are devoid of or contain very few chloroplasts. The water-storing cells have a large vacuole which contains somewhat mucilaginous sap. The mucilaginous substances apparently increase the water-holding capacity of the cell, and they are also present in the cytoplasm and wall.

The parenchyma cells of different storage organs contain water as well as reserve substances, as, for instance, in the potato tuber which supplies water to the developing parts of the plant at the start of the sprouting process.

Many parenchyma cells contain tannins, and such cells may be scattered throughout the plant or they may form continuous systems. Most of the tannins are found in the vacuoles. Tannin-containing cells retain the ability to divide and grow as do parenchyma cells devoid of tannins. Mineral substances can be found in various crystalline forms in parenchyma cells. Some such cells may remain viable after the formation of the crystals but others die.

Idioblastic parenchyma cells may contain various substances such as the enzyme myrosin (Cruciferae, Capparidaceae, Resedaceae, etc.), oily substances (Lauraceae, Simarubaceae, Calycanthaceae, etc.), mucilaginous substances (many monocotyledons and the Cactaceae, Portulacaceae, Malvaceae, Tiliaceae, etc.) and resinous substances (Meliaceae, some Rutaceae, Rubiaceae, etc.) (Fohn, 1935; Sperlich, 1939; Metcalfe and Chalk, 1950, 1983; Mauseth, 1980; Trachtenberg and Fahn, 1981; Baas and Gregory, 1985).

Many elements which cannot be classified in any other tissue are often included in the parenchyma although their characteristics vary from those described in this chapter. Thus elongated rectangular cells with secondary walls, which often occur below the epidermis of many conifer leaves or in the xylem, can be considered as parenchymatous elements. For these cells the term *sclerified parenchyma cells* is suggested. Writing about "sklerotische Zellen", de Bary (1877) refers, at least in part, to such cells.

REFERENCES

BAAS, P. and GREGORY, M. (1985) A survey of oil cells in the dicotyledons with comments on their replacement by and joint occurrence with mucilage cells. *Israel J. Bot.* **34:** 167–186.

CARR, D. J. and CARR, S. G. M. (1970) Oil glands and ducts in *Euclayptus* L'Hérit.: II. Development and structure of oil glands in the embryo. *Aust. J. Bot.* **18:** 191–212.

DE BARY, A. (1877) *Vergleichende Anatomie der Vegetationsorgane der Phanerogamen und Farne.* W. Englemann, Leipzig.

FAHN, A. (1954) The anatomical structure of Xanthorrhoeaceae Dumort. *J. Linn. Soc., Bot.* **55:** 158–84.

FAHN, A. (1979) *Secretory Tissues in Plants.* Academic Press, London, New York, San Francisco.

FAHN, A. and RACHMILEVITZ, T. (1970) Ultrastructure and nectar secretion in *Lonicera japonica.* In: *New Research in Plant Anatomy* (ed. N. K. B. Robson, D. F. Cutler, and M. Gregory). Academic Press, London. *Linn. Soc., Bot.* **63,** suppl. 1: 51–56.

FOHN, M. (1935) Zur Entstehung und Weiterbildung der Exkreträume von *Citrus medica* L. und *Eucalyptus globulus* Lab. *Öst. bot. Z.* **84:** 198–209.

GEESTERANUS, R. A. M. (1941) On the development of the stellate form of the pith cells of *Juncus* species. *Proc. Nederl. Akad. Wetensch.* **44:** 489–501, 648–53.

GUNNING, B. E. S. and PATE, J. S. (1969) "Transfer cells". Plant cells with wall ingrowths, specialized in relation to short distance transport of solutes—their occurrence, structure and development. *Protoplasma* **68:** 107–33.

GUNNING, B. E. S. and PATE, J. S. (1974) Transfer cells. In: *Dynamic Aspects of Plant Ultrastructure* (ed. A. W. Robards). McGraw-Hill, London, pp. 441–80.

GUNNING, B. E. S., PATE, J. S., and BRIARTY, L. G. (1968) Specialized "transfer cells" in minor veins of leaves and their possible significance in phloem translocation. *J. Cell Biol.* **37:** C7–12.

HIGINBOTHAM, N. (1942) The three-dimensional shape of undifferentiated cells in the petiole of *Angiopteris evecta. Am. J. Bot.* **29:** 851–8.

HULBARY, R. L. (1948) Three-dimensional cell shape in the tuberous roots of *Asparagus* and in the leaf of *Rhoeo. Am. J. Bot.* **35:** 558–66.

JOEL, D. M. and FAHN, A. (1980) Ultrastructure of the resin ducts of *Mangifera indica* L. (Anacardiaceae). 1. Differentiation and senescence of the shoot ducts. *Ann. Bot.* **46:** 225–233.

LERSTEN, N. R. (1986) Re-investigation of secretory cavity development in *Lysimachia* (Primulaceae). *New Phytol.* **102:** 193–197.

LEWIS, F. T. (1944) The geometry of growth and cell division in columnar parenchyma. *Am. J. Bot.* **31:** 619–29.

MARVIN, J. W. (1944) Cell shape and cell volume relations in the pith of *Eupatroium perfoliatum* L. *Am. J. Bot.* **31:** 208–18.

MATZKE, E. B. (1946) The three-dimensional shape of bubbles of foam—an analysis of the role of surface forces in three-dimensional cell shape determination. *Am. J. Bot.* **33:** 58–80.

MAUSETH, J. D. (1980) A stereological morphometric study of the ultrastructure of mucilage cells in *Opuntia polyacantha* (Cactaceae). *Bot. Gaz.* **141:** 374–378.

METCALFE, C. R. (1979) Some basic types of cells and tissues. In: *Anatomy of the Dicotyledons.* 2nd ed. Vol. 1 (eds. C. R. Metcalfe and L. Chalk). Clarendon Press, Oxford.

METCALFE, C. R. and CHALK, L. (1950) *Anatomy of the Dicotyledons.* Clarendon Press, Oxford.

METCLAFE, C. R. and CHALF, L. (1983) *Anatomy of the Dicotyledons.* 2nd ed. Vol. II. Clarendon Press, Oxford.

SCHNEPF, E. (1965) Die Morphologie der Sekretion in pflanzlichen Drüsen. *Ber. dt. bot. Ges.* **78:** 478–83.

SHIMONY, C. and FAHN, A. (1968) Light- and electron-microscopical studies on the structure of salt glands of *Tamarix aphylla* L. *J. Linn. Soc. Bot.* **60:** 283–8.

SPERLICH, A. (1939) Das trophische Parenchym. B. Exkretionsgewebe. In: K. Linsbauer, *Handbuch der Pflanzenanatomie,* Band 4, Lief. 38. Gebr. Borntraeger, Berlin.

STEWARD, F. C. (1963) The control of growth in plant cells. *Scient. Am.* **209:** 104–13.

STEWARD, F. C., MAPES, M. O., and MEARS, K. (1958) Growth and organized development of cultured cells: II. Organization in cultures grown from freely suspended cells. *Am. J. Bot.* **45:** 704–8.

STEWARD, F. C., MAPES, M. O., KENT, A. E., and HOLSTEN, R. D. (1964) Growth and development of cultured plant cells. *Science* **143:** 20–27.

STEWARD, F. C., AMMIRATO, P. V., and MAPES, M. O. (1970) Growth and development of totipotent cells: some problems, procedures and perspectives. *Ann. Bot.* **34:** 761–87.

TRACHTENBERG, S. and FAHN, A. (1981) The mucilage cells of *Opuntia ficus-indica* (L.) Mill.—Development, ultrastructure and mucilage secretion. *Bot. Gaz.* **142:** 206–213.

VANCE, C. P., KIRK, T. K. and SHERWOOD, R. T. (1980) Lignification as a mechanism of disease resistance. *Ann. Rev. Phytopathol.* **18:** 259–288.

CHAPTER 5

COLLENCHYMA

THE supporting tissues of the plant, i.e. the collenchyma and the sclerenchyma, are designated from the function point of view by the term *stereome* (Haberlandt, 1918).

Ontogenetically, collenchyma develops from elongated cells which resemble procambium and which appear in the very early stages of the differentiation of the meristem or from more or less isodiametric cells of the ground meristem. Collenchyma consists of living, more or less elongated cells which, generally, have unevenly thickened walls. Collenchyma functions as a supporting tissue in young growing organs and, in herbaceous plants, even in mature organs (Ambronn, 1881; Müller, 1890; Anderson, 1927; Esau, 1936). Collenchyma is plastic and it stretches irreversibly with the growth of the organ in which it occurs. Mature collenchyma is less plastic, harder, and more brittle than young collenchyma. There is a physiological and morphological relationship between collenchyma and parenchyma, and, in places where the two tissues occur side by side, transitional forms can be found between typical collenchyma and typical parenchyma.

Collenchyma, like parenchyma, may contain chloroplasts. Chloroplasts occur in larger numbers in less specialized collenchyma cells which resemble parenchyma and in smaller numbers, or not at all, in the most specialized collenchyma, which consists of elongated narrow cells. Collenchyma cells may also contain tannins.

In a freshly made cross-section of collenchyma the cell walls appear *nacré*-like. In plants exposed to wind the walls of collenchyma cells become thicker. It has been shown that mechanical shaking of a plant causes an increase in the amount of wall thickening in collenchyma (Walker, 1957, 1960). Collenchyma may become lignified and the walls may thicken, thus resulting in the formation of sclerenchyma. The thickened walls of the collenchyma may, secondarily, become thin and then the cells may again become meristematic and start to divide as, for instance, where the phellogen is formed in collenchyma tissue. Primary pit fields can be distinguished in the walls of collenchyma cells.

POSITION OF COLLENCHYMA IN THE PLANT

Collenchyma may occur in stems, leaves, floral parts, fruits, and even in roots, e.g. in *Diapensia* and *Vitis* (Duchaigne, 1955; Metcalfe, 1979). In roots, the collenchyma is mainly developed when they are exposed to light (van Fleet, 1950). Collenchyma is absent in the stems and leaves of many monocotyledons where sclerenchyma develops at an early age. Collenchyma usually forms immediately below the epidermis but, in certain cases, one or two layers of parenchyma occur between the collenchyma and the epidermis. When the collenchyma is found directly beneath the epidermis, the inner walls, or sometimes the entire wall, of the epidermal cells become thickened similarly to the walls of the collenchyma cells. In stems the collenchyma may occur as a complete cylinder or in longitudinal strips. In leaves the collenchyma occurs on one or both sides of the veins and along the margins of the blade. In many plants groups of elongated parenchyma cells which become collenchyma-like occur on the outside of the phloem strands and also on the inside of the xylem, or even as a sheath around the entire vascular bundle. When these groups are only on the sides of the xylem or phloem they appear dome-shaped in cross-section (Fig. 99, no. 2).

85

STRUCTURE AND ARRANGEMENT
OF COLLENCHYMA CELLS

The size and shape of the collenchyma cells varies. The cells may be short prisms and resemble the neighbouring parenchyma cells, or long and fibre-like with tapered ends, but all intermediate shapes and sizes occur. The longest collenchyma cells are found in the central portions of the strands of collenchyma, and the shorter ones on the periphery. This can be explained as follows: the collen-

chyma strands are formed by a series of longitudinal divisions which start in the centre of the strand; the cells continue to elongate after division, thus the central cells are the longest as they are the first to be formed and to reach the maximum length. During the development of the collenchyma strands horizontal divisions also take place.

According to the type of wall thickening three main types of collenchyma have been distinguished (Müller, 1890).

Angular collenchyma (Fig. 47, nos. 1, 5) in which

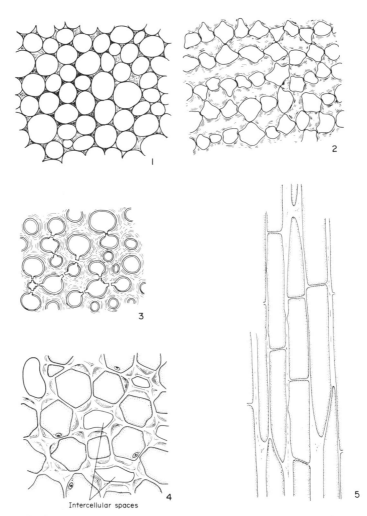

Intercellular spaces

FIG. 47. Different types of collenchyma. 1, Angular collenchyma as seen in a cross-section of a leaf petiole of *Begonia*. 2, Lamellar collenchyma as seen in a cross-section of the cortex of a young branch of *Sambucus*. 3, Annular collenchyma as seen in cross-section of the main vein of a leaf of *Nerium oleander*. 4, Lacunar collenchyma as seen in cross-section of a leaf petiole of *Petasites officinalis*. 5, Angular collenchyma as seen in longitudinal section. (No. 4 adapted from Müller, 1890; no. 5 adapted from Haberlandt, 1918.)

the thickening of the cell wall is longitudinal in the angles of the cells. In cross-section these thickenings are seen to be in those places where three or more cells meet. Examples of such collenchyma are found in the petioles of the leaves of *Vitis, Begonia, Coleus, Cucurbita, Morus, Beta,* and in the stems of *Solanum tuberosum* and *Atropa belladonna*.

Lamellar (or *tangential*) *collenchyma* (Fig. 47, no. 2), in which the thickenings are mainly on the tangential walls of the cells. Examples of this type of collenchyma are found in the stem cortex of *Sambucus nigra* and *Rhamnus* and the petiole of *Cochlearia armoracia*.

Lacunar collenchyma (Fig. 47, no. 4), in which the thickenings appear in those parts of the cell wall which face intercellular spaces. Such collenchyma can be seen in the petioles of species of the Compositae, *Salvia, Malva, Althaea,* and *Asclepias*. However, as intercellular spaces can be distinguished in other types of collenchyma, this does not seem to be a valid criterion for classification of special types.

Duchaigne (1955) distinguished an additional type, i.e. *annular collenchyma* (Fig. 47, no. 3) (cf. "Knorpelcollenchym" in Müller, 1890), in which the cell lumen is circular, or almost so, in cross-section. However, from observations made on the maturation of angular collenchyma, it was seen that, with the continued thickening of the cell wall, the lumen loses its angular appearance.

A special type of collenchyma cell was observed in the pulvinar region of *Lavatera cretica*. The walls of these collenchyma cells have on their inside, a series of transverse, ridge-like thickenings, rather than being smoothly thickened. It has been suggested that the specialized wall structure endow the pulvinar collenchyma with the flexibility required to accommodate the changes in volume occurring in the pulvinus, which are involved in reversible leaf bending (Werker and Koller, 1987).

According to many authors the walls of collenchyma cells consist of alternating layers that are rich in cellulose and poor in pectic compounds, and those which are poor in cellulose and rich in pectic compounds. In fresh material the water content of the entire wall is about 67%. Roelofsen (1959) states that in *Petasites* the collenchyma cell walls contain 45% pectin, 35% hemicellulose, and about 20% cellulose. Majumdar and Preston (1941) and

Preston and Duckworth (1946), working on *Petasites* and *Heracleum*, found that the walls of collenchyma cells, or at least the angular thickenings, consisted of 7–20 lamellae which were alternately rich and poor in cellulose but which became richer in cellulose as they approached the cell lumen.

According to Czaja (1961) transverse lamellation in the wall thickenings of the collenchyma cells of many plants can also be detected with the aid of the polarizing microscope. Chafe (1970) has observed that the orientation of the cellulose microfibrils in the successive lamellae of the wall is alternately transverse and longitudinal. He suggested, therefore, that lamellation seen in the wall at optical level may reflect not primarily a difference in composition of the successive lamellae, but a different arrangement of the cellulose microfibrils.

According to some authors (Duchaigne, 1955; Beer and Setterfield, 1958) the additional layers of microfibrils that appear during the development of the characteristic wall thickenings seem to arise both on the outside and on the inside of those layers that are continuous around the entire cell. In very thick walls the additional layers extend around the cell. In this type of wall pits can be seen. Collenchyma cells are apparently the only cells in which it is not known which part of the wall is laid down during the period of longitudinal growth and which part after the cells have reached maximum length. It is, therefore, impossible to delimit the primary and secondary wall layers in these cells.

In many dicotyledons, e.g. in the petioles and stems of *Medicago sativa, Eryngium maritimum, Viscum album,* and *Salvia officinalis*, the collenchyma may become sclerified. This sclerification is brought about, according to Duchaigne (1955), by a process of centripetal and centrifugal lamellation. The inner lamellae, during the process of growth, together form a layer rich in cellulose which later becomes impregnated with lignin. New concentric, lignified lamellae appear centrifugally around the first such layer. As a result of this centrifugal development of lignified lamellae, the pecto-cellulosic substance of the collenchyma walls, progressively disappears. Often, however, part of this substance remains even after the walls become fully sclerified. Later, additional lamellae develop centripetally and so the cell lumen is gradually reduced. The greatest concentration of lignin is finally found in the outermost wall layers. Simple pits are also present here as in sclerenchyma.

Generally, we can conclude that typical collenchyma is a juvenile supporting tissue, and when it is present in an organ which persists for a long period it becomes sclerified.

The peculiar manner in which the walls of collenchyma cells become thickened, i.e. by intussusception of cellulose microfibrils into the cell wall (Beer and Setterfield, 1958; Setterfield and Bailey, 1958), is a very interesting phenomenon, and further research is necessary to clarify how microfibrils are produced in the wall thickenings of the collenchyma. This, probably, will also lead to a better understanding of wall growth in general.

REFERENCES

AMBRONN, H. (1881) Über die Entwicklungsgeschichte und die mechanischen Eigenschaften des Collenchyms. Ein Beitrag zur Kenntnis des mechanischen Gewebesystems. *Jb. wiss, Bot.* **12:** 473–541.

ANDERSON, D. (1927) Über die Struktur der Kollenchymzellwand auf Grund mikrochemischer Untersuchungen. *S. Akad. Wiss, Wien, Math.-Naturw. Kl.* **136:** 429–40.

BEER. M. and SETTERFIELD, G. (1958) Fine structure in thickened primary walls of collenchyma cells of celery petioles. *Am. J. Bot.* **45:** 571–80.

CHAFE, S. C. (1970) The fine structure of the collenchyma cell wall. *Planta* **90:** 12–21.

CZAJA, A. T. (1961) Neue Untersuchungen über die Struktur der partiellen Wandferdickungen von faserförmigen Kollenchymzellen. *Planta* **56:** 109–24.

DUCHAIGNE, A. (1955) Les divers types de collenchymes chez les Dicotylédones: leur ontogénie et leur lignification. *Annls Sci. nat., Bot ser.* 11, **16:** 455–79.

ESAU, K. (1936) Ontogeny and structure of collenchyma and of vascular tissues in celery petioles. *Hilgardia* **10:** 431–76.

MAJUMDAR, G. P. and PRESTON, R. D. (1941) The fine structure of collenchyma cells in *Heracleum sphondylium* L. *Proc. Roy. Soc. (Lond.)* B, **130:** 201–17.

METCALFE, C. R. (1979) Some basic types of cells and tissues. In: *Anatomy of the Dicotyledons.* 2nd ed. Vol. 1 (eds. C. R. Metcalfe and L. Chalk). Clarendon Press, Oxford.

MÜLLER, C. (1890) Ein Beitrag zur Kenntnis der Formen des Collenchyms. *Ber. dt. bot. Ges.* **8:** 150–66.

PRESTON, R. D. and DUCKWORTH, R. B. (1946) The fine structure of the walls of collenchyma cells in *Petasites vulgaris* L. *Proc. Leeds Phil. Lit. Soc.* **4** (5): 343–51.

ROELOFSEN, P. A. (1959) The plant cell-wall. In: K. Linsbauer, *Handbuch der Pflanzenanatomie*, Abt. 1, Cytologie. Bd. 3, T. 4. Gebr. Borntraeger, Berlin.

SETTERFIELD, G. and BAILEY, S. T. (1958) Deposition of wall material in thickened primary walls of elongating plant cells. *Expl Cell Res.* **14:** 622–5.

VAN FLEET, D. S. (1950) A comparison of histochemical and anatomical characteristics of the hypodermis with the endodermis in vascular plants. *Am. J. Bot.* **37:** 721–5.

WALKER, W. S. (1975) The effect of mechanical stimulation on the collenchyma of *Apium graveolens* L. *Proc. Iowa Acad. Sci.* **64:** 177–86.

WALKER, W. S. (1960) The effect of mechanical stimulation and etiolation on the collenchyma of *Datura stramonium. Am. J. Bot.* **47:** 717–24.

WERKER, E. and KOLLER, D. (1987) Structural specialization of the site of response to vectorial photo-excitation in the solar-tracking leaf of *Lavatera cretica. Am. J. Bot.* **74:** 1339–1349.

CHAPTER 6

SCLERENCHYMA

SCLERENCHYMA is a tissue composed of cells with thickened secondary cell walls, lignified or not, whose principal function is support and sometimes protection. Sclerenchyma cells exhibit elastic properties, unlike collenchyma cells which exhibit plastic properties.

Sclerenchyma cells may differ in shape, structure, origin, and development. Many transitional forms exist between the various cell shapes, and thus it is difficult to classify the different types of sclerenchyma. Generally, sclerenchyma is divided into *fibres* and *sclereids*. Fibres are usually defined as long cells and sclereids as short cells. This definition is not sufficient, as very long sclereids exist and relatively short fibres can be found. It has been suggested that the pits are very narrow, have round apertures, and are more numerous in sclereids. The pit cavities of the sclereids very often have the form of branching canals which is a result of the increase in thickening of the wall.

An attempt was made to distinguish between fibres and sclereids on the basis of the origin of the elements. Sclereids develop from parenchyma cells whose walls become secondarily thickened, whereas fibres develop from meristematic cells and so they are determined from their origin. Other research, however, has shown that these definitions are also insufficient due to their inconstancies. Not only is it difficult to distinguish between the different types of sclerenchyma cells because of the existing transitional forms, but it is also somewhat difficult to distinguish between sclerenchyma and parenchyma as there are parenchyma cells with thick secondary walls, such as the xylem parenchyma.

FIBRES

Fibres occur in different parts of the plant body. They may occur singly as idioblasts (e.g. in the leaflets of *Cycas*), but more usually they form bands or a network or an uninterrupted hollow cylinder (Fig. 48). Fibres are most commonly found among the vascular tissues but in many plants they are also well developed in the ground tissues. According to their position in the plant body, fibres are classified into two basic types—*xylary* and *extraxylary* fibres.

Xylary fibres constitute an integral part of the xylem and they develop from the same meristematic tissues as do the other xylem elements. These fibres are of varied shape in spite of their common origin. Two main types of xylary fibres, i.e. *libriform fibres* and *fibre-tracheids*, are distinguished on the basis of wall thickness and type and amount of pits (Fig. 49, nos 1–3). Libriform fibres resemble phloem fibres (liber = inner bark) and they are usually longer than the tracheids of the plant in which they occur. These fibres have extremely thick walls and simple pits. Fibre-tracheids are forms intermediate between tracheids and libriform fibres. Their walls are of medium thickness—not as thick as those of the libriform fibres but thicker than those of the tracheids. The pits in fibre-tracheids are bordered but their pit chambers are smaller than those of tracheids. In fibre-tracheids and sometimes also in libriform fibres the pit canal is elongated and the inner pit aperture usually becomes slit-like (Fig. 29, no. 1) as a result of the thickening of the wall. In fibre-tracheids, therefore, the length of the pit aperture usually exceeds the diameter of the pit chamber. In both libriform fibres and fibre-tracheids, the

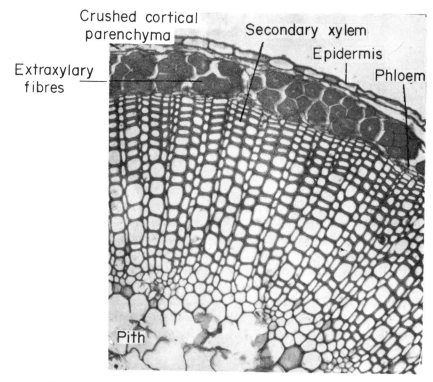

Crushed cortical parenchyma

Extraxylary fibres

Secondary xylem

Epidermis

Phloem

Pith

FIG. 48. Micrograph of portion of a stem cross-section of *Linum usitatissimum* in which the extraxylary fibres can be distinguished. × 150.

inner pit apertures of a pit-pair are usually at right angles to each other.

Another type of fibre present in the secondary xylem of dicotyledons is the *gelatinous* or *mucilaginous fibre* (Fig. 197, no. 1). In such fibres the innermost layer of the secondary wall contains much α-cellulose and is poor in lignin. This layer, termed the *G-layer*, absorbs much water and may swell so as to fill the entire lumen of the fibre. On drying, these layers shrink irreversibly (Dadswell and Wardrop, 1955). The G-layers were found to be relatively porous and less compact than the adjacent outer layers (Côté and Day, 1962). Gelatinous fibres are characteristic of tension wood (see Chapter 15).

Extraxylary fibres occur elsewhere in the plant other than among the xylem elements. They occur, for instance, in the cortex or they may be closely related to the phloem elements. In the stems of many monocotyledons, the extraxylary fibres occur in an uninterrupted hollow cylinder in the ground tissue, and they may be situated at various distances inside the epidermis and may even surround the outermost vascular bundles. Commonly in the monocotyledons, the fibres form sheaths around the vascular bundles. Such fibres develop partly from the procambium and partly from the ground tissue (Esau, 1943).

In the stems of climbing and certain other dicotyledonous plants, such as *Aristolochia* and *Cucurbita*, fibres are found on the inside of the innermost cortical layer and on the periphery of the central cylinder. These fibres are not developmentally connected with the phloem and so they were termed, by many workers, *pericyclic fibres*. As a result of ontogenetic studies in stems of some plants (*Nicotiana, Linum, Ricinus, Apocynum,* and *Nerium*) it has been concluded that the so-called pericyclic fibres of the investigated species develop from the procambium and thus represent primary phloem fibres (Esau, 1938, 1943; Blyth, 1958, Zhenghai and Lanxing, 1980).

The above classification into xylary and extraxylary fibres is not always applicable as there are

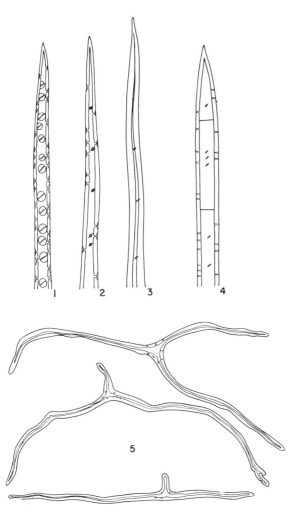

FIG. 49. 1–3, Tips of elements from the secondary xylem of *Quercus ithaburensis*. 1, Tracheid. 2, Fibre-tracheid. 3, Libriform fibre. 4, Septate fibre of *Vitis*. 5, Isolated sclereids from the leaf blade of *Olea*. (No. 5 adapted from Arzee, 1953a.)

fibre tips run out as pointed ends when viewed in longitudinal section of the fibre. The septum consists of a middle lamella and two primary wall-like layers interrupted by numerous plasmodesmata. Septate fibres may contain starch and oils and, therefore, are thought to have a storage function. They may also contain resins and sometimes crystals of calcium oxalate. Septate fibres are present in the secondary xylem of many dicotyledons (Metcalfe and Chalk, 1950; Butterfield and Meylan, 1976). Non-vascular septate fibres occur in some monocotyledons, e.g. in Palmae (Tomlinson, 1961) and in Bambusoideae (Parameswaran and Liese, 1977).

Mature fibres have well-developed, usually lignified secondary walls, which are sometimes so thick as to obscure the lumen of the fibre. Lamellae can be distinguished in these walls. In *Linum*, for example, each lamella is 0.1–0.2 μm thick as seen in a cross-section of the fibre.

Mention should be made here of those elongated cells that sometimes occur in the secondary xylem and whose secondary walls are equal in thickness to those of the xylem parenchyma. These cells contain living protoplasts, and according to Haberlandt (1918), they were termed by Sanio, *substitute fibres* (*Ersatzfasern*). It appears, however, that these cells should be included among the xylem parenchyma and that they should not be confused with the living libriform fibres and fibre-tracheids (Fahn and Arnon, 1963; Fahn and Leshem, 1963) which are discussed later in this chapter.

Form and length of fibres

Fibres are usually very long and narrow cells with tapered and sometimes branched ends. The length of fibres varies very greatly and generally extraxylary fibres are longer than xylary fibres. In *Cannabis sativa* (hemp) the fibres are 0.5–5.5 cm long, in *Linum usitatissimum* (flax) from 0.8 to 6.9 cm, and in *Boehmeria nivea* (ramie) Aldaba (1927) showed, by means of a special maceration, that the fibres may reach a length of 55 cm. These ramie fibres are among the longest cells in the higher plants.

fibres, such as the *septate fibres* (Fig. 49, no. 4), which are found in the xylem and the phloem even of the same species, e.g. in *Vitis* where they are very common (Vestal and Vestal, 1940; Spackman and Swamy, 1949). These fibres are characterized by the presence of internal septa and, usually, of a living protoplast. As seen in septate wood fibres of *Ribes sanguineum* (Parameswaran and Liese, 1969) the internal septa result from mitosis in lignified cells. The septum does not fuse with the fibre wall but becomes broadened, at the place of contact with it. The

Development of fibres

Ontogenetically fibres develop from different meristems, such as the procambium, cambium, ground meristem, and even from the proto- derm, as in certain species of the Gramineae and Cyperaceae. Fibres may also develop from parenchyma cells, e.g. in the protophloem of many dicotyledons. The fibres formed by the cambium develop from fusiform initials and elongate only little or not at all during their maturation.

Fibres that arise from short initials, as in *Linum* (flax) and *Boehmeria nivea* (ramie), must necess- arily elongate greatly in the course of their matur- ation. In ramie, according to Aldaba (1927), the initials of the primary phloem fibres are 20 μm long while the mature fibres can be up to 55 cm (550,000 μm) long. The elongation is very gradual and may take some months. This gradual elonga- tion of primary phloem fibres involves a very complicated development of the secondary wall. While the fibre still grows symplastically, the wall remains thin. Later, when the ends begin to grow by intrusive growth, only the cell walls of the ends remain thin and secondary wall formation com- mences from the middle of the fibre in those parts of wall which have ceased to elongate. In *Linum* and ramie it has been found that this process is gradual so that new lamellae of the secondary wall are added centripetally in the form of cylinders which are open at both ends. At the same time the first-formed lamellae continue to elongate towards the fibre ends which they reach only when the fibre ceases to elongate (Fig. 50, nos. 1–6). According to Kundu and Sen (1960) the upper ends of ramie fibres continue to grow for a longer period than the basal ends. Sometimes not all the lamellae reach the actual fibre end and in some fibres chambers may be formed in the ter- minal portions by the ingrowth, towards the cell lumen, of these lamellae. The lamellae of the primary phloem fibres, or at least of the immature fibres, are often not strongly attached to one another. This feature is easily demonstrated dur- ing the cross-sectioning of such material when the different layers become torn one from the other. In short fibres, such as those found in *Agave*, *San- sevieria*, and *Musa textilis*, whose total length is not more than a few millimetres, all portions of the cell wall grow at the same rate.

When fibres are treated with reagents that cause swelling of cellulose they attain the form of a chain of beads. This phenomenon is referred to as ballooning of the fibres. The reason for this are differences in swelling of the various wall layers. The primary wall, or in some cases the outermost layer of the secondary wall, does not swell or swells to a different degree or in a different way. On application of swelling reagents the inner layers containing microfibrils oriented in a steep spiral swell strongly in the radial direction. The primary wall, unable to withstand the pressure exerted by the rapid swelling of the inner wall layers, breaks up into a number of constricting rings and the swelling inner wall layers expand forming balloons (Fig. 50, no. 7) (Kundu and Rao, 1975; Frey-Wyssling, 1976).

Differences exist in the manner of growth of the fibres in the primary body and of those in the secondary body. The initials of the primary fibres appear early, before the organ in which they occur has elongated, and so they may grow in length symplastically together with the neigh- bouring cells which continue to divide. The sym- plastic growth is augmented by intrusive and glid- ing growth of the ends which thus penetrate between the surrounding cells. The initials of the secondary fibres develop in organs that have ceased to elongate and therefore the growth of secondary fibres can be intrusive only. This is apparently the reason why the primary fibres are usually longer than the secondary fibres of the same plant. Thus it was found in ramie that the average length of the primary phloem fibres is 164.6 mm while that of the secondary phloem fibres is 15.5 mm.

Results of electron microscopical studies of fibre development suggested that ER, Dictyosomes, plasmalemma, lomasomes (multilamellar or multi- vesicular structures occurring between the plasma- lemma and cell wall) and microtubuli are impli- cated in the formation of the cell wall (Juniper *et al.*, 1981).

It has been suggested that a combination of auxin and gibberellin is required for the induction and differentiation of fibres (Aloni, 1987).

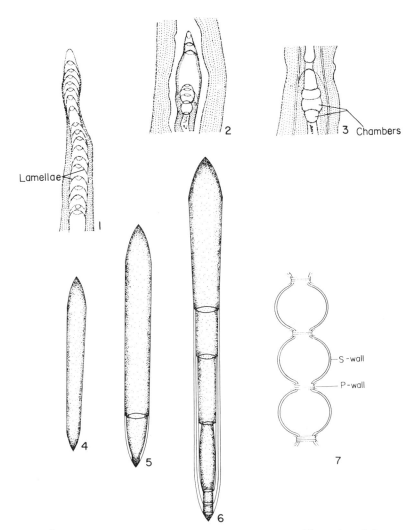

FIG. 50. Stages of the ontogeny of the extraxylary fibres of *Boehmeria nivea*. 1, Elongation of the upper end, showing a series of young lamellae one within the other; each lamella is open at its tip. 2 and 3, Development of chambers by inner lamellae. 4–6, Diagrammatic representation of the differentiation of a phloem fibre, in which the centripetal development of the lamellae of the secondary wall is shown. 7, Diagrammatic representation of a ballooning fibre. (Nos. 1–6 adapted from Aldaba, 1927.)

Fibre protoplasts

During the development of primary phloem fibres of *Nicotiana* and *Linum*, Esau (1938, 1943) observed that the protoplast was multinucleate. The protoplast in developing secondary fibres usually has a single nucleus.

Mature libriform fibres and fibre-tracheids were usually regarded as being dead supporting structures. In mature fibres the presence of a living protoplast and nucleus had been described only in phloem fibres (Kallen, 1882) and in septate fibres (Spackman and Swamy, 1949). According to Bailey (1953) libriform fibres sometimes retain their living contents subsequent to the formation of the thick, lignified secondary wall, thus enabling these cells to assume a storage function in addition to that of support.

Recently, however, living protoplasts and nuclei have been identified in libriform fibres of many species, and even in fibre-tracheids (Fig. 51, nos. 1, 3, 4, 6, 7). Such living fibres were found to occur in the wood of *Tamarix* spp., in many woody species of the Chenopodiaceae, and in trees, shrubs, and subshrubs of many other dicotyledonous families (Fahn and Arnon, 1963; Fahn and Leshem, 1963; Czaninski, 1964; Wolkinger, 1969, 1970; Dumbroff and Elmore, 1977; Gregory, 1978). Living protoplasts have also been found in many monocotyledonous fibres. In fibres with long, narrow lumina the nuclei are usually elongated (Fig. 51, nos. 1, 3, 4, 7). The life span of the wood fibres of *Tamarix aphylla* is about 20 years.

Evolution of xylary fibres

As has been mentioned above, the xylary fibres differ in shape, size, thickness of wall, type, and amount of pits. It is assumed, from the evolutionary point of view, that fibres have developed from tracheids. This assumption is supported by the fact that many transitional forms between these two types of elements are found in some angiosperms, as, for example, *Quercus* spp. From the many transitional forms that have been distinguished it appears that the following changes have taken place during the course of the evolution of fibres from tracheids. The wall has become thickened, the number of pits and the size of the pit chamber has been reduced, leading to the eventual disappearance of the bordered pit, and the cells have become shortened. This assumed shortening of the fibres refers to the shortening of the initials of the fibres in the cambium and not to the mature fibres. In the mature tissues of one plant, the libriform fibres are usually longer than the tracheids, and this increased length is secondary and is the result of the additional growth of the ends of the fibres.

Structure and use of commercial fibres

The term *fibre*, as used in industry, does not generally have the same meaning as that defined by botanists. For instance, the commercial fibres of *Linum*, *Boehmeria*, and *Corchorus* are, in reality, a bundle of fibres, and those from monocotyledonous leaves, such as from *Agave*, *Musa textilis*, and others, are usually the vascular bundles with the surrounding sheaths of fibres. From some plants the commercial fibres comprise the vascular system of the root, e.g. *Muhlenbergia*, or of the entire plant, e.g. *Tillandsia*. The commercial fibres of *Gossypium* (cotton) are the epidermal hairs of the seeds. Kapok fibres are hairs produced on the inner surface of the capsule of *Ceiba pentandra*.

Commercial fibres are divided into two types—hard fibres and soft fibres. Hard fibres are those which have a high lignin content in the walls, and are of a stiff texture. Hard fibres are obtained from monocotyledons. Soft fibres may or may not contain lignin, they are flexible and elastic, and are of dicotyledonous origin. The best-known plants from which hard fibres are produced are different species of *Agave*, especially *A. sisalana* (sisal), *Tillandsia usneoides* (Spanish moss), *Musa textilis* (abaca), *Furcraea gigantea* (Mauritius hemp), and *Phormium tenax*. Soft fibres are mainly produced from *Linum usitatissimum* (flax) (Fig. 52, no. 1), *Cannabis sativa* (hemp), *Boehmeria nivea* (ramie), *Corchorus capsularis* (jute), *Hibiscus cannabinus* (kenaf), and *Ceiba pentandra* (kapok). (See Bailey *et al.*, 1963.)

The fibres of cotton, which are produced from the indumentum of seeds (Fig. 52, no. 2), represent the most important commercial fibres in use today.

Fibres are also classified according to their use (Schery, 1954): (a) textile fibres which are used in the manufacture of fabrics; (b) cordage fibres; (c) brush fibres such as are used in the manufacture of brushes and brooms; and (d) filling fibres such as those used for stuffing upholstery, mattresses and life-belts, caulking (barrels, plumbing), and reinforcing (wall plates, plastics).

In the textile industry the principal fibre used is cotton and, in smaller amounts, flax, ramie, and hemp. For coarser fabrics, such as sacking and bagging, jute is principally used, and cotton, flax, hemp, and a few other hard fibres are used to a lesser extent. For the manufacture of twine, jute, cotton, hemp and, to a lesser extent, flax and several hard fibres are used. Ropes and binder twines are manufactured from hard fibres, such as those of *Musa textilis* (abaca) and *Agave* spp. (sisal), and to a small extent from cotton and other soft fibres. Brushes and brooms are made

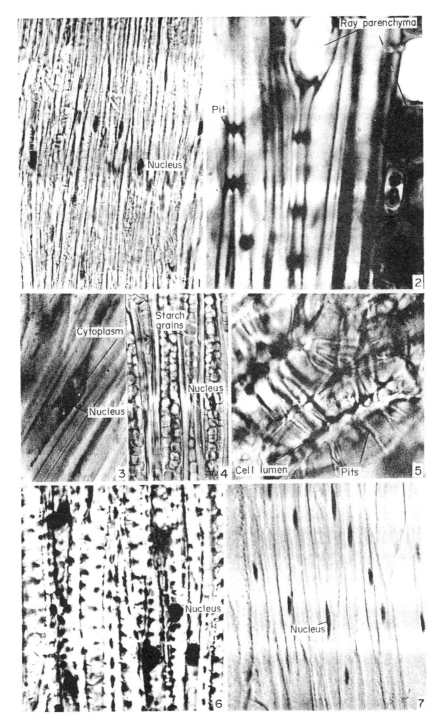

FIG. 51. 1, Micrograph of a longitudinal section of the secondary xylem of *Noëa mucronata* showing nuclei in the fibres. × 450. 2, Tangential longitudinal section of the secondary xylem of *Eucalyptus camaldulensis* in which the common wall between two fibres and pits characteristic of fibre-tracheids can be distinguished. × 1050. 3, Portion of a longitudinal section of the secondary xylem of *Tamarix aphylla* in which a fibre containing cytoplasm and nucleus can be distinguished. × 750. 4, Portion of a longitudinal section of the secondary xylem of *Calligonum comosum* showing fibres filled by starch grains. × 180. 5, Brachysclereids as seen in a cross-section of the stem cortex of *Hoya carnosa*. × 650. 6, Portion of a longitudinal section of the secondary xylem of *Teucrium polium* showing fibre-tracheids with nuclei. × 700. 7, Portion of a longitudinal section of the secondary xylem of *Rubia velutina* showing fibres with nuclei. × 230.

95

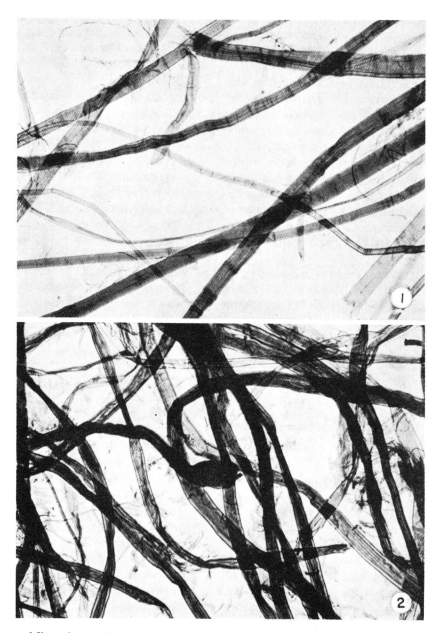

FIG. 52. 1, Processed fibres of *Linum*; fibres of cylindrical shape. × 150. 2, Processed fibres of *Gossypium*; fibres are flat. × 150. (From E. Liebert, in Handbuch der Mikroskopie in der Technik, Umschau Verlag, 1951.)

from *Agave* fibres, fibres from the stems and leaves of the Palmae and the inflorescences of *Sorghum vulgare*, among others. As filling fibres, the fibres of *Ceiba pentandra* (kapok), cotton, jute, the fibres of *Tillandsia usneoides*, several hard fibres, and others are used. For caulking fibres, hemp, jute and sisal are used. Fibres are also used in the paper industry; depending on their physical and chemical properties different types of paper may be made (Clark, 1965).

From a technological point of view, the shape of the fibre cell, its length, and wall structure are of importance in the fibre industry. Special attention is paid to the length of the fibre, the extent to which neighbouring fibres overlap, and how they are joined to one another. The orientation of the cellulose units in the wall has an important effect on the physical properties of wood and commercial fibres. Elasticity and heat conductivity increase as the degree of orientation parallel to the length of the fibre increases. Swelling is much greater in a direction perpendicular to the cellulose molecules than it is in a direction parallel to them; this is the case even in water, where swelling is almost exclusively intermicrofibrillar (Preston, 1963).

Commercially, fibres are ranked according to durability, tensile strength, length of the strands, fineness, uniformity, and elasticity. On the basis of the above features some of the most important commercial fibres may be divided into the following four ranks:

the sclereids may either be irregularly diffused (*diffuse sclereids*), e.g. in *Trochodendron, Olea, Osmanthus,* and *Nymphaea*; or the sclereids may be confined to the vein endings (*terminal sclereids*), e.g. in *Mouriri, Boronia, Arthrocnemum* (Fig. 53, no. 2), *Capparis spinosa* and *Salvadoraceae* species. The elongated sclerenchyma cells occurring in the mesophyll of the Myrtaceae and some other families are often referred to as fibres (Carr *et al.*, 1971).

In some plants, e.g. the Magnoliaceae, sclereids are found in almost all organs: in the cortex and pith of the stem, leaves, stipules, floral appendages, receptacles, fruits, and rarely in roots (Tucker, 1977).

Among the types of idioblastic sclereids there is also a type of cells that simulate tracheids, *tracheoid idioblasts* (Rao and Das, 1979b). Such tracheoids are found, for instance, in *Salicornia*

	Rank 1	Rank 2	Rank 3	Rank 4
Hard fibres	*Musa textilis*	*Agave* spp.	*Phormium tenax*	*Furcraea gigantea*
Soft fibres	*Linum usitatissimum*	*Boehmeria nivea*	*Cannabis sativa*	*Corchorus capsularis*

SCLEREIDS

Form and localization of sclereids

Sclereids occur in many different places in the plant body. In many plants they occur as hard masses of cells within soft parenchyma tissue. Certain organs, such as the shell of walnuts and many other stone and seed coats, are built entirely of sclereids. In many plants sclereids appear as idioblasts, i.e. as cells which are readily distinguished from the surrounding cells of the tissue by their size, shape, and the thickness of their wall. Idioblastic sclereids are very variable in shape. Sclereids of peculiar shape are found in the leaves of various plants, e.g. in *Gnetum, Camellia, Trochodendron, Nymphaea, Cyathocalyx, Desmos, Phaeanthus, Horsfieldia, Salvadora, Monstera, Olea* (Fig. 54) *Fibraurea* and species of Olacaceae (Foster, 1944, 1945, 1946, 1955; Arzee, 1953a, b; Fahn and Arzee, 1959; Gaudet, 1960; Govindarajalu and Parameswaran, 1967; Rao and Das, 1979a, Baas *et al.*, 1982; Wilkinson, 1986) and in aerial roots of *Monstera*. In the leaf mesophyll

species (Fig. 130) (Foster, 1956; Fahn and Arzee, 1959; Lersten and Bender, 1976) and in inflorescence stalks of some orchids (Neubauer, 1978).

Tschirch (1889) suggested the division of sclereids into four types: (1) *brachysclereids* or *stone cells*, which are more or less isodiametric in form; such sclereids are usually found in the phloem, the cortex and the bark of stems and in the flesh of such fruits as pears (*Pyrus communis*) and quinces (*Cydonia oblonga*) (Fig. 51, no. 5); (2) *macrosclereids*, which are rod-shaped sclereids; such sclereids often form a continuous layer in the testa of seeds, e.g. in the seeds of the Leguminosae (Fig. 53, no. 1); (3) *osteosclereids*, which are bone- or spool-shaped sclereids, the ends of which are enlarged, lobed, and sometimes even somewhat branched; such sclereids are mainly found in seed coats and sometimes also in the leaves of certain dicotyledons (Fig. 53, no. 1); (4) *astrosclereids*, which are variously branched and often star-shaped; such sclereids are mainly found in leaves (Fig. 49, no. 5). In *Fibraurea* the hypodermis of the leaf lamina is composed of a horizontal network of astrosclereids with interwoven arms (Wilkinson,

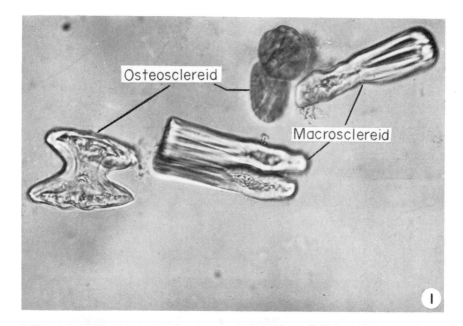

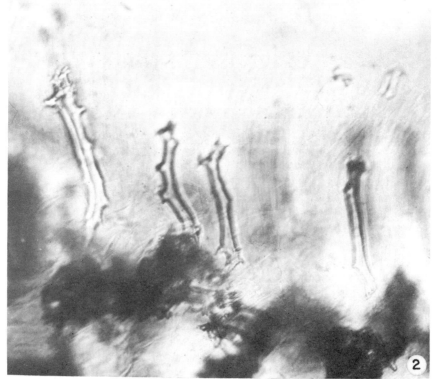

FIG. 53. 1, Isolated osteosclereids and macrosclereids from a macerated seed coat of *Pisum sativum.* × 650. 2, Sclereids with projections as seen in the stem of *Arthrocnemum glaucum* cleared by treatment with lactic acid. × 100.

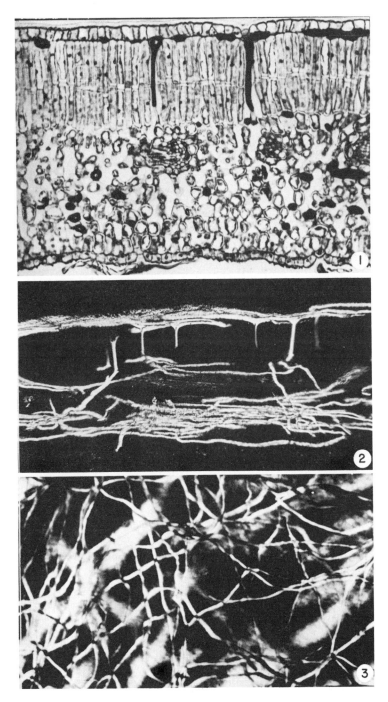

FIG. 54. Sclereids in the leaf blade of *Olea europaea*. 1, Portion of a cross-section of the blade in which parts of the sclereids (darkly stained) can be seen. ×190. 2, Portion of a relatively thick, cleared cross-section of the leaf blade photographed in polarized light in which the sclereids appear white. ×110. 3, Surface view of portion of a cleared leaf, photographed in polarized light, showing the arrangement of the sclereids in the spongy parencyma. ×130. (From Arzee, 1953a.)

1986). A fifth type of sclereids, termed *tricho-sclereids*, was suggested by Bloch (1946). These are very elongated, somewhat hair-like, and regularly once-branched sclereids. Additional terms for sclereid types, such as *filiform, fibriform, columnar*, and *polymorphic*, which occur in leaves, are used by various authors (e.g. Rao and Chin, 1966).

The occurrence of leaf sclereids, their position in the leaf and the type of sclereid are of taxonomic value (Rao and Das, 1979a, b, 1981a, b; Rao *et al.*, 1980).

Ontogeny of sclereids

Typical brachysclereids develop from parenchyma cells by secondary thickening of the cell wall. This secondary wall is very thick, and numerous concentric layers and branched pits can usually be distinguished in it. During the process of wall thickening the inner surface of the wall decreases and pits which start to develop on the outside of the secondary wall are brought together (Fig. 51, no. 5). The physiological reasons for the sclerification of the parenchyma cells are not known, but Bloch (1944) thought the fact that stone cells often appear close to wound tissues suggests that they develop in response to some physiological disturbances. In the bark the change of many parenchyma cells to sclereids suggests that, in this case, the cause is the ageing of the tissue.

Recent observations revealed that although in the wound meristem of injured leaves certain cells develop into sclereids, wounds have an inhibitory influence on the differentiation of sclereids in lamina of developing leaves which normally produce sclereids (Rao, 1969; Rao and Vaz, 1970).

Another interesting example of the development of stone cells is in the continuous cylinders of the phloem fibres. When stems of plants with such cylinders grow in width, neighbouring parenchyma cells penetrate into the spaces formed between the fibres of the cylinder. The parenchyma cells then divide and become sclerified and so close the gaps formed in the cylinder (Haberlandt, 1918).

The manner by which the peculiar structure of the osteosclereids of the seed coat of *Pisum sativum* is obtained was described by Harris (1984). During the early stages of development of these sclereids, the middle portion of the lateral walls become heavily thickened preventing further expansion. The end walls of the cells retain their thin walls and continue to expand, thus creating the typical bone-shaped structure of the osteosclereids. In the synthesis of the cell wall, coated and smooth vesicles of Golgi origin and vesicles with fibrilar contents originating from the ER, are involved.

As a result of investigation of the ontogenetic development of the branched type of sclereids in the leaves of *Trochodendron aralioides* and *Mouriria huberi* (Foster, 1944, 1945, 1947), of *Memecylon* spp. (Rao, 1957), of *Olea* (Arzee, 1953a, b), of *Nymphaea odorata* (Gaudet, 1960) and in the aerial roots of *Monstera* (Bloch, 1946), the following histogenetic facts have been realized. In all the above examples the sclereids develop from small initials with thin walls. These initials are first distinguished from the neighbouring cells by their larger nucleus and nucleolus. Already in the early stages of development they begin to branch and so acquire the form of the mature sclereid. The branches or projections of the sclereid penetrate into the intercellular spaces, but intrusive growth of these branches, between the joined walls of neighbouring cells, is also common (Fig. 55, nos. 1–3). The degree of pitting in these sclereids is not constant.

Sclereids are usually described as non-living cells when mature, but it has been seen that the protoplasts may remain viable throughout the life of the organ in which the sclereids are found. In non-deciduous leaves and in certain stems the life of the sclereids may sometimes be 4–5 years (Puchinger, 1923). The protoplast in the stone cells in the fruits of the pear and quince also remains alive for relatively long periods. According to Alexandrov and Djaparidze (1927), during the ripening of the quince fruit, the stone cells undergo a process of delignification, which they believe is an indication of the enzymatic activity of the protoplast of the stone cell itself.

The structure of the fibre wall has been investigated comparatively thoroughly and emphasis laid on the wall structure of fibres of economic value. Attention has also been paid to the ontogenetic and phylogenetic development of fibres. Cell

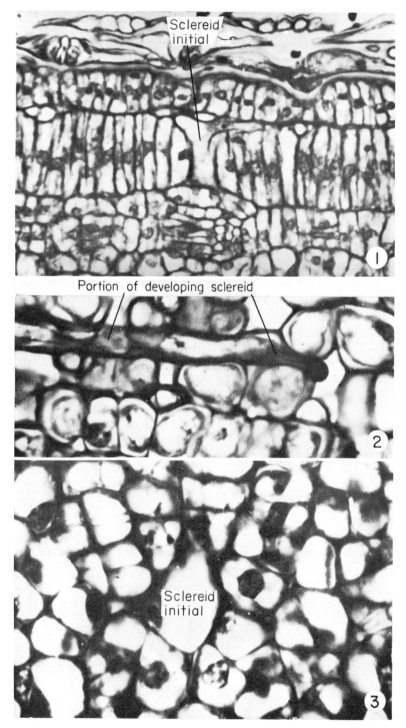

FIG. 55. Stages of development of sclereids in the leaf blade of *Olea europaea*. 1, Portion of a cross-section of a young leaf in which a sclereid initial can be distinguished in the as yet single row of palisade cells. × 420. 2, Portion of a cross-section of a leaf blade, showing the intrusive growth of an arm of a developing sclereid. × 1100. 3, Portion of a section cut parallel to the blade surface in which a sclereid initial can be seen. × 240. (From Arzee, 1953b.)

growth, especially intrusive growth, can be well studied in the course of fibre development.

As was mentioned in the chapter dealing with collenchyma, collenchyma cells may often become sclerified during the maturation of the organs in which they occur. This fact emphasizes the view of the close relationship between these two tissues.

Because of the great variability in the form of fibres and because of the existence of many transitional forms, fibres serve as favourable material for the study of the evolution of an element or part of it, as, for instance, the evolution of the pit.

Substitute and septate fibres have, during the course of evolution, become strikingly different from the typical fibre form and should, actually, be classified as parenchyma cells with secondarily thickened walls. A substitute fibre resembles an elongated parenchyma cell and a septate fibre a longitudinal series of parenchyma cells derived from a single mother cell in which secondary wall lamellae develop before the cell divisions are completed.

The libriform fibres and the fibre-tracheids have, till recently, generally been described as non-living cells devoid of protoplasts and were regarded as having only mechanical function or, at the most, as playing a small role in water conduction in addition to the tracheary elements. However, it now appears, in the light of recent research, that the libriform fibres and even fibre-tracheids of the sap wood of many woody plants contain living protoplasts. Therefore we should begin to consider fibres not only as supporting elements but also as elements that doubtless fulfil various other important physiological functions. This aspect, the investigation of which has been initiated in our laboratory, awaits still further research.

It is possible that the retention of living protoplasts in fibres is more characteristic for certain life forms (e.g. shrubs and subshrubs) or for woody plants of certain habitats, such as xeric ones. The evolutionary and ecological investigation of these assumptions may possibly bring to light some interesting results.

It is also worth mentioning that the appearance of the living protoplasts in libriform fibres and fibre-tracheids represents a further example of the indistinct limits between the various elements that form the highly differentiated tissues of the higher plant body. This and similar phenomena are of great importance in our understanding of the evolution of the various elements.

The appearance of idioblastic sclereids in the leaves of plants that belong to diversified taxonomic and ecological groups makes it difficult to understand both their evolutionary and functional significance.

REFERENCES

ALDABA, V. C. (1927) The structure and development of the cell wall in plants: I, Bast fibers of *Boehmeria* and *Linum*. *Am. J. Bot.* **14:** 16–24.

ALEXANDROV, W. G. and DJAPARIDZE, L. I. (1927) Über das Entholzen und Verholzen der Zellhaut. *Planta* **4:** 467–75.

ALONI, R. (1987) Differentiation of vascular tissues. *Ann. Rev. Plant Physiol.* **38:** 179–204.

ARZEE, T. (1953a) Morphology and ontogeny of foliar sclereids in *Olea europaea*: I, Distribution and structure. *Am. J. Bot.* **40:** 680–7.

ARZEE, T. (1953b) Morphology and ontogeny of foliar sclereids in *Olea europaea*: II, Ontogeny. *Am. J. Bot.* **40:** 745–52.

BAAS, P., VAN OOSTERHOUD, E. and SCHOLTES, C. J. L. (1982) Leaf anatomy and classification of the Olacaceae, *Octoknema,* and *Erythropalum*. Allertonia, A Series of Occasional Papers. Lawai, Kauai, Hawaii. September, 1982. pp. 155–210.

BAILEY, I. W. (1953) Evolution of the tracheary tissue of land plants. *Am. J. Bot.* **40:** 4–8.

BAILEY, T. L. W., JR., TRIPP, V. M., and MOORE, A. T. (1963) Cotton and other vegetable fibres. In: *Fibre Structure* (eds. J. W. S. Hearle and R. H. Peters), The Textile Institute, Butterworths, Manchester, pp. 422–54.

BLOCH, R. (1944) Developmental potency, differentiation and pattern in meristems of *Monstera deliciosa. Am. J. Bot.* **31:** 71–77.

BLOCH, R. (1946) Differentiation and pattern in *Monstera deliciosa.* The idioblastic development of the trichosclereids in the air root. *Am. J. Bot.* **33:** 544–51.

BLYTH, A. (1958) Origin of primary extraxylary stem fibers in Dicotyledons. *Univ. Calif. Publ. Bot.* **30:** 145–232.

BUTTERFIELD, B. G. and MEYLAN, B. A. (1976) The occurrence of septate fibres in some New Zealand woods. *NZ J. Bot.* **14:** 123–30.

CARR, S. G. M., CARR, D. J., and MILKOVITS, L. (1971) Mesophyll fibres in *Eucalyptus* L'Hérit. and *Angophora* Cav. *Ann. Bot.* **35:** 143–9.

CLARK, T. F. (1965) Plant fibres in paper industry. *Economic Bot.* **19:** 394–405.

CÔTÉ, W. A. and DAY, A. C. (1962) The G-layer in gelatinous fibers—electron microscopic studies. *Forest Prod. J.* **12:** 333–8.

CZANINSKI, Y. (1964) Variations saisonnières du chondriome et de l'amidon dans les fibres libriforms du xylème du *Robinia pseudo-acacia. CR Acad. Sci. (Paris)* **258:** 5945–8.

DADSWELL, H. E. and WARDROP, A. B. (1955) The structure and properties of tension wood. *Holzforschung* **9:** 97–104.

DUMBROFF, E. B. and ELMORE, H. W. (1977) Living fibers are a principal feature of the xylem in seedlings of *Acer saccharum* Marsh. *Ann. Bot.* **41**: 471–2.

ESAU, K. (1938) The multinucleate condition in fibers of tobacco. *Hilgardia* **11**: 427–34.

ESAU, K. (1943) Vascular differentiation in the vegetative shoot of *Linum*: III, The origin of the bast fibers. *Am. J. Bot.* **30**: 579–86.

FAHN, A. and ARNON, N. (1963) The living wood fibres of *Tamarix aphylla* and the changes occurring in them in transition from sapwood to heartwood. *New Phytol.* **62**: 99–104.

FAHN, A. and ARZEE, T. (1959) Vascularization of articulated Chenopodiaceae and the nature of their fleshy cortex. *Am. J. Bot.* **46**: 330–8.

FAHN, A. and LESHEM, B. (1963) Wood fibres with living protoplasts. *New Phytol.* **62**: 91–98.

FOSTER, A. S. (1944) Structure and development of sclereids in the petiole of *Camellia japonica* L. *Bull. Torrey Bot. Club* **71**: 302–26.

FOSTER, A. S. (1945) Origin and development of sclereids in the foliage leaf of *Trochodendron aralioides* Sieb & Zucc. *Am. J. Bot.* **32**: 456–68.

FOSTER, A. S. (1946) Comparative morphology of the foliar sclereids in the genus *Mouriria* Aubl. *J. Arnold Arb.* **27**: 253–71.

FOSTER, A. S. (1947) Structure and ontogeny of the terminal sclereids in the leaf of *Mouriria huberi* Cong. *Am. J. Bot.* **34**: 501–14.

FOSTER, A. S. (1955) Structure and ontogeny of terminal sclereids in *Boronia serrulata*. *Am. J. Bot.* **42**: 551–60.

FOSTER, A. S. (1956) Plant idioblasts: remarkable examples of cell specialization. *Protoplasma* **46**: 184–93.

FREY-WYSSLING, A. (1976) *The Plant Cell Wall*. Borntraeger, Berlin and Stuttgart.

GAUDET, J. (1960) Ontogeny of foliar sclereids in *Nymphaea odorata*. *Am. J. Bot.* **47**: 525–32.

GOVINDARAJALU, E. and PARAMESWARAN, N. (1967) On the morphology of foliar sclereids in the Salvadoraceae. *Beitr. Biol. Pfl.* **43**: 41–57.

GREGORY, R. A. (1978) Living elements of the conducting secondary xylem of sugar maple (*Acer saccharum* Marsh.). *IAWA Bull.* **1978** (4): 65–70.

HABERLANDT, G. (1918) *Physiologische Pflanzenanatomie*, 5th edn. W. Engelmann, Leipzig.

HARRIS, W. M. (1984) On the development of osteosclereids in seed coats of *Pisum sativum* L. *New Phytol.* **89**: 135–141.

JUNIPER, B. E., LAWTON, J. R. and HARRIS, P. J. (1981) Cellular organelles and cell-wall formation in fibres from the flowering stem of *Lolium temulentum* L. *New Phytol.* **89**: 609–619.

KALLEN, F. (1882) Verhalten des Protoplasma in den Geweben von *Urtica urens* entwicklungsgeschichtlich dargestellt. *Flora* **65**: 65–80, 81–96, 97–105.

KUNDU, B. C. and RAO, N. S. (1975) Fine structure of jute fibre. *J. Ind. Bot. Soc.* **54**: 85–94.

KUNDU, B. C. and SEN, S. (1960) Origin and development of fibres in ramie (*Boehmeria nivea* Gaud.). *Proc. Natn. Inst. Sci. India* **26** B (Suppl.): 190–8.

LERSTEN, N. R. and BENDER, C. G. (1976) Tracheoid idioblasts in Chenopodiaceae: A review and new observations on *Salicornia virginica*. *Proc. Iowa Acad. Sci.* **82**: 158–62.

METCALFE, C. R. and CHALK, L. (1950) *Anatomy of the Dicotyledons*, Vols. I and II. Clarendon Press, Oxford.

NEUBAUER, H. F. (1978) Über kortikale Idioblasten in Infloreszenzstielen von tropischen Orchideen. *Flora* **167**: 121–5.

PARAMESWARAN, N. and LIESE, W. (1969) On the formation and fine structure of septate wood fibres of *Ribes sanguineum*. *Wood Sci. Tech.* **3**: 272–86.

PARAMESWARAN, N. and LIESE, W. (1977) Structure of septate fibres in bamboo. *Holzforschung* **31**: 55–57.

PRESTON, R. D. (1963) Observed fine structure in plant fibres. In: *Fibre Structure* (eds. J. W. S. Hearle and R. H. Peters), The Textile Institute, Butterworths, Manchester, pp. 235–68.

PUCHINGER, H. (1923) Über die Lebensdauer sclerotinierten Zellen. *Sber. Akad. Wiss. Wien, Math-Naturw. Kl.* I. **131**: 47–57.

RAO, A. N. (1969) Effect of injury on the foliar sclereid development in *Fagraea fragrans*. *Experientia* **25**: 884–5.

RAO, A. N. and VAZ, S. J. (1970) Morphogenesis of foliar sclereids: II, Effects of experimental wounds on leaf sclereid development and distribution in *Fagraea fragrans*. *J. Singapore. Natn. Acad. Sci.* **1** (3) 1–7.

RAO, A. N. and WEE YEOW CHIN (1966) Foliar sclereids in certain members of Annonaceae and Myristicaceae. *Flora*, Abt. B, **156**: 220–31.

RAO, T. A. (1957) Comparative morphology and ontogeny of foliar sclereids in seed plants: I, *Memecylon* L. *Phytomorphology* **7**: 306–30.

RAO, T. A. and DAS, S. (1979a) Leaf sclereids—occurrence and distribution in the angiosperms. *Bot. Notiser* **132**: 319–24.

RAO, T. A. and DAS, S. (1979b) Typology of foliar tracheoids in angiosperms. *Proc. Indian Acad. Sci.* **88B**, Part II: 331–45.

RAO, T. A. and DAS, S. (1981a) Typology and taxonimic value of foliar sclereids in Proteaceae. 1. *Isopogon* R.Br. *Proc. Indian Acad. Sci.* (*Plant Sci.*) **90**: 31–43.

RAO, T. A. and DAS, S. (1981b) Comparative morphology and taxonomic value of foliar sclereids in *Limonium* Tour. (Limoniaceae). *Proc. Indian Acad. Sci.* (*Plant Sci.*) **90**: 153–162.

RAO, T. A., BREMER, K. and CHAKRABORTI, S. (1980) Foliar sclereids in Sri Lanka (Ceylonese) species of *Memecylon* (Melastomataceae). *Bot. Notiser* **133**: 397–401.

SCHERY, R. W. (1954) *Plants for Man*. Allen & Unwin, London.

SPACKMAN, W. and SWAMY, B. G. L. (1949) The nature and occurrence of septate fibres in dicotyledons. *Abst. Am. J. Bot.* **36**: 804.

TOMLINSON, P. B. (1961) *Anatomy of the Monocotyledons*. Vol. II. Palmae. Clarendon Press, Oxford.

TSCHIRCH, A. (1889) *Angewandte Pflanzenanatomie*. Urban & Schwarzenberg, Wien.

TUCKER, S. C. (1977) Foliar sclereids in the Magnoliaceae. *J. Linn. Soc. Bot.* **75**: 325–56.

VESTAL, P. A. and VESTAL, M. R. (1940) *The Formation of Septa in the Fibertracheids of Hypericum androsaemum* L. Bot. Mus. Leaf. I, Harv. Univ. **8**: 169–88.

WILKINSON, H. P. (1986) Leaf anatomy of *Tinomiscium* and *Fibraurea* (*Menispermaceae* tribe *Fibraureae*) with special reference to laticifers and astrosclereids. *Kew Bull.* **41**: 153–169.

WOLKINGER, F. (1969) Morphologie und systematische Verbreitung der lebenden Holzfasern bei Sträucher und Bäumen: I, Zur Morphologie und Zytologie. *Holzforschung* **23**: 135–44.

WOLKINGER, F. (1970) Das Vorkommen lebender Holzfasern in Sträuchern und Bäumen. *Phyton* (*Austria*) **14**: 55–67.

ZHENGHAI, H. and LANXING, T. (1980) A study on formation of bast fibers in the stem of *Apocynum lancifolium* Rus. *Kexue Tonobao* **25**: 773–778.

CHAPTER 7

XYLEM

THE vascular system of the sporophytes of the higher plants consists of *xylem*, the main function of which is the transport of water and solutes, and *phloem*, which mainly transports the products of photosynthesis.

On the basis of its physiological and phylogenetic importance, the vascular system, and especially the xylem, has been used for the classification of a large group of plants. The term *vascular plants* was first used in 1917 by Jeffrey. Later, the term *Tracheophyta* has been introduced to cover this group of plants which comprises the Pteridophyta and Spermatophyta. The term *Tracheophyta* has been derived from the xylem, and not the phloem, because of the firm and enduring structure of the tracheary elements. These elements have thick, hard walls and so can be distinguished more easily than the phloem elements. Also the xylem is more readily preserved in fossils and so can be identified more easily.

It should be mentioned that specialized cells conducting water, termed *hydroids*, occur in bryophytes (see Hebant, 1977).

Xylem is a complex tissue as it consists of several types of cells. The most important cells are the *tracheary elements* which are the non-living cells that are principally concerned with the transport of water and which also, to a certain degree, have a supporting function. Fibres are present in the xylem where they are mainly concerned with the strengthening of the plant body. Sometimes sclereids may also be present. Parenchyma cells which have storage and other functions also occur in the xylem. The xylem of some plants contains resin ducts (see Chapter 9).

The xylem and phloem elongate in developing organs by the continual differentiation of new elements produced by the procambium, which itself is continuously produced by the apical promeristem. The xylem produced by the procam-

bium in the primary body is called the *primary xylem*. In many plants, after the completion of the formation of the primary body, secondary tissues are developed. The xylem that is produced as a result of the activity of the vascular cambium is called the *secondary xylem*.

In the primary xylem the elements that are completed early, i.e. the *protoxylem*, are distinguished from those completed later, i.e. the *metaxylem*.

TRACHEARY ELEMENTS

Two basic types of tracheary elements are distinguished—*tracheids* and *vessel members*. The term *tracheid* was introduced in 1863 by Sanio who discussed the similarity and differences between this element and the vessel member. Since then much work has been devoted to the investigation of the structure, shape, function, ontogeny, and phylogeny of these elements.

The main difference between tracheids and vessel members is that the former are not perforated while the end walls of the latter are perforated (Fig. 58). A vessel, which is also termed a *trachea*, is built of numerous vessel members that are joined one to the other by their end walls. Vessels are terminated by a vessel member of which the proximal end wall is perforated, whereas the distal end wall is not, i.e. the distal parts of a vessel are tracheid-like.

The conductance of water by tracheary elements is related to their structure and dimensions (see Zimmermann, 1982, 1983).

Structure and shape of the secondary wall of tracheary elements

In a radial longitudinal section of vascular bundles it can be seen that the tracheary elements

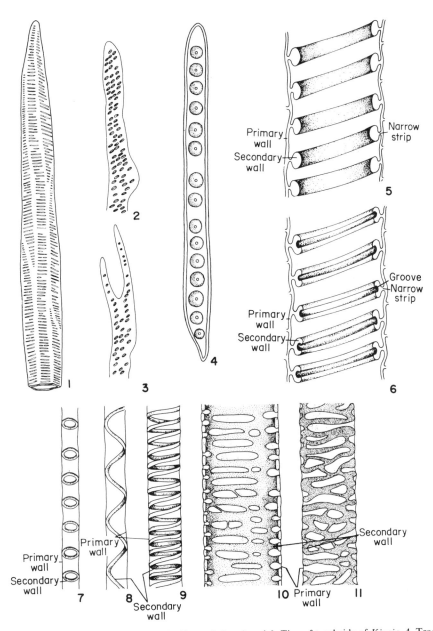

FIG. 56. 1, Tip of tracheid of *Dryopteris*; with scalariform pitting. 2 and 3, Tips of tracheids of *Kingia*. 4, Tracheid of *Pinus*. 5, Portion of a longitudinally sectioned tracheary element showing the helical wall thickenings and the strips by which they are joined to the primary wall. 6, As in no. 5, but in which the helical thickening is deeply grooved. 7–11, Different types of wall thickening in tracheary elements. 7, Annular thickening. 8, Helical thickening. 9, Dense helical thickening. 10, Scalariform thickening. 11, Reticulate thickening.

differ one from the other in the shape and structure of the secondary wall. In many plants the secondary wall thickening of the first-formed xylem (protoxylem) is *annular* or *helical* (Fig. 56, nos. 7–9; Fig. 57, no. 1). The helical thickening may be single, or more than one helix may be present in a single element. The rings or helices may be arranged in a loose or a dense manner. From an ontogenetic viewpoint, the annular elements precede the helical elements. In later-formed

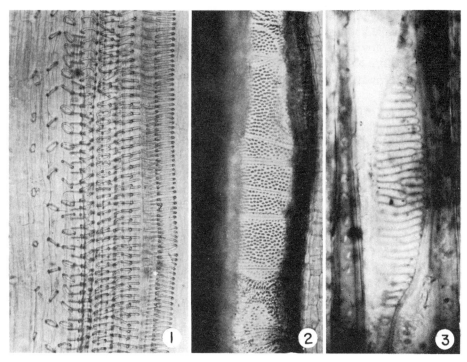

FIG. 57. 1 and 2. Micrographs of longitudinal sections of the young stem of *Cucurbita*. × 160. 1, Protoxylem elements with annular and helical thickening. 2, Pitted metaxylem vessel. 3, Micrograph of a radial longitudinal section in the secondary xylem of *Viburnum tinus* showing a scalariform perforation plate. × 500.

tracheary elements (in early metaxylem) the helical bands become joined in certain areas, giving rise to a ladder-like type thickening; such thickening is termed *scalariform thickening* (Fig. 56, no. 10). In tracheary elements formed at a still later ontogenetic stage the wall thickening is in the form of a network, i.e. *reticulate thickening* (Fig. 56, no. 11). When the openings in the secondary wall of such a network are elongated in a direction perpendicular to the longitudinal axis of the element, the thickening is termed *scalariform-reticulate*. In the ontogenetically most advanced elements, the secondary cell wall is interrupted only at the pits; such elements are termed *pitted elements* (Fig. 57, no. 2). Pitted elements are characteristic of the late primary and of the secondary xylem. Not all the above types are always found in a single plant. On the other hand, intermediate types not mentioned above can be found, as well as combinations of more than one form of thickening which may occur in a single element. The annular and helical wall thickenings may vary in

thickness and certain helices are so deeply grooved on their inner surfaces as to appear double (Fig. 56, no. 6). In some cases the helical thickening is joined by a narrow strip to the primary wall (Fig. 56, nos. 5, 6). In general, it has been shown that the range of variation in the pattern of wall thickening of the tracheary elements is very extensive (Bierhorst, 1960).

The pits in pitted tracheary elements are bordered. The well-developed bordered pit-pairs which are usually present between two tracheary elements are termed *intervascular pits* (see pp. 41–44). Between tracheary elements and fibres there may be only a few small pits or even none at all. Between tracheary elements and parenchyma cells the pit-pairs are mostly half-bordered, i.e. bordered on the side of the tracheary element and simple on the side of the parenchyma cell.

When the bordered pits are transversely elongated and are arranged in longitudinal rows along the element, the pitting is termed *scalariform pitting* (Fig. 29, no. 2). *Circular* and *elliptical* pits are arranged in horizontal or diagonal rows. The former arrangement is called *opposite pitting* (Fig. 29, no. 4) and the latter *alternate pitting* (Fig. 29, no. 5). On the inside surface of a pitted secondary wall a helical thickening may develop (Fig. 192; Fig. 201, no. 1).

In the Ophioglossales, the Ginkgoales, the Coniferales, and the Gnetales no scalariform pitted elements are found. In plants of these orders bordered pits, which are similar to those found in the secondary tracheary elements of the same plant, are found on the reticulate and helical thickenings of the primary xylem (Fig. 59, no. 10).

Wall development

The following facts are known about the formation of the special wall thickenings of the tracheary elements.

Crüger (1855) observed that in the positions where the thickenings of the secondary wall will develop, strips of actively streaming cytoplasm appear. Similar conclusions were reached by Barkley (1927) who believed that the position of the cytoplasmic strips is determined by the position of rows of vacuoles.

A similar phenomenon was also observed by Sinnott and Bloch (1945) who studied the development of tracheary elements from parenchyma cells during the regeneration in the vascular bundles of *Coleus*. Several workers have suggested that dictyosomes are involved in the formation of the wall thickenings (Wooding and Northcote, 1964; Esau et al., 1966a, b). There are indications that the endoplasmic reticulum may also be concerned with this process.

Recently it has been observed (Pickett-Heaps, 1968) that preceding the formation of the wall thickenings of the tracheary elements microtubules become grouped in bands between small, regular corrugations in the wall. In later stages many vesicles, probably derived from dictyosomes, collect at the regions occupied by the microtubules. Labelling indicated that both the endoplasmic reticulum and dictyosomes were associated with the wall incorporation of lignin precursors. Cronshaw (1965) has noted that microtubules, which are often seen in the cytoplasm close to the cell surface and, according to Ledbetter and Porter (1963),

may govern cytoplasmic streaming or exert an influence over the disposition of cell-wall material, are oriented in the same direction as the cellulose microfibrils of the developing wall thickenings of tracheary elements of the protoxylem of the *Avena* coleoptile. Herth (1985) observed in developing vessel elements of *Lepidium sativum* that "rosettes" (see p. 31) on the plasmalemma were restricted to regions of secondary wall thickenings. Barnett (1977, 1979), who studied tracheid differentiation in *Pinus radiata*, did not find significant numbers of microtubules during the main phase of secondary wall formation. Their number, however, increased much in the final stage of secondary wall formation, when lignification of the secondary wall begun.

After completion of the elaborate secondary wall thickenings and their lignification, the protoplast and certain parts of the cell wall enter the stage of lysis. It has been suggested that autolysis of the cytoplasm occurs when the latter is exposed to the vacuolar sap containing hydrolytic enzymes after a rupture of the tonoplast (Wodzicki and Brown, 1973). Hydrolysis affects also the non-cellulosic components of those parts of the primary wall that are not covered by lignified secondary wall (i.e. wall regions between helical wall thickenings or pit membranes). The residual material of the hydrolysed primary wall forms a coarse net of fibrils (Esau and Charvat, 1978). According to Czaninski (1968), in *Robinia* at the end of the differentiation of the vessel element a tertiary wall lamella may be deposited (see wart structure, p. 45). Scott et al. (1960) described a suberization of the vessel wall in *Ricinus* after the death of the protoplast.

The tracheary elements which mature later than the helical ones go through a two-phase wall deposition process. In the first phase the helical framework is formed and during the second phase secondary wall material is deposited between the gyres of the helix (Bierhorst and Zamora, 1965; Falconer and Seagull, 1988).

The functional significance of the different types of wall thickenings in the tracheary elements is not clear. It is possible that the exclusive appearance of annular and spiral thickening in elements in those organs that are still elongating has some connection with the rapid increase in length of the organ. Investigations using X-rays together with the regulation of light or osmotic regulation to alter the rate of stem elongation proved this assumption. Goodwin (1942), Smith and Kersten (1942) and Brower and Hepler (1972) observed that if stem elongation is inhibited the production of annular and spiral vessels is reduced or stopped and pitted vessels develop.

The development of vessel members with wall thickenings that are intermediate between protoxylem and metaxylem, from a file of procambial cells which is continuous with typical protoxylem elements, occurs at the end of intercalary growth of the internode of *Cyperus alternifolius*. This supports the view that the extent of elongation of an organ determines the wall pattern of the tracheary elements maturing in the organ (Fisher, 1970).

Vessels and the structure of perforation plates

As has been mentioned previously, two main types of tracheary elements can be distinguished —tracheids and vessel members. Tracheids are non-perforated cells in which only bordered pit-pairs are found in the areas of contact between them, while vessel members are perforated at their ends. By these perforations the vessel members become joined to form a tube-like series of cells which is termed a *vessel* or *trachea*. Vessels are limited in length and those vessel members which terminate a vessel are perforated at one end only, i.e. the terminating end is not perforated. Therefore the passage of water from vessel to vessel takes place via the pits as from tracheid to tracheid. It is difficult to measure the length of vessels but this was done successfully by Handley in 1936 who made use of the fact that, although water and solution pass through the pits, gas does not. Handley forced coal gas into one end of a cut branch and attempted to light it at the other end. Zimmermann and Jeje (1981) and Zimmermann *et al.* (1982) calculated vessel length from measurements of air-volume flow and of latex-paint particles (in suspension) penetration through the xylem. The length of vessels was found to vary from a few centimetres to 5 m or even more. In general, vessel length was found to be correlated with vessel diameter, wide vessels being longer.

A more accurate analysis of length as well as of distribution of all vessels in a plant organ can be achieved by the use of motion picture films assembled from images of serial transverse sections (Zimmermann, 1978, 1983).

The vessel members are usually perforated on the end walls, but sometimes the perforations are formed on the side walls. Those parts of the cell wall that bear perforations are called *perforation plates*. The perforation plate may contain one large perforation and then it is termed a *simple perforation plate* (Fig. 58, nos. 4–6), or it may contain numerous perforations and form a *multi perforation plate*. In the latter case there are several possible ways in which the perforations can be arranged. When the perforations are elongated and are arranged in a parallel series the plate is termed a *scalariform perforation plate* (Fig. 57, no. 3; Fig. 58, nos. 1, 2), when in a reticulate manner *reticulate perforation plate* (Fig. 59, nos. 2, 4) and when the perforations are almost circular the plate is termed a *foraminate* (or *ephedroid*) *perforation plate* (Fig. 59, no. 11).

The scalariform perforation plates may sometimes be very long and then they contain hundreds of perforations. In such cases the end wall bearing the plate is very long and oblique so that it is sometimes difficult to decide whether it is a vessel member or a tracheid (Fahn, 1953). The identity of these elements can be established by passing a carbon suspension through sectioned portions of branches. The suspended particles can pass only through perforations as the pit membrane prevents their passage through the pits. However, sometimes even the above method is not reliable and a further method in which very fine longitudinal sections of the perforation plate are cut is used in order to discover if the primary wall is present or not. Transmission and scanning electron microscopy can of course be of great help in this matter.

In many dicotyledonous species the middle portion of the vessel members of the secondary xylem widens during ontogenetic development while the tips remain narrow and elongated. These tips are not perforated and they appear as projections that overlap the walls of the neighbouring vessel members; they have been termed *tails* (Chalk and Chattaway, 1934, 1935). The per-

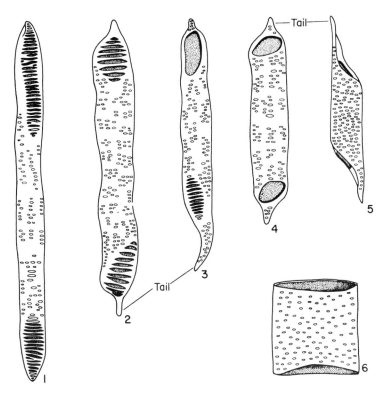

FIG. 58. Dicotyledonous vessel members. 1 and 2, Vessel members in which the perforation plates at both ends are scalariform. 3, Vessel member with one scalariform and one simple perforation plate. 4–6, Vessel members with simple perforation plates. "Tails", the narrow elongated tips of the vessel members, can be seen in nos. 2–5. (Courtesy of I. W. Bailey.)

forations are present at the end of the widened part of the element, i.e. near the base of the tails (Fig. 58, nos. 2–5).

When the vessels are wide, in ring-porous and sometimes in diffuse-porous wood, their growth in width separates the neighbouring cells from one another. In this way the vessel is brought into contact with new cells (Fig. 60, nos. 6, 7). In many cases it is possible to observe that where the position of the separating cells is shifted relative to the widening vessel, the cells retain, at least partially, their original attachments in those positions where there are pits. This occurs by the extension of the cell wall to form bridge-like connections in the region of the pits (Fig. 60, no. 8). According to Priestley *et al.* (1935) this phenomenon is apparently made possible because of the greater plasticity of the pits themselves or their margins. These workers state that in ring-porous wood the cells surrounding the vessels differentiate prior to the vessels or at the latest simultaneously with them, and are therefore separated

by vessel growth. In diffuse-porous wood the walls of the cells next to the vessel become thickened and lignified somewhat later than those of the vessel members, and are able to expand during the growth of the vessel.

Development of the perforation

Vessels develop from meristematic cells—procambial cells in the primary xylem and cambial cells in the secondary xylem. The vessel members may or may not elongate prior to the thickening of the wall but they usually widen in this stage of development. During the stage of the growth of the meristematic vessel element, the nucleus becomes polyploid and increases in volume (Innocenti and Avanzi, 1971).

Much attention has been paid by workers studying the ontogeny of vessels to the end walls in which the perforations develop. The end wall occupying the future site of the perforation does not become covered with a secondary wall layer,

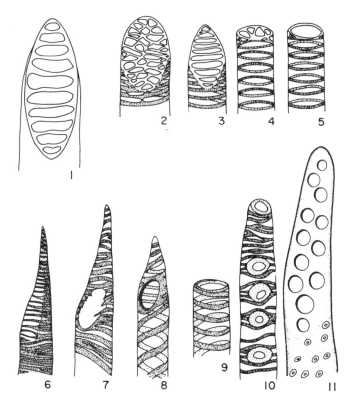

FIG. 59. 1–5, Perforation plates of vessel members in the primary xylem of monocotyledons. 1, Scalariform perforation plate from the stem of *Phoenix dactylifera*. × 70. 2, Reticulate perforation plate from the root of *Hymenocallis caribaea*. × 200. 3–5, Vessel members from the stem of *Rhoeo discolor*. × 150. 3, Scalariform perforation plate of a helically thickened vessel member. 4, Reticulate perforation plate of an annularly thickened vessel member. 5, Simple perforation plate. 6–9, Ends of vessel members with helical thickening from dicotyledonous primary xylem. 6, Scalariform perforation plate. 7, Transitional form between a scalariform and simple perforation plate. 8 and 9, Simple perforation plates. 10, Tracheid of *Gnetum* with helical thickening and circular bordered pits. 11, Vessel member end of *Ephedra* with a foraminate perforation plate. (Nos. 1–5 adapted from Cheadle, 1953; nos. 6–10 adapted from Bailey, 1944.)

but it does become thicker than the primary wall elsewhere. On the basis of a cytochemical study of the differentiation of the end wall of secondary vessels of *Populus italica* and *Dianthus caryophyllus* it was suggested that this wall consists primarily of non-cellulosic polysaccharides and of only very little cellulose (Fig. 60, nos. 1–5; Fig. 61) (Benayoun *et al.*, 1980).

Czaninski (1968), who made an electron microscope study of the vessel members of *Robinia*, has observed that the plasmalemma bordering the end wall of the developing element forms many vesicles. This suggests an exchange of numerous substances between the cytoplasm and the end wall. At this stage the cytoplasm adjacent to the end walls contains mitochondria, dictyosomes,

endoplasmic reticulum, and polyribosomes, as does the cytoplasm bordering the side walls of the vessel member. At a later stage an electron dense layer between the electron translucent layer of the primary wall and the plasmalemma becomes visible. This phenomenon occurs concurrently with the completion of the degeneration of the cytoplasm. At the same time the whole end wall becomes sinuous. Finally, the end wall disappears.

The manner of removal of the end wall may differ in different species and in vessel elements of the primary or secondary xylem. So, for instance, it was reported that in the central vessel of the root of *Hordeum vulgare* lysis begins along the margin of the end wall (Sassen, 1965). In the

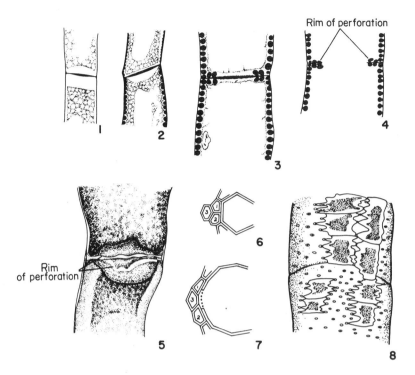

FIG. 60. 1–4, Development of a perforation plate in vessel members of *Apium graveolens*, after Esau, 1936. The development of helical secondary thickening on the side walls and the presence of the primary end wall can be seen in nos. 1–3. The end wall has disappeared in no. 4. 5, End portions of two adjacent vessel members of *Fraxinus* in which it is possible to distinguish the perforation rim and the two protoplasts that have separated from one another. × 125. 6 and 7, Diagrams of cross-sections of a vessel and neighbouring cells showing how the vessel, during enlargement, comes into contact with new cells. Neighbouring cells indicated by numerals. 8, Drawing of portion of a vessel of *Ulmus* in surface view, showing how the cells around the vessel tear away from each other as a result of the widening of the vessel. The neighbouring cells retain their original attachments to the vessels where there are pits, resulting in the formation of bridge-like extensions. × 95. (Nos. 5–8 adapted from Priestley *et al.*, 1935.)

secondary vessel elements of *Knightia excelsa* small holes develop in the end wall before final disappearance (Meylan and Butterfield, 1972). A similar process occurs in *Quercus rubra* (Murmanis, 1978). In *Pisum sativum* and in *Dianthus caryophyllus* lysis of the end wall begins in the centre (Niedermeyer, 1974; Benayoun *et al.*, 1980). In the vessels of wood of *Laurelia novae-zelandiae* the degradation of the end wall in the scalariform plates leaves a network of fibrils which later disappears, possibly by the action of the transpiration stream (Butterfield and Meylan, 1972).

Phylogenetic development of tracheary elements

The xylem holds an important position in the study of plant tissues as the structure of its elements is of extreme importance in taxonomy and phylogeny. More attention has been paid to the phylogenetic development of xylem than to any other tissue. The structure of the tracheary elements has been studied in special detail. The research has been aided by statistical methods which emphasize the differences in structure and shape of the tracheary elements and which have explained their phylogenetic significance.

It has been obvious for a long time that the tracheid is a more primitive element than the vessel member. The tracheid is the only tracheary element found among Pteridospermae, in fossil Spermatophyta, in most of the lowest vascular plants of today, and in nearly all the Gymnospermae. It is commonly accepted that vessel members have developed from tracheids. Vessel members occur in the following diverse groups of plants: in the most advanced gymnosperms, the

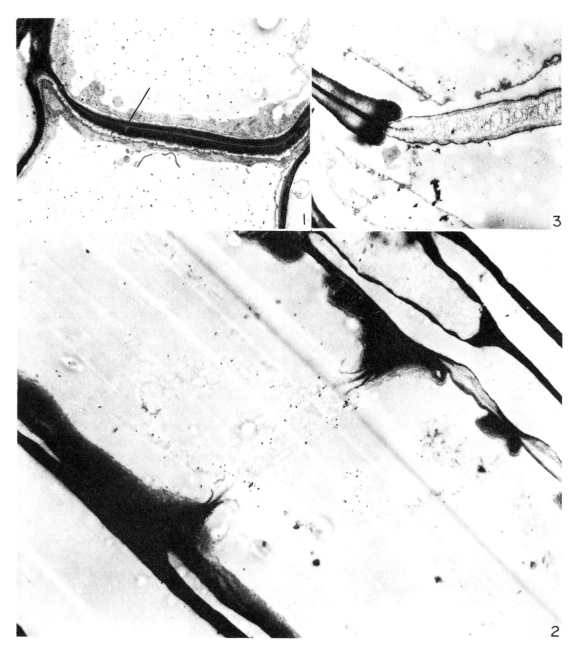

FIG. 61. Electron micrographs of changes occurring in the end wall of differentiating vessel members of *Dianthus caryophyllus* (nos. 1 and 2) and *Populus italica* (no. 3). 1, Early stage of differentiation. The thick end wall consisting of the darkly stained primary walls of the two contiguous vessel members and of the lightly stained narrow middle lamelia. × 8,500. 2, Vessel members at the stage of secondary wall deposition stained for lignin (with CIEtAg) and contrasted with a polysaccharide reagent (PATAg). The fragile end wall is partly disrupted by the treatment. × 8,000. 3, As in no. 2 but after extraction of pectin (with EDTA). × 8,500. (From Benayoun *et al.*, 1981.)

Gnetales; in dicotyledons except for the lower-most taxonomic groups; in monocotyledons; in the fern, *Pteridium*; in certain species of *Selaginella*; in *Equisetum* (Bierhorst, 1958), and in the root of some species of *Marsilea* and *Regnellidium* (Mehra and Soni, 1971; Tewari, 1975). From the above it can be assumed that vessels developed independently, by parallel evolution, in each of these groups.

Origin and phylogenetic development of vessel members in angiosperms

In order to understand the problems of plant phylogeny, fundamental methods of logic have been used. Frost (1930a, b, 1931) clearly defined some of these fundamental logical assumptions while trying to establish the origin of the vessels in the dicotyledons. The following are the principal assumptions used by Frost.

1. *The association method.* This method states that if it is possible to determine which of two structures is the more primitive, and if it is assumed that the two structures have a direct genetic relationship, it will be possible to conclude that the primitive condition of the more advanced structure will be similar to the general condition of the primitive structure. If there is not much similarity then the assumption of direct genetic relationship is not correct or the elements in question are, apparently, so far separated in the scale of evolution that the primitive form of the advanced element has been lost. Therefore, with reference to tracheary elements, if it is assumed that the tracheids are more primitive than the vessel members and that the two elements have a direct genetic relationship, then it must be concluded that the most primitive vessel members will be those that are most similar to tracheids.

2. *The correlation method.* By this method it is assumed that in a certain homogeneous tissue, as, for instance, the secondary xylem, there will exist a statistically significant correlation between the degrees of specialization of the main characteristics of a structure in a large random sample (many species), i.e. the various features have undergone evolutionary changes simultaneously. Therefore, features occurring together with those features that are defined as being primitive by the association method are themselves primitive, and those that occur together with features defined as being advanced are advanced. It is necessary to bear in mind that such correlations express only the general trends of development and that exceptions exist. The development of some features may be delayed and of others advanced. The investigations of these exceptions can indicate the lines of secondary specialization which only become clear after the principal lines of development have been determined. In relation to the vessel members, if great length is a primitive feature (as is derived by the association method) then all other features that are found in correlation with great length are also primitive features.

3. *The sequence method.* This method deals with the reconstruction of the evolutionary variability on the basis of the variation, as seen in living forms. These variations can be seen ontogenetically or by the comparison of different plants belonging to a single taxonomic group. The contribution of this method to the problem of the origin of vessels has been the determination of the origin of vessel members from tracheids with scalariform pitting. Typical tracheids of this type appear only in the secondary xylem of those dicotyledonous genera that have no vessels, e.g. genera of the Winteraceae, Chloranthaceae, and Tetracentraceae, but are completely absent from the secondary xylem of angiosperms that contain vessels (Bailey, 1944). In some trees and large shrubs of various primitive dicotyledonous families, the vessel members of the secondary xylem are similar in size, angular cross-section, pitting, and thin secondary walls to tracheids with scalariform pitting. It is important to mention that the scalariform pitted tracheids have served as the origin not only of vessel members but also of tracheids with circular bordered pits and apparently also, indirectly, of fibre-tracheids and libriform fibres (Tippo, 1946; Bailey, 1936, 1953; Bailey and Tupper, 1918). The complete or almost complete absence of primitive tracheids with scalariform pitting in the Angiospermae is related to their development, in the process of phylogeny, into vessel members or into tracheids with a more advanced form of pitting.

The following structural features of the angiosperm tracheary elements are those that are used as a basis for the study of their evolution.

1. *The length of the element.* The tracheids are long cells whose average length reaches 6.29 mm in *Bellidium gracile,* a vessel-less dicotyledon (Carlquist, 1983b). Their average length as calculated from many hundreds of measurements made in monocotyledons is 5.07 mm (Cheadle, 1943a). In monocotyledons, according to Cheadle, vessels can be divided into four groups according to their degree of specialization. The average lengths of the vessel members in these four groups are 3.96 mm, 2.58 mm, 1.47 mm, and 0.76 mm. As the vessel members are shorter than the tracheids the shorter the vessel member, the more advanced it is considered to be.

2. *The diameter of the element.* The diameter of the tracheid is smaller than that of the vessel member.

3. *The thickness of the wall.* The wall of a typical tracheid is thin and is of equal thickness over the entire circumference. This feature is also seen in primitive vessel members.

4. *The perforation plates.* Those scalariform perforation plates that are long, oblique and with numerous perforations are considered the most primitive and the simple, horizontal perforation plates the most advanced.

5. *The shape of the element in cross-section.* The shape of the tracheids and the primitive vessels in cross-section is angular, while that of advanced vessel members is circular or nearly so.

6. *The type of pitting.* In the dicotyledons scalariform pitting in vessel members is considered to be primitive. The structure and arrangement of pits developed from scalariform pitting, through intermediate forms in which scalariform pits occur together with circular or elliptical pits (Fig. 29, no. 3), to forms with only circular or elliptical pits. Of this advanced type of pitting, that in which the pits are arranged in parallel rows, i.e. opposite pitting, is more primitive than alternate pitting, in which the circular or elliptical pits are arranged along more or less helical lines (Fig. 29, nos. 4, 5). The appearance of the spiral thickenings on the inside of the secondary wall of the

tracheary elements is evidence of advanced development.

The phylogenetic development of the side walls of the tracheary elements was prior to that of the perforation of the end walls.

Summary of suggested origin and specialization of vessels

From investigations based on the methods and facts that have been mentioned above the present knowledge of the evolutionary development of the mono- and dicotyledonous vessel members can be summarized, after Cheadle (1953), as follows:

Dicotyledons

1. Ten woody genera are known that completely lack vessels. These genera belong to the following five families: Chloranthaceae, Winteraceae, Tetracentraceae, Trochodendraceae, and Amborellaceae.

2. There are 52 out of 147 families that consist of woody plants only and that contain one or more species that have only scalariform-perforated vessel members. The following are some of these families: Aquifoliaceae, Betulaceae, Buxaceae, Celastraceae, Magnoliaceae, Myrtaceae, and Styracaceae.

3. Of 82 families that contain both woody and herbaceous species, only 7 families contain one or more species with exclusively scalariform-perforated vessel members.

4. Of the herbaceous plants, the internal structure of which has been adequately studied, only *Paeonia* of the Ranunculaceae, *Pentaphragma* of the Campanulaceae, and a few other species of three other families have exclusively scalariform-perforated vessel members. However, in these examples the perforation plate is mostly not of the very primitive type as the plate is short and has only a few perforations.

5. Of the remaining herbaceous families, in 61 families only vessel members with simple perforation plates are found and in 20 families the perforation is mainly simple but a few scalariform perforation plates (usually short) can be found.

From all the above facts it appears that in the dicotyledons the vessels arose first in woody plants. Apparently they developed independently a number of times as vessel-less species are found in different families. Because of the advanced character of the vessels in herbaceous plants it cannot be suggested that the woody plants have been derived from the herbaceous plants.

As a result of the data that have accumulated, it has been concluded that the vessels arose first in the secondary xylem and later in the metaxylem. The specialization has also gradually advanced from the secondary to primary xylem.

It can also be assumed that the herbaceous plants have developed from the woody plants by reduction of cambial activity only after obvious development of the vessel members had taken place in the woody ancestral plants.

On the basis of cladistic analysis, Young (1981) suggested that the absence of vessels in dicotyledons is a derived feature. Carlquist (1983a) rejected this view, and brought further evidence to support the widely accepted hypothesis that the original angiosperms were vessel-less. Carlquist stated that secondary vessel-less can be hypothesized only for aquatic plants in which no secondary xylem is present.

In some specialized dicotyledons, such as in certain representatives of the Cactaceae, vessels in the secondary xylem are replaced by so-called vascular tracheids (short, imperforated cells with annular and helical thickenings). However, in such plants the reduction of vessels cannot be cited as support for evolutionary reversal to primitive vessel-less conditions (Bailey 1957, 1966).

Monocotyledons

1. From a phylogenetic point of view, vessels in the monocotyledons first appeared in the roots and later in the stems and leaves. The specialization of the vessels followed the same pattern (Cheadle, 1943a, b).

2. Phylogenetically, the vessels first appeared and became specialized in the late-formed metaxylem and progressed gradually into the early formed metaxylem and finally into the protoxylem (Cheadle, 1944).

3. Monocotyledons exist today that have, in the last-formed metaxylem of their roots, only the most primitive vessels the perforation plates of

which are scalariform and which contain more than 100 parallel perforations.

4. A few monocotyledonous families with only aquatic species are known to include plants that lack vessels completely in all their organs. This feature, however, may be a secondary one.

The tracheary elements have developed during the evolution of the land plants. As has been pointed out by Bailey (1953), two main functional trends have become evident during the course of the morphological evolution of these elements, i.e. the development of those structures that enhance rapid conduction, on the one hand, and of those that strengthen the elements, on the other hand. These two trends are antagonistic to a great extent because certain structures that increase the efficiency of conduction tend to weaken the cells and vice versa. However, during the course of evolution, structures have been developed that have, to various extents, resolved these two trends.

Pitted tracheary elements, in addition to those with annular and helical wall thickenings, are found in most of the Tracheophyta, with the exception of certain lower Devonian plants and some hydrophytes. Elements with such wall thickenings give support to the mature stem. The absence of living protoplasts in the tracheary elements, the development of elongated tracheids and the occurrence of vessel members are all features that increase the efficiency of water conduction. The bordered pit-pairs which are characteristic of the tracheary elements, are, as has been shown by Bailey, well adapted to their function and they combine the two above-mentioned trends. On the one hand, the area of the pit membrane is comparatively large and so the passage of water is fairly easy and, on the other hand, the extent of the development of the secondary wall is maximal because the secondary wall overarches the pit membrane in such a manner that the pit membrane remains comparatively large whereas the pit aperture is very small. This feature greatly strengthens the tracheary elements.

In tracheids more rapid conduction is obtained by the elongation of the cells, the increase in diameter of the lumen and in the number of pits and the reduction of wall thickness. Strengthening of the tissue is brought about by the shortening of the cells, narrowing of the lumen, increase in wall thickness and the reduction in the number of pits. In the secondary xylem of conifers, for instance, the early wood is more adapted for efficient water conduction, and the tracheids of the late wood for support.

Conduction is further facilitated by the complete disappearance of the pit membranes in certain areas, resulting in the formation of vessel members. In the secondary xylem of certain primitive dicotyledons primitive vessel members, resembling scalariform pitted tracheids, and thick-walled, narrow tracheids with a few round bordered pits have been observed to occur side by side. This phenomenon proves that the evolution of tracheids, in relation to function, was dichotomous, i.e. the scalariform tracheids evolved into vessel members which were better adapted to conduction, whereas the tracheids with round bordered pits are modified to give better support and, through various intermediate forms, give rise to the libriform fibres. Fibre-tracheids, which are an intermediate form, and libriform fibres functionally differ greatly from the tracheary elements, and in many plants they contain even living protoplasts and store reserve materials.

From the large amount of data that has been accumulated from the study of the various angiosperm groups it is possible to build a clear picture of the trend of evolution that has taken place in the development of the tracheary elements. The fact that this trend is unidirectional, irreversible, and cannot be interpreted in the reverse direction is important and should be emphasized. The structural evolution of the tracheary elements presents one of the most convincing examples of evolutionary development. However, although it is obvious from the great amount of data that this structural evolution has been accompanied by functional specialization, as yet only little is known about whether or not a correlation exists between the types of tracheary elements and ecological conditions. This is a problem that still needs to be investigated.

The sequence of the different types of tracheary elements and the numerous transitional forms are important features in the study of the origin of the Angiospermae and the phylogeny of the various taxonomic groups among them. It is still necess-ary, however, because of parallel and convergent evolution, to accumulate more data concerning the tracheary elements of the various species and to use such data together with morphological and structural data concerning other elements and tissues before any definite conclusions can be drawn (Fahn, 1954).

ECOLOGICAL ASPECTS OF XYLEM EVOLUTION

Much attention has recently been paid to the ecological aspects of xylem evolution (Baas, 1973, 1976, 1980, 1986; Baas et al., 1983; Carlquist, 1975, 1977, 1980, 1982; Fahn et al., 1986). The structural diversity in xylem of presently living plants is suggested to be a result of evolutionary changes which are functionally adaptive to the various habitats. The adapted changes were brought about through selective pressure in the different environments in which the various plant taxa evolved. It is thus expected that the adaptation of anatomical features of wood are related to moisture availability, transpiration, and requirements for mechanical strength. Carlquist (1975) considers, for instance, that great length of vessel members with scalariform perforation plates reduces resistance to water flow. Short and narrow vessel members with thick walls would increase the strength of a vessel and make it resistant to high negative pressure which occurs in xeric habitats. Scalariform perforation plates are regarded as functionally disadvantageous as compared to simple perforations plates, because of resistance to water flow. Wood with scalariform perforations is rare in plants of xeric habitats and has not been found in plants growing in deserts. In highly mesic conditions wood with scalariform perforations seems not to be a limiting factor for evolutionary success because of the consistently low flow of water in the vessels. Vines and lianas, with a requirement for efficient water conduction, possess exceptionally wide vessels causing minimal friction to water flow. The vessel members in these plants are relatively short, apparently enhancing resistance to negative pressure. The number of vines and lianas with scalariform perforation plates is very small.

Baas (1986) distinguished two partly parallel

ecological trends in vessel characteristics. On the basis of floristic analysis as well as from studies of anatomical variation within widely distributed genera and families he summarized these trends as follows:

In mesic floras or taxa, vessel member length and vessel diameter decrease, and the number of vessels per area of cross-section and incidence of scalariform perforations increases, with increasing latitude or altitude (in cooler environment).

If mesic floras or taxa are compared with xeric ones, woods from drier habitats tend to have shorter vessel members and the incidence of scalariform perforations is less, and may even be nill in desert floras. Many xeric species have two vessel size classes: numerous very narrow vessels in addition to wide vessels. The incidence of helical (or spiral) thickenings on the inner vessel-wall surface, in extremely arid floras is fairly low, but is very high in moderately arid Mediterranean or sclerophyllous vegetation.

REFERENCES

BAAS, P. (1973) The wood anatomy of *Ilex* (Aquifoliaceae) and its ecological and phylogenetic significance. *Blumea* **21**: 193–258.

BAAS, P. (1976) Some functional and adaptive aspects of vessel member morphology. *Leiden Botanical Series*, No. 3: 157–81.

BAAS, P. (1980) Further concepts in ecological wood anatomy with comments on recent work in wood anatomy and evolution. *Aliso* **9**: 499–553.

BAAS, P. (1986) Ecological patterns in xylem anatomy. In: *On the Economy of Plant Form and Function* (ed. T. J. Givnish). Cambridge University Press, Cambridge, pp. 327–352.

BAAS, P., WERKER, E. and FAHN, A. (1983) Some ecological trends in vessel characters. *IAWA Bull.*, n.s. **4**: 141–159

BAILEY, I. W. (1936) The problem of differentiating and classifying tracheids, fiber-tracheids and libriform fibers. *Trop. Woods.* **45**: 18–23.

BAILEY, I. W. (1944) The development of vessels in angiosperms and its significance in morphological research. *Am. J. Bot.* **31**: 421–8.

BAILEY, I. W. (1953) Evolution of the tracheary tissue of land plants. *Am. J. Bot.* **40**: 4–8.

BAILEY, I. W. (1957) Additional notes on the vesselless dicotyledon *Amborella trichopoda* Baill. *J. Arnold Arb.* **38**: 374–8.

BAILEY, I. W. (1966) The significance of the reduction of vessels in the Cactaceae. *J. Arnold Arb.* **47**: 288–92.

BAILEY, I. W. and TUPPER, W. W. (1918) Size variation in tracheary cells: A comparison between the secondary xylem of vascular cryptogams, gymnosperms, and angiosperms. *Am. Acad. Arts Sci. Proc.* **54**: 149–204.

BARKLEY, G. (1927) Differentiation of vascular bundle of *Trichosanthos anguina. Bot. Gaz.* **83**: 173–84.

BARNETT, J. R. (1977) Tracheid differentiation in *Pinus radiata. Wood. Sci. Tech.* **11**: 83–92.

BARNETT, J. R. (1979) Current research into tracheary element formation. *Current Advances in Plant Science* **33**: 1–13.

BENAYOUN, J., CATESSON, A. M. and CZANINSKI, Y. (1981) A cytochemical study of differentiation and breakdown of vessel end walls. *Ann. Bot.* **47**: 687–698.

BIERHORST, D. W. (1958) Vessels in *Equisetum. Am. J. Bot.* **45**: 534–7.

BIERHORST, D. W. (1960) Observations on tracheary elements. *Phytomorphology* **10**: 249–305.

BIERHORST, D. W. and ZAMORA, P. M. (1965) Primary xylem elements and associations of angiosperms. *Am. J. Bot.* **52**: 657–710.

BROWER, D. L. and HEPLER, P. K. (1976) Microtubules and secondary wall deposition in xylem: the effects of isopropyl N-phenylcarbamate. *Protoplasma* **87**: 91–111.

BUTTERFIELD, B. G. and MEYLAN, B. A. (1972) Scalariform perforation plate development in *Laurelia novae-zelandiae* A. Cunn.: a scanning electron microscope study. *Aust. J. Bot.* **20**: 253–9.

CARLQUIST, S. (1975) *Ecological Strategies of Xylem Evolution.* University of California Press, Berkeley.

CARLQUIST, S. (1977) Ecological factors in wood evolution, a floristic approach. *Am. J. Bot.* **64**: 887–896.

CARLQUIST, S. (1980) Further concepts in ecological wood anatomy, with comments on recent work in wood anatomy and evolution. *Aliso* **9**: 499–553.

CARLQUIST, S. (1982) Wood anatomy of *Illicium* (Illiciaceae). Phylogenetic ecological and functional interpretations. *Am. J. Bot.* **69**: 1587–1598.

CARLQUIST, S. (1983a) Wood anatomy of *Bubbia* (Winteraceae), with comments on origin of vessels in dicotyledons. *Am. J. Bot.* **70**: 578–590.

CARLQUIST, S. (1983b) Wood anatomy of *Belliolum* (Winteraceae) and a note of flowering. *J. Arnold Arb.* **64**: 161–169.

CHALK, L. and CHATTAWAY, M. M. (1934) Measuring the length of vessel members. *Trop. Woods* **40**: 19–26.

CHALK, L. and CHATTAWAY, M. M. (1935) Factors affecting dimensional variations of vessel members. *Trop. Woods* **41**: 17–37.

CHEADLE, V. I. (1943a) The origin and certain trends of specialization of the vessel in the Monocotyledoneae. *Am. J. Bot.* **30**: 11–17.

CHEADLE, V. I. (1943b) Vessel specialization in the late metaxylem of the various organs in the Monocotyledoneae. *Am. J. Bot.* **30**: 484–90.

CHEADLE, V. I. (1944) Specialization of vessels within the xylem of each organ in the Monocotyledoneae. *Am. J. Bot.* **31**: 81–92.

CHEADLE, V. I. (1953) Independent origin of vessels in the monocotyledons and dicotyledons. *Phytomorphology* **3**: 23–44.

CRONSHAW, J. (1965) Cytoplasmic fine structure and cell development in differentiating xylem elements. In: *Cellular Ultrastructure of Woody Plants* (ed. W. A. Côté, Jr.), Syracuse Univ. Press, Syracuse, NY, pp. 99–124.

CRÜGER, H. (1855) Zur Entwicklungsgeschichte der Zellwand. *Bot. Ztg.* **13**: 601–13; 617–29.

CZANINSKI, Y. (1968) Étude cytologique de la différentiation cellulaire du bois de Robinier. *J. Microscopie* **7**: 1051–68.

ESAU, K. and CHARVAT, I. (1978) On vessel member differentiation in the bean (*Phaseolus vulgaris* L.). *Ann. Bot.* **42**: 665–77.

ESAU, K., CHEADLE, V. I. and GILL, R. H. (1966a) Cytology of differentiating tracheary elements: I. Organelles and membrane systems. *Am. J. Bot.* **53:** 756–64.

ESAU, K., CHEADLE, V. I. and GILL, R. H. (1966b) Cytology of differentiating tracheary elements: II, Structures associated with cell surfaces. *Am. J. Bot.* **53:** 765–71.

FAHN, A. (1953) Metaxylem elements in some families of the Monocotyledoneae. *New Phytol.* **53:** 530–40.

FAHN, A. (1954) The anatomical structure of the Xanthorrhoeaceae Dumort. *J. Linn. Soc. Lond., Bot.* **55:** 158–84.

FAHN, A., WERKER, E. and BAAS, P. (1986) *Wood Anatomy and Identification of Trees and Shrubs from Israel and Adjacent Regions.* Israel Academy of Sciences, Jerusalem.

FALCONER, M. M. and SEAGULL, R. W. (1988) Xylogenesis in tissue culture III: Continuing wall deposition during tracheary element development. *Protoplasma* **144:** 10–16.

FISHER, J. B. (1970) Xylem derived from the intercalary meristem of *Cyperus alternifolius. Bull. Torrey Bot. Club* **97:** 58–66.

FROST, F. H. (1930a) Specialization in secondary xylem of dicotyledons: I, Origin of vessel. *Bot. Gaz.* **89:** 67–94.

FROST, F. H. (1930b) Specialization in secondary xylem of dicotyledons: II, Evolution of end walls of vessel segment. *Bot. Gaz.* **90:** 198–212.

FROST, F. H. (1931) Specialization in secondary xylem of dicotyledons: III, Specialization of lateral wall of vessel segment. *Bot. Gaz.* **91:** 88–96.

GOODWIN, R. H. (1942) On the development of xylary elements in the first internode of *Avena* in dark and light. *Am. J. Bot.* **29:** 818–28.

HANDLEY, W. R. C. (1936) Some observations on the problem of vessel length determination in woody dicotyledons. *New Phytol.* **35:** 456–71.

HÉBANT, C. (1977) *The Conducting Tissues of Bryophytes.* J. Cramer, Vaduz.

INNOCENTI, A. M. and AVANZI, S. (1971) Some cytological aspects of the differentiation of metaxylem in the root of *Allium cepa. Caryologia* **24:** 283–92.

JEFFREY, E. C. (1917) *The Anatomy of Woody Plants.* Univ. Chicago Press, Chicago.

LEDBETTER, M. C. and PORTER, K. R. (1963) A "microtubule" in plant cell fine structure. *J. Cell Biol.* **19:** 239–50.

MEHRA, P. N. and SONI, S. L. (1971) Morphology of tracheary elements in *Marsilea* and *Pteridium. Phytomorphology* **21:** 68–71.

MEYLAN, B. A. and BUTTERFIELD, B. G. (1972) Perforation plate development in *Knightia excelsa* R. Br.: a scanning electron microscope study. *Aust. J. Bot.* **20:** 79–86.

MURMANIS, L. L. (1978) Breakdown of end walls in differentiating vessels of secondary xylem in *Quercus rubra* L. *Ann. Bot.* **42:** 679–82.

NIEDERMEYER, W. (1974) Auflösung der Endwände in differenzierenden Gefässzellen. *Ber. dt. bot. Ges.* **86:** 529–36.

PICKETT-HEAPS, J. D. (1968) Xylem wall deposition: radioautographic investigations using lignin precursors. *Protoplasma* **65:** 181–205.

PRIESTLEY, J. H., SCOTT, L. I., and MALLINS, M. E. (1935) Vessel development in the angiosperms. *Proc. Leeds Phil. Lit. Soc.* **3:** 42–54.

SANIO, C. (1863) Vergleichende Untersuchungen über die Elementarorgane des Holzkörpers. III. Tracheales System. *Bot. Ztg.* **21:** 113–18; 121–8.

SASSEN, M. M. A. (1965) Breakdown of plant cell wall during the cell-fusion process. *Acta bot. neerl.* **14:** 165–96.

SCOTT, F. M., SJAHOLM, V. and BOWLER, E. (1960) Light and electron microscope studies of the primary xylem of *Ricinus communis. Am. J. Bot.* **47:** 162–73.

SINNOTT, E. W. and BLOCH, R. (1945) The cytoplasmic basis of intercellular patterns in vascular differentiation. *Am. J. Bot.* **32:** 151–6.

SMITH, G. F. and KERSTEN, H. (1942) The relation between xylem thickenings in primary roots of *Vicia faba* seedlings and elongation as shown by soft X-ray irradiation. *Bull. Torrey Bot. Club* **69:** 221–34.

TEWARI, R. B. (1975) Structure of vessels and tracheids of *Regnellidium diphyllum* Lindman (Marsileaceae). *Ann. Bot.* **39:** 229–231.

TIPPO, O. (1946) The role of wood anatomy in phylogeny. *Am. Midl. Nat.* **36:** 362–72.

WODZICKI, T. J. and BROWN, C. L. (1973) Organization and breakdown of the protoplast during maturation of pine tracheids. *Am. J. Bot.* **60:** 631–40.

WOODING, F. B. P. and NORTHCOTE, D. H. (1964) The development of the secondary wall of the xylem in *Acer pseudoplatanus. J. Cell Biol.* **23:** 327–37.

YOUNG, D. A. (1981) Are the angiosperms primitively vesselless? *Systematic Bot.* **6:** 313–330.

ZIMMERMANN, M. H. (1978) Vessel ends and the disruption of water flow in plants. *Phytopathology* **68:** 253–5.

ZIMMERMANN, M. H. (1982) Functional xylem anatomy of angiosperm trees. In: *New Perspectives in Wood Anatomy*, (ed. P. Baas). Martinus Nijhoff/Dr. W. Junk Publishers, The Hague.

ZIMMERMANN, M. H. (1983) *Xylem Structure and the Ascent of Sap.* Springer-Verlag, Berlin.

ZIMMERMANN, M. H. and JEJE, A. A. (1981) Vessel-length distribution in stems of some American woody plants. *Can. J. Bot.* **59:** 1882–1892.

ZIMMERMANN, M. H., McCUE, K. F. and SPERRY, J. S. (1982) Anatomy of the palm *Rhapis excelsa*. VIII. Vessel network and vessel-length distribution in the stem. *J. Arnold Arbor.* **63:** 83–95.

CHAPTER 8

PHLOEM

The phloem, together with the xylem, constitute the conducting system of vascular plants. The xylem functions principally in the conduction of water, and the phloem of products of photosynthesis. Similarly to the xylem the phloem also is a compound tissue. The important cells of the phloem are the *sieve elements* which serve for the conduction of the photosynthetic products. Additional to these elements phloem contains typical parenchyma cells in which reserve substances are stored as well as specialized parenchyma cells, i.e. the *companion cells* and *albuminous cells*, which are connected with functioning of the sieve elements. Fibres, sclereids, and sometimes laticifers and resin ducts may also be found in phloem tissue.

The primary phloem, similarly to the primary xylem, develops from the procambium. The primary phloem is divided into the *protophloem*, which develops from the procambium during an early ontogenetic stage and matures before the organ in which it is located has ceased its longitudinal extension growth, and the *metaphloem* which also develops from the procambium, but at a later stage of development.

The sieve elements were first discovered by Hartig in 1837, and the term *phloem* was coined from the Greek word for bark by Nägeli in 1858.

The phloem in the stem is usually external to the xylem, but in many ferns and in different species of numerous dicotyledonous families, e.g. Asclepiadaceae, Cucurbitaceae, Myrtaceae, Apocynaceae, Convolvulaceae, Compositae, and Solanaceae, phloem is also present on the inside of the xylem. Phloem on the inside of the xylem is called *internal* or *intraxylary phloem* (Fig. 99, no. 3) and it develops a little later than the external phloem. In certain families, such as the Chenopodiaceae, Amaranthaceae, Nyctaginaceae, Salvadoraceae,

and others, phloem is also present within the secondary xylem. This type of phloem is called *interxylary phloem* or *included phloem* (Fig. 223, nos. 1, 3). The extensive literature on phloem was summarized by Esau (1969).

SIEVE ELEMENTS

The most characteristic features of sieve elements are the *sieve areas* in the walls and the disappearance of the nucleus from the protoplast.

The sieve areas are interpreted as being modified primary pit fields and they appear as depressions in the wall in which groups of pores are located. *Connecting strands*, the contents of pores, connect the protoplasts of the neighbouring sieve elements (Fig. 67 no. 1). Sieve areas can be distinguished from primary pit fields by the following two features: (a) in the sieve areas the connecting strands are much thicker than the plasmodesmata that occur in the primary pit fields; (b) in the sieve area each pore usually contains a small cylinder of *callose* which surrounds the connecting strand (Fig. 62, nos. 2–4; Fig. 66; Fig. 67, no. 1). The diameter of the pores of the more specialized sieve areas, which are present in the sieve plates (see further), varies in the different species from a fraction of a micron to 14 μm. In *Spiraea vanhouttei* the diameter of the pores is less than 1 μm, while in *Pyrus malus* and *Pyrus communis* the maximum diameter is only a little larger than 1 μm, in *Cucurbita* spp. it is 10.3 μm, and in *Ailanthus altissima,* 14.3 μm (Esau and Cheadle, 1959). The presence of callose can easily be demonstrated by staining with aniline blue or resorcin blue (Esau, 1948). In ultraviolet light even minute amounts of callose stained with aniline blue give a lemon-yellow fluorescence. Cal-

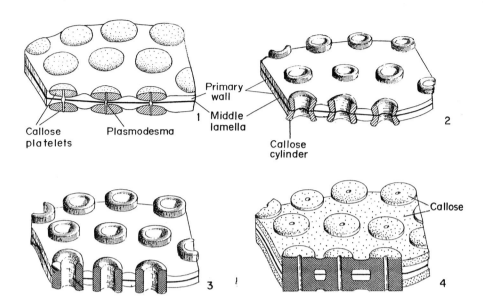

FIG. 62. Diagrams showing development and maturation of a sieve plate. 1. Early stage, showing appearance of callose platelets on the two opposite ends of the plasmodesmata. 2. The callose replaces the cellulosic wall in the developing pore. The middle lamella has not yet disappeared. 3. The mature pores are lined by relatively thick callose cylinders. 4. Sieve plate with callose over their whole surface.

lose is an unbranched β-1,3-linked glucan, it is not birefringent and shows no structure under the electron microscope. Callose also occurs on the walls of fungal cells, in the germination tube of pollen grains, in cystoliths, and has even been shown to be present in the primary pit fields of epidermal cells (Currier and Strugger, 1956).

Callose is also found on those portions of the wall between the callose cylinders surrounding the connecting strands. In old elements large amounts of callose accumulate to form continuous thick layers which were termed *callus* by Hanstein in 1864. This term callus should not be confused with that referring to wound tissue.

There is evidence that in response to injury, callose is formed and deposited very rapidly. In view of this phenomenon some workers have questioned the presence of callose in active sieve elements in intact plants (Eschrich, 1963; Evert, 1984; Evert and Derr, 1964). In any case, callose is generally present in active sieve elements of material fixed according to the usual methods and serves as a diagnostic feature of sieve elements.

In sieve areas that are not highly specialized the connecting strands are extremely thin and are nearly indistinguishable from plasmodesmata. In highly specialized sieve areas the connecting strands are very thick and they stain intensely. When the sieve element is young, the sieve area is thinner than the other portions of the cell wall and it appears as a depression on the inner surface of the wall. As the element matures additional callose is laid down not only on the cylinders, which line the pores, but also on the surface of the sieve area between the pores (Fig. 62, no. 4), so that, finally, in old elements, the sieve areas do not appear as depressions but as raised portions above the surface. When the element ceases to function the connecting strands become very thin and may even disappear. When sieve areas develop close to one another the callose masses from each sieve area may fuse to form a single mass. Large masses of callose which develop with the cessation of function of the elements are called *definitive callose* (Fig. 63, no. 1). Usually, with the complete disintegration of the protoplast

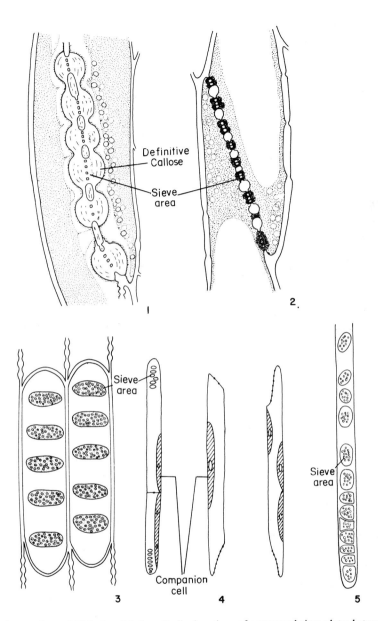

FIG. 63. 1–4, Sieve-tube members of *Vitis*. 1 and 2. Longitudinal sections of compound sieve plates between two elements. 1, Elements in dormant state in which the plate is covered by a thick layer of callose. 2, Elements reactivated after the removal of the callose. The slime which fills the sieve areas is indicated by heavy stippling. 3, Surface view of two compound sieve plates. 4, Sieve-tube elements with companion cells. 5, Portion of a sieve cell of *Pinus*. (Nos. 1–4 adapted from Esau, 1948.)

of the element the callose peels away from the sieve areas. In most dicotyledons the sieve elements function during a single growing season, but in certain plants, such as *Suaeda, Tilia,* and *Vitis,* the sieve elements function for two or more years. In woody stems of *Atriplex* and other plants with included phloem the sieve elements remain active in the whole sapwood (Fahn and Shchori, 1967). In palm stems the sieve tubes in the metaphloem function throughout the life of the plant and no conspicuous structural features which might account for the long life of the sieve elements can be discerned (Parthasarathy and Tomlinson, 1967). In *Tilia* no distinct changes can be seen in the sieve elements with the start of the resting season, while in *Vitis* large amounts of callose accumulate in the autumn, disintegrate in the spring before the commencement of cambial activity, and so the sieve elements start to function for a second year (Esau, 1948; Bernstein and Fahn, 1960).

The density and arrangement of the sieve areas on the sieve elements exhibit the same degree of variation as does the pitting on the walls of the tracheary elements. On the basis of the thickness of the connecting strands and the degree of development of the callose cylinders, sieve areas with different stages of specialization can be distinguished. In certain plants, such as the Coniferales, all the sieve areas of an element are equal, while in other plants, for instance, most of the angiosperms, some of the sieve areas are more specialized, i.e. they have more well-developed connecting strands and callose cylinders than those found in other areas. These more specialized sieve areas are usually situated on the end walls of the elements which are horizontal or oblique to the longitudinal axis of the element. Those parts of the cell wall that bear such specialized sieve areas are called *sieve plates* (Fig. 63, no. 3; Fig. 64).

At the early stage of sieve plate development callose, in the form of platelets, is deposited above the cellulosic wall, at the two opposite ends of the plasmodesmata which will develop into pores (Fig. 62, no. 1). At a later stage, callose replaces the cellulosic wall in the developing pore. The middle lamella may not disappear completely until just before opening of the pore, when a complete callose cylinder is formed (Fig. 62, nos. 2–4). Esau and Thorsch (1985) suggested that since callose is more easily digested enzymatically than cellulose, the deposition of callose may facilitate the opening of the pores. The plasmodesmata in the developing pores increase in diameter, thus forming wide cytoplasmic connections (*connecting strands*) between the sieve tube members. The desmotubules of the original plasmodesmata disappear, and only the plasmalemma, now surrounding the connecting strands, remains.

Those sieve elements that have unspecialized sieve areas that are similar throughout the element are called *sieve cells* (Fig. 63, no. 5). Sieve cells, therefore, do not contain sieve plates. These cells are usually elongated with tapering ends or their end walls are very oblique. In the positions where sieve cells overlap one another the sieve areas are more numerous. The sieve elements of the gymnosperms and vascular cryptogams are sieve cells. They are narrower and much longer than sieve tube members. In some species of the lower vascular plants the sieve elements may reach a length of 30–40 mm (Lamourex, 1961, cited by Evert, 1984). In the carboniferous pteridosperm *Medullosa*, Smoot (1984) found sieve cells which reached the length of up to 4.2 mm.

Elements in which sieve plates can be distinguished are called *sieve-tube members* (Fig. 63, no. 4). Sieve plates are usually found on the end walls which may be very oblique or horizontal or in intermediate planes. In certain elements, e.g. those of *Vitis* and *Pyrus malus,* the sieve plate contains several sieve areas; while in other elements, e.g. those of *Cucurbita,* only one sieve area may be present. The former type of sieve plate is termed a *compound sieve plate* (Fig. 63, nos. 1–4) and the latter, a *simple sieve plate* (Fig. 64, no. 2). The sieve-tube members are connected one to the other by the walls that contain the sieve plates and so form *sieve tubes*. Sieve plates are found only very occasionally on the longitudinal walls of the sieve-tube members. On these walls unspecialized sieve areas develop.

The cell wall

The walls of sieve elements are usually only primary and consist mainly of cellulose and pectin. The thickness of the wall of the sieve elements varies in different species; in some species the cell wall is 1 μm thick while in other species the wall nearly fills the cell lumen. Esau and Cheadle (1958) also found differences in the structure of the wall. In certain species they found that the wall is homogeneous, while in others the wall is composed of two layers—a thin layer close to the middle lamella and a thicker layer next to the

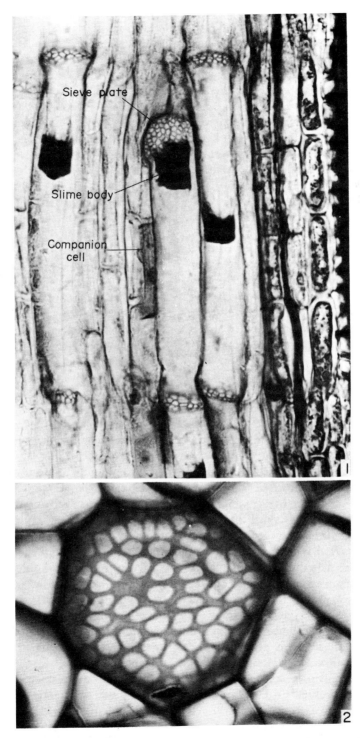

FIG. 64. 1, Micrograph of a longitudinal section of phloem of *Robinia pseudacacia*. ×500. 2, Face view of a sieve plate of *Curcubita*. ×960. (No. 1 courtesy of R. F. Evert.)

cytoplasm. The inner layer as seen in cross-sections of fresh material has a sheen similar to mother-of-pearl, and therefore has been termed the *nacreous* layer. The thickness of the wall of the sieve elements usually decreases with the ageing of the element. Thick nacreous walls can be seen, for example, in *Magnolia, Laurus, Rhamnus,* and *Persea* but they are not present in *Casuarina, Crataegus, Fraxinus, Morus, Populus, Salix,* and *Passiflora,* among others. In considerably thickened walls the inner layers are sometimes interpreted as secondary walls (Esau, 1969). The nacreous wall layer in *Cucurbita* sieve tube members was found by Deshpande (1976) to be polylamellate, the microfibrils being very densely packed and the lammellae not separating readily. The orientation of the microfibrils is very close to transverse. In *Pinus strobus* Chafe and Doohan (1972) reported that the thickened sieve cell wall comprises a crossed-helical polylamellate structure in which the microfibrils are oriented in relation to the cell axis in an angle predominantly greater than 45°.

In differentiating protophloem elements of *Triticum aestivum,* Eleftheriou (1987) observed that the number of microtubules increases toward the initiation of wall thickening. The microtubuli were found to be aligned only on the thicker region of the wall and not on the thinner future places of the plasmodesmata, which remained as narrow depressions.

The protoplast

A common property of the protoplast of the mature and active sieve element is the lack of a nucleus. The structure of immature sieve elements resembles that of the procambial or cambial cells from which they develop. In this stage the protoplast contains a large nucleus, vacuoles, and the other usual cell organelles. With specialization of the sieve elements, degradation of the nucleus takes place. In sieve tube members the nucleus may disappear completely (Hoefert, 1980; Danilova and Telepova, 1978, 1981; Thorsch and Esau, 1981) or remnants of the nuclear envelope and/or chromatine may persist (Eleftheriou, 1986; Eleftheriou and Tsekos, 1982; Walsh and Evert, 1975). The nuclear degeneration may occur *chromato,* by gradual loss

of stainable content, or *picno,* by the nucleus becoming very dark (dense mass of chromatin) before the rupture of the envelope. Chromatolitic type of degeneration is characteristic of most angiosperms and the pycnotic type of the gymnosperms (Behnke, 1986; Evert, 1977; Hoefert, 1980). The nuclear degeneration in the sieve elements of the lower vascular plants may be either picnotic or chromatolitic, or by an intermediate process (Evert, 1984).

The nuclei of some sieve tube members (e.g. in *Echium*) contain proteinaceous inclusions (Esau and Thorsch, 1982). During nuclear degeneration, these cyrstalloids are released into the cell lumen where they remain intact or fragmented.

A more or less viscous substance, which stains readily with cytoplasmic stains, is present in the sieve elements. This substance, which was called *slime,* is now recognized as proteinaceous and has been re-named P-protein (Figs. 66, 67) (Esau and Cronshaw, 1967).

With the aid of the electron microscope the P-protein has been observed to vary from species to species and to appear in several morphological forms—tubular, filamentous or fibrillar, granular, and crystalline. Changes from one type of P-protein to another may take place within a differentiating sieve-tube member (Cronshaw and Esau, 1967; Parthasarathy and Mühlethaler, 1969; Evert *et al.* 1973; Cronshaw, 1974; Evert, 1977; Esau, 1978a, b; Hoefert, 1979, 1980). With few exceptions, P-protein was found in sieve-tube members of all dicotyledonous species examined. P-protein is not common in sieve-tube members of monocotyledons and in sieve cells of gymnosperms and vascular cryptophytes. It has been reported lacking in many palms, grasses, pines and pteridophytes (Evert 1976, 1977; Walsh and Evert, 1975; Neuberger and Evert, 1976). It was, however, found in *Ephedra* (Alosi and Alfieri, 1972).

The P-protein is produced in the cytoplasm in the form of discrete bodies. It spreads out in the cytoplasm concomitantly with the disintegration of the nucleus. When special care was taken not to disturb the phloem while preparing sections, it was found that after the tonoplast disappears the dispersed P-protein occupies a parietal position (Figs. 66, 67) (Evert and Deshpande, 1969; Evert *et*

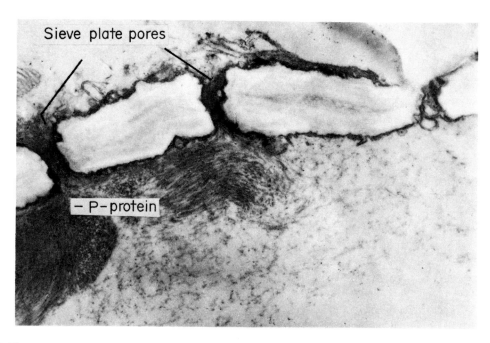

Sieve plate pores

– P-protein

FIG. 65. Electron micrograph of a sectioned sieve plate of *Anabasis articulata*, showing P-protein and some membranous structure next to and inside the pores. × 28,000.

al., 1969, 1973). When such precautions are not taken the P-protein is found throughout the cell lumen and plugs the sieve pores. Accumulated P-protein near the sieve plates has been termed *slime plug* (Fig. 64, no. 1).

During the final stages of differentiation of the sieve-tube member, all cellular components become disorganized and/or disappear, except the plasmalemma, mitochondria, plastids, P-protein, and at least some endoplasmic reticulum. The mitochondria undergo the least amount of structural change during the sieve-tube member differentiation. In the mature active sieve elements the thin cytoplasmic layer lining the inner surface of the cell walls retains its outer membrane, the plasmalemma, but lacks the tonoplast, the membrane bordering the large central vacuole (Esau and Cheadle, 1962).

In young sieve elements still with intact nuclei the endoplasmic reticulum cisternae become aggregated into stacks which are retained after the disintegration of the nucleus. Before the aggregation the ER is associated with ribosomes; later, however, it becomes smooth (Esau and Gill, 1971).

Special attention was paid to the nature of the connecting strands which pass through the pores of the sieve areas. It is commonly accepted that the pores are lined by plasmalemma. According to some authors the pores are filled with P-protein filaments (**Fig. 65**) (Buvat, 1963; Cronshaw and Esau, 1967; Johnson, 1968; Spanner, 1978). However, Evert *et al.* (1973) showed that in *Cucurbita maxima* the contents of the pores are essentially similar in composition and distribution to the contents of the lumina of the sieve-tube elements. The pores are lined with the plasmalemma, portions of the parietal network of ER, and P-protein (**Figs. 66, 67**). They lack only plastids and mitochondria, which occur in the lumina of the elements. Parthasarathy and Klotz (1976), who studied sieve elements of palms, also found that the sieve plate pores may be free of P-protein occlusions in intact sieve elements. Deshpande (1984) suggested that the components of the meshwork of ER—P-protein in *Cucurbita* may change with the state of activity of the sieve elements, and that unobstructed or arbitrarily obstructed sieve plate pores represent a state of rest or low transport.

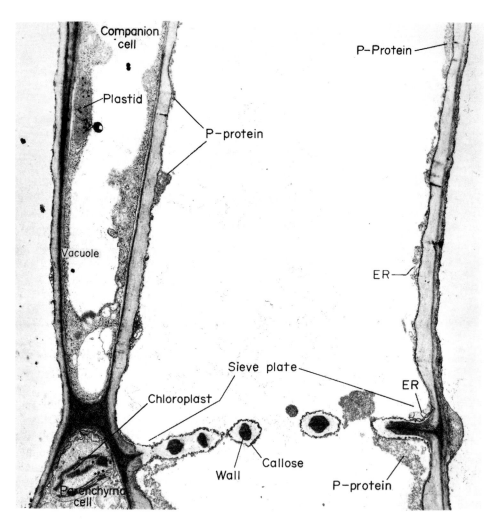

FIG. 66. Electron micrograph of a longitudinal section of *Cucurbita maxima* sieve-tube members, showing sieve plate. Companion cells on left, above × 10,000. (Courtesy of R. F. Evert.)

The plastids occur in two major types: S-type plastids, which store starch and P-type plastids, which contain protein in the form of fibrils or crystals but may also contain starch grains. Only P-type plastids have been recorded in monocotyledons. Ultrastructural differences in sieve-tube elements are useful taxonomic characters (Behnke, 1972, 1975, 1976, 1978, 1984, 1986a). The sieve-tube starch stains brownish red with iodine and not blue-black as does ordinary starch.

As mentioned above, it is difficult to get an accurate picture of the structure of the protoplast in a mature sieve-tube member from the study of microscope sections because of the changes in the position of the protoplasmic constituents that take place during the preparation for sectioning. The cell contents are pushed in the direction of the cuts made before the material is fixed and a portion may even be extruded on the surface by the pressure that exists in the mature sieve element. P-protein accumulates near those sieve plates which are close to the cuts, and it appears

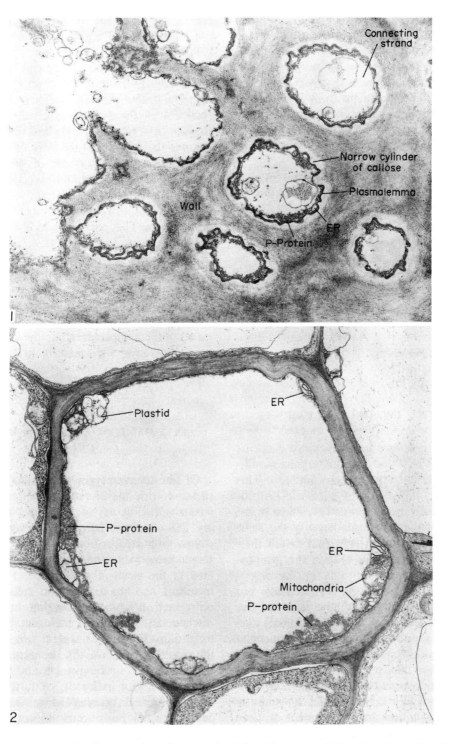

FIG. 67. Electron micrographs of cross-sections of a sieve tube of *Cucurbita maxima*. 1, Face view of a portion of a sieve plate. × 26,000. 2, Cross-section of a mature sieve-tube member. × 7800. (Courtesy of R. F. Evert.)

as if this substance passes through the sieve areas.

The plugging of the sieve plate prevents continuous exudation from the sieve tubes. Callose is formed rapidly as a result of the injury exerted to the phloem and blockage of pores takes place.

In sieve cells of vascular cryptogams, with apparent exception of the Lycophytina, dense, membrane-bound, proteinaceous bodies occur, which are termed *refractive spherules* (Evert and Eichhorn, 1974; Fisher and Evert, 1979).

Cells which resemble sieve elements occur in bryophytes. They were termed *leptoids*. The nucleus persists in mature leptoids but it appears that it is degenerated at least in some mosses (Hébant, 1974, 1977).

Different and contrary concepts exist as to the mechanism of translocation of the photosynthetic products in the sieve elements (Esau *et al.*, 1957; Esau, 1969; MacRobbie, 1971; Evert *et al.*, 1973; Zimmermann and Milburn, 1975; Spanner, 1979, Evert, 1982). The most widely accepted concept today is the osmotically generated pressure flow hypothesis originally proposed by Münch (1930). The supporters of this theory assume that the sieve plate pores are "open". The researchers who have put forward the hypothesis of electro-osmotic mechanism of transport presume that the pores are occluded with P-protein or other material.

Several workers (Thaine, 1969; Fenson, 1972) have suggested that P-protein is a contractile protein and that it plays an active role in the transport of photosynthetic products through the sieve tube. As an argument against this hypothesis the fact that in the sieve elements of many vascular plants P-protein is lacking can be used (Evert *et al.*, 1973).

Loading sugars into the complex of companion cells and sieve element from the apoplast is an active, energy requiring process (Evert, 1982; Giaquinta, 1983; Russin and Evert, 1985).

On the basis of the study of a number of species, it has been suggested that the unloading of translocated sucrose from the sieve elements to the cortical root cells is predominantly via the symplast (Giaquinta *et al.*, 1983; Warmbrodt, 1985a, b).

It is known that the passage of photosynthates in the sieve elements is much faster than that in ordinary parenchyma cells. The velocity of long-distance transport in the phloem varies from 10 to 100 cm per hour or even more (Zimmermann,

1961, 1963, 1964). Velocity of movement can be measured by the amount of exudate collected from an aphid proboscis inserted into a sieve element after removing the aphid body from the inserted stylet of the proboscis. Zimmermann (1964) has estimated that a sieve element is emptied 3–10 times per second. The velocity of the transport of assimilates in angiosperms, where the sieve-plate pores are not occluded, is typically 50–100 cm per hour. In gymnosperms, where the sieve areas are largely occluded with tubular ER elements, the velocity was found to be lower (Evert, 1984).

PHYLOGENY OF SIEVE ELEMENTS

In the most primitive form the sieve elements are parenchyma cells that have undergone modifications in connection with their function. This was followed by the loss of the nucleus. This protoplasmic specialization apparently resulted in the development of the interdependence of the sieve elements and parenchyma cells that retain their nucleus, i.e. the albuminous cells in the gymnosperms and the companion cells in the angiosperms. In the angiosperms the sieve elements and companion cells develop from the same mother cell. The specialization of the sieve elements also involved the development of thick connecting strands. In the lower vascular plants with few exceptions (Evert, 1984) and in the gymnosperms these strands are thin and resemble plasmodesmata, while in the angiosperms they are thick and conspicuous. The evolutionary trends in the shape of the sieve elements and the arrangement of sieve areas have been more thoroughly studied in the monocotyledons (Cheadle and Whitford, 1941; Cheadle, 1948; Cheadle and Uhl, 1948; Parthasarathy, 1968). The specialization, in this group of plants, has taken the following courses: (1) gradual localization of highly specialized sieve areas to the end walls of the elements; (2) gradual changes in the position of the end wall from very oblique to horizontal; (3) gradual change from compound sieve plates to simple ones; (4) gradual reduction of the sieve areas on the side walls of the elements. In monocotyledons the above investigators also found that the sieve tubes first developed in the aerial portions of the plants from where the development and specialization spread to the roots. This direction of development is opposite to that of the development of the vessels in the xylem. This phenomenon is understandable from a functional point of view. Features of the dicotyledonous sieve-tube members suggest that the phylogenetic development in this group of plants is similar to that of the monocotyledons, but as yet no conclusive research has been done. According to Zahur (1959), the sieve-tube elements in the angiosperms, like the vessel elements, have undergone a decrease in length during the course of evolution.

COMPANION AND ALBUMINOUS CELLS

Of the different types of parenchyma cells associated with the sieve-tube members the *companion cells* (Fig. 63, no. 4; Fig. 64, no 1; Fig. 66) are the most specialized. However, the parenchyma cells are not always unequivocally distinguishable from one another even when examined in the electron microscope. The sieve-tube members and the companion cells, are related ontogenetically as they develop from the same meristematic cell. Such a meristematic cell divides longitudinally once or several times and one of the resulting cells, usually the largest, specializes to form the sieve-tube member and the others develop directly or indirectly, by further transverse or longitudinal divisions, into the companion cells. One or more companion cells may accompany a sieve-tube member. Companion cells vary in size—they may be as long as the sieve-tube member to which they are related or they may be shorter. Companion cells may develop on various sides of the sieve tube or they may form longitudinal rows on one side only. Companion cells are strongly attached to the sieve-tube members from which they usually cannot be separated even by maceration. The walls between sieve-tube members and the companion cells are thin or possess many thin areas which, apparently, are sieve areas on the side of the sieve-tube member and primary pit fields on the side of the

companion cell. The plasmodesmata on the companion-cell side are usually branched. The length of life of the companion cells is usually the same as that of the sieve-tube member to which they are attached. This feature proves not only the ontogenetic connection but also the functional connection that exists between these cells. The companion cells and also the phloem parenchyma play an important part in the maintenance of a pressure gradient in the sieve tubes (Esau, 1961).

As the companion cell differentiates, its protoplast increases in density. The mature companion cell retains its nucleus, is very rich in ribosomes, contains numerous mitochondria, rough ER, and plastids. Some companion cells contain P-protein (Evert et al., 1966; Esau, 1947, 1948, 1969; Deshpande, 1974).

In most parts of the plant the sieve elements are usually considerably wider than the companion cells. However, in the minor leaf-veins the sieve elements are often conspicuously smaller than their associated parenchyma cells. These parenchyma cells (companion cells or cells not ontogenetically related to the sieve elements) were called by Fischer (1884) "intermediary cells", because he believed that they mediate in the transfer of photosynthates between the mesophyll cells and sieve elements.

In the minor veins of many herbaceous dicotyledons the intermediary cells were found to develop wall ingrowths. Such cells were termed *transfer cells* (Fig. 135, no. 4) (Gunning et al., 1968; Gunning and Pate, 1969; Pate and Gunning, 1969, 1972).

In the small and intermediate vascular bundles of *Zea mays*, Evert (1980) did not find phloem transfer cells. He distinguished, however, two types of sieve tubes: relatively thin-walled ones that have numerous cytoplasmic connections with companion cells, and thick-walled ones that lack companion cells but possess numerous connections with vascular parenchyma cells. The vascular parenchyma cells contain tubular extensions of the plasmalemma which enlarge to a great extent the plasmalemma surface.

In gymnosperms companion cells, as described above, do not occur, but cells which stain intensely with cytoplasmic stains are present. These cells are apparently connected physiologically and morphologically to the sieve cells and have been termed *albuminous cells.* Ontogenetically these cells develop from the phloem parenchyma or from cells of the phloem rays. However, Alosi and Alfieri (1972) found that many of the albuminuous cells in the secondary phloem of *Ephedra* are ontogenetically related to the sieve cells, as are companion cells.

The albuminous cells can be distinguished from the other parenchyma cells of the phloem by the presence of connections with the sieve cells and by the usual absence of starch (Srivastava, 1963). The albuminous cells are referred to by many authors as *Strasburger cells* (Parameswaran and Liese, 1970; Sauter *et al.,* 1976).

It was found that high respiratory and acid phosphatase activities occur in the albuminous cells associated with fully differentiated sieve cells. The increased activity is restricted to those periods during which loading and unloading of the sieve cells is taking place (Sauter and Braun, 1972; Sauter, 1974, 1980).

PROTOPHLOEM AND METAPHLOEM

The primary phloem, as described previously, consists of protophloem and metaphloem. The protophloem, together with protoxylem, constitute the vascular tissue of the young elongating parts of the plant.

The description of the phloem elements given above refers only to metaphloem and secondary phloem. Sieve areas cannot be distinguished in the protophloem elements of gymnosperms. In angiosperms there are sieve-tube members in the protophloem, but in many plants there are no companion cells. These sieve-tube members are long and narrow, and sieve areas can be distinguished only with difficulty. The walls are somewhat thick and the cell contents stain only slightly. The sieve tubes of the protophloem are apparently active for a short period only. As the sieve-tube members have no nucleus they cannot divide and grow with the elongating organ, and so they become obliterated by the surrounding cells. The remnants of these obliterated cells may completely disappear in time. In many dicoty-

ledonous stems the parenchyma of the protophloem remains after the obliteration of the sieve-tube members, and then these cells become fibres. In leaves they form elongated collenchyma cells.

In contrast to those sieve elements of the protophloem which only function for a short period and which are early obliterated, the sieve elements of the metaphloem of Pteridophyta and long-living monocotyledons, such as the Palmae, function for many years.

Recently a detailed ultrastructural study on the differentiation of protophloem sieve elements in adventitious roots of *Salix* has been carried out by Fjell (1987). Basically the manner of development of the protophloem sieve elements is similar to that of the sieve elements of meta- and secondary phloem.

When fibres occur in the primary phloem of dicotyledonous plants they are always restricted to the protophloem even in those cases where fibres develop later in the secondary phloem of the same plant. In herbaceous species, however, parenchyma cells of the old metaphloem may become sclerified.

On the basis of experimental studies it has been shown that the differentiation of primary phloem fibres in stems is induced by stimuli originating in the leaves (Sachs, 1972; Aloni, 1976, 1978).

REFERENCES

ALONI, R. (1976) Polarity of induction and pattern of primary phloem fiber differentiation in *Coleus. Am. J. Bot.* **63:** 877–89.

ALONI R. (1978) Source of induction and sites of primary phloem fibre differentiation in *Coleus blumei. Ann. Bot.* **42:** 1261–9.

ALOSI, M. C. and ALFIERI, F. J. (1972) Ontogeny and structure of the secondary phloem in *Ephedra. Am. J. Bot.* **59:** 818–27.

BEHNKE, H.-D. (1972) Sieve-tube plastids in relation to angiosperm systematics—an attempt towards a classification by ultrastructural analysis. *Bot. Rev.* **38:** 155–97.

BEHNKE, H.-D. (1975) P-type sieve-element plastids: a correlative ultrastructural and ultrahistochemical study on the diversity and uniformity of a new reliable character in seed plant systematics. *Protoplasma* **83:** 91–101.

BEHNKE, H.-D. (1976) Ultrastructure of sieve-element plastids in Caryophyllales (Centrospermae), evidence for the delimitation and classification of the order. *Plant Syst. Evol.* **126:** 31–54.

BEHNKE, H.-D. (1978) Elektronenoptische Untersuchungen am Phloem sukkulenter Centrospermen (incl. Didiereaceen). *Bot. Jahrb. Syst.* **99**: 341–52.

BEHNKE, H.-D. (1984) Ultrastructure of sieve-element plastids of Myrtales and allied groups. *Ann. Missouri Bot. Gard.* **71**: 824–831.

BEHNKE, H.-D. (1986a) Contribution to the knowledge of P-type sieve-element plastids in dicotyledons. IV. Acanthaceae. *Bot. Jahrb. Syst.* **106**: 499–510.

BEHNKE, H.-D. (1986b) Sieve element characters and the systematic position of *Austrobaileya*, Austrobaileyaceae—with comments to the distinction and definition of sieve cells and sieve-tube members. *Pl. Syst. Evol.* **152**: 101–121.

BERNSTEIN, Z. and FAHN, A. (1960) The effect of annual and bi-annual pruning on the seasonal changes in xylem formation in the grapevine. *Ann. Bot.* **24**: 159–71.

BUVAT, R. (1963) Les infrastructures et la différentiation de cellules criblées de *Cucurbita pepo* L. *Port. Acta biol.,* sér. A **7**: 249–99.

CHAFE, S. C. and DOOHAN, M. E. (1972) Observations on the ultrastructure of the thickened sieve cell wall in *Pinus strobus* L. *Protoplasma* **75**: 67–78.

CHEADLE, V. I. (1948) Observations on the phloem in the Monocotyledoneae. II, Additional data on the occurrence and phylogenetic specialization in structure of the sieve tubes in the metaphloem. *Am. J. Bot.* **35**: 129–31.

CHEADLE, V. I. and UHL, N. W. (1948) The relation of metaphloem to the types of vascular bundles in the Monocotyledoneae. *Am. J. Bot.* **35**: 578–83.

CHEADLE, V. I. and WHITFORD, N. B. (1941) Observations on the phloem in the Monocotyledoneae: I, The occurrence and phylogenetic specialization in structure of the sieve tubes in the metaphloem. *Am. J. Bot.* **28**: 623–7.

CRONSHAW, J. (1974) P-proteins. In: *Phloem Transport* (ed. S. Aronoff). NATO Advanced Study Institutes, P. a: *Life Sciences,* Vol. 4, Plenum Press, New York, pp. 79–115.

CRONSHAW, J. and ESAU, K. (1967) Tubular and fibrillar components of mature and differentiating sieve elements. *J. Cell Biol.* **34**: 801–16.

CURRIER, H. B. and STRUGGER, S. (1956) Aniline blue and fluorescence microscopy of callose in bulb scales of *Allium cepa* L. *Protoplasma* **45**: 552–9.

DANILOVA, M. F. and TELEPOVA, M. N. (1978) Differentiation of protophloem sieve elements in seedling roots of *Hordeum vulgare. Phytomorphology* **28**: 418–431.

DANILOVA, M. F. and TELEPOVA, M. N. (1981). Differentiation of proto- and meta-phloem sieve elements in roots of *Hordeum vulgare* (Poaceae). *Bot. Zh.* Leningrad) **66**: 169–178.

DESHPANDE, B. P. (1974) On the occurrence of spiny vesicles in the phloem of *Salix. Ann. Bot.* **38**: 865–8.

DESHPANDE, B. P. (1976) Observations on the fine structure of plant cell walls. III. The sieve-tube wall in *Cucurbita. Ann. Bot.* **40**: 443–6.

DESHPANDE, B. P. (1984) Distribution of P-protein in mature sieve elements of *Cucurbita maxima* seedling subjected to prolonged darkness. *Ann. Bot.* **53**: 237–247.

ELEFTHERIOU, E. P. (1987) Microtubules and cell wall development in differentiating protophloem sieve elements of *Triticum aestivum* L. *J. Cell Sci.* **87**: 595–607.

ELEFTHERIOU, E. P. and TSEKOS, I. (1982) Development of protophloem in roots of *Aegilops comosa* var. *thessalica.* II. Sieve-element differentiation. *Protoplasma* **113**: 221–233.

ESAU, K. (1947) A study of some sieve-tube inclusions. *Am. J. Bot.* **34**: 224–33.

ESAU, K. (1948) Phloem structure in the grapevine and its seasonal changes. *Hilgardia* **18**: 217–96.

ESAU, K. (1950) Development and structure of the phloem tissue: II, *Bot. Rev.* **16**: 67–114.

ESAU, K. (1961) *Plants, Viruses and Insects.* Harvard Univ. Press, Cambridge, Mass.

ESAU, K. (1969) *The Phloem.* In: K. Linsbauer, *Handbuch der Pflanzenanatomie,* Bd. V, T. 2. Gebr. Borntraeger, Berlin.

ESAU, K. (1978a) Developmental features of the primary phloem in *Phaseolus vulgaris* L. *Ann. Bot.* **42**: 1–13.

ESAU, K. (1978b) The protein inclusions in sieve elements of cotton (*Gossypium hirsutum* L.). *J. Ultrastruct. Res.* **63**: 224–35.

ESAU, K. and CHEADLE, V. I. (1958) Wall thickening in sieve elements. *Proc. Natn. Acad. Sci.* **44**: 546–53.

ESAU, K. and CHEADLE, V. I. (1959) Size of pores and their contents in sieve elements of dicotyledons. *Proc. Natn. Acad. Sci.* **45**: 156–62.

ESAU, K. and CHEADLE, V. I. (1962) An evaluation of studies on ultrastructure of tonoplast in sieve elements. *Proc. Natn. Acad. Sci.* **48**: 1–8.

ESAU, K. and CRONSHAW, J. (1967) Tubular components in cells of healthy and tobacco mosaic virus-infected *Nicotiana. Virology* **33**: 26–35.

ESAU, K. and GILL, R. H. (1971) Aggregation of endoplasmic reticulum and its relation to the nucleus in a differentiating sieve element. *J. Ultrastruct. Res.* **34**: 144–58.

ESAU, K. and THORSCH, J. (1982) Nuclear crystalloids in sieve elements of species of *Echium* (Borraginaceae). *J. Cell Sci.* **54**: 149–160.

ESAU, K. and THORSCH, J. (1985) Sieve plate pores and plasmodesmata, the communication channels of the symplast: ultrastructural aspects and developmental relations. *Am. J. Bot.* **72**: 1641–1653.

ESAU, K., CURRIER, H. B. and CHEADLE, V. I. (1957) Physiology of phloem. *A. Rev. Pl. Physiol.* **8**: 349–74.

ESAU, K., CHEADLE, V. I., and RISLEY, E. B. (1962) Development of sieve-plate pores. *Bot. Gaz.* **123**: 233–43.

ESCHRICH, W. (1963) Beziehungen zwischen dem Auftreten von Callose und der Feinstruktur des primären Phloems bei *Cucurbita ficifolia. Planta* **59**: 243–61.

EVERT, R. F. (1976) Some aspects of sieve-element structure and development in *Botrychium virginianum. Israel J. Bot.* **25**: 101–26.

EVERT, R. F. (1977) Phloem structure and histochemistry. *Ann. Rev. Pl. Physiol.* **28**: 199–222.

EVERT, R. F. (1980) Vascular anatomy of angiospermous leaves, with special consideration of maize leaf. *Ber. dt. bot. Ges.* **93**: 43–55.

EVERT, R. F. (1982) Sieve-tube structure in relation to function. *BioScience* **32**: 789–795.

EVERT, R. F. (1984) Comparative structure of phloem. In: *Contemporary Problems in Plant Anatomy* (eds. R. A. White and W. C. Dickison). Academic Press, Orlando, Fla. pp. 145–234.

EVERT, R. F. (1986) Phloem loading in Maize. In: *Regulation of Carbon and Nitrogen Reduction in Maize* (eds. J. C. Shannon, D. P. Knievel and C. D. Boyer). The American Society of Plant Physiologists, pp. 67–81.

EVERT, R. F. and DERR, W. F. (1964) Callose substance in sieve elements. *Am. J. Bot.* **51**: 552–9.

EVERT, R. F. and DESHPANDE, B. P. (1969) Electron microscope investigation of sieve-element ontogeny and structure in *Ulmus americana. Protoplasma* **68**: 403–32.

EVERT, R. F. and EICHHORN, S. E. (1974) Sieve-element ultra-structure in *Platycerium bifurcatum* and some other polypodiaceous ferns: the refractive spherules. *Planta* **119**: 319–34.

EVERT, R. F., MURMANIS, L., and SACHS, I. B. (1966) Another view of the ultrastructure of *Cucurbita* phloem. *Ann. Bot.* **30**: 563–85.

EVERT, R. F., TUCKER, C. M., DAVIS, J. D. and DESHPANDE, B. P. (1969) Light microscope investigation of sieve-element ontogeny and structure in *Ulmus americana. Am. J. Bot.* **56**: 999–1017.

EVERT, R. F., ESCHRICH, W., and EICHHORN, S. E. (1973) P-protein distribution in mature sieve elements of *Cucurbita maxima. Planta* **109**: 193–210.

FAHN, A. and SHCHORI, Y. (1967) The organization of the secondary conducting tissues in some species of the Chenopodiaceae. *Phytomorphology* **17**: 147–54.

FENSON, D. S. (1972) A theory of translocation in phloem of *Heracleum* by contractile protein microfibrillar material. *Can. J. Bot.* **50**: 479–97.

FISCHER, A. (1884) *Untersuchungen über das Siebröhren-System der Cucurbitaceen.* Gebr. Borntraeger, Berlin.

FISHER, D. G. and EVERT, R. F. (1979) Endoplasmic reticulum-dictyosome involvement in the origin of refractive spherules in sieve elements of *Davallia fijiensis* Hook. *Ann. Bot.* **43**: 255–8.

FJELL, I. (1987) Ultrastructural features of differentiating protophloem sieve elements in adventitious roots of *Salix viminalis. Nord. J. Bot.* **7**: 135–151.

GIAQUINTA, R. T. (1983) Phloem loading of sucrose. *Ann. Rev. Plant Physiol.* **34**: 347–387.

GIAQUINTA, R. T., SADLER, L. N. and FRANCESCHI, V. R. (1983) Pathway of phloem unloading of sucrose in corn roots. *Plant Physiol.* **72**: 362–367.

GUNNING, B. E. S. and PATE, J. S. (1969) "Transfer cells"—Plant cells with wall ingrowths, specialized in relation to short distance transport of solutes—their occurrence, structure and development. *Protoplasma* **68**: 107–33.

GUNNING, B. E. S., PATE, J. S., and BRIARTY, L. G. (1968) Specialized "transfer cells" in minor veins of leaves and their possible significance in phloem translocation. *J. Cell Biol.* **37**: C 7–12.

HANSTEIN, J. (1864) *Die Milchsaftgefässe und die verwandten Organe der Rinde.* Wiegandt und Hempel, Berlin.

HARTIG, T. (1837) Vergleichende Untersuchungen über die Organisation des Stammes der einheimischen Waldbäume. *Jahresber. Forsch. Forstwissensch. und Forstl. Naturkunde* **1**: 125–168.

HÉBANT, C. (1974) Polarized accumulation of endoplasmic reticulum and other ultrastructural features of leptoids in *Polytrichadelphus magellanicus* gametophytes. *Protoplasma* **81**: 373–82.

HÉBANT, C. (1977) *The Conducting Tissues of Bryophytes.* J. Cramer, Vaduz.

HOEFERT, L. L. (1979) Ultrastructure of developing sieve elements in *Thlaspi arvense* L. I. The immature state. *Am. J. Bot.* **66**: 925–32.

HOEFERT, L. L. (1980) Ultrastructure of developing sieve elements in *Thalspi arvense* L. II. Maturation. *Am. J. Bot.* **67**: 194–201.

JOHNSON, R. P. C. (1968) Microfilaments in pores between frozen-etched sieve elements. *Planta* **81**: 314–32.

LAMOUREX, C. H. (1961) Comparative studies on phloem of vascular cryptogams. Ph.D. Diss. University of California, Davis.

MACROBBIE, E. A. C. (1971) Phloem translocation: facts and mechanism: a comparative survey. *Biol. Rev.* **46**: 429–81.

MÜNCH, E. (1930) *Die Stoffbewegungen in der Pflanze.* G. Fischer, Jena.

NÄGELI, C. W. (1858) Das Wachstum des Stammes und der Wurzel bei den Gefässpflanzen und die Anordnung der Gefässtränge in Stengel. *Beitr. z. Wiss. Bot.* **1**: 1–156.

NEUBERGER, D. S. and EVERT, R. F. (1976) Structure and development of sieve cells in the primary phloem of *Pinus resinosa. Protoplasma* **87**: 27–37.

PARAMESWARAN, N. and LIESE, W. (1970) Zur Cytologie der Strasburger-Zellen in Coniferennadeln. *Die Naturwissenschaften* **57**: 45–46.

PARTHASARATHY, M. V. (1968) Observations on metaphloem in the vegetative parts of palms. *Am. J. Bot.* **55**: 1140–68.

PARTHASARATHY, M. V. and KLOTZ, L. H. (1976) Palm "Wood". II. Ultrastructural aspects of sieve elements, tracheary elements and fibres. *Wood Sci. Tech.* **10**: 247–71.

PARTHASARATHY, M. V. and MÜHLETHALER, K. (1969) Ultrastructure of protein tubules in differentiating sieve elements. *Cytobiologie* **1**: 17–36.

PARTHASARATHY, M. V. and TOMLINSON, P. B. (1967) Anatomical features of metaphloem in stems of *Sabal, Cocos,* and two other palms. *Am. J. Bot.* **54**: 1143–51.

PATE, J. S. and GUNNING, B. E. S. (1969) Vascular transfer cells in angiosperm leaves—a taxonomic and morphological survey. *Protoplasma* **68**: 135–56.

PATE, J. S. and GUNNING, B. E. S. (1972) Transfer cells. *Ann. Rev. Pl. Physiol.* **23**: 173–196.

RUSSIN, W. A. and EVERT, R. F. (1985) Studies on the leaf of *Populus deltoides* (Salicaceae): ultrastructure, plasmodesmatal frequency, and solute concentrations. *Am. J. Bot.* **72**: 1232–1247.

SACHS, T. (1972) The induction of fibre differentiation in peas. *Ann. Bot.* **36**: 189–97.

SAUTER, J. J. (1974) Structure and physiology of Strasburger cells. *Ber. dt. bot. Ges.* **87**: 327–36.

SAUTER, J. J. (1980) The Strasburger cells–equivalents of companion cells. *Ber. dt. bot. Ges.* **93**: 29–42.

SAUTER, J. J. and BRAUN, H. J. (1972) Cytochemische Untersuchungen der Atmungsaktivität in den Strasburger-Zellen von *Larix* und ihre Bedeutung für den Assimilattransport. *Z. Pflanzenphysiol.* **66**: 440–58.

SAUTER, J. J., DÖRR, I., and KOLLMANN, R. (1976) The ultrastructure of Strasburger cells (=albuminous cells) in the secondary phloem of *Pinus nigra* var. *austriaca* (Hoess) Badoux. *Protoplasma* **88**: 31–49.

SMOOT, E. L. (1984) Phloem anatomy of the carboniferous pteridosperm *Medullosa* and evolutionary trends in gymnosperm phloem. *Bot. Gaz.* **145**: 550–564.

SPANNER, D. C. (1978) Sieve-plate pores, open or occluded? A critical review. *Plant, Cell and Environment* **1**: 7–20.

SPANNER, D. C. (1979) The electroosmotic theory of phloem transport: a final restatement. *Plant Cell Environ.* **2**: 107–121.

SRIVASTAVA, L. M. (1963) Secondary phloem in the Pinaceae. *Univ. Calif. Publ. Bot.* **36**: 1–142.

THAINE, R. (1969) Movement of sugars through plants by cytoplasmic pumping. *Nature* **222**: 873–5.

THORSCH, J. and ESAU, K. (1981) Nuclear degeneration and the association of endoplasmic reticulum with the nuclear envelope and microtubules in maturing sieve elements of *Gossypium hirsutum. J. Ultrastruct. Res.* **74**: 195–204.

WALSH, M. A. and EVERT, R. F. (1975) Ultrastructure of metaphloem sieve elements in *Zea mays*. *Protoplasma* **83:** 365–88.

WARMBRODT, R. D. (1985a) Studies on the root of *Zea mays* L.—Structure of the adventitious roots with respect to phloem unloading. *Bot. Gaz.* **146:** 169–180.

WARMBRODT, R. D. (1985b) Studies on the root of *Hordeum vulgar* L. Ultrastructure of the seminal root with special reference to the phloem. *Am. J. Bot.* **72:** 414–432.

ZAHUR, M. S. (1959) Comparative study of secondary phloem of 423 species of woody dicotyledons belonging to 85 families. *Mem. Cornell Univ. Agric. Exp. Sta.*, no. 358.

ZIMMERMANN, M. H. (1961) Movement of organic substances in trees. *Science* **133:** 73–79.

ZIMMERMANN, M. H. (1963) How sap moves in trees. *Scient. Am.* **208:** 132–42.

ZIMMERMANN, M. H. (1964) Sap movements in trees. *Biorheology* **2:** 15–27.

ZIMMERMANN, M. H. and MILBURN, J. A. (1975) *Transport in Plants* I, *Phloem Transport, Encyclopedia of Plant Physiology*, NS, Vol. I, Springer, Berlin.

CHAPTER 9

SECRETORY DUCTS AND LATICIFERS

SECRETION in plants is a common phenomenon. The formation of the cell wall and cuticle, suberization, wax deposition, and the migration of specific substances from the cytoplasm into the vacuoles represent processes of secretion. Some of these types of secretion occur in all plant cells, and some are common to certain tissues of most plants.

In addition to the secretion of the above-mentioned type of materials, there are cells, groups of cells, or more complicated structures which secrete specific substances. These secretory structures occur in various plants on or in different organs and tissues.

Under the influence of studies on the processes of secretion in the animal kingdom, attempts were made to distinguish between different types of secretion in plants. Thus the terms *excretion* (elimination of end products of the metabolism) and *secretion* (elimination of substances which may still take part in the metabolic processes) were usually used. Frey-Wyssling (1935) suggested also the use of the term *recretion* for the elimination of salts, a process which regulates the ion content of the cell. For many reasons it appears that in plants no clear distinction can be made between secretion and excretion. Therefore one term should be used for all types of elimination of substances from the active part of the cell. In this book the term *secretion* is used.

Many secretory structures, e.g. certain secretory idioblasts, glandular trichomes and nectaries, are discussed in other chapters. In this chapter some examples of anatomically complicated structures which form organized systems inside the plant organs will be presented.

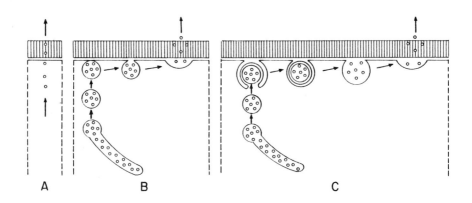

FIG. 68. Diagram showing possible mechanisms of secretion: A, membrane transport of individual particles (*eccrine secretion*); B and C, Transport of masses of particles (*granulocrine secretion*). B, Membrane of the vesicle fuses with the plasmalemma. C, Vesicle with the secreted substance is eliminated by invaginations of the plasmalemma. (From Fahn, 1979.)

TYPES OF SECRETION BASED ON SITE OF DEPOSITION OF THE SECRETED MATERIAL

According to the site of deposition of the secreted material a distinction is made between secretion into membrane bound compartments within the cell (*intracellular secretion*) and secretion to the outside of the cell (*extracellular secretion*). Within the extracellular type of secretion a distinction is made between *endogenous secretion* and *exogenous secretion*. The former is characterized by the accumulation of the secreted material in intercellular spaces, in the second the secreted substance is discharged to the outside of the plant and occurs in various types of secretory epidermal structures.

Intracellular secretion occurs in idioblasts with various secreted substances (myrosin, tannins, mucilages, crystals, essential oils, resins, etc.). Secretory idioblasts differing in size only slightly from the neighbouring cells may occur singularly or be arranged in long rows (e.g. articulated laticifers). Very large cells, sometimes as long as the whole plant (e.g. non-articulated laticifers), also occur.

The secreted material in *extracellular secretion* collects in special intercellular spaces. These spaces may be of various forms (Metcalfe and Chalk, 1950, 1983) and origin. They may be more or less spherical, e.g. in *Citrus, Eucalyptus,* and termed *secretory cavities,* or they may be elongated and called *ducts* or *canals,* e.g. in the Compositae (Moraes Castro, 1987; Lersten and Curtis, 1988), Anacardiaceae (Joel and Fahn, 1980), and Umbelliferae. Ducts may anastomose to some extent and form two-dimensional systems according to a more or less constant pattern, e.g. in the secondary xylem of *Pinus*. Ducts and cavities may develop schizogenously, lysigenously or schizo-lysigenously (Fahn, 1979). Ducts, for instance, were reported to develop schizogenously in the Umbelliferae and Pinaceae, or lysigenously, e.g. in *Copaifera* (Caesalpiniaceae), *Mangifera* (Anacardiaceae) and *Dipterocarpus* (Dipterocarpaceae). Secretory cavities and ducts may occur in various organs and tissues.

Mechanism of secretion

The main ways in which substances may be secreted are: A. The compounds may leave the secretory cell as a result of its disintegration—*holocrine secretion*. B. The secreted substances may be eliminated from the cell, which however remain intact—*merocrine secretion*.

In the latter type of secretion the compound leaves the cytoplasm either as individual molecules by a mechanism of active transport through membranes—*eccrine secretion,* or they accumulate in membrane bound vesicles, which fuse with the plasmalemma (or tonoplast) and then discharge the material from the cytoplasm—*granulocrine secretion* (Fig. 68).

For more information on secretory tissues see Fahn (1979, 1988).

Resin ducts

Resin ducts are a common feature in conifers. In the primary body they seem to be quite a normal feature (Hanes, 1927). In the secondary body they are more liable to be influenced by external factors. In the Pinaceae it has usually been stated that in genera such as *Abies, Tsuga, Cedrus,* and *Pseudolarix* the resin ducts are produced only as a result of injury (*traumatic ducts*), while in other genera such as *Pinus, Picea, Larix,* and *Pseudotsuga* they appear as a normal feature of the wood. The Pinaceae have been classified in accordance to this feature by Jeffrey in 1905 (Bannan, 1936).

Some workers suggest that even in the second group the ducts are much more closely correlated with injury than was thought earlier. In the first group of Pinaceae mentioned above, the ducts are generally cyst-like and confined to tangential series at wounds. In the second group they are longer and often scattered in distribution, becoming further dispersed at a distance from the centres of injury. The more general occurrence of ducts in this second group as compared with the first one is correlated with the lengthening and scattering of ducts produced subsequently to injury rather than to normal occurrence independent of wounding as has been commonly supposed. The genus *Pinus* might be considered to have the ducts which are least controlled by external factors (Thomson and Sifton, 1925; Bannan, 1933, 1936; Mirov, 1967).

In normal growing trees of *Pinus halepensis* vertical ducts do not appear immediately after the commencement of cambial activity, in early spring, but only later in the season close to the formation of the late wood. Auxins were found to induce the formation of vertical resin ducts as well as an enhanced radial growth of the wood. The formation of induced vertical ducts did not start immediately after the application of auxins but only about a month later, resembling the time gap between commencement of cambial activity and duct formation in nature. Wounding and pressure also stimulate the formation of resin ducts only after a considerable amount of ductless wood has been produced (Fahn and Zamski, 1970).

Many investigators have dealt with the secondary resin ducts of conifers because of their importance in wood technology and resin extraction.

The resin duct is an elongated structure built throughout its length of cells surrounding a cavity. In the genus *Pinus* the cavity is surrounded by a layer of thin-walled and unlignified cells which are termed *epithelial cells*. Outside these are one or more layers of cells with relatively thick, unlignified walls, which are termed *sheath cells*. The walls of these cells are apparently very rich in pectic substances. Among the sheath cells there may be some dead cells which may form a cylinder around the epithelial cells. The inner lamella of the wall of these cells consists of suberin. These cells become crushed by further expansion of the duct during growth. There is a gradual increase in the height of the duct cells from the duct cavity outwards (Fig. 70, nos. 1, 2).

The primary and secondary systems of the resin ducts of *Pinus halepensis* according to Werker and Fahn (1969) are described below.

Resin ducts of the primary body

The root-hypocotyl. The number of resin ducts in the primary body of the root was found to correspond with the number of primary xylem strands (Fig. 69, no. 1). The ducts extend from a region adjacent to the root apex, along the axis, into the hypocotyl. Some of the hypocotyl ducts may branch. The ducts end at the upper portion of the hypocotyl, at the base of the cotyledons, or they extend to the lower portions of the cotyledons.

The shoot. In the region of the cotyledonary node new resin ducts develop. These ducts, which continue upwards into the cortex of the stem, have no connections with the root-hypocotyl ducts.

In the seedling only juvenile leaves are present. When the seedling grows these leaves become more and more scale-like, and during the second year of the sapling brachyblasts bearing needles appear in the axils of these leaves. In the stem, each vascular bundle which diverges into a juvenile leaf passes between two cortical resin ducts (Fig. 69, no. 2). Each of these stem ducts splits periclinally into two ducts. One branch continues along the stem in the cortex and the second one enters a juvenile leaf. Thus each juvenile leaf has two longitudinal resin ducts along its abaxial side. A high correlation exists between the vascular and resin duct systems and both are correlated with the phyllotaxis. The same pattern prevails in the primary body of the shoot throughout the plant from a seedling to a mature tree. While the juvenile leaves change to scales no change occurs in the shoot duct system. The system becomes only more complicated by the emergence of lateral shoots. The differentiation of the stem ducts is acropetal.

Needles. In the needles of the brachyblasts, as in all other organs of *Pinus halepensis* that have secondary thickening, primary and secondary ducts are present. The primary ducts appear in the mesophyll parallel to the longitudinal axis of the leaf (Fig. 142, no. 2).

In the lower portion of the leaf two lateral and 2–7 abaxial ducts can usually be distinguished. These ducts are of varying length.

The secondary ducts of the needles are continuous with the ducts of the secondary xylem of the brachyblast's axis. The secondary ducts of the needle end a short distance above the needle's base.

Resin ducts of the secondary body

In the secondary xylem of pines vertical and

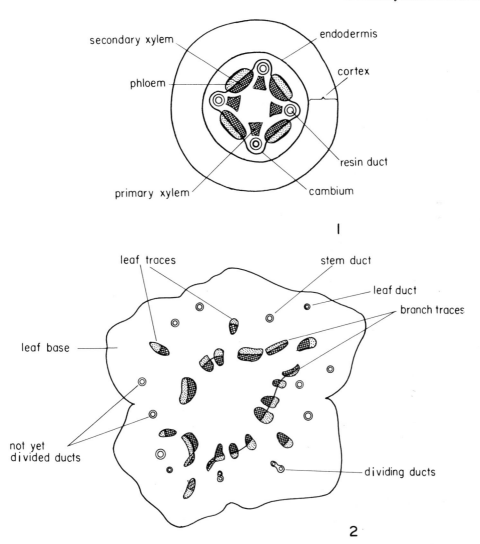

FIG. 69. 1, Diagram of a cross-section of a young seedling of *Pinus halepensis* showing a tetrarch root with four primary resin ducts. 2, As no. 1 but of a stem showing that each trace of a juvenile leaf is accompanied by two resin ducts. (From Werker and Fahn, 1969.)

horizontal resin ducts are present (Fig. 70, nos. 1, 3). The vertical ducts of each growth ring usually appear in the outer part of the early wood and in the part of the late wood which is formed first. In the first growth-ring of tree branches, however, vertical resin ducts usually develop from the cambium immediately after the formation of 0–4 layers of tracheids. At various levels these inner secondary ducts are tangentially split and their number increases.

The inner end of each radial resin duct is connected to a vertical duct of the secondary xylem, and the lumina of the two types of ducts are continuous. Each xylary radial duct extends to the cambium, and from the cambium outwards it continues as a phloic radial duct. The outer end of each radial duct of the phloem is enlarged into a cyst-like vesicle. There are no vertical ducts in the phloem. In the cambial region the lumen of the radial duct is closed.

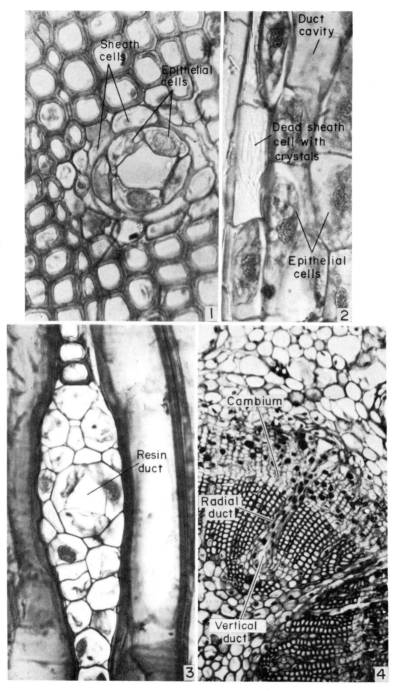

FIG. 70. Resin ducts of *Pinus halepensis*. 1, Cross-section of a vertical duct of the secondary xylem. × 480. 2, Portion of a longitudinal section of a vertical duct of the secondary xylem. × 500. 3, Tangential section of secondary xylem showing a radial resin duct. × 480. 4, Portion of a cross-section of a young stem showing a radial duct connected to a vertical duct. × 110. (From Werker and Fahn, 1969.)

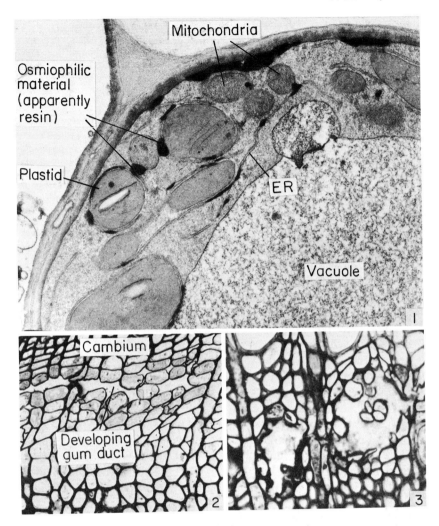

FIG. 71. 1, Electron micrograph of a portion of an epithelial cell of a primary resin duct of *Pinus halepensis*, × 28,000. 2 and 3, Gum ducts of *Citrus*. 2, Young duct in cambial region, × 500. 3, Ducts embedded in the secondary xylem, × 500. (No. 1 from Benayoun and Fahn, 1979; nos. 2 and 3 from Gedalovich and Fahn, 1985.)

Radial resin ducts come in contact with vertical resin ducts situated on the same radial plane, and connection may occasionally result between the lumina of the two types of duct. The radial ducts are found within rays which have a special fusiform shape (Fig. 70, nos. 3, 4).

Connections between different vertical ducts are present only on the same radial plane, and these occur through the radial resin ducts. This means that the resin duct system cannot be considered as a spatial network. The lack of tangential connections was found to be true for all the secondary resin ducts except for those formed in the innermost portion of the first growth ring which branch and unite in a tangential direction.

The vertical ducts of *Pinus halepensis* were found to be 4–10 cm in length. Apparently in the young branches the proportion of the short ducts is higher. According to Münch (1919), Bannan (1936), and Reid and Watson (1966), vertical ducts in old trees of other pine species may be much longer.

The lack of tangential connections between the ducts of each radial plane in the secondary body and the limited length of the vertical ducts, the lack of connections between the resin systems of the primary body in the different organs and between it and the duct system of the secondary body, may explain how differences in composition of the terpene fraction of resin could be detected at different heights of the trunk of conifer trees and between different organs and tissues of the plant (Zavarin, 1968a, b; Roberts, 1970).

Development of ducts

The resin ducts throughout the plant, in both primary and secondary tissues, were found to develop in a similar manner. The duct initials can be recognized very early in the plant's life, sometimes even in the seed and in the early stages of developing organ and tissue. In the primary body they can be distinguished close to the root and shoot apices, in both a seedling and the mature plant. In the secondary body they are discernible in the daughter cells of the fusiform initials of the cambium and form the vertical ducts, while in the ray cambial initials they form the radial ones. As seen in cross-section a group of a few initials develops from the meristematic tissue, forming a rosette in a cross-section. A few cells undergo divisions in various directions, the main direction being periclinal to the future duct cavity. In this manner the rosette becomes more distinct from the neighbouring cells. The duct cavity develops between two cells which are situated more or less in the middle of the rosette. The intercellular space thus formed is surrounded by the innermost two cells and usually two other neighbouring cells. These four cells and the cells surrounding them may undergo further divisions periclinally to the developing cavity. In this manner a 2–3 cell layered sheath is formed around the epithelial cells. Further from this region the intercellular space expands and penetrates between the radial walls of some epithelial cells and even between cells of the sheath. Thus some of the sheath cells become also part of the final epithelial layer.

Prior to the separation which occurs during the formation of the duct cavity, the walls swell and become lens-shaped. This swelling apparently takes place in the middle lamella and precedes its disintegration.

The main ultrastructural feature of the secreting epithelial cells is the occurrence of many plastids (Fig. 71, no. 1). For more information on the ultrastructure and the process of resin secretion, see Fahn (1979, 1988; Fahn and Benayoun, 1976).

Mucilage ducts and cavities

The occurrence of mucilage cavities (isodiametric structures) or ducts (elongated structures) has been reported in several families, e.g. Malvaceae, Sterculiaceae, and Tiliaceae (Metcalfe and Chalk, 1950, 1983).

In *Sterculia urens* the ducts were found to occur in the cortex and pith of the stem, in the petiole and midvein of the leaf, and in the cortex, phloem parenchyma, and xylem parenchyma of the root (Shah and Setia, 1976).

The ducts, for example of *Sterculia* and *Brachychiton*, consist of a lumen filled with mucilage and cell remnants and a border formed by flattened cells. In the mucilage duct cells, the dictyosomes are involved in secretion of the mucilage (see Fahn, 1979, 1988).

Gum production in ducts of stems of many plants, e.g. in species of the Prunoideae (Fig. 72), *Acacia, Citrus* and many others, is usually referred to as *gummosis*. Different views have been expressed regarding the development of these ducts. Gummosis in the Prunoideae, for instance, was considered by some authors to be primarily a response to injury or to pathogen attack, while others suggested that it is a natural phenomenon which is intensified by injury or pathogen infection (Grosclaude, 1966, Morrison and Polito, 1985).

Gum ducts in *Citrus* trees develop in response to fungal and viral diseases. The well-known "brown rot" gummosis of *Citrus* trees is caused by the fungus *Phytophthora citrophthora*. When a tree is infected with this fungus, gum ducts start to develop schizogenously in the cambium. With continuing development of the cambium and differentiation xylem, the gum ducts become embedded in the latter (Fig. 71, nos. 2, 3) and the epithelial cells cease to secrete. The walls of many cells rupture and the gum still present in the cells is released (Gedalovich and Fahn, 1985).

Different views have been expressed regarding the way the gum is produced during gummosis.

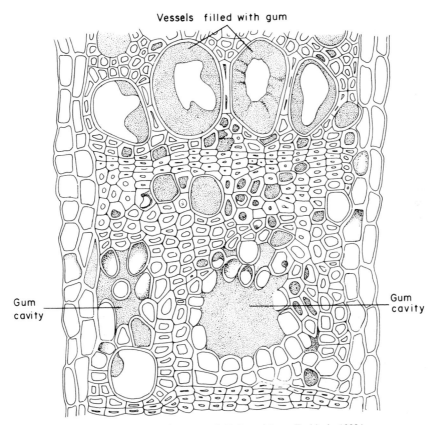

Vessels filled with gum

Gum cavity

Gum cavity

FIG. 72. Gummosis in cherry wood. (Adapted from Tschirch, 1889.)

Some authors attribute gum formation to cell wall decomposition (Tschirch, 1889; Butler, 1911; Groom, 1926; Vander Mollen *et al.*, 1977). However, in *Citrus* and some other plants gum production has been shown to result from synthetic activity of secretory cells (Catesson and Moreau, 1985; Gedalovich and Fahn, 1985; Morison and Polito, 1985). In all cases the dictyosomes were found to take part in the synthesis of the gum.

Another type of gummosis is the occlusion of vessels with gum-like material. This may occur in response to physiological stresses arising from wounding, e.g. in cassava roots (Rickard and Gahan, 1983) or infection. In a number of plants, e.g. *Dianthus caryophyllus, Ulmus campestris* and *Ailanthus excelsa,* vascular occlusion develops in response to infection by fungi (Catesson and Moreau, 1985; Shah and Babu, 1986). In *Dianthus*

it has been shown that as a result of infection, living cells in contact with vessels acquire secretory function. They secrete carbohydrates, glycoproteins and polyphenols into the damaged vessels and occlude them.

Kino veins

A special type of duct, termed a *kino vein,* occurs in the genus *Eucalyptus* (Fig. 73, no. 1). Kino differs from gum in containing polyphenols. The kino veins vary considerably in size: from a few isolated veins to a dense anastomosing mass completely encircling a tree, and 1.5 to 5 mm in width. The length of the veins varies from a few centimetres to 3 m.

Kino veins form in the cambial region as a result of injury. When a patch of veins of moderate size is formed, as seen in tangential view, the veins extend for a greater distance upwards from the wound than downwards.

Kino veins develop in the zone of traumatic parenchyma, which is produced by the cambium shortly after the stimulus to vein formation is given and in which groups of cells accumulate large quantities of polyphenols. At particular foci, groups of cells containing polyphenols break down and form ducts into which the contents of the kino-producing cells are released. At about the same time the cells surrounding the future veins divide and form a peripheral "cambium". The derivatives of this "cambium" grow, accumulate polyphenols, break down, and enlarge the quantity of kino already present in the ducts. In the final stage the peripheral "cambium" produces a layer of derivatives which becomes suberized in the form of a typical periderm (Skene, 1965).

The kino veins may occur in the secondary xylem or in the secondary phloem. If the traumatic parenchyma bands (as seen in cross-section) are produced on the xylem side of the cambium, the kino veins become included in the xylem; if they are formed on the phloem side of the cambium, the veins become included in the phloem. According to Tippett (1986), some *Eucalyptus* species exhibit kino veins in the xylem, while other species have phloem veins.

Laticifers

Latex, a suspension or, in certain cases an emulsion, of many small particles in a liquid with a varying refractive index, is known to occur in many angiosperms. This liquid may occur either in series of cells or in single long cells; both these specialized structures have been termed *laticifers*. However, latex may also occur in unspecialized parenchyma cells.

Laticifers are present in a large number of species and genera belonging to about twenty families (Metcalfe, 1967). These belong mostly to the Dicotyledons, but laticifers were also found in a few monocotyledonous families, in the genus *Regnellidium* of the Marsiliaceae (Pteridophyta) and in *Gnetum* (Behnke and Herrmann, 1978). The secretory cells (not those of the resin ducts) occurring in the conifers are also regarded as laticifers (see Werker, 1970). The fact that laticifers are restricted to a small number of families, between many of which there is no evidence of close taxonomic relationship, suggests that the capacity to produce latex has evolved more than once (Metcalfe, 1967).

The chemical composition of the suspended or emulsified particles of the latex differs in the different species. Among the suspended material rubber particles [$(C_5H_8)n$], waxes, resins, proteins, essential oils, mucilages, and in certain *Euphorbia* species, variously shaped starch grains can be found (Fig. 73, no. 4). In the latex of *Ficus carica* proteolytic enzymes occur (Zuckerman-Stark, 1967). Latex, like the cell sap, also contains salts, organic acids, and other substances in true solution. Certain plants contain, in the latex, sugars (the Compositae), tannins (*Musa*), alkaloids (*Papaver somniferum*), and in *Carica papaya,* a proteolytic enzyme, papain, is present. The colour of latex varies in the different plant species—it may be white and milky (*Euphorbia, Lactuca, Asclepias*), yellow-brown (*Cannabis*), yellow to orange (*Chelidonium*) or colourless (*Morus*).

Heinrich (1967, 1970), who studied the laticifers of *Taraxacum bicorne* and *Ficus elastica* with the aid of the electron microscope, expressed the view that rubber particles are synthesized in the cytoplasm and not in the plastids as thought before (cf. Frey-Wyssling, 1935). In *Ficus* the rubber particles were surrounded by an envelope, possibly of membranous nature. As seen in *Taraxacum*, the particles remain in the cytoplasm and only in old laticifers, when the tonoplast disintegrates, are they freed. Many vacuoles appear in the initials of the laticifers. Thureson-Klein (1970), who studied the laticifers of *Papaver somniferum*, has noticed in the cytoplasm the appearance of many vesicles with dense material which may contain alkaloids. In very young laticifers of *Euphorbia marginata*, Schulze et al. (1967) found rubber drops first in the cytoplasm but later almost only in vacuoles. In later developmental stages a large central vacuole bounded by a tonoplast is formed. The small vacuoles with rubber particles join the central vacuole which becomes more and more filled with rubber. A tonoplast was also reported by Schnepf (1964) to occur in mature laticifers of *Euphorbia pulcherrima*. Schulze et al. concluded that in the mature articulated-branched laticifers of *Taraxacum, Papaver,* and *Hevea* the rubber particles

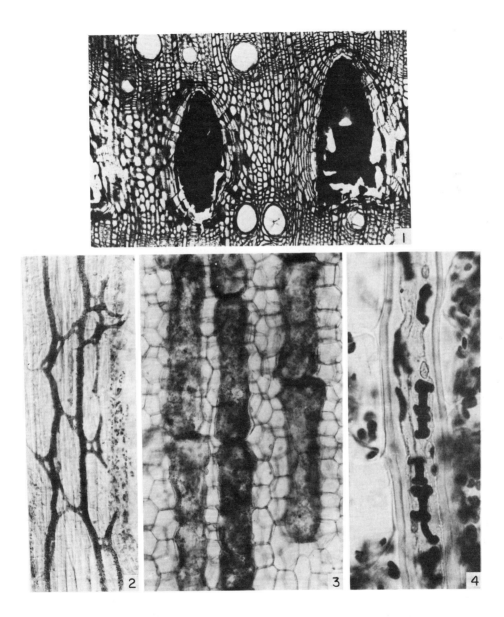

FIG. 73. 1, A cross-section of *Eucalyptus regnans* wood with kino veins. × 90. 2, Micrograph of a stem of *Sonchus oleraceus*, cleared with lactic acid, to show the branched, articulated laticifers. × 180. 3, Tangential sections at a depth of two cells below the abaxial epidermis of a bulb scale of *Allium cepa* in which simple articulated laticifers can be distinguished. × 150. 4, Laticifer with osteoid starch grains in *Monadenium ellenbeckii* (Euphorbiaceae), × 600. (No. 1 courtesy of the Forestry and Forest Products Laboratory, CSIRO, Melbourne, Australia; No. 4 courtesy of Paula J. Rudall.)

appear as single drops in the cytoplasm, whereas in the non-articulated laticifers of *Euphorbia* they appear assembled in central vacuoles. In developing laticifers of *Euphorbia pulcherrima*, Fineran (1983) observed tubular and spherical vacuoles which contained latex particles. The latter were first seen in the narrow tubular vacuoles. In *Hevea*, vesicules 1–5 μm in diameter, called *lutoids*, are present. These vesicles contain a fluid within which there are usually numerous small bodies (see Fahn, 1979).

Studies of *Ficus* and *Taraxacum* showed that in the cytoplasm of developing laticifers a relatively well-developed endoplasmic reticulum and many ribosomes occur. The mitochondria are small and degenerate when the rubber particles start to appear. Dictyosomes appear only in the growing portions of the laticifers. In old laticifers the disintegration of the organelles, as seen in *Taraxacum*, proceeds in the following sequence: dictyosomes, ribosomes and mitochondria, the ER, the nucleus, and plastids. In *Ficus* mitotic activity of the nuclei seems to be confined to apical zones. In the exuded latex only degenerated nuclei and plastids, and more rarely mitochondria, appear. In mature laticifers of *Ficus*, filaments similar to those in sieve elements were observed (Heinrich, 1970).

The cell wall of the laticifers is entirely primary and may be as thick as or thicker than the neighbouring parenchyma cells. The thick walls contain cellulose and a high proportion of pectic substances and hemicelluloses (Moor, 1959). These walls are highly hydrated. Both the thick walls and the thin ones, which do not differ from the walls of the neighbouring parenchyma cells, are very elastic. In the cell wall of a number of Convolvulaceae species an impregnated suberized layer was found to be present (Fineran *et al.*, 1988). There is controversy as the occurrence of primary pit fields.

In the laticifers of some plants the protoplasts are reported to remain intact even after maturation; in others the cytoplasm, including the organelles, may undergo degeneration (Heinrich, 1967; Neumann and Müller, 1972). In some plants the mature laticifers were reported to contain large central vacuoles bounded by a tonoplast, e.g. in *Euphorbia marginata* (Schulze *et al.*, 1967). In *Tar-*

axacum bicorne the tonoplast disappears (Heinrich, 1967).

In addition to the usual components of the cytoplasm, electron-dense globules and small vacuoles or vesicles occur in laticifers. In mature regions of the laticifers of *Ficus carica* vesicular structures fill the vacuolar space and are present also in the cytoplasm. The endoplasmic reticulum and the dictyosomes have been suggested as possible sources of the vesicular structures occurring in the laticifers of *Ficus carica* and *Nerium oleander* (Rachmilevitz and Fahn, 1982, Stockstill and Nessler, 1986). In the laticifer initials of *Papaver somniferum* the ER was found to be abundant (Nessler and Mahlberg, 1977). In the initials of the *Taraxacum bicorne* laticifers dictyosomes were observed to be rare and to degenerate before the appearance of the first rubber particle. The relatively numerous mitochondria remain small and their inner structures undergo early lysis. The degeneration of the mitochondria and ribosomes occurs long before the dissolution of the end wall. The nuclei and plastids are the longest persisting organelles (Heinrich, 1967). In *Hevea* bits of endoplasmic reticulum and several mitochondria may still be found in mature laticifers (Gomez, 1975; Fahn, 1979).

Several different views have been expressed regarding the function of the laticifers (de Bary, 1877; Parkin, 1900; Tschirch, 1906; Haberlandt, 1918; Frey-Wyssling, 1935; Sperlich, 1939; Bonner and Galston, 1947; Benedict, 1949). Because of their distribution in the plant body and their liquid content that flows out promptly when the plant is cut, the early botanists called them *Lebenssaftgefässe* (vital sap vessels) and compared them with the blood vessels of animals. Their association with the vascular bundles, especially with the phloem (in both the pressure is high), brought the researchers later to the assumption that the laticifers take part in translocation of assimilates. This, however, is incorrect. The laticifers at the separation layer of leaves of *Hevea* which have completed their development were found to be occluded by callose (Spencer, 1939), This would prevent translocation of substances from leaves to the stem via laticifers. The advantage of association with phloem may be an efficient supply of initial latex precursors from the assimilating tissues.

Haberlandt (1918) believed the latex to be of nutritional value in many plants. The laticifers have also been considered to participate in storage of food materials (Sperlich, 1939). The results of experiments also contradicted this assumption. The storage materials present in the latex are not readily mobilized when the plant is grown under conditions unfavourable for carbohydrate synthesis (Traub, 1946; Benedict, 1949). Although α-amylase has been found in the latex of *Euphorbia,* its starch grains are not susceptible to digestion. It has been suggested that the starch grains of the laticifers may have evolved to perform a function in plugging and rapid wound closure (Biesboer and Mahlberg, 1981; Spilatro and Mahlberg, 1985).

Most of the substances produced and stored by laticifers do not re-enter the plant metabolism. Rubber, for instance, like the essential oils cannot revert to nutritional substrates as no enzymes capable of breaking down rubber are present in the plant (Bonner and Galston, 1947). The most probable main function of the laticifers seems to be protection. The latex may play a role in covering wounds, as a defence against herbivores, and perhaps also as a defence against microorganisms.

In the laticifers latex is under pressure and therefore when the laticifers are cut the latex is exuded. This is a pressure flow. The dynamics of latex flow have been dealt with in detail by Frey-Wyssling (1952).

TYPES OF LATICIFERS

According to de Bary (1877), the laticifers are divided into two main types: *non-articulated* and *articulated.* This classification has no relationship to taxonomic groups and thus different types of laticifers may be found in different species of one family.

The non-articulated laticifers develop from a single cell which greatly elongates with the growth of the plant and which is sometimes branched. Such laticifers are also termed *laticiferous cells.* Articulated laticifers consist of simple or branched series of cells which are usually elongated. The end walls of such cells remain entire or become porous or disappear completely. Such laticifers are also termed *laticiferous vessels.*

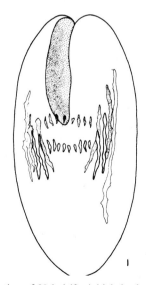
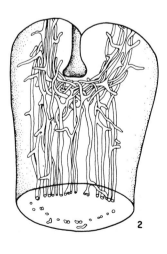

FIG. 74. 1, Reconstruction of 28 laticifer initials in the cotyledonary node of an immature embryo (550 μm long) of *Nerium oleander.* The initials are situated on the periphery of the provascular system. 2, As above, but cotyledonary node of a mature embryo (5 mm long). (Adapted from Mahlberg, 1961.)

Non-articulated laticifers

Non-articulated laticifers are characteristic of various species of the following families: Apocynaceae, Asclepiadaceae, Euphorbiaceae, Moraceae, and Urticaceae. Simple (unbranched) non-articulated laticifers are found, for example, in *Vinca, Urtica,* and *Cannabis,* and branched non-articulated laticifers in *Euphorbia, Nerium* (Fig. 74, nos. 1, 2), *Ficus,* and *Asclepias.*

As has already been mentioned, different forms of non-articulated laticifers exist and in certain mature plants laticiferous cells may develop into very large systems which extend throughout the different shoot and root tissues. Scharffstein (1932) and Rosowski (1968) who worked on some *Euphorbia* species express the view that their entire laticifer system is derived from a few initials that are already present in the embryo. Mahlberg (1961) found in *Nerium* that the number of initials is constant and that they can all be distinguished already in the embryo where they appear in the cotyledonary node from where they send branches into the cotyledons, the hypocotyl and the radicle (Fig. 74, nos. 1, 2). In the course of development of certain species of *Euphorbia* laticiferous cells are found on the circumference of the central cylinder and they branch into the leaves and pith. In the leaves of certain species of the Euphorbiaceae the laticiferous cells reach the epidermis where they may even come into contact with the cuticle. Blaser (1945) working on *Cryptostegia grandiflora* (Asclepiadaceae) found that the early formed laticiferous cells in the cortex branch radially in the position of the leaf-gaps and penetrate into the pith. After a period of cambial activity these branches of the laticiferous cells become surrounded by the secondary phloem and xylem. Laticifers were observed in the rays of the secondary xylem of some plants (Fahn *et al.,* 1986). In a number of species of *Artocarpus* (Moraceae) non-articulated laticifers branch in the wood both in the radial and in the axial directions (Topper and Koek-Noorman, 1980). The growth and expansion of the branched laticiferous cells is continuous throughout the life of the plant and their ends penetrate into the new buds, developing leaves and growing regions of the root. Such development of laticiferous cells have been observed, for example, in *Nerium* (Mahlberg, 1963). The growth of these branched laticiferous cells is by a combination of intrusive and symplastic growth (Mahlberg, 1959b). It is of interest that in certain plants even the old portions of these cell walls retain the ability to grow and produce new branches, as was shown by grafting experiments carried out by Scharffstein on *Euphorbia esculenta.* Numerous nuclei have been observed in the branched non-articulated laticifers of different plants. Scharffstein (1932) and Mahlberg (1959a) drew attention to the fact that the multiplication of the nuclei takes place early in the ontogeny of the laticiferous cell.

Mahlberg and Sabharwal (1967) found that in the developing embryo of *Euphorbia marginata* successive stages of mitosis in the nuclei of the laticifers appear in waves and that the mitotic stimulus does not start simultaneously in all laticifers. The mitotic waves originate in the laticifer distally from the meristems and move along its longitudinal axis. Polyploidization was not observed in these laticifers.

The growth of the simple non-articulated laticifers is simpler than that of the branched non-articulated ones. The initials of the former have not been observed in the embryo but only in the developing shoot, as, for instance, in *Vinca* and *Cannabis* (Zander, 1928; Scharffstein, 1932). The initials appear below the apical meristem and they develop into long, unbranched ducts which grow by a combination of intrusive and symplastic growth. These laticiferous cells may penetrate from the stem into the leaves (*Vinca*) or they may develop independently in these organs (*Cannabis*). In certain species multiplication of the nuclei also takes place during the development of simple non-articulated laticifers.

Articulated laticifers

Articulated laticifers are characteristic of different species of the Compositae, Convolvulaceae, Papaveraceae, Euphorbiaceae, Caricaceae, Sapotaceae, Liliaceae, and Musaceae.

The occurrence of articulated and non-articulated laticifers in the Euphorbiaceae has been discussed recently by Rudal (1987) and Mahlberg *et al.* (1987).

Simple non-anastomosing articulated laticifers occur in *Musa* (Fig. 76, nos. 1–3), *Allium* (Fig. 73, no. 3), *Convolvulus*, and *Ipomoea*. Anastomosing articulated laticifers are found in *Sonchus* (Fig. 73, no. 2), *Cichorium, Lactuca, Taraxacum, Tragopogon, Scorzonera, Carica, Manihot, Papaver*, and *Hevea*. Like the non-articulated laticifers, the articulated laticifers appear in the early ontogenetic stages. In *Taraxacum kok-saghyz* the differentiation of the laticifers commences with the uptake of water at the time of germination of the seed. The primary laticifers appear in the pericycle in close connection with the phloem, whereas the secondary laticifers differentiate in the secondary phloem very close to the cambium (Rudenskaja, 1938; Artschwager and McGuire, 1943; Krotkov, 1945). Scharffstein (1932), Sperlich (1939), Artschwager and McGuire (1943), and other workers observed the process of the formation of the articulated laticifers from single cells in the embryos of *Tragopogon, Scorzonera, Taraxacum*, and *Hevea brasiliensis*. Articulated laticiferous vessels develop in many plants in the phloem or pericycle of the stem and root. They also occur in the leaf mesophyll. Apart from the connections formed between the cells of one laticifer by the entire or partial dissolution of the end walls, connection may also be made between adjacent laticiferous vessels in species with branched articulated laticifers. The process of degradation of the end walls of the laticifers of *Lactuca sativa* has been described by Giordani (1980, 1981) and Nessler (1982). Articulated laticifers may also become connected by horizontally or diagonally orientated elements. Reticulate systems may also develop as a result of the formation of branches which arise on the side walls of the laticifers. Sometimes parenchyma cells between two laticiferous vessels redifferentiate and so connect the two vessels.

The laticifer system in the capsules (which supply the opium) of *Papaver somniferum* was studied by several workers (see Fahn, 1979; Thureson-Klein, 1970). The laticifers in this plant differentiate soon after seed germination. They appear in all organs but they are most abundant in the ovary. Here they produce a peripheral network the main strands of which are parallel to the vascular bundles and are interconnected by anastomoses. The network becomes denser with the development of the capsule.

The laticifers of *Hevea brasiliensis* appear mainly in the bark, but they occur also in the pith

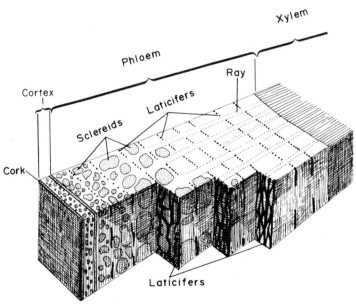

FIG. 75. Three-dimensional diagram of the bark of *Hevea brasiliensis*, showing arrangement of laticifers in the secondary phloem. (Adapted from Vischer, 1923.)

and leaves. The most important laticifers as a source of rubber are those which appear in the inner bark. These are derived from the cambium. The laticifers are arranged in concentric zones, with lateral tangential anastomoses (Fig. 75). Radial connections apparently occur only as a result of disturbances in the cambial activity. The number of rings of laticifers, the density of the network of each zone and the diameter of the individual laticifers vary from one clone to another (Vischer, 1923; Frey-Wyssling, 1930, cited in Sperlich, 1939). Correlation between the various anatomical features of the laticifers and rubber yield is restricted to clones and cannot be drawn for populations. Grafting experiments showed that laticifers become united (Metcalfe, 1967).

Bouychou (1952, according to Metcalfe, 1967) found that laticifers in *Hevea* were produced from the cambium of tissue grown *in vitro* and the segments of laticifers were still capable of producing rubber.

The simple (unbranched) laticifers of *Musa* usually accompany the vascular tissues (Skutch, 1932; Tomlinson, 1959). In *Musa* the central portion of the end walls of two adjacent cells becomes torn away, but usually it remains attached to one side of the cell where it appears as a loose flap (Fig. 76). In *Allium* the laticiferous vessels are in no way connected with the vascular tissue and they are found in the third layer of leaf mesophyll or in the third layer below the abaxial epidermis of the bulb scales. The end walls of the cells of the laticifers of *Allium* are not perforated, but well-developed primary pit fields appear in them.

Rubber plants

There are many plant species from which rubber can be obtained, but of them, *Hevea brasiliensis* in the Euphorbiaceae, which is known as the Para rubber tree, is the most important in world economics. This plant is today grown in Central America, the West Indies, Brazil, Liberia, Ceylon, Malayan Archipelago, Sumatra, Java, and eastern India. The latex of *H. brasiliensis* contains about 30% rubber.

Other rubber plants of secondary economic im-

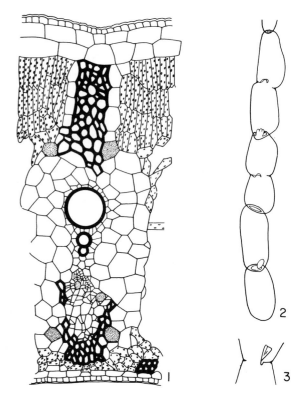

FIG. 76. Laticifers in *Musa*. 1, Portion of a cross-section of a leaf showing laticifers accompanying the vascular bundle. Laticifers, stippled. 2, Diagram of a portion of laticifer showing the articulation. 3, Diagram showing opening and flap between two adjacent cells of the laticifer.

portance (Schery, 1954) are *Castilla* (Panama rubber) in the Moraceae, *Manihot* (Ceara rubber) in the Euphorbiaceae, *Parthenium argentatum* (guayule) and *Taraxacum kok-saghyz* in the Compositae, *Hancornia* and *Landolphia* in the Apocynaceae, and *Cryptostegia* in the Asclepiadaceae. *Parthenium argentatum* was investigated thoroughly in the United States, during World War II, when rubber was unobtainable from the *Hevea* plantations in Asia.

The analogy in the structural organization of the laticifers with that of the vascular system is of interest. On this basis, the non-articulated laticifers can be compared to tracheids and sieve cells. However, laticifers, which have no conducting function, do not necessarily have to be mature throughout their entire length and so their ends can grow continuously at the plant apices. It is possible that if tracheids and sieve cells did not necessarily have to be of so specialized a structure in order to function soon after their initiation, they too would be capable of such extensive growth as is exhibited by these laticiferous cells. The articulated laticifers can be compared with vessels and sieve tubes as they are similarly built of series of many cells. This similarity was probably one of the reasons why early botanists considered the laticifers to be structures concerned with the conduction of nutrients. It may also be suggested that groups of idioblasts containing secretory substances which are arranged in rows may represent transitional forms between idioblasts and laticifers.

It is commonly held that resin and latex are end- or byproducts of metabolism. Some facts, however, suggest that this may not be so, and that resin and latex are not excretions in the meaning of animal physiology. (1) Laticifers in Papaveraceae, for instance, contain such specialized chemical substances as alkaloids at a very early stage of their development (Metcalfe, 1967). (2) Latex can be collected in *Hevea* from the same region of the bark for 8–9 years by repeated opening of the laticifers (Vischer, 1923; Sperlich, 1939). (3) Resin ducts of *Pinus* were found to secrete similar amounts of resin when first opened and re-opened after 10 days, 21 days, or after a year. If the latex and resin are merely byproducts, it is difficult to explain why they regenerate so quickly and to the same amount at various time intervals.

REFERENCES

Artschwager, E. and McGuire, R. C. (1943) Contribution to the morphology and anatomy of the Russian dandelion (*Taraxacum kok-saghyz*). *USDA Tech. Bull. US Dep. Agric.* no. 843.

Bannan, M. W. (1933) Factors influencing the distribution of vertical resin canals in the wood of the larch, *Larix laricina* (Du Roi) Koch. *Trans. R. Soc. Can.* **5:** 203–18.

Bannan, M. W. (1936) Vertical resin ducts in the secondary wood of the Abietineae. *New Phytol.* **35:** 11–46.

Behnke, H.-D. and Herrmann, S. (1978) Fine structure and development of laticifers in *Gnetum gnemon* L. *Protoplasma* **95:** 371–384.

Benayoun, J. and Fahn, A. (1979) Intracellular transport and elimination of resin from epithelial duct-cells of *Pinus halepensis*. *Ann Bot.* **43:** 179–181.

Benedict, H. M. (1949) A further study of the non-utilization of rubber as a food reserve by guayule. *Bot. Gaz.* **111:** 36–43.

Biesboer, D. and Mahlberg, P. (1981) A comparison of alpha-amylases from the latex of three selected species of *Euphorbia* (Euphorbiaceae). *Am. J. Bot.* **68:** 498–506.

Blaser, H. W. (1945) Anatomy of *Cryptostegia grandiflora* with special reference to the latex. *Am. J. Bot.* **32:** 135–41.

Bonner, J. and Galston, A. W. (1947) The physiology and biochemistry of rubber formation in plants. *Bot. Rev.* **13:** 543–96.

Butler, O. (1911) A study of gummosis of *Prunus* and *Citrus,* with observations on squamosis and exanthema of *Citrus*. *Ann. Bot.* **25:** 107–153.

Catesson, A. M. and Moreau, M. (1985) Secretory activity in vessel contact cells. *Israel J. Bot.* **34:** 157–165.

de Bary, A. (1877) *Vergleichende Anatomie der Vegetationsorgane der Phanerogamen und Farne*. W. Engelmann, Leipzig.

Fahn, A. (1979) *Secretory Tissues in Plants*. Academic Press, London.

Fahn, A. (1988) Secretory tissues in vascular plants. *New Phytol.* **108:** 229–257.

Fahn, A. and Benayoun, J. (1976) Ultrastructure of resin ducts in *Pinus halepensis*: development, possible sites of resin synthesis, and mode of its elimination from the protoplast. *Ann. Bot.* **40:** 857–863.

Fahn, A. and Zamski, E. (1970) The influence of pressure, wind, wounding and growth substances on the rate of resin duct formation in *Pinus halepensis* wood. *Israel J. Bot.* **19:** 429–46.

Fineran, B. A. (1983) Differentiation of non-articulated laticifers in Poinsettia (*Euphorbia pulcherrima* Willd.) *Ann. Bot.* **52:** 279–293.

Fineran, B. A., Condon, J. M. and Ingerfeld, M. (1988) An impregnated suberized wall layer in laticifers of the Convolvulaceae. *Protoplasma* **147:** 42–54.

Frey-Wyssling, A. (1935) *Die Stoffausscheidung der höheren Pflanzen*. Springer-Verlag, Berlin.

Frey-Wyssling, A. (1952) Latex flow. In: *Deformation and Flow in Biological Systems* (ed. A. Frey-Wyssling), North-Holland, Amsterdam, pp. 322–43.

Gedalovich, E. and Fahn, A. (1985) The development and ultrastructure of gum ducts in *Citrus* plants formed as a result of brown-rot gummosis. *Protoplasma* **127;** 73–81.

Giordani, R. (1980) Dislocation du plasmalemme et libération de vésicules pariétales lors la dégradation des parois terminales durant la différenciation des laticifères articulés. *Biol. Cellulaire* **38:** 231–236.

Giordani, R. (1981) Activités hydrolasiques impliquées dans le processus de dégradation pariétal durant la différenciation des laticifères articulés. *Biol. Cell.* **40:** 217–224.

Gomez, J. B. (1975) Comparative ultracytology of young and mature latex vessels in *Hevea brasiliensis*. *International Rubber Conference 1975, Kuala Lumpur*.

Groom, P. (1926) Excretory systems in secretory xylem or Meliaceae. *Ann. Bot.* **40:** 633–649.

Grosclaude, C. (1966) La gommose des arbres fruitiers. *Annales des Epiphyties* **17:** 129–137.

Haberlandt, G. (1918) *Physiologische Pflanzenanatomie*, 5th edn. W. Engelmann, Leipzig.

Hanes, C. S. (1927) Resin canals in seedling conifers. *J. Linn. Soc. Bot.* **47:** 613–36.

Heinrich, G. (1967) Licht- und elektronenmikroskopische Untersuchungen der Milchröhren von *Taraxacum bicorne*. *Flora, Abt. A,* **158:** 413–20.

Heinrich, G. (1970) Elektronenmikroskopische Untersuchung der Milchröhren von *Ficus elastica*. *Protoplasma* **70:**

317–23.

JEFFREY, E. C. (1905) The comparative anatomy and phylogeny of Coniferales: II, The Abietineae. *Bull. Boston Soc. Nat. Hist.* **6**: 1–37.

JOEL, D. M. and FAHN, A. (1980) Ultrastructure of resin ducts of *Mangifera indica* L. (Anacardiaceae). 1. Differentiation and senescence of the shoot ducts. *Ann. Bot.* **46**: 225–233.

KROTKOV, G. (1945) The review of literature on *Taraxacum kok-saghyz* Rod. *Bot. Rev.* **11**: 417–61.

LERSTEN, N. R. and CURTIS, J. D. (1988) Secretory reservoirs (ducts) of two kinds in giant ragweed. (*Ambrosia trifida*; Asteraceae). *Am. J. Bot.* **75**: 1313–1323.

MAHLBERG, P. G. (1959a) Karyokinesis in non-articulated laticifers of *Nerium oleander* L. *Phytomorphology* **9**: 110–18.

MAHLBERG, P. G. (1959b) Development of the non-articulated laticifer in proliferated embryos of *Euphorbia marginata* Pursh. *Phytomorphology* **9**: 156–62.

MAHLBERG, P. G. (1961) Embryogeny and histogenesis in *Nerium oleander*: II, Origin and development of the non-articulated laticifer. *Am. J. Bot.* **48**: 90–99.

MAHLBERG, P. G. (1963) Development of non-articulated laticifers in seedling axis of *Nerium oleander*. *Bot. Gaz.* **124**: 224–31.

MAHLBERG, P. G. and SABHARWAL, P. (1967) Mitosis in the non-articulated laticifer of *Euphorbia marginata*. *Am. J. Bot.* **54**: 465–72.

MAHLBERG, P. G., DAVIS, D. G., GALITZ, D. S. and MANNERS, G. D. (1987) Laticifers and the classification of *Euphorbia*: the chemotaxonomy of *Euphorbia esula* L. *Bot. J. Linn. Soc.* **94**: 165–180.

METCALFE, C. R. (1967) Distribution of latex in plant kingdom. *Economic Bot.* **21**: 115–25.

METCALFE, C. R. and CHALK, L. (1950) *Anatomy of the Dicotyledons*. Clarendon Press, Oxford.

METCALFE, C. R. and CHALK, L. (1983) *Anatomy of the Dicotyledons*. 2nd ed. Vol. II. Clarendon Press, Oxford.

MIROV, N. T. (1967) *The Genus Pinus*. Ronald Press, New York.

MOOR, H. (1959) Platin Kohle-Abdruck-Technik angewandt auf Feinbau der Milchröhren. *J. Ultrastruct. Res.* **2**: 393–422.

MORAES CASTRO, M. DE (1987) Estrudas Secretoras Em Folhas de Especies da Familia Asteraceae: Aspectos Estruturais e Histoquimicos. D.Sc. Dissertation of the Institute of Biosciences of The University of São Paulo.

MORRISON, J. C. and POLITO, V. S. (1985) Gum duct development in almond fruit, *Prunus dulcis* (Mill.) D. A. Webb. *Bot. Gaz.* **146**: 15–25.

MÜNCH, E. (1919–21) Naturwissenschaftliche Grundlagen der Kiefernharznutzung. *Arb. biol. BundAnst. Land- u. Forstw.* **10**: 1–140.

NESSLER, C. L. (1982) Ultrastructure of laticifers in seedlings of *Glaucium flavum* (Papaveraceae). *Can. J. Bot.* **60**: 561–567.

NESSLER, C. L. and MAHLBERG, P. G. (1977) Ontogeny and cytochemistry of alkaloidal vesicles in laticifers of *Papaver somniferum* L. (Papaveraceae). *Am. J. Bot.* **64**: 541–51.

NEUMANN, D. and MÜLLER, E. (1972) Beiträge zur Physiologie der Alkaloide. III. *Chelidonium majus* L. und *Sanguinaria canadensis* L.: Ultrastruktur der Alkaloidbehälter Alkaloidaufnahme und Verteilung. *Biochem. Physiol. Pflanzen* **163**: 375–91.

PARKIN, J. (1900) Observations of latex and its functions. *Ann. Bot.* **14**: 193–214.

RACHMILEVITZ, T. and FAHN, A. (1982) Ultrastructure and development of the laticifers of *Ficus carica* L. *Ann. Bot.* **49**:

13–22.

REID, R. W. and WATSON, J. A. (1966) Sizes, distributions and numbers of vertical resin ducts in lodgepole pine. *Can. J. Bot.* **44**: 519–25.

RICKARD, J. E. and GAHAN, P. B. (1983) The development of occlusions in cassava (*Manihot esculenta* Crantz) root xylem vessels. *Ann. Bot.* **52**: 811–821.

ROBERTS, D. R. (1970) Within-tree variation of monoterpene hydrocarbon composition of slash pine oleoresin. *Phytochemistry* **9**: 809–15.

ROSOWSKI, J. R. (1968) Laticifer morphology in the mature stem and leaf of *Euphorbia supina*. *Bot. Gaz.* **129**: 113–20.

RUDAL, P. J. (1987) Laticifers in Euphorbiaceae—a conspectus. *Bot. J. Linn. Soc.* **94**: 143–163.

RUDENSKAJA, S. J. (1938) Development of the latex vessel system as a factor of rubber accumulation in kok-saghyz roots. *Doklady (Proc.) Acad. Sci. USSR* **20**(5): 399–403.

SCHARFFSTEIN, G. (1932) Untersuchungen an ungegliederten Milchröhren. *Beih. bot. Zbl.* **49**(1) 197–220.

SCHERY, R. W. (1954) *Plants for Man*. Allen & Unwin, London.

SCHNEPF, E. (1964) Zur Cytologie und Physiologie pflanzlicher Drüsen. 5. Teil: Elektronenmikroskopische Untersuchungen an Cyathialnektarien von *Euphorbia pulcherrima* in verschiedenen Funktionszuständen. *Protoplasma* **58**: 193–219.

SCHULZE, CH., SCHNEPF, E., and MOTHES, K. (1967) Über die Lokalisation der Kautschukpartikel in verschiedenen Typen von Milchröhren. *Flora*, Abt. A, **158**: 458–60.

SHAH, J. J. and BABU, A. M. (1986) Vascular occlusions in stem of *Ailanthus excelsa* Roxb. *Ann. Bot.* **57**: 603–611.

SHAH, J. J. and SETIA, R. C. (1976) Histological and histochemical changes during the development of gum canals in *Sterculia urens*. *Phytomorphology* **26**: 151–8.

SKENE, D. S. (1965) The development of kino veins in *Eucalyptus obliqua* L'Hérit. *Aust. J. Bot.* **13**: 367–78.

SKUTCH, A. F. (1932) Anatomy of the axis of the banana. *Bot. Gaz.* **93**: 233–58.

SPENCER, H. J. (1939) On the nature of the blocking of the laticiferous system at the leaf-base of *Hevea brasiliensis*. *Ann. Bot.*, **3**: 231–5.

SPERLICH, A. (1939) Das trophische Parenchym. B. Exkretionsgewebe. In: K. Linsbauer (ed.), *Handbuch der Pflanzenanatomie*, Band 4, Lief. 38, Gebr. Borntraeger, Berlin.

SPILATRO, S. R. and MAHLBERG, P. G. (1985) Composition and structure of nonutilizable laticifer starch grains of *Euphorbia pulcherrima* Wild. *Bot. Gaz.* **146**: 26–31.

STOCKSTILL, B. L. and NESSLER, C. L. (1986) Ultrastructural observations on the non-articulated, branched laticifers in *Nerium oleander* L. (Apocynaceae). *Phytomorphology* **36**: 347–355.

THOMSON, R. B. and SIFTON, H. B. (1925) Resin canals in the Canadian spruce, *Picea canadensis*. *Phil. Trans. R. Soc. B*, **214**: 63–111.

THURESON-KLEIN, Å. (1970) Observations on the development and fine structure of the articulated laticifers of *Papaver somniferum*. *Ann. Bot.* **34**: 751–9.

TIPPETT, J. T. (1986) Formation of kino veins in *Eucalyptus* L'Herit. *IAWA Bull.* **7**: 137–143.

TOMLINSON, P. B. (1959) An anatomical approach to the classification of the Musaceae. *J. Linn. Soc., Bot.* **55**: 779–809.

TOPPER, S. M. C. and KOEK-NOORMAN, J. (1980) The occurrence of axial latex tubes in the secondary xylem of some species of *Artocarpus* J. R. & G. Forster (Moraceae). *IAWA Bull.* NS, **1**: 113–19.

TRAUB, H. P. (1946) Concerning the function of rubber hydrocarbon (caoutchouc) in the guayule plant, *Parthenium argentatum* A. Gray. *Pl. Physiol.* **21**: 425–44.

TSCHIRCH, A. (1889) *Angewandte Pflanzenanatomie.* Wien und Leipzig.

TSCHIRCH, A. (1906) *Die Harze und die Harzbehälter mit Einschluss der Milchsäfte* 2. Aufl., Gebr. Bornträger, Leipzig.

VANDER MOLLEN, G. E., BECKMAN, C. H. and RODEHORST, E. (1977) Vascular gelation, a general response phenomenon following infection. *Physiol. Plant Pathol.* **11**: 95–100.

VISCHER, W. (1923) Über die Konstanz anatomischer und physiologischer Eigenschaften von *Hevea brasiliensis* Müller Arg. (Euphorbiaceae). *Verh. naturf. Ges. Basel* **35**: 174–85.

WERKER, E. (1970) The secretory cells of *Pinus halepensis* Mill. *Israel J. Bot.* **19**: 542–57.

WERKER, E. and FAHN, A. (1969) Resin ducts of *Pinus halepensis* Mill.: their structure, development and pattern of arrangement. *J. Linn. Soc., Bot.* **62**: 379–411.

ZANDER, A. (1928) Über Verlauf und Entstehung der Milchröhren des Hanfes (*Cannabis sativa*). *Flora* **23**: 191–218.

ZAVARIN, E. (1968a) Monoterpenoids of Coniferales. *Bull. Int. Ass. Wood Anat.* **1**: 3–12.

ZAVARIN, E. (1968b) Chemotaxonomy of the genus *Abies*: II, Within tree variation of the terpenes in cortical oleoresin. *Phytochemistry* **7**: 99–107.

ZUCKERMAN-STARK, SH. (1967) On proteolytic enzymes from the fig. *Enzymology* **32**: 380–382.

CHAPTER 10

EPIDERMIS

THE epidermis constitutes the outermost layer of cells of the leaves, floral parts, fruits, and seeds, and of stems and roots before they undergo considerable secondary thickening. Functionally and morphologically the epidermal cells are not uniform and among them, apart from the ordinary cells, many types of hairs, stomatal guard cells, and other specialized cells are found. Topographically, however, and to a certain extent also ontogenetically, the epidermis constitutes a uniform tissue.

The earliest stages of the ontogenetic development of the epidermis differ in the root and shoot (see Chapter 3). This fact caused certain investigators to coin special terms, *epiblem* and *rhizodermis*, for the outermost layer of the root (Linsbauer, 1930; Guttenberg, 1940; and others). However, if the development of the epidermis from the protoderm is traced ignoring the problem of the origin of this meristematic tissue, it is possible to apply the term *epidermis* to all the organs of the various groups of vascular plants.

The epidermis usually exists throughout the entire life of those organs that have no secondary thickening. In a few plants, such as long-lived monocotyledons with no secondary thickening, the epidermis is replaced by a cork tissue as the organs age. The duration of the epidermis in organs with secondary growth differs; usually in stems and roots the epidermis is replaced by the periderm during the plant's first year, but there are certain trees, e.g. *Acer striatum*, in which the periderm develops only after several years of secondary growth of the organ (de Bary, 1877). In such cases the epidermal cells continue to divide anticlinally and to enlarge tangentially (Fig. 213, nos. 3–6).

UNISERIATE AND MULTISERIATE EPIDERMIS

In most spermatophytes the epidermis consists of a single layer of cells, but in certain plants one or several cell layers, which are morphologically and physiologically distinct from the inner ground tissue, are found on the inside of the surface layer. These layers may develop ontogenetically from two different meristematic tissues, i.e. from the meristem of the ground tissue or from the protoderm. In the former case, these layers are termed a *hypodermis* and, in the latter case, the tissue formed is regarded as being a *multiseriate epidermis*. A multiseriate epidermis develops as a result of periclinal divisions of the protodermal cells. These divisions are relatively delayed and occur in late ontogenetic stages, e.g. in the leaves of *Ficus elastica* the epidermis remains single-layered till the stage when the leaf begins to expand in the bud and the stipules are shed (Fig. 77, nos. 1–6). Multiseriate epidermis occurs in the Moraceae, certain species of the Begoniaceae and Piperaceae, and some articulated Chenopodiaceae. In *Anabasis articulata* the multiseriate epidermis develops in the lower region of each internode (Figs. 91, 92).

The *velamen*, the special absorbing tissue of orchid aerial roots, is also a multiseriate epidermis (Fig. 96, nos. 4, 5). The cells of the innermost layer of the multiseriate epidermis of leaves usually functions as a water-storing tissue (de Bary, 1877).

EPIDERMAL CELLS

Various types of epidermal cells can be dis-

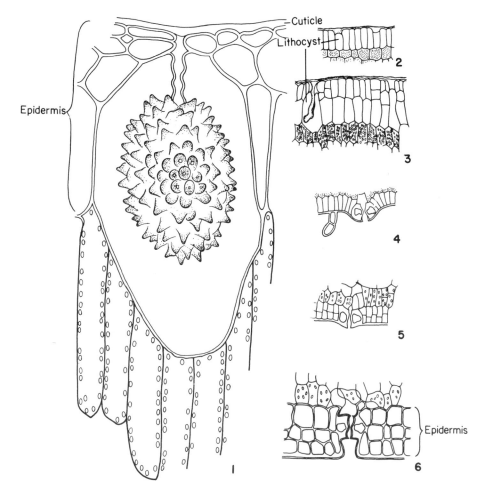

FIG. 77. Portions of cross-sections of the leaf blade of *Ficus elastica*. Approx. × 500. 1, Adaxial side of the blade in which the multiseriate epidermis, the cells of which include a lithocyst (a cystolith-containing cell), can be distinguished. 2 and 3, Two early stages in the development of the adaxial multiseriate epidermis and a lithocyst. 4–6, Stages in the development of the multiseriate abaxial epidermis. (Adapted from de Bary, 1877.)

tinguished in different plants: the ordinary cells of the epidermis; single cells or groups of cells with special structure, form or content; cells connected with stomata; and epidermal appendages termed *trichomes.*

The ordinary cells of the epidermis

The ordinary cells of the epidermis vary in shape, size, and arrangement, but they are always closely attached to form a compact layer devoid of intercellular spaces. In the epidermis of petals, air spaces may sometimes occur but they are always covered by the cuticle (Eames and Mac-

Daniels, 1947). Many epidermal cells are tubular, and in the leaf blade of dicotyledons the anticlinal walls are mostly sinuous. In the stems, and especially in the leaves of many monocotyledons, the epidermal cells are elongated. In the epidermis of certain seeds (in species of the Leguminosae and in *Punica*) the cells are relatively very elongated in a radial direction and are rod-shaped. In certain plants, for instance *Aloë aristata*, the epidermal cells appear to be hexagonal in surface view but actually they are polyhedral, and according to Matzke (1947), the average number of faces is 10.885. The external wall of the epidermal cells of certain leaves and petals is raised in the form of papillae. Haberlandt (1918) believed that the papillae may have a function in concentrating light which is limited for plants growing in the shade. In certain pteridophytes such papillae are found on the epidermal cell wall facing the mesophyll.

Wall structure

The epidermal cell wall differs in thickness—some cells are thin-walled while in others the outer or the outer and inner periclinal walls are thicker than the anticlinal walls. In seeds, scales, and certain foliage leaves, e.g. the leaves of conifers, the epidermal cell walls are very thick and lignified (Fig. 142, no. 2; Fig. 143, no. 2). Outside the outer cellulose wall there is usually a layer of more or less pure pectin. This layer enables the isolation of the cuticle from many leaves with the aid of pectinases or by other means. In many secretory trichomes the cuticle becomes detached during the process of secretion (Fig. 269, no. 1).

In leaf petioles and epicotyles of some plants in which the periclinal walls are thicker than the anticlinal walls, the periclinal walls were found to consist of lamellae in which the cellulose microfibrills are oriented alternately, parallel or transverse to the longitudinal cell axis (Chafe and Wardrop, 1972; Takeda and Shibaoka, 1978). This arrangement of fibrils resembles that of collenchyma cell walls mentioned above.

Primary pit fields and plasmodesmata are often found in the walls, especially on the radial and inner walls of the epidermal cells. In epidermal cells of the aerial organs there occur bundles of specific interfibrillar spaces, mostly in their outer walls. These bundles of spaces in the cellulosic wall obtained the name *ectodesmata*, and can be demonstrated by specific reagents. In cross-section of the wall they form thread-shaped, ribbon-like or cone-like structures. Occasionally mushroom-like forms, the broader part of which faces the cell lumen, also occur. It has been shown that ectodesmata in contrast to plasmodesmata are not plasmatical structures. A suggestion was made that they may serve as pathways of penetration of solutions to and from the protoplast (Franke, 1971a). In order to distinguish sharply between ectodesmata and plasmodesmata Franke (1971b)

has proposed to use the term *teichodes* (teichos = wall, hodos = path) instead of ectodesmata.

In the epidermal cells of certain leaves, petals, and dry fruits, partial partitions are found on the inside of the outer wall. These partitions sometimes almost reach the inner epidermal wall (Fig. 282, no. 1).

Cutin, a complex polymer of lipid derivatives is usually present in the outer walls of the epidermal cells. This substance is found within the cell wall, i.e. in the interfibrillar and intermicellar spaces of the cellulose, and it also constitutes a special layer—the *cuticle*—on the outer surface of the cell wall (Fig. 128, no. 1). Cutin stains red with Sudan IV. All the parts of the herbaceous stem, the leaves, and, to a certain extent, the maturing portions of the root, are covered with a cuticle. The cuticle is usually absent from the actively growing parts of the roots. The cuticle possesses in some genera (e.g. *Eucalyptus*) a complex and species-specific ornamentation which can be used for diagnostic purposes (Carr *et al.*, 1971).

In certain plants, which have been preserved from early geological eras, the cuticle has retained its shape, and the structure of the epidermal cells can be learnt from it. These preserved fragments of cuticle are often used to classify these early plants (Fig. 85, no. 1).

The cuticle is of varying thickness in different plants, and it is usually thicker in plants growing in dry habitats (Fig. 128, no. 1). The surface of the cuticle may be smooth, rough, ridged, or furrowed (Fig. 267, nos. 2, 3). Priestley (1943) states that very thin layers of cutin can be distinguished on those mesophyll cell walls that border the intercellular spaces. These spaces form a continuous system connected with the stomata, and therefore these cutin layers constitute a continuation of the cuticle on the outer surface of the epidermis. The cuticle on the surface of the epidermis often penetrates, to a certain extent, between the radial walls of the epidermal cells (Fig. 143, no. 2) The cuticle consists of two layers; an outer layer built of cutin only—the *cuticle proper*, and a layer below it—the *cuticular layer* which consists of cutin and wall materials. The cuticle proper is formed by the secretion of cutin or its precursors to the surface of the cell wall. This process is called *cuticularization*. The cuticular layer is

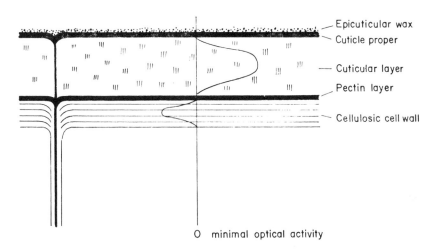

Epicuticular wax
Cuticle proper

Cuticular layer

Pectin layer

Cellulosic cell wall

O   minimal optical activity

FIG. 78. Diagram of an outer epidermal wall.

formed by deposition of cutin between the cellulose microfibrils of the outermost wall layers, where pectin and hemicellulose may also be present. This process of cutin deposition is called *cutinization* (Fritz, 1935, 1937; Roelofsen, 1959). Between the outer periclinal cellulose wall and the cuticular layer there is in many plants a layer very rich in pectin—the *pectin layer*. This layer is believed to be continuous with the middle lamella of the anticlinal walls of the epidermal cells. In the cuticular layer birefringent wax may be embedded (Fig. 78; Fig. 79, no. 1). In the stems of the xerophyte *Monttea aphylla* this layer may reach a thickness of 140–180 μm (Böcher and Lyshede, 1968). Electron microscopical studies suggested that the cuticle appears first as a thin, lamellate layer (the cuticle proper). Later another cutin layer containing a reticulum of electron dense material, apparently polyuronides (O'Brien, 1967) becomes visible (the cuticular layer). This layer eventually becomes the largest part of the cuticle (Jarvis and Wardrop, 1974). In *Hakea suaveolens* the reticulate cuticular layer was found to develop long before the final expansion of the cells and parallel with the late phase of lamellae differentiation (Heide-Jørgensen, 1978).

The outer layers of the cuticle of many *Eucalyptus* species consist largely of rubber. In young leaves these layers are elastic and can be pealed off (Carr and Carr, 1987).

Deposition of wax on the surface of the cuticle (*epicuticular wax*) may appear in various morphological forms (Metcalfe and Chalk, 1979; Barthlott and Frolich, 1983). It may occur in granules, as in *Brassica, Dianthus,* or as rods, as in *Saccharum* (Fig. 80, no. 2), or as continuous layers, as in *Thuja orientalis*, are often found on the surface of the cuticle. Wax may also be deposited in the form of platelets or scales (Fig. 80, nos. 1, 3). The wax gives the "bloom" of many leaves and fruits, and is important in reducing the wettability of surfaces. Some wax mixtures fail to crystallize and may form an oily layer of irregular flat plates above the cuticle as, for instance, in many apple fruit surfaces (Amelunxen *et al.*, 1967). As seen under the scanning microscope, waxes in crystalline state may form very elaborate structures (Fig. 81). In the wax palm (*Copernicia cerifera*) and in *Ceroxylon andicola* the wax layer may be 5 mm thick (Martin and Juniper, 1970). The structure and amount of the epicuticular wax affects the cuticular permeability and the degree to which a surface can be wetted. This phenomenon represent one of the factors which govern the selective effect of herbicides on some crop plants (Price, 1982). The epicuticular wax plays also a role in reducing the damage to photosynthesis and heat load of leaves by reflecting of light (McClendon, 1984). In the leaves of *Agave* a continuous layer of wax is also found below the cuticle (Schieferstein and Loomis, 1959).

In certain plants platelets of wax have been found within the cutin (*intracuticular wax*) of outer epidermal cell walls (Roelofsen, 1952).

Cuticles develop during the early stages of growth of the organs. The precursor of cutin mi-grates in the form of minute droplets through the matrix of the epidermal wall (Frey-Wyssling and Mühlethaler, 1965). The precursors seem to be unsaturated fatty acids (Bolliger, 1959). The cuticle is built up centripetally, its outer zone being

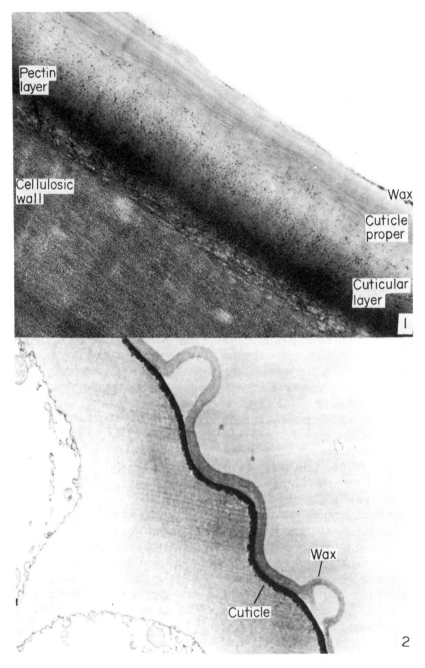

FIG. 79. Electron micrographs of outer epidermal walls. 1, Epidermis of the stem of *Anabasis articulata*. × 37,000. 2, Epidermis of a juice vesicle of a grapefruit. × 13,500. (No. 1 from Lyshede, 1977; no. 2 from Fahn *et al.*, 1974.)

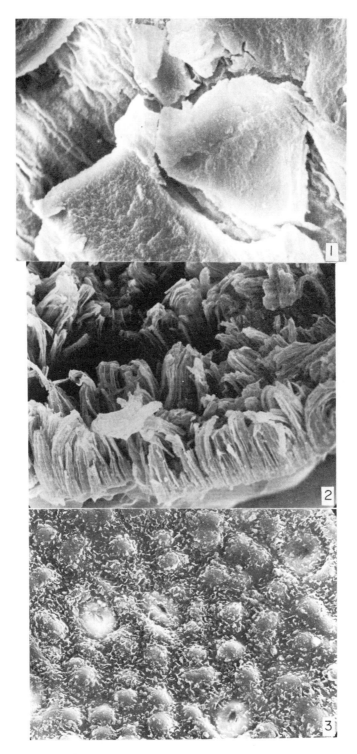

FIG. 80. Scanning electron micrographs showing various forms of epicuticular wax. 1, Platelets on a juice vesicle of a Shamouti orange. ×2500. 2, Rods or hairs on *Saccharum officinarum* stem. ×1000. 3, Small scales on *Eucalyptus camaldulensis* leaf. ×390. (No. 1 from Fahn *et al.*, 1974.)

formed first. The cuticular material hardens gradually, due to a continuous process of oxidation and polymerization.

Various theories have been proposed to explain this outward movement of the cutin (Lyshede, 1982). According to some investigators, special channels are present in the outer epidermal walls through which precursors of cutin pass (Scott *et al.*, 1957); other investigators believe that the porous nature of the wall suffices (see also observations and discussion by Heide-Jørgensen, 1978). Miller (1983) observed the occurrence of canals in the cuticle of many fruits.

The way in which the epicuticular wax is extruded is still disputed. Some workers suggest that the exodesmata or special canals in the wall and cuticle are involved in this process (Hall, 1967a, b; Lyshede, 1978). Other workers, on examination of a wide range of plant cuticles, have not detected canals (cf. Martin and Juniper, 1970). Schieferstein and Loomis (1956, 1959) doubt the existence of true canals. They suggest that the wax is extruded only through the fragile cuticle of young leaves and that later extrusion is prevented by the thickening of the cuticular layer.

Deposits of salts in the form of crystal, e.g. in *Tamarix* and *Plumbago capensis* (Fig. 82, no. 2), of caoutchouc, e.g. *Eucalyptus*, or of oils and resins, sometimes occur on the surface of cuticle or within it. Deposits of silicon salts are found in the epidermal cell walls of many plants, as, for example, *Equisetum*, the Gramineae, many species of the Cyperaceae, the Palmae and certain species of the Moraceae, the Aristolochiaceae, the Compositae and the Magnoliaceae (Metcalfe and Chalk, 1983; Parry *et al.*, 1986).

Lignin is rarely found in the epidermal cell walls. When it is present it may be found in all the walls or only in the outer wall. Lignified epidermal walls are found in the leaves of the Cycadaceae, in the needles of conifers, in the rhizomes of the Gramineae, in the strips of epidermis above the bundles of sclerenchyma in the leaves of the Gramineae, Juncaceae, and Cyperaceae, in the leaves of certain species of *Eucalyptus* and *Quercus*, and in *Laurus nobilis* and *Nerium oleander*.

Parts of the epidermal cell walls of groups of cells or of single cells may become mucilaginous

in certain dicotyledonous families, such as the Moraceae, Malvaceae, Rhamnaceae, Thymelaeaceae, and Euphorbiaceae. In certain seeds, such as those of *Linum usitatissimum* (Fig. 298, no. 2) and species of *Alyssum*, the outer walls of the epidermal cells become mucilaginous. In the nectaries of certain plants the epidermal cells also become mucilaginous at the time of nectar secretion (Fahn, 1952).

Protoplast

The epidermal cells are usually highly vacuolated. However, those of *Eucalyptus papuana* leaves were observed to have only few small vacuoles (Hallam, 1967, cited in Martin and Juniper, 1970). It is generally assumed that the epidermal cells contain leucoplasts, but these could not be seen in several species studied, e.g. species of *Eucalyptus*, *Ligustrum*, and *Phaseolus* (Martin and Juniper, 1970). The leucoplasts of the epidermal cells of *Origanum dictamnus* possess membrane bound amorphous protein bodies (Bosabalidis, 1987). Chloroplasts are found in certain pteridophytes, hydrophytes, and in some shade plants. Numerous mitochondria, ER, spherosomes, and dictyosomes are usually present. Anthocyanins are found in the vacuoles of the epidermal cells of the petals of numerous flowers, in the leaves of *Zebrina pendula* and red cabbage, in the stems and petioles of *Ricinus*, and in different organs of many other plants. Tannins, mucilage, and crystals may be present in epidermal cells.

Epidermal cells with special structure or content

In certain pteridophytes, in the gymnosperms, in many species of the Gramineae, and certain dicotyledons, fibre-like epidermal cells are found.

In the Gramineae, between the elongated epidermal cells, i.e. *long cells*, above the veins, there are *short cells* which are of two types—*silica cells* and *cork cells*. The latter two types of cells often occur successively in pairs, throughout the length of the leaf. The fully developed silica cells contain silica bodies which are isotropic masses of silica in the centre of which are usually minute gran-

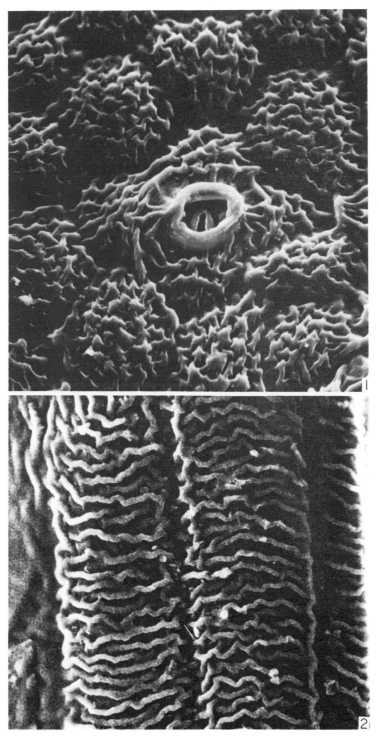

FIG. 81. Scannng electron micrographs of leaf surfaces showing shapes of epicuticular wax. 1, *Haworthia altilinea*; a sunken stoma is seen in the centre. × 525. 2, *Bulbine alooides*. × 1000. (Courtesy of D. F. Cutler.)

ules. In surface view the silica bodies may be circular, elliptic, dumb-bell, or saddle-shaped (Fig. 83, no. 5; Fig. 84, no. 1). The silicon was reported to be present, if at all, only in small amounts in the young silica cells, its accumulation being rapid in senescing cells (Lawton, 1980). The walls of the cork cells are impregnated with suberin and many of them contain solid organic substances. The above short cells sometimes bear papillae, setae, spines, or hairs. Metcalfe (1960) draws attention to the fact that the cork cells in many plants contain silica bodies, and that in certain grasses silica bodies also occur in some elongated cells. Silica bodies also occur in specialized epi-

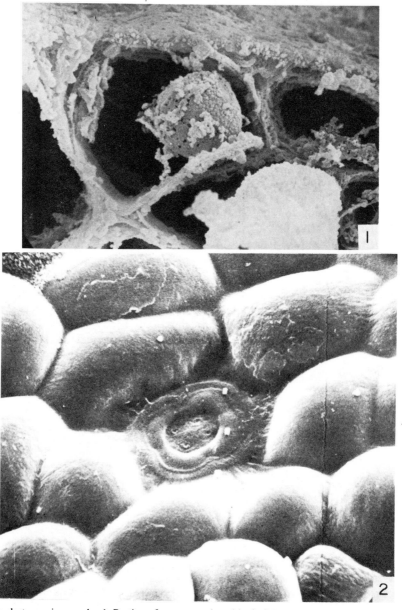

FIG. 82. Scanning electron micrographs. 1, Portion of a cross-sectioned leaf of *Pennisetum clandestinum* (Gramineae) showing a silica body in an epidermal cell. × 3900. 2, A chalk gland on the leaf surface of *Plumbago capensis*. × 1000. (No. 2 from Fahn, 1979.)

dermal cells of the Cyperaceae and some other monocotyledons (Metcalfe, 1963).

In the Gramineae and many other monocotyledons, with the exception of the Helobiae, *bulliform cells* are found in the epidermis. These cells are larger than the typical epidermal cells and they are thin-walled and have a large vacuole. The bulliform cells may constitute the entire adaxial epidermis of the leaf or they form isolated

parallel strips in the area between the veins. In a cross-section of the leaf these cells appear in a fan-like arrangement in which the central cell is the tallest. In certain plants bulliform cells are also found on the abaxial surface of the leaf. These cells are sometimes accompanied by similar mesophyll cells. Bulliform cells contain much water and are devoid, or nearly so, of chloroplasts. Their wall consists of cellulose and pectic

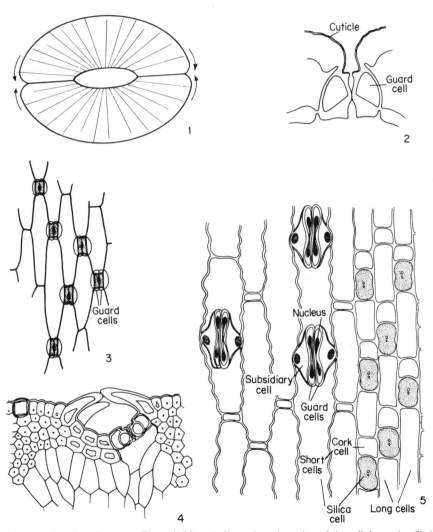

Fig. 83. 1, A schematic drawing of a stoma. The radial lines indicate the orientation of the cellulose microfibrils in the walls of the guard cells. The arrows indicate the direction of elastic stretching. 2, Cross-section of a stoma of *Allium cepa*. 3, Surface view of the epidermis of *Iris* in which the stomata can be seen to be arranged in longitudinal rows. The stomata are sunken and the rectangular depressions formed by the overarching of the neighbouring epidermal cells are shown as shaded areas in the diagram. 4, Portion of a cross-section of the leaf of *Xanthorrhoea gracilis* in which a sunken stoma overarched by trichomes can be seen. 5, Surface view of the epidermis of *Pennisetum clandestinum* showing epidermal cells characteristic to the Gramineae. (No. 2 adapted from Haberlandt, 1918.)

substances, and the outermost wall contains cutin and is covered by cuticle (Fig. 84, no. 2).

Different opinions exist as to the function of the bulliform cells. According to one opinion they function in the opening of the rolled leaf as present in the bud. According to a second view they bring about the rolling or unrolling of mature leaves as a result of their loss or uptake of water. An investigation carried out by Shields (1951) on twelve xerophytic grass species suggests

that, in the opening of the young leaves from the bud and in the hygrochastic movements (opening due to water absorption) of the mature leaves, elements other than the bulliform cells may actively participate. According to Metcalfe (1959), the bulliform cells often become filled with large masses of silica and their outer walls often become thick and cutinized.

Other specialized epidermal cells are the lithocysts (Fig. 77; Fig. 128, no. 2; see also Chapter 2),

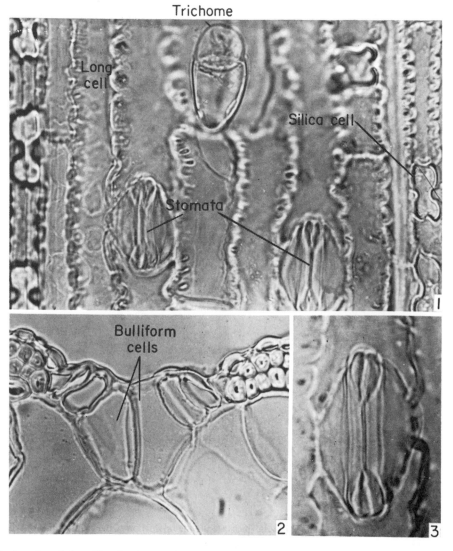

FIG. 84. 1, Surface view of the epidermis of a grass leaf (*Crypsis schoenoides*). × 780. 2, Portion of a cross-section of a grass leaf (*Dactyloctenium robecchi*) in which bulliform cells can be distinguished in the epidermis. × 140. 3, Surface view of a stoma of *C. schoenoides*. × 1300.

which are found in the Acanthaceae, Moraceae, Urticaceae, and Cucurbitaceae. In the Cruciferae *myrosin cells* are sometimes found in the epidermis. These cells are large secretory cells which contain the enzyme myrosin, and they stain red in the Millon test, or violet with orcein solution and concentrated hydrochloric acid.

Large idioblastic cells containing mucilage

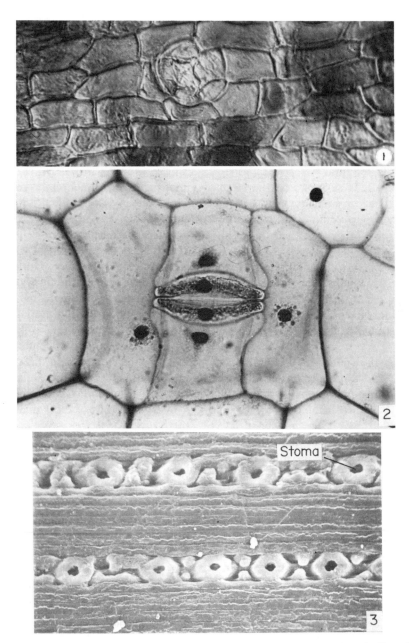

FIG. 85. 1, Micrograph of the cuticle of the fossil *Cupressinocladus* from which the shape of the epidermal cells can be determined. × 820. 2, Micrograph of a surface view of the abaxial epidermis of *Zebrina pendula* in which guard cells and subsidiary cells can be distinguished. The nuclei of the subsidiary cells are surrounded by leucoplasts. × 540. 3, Scanning electron micrograph of the surface of a *Pinus pinea* leaf, showing circular rims, formed by the neighbouring epidermal cells, above the stomata (Florin rings). × 220. (No. 1 from Chaloner and Lorch, 1960).

occur in the epidermis of Lythraceae and very large sac-like cells which intrude into the mesophyll were observed in *Cleome aspera* (Rajagopal and Ramayya, 1968). Secretory cavities may sometimes develop from epidermal cells, e.g. in *Psoralea* (de Bary, 1877). Long pipe-like cells filled with tannic substances appear in the epidermis of some *Saxifraga* species.

Stomata

The continuity of the epidermis is interrupted by minute openings. These are intercellular spaces each of which is limited by two specialized cells termed the *guard cells* (Fig. 85, no. 2). The guard cells, together with the opening between them, constitute the *stoma*. In many plants *subsidiary* or *accessory cells* can be distinguished. These cells differ morphologically from the typical epidermal cells and they constitute two or more cells bordering the guard cells to which they are apparently functionally connected. The stoma, together with the subsidiary cells if present, is termed the *stomatal apparatus* or the *stomatal complex* (Fig. 85, no. 2). The subsidiary cells usually develop from protodermal cells adjacent to the stomatal mother cells, but they may also develop from sister cells of the mother cell (de Bary, 1877). Based on the ontogenetic relationship between the guard and subsidiary cells, stomata can be divided into three types: *mesogenous* stomata, in which the subsidiary cells have a common origin with the guard cells; *perigenous* stomata, in which the subsidiary cells develop from protodermal cells adjacent to the stomatal mother cell; and *mesoperigenous* stomata, in which the cells surrounding the stoma are of dual origin, one or sometimes more of the subsidiary cells having a common origin with a guard cell, while the other or others have not (Pant, 1965).

On the basis of the relationship between the cells of the stomatal complex and the meristemoid (see Ontogeny of stomata, p. 170), and the orientation of the dividing walls, several additional terminologies and classifications of stomatal development have been proposed (Payne, 1979; Kidwai, 1981; Rasmussen, 1981, 1983).

Stomata are usually found on the aerial portions of the plant and especially on leaves, ordinary stems, and rhizomes. Stomata usually do not occur on roots. They have, however, been observed to appear on seedling roots of some plants, e.g. *Ceratonia siliqua* and *Pisum arvense* (Christodoulakis and Psaras, 1987; Tarkowska and Wacowska, 1988). Stomata are absent from the entire plant body of certain parasitic plants that lack chlorophyll, such as *Monotropa* and *Neottia*. In *Orobanche*, however, stomata are found on the stem although this genus is also devoid of chlorophyll. Stomata are present in some submerged water plants but not in most others. Stomata may be found on petals, staminal filaments (e.g. *Colchicum*), carpels, and seeds, but these stomata are usually non-functional.

In photosynthesizing leaves stomata may be found on both sides of the leaf or on one side only. In certain water plants with leaves that float on the surface of the water, e.g. *Nymphaea*, stomata are found only on the upper surface of the leaf which is exposed to the atmosphere. Stomata on land plants may be present either only on the lower (abaxial) surface, or on both surfaces. Stomata on the lower as well as on the upper (adaxial) leaf surface are more common in plants with a high photosynthetic capacity, plants living in full-sun environments (Mott *et al.*, 1982). The number of stomata per square millimetre is different in different plants; for example, on the abaxial leaf surface of *Oxalis acetosella* there are 37 stomata per square millimetre (Martin and Juniper, 1970); of *Pistacia palaestina*, 176; of *Pistacia lentiscus*, 255; of *Styrax officinalis*, 261; of *Quercus calliprinos*, 402; of *Olea europaea*, 545; and of *Quercus lyrata*, 1,198 (Meyer and Meola, 1978). Experiments with leaves of *Iris* grown under different light intensities showed that the stomatal frequency decreases with the decrease of light intensity (Pazourek, 1970).

In leaves with reticulate venation the stomata are distributed in no particular order, while in leaves in which the majority of veins are parallel, as in the Gramineae, the stomata are arranged in parallel rows.

The guard cells of the stomata may be level, sunken (Fig. 83, no. 2; Fig. 86, no. 2), or raised relative to the other epidermal cells. In cases where the epidermis is multiseriate, e.g. as in species of *Anabasis* and *Haloxylon* and in *Ficus elas-*

tica, the guard-cell mother cells are differentiated at that stage of development when the protoderm is as yet a single layer. During the process of further development the surrounding protodermal cells undergo several periclinal divisions resulting in the raising of the multiseriate epidermis above the level of the guard cells (Fig. 91, nos. 1–6; Fig. 92, nos. 1, 2) (Fahn and Dembo, 1964). In *Pinus* the cells encircling the stomata produce a raised ring termed *Florin ring* (Fig. 85, no. 3) (Yoshie and Sakai, 1985).

Below the stomata and directed inwards to the mesophyll are large intercellular spaces which are termed *substomatal chambers* (Fig. 142, no. 2).

The guard cells of most plants, except those of the Gramineae, Cyperaceae, and some others, are kidney-shaped in outline. The size of the aperture between the guard cells increases or decreases as a

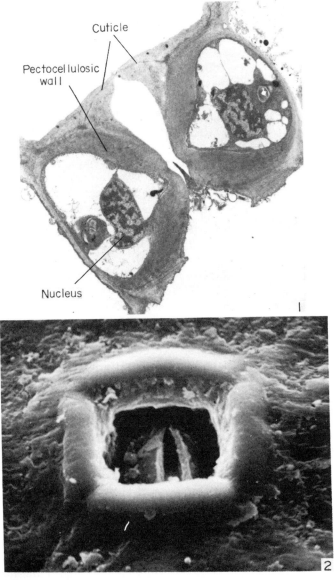

Cuticle

Pectocellulosic wall

Nucleus

FIG. 86. 1, Electron micrograph of a cross-section of a stoma of *Lathyrus latifolius.* × 4150. 2, A scanning electron micrograph of a stoma of *Aloë megalacantha* × *A. pubescens.* × 1775. (No. 2 courtesy of D. F. Cutler.)

result of turgor changes in the guard cells. The guard cells tend to accumulate potassium; apparently from the adjacent cells. A relationship between the extent of potassium accumulation and the extent of stomatal opening was found to exist (Dayanandan and Kaufman, 1975; Ting, 1982). In *Vicia faba* it was found that the average guard cell volumes were 4.8×10^{-12} litres per stomatal apparatus in open stomata and 2.6×10^{-12} litres for the ones in closed stomata (Humble and Raschke, 1971). As a result of unequal thickness and consistency of the walls, the guard cells undergo changes in form with the increase in volume. These changes cause the opening of the stoma. In most cases the thinnest wall is that closest to the subsidiary cells (Fig. 83, no. 2), and is termed the back or dorsal wall. With the increase of turgor this thin, flexible back wall expands more than the other walls, which are more resistant to pressure. As a result the shape of the guard cells changes in such a way that the stomatal pore opens. According to another hypothesis the radial arrangement of the cellulose microfibrils (radial micellation) in the guard cell walls (Fig. 83, no. 1) plays a major role in stomatal movement (Ziegenspeck, 1938, 1944; Aylor *et al.,* 1973; Palevitz and Hepler, 1976). In guard cells with very thick periclinal walls, e.g. in *Araucaria, Agathis,* and *Anabasis articulata,* the additional wall layers consist of numerous lamellae in which the cellulose microfibrils alternate in their orientation (Vassilyev and Vassilyev, 1976; Gedalovich and Fahn, 1981). For theories of the mechanism of stomatal movement, see Levitt (1974) and Raschke (1975, 1979; Wu *et al.,* 1985).

Protuberances of the guard-cell wall may be present above, or both above and below, the stomatal aperture (Fig. 86, no. 1; Fig. 87, no. 7). In cross-section these protuberances appear as horn-shaped ledges. The outer ledge delimits the front cavity above the aperture, and the inner ledge delimits the back cavity which abuts on the substomatal chamber. In some *Eucalyptus* species the cuticle forms special ridges termed *stomatal bars* (Carr and Carr, 1978, 1979).

The guard cells in the Gramineae and Cyperaceae are different in form from that described above. They are elongated and dumb-bell shaped. The ends of these guard cells are expanded and thin-walled, while the middle portions· are elon-

gated and thick-walled and the cell lumen is narrow (Fig. 83, no. 5). As a result of turgor increase in this type of guard cell, the expanded ends swell and so push apart the middle elongated portions of the cells. Because of the above structure the nuclei in the guard cells of grasses appear as two elipses connected by a narrow thread. According to Flint and Moreland (1946) the two parts of the nucleus may become completely separated. In the stomata of the Gramineae, as seen with the aid of the electron microscope, the wall between the ends of the guard cells is incomplete and the protoplasts are confluent (Brown and Johnson, 1962; Pickett-Heaps, 1967; Ziegler *et al.,* 1974). Perforations between the guard cells at their polar ends were observed, with the aid of a light microscope, in some species of *Dicranopygium* (Cyclanthaceae) (Wilder, 1985).

The chemical composition of the guard cell wall is the same as that of the ordinary epidermal cells of the same plant. They are usually covered by cuticle which generally continues on that wall that faces the aperture and which also reaches the cells abutting on the substomatal chamber. In *Citrus,* cuticle is absent from the cell wall facing the stomatal aperture (Turrell, 1947).

Apart from the types of guard-cell structure described above many other structural variations exist in the mono- and dicotyledons, e.g. in species of *Haloxylon* (Fig. 88, nos. 1–3) and *Anabasis,* in the Palmae (Fig. 87, nos. 5–7) and in others (Tomlinson, 1961; Fahn and Dembo, 1964; Napp-Zinn, 1966, 1973; Gedalovich and Fahn, 1983).

Guard cells of many vascular cryptogams, gymnosperms (Fig. 87, nos. 1–4) and some angiosperms possess lignified wall thickenings (Kaufman, 1927; Florin, 1931, 1933; Boulter, 1970). Kaufman suggested that the lignified wall areas are related to the opening mechanism of the stomata.

Electron microscope studies have shown that the guard cells contain numerous mitochondria, elements of ER, dictyosomes, and vacuoles of various sizes. Although the plastids contain few grana and frets, the total amount of photosynthesis occurring in the guard-cell plastids is, according to Thomson and de Journett (1970), sufficient to support the function of these cells. However, in some investigated species of *Paphio-*

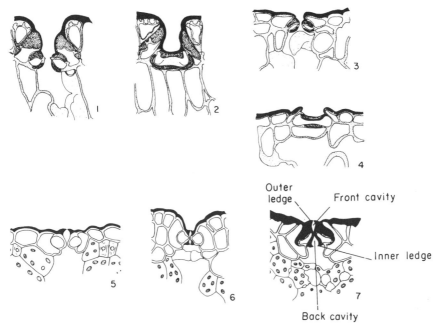

FIG. 87. Various types of stomata. 1 and 2, *Abies pinsapo.* 3 and 4, *Juniperus chinensis.* 5, *Synechanthus fibrosus.* × 280. 6, *Corypha talieri.* × 175. 7, *Chamaerops humilis.* × 280. Nos. 2 and 4, longitudinal sections; the rest cross-sections. Cuticle represented by solid black, lignified areas by dark shading. (Nos. 1–4 adapted from Florin, 1931; nos. 5–7 adapted from Tomlinson, 1961.)

pedilum (Orchidaceae) internal lamellar structures and chlorophyll are completely lacking in the guard-cell plastids although the stomata are functional (Rutter and Willmer, 1979). Starch grains occur in the plastids. Complete plasmodesmatal connections between guard cells and subsidiary cells were not observed to occur in mature stomata of most plants investigated with the aid of the electron microscope. However, in developing stomatal complexes, plasmodesmata were found between guard cells and subsidiary cells and also between sister guard cells (Peterson and Hambleton, 1978; Willmer and Sexton, 1979). In mature stomata of *Phaseolus vulgaris* Willmer and Sexton (1979) reported that aborted incomplete plasmodesmata occasionally occur in the common guard–subsidiary cell wall. The plasmalemma shows numerous invaginations. Microinjection of Lucifer yellow into stomatal cells, showed that the dye does not move out of injected mature guard cells, and it does not enter them when adjacent epidermal or subepidermal cells are injected. The dye rapidly moves into and out of young guard cells (Palevitz and Hepler, 1985).

According to the relation of stomata to the neighbouring epidermal cells, the stomata have been classified into various types. This classification is separate from the one based on development (see p. 164). Although different types may occur in one and the same family (Metcalfe and Chalk, 1950), or even in a leaf of the same species (Pant and Banerji, 1965; Pant and Kidwai, 1967), the structure of the stomatal apparatus can be of use in taxonomic studies.

Morphologically, four main types of stomata have been distinguished in the dicotyledons on the basis of the arrangement of the epidermal cells neighbouring the guard cells (Metcalfe and Chalk, 1950).

1. *Anomocytic type* (ranunculaceous) (Fig. 89, no. 4) in which the guard cells are surrounded by a certain number of cells that do not differ in size and shape from the other epidermal cells. This type is common in the Ranunculaceae, Geraniaceae, Capparidaceae, Cucurbitaceae, Malvaceae, Scrophulariaceae, Tamaricaceae, and Papaveraceae.

2. *Anisocytic type* (cruciferous) (Fig. 89, no. 2) in

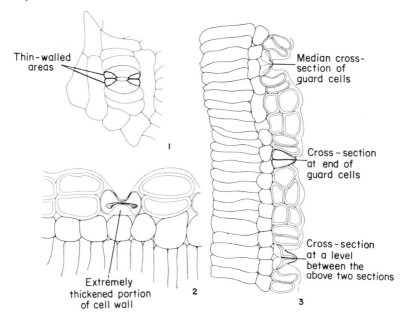

FIG. 88. Structure of stomata of *Haloxylon articulatum*. 1, Surface view showing thin-walled areas at the ends of the guard cells. 2, Portion of a cross-section of the stem, showing a guard cell in longitudinal section. The cell-lumen is dumb-bell shaped; nucleus heavily stippled. 3, Portion of longitudinal section of stem showing the biseriate epidermis and sunken stomata, sectioned transversely at various levels. × 500.

which the guard cells are surrounded by three unequally sized subsidiary cells. This type is common in the Cruciferae, in *Nicotiana*, *Solanum*, *Sedum*, and others.

3. *Paracytic type* (rubiaceous) (Fig. 89, no. 1) in which each guard cell is accompanied by one or more subsidiary cells, the longitudinal axes of which are parallel to that of the guard cells and aperture. This type is common in the Rubiaceae, Magnoliaceae, most species of the Convolvulaceae and Mimosaceae, some genera of the Papilionaceae such as *Ononis*, *Arachis*, *Phaseolus*, and *Psoralea*, and various species of other families.

4. *Diacytic type* (caryophyllaceous) (Fig. 89, no. 3) in which each stoma is surrounded by two subsidiary cells, the common wall of which is at right angles to the longitudinal axis of the stoma. This type is common in the Caryophyllaceae, Acanthaceae, and others.

5. *Actinocytic type* (Fig. 89, no. 5) in which the stomata are surrounded by a circle of radiating cells. This type is rather uncommon.

It should be mentioned that modifications of these types and additional types may occur in species of various families (see Payne, 1970; Wilkinson, 1979). Stomata of more than one type may sometimes occur together on the same organ (Metcalfe and Chalk, 1950; Rao and Ramayya, 1977; Kidwai, 1981; Baranova, 1987).

In the monocotyledons, Stebbins and Khush (1961) have distinguished the following types of stomatal complexes:

1. The guard cells are surrounded by four to six subsidiary cells (Fig. 90, no. 8). This type is common in many species of the Araceae, Commelinaceae, Musaceae, Strelitziaceae, Cannaceae, and Zingiberaceae.

2. The guard cells are surrounded by four to six subsidiary cells of which two are roundish, smaller than the rest, and are situated at the ends of the guard cells (Fig. 90, no. 10). This type is found in many species of the Palmae, Pandanaceae, and Cyclanthaceae.

3. The guard cells are accompanied laterally by two subsidiary cells—one on each side (Fig. 90,

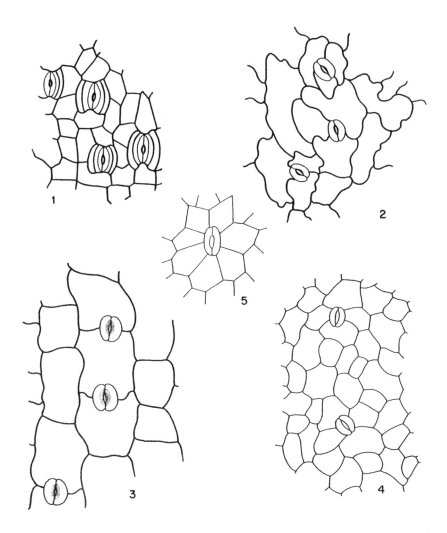

FIG. 89. Different types of arrangement, as seen in surface view of the leaf, of the subsidiary cells relative to the stoma. 1, *Acacia*; rubiaceous or paracytic type. 2, *Brassica*; cruciferous or anisocytic type. 3, *Dianthus*; caryophyllaceous or diacytic type. 4, *Pelargonium*; ranunculaceous or anomocytic type 5, Actinocytic type.

no. 9). This type is found in many species of the Pontederiaceae, Flagellariaceae, Butomales, Alismatales, Potamogetonales, Cyperales, Xyridales, Juncales, Graminales, and others.

4. The guard cells are not associated with any subsidiary cells (Fig. 83, no. 3). This type can be seen in many species of the Liliales (with the exception of the Pontederiaceae), the Dioscoreales, Amaryllidales, Iridales, Orchidales, and others.

The above monocotyledons have been modified by Paliwal (1969), who suggested a special terminology for the five main types which he distinguished.

Van Cotthem (1970), who stated that "the type of adult stomata not only has a diagnostic value, but can be used in many cases as an indicator of natural taxonomic affinity too", has distinguished 15 main types of stomata in the ferns, gymnosperms, and angiosperms on the basis of surface appearance only.

Ontogeny of stomata

The stomata develop from the protoderm (Fig. 90, nos. 1–7; Fig. 91, nos. 1–6; Fig. 92, nos. 1, 2). The guard mother cell differentiates from a *meristemoid* (Fig. 90, no. 3) which is usually the smaller of the two cells that result from an unequal division of a protodermal cell (Bünning and Biegert, 1953; Bonnett, 1961; and others) (see p. 164). The guard mother cell divides to form two cells which differentiate into the guard cells. At first these cells are small and have no special shape, but, as they develop, they enlarge and become characteristically shaped. During their development the middle lamella between the two guard cells swells and becomes lens-shaped shortly before the time when it disintegrates to form the stomatal aperture (Ziegenspeck, 1944; Galatis and Mitrakos, 1980). Stevens and Martin (1978) suggested that, whilst pore formation is initiated enzymatically, the separation of the two guard cells is brought about by osmotic force derived from starch hydrolysis. Even in those cases in which the mature guard cells are sunken or raised relative to the ordinary cells of the uniseriate epidermis, the guard cell mother cells and the guard cells are level with the other epidermal cells immediately after their formation. The sinking and the raising is brought about during the maturation of the guard cells. The development of stomata in the leaf continues for a relatively long period during the growth of the leaf. Arrested stomatal development and abnormal stomata with a single guard cell have been reported (Pant and Kidwai, 1967).

The early stages of development of the stomata complex in grass leaves and some other monocotyledons were described by Stebbins and Jain (1960), Stebbins and Shah (1960), and in wheat by Pickett-Heaps and Northcote (1966). The first step in the differentiation process is the asymmetric division of certain protodermal cells. Before the cell divides, the nucleus becomes displaced in one direction and the vacuoles occupy the other end of the cell. The nucleus then divides. One of the daughter nuclei which is closer to the vacuoles becomes larger and more weakly stain-

ing than the other at the denser end of the dividing cell. The smaller distal nucleus divides later to form the two guard cells, but before doing so it induces division of neighbouring epidermal cells. In these cells also there is denser cytoplasm adjacent to the guard mother cell, and the induced divisions are also asymmetrical. The daughter nucleus in the centre of the cell becomes larger and more weakly staining than that near the guard mother cell which becomes the nucleus of a subsidiary cell (Fig. 90, no. 11). After the formation of the subsidiary cells the guard mother cell divides to form the guard cells.

It has been observed (Pickett-Heaps and Northcote, 1966) that prior to each of the asymmetric divisions a band of microtubules appears in the peripheral cytoplasm. This band indicates the location on the wall of the dividing cell where the cell plate will join it at the final stage of division. Thus the future plane of cell division is indicated by these microtubules at preprophase.

Stebbins et al. (1967), by using a substance that interferes with spindle formation or by removing the sheaths from culm leaves of barley plants, obtained stomata lacking subsidiary cells and a change in orientation of the division of some guard cells.

On the basis of the order of the appearance of the stomata on the photosynthesizing organ, two main types of development can be distinguished: (1) that in which the stomata appear gradually in a basipetal sequence, i.e. from the tip of the organ to its base, as is the case in leaves with parallel venation and in the internodes of the articulated species of the Chenopodiaceae; (2) that in which there is no regularity in the appearance of the stomata in the various regions of the growing organ, as is the case in leaves with reticulate venation.

Stomata are distributed at more or less equal distances, the extent of which is specific to the plant species and leaf side. Several theories were suggested to explain the presence of the stomata-free regions surrounding each stoma: (1) inhibition of additional stomata by already differentiated ones; (2) formation of the stoma together with the surrounding cells as part of the same developmental pattern; (3) induction of stomata pattern by the pattern of the underlying tissue, i.e.

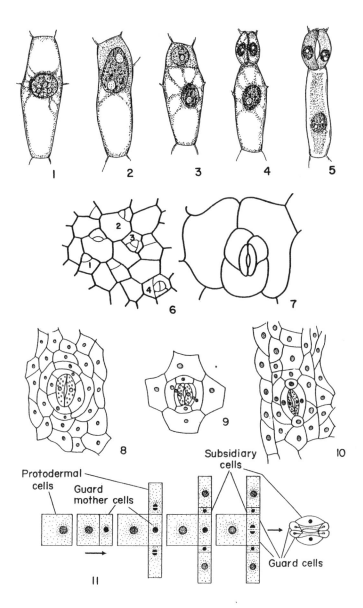

FIG. 90. 1–5, Ontogeny of stomata of *Allium cepa*; 1 and 2, Elongated epidermal cells before unequal division; 3, Showing that the smaller cell (the meristemoid) resulting from this division is richer in protoplasm. It is this cell that gives rise, by a longitudinal division, to the two guard cells which in no. 4 are not yet separated by the stomatal aperture. In no. 5 the stomatal aperture has developed. 6 and 7, Epidermis of *Sedum pubescens*. 6, Early stages in the development of stomata up to the stage where three subsidiary cells and two guard cells, but no aperture, can be distinguished. Numerals indicate ontogenetic stages. 7, Portion of epidermis with mature stoma. 8–10, Types of monocotyledonous stomata. 8, *Strelitzia nicolei*. 9, *Commelina communis*. 10, *Pandanus haerbachii*. 11, Diagram showing the manner of formation of the guard cells and subsidiary cells in *Hordeum*. The arrows indicate the direction of the successive stages. (Nos. 1–5 adapted from Bünning and Biegert, 1953; nos. 8–10 adapted from Stebbins and Khush, 1961; no. 11 adapted from Stebbins and Jain, 1960.)

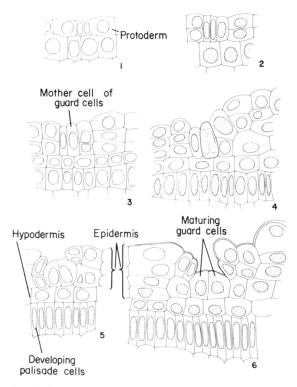

FIG. 91. Ontogeny of the stomata and multiseriate epidermis in *Anabasis articulata*. Base of internode on the left. × 500.

the mesophyll. Sachs (1978, 1979, 1984) suggested that the role of mutual inhibition between developing stomata and of inductive effects of sub-epidermal tissues, is of minor importance in the establishment of stomatal patterns.

Epidermal appendages

All unicellular and multicellular appendages of the epidermis are designated by the term *trichome*. More massive structures, such as warts, some secretory structures, and spines (e.g. the thorn of *Rosa*), which consist of epidermal as well as sub-epidermal tissues, are termed *emergences* (see Ramayya, 1964). In some cases it is difficult to distinguish clearly between these two types of appendage without an ontogenetic study. For practical reasons, the appendages discussed below, either of epidermal or epidermal and sub-

epidermal origin, will be termed trichomes. The cells of the trichomes may develop secondary walls, which in some cases may become lignified. Some trichomes may lose their living protoplasts.

The use of trichomes in taxonomy is well known. Some families can be easily identified by the presence of a particular type or types of hair. In other cases the hairs are important in the classification of genera and species and in analysing interspecific hybrids (Metcalfe and Chalk, 1950; Metcalfe, 1963; Rollins, 1944; Carlquist, 1961; Hummel and Staesche, 1962).

Trichomes can be classified into several types

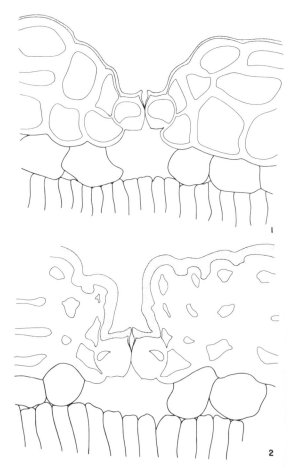

FIG. 92. Median cross-sections of maturing stomata of *Anabasis articulata*. 1, × 450. 2, × 550.

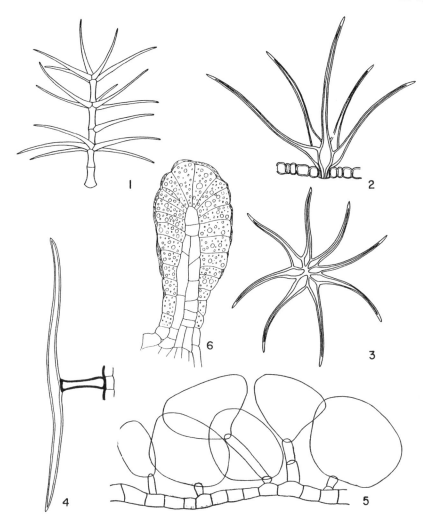

FIG. 93. Different types of trichomes. 1, Candelabrum-like, branched, multicellular trichome from the leaf of *Verbascum*. 2, Stellate multicellular trichome from the leaf of *Styrax officinalis*; lateral view showing how the trichome arises from between ordinary epidermal cells. 3, As in no. 2, but surface view. 4, A T-shaped trichome of *Corokia buddleioides*. 5, Vesiculate hairs of *Atriplex portulacoides*. 6, Colleter on a stipule of *Viola*. (No. 6 adapted from Strasburger, 1911.)

(Solereder, 1908; Foster, 1950; Metcalfe and Chalk, 1950, 1979; Uphof, 1962). A glossary of the very rich terminology of plant trichomes has been compiled by Payne (1978).

1. Non-glandular trichomes

(a) Simple unicellular or multicellular uniseriate, non-flattened hairs, are common, for example, in the Lauraceae, Moraceae, *Triticum, Hordeum, Pelargonium*, and *Gossypium*. In *Gossypium* the fibres used in commerce constitute unicellular epidermal hairs, which may be up to 6 cm long and which are located on the seed coat. This group includes papillae and bladders, which are also known as *vesicular hairs* (e.g. in Crassulaceae).

(b) Squamiform hairs which are usually flattened

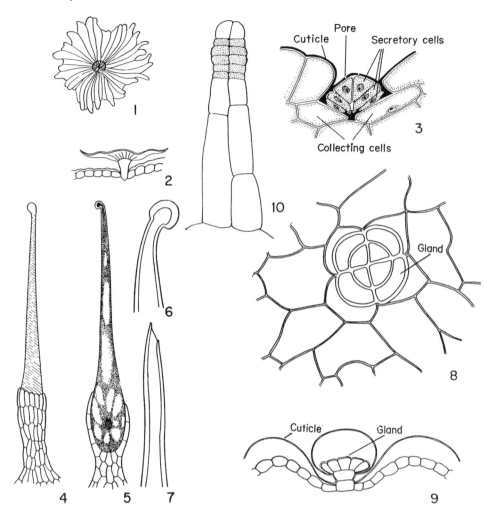

FIG. 94. Trichomes and epidermal glands. 1 and 2, Peltate hair of *Olea europaea*. 1, Surface view showing in the centre the stalk cell around which the "shield" develops. 2, Lateral view. 3, Portion of cross-section of the leaf of *Tamarix* showing a multicellular salt gland. 4–7, Stinging hair of *Urtica dioica*. 4, As seen under a microscope with the focus on the surface of the trichome. 5, As in no. 4, but with focus on centre of the trichome. 6, Intact tip. 7, Trichome with broken tip. 8, Chalk gland of *Plumbago capensis* in surface view of the epidermis. 9, Portion of a cross-section of the leaf of *Thymus capitatus* showing a secretory gland. 10. A stalked glandular hair of *Inula viscosa* (photosynthesizing cells are dotted). (Nos. 4–7 adapted from Troll, 1948.)

and multicellular. These may be sessile and are then termed scales, or stalked, and known as peltate hairs, e.g. in *Olea* (Fig. 94, nos. 1, 2) or dendroid or dendritic, e.g. in Cruciferae (Rollins and Banerjee, 1975).

(c) Multicellular hairs which may be stellate, e.g. in *Styrax* (Fig. 93, nos. 2, 3), branched candelabrum-like, e.g. in *Platanus* and *Verbascum*

(Fig. 93, no. 1).

(d) T-shaped hairs (Fig. 93, no. 4), consisting of long, more or less horizontally oriented terminal cell and a stalk of one cell or of a row of a few cells. Such trichomes occur, for instance, in *Astragalus canadensis, Corokia buddleioides* and *Olearia rotundifolia* (Fahn, 1986).

Many non-glandular trichomes of xeromorphic

plants possess at their base endodermal cells which prevent apparently apoplastic water leakage (Fahn, 1986).

In some species the hairs may show movements. This may be brought about in two ways: either by hygroscopic mechanisms, i.e. by the differential swelling and shrinking of the cell walls e.g. as on the seed of *Tamarix*); or by the action of living cells which may comprise the hair itself or which may be present only at the base of the hair or close to it (Uphof, 1962).

2. Glandular trichomes

Glandular trichomes, often called glands, are involved in the secretion of various substances, e.g. salt solution, sugar solution (nectar), lipids, and gums (polysaccharides).

Almost all glandular trichomes possess *endodermal cells* below the secretory cells (Fig. 95). The endodermal cells prevent the secreted solution from flowing back into the plant through the apoplast. The secretory cells usually contain numerous mitochondria. The frequency of other organelles varies according to the material secreted (Fahn, 1979, 1988; Joel, 1986).

Salt-secreting trichomes. (1) Bladder-like hairs consisting of a large secretory cell on top of a narrow stalk consisting of one or sometimes a few cells and a basal cell as seen in *Atriplex* (Fig. 93, no. 5). The salt is here secreted by the cytoplasm into the large vacuole (Osmond *et al.*, 1969). The secretory cell dries out with the ageing of the leaf, and the salt content remains on the leaf surface as a white, powdery layer. (2) Multicellular glands consisting of several secretory cells and basal collecting cells. In some cases a stalk cell may also be present. To this group belong the chalk glands of *Plumbago capensis* (Fig. 94, no. 8) and salt glands of *Limonium, Avicennia* and *Tamarix* (Fig. 94, no. 3; Fig. 95, no. 5; Fig. 96, nos. 1–3). In these glands the cytoplasm is dense, rich in mitochondria, ER and dictyosomes and has many vesicular structures. The salt solution is actively secreted onto the surface of the secretory cells (Volkens, 1887; Brunner, 1909; Shimony and Fahn, 1968; Shimony *et al.*, 1973). Pores are present in the cuticle covering the secretory cells of these glands.

Trichomes that secrete an aqueous solution containing some inorganic and organic substances were termed *trichome-hydathodes.* Such glandular trichomes occurring in young leaves and stems of *Cicer arietinum* (Fig. 95, no. 2) consist of a uniseriate stalk and a multicelled oval head. Between the cellulose layer of the wall and the cuticle at the top of the gland a subcuticular space is formed during secretion. When pressure reaches a certain value, pores in the cuticle open, and droplets appear on the surface. The occurrence of many mitochondria in the cells of these trichomes indicates a process of active secretion (Schnepf, 1965). Perrin (1970), who worked on the ultrastructure of the trichome-hydathodes of *Phaseolus multiflorus*, came to a similar conclusion concerning the active nature of the secretion. As these trichomes secrete actively, the term trichome-hydathode may not be quite appropriate.

Nectar-secreting trichomes, e.g. in the calyx of *Abutilon*, in the corolla of *Lonicera japonica* and *Tropaeolum majus* (Fig. 267; Fig. 268) and on the stipules of *Vicia faba*. The cytoplasm at the stage of secretion is very dense and especially rich in ER. It has been suggested that vesicles mainly of ER origin are involved in the secretion of the nectar. In some plants dictyosomes take part also in nectar secretion (see also Nectaries in Chapter 19).

Epidermal cells not in the form of hair, may also become glandular. Such a secretory epidermis occurs on special outgrowths (Fig. 97, no. 1), on the teeth of leaf margins (*Prunus amygdalus, Ailanthus altissima*) or on different parts of the floral organs. Structure of nectaries and mode of secretion will be described in more detail in Chapter 19.

Mucilage-secreting glands, for example, on the membraneous sheath arising from the leaf base (ochrea) of *Rumex* (Fig. 95, no. 3) and *Rheum*. The secreted mucilage is mainly a polysaccharide. Golgi vesicles are involved in its secretion. The extruded mucilage accumulates in spaces between the cell wall and the cuticle. The cuticle may rupture and the mucilage thus reaches the surface. Pores filled with mucilage were observed in the cuticle (Schnepf, 1969).

Glands of carnivorous plants (Fig. 95, nos. 4, 6, 7). The capture organs of these plants are usually modified leaves. The prey, insects, and

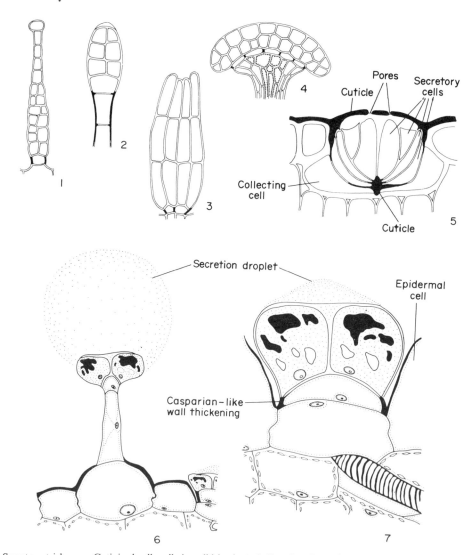

Fig. 95. Secretory trichomes. Cutinized cell walls in solid back. 1–4, Showing Casparian-like wall thickenings (solid black) in basal stalk cells. 1, Nectar-secreting trichome of the calyx of *Abutilon* sp. 2, Hydathode-trichome of *Cicer arietinum*. 3, Mucilage-secreting trichome of *Rumex maximus*. 4, A stalked gland of *Drosophyllum*. 5, Salt gland of *Limonium latifolium*. 6, A stalked gland, and 7, a sessile gland of *Pinguicula grandiflora*. (Nos. 1–5, adapted from Schnepf, 1969; nos. 6 and 7 adapted from Heslop-Harrison and Knox, 1971.)

often other small animals, are attracted to the capture organs by attractive colours, odour, and nectar secretion. The animals are captured by the organs in various ways; by mucilages secreted by special trichomes (*Drosera*, *Pinguicula*); by rapid closure of the two halves of the leaves of *Dionaea*, by attraction to pitfalls (*Nepenthes*, *Sarracenia*); or by a complicated trap ("mousetrap") as in *Utricularia* (see Lloyd, 1942; Schmucker and Linnemann, 1959; Juniper *et al.*, 1989). Nutritional benefit is derived from the captured prey as a result of secretion of proteolytic enzymes by the capture organs and resorption of the digestive products. This is done by the aid of glandular trichomes (Gilchrist and Juniper, 1974). Much research has been done on these glands both from the physiological and the anatomical points of view. Many studies with the aid of the electron microscope have

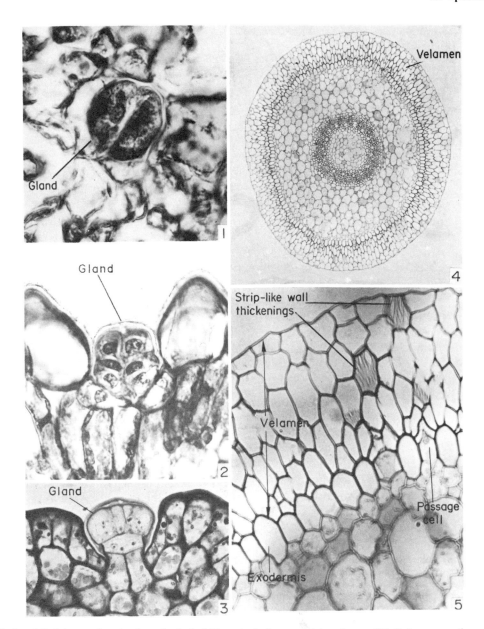

FIG. 96. 1 and 2, Salt-secreting gland on the leaf of *Tamarix*. 1, In tangential section. ×580. 2, In cross-section. ×670. 3, Portion of a leaf cross-section of *Avicennia marina*, showing a salt-secreting gland. ×640. 4, Cross-section of the aerial root of an epiphytic orchid, in which it is possible to distinguish the velamen. ×40. 5, As in no. 4, but outer portion, consisting mainly of velamen; in some of the cells the walls, with strip-like thickenings, which strengthen the cells, can be seen. ×225.

been carried out (see Lüttge, 1967; Schwab *et al.*, 1969; Juniper *et al.*, 1989).

Heslop-Harrison and Knox (1971) investigated *Pinguicula* in which two types of glands are present—stalked and sessile (Fig. 95, nos. 6, 7). The stalked glands secrete a muco-polysaccharide and are concerned with capture of the prey. The sessile glands are concerned both with the secretion of proteolytic enzymes and resorption of the digestion products.

Trichomes secreting lipophilic material (see Kisser, 1958; Fahn, 1979). (1) Glandular hairs, e.g.

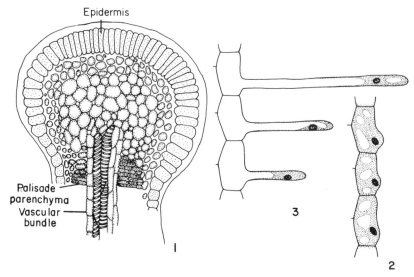

FIG. 97. 1, Longitudinal section through a gland present on the petiole of *Prunus amygdalus* showing a palisade-like secretory epidermis. 2 and 3, Development of root hairs. 2, Epidermal cells showing the beginning of a protuberance at the apical end of the cell. 3, Maturing root hairs that develop from the above protuberances as the cells become further distant from the root apex. (Nos. 2 and 3 adapted from Troll, 1948.)

the essential oil-secreting glands of the Labiatae (Fig. 94, no. 9). These consist of a basal cell, a uniseriate stalk one or several cells long, and a head of one or several secretory cells. The cell wall around the secretory cells is differentiated into a cuticle, cuticular layer, a pectic layer, and a cellulotic layer. Under the electron microscope the vacuoles are seen to contain osmiophilic substances. At the stage of secretion there is an increase in the amount of dictyosomes and ER, the cisternae of which become dilated. The cytoplasm retracts irregularly from the cell wall. The cell wall becomes split between the pectic and cuticular layers, resulting in the characteristic subcuticular space. During the process of secretion the vacuoles lose their content. In old cells they are empty, whereas the extra plasmatic space is much enlarged. The secreted substance collects in the subcuticular space. In some cases this space remains small apparently due to the presence of pores in the cuticle. In others the secretion process terminates with the death of the cells (Amelunxen, 1964, 1965). (2) Glandular shaggy hairs. These hairs consists of multiseriate stalks and heads, e.g. in *Cleome* (Ramayya and Gopalacharyulu, 1968; Amelunxen and Arbeiter,

1969). Electron microscopic studies of *Cleome spinosa* showed that the essential oil at first appears as many small droplets. The oil, which is only formed in small quantities, remains in the cytoplasm.

Some Compositae species bear multicellular, biseriate glandular trichomes (Werker and Fahn, 1981; Ascensao and Pais, 1987). In *Inula viscosa* there are sessile and stalked glandular trichomes. Both types of trichome possess a secretory head, consisting of three types of cells: one pair of summit cells, one pair of cells below the summit cells, and three pairs of photosynthetic cells (Fig. 94, no. 10). All head cells secrete lipophilic substances, polysaccharides and protein (Werker and Fahn, 1981).

Colleters. Trichomes secreting sticky substances (Hanstein, 1868; Kisser, 1958; Ramayya and Bahadur, 1968; Lersten, 1974). Glandular trichomes consisting usually of a multicellular head and a stalk which may sometimes be lacking (Fig. 93, no. 6). All external and often also the neighbouring epidermal cells have the ability to secrete. The secreted sticky substance, frequently a mixture of terpenes and mucilage, reaches the gland surface usually by rapid rupture of the cuticle. The secretion continues for a long period. Colleters appear mostly on bud scales (e.g. *Syringa, Rosa, Aesculus, Alnus, Coffea*) but may also be found on other organs. In *Ononis spinosa* the cuticle apparently does not rupture during secretion.

Stinging hairs. The stinging hairs of *Urtica* are highly specialized glandular trichomes. They consist of a single, long cell which has a broad, bladder-like base and narrow, needle-like upper part (Fig. 94, nos. 4–7). The broad base is surrounded by epidermal cells which are raised above the level of the other epidermal cells. The wall of the distal needle-like part of the secreting cell is impregnated with silica at the tip and with calcium somewhat lower. The very tip is spherical and breaks off, along a predetermined line, when the hair is touched. The broken tip resembles the tip of a syringe and thus easily penetrates the skin, into which the poisonous, irritating cell contents (histamine and acetylcholine) are injected.

Stinging emergences also occur in other plants, e.g. in *Cnidoscolus* (Euphorbiaceae) and *Loasa* (Loasaceae). Very little research work has been done on stinging hairs although they present many intriguing unsolved problems (Thurston and Lersten, 1969).

In many secretory trichomes one or both of the two following types of wall structure are present: (1) an inner wall layer, consisting of protuberances of varying and complex form (Fig. 269), which may have a function in transport of the secreted substances through the wall to the gland surface, and (2) Casparian-like wall thickenings, impregnated with cutin and/or suberin, in the side walls surrounding each of the basal stalk cells of the gland (Fig. 95, nos. 1–5, 7). These endodermis-like cells may force the secreted solution to migrate outwards through the unimpregnated walls of the secretory cells by preventing the flow back of the solution through the apoplast (cf. Schrödter, 1926; Schnepf, 1969; Shimony *et al.*, 1973).

In the terpene-secreting trichomes the cuticle at the top of the gland is raised. A subcuticular space is thus formed. Since the Casparian-like wall thickenings and the cuticle are tightly connected, the detachment of the cuticle is enabled at the upper part of the gland only (Amelunxen, 1965).

3. Root hairs

Root hairs are tubular elongations of epidermal cells. They are branched in only very few plants. Root hairs are 80–1500 μm long and 5–17 μm in diameter (Dittmer, 1949). The root hairs have large vacuoles and they are usually thin-walled. On the aerial, adventitious roots of *Kalanchoë fedtschenkoi* multicellular root hairs have been found (Popham and Henry, 1955).

Electron microscopic studies (Leach *et al.*, 1963) have suggested that the outer epidermal wall of the root in the region of the root hairs consists of two layers—a thin inner layer which stains densely, and a wide outer layer which is less dense. The wall of the developing hair appears to be a continuation of the inner layer only.

Root hairs begin to form beyond the meristematic zone of the young roots in regions where the epidermal cells can still elongate. The root hairs usually first appear as small protuberances near the apical end of the epidermal cell. If the epidermal cell continues to elongate after the appearance of the protuberance, the root hair is found somewhat distant from the apical end of the mature epidermal cell (Fig. 97, nos. 2, 3). Root hairs elongate at their tips where the wall is thinner, softer, and more delicate. Sievers (1963a, b) has shown that at the tip of the root hairs there is a concentration of cytoplasmic organelles (ER, ribosomes, mitochondria), especially of many active dictyosomes. The nucleus is usually located close to the growing tip of the root hair. The epidermal cells which give rise to root hairs elongate less than the other epidermal cells. (See Cormack, 1949, 1962, for further anatomical and physiological details.)

In some plants only certain of the root epidermal cells, termed *trichoblasts* or *piliferous cells*, can produce root hairs. These are small cells which result from unequal divisions of epidermal cells.

Cutter and Feldman (1970a, b) have studied the trichoblasts of *Hydrocharis*. They have shown that during development the trichoblasts, their nuclei, and nucleoli increase in volume. The trichoblasts contain more nucleohistone, total protein, RNA, and nuclear DNA than the neighbouring cells. The trichoblasts do not divide, and their nuclei become increasingly polyploid with distance from the root tip. This is a result of the delay in maturation of the developing root hair. According to Cutter and Feldman the delay in maturation is probably an essential factor in the differentiation of root hairs.

Root hairs are usually viable for only a short period, generally only a few days. With the death of the root hairs and if the cells are not sloughed, the walls of the epidermal cells become suberized and lignified. In some plants root hairs have been found that remain permanently on the plant. The walls of such root hairs become thick and then apparently lose their ability to take up water from the soil (Artschwager, 1925; Cormack, 1949).

Both ontogenetically and functionally the epidermis may be considered as a separate tissue. The epidermis develops from the protoderm by continuous anticlinal cell divisions. As it is a compact tissue, devoid of intercellular spaces and covered with a cuticle, the epidermis provides protection to all those plant organs that consist entirely, or almost so, of primary tissues.

The specialized cells of the epidermis are of great interest as they usually are of characteristic structure, ontogenetic development, and function. They are also of great value in the study of taxonomic and evolutionary problems. Specialized cells, such as the trichoblasts and guard cells, are the products of unequal cell divisions, and they arise from the smaller of the two cells thus formed. Bünning (1952, 1953) has already drawn attention to this phenomenon which is characteristic not only for the epidermis but also for idioblasts, such as raphide-containing cells and idioblastic sclereids, which develop in inner tissues. Bünning states that the small embryonic cells, which are capable of varied development, are at first qualitatively equal and that their consequent development is, therefore, controlled by factors which as yet have not been analysed.

Till now no satisfactory explanation has been given of the phylogeny and functional significance of certain of the specialized cells, such as the bulliform, silica- and cork-cells of grass leaves.

Several types of arrangement of the subsidiary cells around the guard cells are recognized in both monocotyledons and dicotyledons. In relation to the dicotyledons, no conclusions have been put forward as to the evolutionary trends among these types. For the monocotyledons, Stebbins and Khush (1961) suggest that the type with four or more subsidiary cells is the most primitive, and that those types with few subsidiary cells or those lacking them entirely have been derived, independently, from this primitive type by reduction. The Stebbins and Khush approach was criticized by Paliwal (1969). The argument for numerous subsidiary cells as the primitive type in monocotyledons was also not supported by the study of the development of the stomatal complex in representatives of 48 families carried out by Tomlinson (1974) and by the work on the subsidiary cells of Williams (1979).

It should also be mentioned that this evolutionary trend does not correlate with that of other anatomical characteristics such as the evolution of the tracheary elements, for instance. It is seen, therefore, that additional research should be made in this field.

As regards the opening and closing of the stomatal aperture, the structure of only a few types of guard cell has been thoroughly investigated. There are indications that the variations in guard-cell structure result in different opening and closing mechanisms. In certain cases, such as in hydathodes and nectaries, the stomata are so modified that their apertures remain open permanently.

In some desert plants the walls of the guard cells become thickened and sometimes cutinized at the end of the summer to such an extent that the cell lumen is almost completely obliterated. Therefore it was suggested (Volkens, 1887; Fahn and Dembo, 1964; Böcher and Lyshede, 1968; Gedalovich and Fahn, 1983) that stomata with such guard cells remain closed during the critical drought period. This is a feature of extreme interest and further study may lead to a better understanding of the anatomical adaptations of desert plants.

REFERENCES

AMELUNXEN, F. (1964) Elektron–mikroskopische Untersuchungen an den Drüsenhaaren von *Mentha piperita* L. *Planta med.* **12:** 121–39.
AMELUNXEN, F. (1965) Elektronenmikroskopische Untersuchungen an den Drüsenschuppen von *Mentha piperita* L. *Planta med.* **13:** 457–73.

AMELUNXEN, F. and ARBEITER, H. (1969) Untersuchungen an den Drüsenhaaren von *Cleome spinosa* L. *Z. Pflanzenphysiol.* **61**: 73–80.

AMELUNXEN, F., MORGENROTH, K. and PICKSAK, T. (1967) Untersuchungen an der Epidermis mit dem Stereoscan-Elektron-mikroskop. *Z. Pflanzenphysiol.* **57**: 79–95.

ARTSCHWAGER, E. (1925) Anatomy of vegetative organs of sugar cane. *J. Agric. Res.* **30**: 197–221.

ASCENSAO, L. and PAIS, M. S. S. (1987). Glandular trichomes of *Artemisia campestris* (ssp. *maritima*): ontogeny and histochemistry of the secretory product. *Bot. Gaz.* **148**: 221–227.

AYLOR, D. E., PARLANGE, J.-Y. and KRIKORIAN, A. D. (1973) Stomatal mechanics. *Am. J. Bot.* **60**: 163–71.

BARANOVA, M. A. (1987) Historical development of the present classification of morphological types of stomat. *Bot. Rev.* **53**: 53–79.

BARTHLOTT, W. and FROLICH, D. (1983) Micromorphologie un Orientierungsmuster epicuticularer Wachs-Kristalloide: Ein neues systematisches Merkmal bei Monokotylen. *Pl. Syst. Evol.* **142**: 171–185.

BÖCHER, T. W. and LYSHEDE, O. B. (1968) Anatomical studies in xerophytic apophyllous plants: I, *Monttea aphylla, Bulnesia retama* and *Bredemeyera colletioides. K. danske Vidensk. Selsk. Biologiske Skr.* **16**: 1–44.

BOLLIGER, R. (1959) Entwicklung und Struktur der Epidermisaussenwand bei einigen Angiospermenblättern. *J. Ultrastruct. Res.* **3**: 105–30.

BONNETT, O. T. (1961) The oat plant: its histology and development. *Bull. Ill. Agr. Exp. Sta.*, no. 672.

BOSABALIDIS, A. M. (1987) Origin, differentiation and cytochemistry of membrane-limited inclusion bodies in leucoplasts of leaf epidermal cells of *Origanum dictamnus* L. *Cytobios* **50**: 77–88.

BOULTER, M. C. (1970) Lignified guard cell thickenings in the leaves of some modern and fossil species of Taxodiaceae (Gymnospermae). *J. Linn. Soc. Biol.* **2**: 41–46.

BROWN, W. V. and JOHNSON, S. C. (1962) The fine structure of the grass guard cell. *Am. J. Bot.* **49**: 110–15.

BRUNNER, C. (1909) Beiträge zur vergleichenden Anatomie der Tamaricaceen. *Jb. wiss. Anst. Hamburg* 1909: 89–162.

BÜNNING, E. (1952) Morphogenesis in plants. *Survey of Biological Progress* **2**: 105–40.

BÜNNING, E. (1953) *Entwicklungs- und Bewegungsphysiologie der Pflanze*, 3rd edn. Springer, Berlin.

BÜNNING, E. and BIEGERT, F. (1953) Die Bildung der Spaltöffnungsinitialen bei *Allium cepa. Z. wiss. Bot.* **41**: 17–39.

CARLQUIST, S. (1961) *Comparative Plant Anatomy.* Holt, Rinehart, & Winston, New York.

CARR, D. J. and CARR, S. G. M. (1978) Origin and development of stomatal microanatomy in two species of *Eucalyptus. Protoplasma* **96**: 127–48.

CARR, S. G. M. and CARR, D. J. (1979) An unusual feature of stomatal microanatomy in certain taxonomically-related *Eucalyptus* spp. *Ann. Bot.* **44**: 239–43.

CARR, D. J. and CARR, S. G. M. (1987) *Eucalyptus II. The Rubber Cuticle, and other studies of the Corymbosae.* Phytoglyph Press, Canberra.

CARR, S. G. M., MILKOVITS, L. and CARR, D. J. (1971) Eucalypt phytoglyphs: the microanatomical features of the epidermis in relation to taxonomy. *Aust. J. Bot.* **19**: 173–90.

CHAFE, S. C. and WARDROP, A. B. (1972) Fine structural observations on epidermis. I. The epidermal cell wall. *Planta* **107**: 269–78.

CHALONER, W. G. and LORCH, J. (1960) An opposite-leaved conifer from the Jurassic of Israel. *Palaeontology* **2**: 226–42.

CHRISTODOULAKIS, N. S. and PSARAS, G. K. (1987) Stomata on the primary root of *Ceratonia siliqua. Ann. Bot.* **60**: 295–297.

CORMACK, R. G. H. (1949) The development of root hairs in angiosperms. *Bot. Rev.* **15**: 583–612.

CORMACK, R. G. H. (1962) Development of root hairs in angiosperms: II, *Bot. Rev.* **28**: 446–64.

CUTTER, E. G. and FELDMAN, L. J. (1970a) Trichoblasts in *Hydrocharis*: I, Origin, differentiation, dimensions and growth. *Am. J. Bot.* **57**: 190–201.

CUTTER, E. G. and FELDMAN, L. J. (1970b) Trichoblasts in *Hydrocharis*: II, Nucleic acids, proteins and a consideration of cell growth in relation to endopolyploidy. *Am. J. Bot.* **57**: 202–11.

DAYANANDAN, P. and KAUFMAN, P. B. (1975) Stomatal movement associated with pottasium fluxes. *Am. J. Bot.* **62**: 221–231.

DE BARY, A. (1877) *Vergleichende Anatomie der Vegetationsorgane der Phanerogamen und Farne.* W. Engelmann, Leipzig.

DITTMER, H. J. (1949) Root hair variations in plant species. *Am. J. Bot.* **36**: 152–5.

EAMES, A. J. and MACDANIELS, L. H. (1947) *Introduction to Plant Anatomy*, 2nd edn. McGraw-Hill, New York.

FAHN, A. (1952) On the structure of floral nectaries. *Bot. Gaz.* **113**: 464–70.

FAHN, A. (1979) *Secretory Tissues in Plants.* Academic Press, London.

FAHN, A. (1986) Structure and functional properties of trichomes of xeromorphic leaves. *Ann. Bot.* **57**: 631–637.

FAHN, A. (1988) Secretory tissues in vascular plants. *New Phytol.* **108**: 229–257.

FAHN, A. and DEMBO, N. (1964) Structure and development of the epidermis in articulated Chenopodiaceae. *Israel J. Bot.* **13**: 177–92.

FAHN, A., SHOMER, I. and BEN-GERA, I. (1974) Occurrence and structure of epicuticular wax on the juice vesicles of citrus fruits. *Ann. Bot.* **38**: 869–72.

FLINT, L. H. and MORELAND, C. F. (1946) A study of the stomata in sugar cane. *Am. J. Bot.* **33**: 80–82.

FLORIN, R. (1931) Untersuchungen zur Stammesgeschichte der Coniferales und Cordaitales. *K. svenska Vetensk Akad. Handl.*, ser. 5, **10**: 1–558.

FLORIN, R. (1933) Studien über die Cycadales des Mesozoikums nebst Erörterungen über die Spaltöffnungsapparate der Bennettitales. *K. svenska Vetensk Akad. Handl.*, ser. 3, **12**: 1–134.

FOSTER, A. S. (1950) *Practical Plant Anatomy.* van Nostrand, New York.

FRANKE, W. (1971a) The entry of residues into plants via ectodesmata (ectocythodes). *Residue Rev.* **38**: 81–115.

FRANKE, W. (1971b) Über die Natur der Ectodesmen und einen Vorschlag zur Terminologie. *Ber. dt. bot. Ges.* **84**: 533–7.

FREY-WYSSLING, A. and MÜHLETHALER, K. (1965) *Ultrastructural Plant Cytology.* Elsevier, Amsterdam.

FRITZ, F. (1935) Über Kutikula von Aloë- und Gasteriaarten. *Jb. wiss. Bot.* **81**: 718–46.

FRITZ, F. (1937) Untersuchungen über die Kutinisierung der Zellmembranen und den rhythmischen Verlauf dieses Vorganges. *Planta* **26**: 693–704.

GALATIS, B. and MITRAKOS, K. (1980) The ultrastructural cytology of the differentiating guard cells of *Vigna sinensis.*

Am. J. Bot. **67**: 1243–61.

GEDALOVICH, E. and FAHN, A. (1983) Ultrastructure and development of the inactive stomata of *Amabasis articulata* (Forsk.) Moq. *Am. J. Bot.* **70**: 88–96.

GILCHRIST, A. J. and JUNIPER, B. E. (1974) An excitable membrane in the stalked glands of *Drosera capensis* L. *Planta* **119**: 143–147.

GUTTENBERG, H. V. (1940) Der primäre Bau der Angiospermenwurzel. In: K. Linsbauer, *Handbuch der Pflanzenanatomie*. Bd. 8, Lief. 39. Gebr. Borntraeger, Berlin.

HABERLANDT, G. (1918) *Physiologische Pflanzenanatomie*, 5th edn. W. Engelmann, Leipzig.

HALL, D. M. (1967a) The ultrastructure of wax deposits on plant leaf surfaces: II, Cuticular pores and wax formation. *J. Ultrastruct. Res.* **17**: 34–44.

HALL, D. M. (1967b) Wax microchannels in the epidermis of white clover. *Science* **158**: 505–6.

HANSTEIN, J. (1868) Über die Organe der Harz- und Schleimabsonderung in den Laubknospen. *Bot. Ztg.* **26**: 697–713, 721–35, 745–61, 769–87.

HEIDE-JØRGENSEN, H. S. (1978) The xeromorphic leaves of *Hakea suaveolens* R. Br. II. Structure of epidermal cells, cuticle development and ectodesmata. *Bot. Tidsskr.* **72**: 227–44.

HESLOP-HARRISON, J. and KNOX, R. B. (1971) A cytochemical study of the leaf-gland enzymes of insectivorous plants of the genus *Pinguicula. Planta* **96**: 183–211.

HUMBLE, G. D. and RASCHKE, K. (1971) Stomatal opening quantitatively related to potassium transport. *Plant Physiol.* **48**: 447–53.

HUMMEL, K. and STAESCHE, K. (1962) Die Verbreitung der Haartypen in den natürlichen Verwandschaftsgruppen. In: K. Linsbauer, *Handbuch der Pflanzenanatomie*, Bd. 4, Tl. 5. Gebr. Borntraeger, Berlin.

JARVIS, L. R. and WARDROP, A. B. (1974) The development of the cuticle in *Phormium tenax. Planta* **119**: 101–12.

JOEL, D. M. (1986) Glandular structures in carnivorous plants: their role in mutual and unilateral exploitation of insects. In: *Insects and Plant Surfaces* (eds. B. E. Juniper and T. R. E. Southwood). Edward Arnold. pp. 219–234.

JOHNSON, H. B. (1975) Plant pubescence: an ecological perspective. *Bot. Rev.* **41**: 233–58.

JUNIPER, B. E., ROBINS, R. J. and JOEL, D. M. (1989) *The Carnivorous Plants*. Academic Press, London.

KAUFMAN, K. (1927) Anatomie und Physiologie der Spaltöffnungsapparate mit verholzten Schliesszellmembranen. *Planta* **3**: 27–59.

KIDWAI, P. (1981) *An Illustrated Glossary of Technical Terms Used in Stomatal Studies. Indian J. Forestry* Additional Series I. Bishen Singh Mahendra Pal Singh. Dehra Dun, India.

KISSER, J. (1958) Der Stoffwechsel sekundärer Pflanzenstoffe. In: W. Ruhland, *Handbuch der Pflanzenphysiologie*. Springer-Verlag, Berlin, **10**: 91–131.

LAWTON, J. R. (1980) Observations on the structure of epidermal cells, particularly the cork and silica cells, from the flowering stem internode of *Lolium temulentum* L. (Gramineae). *J. Linn. Soc. Bot.* **80**: 161–77.

LEACH, J. H., MOLLENHAUER, H. H. and WHALEY, W. G. (1963) Ultrastructural changes in root apex. *Symp. Soc. Exp. Biol.* **17**: 74–84.

LERSTEN, N. R. (1974) Morphology and distribution of colleters and crystals in relation to the taxonomy of bacterial leaf nodule symbiosis of *Psychotria* (Rubiaceae). *Am. J. Bot.* **61**:

973–981.

LEVIN, D. A. (1973) The role of trichomes in plant defence. *Q. Rev. of Biol.* **48**: 3–15.

LEVITT, J. (1974) The mechanism of stomatal movement—once more. *Protoplasma* **82**: 1–17.

LINSBAUER, K. (1930) Die Epidermis. In: K. Linsbauer, *Handbuch der Pflanzenanatomie*, Bd. 4, Lief. 27, Gebr. Borntraeger, Berlin.

LLOYD, F. D. (1942) *The Carnivorous Plants*. Chronica Botanica, 9. Waltham, Mass.

LÜTTGE, U. (1967) Drüsenfunktionen bei fleischfressenden Pflanzen. *Umschau* **67**: 181–6.

LYSHEDE, O. B. (1977) Structure of the epidermal and subepidermal cells of some desert plants of Israel. *Anabasis articulata* and *Calligonum comosum. Israel J. Bot.* **26**: 1–10.

LYSHEDE, O. B. (1978) Studies on outer epidermal cell walls with microchannels in a xerophytic species. *New Phytol.* **80**: 421–6.

LYSHEDE, O. B. (1982) Structure of the outer epidermal wall in xerophytes. In: *The Plant Cuticle* (eds. D. F. Cutler, K. L. Alvin and C. E. Price). Academic Press, London.

MARTIN, J. T. and JUNIPER, B. E. (1970) *The Cuticles of Plants*. E. Arnold, London.

MATZKE, E. B. (1947) The three-dimensional shape of epidermal cells of *Aloë aristata. Am. J. Bot.* **34**: 182–95.

McCLENDON, J. H. (1984) The micro-optics of leaves. I. Patterns of reflection from epidermis. *Am. J. Bot* **71**: 1391–1397.

METCALFE, C. R. (1959) A vista in plant anatomy. In: *Vistas in Botany* (ed. W. B. Turrill), Pergamon Press, London, Vol. I, pp. 76–99.

METCALFE, C. R. (1960) *Anatomy of the Monocotyledons. I. Gramineae*. Clarendon Press, Oxford.

METCALFE, C. R. (1963) Comparative anatomy as a modern botanical discipline. In: *Adv. Bot. Res.* (ed. R. D. Preston), Academic Press, London, Vol. I, pp. 101–47.

METCALFE, C. R. and CHALK, L. (1979) *Anatomy of the Dicotyledons*. Vol. I, 2nd edn. Clarendon Press, Oxford.

METCALFE, C. R. and CHALK, L. (1983) Anatomy of the Dicotyledons, 2nd ed. Vol. II. Clarendon Press, Oxford.

MEYER, R. E. and MEOLA, S. M. (1978) *Morphological Characteristics of Leaves and Stems of Selected Texas Woody Plants*. USDA, Technical Bull. No. 1564.

MILLER, R. H. (1983) Cuticular pores and transcuticular canals in diverse fruit varieties. *Ann. Bot.* **51**: 697–709.

MOTT, K. A., GIBSON, A. C., and O'LEARY, J. W. (1982) The adaptive significance of amphistomatic leaves. *Plant Cell Environ.* **5**: 455–460.

NAPP-ZINN, K. (1966) Anatomie des Blattes. I. Blattanatomie der Gymnospermen. In: *Handbuch der Pflanzenanatomie* (eds. W. Zimmermann, P. Ozenda and H. D. Wulff). Gebr. Borntraeger, Berlin-Nikolassee, Part 1, Vol. VIII.

NAPP-ZINN, K. (1973) Anatomie des Blattes. II. Blattanatomie der Angiospermen. A. Entwicklungsgeschichtliche und topographische Anatomie des Angiospermenblattes. In: *Handbuch der Pflanzenanatomie* (eds. S. Carlquist, P. Ozenda, and H. D. Wulff). Gebr. Borntraeger, Berlin, Stuttgart, Part 2A, Vol. VIII.

O'BRIEN, T. P. (1967) Observations of the fine structure of the root coleoptile. I. The epidermal cells of the extreme apex. *Protoplasma* **63**: 385–416

OSMOND, C. B., LÜTTGE, U., WEST, K. R., PALLAGHY, C. K. and SHACHER-HILL, B. (1969) Ion absorption in *Atriplex* leaf tissue: II, Secretion of ions to epidermal bladders. *Aust. J. Biol. Sci.* **22**: 797–814.

PALEVITZ, B. A. and HEPLER, P. K. (1976) Cellulose microfibril orientation and cell shaping in developing guard cells of *Allium*: the role of microtubules and ion accumulation. *Planta* **132**: 71–93.

PALEVITZ, B. A. and HEPLER, P. K. (1985) Changes in dye coupling of stomatal cells of *Allium* and *Commelina* demonstrated by microinjection of Lucifer yellow. *Planta* **164**: 473–479.

PALIWAL, G. S. (1969) Stomatal ontogeny and phylogeny: I, Monocotyledons. *Acta bot. neerl.* **18**: 654–68.

PANT, D. D. (1965) On the ontogeny of stomata and other homologous structures. *Plant Sci. Ser. (Allahabad)* **1**: 1–24.

PANT, D. D. and BANERJI, R. (1965) Epidermal structure and development of stomata in Convolvulaceae. *Senckenberg. Biol.* **46**: 155–73.

PANT, D. D. and KIDWAI, P. F. (1967) Development of stomata in some Cruciferae. *Ann. Bot.* **31**: 513–21.

PARRY, D. W., O'NEILL, C. H. and HODSON, M. J. (1986) Opaline silica deposits in the leaves of *Bideus pilosa* L. and their possible significancer in cancer. *Ann. Bot.* **58**: 641–647.

PAYNE, W. W. (1970) Helicocytic and allelocytic stomata: unrecognized patterns in the Dicotyledonae. *Am. J. Bot.* **57**: 140–7.

PAYNE, W. W. (1978) A glossary of plant hair terminology. *Brittonia* **30**: 239–55.

PAYNE, W. W. (1979) Stomatal patterns in embryophytes: their evolution, ontogeny and interpretation. *Taxon* **28**: 117–32.

PAZOUREK, J. (1970) The effect of light intensity on stomatal frequency in leaves of *Iris hollandica* hort. var. *Wedgwood*. *Biol. Plant.* **12**: 208–15.

PERRIN, A. (1970) Organisation infrastructurale, en rapport avec les processus de sécrétion, des poils glandulaires (trichome-hydathodes) de *Phaseolus multiflorus*. *CR Acad. Sc. (Paris)* **270**: 1984–7.

PETERSON, R. L. and HAMBLETON, S. (1978) Guard cell ontogeny in leaf stomata of the fern *Ophioglossum petiolatum*. *Can. J. Bot.* **56**: 2836–52.

PICKETT-HEAPS, J. D. (1967) Further observations on Golgi apparatus and its functions in cells of the wheat seedling. *J. Ultrastruct. Res.* **18**: 287–303.

PICKETT-HEAPS, J. D. and NORTHCOTE, D. H. (1966) Cell division in the formation of the stomatal complex of the young leaves of wheat. *J. Cell Sci.* **1**: 121–8.

POPHAM, R. A. and HENRY, R. D. (1955) Multicellular root hairs on adventitious roots of *Kalanchoë fedtschenkoi*. *Ohio J. Sci.* **55**: 301–7.

PRICE, C. E. (1982) A review of factors influencing the penetration of pesticides through plant leaves. In: *The Plant Cuticle* (eds. D. F. Cutler, K. L. Alvin and C. E. Price). Academic Press, London. pp. 237–252.

PRIESTLEY, J. H. (1943) The cuticle in angiosperms. *Bot. Rev.* **9**: 593–616.

RAJAGOPAL, T. and RAMAYYA, N. (1968) Occurrence of idioblastic "cell-sacs" in the leaf epidermis of *Cleome aspera* Koen. ex DC with observations on their taxonomic significance, structure and development. *Curr. Sci.* **37**: 260–2.

RAMAYYA, N. (1964) Morphology of the emergences. *Curr. Sci.* **33**: 577–80.

RAMAYYA, N. and BAHADUR, B. (1968) Morphology of the "squamellae" in the light of their ontogeny. *Curr. Sci.* **37**: 520–2.

RAMAYYA, N. and GOPALACHARYULU, M. (1968) Morphology of the shaggy glands of *Cleome viscosa* L. *Curr. Sci.* **37**:

457–9.

RAO, S. R. S. and RAMAYYA, N. (1977) Stomatogenesis in the genus *Hibiscus* L. (Malvaceae). *J. Linn. Soc. Bot.* **74**: 47–56.

RASCHKE, K. (1975) Stomatal action. *A. Rev. Pl. Physiol.* **26**: 309–40.

RASCHKE, K. (1979) Movements of stomata. In: *Physiology of Movements*. *Encyclopedia of Plant Physiology*, N.S. (eds. W. Haupt and M. E. Feinleib). Springer-Verlag, Berlin, **7**: 383–441.

RASMUSSEN, H. (1981) Terminology and classification of stomata and stomatal development—a critical survey. *Bot. J. Linn. Soc.* **83**: 199–212.

RASMUSSEN, H. (1983) Stomatal developmental in families of Liliales. *Bot. Jahrb. Syst.* **104**: 261–287.

ROELOFSEN, P. A. (1952) On the submicroscopic structure of cuticular cell walls. *Acta bot. neerl.* **1**: 99–114.

ROELOFSEN, P. A. (1959) The plant cell-wall. In: K. Linsbauer, *Handbuch der Pflanzenanatomie*, Bd. 3, T. 4. Gebr. Borntraeger, Berlin.

ROLLINS, R. C. (1944) Evidence for natural hybridity between guayule (*Parthenium argentatum*) and mariola (*P. incanum*). *Am. J. Bot.* **31**: 93–99.

ROLLINS, R. C. and BANERJEE, U. C. (1975) *Atlas of the Trichomes of* Lesquerella (*Cruciferae*). The Bussey Institution of Harvard.

RUTTER, J. C. and WILLMER, C. M. (1979) A light and electron microscopy study of the epidermis of *Paphiopedilum* spp. with emphasis on stomatal ultrastructure. *Plant, Cell and Environment* **2**: 211–9.

SACHS, T. (1978) Patterned differentiation in plants. *Differentiation* **11**: 65–73.

SACHS, T. (1979) Cellular interactions in the development of stomatal patterns in *Vinca major* L. *Ann. Bot.* **43**: 693–700.

SACHS, T. (1984) Controls of cell patterns in plants. In: *Pattern Formation. A Primer in Developmental Biology* (eds. G. M. Malacinski and S. V. Bryant). Macmillan Publishing Co., New York. pp. 367–391.

SCHIEFERSTEIN, R. H. and LOOMIS, W. E. (1956) Wax deposits on leaf surfaces. *Plant Physiol.* **31**: 240–7.

SCHIEFERSTEIN, R. H. and LOOMIS, W. E. (1959) Development of the cuticular layer in angiosperm leaves. *Am. J. Bot.* **46**: 625–35.

SCHMUCKER, TH. and LINNEMANN, G. (1959) Carnivorie. In: W. Ruhland (ed.), *Handbuch der Pflanzenphysiologie*. Springer-Verlag, Berlin; Bd. XI: 198–283.

SCHNEPF, E. (1965) Licht- und elektron–mikroskopische Beobachtungen an den Trichom-Hydathoden von *Cicer arietinum*. *Z. Pflanzenphysiol.* **53**: 245–54.

SCHNEPF, E. (1969) *Sekretion und Exkretion bei Pflanzen*. In: *Protoplasmatologia*, Bd. 8, Springer-Verlag, Wien.

SCHRÖDTER, K. (1926) Zur physiologischen Anatomie der Mittelzellen drüseger Gebilde. *Flora* **120**: 19–86.

SCHWAB, D. W., SIMMONS, E. and SCALA, J. (1969) Fine structure changes during function of the digestive gland of Venus's flytrap. *Am. J. Bot.* **56**: 88–100.

SCOTT, F. M., HAMMER, K. C., BAKER, E. and BOWLER, E. (1957) Ultrasonic and electron microscope study on onion epidermal wall. *Science* **125**: 399–400.

SHIELDS, L. M. (1951) The involution mechanism in leaves of certain xeric grasses. *Phytomorphology* **1**: 225–41.

SHIMONY, C. and FAHN, A. (1968) Light- and electron-microscopical studies on the structure of salt glands of *Tamarix aphylla* L. *J. Linn. Soc. Bot.* **60**: 283–8.

SHIMONY, C., FAHN, A. and REINHOLD, L. (1973) Ultrastructure and ion gradients in the salt glands of *Avicennia marina* (Forssk.) Vierh. *New Phytol.* **72**: 27–36.

SIEVERS, A. (1963a) Beteiligung des Golgi-Apparatus bei der Bildung der Zellwand von Wurzelhaaren. *Protoplasma* **56**: 188–92.

SIEVERS, A. (1963b) Über die Feinstruktur des Plasmas wachsender Wurzelhaare. *Z. Naturf.* **18b**: 830–6.

SOLEREDER, H. (1908) *Systematische Anatomie der Dikotyledonen.* Ergänzungsbd. F. Enke, Stuttgart.

STEBBINS, G. L. and JAIN, S. K. (1960) Developmental studies of cell differentiation in the epidermis of monocotyledons: I, *Allium*, *Rhoeo* and *Commelina*. *Devl. Biol.* **2**: 409–26.

STEBBINS, G. L. and KHUSH, G. S. (1961) Variation in the organization of the stomatal complex in the leaf epidermis of monocotyledons and its bearing on their phylogeny. *Am. J. Bot.* **48**: 51–59.

STEBBINS, G. L. and SHAH, S. S. (1960) Developmental studies of cell differentiation in the epidermis of monocotyledons: II, Cytological features of stomatal development in the Gramineae. *Devl. Biol.* **2**: 477–500.

STEBBINS, G. L., SHAH, S. S., JANIN, D. and JURA, P. (1967) Changed orientation of the mitotic spindle of stomatal guard cell divisions in *Hordeum vulgare*. *Am. J. Bot.* **54**: 71–80.

STEVENS, R. A. and MARTIN, E. S. (1978) Structural and functional aspects of stomata. I. Development studies in *Polypodium vulgare*. *Planta* **142**: 307–16.

STRASBURGER, E. (1911) *Lehrbuch der Botanik.* Gustav Fischer, Jena.

TAKEDA, K. and SHIBAOKA, H. (1978) The fine structure of the epidermal cell wall in Azuki bean epicotyl. *Bot. Mag. Tokyo* **91**: 235–45.

TAROWSKA, J. A. and WACOWSKA, M. (1988) The significance of the presence of stomata on seedling roots. *Ann. Bot.* **61**: 305–310.

THOMSON, W. W. and DE JOURNETT, R. (1970) Studies on the ultrastructure of the guard cells of *Opuntia*. *Am. J. Bot.* **57**: 309–16.

THURSTON, E. L. and LERSTEN, N. R. (1969) The morphology and toxicology of plant stinging hairs. *Bot. Rev.* **35**: 393–412.

TING, I. P. (1982) *Plant Physiol.* Addison-Wesley, Reading, Massachusetts.

TOMLINSON, P. B. (1961) Palmae. In: *Anatomy of the Monocotyledons* (ed. C. R. Metcalfe), Vol. II. Clarendon Press, Oxford.

TOMLINSON, P. B. (1974) Development of the stomatal complex as a taxonomic character in the monocotyledons. *Taxon* **23**: 109–28.

TROLL, W. (1948) *Allgemeine Botanik.* F. Enke, Stuttgart.

TURRELL, F. M. (1947) Citrus leaf stomata: structure, composition and pore size in relation to penetration of liquids. *Bot. Gaz.* **108**: 476–83.

UPHOF, J. C. T. (1962) Plant hairs. In: K. Linsbauer, *Handbuch der Pflanzenanatomie*, Bd. 4, T. 5. Gebr. Borntraeger, Berlin.

VAN COTTHEM, W. R. J. (1970) A classification of stomatal types. *J. Linn. Soc., Bot.* **63**: 235–46.

VASSILYEV, A. E. and VASSILYEV, G. V. (1976) The ultrastructure of the stomatal apparatus in gymnosperms (with special reference to the stomatal movements). *Bot. Zh. SSSR* **61**: 449–65 (in Russian).

VOLKENS, G. (1887) *Die Flora der aegyptisch-arabischen Wüste auf Grundlage anatomisch physiologischer Forschungen.* Gebr. Borntraeger, Berlin.

WERKER, E. and FAHN, A. (1981) Secretory hairs of *Inula viscosa* (L.) Ait.—development, ultrastructure, and secretion. *Bot. Gaz.* **142**: 461–476.

WILDER, G. J. (1985) Anatomy of noncostal portions of lamina in the Cyclanthaceae (Monocotyledoneae). I. Epidermis. *Bot. Gaz.* **146**: 82–105.

WILKINSON, H. P. (1979) The plant surface (mainly leaf). In: C. R. Metcalfe and L. Chalk, *Anatomy of Dicotyledons*, 2nd edn. Clarendon Press, Oxford, pp. 97–165.

WILLIAMS, N. H. (1979) Subsidiary cells in the Orchidaceae: their general distribution with special reference to development in the Oncidieae. *J. Linn. Soc. Bot.* **78**: 41–66.

WILLMER, C. M. and SEXTON, R. (1979) Stomata and plasmodesmata. *Protoplasma* **100**: 113–24.

WU, H., SHARPE, P. J. H. and SPENCE, R. D. (1985) Stomatal mechanics. III. Geometric interpretation of the mechanical advantage. *Plant Cell Environ.* **8**: 269–274.

YOSHIE, F. and SAKAI, A. (1985) Types of Florin rings, distributional patterns of epicuticular wax, and their relationships in the genus *Pinus*. *Can. J. Bot.* **63**: 2150–2158.

ZIEGENSPECK, H. (1938) Die Micellierung der Turgeszenzmechanismen. T. I: Die Spaltöffnungen (mit phylogenetischen Ausblicken). *Bot. Arch.* **39**: 268–309, 332–72.

ZIEGENSPECK, H. (1944) Vergleichende Untersuchungen der Entwicklung der Spaltöffnungen von Monokotyledonen und Dikotyledonen im Lichte der Polariskopie und Dichroskopie. *Protoplasma* **38**: 197–224.

ZIEGLER, H., SHMUELI, E. and LANGE, G. (1974) Structure and function of the stomata of *Zea mays*. I. The development. *Cytobiologie* **9**: 162–8.

PRIMARY VEGETATIVE BODY OF THE PLANT

THE previous chapters considered the structure of the tissues and the types and characteristics of cells of which they consist. In the following three chapters the arrangement of an interrelationship between these tissues in the different vegetative organs of the primary body of the sporophyte will be discussed. An attempt will also be made to relate the structure to the function of the different organs and to consider it from a phylogenetic standpoint.

The vegetative organs to be discussed are the stem (*caulome*), the leaf (*phyllome*), and the *root*. The stem and leaves constitute the *shoot*. It is difficult, however, to distinguish between the stem and the leaves—these organs develop from a common apical meristem of the shoot apex. The connection between and mutual dependence of these two organs exist throughout the entire growth period of the plant (Wetmore, 1943; Wardlaw, 1960; Steeves and Sussex, 1972). Arber (1950) claims that the leaf is actually a stem-like structure which has secondarily become flattened. In certain ferns leaf primordia were induced to become shoots when their sites were isolated from the shoot apices by deep incisions (Cutter, 1956; Wardlaw, 1968; Hicks and Steeves, 1969), or when they were isolated in culture (Steeves, 1961). In angiosperms, Sussex (1955) isolated a young leaf primordium of *Solanum tuberosum* from the remainder of the meristem, and obtained an organ that was determined in growth but radially symmetrical, a "centric leaf".

The acceptance of this relationship and connection between the stem and the leaf enables a more thorough understanding of the primary structure of the stem. Sattler (1974) found the classical model, according to which the shoot consists of the categories caulome and phyllome, too rigid, and proposed that intermediate structures should be recognized. Sattler thus suggests that the five categories of the classical morphological model, i.e. shoot, caulome, phyllome, root, and emergence, are not mutually exclusive. Various positional relationships between the different organs are possible; caulome, phyllome, and roots may be inserted either on caulome or on phyllome; intermediate forms may be inserted on other intermediate forms or on caulome or phyllome (see also Dickinson, 1978, on epiphylly in angiosperms).

CHAPTER 11

THE STEM

ONTOGENETIC DEVELOPMENT OF THE STEM

THE axis of the embryo in the seed consists of a *hypocotyl* and *radicle*. At the tip of the hypocotyl one or more cotyledons and the bud of the shoot, i.e. the *plumule*, are found. At the tip of the radicle is the root cap.

The bud of the shoot usually consists of an axis, the *epicotyl*, containing a few internodes, which have not elongated, and some leaf primordia. With the germination of the seed the embryo enlarges and starts to grow, the apical meristem of the young shoot adds further leaf primordia, and the internodes between the lower primordia, which in the meantime have become distant from the apex, elongate. Buds develop in the axils of the developing leaves giving rise to a branched shoot.

In mature plants the development of leaf primordia at the shoot apex and the elongation of the internodes below it are the same as that in the growing embryo of the germinating seed. The order of appearance and the arrangement of leaves on the stem are more or less characteristic of each species.

That part of the stem from which a leaf or leaves develop is called the *node* and that portion of the stem between two such nodes, the *internode*. The length of the internodes varies in the different species. In certain plants, such as *Chichorium*, *Thrincia*, and others, where the leaves are arranged in a basal rosette, the internodes hardly elongate at all, but in most spermatophytes the internodes elongate to differing extents. At each node one, two, or more leaves may be found. The arrangement of the leaves on the stem is termed *phyllotaxy* or *phyllotaxis*. When there are more than two leaves at one node the ar-

rangement is termed *whorled*. When there are two leaves at each node the leaves are said to be *opposite*; in this type of arrangement the leaves of the successive nodes may be at right angles to each other and then the arrangement is termed *decussate*; or the leaves may form two parallel ranks along the stem, i.e. *distichous*. The straight vertical line along which the leaves occur is termed *orthostichy*. When there is a single leaf at each node and the leaves are arranged spirally on the stem, the phyllotaxy is said to be *alternate*. The helix along which the leaves occur is termed *parastichy*. The space, on the circumference of the stem, between two successive leaves, whether they arise at a single node or whether they are arranged spirally on the stem, is constant, i.e. two successive leaves are separated by a constant portion of the perimeter of the stem (Fig. 98, no. 1). If internodes are short, as is always the case in the apical region, two series of parastichies can usually be recognized winding in opposite directions (see Snow, 1955). In most species the leaves along the parastichies are in contact at their bases (*contact parastichies*) (Fig. 98, no. 2) (Rutishauser, 1982).

The position of the leaf primordia on the stem apex is determined before it is possible to distinguish any feature which indicates that such development has begun. Therefore it appears that the factors determining the position of the primordia on the stem apex are internal and, in general, they are identical with those factors that control the distribution of the growth potential in the apical meristem. The determination of leaf arrangement has been ascribed either to interactions within the apex (Snow and Snow, 1947; Wardlaw, 1948, 1950, 1955) or to the influence of the mature tissues below the apex, via the procambium (Crafts, 1943; Gunckel and Wetmore, 1946a, b; Philipson, 1949; Esau, 1965a; Larson,

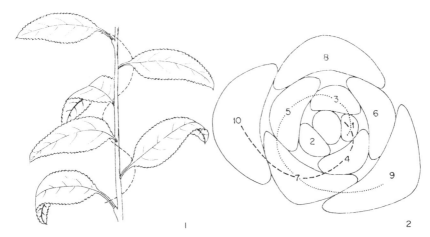

FIG. 98. 1, Portion of a branch of *Prunus*. The broken line which passes from leaf to leaf indicates the phyllotaxis which, in this case, is two-fifths. 2, Cross-section of an apical bud of *Jasminum fruticosum*; one of each of the two sets of the contact parastichies (2 and 3) was drawn starting at leaf 1. (No. 2 adapted from Snow, 1955.)

1975). The evidence in favour of local determination within the meristematic tissues comes from studies of regeneration after wounds (Snow and Snow, 1947; Ball, 1948; Clowes, 1961). Three main theories have been the basis of much research concerning the possible nature of these local interactions within the apex: (1) the theory of the *first available space*, according to which a leaf primordium arises in the first space which attains a minimum width and a minimum distance below the tip of the shoot apex (Snow and Snow, 1947); (2) the theory of *leaf fields* or *primordial fields*, according to which the existing primordia together with portions of the apical meristem form physiological units, new primordia being formed at specific points with reference to these fields (Wardlaw, 1968); (3) the theory of multiple foliar helices, according to which special mitotic properties are transmitted acropetally along foliar parastichies which terminate in leaf-generating centres that function in the "anneu initial" (Plantefol, 1947) (see Chapter 3).

To the author of this book it appears that the first two theories, those of minimum free space and leaf fields, are complementary and together provide the most satisfying explanation of phyllotaxis.

The association between leaf and axis, suggested by the theory of leaf fields is expressed in the mature shoot by the relationship between phyllotaxis and the vascularization.

ARRANGEMENT OF PRIMARY TISSUES IN THE STEM

The primary body develops from the protoderm, procambium, and ground meristem. The arrangement and structure of the primary tissues are as follows.

The epidermis

Externally, the stem is bounded by the epidermis which contains, apart from the typical epidermal cells, guard cells, idioblasts, and different types of trichomes (see Chapter 10).

The stem cortex

The stem cortex is that cylindrical region between the epidermis and the vascular cylinder (Fig. 99; Fig. 114, no. 2). It may comprise various cell types. In the simplest case, the cortex consists entirely of thin-walled parenchyma tissue. In many stems, as for instance *Pelargonium*, *Retama*, and *Salicornia*, this parenchyma may have a photosynthetic function in addition to that of the temporary storage of starch and other metabolites. In other cases the outer region of the cortex, which borders on the epidermis, may include collenchyma or fibres, and the inner region, par-

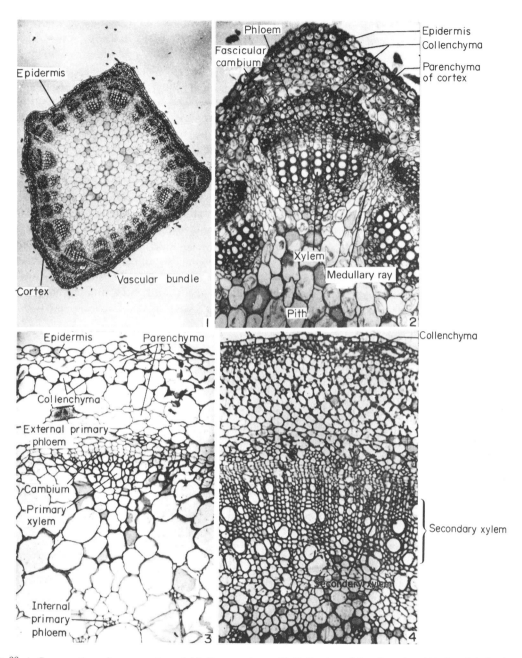

FIG. 99. 1, Cross-section of a young stem of *Medicago sativa*. × 35. 2, Portion of 1, enlarged. × 140. 3 and 4, Portions of cross-sections of a stem of *Lycopersicon esculentum*. 3, A young stem in which internal phloem can be distinguished. × 110. 4, A mature stem in which it is possible to distinguish that the cortical parenchyma below the epidermis has developed into chlorenchyma and that a fair amount of secondary xylem has been produced. × 115.

enchyma. The collenchyma or fibres may form a continuous cylinder or they may be present in the form of separated strips. In some plants, e.g.

Penthorum (Saxifragaceae) the cortex is lacunate (Haskins and Hayden, 1987). The stem cortex may contain sclereids, secretory cells, and laticifers.

The endodermis

In the stem, difficulty exists in determining the anatomical border between the cortex and the stele (Esau, 1950) as in stems a distinct *endodermis*, i.e. a layer of specialized cells which delimits the cortex from the vascular cylinder, is not usually developed. Whether the endodermis is ontogenetically related to the cortex or to the stele has not yet been clarified.

Morphologically, a well-developed endodermis is a compact layer of living cells which form a hollow cylinder. The walls of the endodermal cells are of characteristic and specialized structure. In the typical cells, bands or strips which contain lignin and suberin, apart from cellulose, are present on the radial and transverse walls; these are called *Casparian strips* or *Casparian bands*

(Fig. 153, nos. 2, 3). In the course of maturation the endodermal cells may undergo changes that involve the addition of suberin lamellae on the entire inner surface of the cell walls. This may be followed by the addition of a secondary layer of cellulose, which may sometimes contain lignin, on the inside of suberin lamellae. (See also Chapter 13.)

The endodermis is most conspicuous in the stems of the lower vascular plants. Here the cell walls have Casparian strips and a suberin lamella, and the endodermis surrounds the vascular tissue. It is sometimes also found between the vascular cylinder and the pith (*Marsilea, Ophioglossum,* and others). In certain ferns, e.g. *Dryopteris,* the endodermis surrounds the individual bundles. In the spermatophytes the endodermis is usually most obvious in the root, but, in some herbaceous

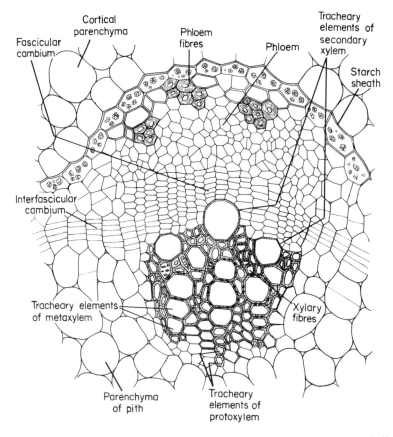

FIG 100. Portion of a cross-section of a well-developed hypocotyl of *Ricinus*. (Adapted from Palladin, 1914.)

plants and water plants (Ancibor, 1979), Casparian strips can also be observed in the endodermis of the stem. In stems, however, a typical endodermis is more usually found in underground stems, such as rhizomes, than in aerial stems. In the herbaceous stems of *Senecio* and *Leonurus* the endodermis is developed only when the plant reaches the flowering stage.

The innermost cortical layer of young dicotyledonous stems usually contains many large starch grains (Fig. 100). This layer has been termed the *starch sheath*, and, because of its position, it is considered to be homologous with the endodermis. In older stem regions the starch disappears from the sheath cells. In some plants, e.g. in *Senecio vulgaris*, these cells develop Casparian strips. In this species the endodermis develops in the axis of seedlings and in the developing stems, but in plants under summer conditions this development is slow until the onset of flowering when it proceeds rapidly (Warden, 1935).

Different investigations have shown that, in some monocotyledons, an endodermis with typical secondary cell walls may be induced to a greater extent under the influence of external factors, such as the lack of nitrogenous salts and a high degree of aeration of the soil, among others (van Fleet, 1942a, b; 1950a, b). In some dicotyledonous stems Casparian strips may develop under conditions that cause etiolation (Priestley, 1926; van Fleet, 1961).

In the stems of the lower vascular plants and in the roots of all vascular plants, the pericycle can readily be located between the endodermis and the vascular tissues. When dealing with stem anatomy of angiosperms, Blyth (1958) suggested to avoid the term pericycle, because in several species she has studied, the so-called pericyclic fibres have been found by her to be primary phloem fibres.

The primary vascular system

Internal to the cortex is the vascular system of the stem. In the gymnosperms and most of the dicotyledons the vascular system consists of a continuous or a split cylinder which encloses the *pith*, i.e. the central portion of the stem (Fig. 99, nos. 1, 2; Fig. 107, nos. 1–4). In this cylinder two types of vascular tissue can be distinguished—the phloem which is usually external, and the xylem which is usually internal. In the case of the split cylinder each strand is termed a *vascular bundle*. A vascular bundle in which the phloem is only external to the xylem is said to be a *collateral bundle* (Fig. 99, no. 2; Fig. 101, nos. 3, 4; Fig. 115, no. 1). In some dicotyledonous families, e.g. the Solanaceae, Cucurbitaceae, Asclepiadaceae, Apocynaceae, Convolvulaceae, and Compositae, internal phloem is also present. The internal or intraxylary phloem may be present as separate strands on the border of the pith, as in *Lycopersicon* (Fig. 99, no. 3), or it may be in close contact with the inner side of the xylem as in the stems of the Cucurbitaceae and Myrtaceae. The internal phloem may differentiate both from procambium and from partly differentiated parenchyma cells (Fukuda, 1967). A vascular bundle in which the intraxylary phloem is in close contact with the xylem is termed a *bicollateral bundle* (Fig. 114, no. 2).

Anastomoses between external and internal phloem of small bicolateral leaf veins have been reported for *Ecbalium elaterium* (Koch, 1884); anastomoses between the phloem of neighbouring vascular stem bundles of many species of the Cucurbitaceae have also been reported (Fischer, 1884; Zimmermann, 1922). Careful observation of cleared stem portions stained for callose has revealed that phloem anastomoses between vascular bundles may be common to other families as well (Aloni and Sachs, 1973). In *Pharbitis nill* (Convolvulaceae), anastomoses between internal and external phloem occur at three locations along the plant axis: the transition region, sites of adventitious roots formation in the hypocotyl, and in leaf gaps (Mikesell and Schroeder, 1984). Transport of labelled compounds in intact tobacco plants showed that translocation in the internal phloem is of lesser importance than that of the outer phloem (Zamski and Tsivion, 1977). Botha and Evert (1978) arrived at a similar conclusion regarding leaf veins of *Cucurbita maxima*. These authors found that most of the stylets of aphids which were feeding on the leaves terminated in the abaxial phloem. In stems of some Asclepiadaceae, however, most of the aphid stylets were observed to end within the internal primary phloem (Botha *et al.*, 1977).

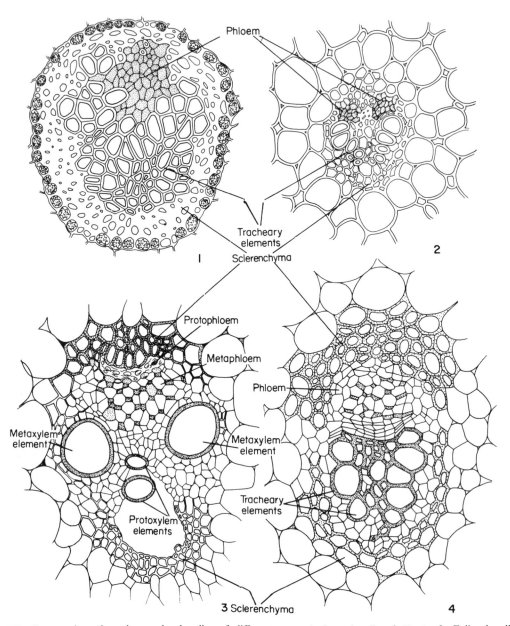

FIG. 101. Cross-sections through vascular bundles of different types. 1, Stem bundle of *Kingia*. 2, Foliar bundle of *Xanthorrhoea*. 3, Stem bundle of *Zea*, showing a lacuna which has developed as a result of separation of the parenchyma from the two protoxylem elements which it surrounded. 4, Stem bundle of *Ranunculus*. (Nos. 3 and 4 adapted from Palladin, 1914.)

In some monocotyledons, as, for example, *Convallaria majalis*, *Acorus*, some genera of the Xanthorrhoeaceae, as well as the secondary bundles of *Aloë arborescens*, *Dracaena*, *Cordyline*, and others, the xylem surrounds the phloem. Such bundles are termed *amphivasal bundles* (Fig. 115, no. 3). Bundles common in the Pteridophyta, in which the phloem surrounds the xylem, are termed *amphicribral bundles*. There are also bundles in which the xylem is seen to be V- or U-shaped in cross-section. In the former, as can be seen in the leaf of *Xanthorrhoea*, for example, the phloem groups are situated at the free ends of the xylem arms (Fig. 101, no. 2). In the case of U-shaped xylem the phloem is surrounded on three sides by xylem; such bundles occur in the stem of *Asparagus aphyllus* and *Kingia australis* (Fig. 101, no. 1).

In most monocotyledons and in a few dicotyledons no distinct vascular cylinder exists. The primary vascular system consists of a large number of bundles which are scattered irregularly, and it is impossible to distinguish clearly the boundary between the cortex, the vascular cylinder and the pith (Fig. 114, no. 1).

The vascular system plays an important role in the attempt to solve phylogenetic problems. It has been thoroughly investigated, by comparative methods, both in plants living today and in fossil forms. Van Tieghem and Douliot (1886) proposed the *stelar concept* to explain the structure of the plant axis. This concept greatly influenced later investigators who worked on the comparative anatomy of the Tracheophyta (Beck *et al.*, 1982; Schmid, 1982). According to the above concept the gross anatomical structure of the root and stem is similar, i.e. in both of them the cortex surrounds a central core which they termed the *stele*. The stele comprises the *pericycle*, i.e. the non-vascular tissue between the phloem and cortex, the *vascular tissues* and the *pith*, when present.

Medullary and cortical bundles, present on the inside and outside of the stele respectively, are found in certain plants (de Bary, 1877; Eames and MacDaniels, 1947; Metcalfe and Chalk, 1950). These bundles are associated with stems of both anomalous and typical structure. Medullary bundles occur in numerous dicotyledonous families, e.g. in the Amaranthaceae, Chenopodiaceae, Orobanchaceae, Berberidaceae, Cucurbitaceae and Phytolaccaceae (Fig. 111) (Kirchoff and Fahn, 1984). Cortical bundles are less common and are known to occur, for example, in the Melastomaceae, Proteaceae, Araliaceae, and Calycanthaceae. So-called "cortical" bundles are often leaf-traces which descend through the cortex for some distance before entering the stele, e.g. as in *Begonia* and *Casuarina*. In many plants with reduced leaves and a fleshy, photosynthetic cortex, branches from the base of the leaf-trace penetrate into the cortex (Eames and MacDaniels, 1947; Fahn, 1963).

Types of stele

The pattern of arrangement of the primary vascular tissues differs in the various plant groups and sometimes even within the same species. It has been found that the different patterns illustrate various stages in the evolution of the primary vascular system. The pattern in each plant is determined in the shoot and root apices from which the vascular tissues differentiate according to its genotype.

The stele of the sporophyte of vascular plants may be divided into two basic types (Smith, 1955; Foster and Gifford, 1959; and others): (1) *protostele* which consists of a solid central cylinder of xylem surrounded by phloem; and (2) *siphonostele* in which there is a cylinder of pith within the xylem. Both ontogenetically and phylogenetically the protostele is the more primitive, and it is thought that the siphonostele has developed phylogenetically from the protostele.

Three types of protostele can be distinguished: (1) *haplostele* (Fig. 102, no. 1) which is the simplest type, in which the xylem appears more or less circular in cross-section, e.g. *Rhynia* and *Selaginella*; (2) *actinostele* (Fig. 102, no. 2) in which the xylem is stellate in cross-section, e.g. *Psilotum*; and (3) *Plectostele* (Fig. 102, no. 3) in which the xylem is split into longitudinal plates of which some are joined and others separate, e.g. *Lycopodium*.

The various types of protostele are characteristic of the Lycopsida (the lower Pteridophyta) and the siphonostele of the Pteropsida (the phylogenetically more advanced Pteridophyta and the Spermatophyta).

Two types of siphonostele are distinguished according to the positions of the phloem and xylem: (1) *ectophloic siphonostele* in which phloem only surrounds the xylem externally (Fig. 102, no. 6); and (2) *amphiphloic siphonostele* (Fig.102, no. 4) in which the phloem surrounds the xylem both externally and internally and where the endodermis appears both outside and inside the vascular tissue on the borders of the cortex and pith, respectively (e.g. *Adiantum* and *Marsilea*).

The siphonostele may consist of a continuous cylinder of vascular tissue or of a network of bundles (Fig. 103, no. 1; Fig. 104). The latter type is the more advanced and the vascular tissues of this type appear in cross-section as a ring of separ-

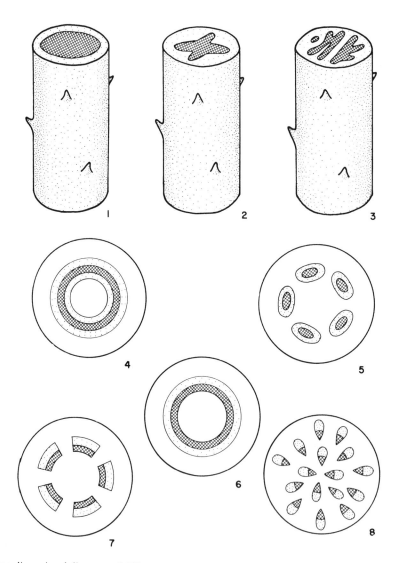

Fig. 102. 1–3, Three-dimensional diagrams of different types of protostele. The diagrams represent the stele alone without the cortex and epidermis. The microphylls appear in those positions where there are protuberances on the surface of the stele. 1, Haplostele. 2, Actinostele. 3, Plectostele. 4–8, Diagrams of cross-sections of stems with siphonosteles showing different stages in evolutionary development. 4, Amphiphloic siphonostele (solenostele). 5, Dictyostele. 6, Ectophloic siphonostele. 7, Eustele. 8, Atactostele. Xylem—hatched; phloem—stippled.

ate bundles. The regions between the bundles are parenchymatous. Those parenchymatous regions that occur in the stele, above the positions where the leaf traces diverge from the stele to the leaves, are termed *leaf gaps* (Fig. 104, no. 1). An amphiphloic siphonostele in which the successive leaf gaps are considerably distant, one from the other, is termed a *solenostele*. An amphiphloic siphonostele with overlapping gaps, i.e. that in which the lower part of one gap is parallel with the upper part of another gap, is termed a *dictyostele* (Fig. 102, no. 5). In this case the bundles are interconnected to form a cylindrical network (Fig. 103, no. 1), and each bundle is of concentric structure consisting of a central strand of xylem surrounded by phloem. Individually such bundles are termed *meristeles*. From the anatomical point of view these are amphicribral bundles.

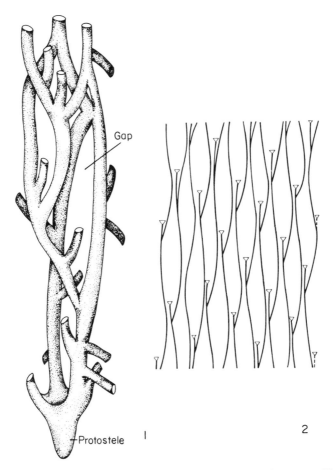

FIG. 103. 1, Three-dimensional diagram of the stele of the rhizome of *Ophioglossum lusitanicum*. The larger part of the stele is dictyostelic and only the small basal portion is protostelic. Both leaf traces (directed upwards) and bundles that enter the roots (directed downwards) arise from the bundles which surround the gaps. 2, Diagram of the primary vascular system of the stem of *Abies concolor*, spread out on one plane. The system is an open one. (No. 2 adapted from Namboodiri and Beck, 1968a.)

During the course of evolutionary development the *eustele* (Fig. 102, no. 7; Fig. 109; Fig. 110), with collateral bundles, developed from the ectophloic siphonostele. Bicollateral vascular bundles in which the xylem strands are accompanied externally and internally by phloem strands and which are found in advanced dicotyledonous families appear to be the result of a secondary specialization and not a relic of the primitive structure characteristic of the Filicinae. The vascular system of eustelic plants composed of bundles that branch and anastomose is referred to as a closed system (Fig. 109). A vascular system characterized by bundles which branch, but still form independent, separate systems, is referred to as an open

system (Fig. 103, no. 2). A compact arrangement of vascular bundles, which may form an open or closed system, seems to correlate with woody habit (Devadas and Beck, 1972).

In some plants, e.g. *Marattia*, *Pteridium*, and *Matonia*, two or more concentric cylinders of vascular tissue are present. Such a stele is termed a *polycyclic stele* (Fig. 105). The individual cylinders in this case are interconnected. In rare cases stems and roots contain more than one stele; such a condition is termed *polystelic* (see Chapter 13, p. 279).

A different interpretation of the above nomenclature exists and has been summarized by Sporne (1962). According to this interpretation

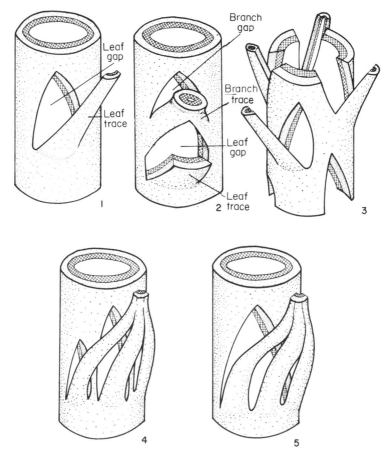

FIG. 104. Diagrams of siphonosteles with different types of arrangement of the leaf and branch traces and gaps. 1, Unilacunar node with one leaf trace. 2, Unilacunar node with associated branch traces and gap. 3, Overlapping gaps so that the stele forms a network of bundles. 4, Trilacunar node with three leaf traces. 5, Unilacunar node with three traces.

the ectophloic siphonostele without leaf gaps, such as is found in the Pteridophyta, is considered as a protostele and is termed a *medullated protostele*. Sporne does not use the term siphonostele for the Pteridophyta.

As in the dictyostele, the bundles of the eustele are usually interconnected. That type of stele in which the bundles are scattered (Fig. 102, no. 8), such as is characteristic of the monocotyledons, is called an *atactostele* (Nast, 1944).

In the siphonostele not all the interruptions in the vascular tissue are leaf gaps as described above. Some interruptions result from the secondary reduction of vascular tissue and the formation of interfascicular parenchyma. Such interruptions are termed *perforations*. When such perforations

occur in a solenostele it may be confused with a dictyostele. The parenchymatous connections between the pith and cortex are termed *medullary rays*.

Differing from the Pteropsida, whose steles form leaf gaps, the steles in microphyllous plants, i.e. the Psilopsida, Lycopsida, and Sphenopsida, are devoid of such parenchymatous regions (Fig. 102, nos. 1–3). In microphyllous plants with a siphonostele the only gaps present are *branch gaps* which are those gaps associated with bundles that depart from the central cylinder to the lateral branches. This type of siphonostele has been termed by some authors *cladosiphonic*, and that found in the Pteropsida *phyllosiphonic* (Jeffrey, 1910).

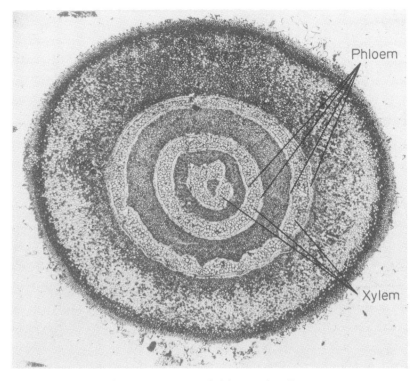

FIG. 105. A cross-section of a rhizome of *Matonia pectinata*, showing a polycyclic stele with three concentric amphiphloic vascular cylinders. × 13.

It appears that the siphonostele developed from the protostele by the transformation of the inner initials of the tracheary elements into parenchyma initials. This view is supported by the fact that tracheary elements may be found in the centre of the stele scattered among the parenchyma cells. This feature is common in relatively primitive plants, both living and fossil. Such steles may be considered as intermediate stages between protosteles and siphonosteles. Research on *Ophioglossum lusitanicum* (Gewirtz and Fahn, 1960), for instance, showed that the stele of the rhizome of the sporophyte, of this species, that developed from the gametophyte (not as a result of vegetative reproduction) was protostelic at the base and siphonostelic (dictyostelic) in its upper portion. Furthermore, it was clearly seen in the transition zone, below the level of the first leaf gap, that the pith undoubtedly originated from the xylem. Here parenchyma cells were seen mingled with tracheids and the number of parenchyma cells was

seen to increase in an upwards direction (Fig. 106, nos. 1, 2).

The vascular systems discussed above are composed entirely of primary tissues, i.e. of protoxylem and metaxylem and protophloem and metaphloem. The order in which successive metaxylem elements mature may be centrifugal or centripetal. When the protoxylem is on the inside and the differentiation of the metaxylem proceeds progressively towards the periphery, as in angiosperm stems (Fig. 100), the xylem is described as *endarch*. When the protoxylem is on the outside and the metaxylem differentiates centripetally, as in angiosperm roots (Fig. 156, no. 1), the xylem is described as *exarch*. In a third arrangement termed *mesarch*, the differentiation of the metaxylem proceeds both centripetally and centrifugally to the protoxylem. The exarch and mesarch types of primary xylem are apparently the more primitive.

The phloem always develops towards the

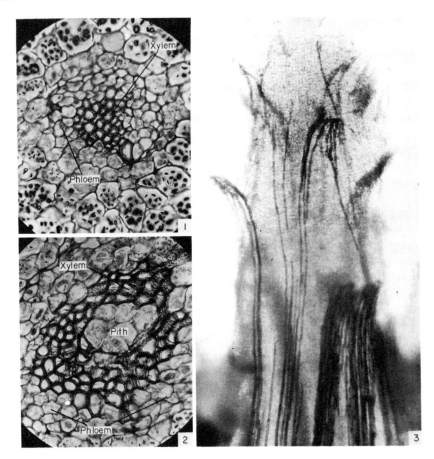

FIG. 106. 1 and 2, Micrographs of the central portion of cross-sections of the erect rhizome of *Ophioglossum lusitanicum* showing the transition from protostele to siphonostele. 1, Protostele. 2, Siphonostele. 3, A branch tip of *Chimonanthus* (a woody genus of the Ranales) cleared by treatment with lactic acid to show the pattern of vascularization in which the nodes are unilacunar with two leaf traces; the traces fuse on entry into the leaf. Nos. 1 and 2, × 200.

xylem. In those cases where the phloem is situated only on the outside of the xylem the development of the phloem is centripetal, and in internal phloem the development is centrifugal.

In the gymnosperms the evolution of the stele proceeded, as suggested by Sporne (1965), as follows: solid protostele with mesarch protoxylem; split tracheary tissue mixed with parenchyma, with protoxylem only in the outermost strands; mesarch xylem strands surrounding a pith; endarch xylem strands surrounding a pith.

The direct evolution of the eustele of the gymnosperms from the protostele by gradual medullation and concurrent separation of the peripheral

conducting tissue into sympodial bundles was suggested also by Namboodiri and Beck (1968b). The closed vascular system of conifers is regarded by these authors as derived from the open sympodial system characteristic of most gymnosperms (Fig. 103, no. 2), by fusion of pairs of leaf traces. Slade (1971) and Devadas and Beck (1972) suggested that the vascular system of the angiosperms has also developed from a protostele similar to that of the gymnosperms.

If the above concepts of stelar evolution in the seed plants were accepted the term *leaf gap* would not have evolutionary significance, though it would have useful descriptive meaning with reference to this group of plants.

Anatomy of the node

In the angiosperms, and especially in the dicotyledons, the primary vascular cylinder is interrupted at each node by the exit of one or more bundles that enter the leaves. The stelar bundles, which are the continuation of the bundles in the leaf bases, are called *leaf traces*. According to the number of leaf gaps per leaf the node is termed *unilacunar, trilacunar* or *multilacunar* (Figs. 104, and 107). The anatomy of the node is an important aspect of taxonomy (e.g. Sugiyama, 1979; Dickison and Endress, 1983) and of the comparative morphology of the stem, leaf, and flower. Sinnott (1914) concluded that the trilacunar node is the primitive type in the angiosperms and that the unilacunar node developed, phylogenetically from it by the loss of the two lateral gaps together with their respective traces, or by the approximation of the lateral traces to the median bundle to form a bundle composed of three traces, which is associated, therefore, with a single gap. Contrary to the reduction in the process of the formation of the unilacunar node, Sinnott states that the multilacunar node is formed by the addition of new gaps and traces.

The assumption that the trilacunar node is the primitive one among the angiosperms has been refuted by later research on nodal anatomy in the gymnosperms and angiosperms (Gunckel and Wetmore, 1946a, b; Marsden and Bailey, 1955; Marsden and Steeves, 1955; Bailey, 1956; Fahn and Bailey, 1957). By these workers and others it has been shown that in many Pteridophyta, in the Cordaitales, Bennettitales, in *Ginkgo* and *Ephedra*, a single gap is found in that position where the leaf traces depart from the stele. Similarly, in many dicotyledons unilacunar nodes with two leaf traces have also been found. Many of the dicotyledonous genera with unilacunar nodes with a double leaf trace belong to the primitive groups of the Ranales and the Chenopodiaceae. Bailey (1956) found in many dicotyledons that the vascular supply to the cotyledons consists of a double leaf trace which arises from a unilacunar node (Fig. 108).

Bailey is of the opinion that the leaves of angiosperms are able to undergo reversible changes in shape and vascularization.

In the light of the above facts it can be assumed: (1) that the unilacunar node of certain genera of the Ranales is primitive and has not changed during its evolution, from that of the lower Pteropsida; (2) in certain other dicotyledonous genera, e.g. genera of the Leguminosae, Anacardiaceae, and others, the unilacunar node has apparently been derived, by reduction, from a trilacunar node; (3) there are positive indications in certain dicotyledonous groups, such as the Epacridaceae and Chloranthaceae, that in the course of evolution the tri- and multilacunar nodes have arisen from the unilacunar node (Bailey, 1956).

The evolutionary development from a unilacunar node with two leaf traces to other types of unilacunar nodes with one, three, or more traces may occur in a single family as can be demonstrated in the Chenopodiaceae (Bisalputra, 1962; Fahn and Broido, 1963). In the Gnetaceae, *Ephedra* has unilacunar nodes with two leaf traces whereas in *Gnetum* the nodes are multilacunar. As stated by Rodin and Paliwal (1970) the nodal anatomy of *Gnetum* represents the highest degree of specialization among the gymnosperms.

In order further to clarify the nature of the leaf traces and the nodal type, it is necessary to study more accurately the nature of the traces and to follow their passage downwards in the stem. Often the picture obtained at the node is not a true reflection of the situation, but lower in the stem the stele is more conservative (Fig. 106, no. 3) and a more correct picture of the arrangement of the leaf traces is obtained (Fahn and Bailey, 1957). The evolutionary trends of the basic types of nodal structure, as suggested by Bailey and his co-workers, are presented in Fig. 107, no. 5. Trends that differ from these have been suggested by some other authors (Neubauer, 1981).

The vascular bundles of the stem are usually classified into *axial bundles* (also *cauline bundles* and *common bundles*), leaf trace bundles, branch trace bundles and sympodia. An *axial bundle* refers to those bundles which do not clearly belong to any one leaf or lateral branch. They form the major vascular system of the stem, and may anastomose and give rise to leaf and branch traces. A *leaf trace* is a bundle that diverges from an axial bundle (or

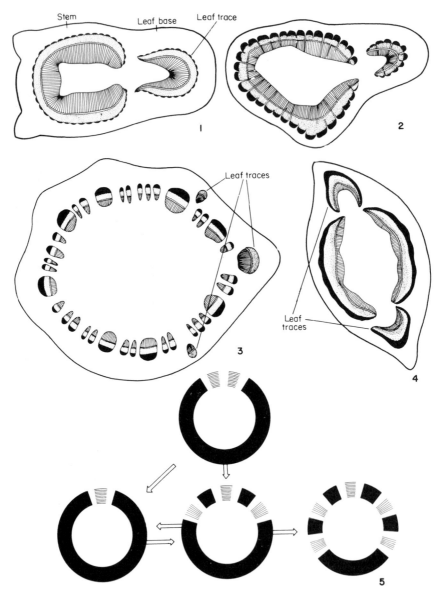

Fig. 107. Diagrams showing the relationship between the vascular systems of the leaf and the stem. 1–4, Cross-sections of the nodes of young stems. 1, *Eucalyptus camaldulensis* in which the node is unilacunar. 2, *Laurus nobilis*, also with unilacunar node. 3, *Chrysanthemum anethifolium* in which the node is trilacunar with three traces. 4, *Dianthus caryophyllus* which has opposite leaves and unilacunar nodes. 5, Diagrams showing the possible ways of development of the nodal vascularization in dicotyledons from the unilacunar node with two traces. (No. 5 adapted from Marsden and Bailey, 1955.)

another leaf trace) and extends into a leaf. *Branch trace* differs from the leaf trace by diverging into a lateral shoot and not into a leaf. A *sympodium* consists of an axial bundle and its associated leaf and branch traces.

Some authors consider the vascular system of the stem to consist entirely of interconnecting leaf traces organized in sympodia (O'Neill, 1961; Balfour and Philipson, 1962; Philipson and Balfour, 1963; Esau, 1965a, b; Fahn, in the

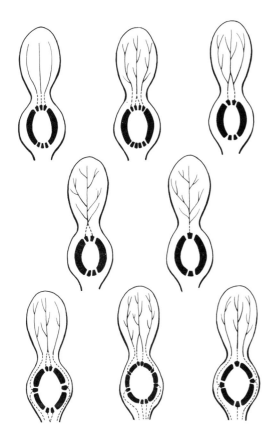

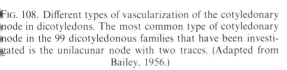

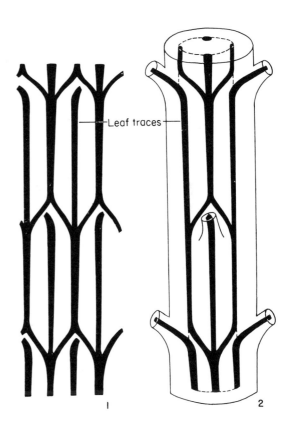

FIG. 108. Different types of vascularization of the cotyledonary node in dicotyledons. The most common type of cotyledonary node in the 99 dicotyledonous families that have been investigated is the unilacunar node with two traces. (Adapted from Bailey, 1956.)

FIG. 109. Diagrams of the primary vascular system, of the stem of *Anabasis articulata* (closed system). 1, Vascular system spread out on one plane. 2, Three-dimensional diagram showing the bundles on the one side of the stem. The number of bundles and their arrangement as seen in cross-section can be seen at the level of the cut.

previous editions of the book). This view is based on morphogenetic studies of the apical region, which suggest that the pattern of the vascular bundles of the shoot is determined by the developing leaf primordia (Sachs, 1981).

The number of bundles at a given level of the stem is determined by the phyllotaxy, the number of axial bundles, the number of leaf and branch traces and the number of internodes the traces traverse (Fig. 106, no. 3; Figs 110 and 111). Inasmuch as the phyllotaxis is more dense and the leaf traces continue down along a larger number of internodes, the number of bundles, as seen in a

cross-section of the stem, will be greater. Changes in number may sometimes occur along the stem of one plant. It is known, for example, that the number of bundles seen in a cross-section of that part of the stem that develops first, i.e. the lowermost portion, is less than that seen in cross-sections of higher portions, and that the number of bundles again decreases in the uppermost portion prior to the development of floral primordia.

Anisophylly and dorsiventral symmetry of shoots may cause pronounced modification of the vascular organization, as has been observed in *Pellionia daveauana* (Dengler and Donnelly, 1987).

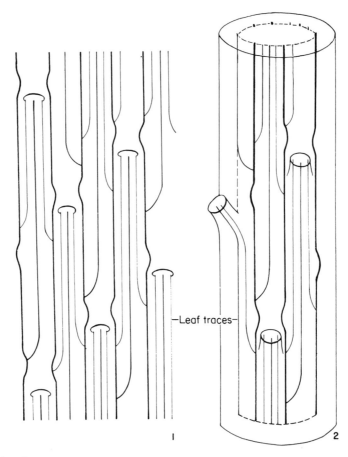

−Leaf traces−

1   2

FIG. 110. Diagrams of the primary vascular system of the stem of *Chenopodium glaucum* (open system). 1, Vascular system spread out on one plane. 2, Three-dimensional diagram showing the bundles on one side of the stem. The number of bundles and their arrangement as seen in a cross-section can be seen at the level of the cut.

Attempts have been made to use nodal anatomy as an aid to taxonomy (e.g. Philipson and Philipson, 1968).

Branch traces

Branches develop from axillary buds and have vascular connections with the main axis. These vascular connections are termed *branch traces*. At the node, the branch traces are situated very close to the leaf traces that enter the leaf in whose axil the branch develops. The branch gap is situated above the leaf gap and together these two gaps appear as one. In the dicotyledons and gymno-sperms the vascular supply to the axillary branches usually consists of two traces (Fig. 104, no. 2). However, in some plants only one trace is present and in others more than two. In the branches themselves, the stele is similar to that of the main axis.

Summary of the arrangement of the vascular system of dicotyledons

The stems of the various dicotyledons differ from one another in the pattern of the primary vascularization (Balfour and Philipson, 1962; Philipson and Balfour, 1963; and others). These differences are apparently connected with evolutionary development.

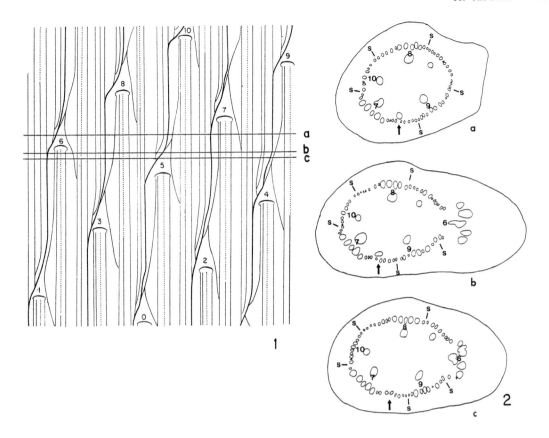

FIG. 111. Primary vascular system of *Phytolacca dioica* with medullary bundles. 1, Simplified diagram of a shoot with 2/5 phyllotaxis. One line represents several vascular strands. Five sympodial bundles are shown by heavy lines. The leaf traces and their branches are represented by thin lines. The course of the medullary bundles are marked by dotted lines. At the point where dotted and solid lines meet, several vascular strands move into the pith to form the medullary bundle. Horizontal lines *a, b,* and *c* indicate the levels of the sections shown in no. 2. 2, Tracings of sections at the levels indicated in no. 1. The numbers in the drawings indicate the vascular supply to the leaves with the same number as in the diagram of no. 1. *s,* regions of sympodial bundles; arrow, region in which vascular strands depart the vascular cylinder to give rise to a medullary bundle. (From Kirchoff and Fahn, 1984.)

The amount of primary vascular tissue, as has been described above, varies from a solid through a hollow uninterrupted cylinder to a small number of narrow separate bundles. It is assumed that during the course of evolution the primary vascular cylinder became thinner, i.e. it underwent reduction in a radial direction and because of the appearance of leaf gaps, branch gaps, and perforations, and, because of further reduction of the vascular tissue in a tangential direction, the cylinder became split into the longitudinal strands such as are seen in most dicotyledons.

The arrangement of secondary vascular tissue

of the gymnosperms and dicotyledons bears no relation to the arrangement of the primary vascular tissue, and may be in the form of an entire cylinder. However, the amount and arrangement of the secondary vascular tissues and especially that of the xylem may also vary from an entire cylinder of various widths, as in trees, to separate strands, as in the herbaceous stems of certain annual dicotyledons, e.g. *Cucurbita,* and in certain species it may even be almost completely absent. It is assumed that the reduction of the secondary vascular tissue is also the result of evolutionary processes.

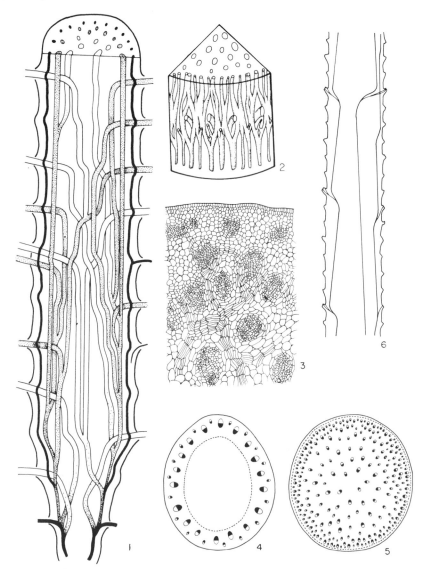

Fɪɢ. 112. 1–3, *Zea mays*. 1, Diagrammatic representation of a longitudinally sectioned stem showing the vascularization. Mid-rib leaf traces—unshaded; large lateral leaf traces—stippled; small lateral traces and very thin leaf bundles which fade out soon after entering the cortex—black. Coleoptile traces and the vascular system of the hypocotyl appear at the base of the diagram and are drawn in solid black. 2, Diagram of a reconstruction of part of a node. 3, Outer portion of cross-section of a young node showing the development of horizontal procambial strands which interconnect the vertical bundles. 4 and 5, Cross-sections of monocotyledonous stems showing two types of arrangement of the vascular bundles. 4, *Secale*, in which the vascular bundles are arranged in two rings. 5, *Zea mays*, in which the vascular bundles are scattered throughout the cross-section. 6, Diagram of the course of vascular bundles in the stem of *Rhapis excelsa*. (Nos. 1–3 adapted from Kumazawa, 1961; nos. 4–5 adapted from Troll, 1948; no. 6 adapted from Zimmermann and Tomlinson, 1965.)

Vascular system of monocotyledons

The vascular system of monocotyledons usually consists of bundles that are scattered throughout the ground tissue of the stem. In cross-sections of such stems it can be seen that the bundles do not form a ring such as is found in cross-sections of most dicotyledons. Among dicotyledons, however, an arrangement of more or less scattered bundles does occur in a small number of plants, e.g. in the

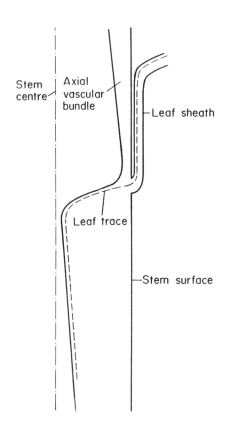

Stem centre

Axial vascular bundle

Leaf sheath

Leaf trace

Stem surface

FIG. 113. Diagram of a departure of a leaf-trace bundle in *Rhapis excelsa*. Solid line of the vascular bundle indicates the presence of metaxylem elements; dashed line, narrow protoxylem elements. (Adapted from Zimmermann and Sperry, 1983.)

Nymphaeaceae, many plants of the Ranunculaceae and the herbaceous genera of the Berberidaceae.

The course of vascular bundles in the stem of the palm *Rhapis excelsa* was investigated by Zimmermann and Tomlinson (1965, 1966) by a cinematographic method. According to these studies essentially "all bundles maintain their individuality and proceed indefinitely up the stem" from the base of the stem to the apex. Each of these vertical bundles—minor, intermediate, and major—give off a leaf trace at intervals. The minor bundles give off leaf traces more frequently than the major ones. Each vertical bundle slowly reaches a more central position, when followed from the base to the apex. At intervals the bundle "turns" sharply towards the periphery whilst it gives off the

leaf trace, satellite bundles which go into the inflorescence and bridge bundles which make connections with neighbouring bundles. Major bundles extend more into the centre of the stem than the minor ones (Fig. 112, no. 6). Because only relatively few bundles extend to the centre of the stem and because all bundles in parts of their course are found in the peripheral region, this region is crowded with bundles. In the central, uncrowded part of the stem, all bundles when traced upwards rotate uniformly and describe a helix in the direction of the phyllotactic spiral. In *Rhapis*, Zimmermann and Sperry (1983) observed that in the area of the leaf trace between the junction of axial bundles and the base of the leaf sheath metaxylem is absent and only narrow protoxylem tracheids are present (Fig. 113). This causes considerable hydraulic constriction in the xylem path between stem and leaf petiole. Zimmermann and Sperry suggested, that this constriction plays a critical role in confining cavitation (embolism, air entrance into tracheary elements) at damaging situations and subsequent conduction failure to further areas of the stem. The occurrence of only narrow protoxylem elements at the leaf insertion may thus insure the functional integrity of the xylem of the stem.

The primary vascular system of arborescent Liliflorae is in principle similar to that of palms (Zimmermann and Tomlinson, 1969). The patterns of organization of the vascular bundles in the Araceae have been studied in detail by French and Tomlinson (1981a–d). In other monocotyledons, e.g. in the Dioscoraceae, the vasculature of the stem is very complicated and differs markedly from that of the palms (Ayensu, 1969, 1970, 1972).

In the Gramineae there are two basic types of arrangement of the vascular bundles: (1) that in which the vascular bundles are arranged in two circles (Fig. 112, no. 4), the outer circle consisting of thin bundles and the inner of thick bundles, e.g. *Triticum, Hordeum, Avena,* and *Oryza*; and (2) that in which the vascular bundles are scattered throughout the entire cross-section of the stem, e.g. *Sorghum, Saccharum,* and *Zea* (Fig. 112, no. 5; Fig. 114, no. 1). The individual bundles of the Gramineae are collateral and are usually surrounded by a sclerenchymatous sheath (Metcalfe, 1960).

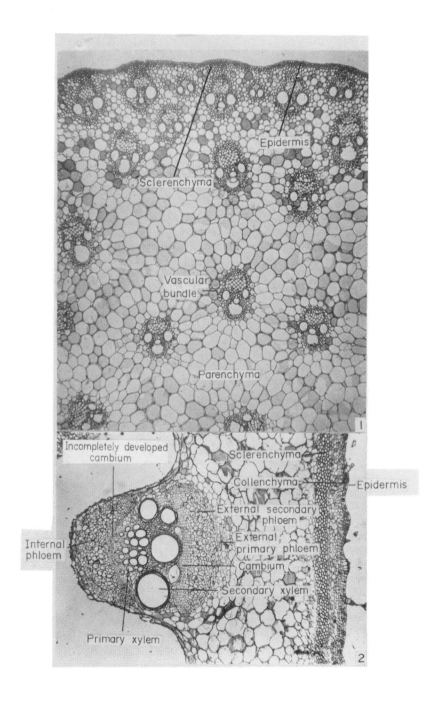

F<small>IG</small>. 114. 1, Micrograph of the outer portion of a cross-section of the stem of *Zea mays* showing the vascular bundles scattered in the ground parenchyma. × 46. 2. Portion of a cross-section of *Cucurbita* in which external and internal phloem can be distinguished. × 32.

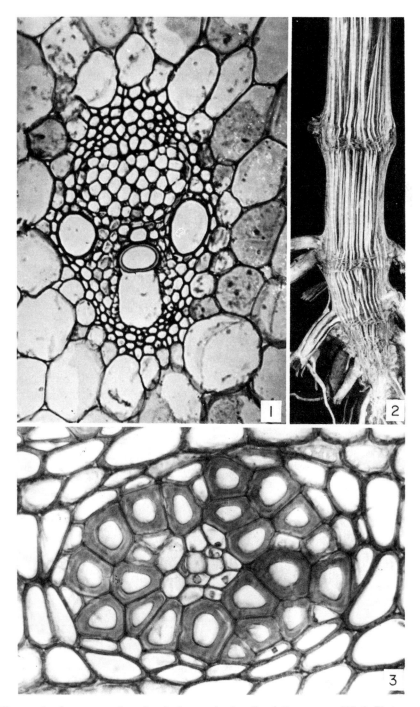

FIG. 115. 1, Micrograph of a cross-section of a single vascular bundle of *Zea mays*. ×230. 2, Photograph of the lower portion of a longitudinally sectioned mature stem of *Zea mays* in which most of the parenchymatous tissues have been destroyed and showing horizontally orientated vascular bundles at the nodes. 3, Micrograph of a cross-section of an amphivasal secondary vascular bundle from the stem of *Dracaena*. ×490. (No. 2 from Sharman, 1942.)

The complicated arrangement of vascular bundles in the monocotyledonous stem, and especially in the Gramineae, is connected with the position of the various bundles of each leaf in the stem. According to Kumazawa (1961), three types of leaf trace bundles are present in the stem of *Zea* (Fig. 112, nos. 1, 2). The first type is represented by those bundles derived from mid-ribs and the larger lateral veins. These penetrate deep into the interior of the stem and then pass to the periphery in lower nodes. The second type is represented by leaf trace bundles derived from smaller lateral veins which, immediately on entering the stem, occupy the position of the outermost peripheral bundles with which they fuse sooner or later. The third type is represented by bundles that are very thin and which soon fade out within the cortex close to the level of the node. In the Gramineae there are special horizontal bundles at the nodes which connect the leaf bundles together (Fig. 112, no. 3; Fig. 115, no. 2).

In the stem of *Triticum* the internodes are hollow and the nodes solid. The thin leaf bundles are continued in the outer circle of the stem, and the thick bundles constitute the inner circle of bundles.

In the nodes of *Triticum* O'Brien and Zee (1971) distinguished three types of vascular bundle: (1) nodal plate bundles which contain phloem transfer cells; (2) large bundles elliptical in cross-section, containing xylem transfer cells; and (3) diffuse bundles with irregular outline, which are apparently a result of fusion of collateral bundles of the internode, and which alternate with the elliptical bundles. These diffuse bundles lack xylem transfer cells but contain phloem transfer cells, especially where they join the nodal plate bundles. (For transfer cells see Chapter 12.)

Ontogeny of the primary vascular system

With the progressive differentiation of the promeristematic cells the three meristems, the protoderm, ground meristem and procambium, are formed. The procambium may be in the form of a solid or hollow cylinder or of strips. The developing procambial strips, as seen in cross-sections, are organized in groups of cells which divide in all

directions, the commonest direction being periclinal. The last gives the procambial strand a concentric appearance (Fig. 172, nos. 1, 2).

The procambium stains more intensely than the neighbouring tissues as a result of the delayed appearance of the vacuoles in it. The development of the procambium is continuous in an acropetal direction, i.e. from below towards the apex. The procambium differentiates below the position where a leaf primordium is formed as a small buttress on the surface of the apex.

The differentiation of procambial cells into cells of the vascular tissue takes place in different plants and organs at different stages of procambial development. For instance, in aerial stems of the spermatophytes the procambial cells undergo differentiation into tracheary elements even at the level where cell division still takes place in the procambium, while in the roots of most spermatophytes and in the stems of most pteridophytes the first differentiation of the procambium takes place only where almost all of the procambium cells have ceased to divide.

The earliest sieve elements (those of the protophloem) usually undergo acropetal differentiation, i.e. from the phloem of the traces of more mature leaves, to the leaf primordia. The development of the phloem commences before the appearance of the xylem elements. In spermatophytes the primary xylem begins to differentiate from the procambium near the base of the developing leaf, from where its further differentiation continues in two directions—acropetally into the leaf, and basipetally into the stem. In the stem the newly formed xylem becomes continuous with the xylem of earlier formed strands.

The above processes of differentiation have been proved only in relation to the development of the protophloem and protoxylem. As yet the pattern of differentiation of the metaphloem and metaxylem along the shoot has not been investigated thoroughly in the spermatophytes.

In the procambium it is possible to distinguish between that part that will differentiate into xylem and that part that will differentiate into phloem by the more intense staining of the latter as well as by the differences in the plane of cell division.

In plants that exhibit secondary thickening that

part of the procambium between the primary xylem and phloem remains meristematic and forms part of the vascular cambium (see Chapter 14).

Experimental work on the control of vascular tissue differentiation has yielded some understanding of the mechanisms determining the pattern of vascular strands. It has been shown, both in plants and in tissue culture, that young leaves cause the differentiation of xylem and this effect can be replaced by auxin (Jacobs, 1952; Wetmore and Rier, 1963; Wangermann, 1967; McArthur and Steeves, 1972; Aloni, 1987, 1988). The differentiation of vascular strands appear to occur along the path of auxin transport, from the shoot to the root (Thompson and Jacobs, 1966; Sachs, 1968a). It has been shown that differentiation into defined strands, rather than into a diffuse system, can be explained by polarization of auxin transport occurring at an early stage in vascular differentiation (Sachs, 1969). The cells which first start to differentiate, therefore, become the preferred channels for auxin transport, and this in turn causes their further differentiation and diverts the auxin from cells which are not within the differentiating strand. Leaf removal and hormone treatment experiments showed that contacts between strands are readily formed only between a new strand and an established strand which does not connect to a young leaf and, therefore, is not loaded with auxin (Sachs, 1968b, 1981).

The pith

The pith is a more or less cylindrical body of tissue in the centre of the axis, enclosed by the vascular tissues. It has long been known under the name of *medulla* which was later abandoned, although the extensions of the pith between the vascular bundles have retained the name *medullary rays*. The pith consists of a rather uniform tissue, mainly parenchymatous, in which the cells are arranged usually loosely. Often thick-walled, lignified parenchyma cells and sclereids are also present. Fibres occur only rarely, and if so, in the peripheral region where they are associated with the primary vascular tissue. In some species secretory structures occur in the pith.

In the ontogeny of the stem, the pith cells in the internodes of many species mature very early and

stop growing, whereas the surrounding tissues are still meristematic and continue to enlarge longitudinally and in circumference. Thus the pith may be torn apart and a hollow pith is formed, with the broken cell walls lining the cavity. This is common in herbaceous plants. In stems of some plants, e.g. *Phytolacca americana*, the pith is chambered: the internodes consist of hollow chambers separated by parenchymatous diaphragms (Mikesell and Schroeder, 1980).

Storage stems

Portions of stems may become thickened and remain prevailingly parenchymatous and store reserve materials. The development of storage organs from stems results from the activity of primary or secondary meristems. In kohlrabi, for instance, this occurs between the third and fifth nodes. The meristematic activity resulting in bulbing of the stem initiates in the central pith, after a normal cambial zone has been established (Selman and Kulasegaram, 1967).

In *Hordeum bulbosum* the bulb is formed from the lowest internode. Bulb initiation begins by rapid cell divisions and elongation. The bulb form is achieved by cell files directed diagonally outward in the upper part of the bulb internode and by a peripheral meristem contributing cells to the middle and lower parts of it (Leshem, 1971).

ADAPTATION OF THE STEM TO DESERT AND SALINE HABITATS

In desert perennials the leaves are usually very much reduced, e.g. as in the articulated Chenopodiaceae, or they are shed at the beginning of the very long dry season. In some plants, e.g. *Artemisia* spp., *Reaumuria* Spp., *Gymnocarpos fruticosus*, and *Atriplex* spp., the shed leaves are replaced, during the dry season by smaller and more xeromorphic leaves (Zohary, 1961). In other plants, e.g. *Zygophyllum dumosum*, the chloroplast-containing petioles are retained after the leaflets are shed. In still others, such as *Retama raetam* and *Calligonum comosum*, in which the function of photosynthesis is taken over by the young green branches, these branches may be shed. In many desert plants even ordinary branches and larger portions of the plants die during the dry season (Orshan, 1953) so reducing the plant body and its requirements to a minimum. However, although most desert perennials appear dead at the end of the dry season they are capable of developing new shoots with the onset of the rainy season.

As little or no foliage actually remains on desert plants during the dry season, the main problem of adaptation is not by what means transpiration is reduced in the leaves, but how the plant remains viable until the onset of the rainy season. The answer to this question must therefore be sought in the axis of the plant.

In some desert and semi-desert plants the function of photosynthesis is taken over by the cortex of the stem (Fahn, 1964; Gibson, 1983). In others, in addition to the presence of photosynthesizing tissues, water-storing parenchyma is developed. The epidermis is often multiseriate and covered by a thick cuticle. In some apophyllous and other xerophytic plants stomata openings of older but still assimilating regions may become plugged by wax or other substances. The stomata may be also blocked from the inside by cells developing from the subsidiary cells or from palisade cells (Volkens, 1887; Fahn and Dembo, 1964; Böcher and Lyshede, 1968, 1972; Böcher, 1972, 1975; Lyshede, 1977a, b, 1979; Heide-Jørgensen, 1978). In older but still assimilating internode portions of *Anabasis articulata* the stomata apparently cease functioning as a result of a large increase in the thickness of the guard-cell walls and the senescence of the guard-cell protoplasts (Gedalovich and Fahn, 1983).

Another commonly observed character of desert shrubs is the occurrence of a split axis. The axes of *Artemisia herba-alba, Peganum harmala, Zygophyllum dumosum, Zilla spinosa, Ambrosia dumosa,* and some *Frankenia* species for instance, become split by various anatomical mechanisms into separate parts "splits" (Ginzburg, 1963; Jones and Lord, 1982; Wahlen, 1987) which may compete with one another; the split in the most favourable microhabit around the mother plant will probably be the one to survive.

Many perennial plants growing in a Medi-

terranean-type of climate, (e.g. species of *Eucalyptus, Adenostoma, Arctostaphylos* , and *Ceanotus*) produce woody outgrowths, called *lignotubers* or *burls*, at the base of their stems. These outgrowths contain dormant buds which sprout after the shoots are mechanically damaged or consumed by fire (Naveh, 1975; Carr *et al.*, 1982; Lacey, 1983; Montenegro *et al.*, 1983; James, 1984).

In primary stems the cortex is generally narrow and the vascular tissues are situated on the periphery of a wide pith. However, it has been observed that the cortex of the primary stem of plants growing in deserts or salt marshes is recognizably thicker than in mesophytes. This feature, which is accompanied by the "contraction" of the vascular strands around a narrow pith, may be an adaptive one by which the vascular tissues are protected from drought or other damage in the early stages of development before the periderm is developed.

It is known that in the articulated Chenopodiaceae, such as *Anabasis* spp., *Haloxylon* spp., and *Arthrocnemum glaucum*, the fleshy photosynthesizing cortex is shed from the mature stems in the summer as the result of the formation of a periderm which develops in the phloem parenchyma (Fahn, 1963). It is also of interest that in some desert plants, such as *Atriplex halimus, Zygophyllum dumosum*, and *Fagonia cretica*, for example, which do not have a fleshy cortex, the first-formed periderm also develops deep within the stem in the pericycle or phloem parenchyma. This is an adaptive feature.

In some desert shrubs, such as *Artemisia* spp. and *Achillea fragrantissima*, interxylary cork rings are produced at the end of each annual wood increment. Moss (1940) has already pointed out the important adaptive value of these interxylary cork tissues which reduce water loss and restrict the upward passage of water to a narrow zone of functioning secondary xylem.

The anatomical structure of the stems of some *xerophytes* (plants growing in arid habitats) and one *halophyte* (plant growing in saline habitat) will be described here.

Retama raetam (Fig. 116, no. 3) may be cited as an example of plants with xeromorphic non-succulent stem (Evenari, 1938). Along the green branches of this plant there are ribs and furrows. The stomata

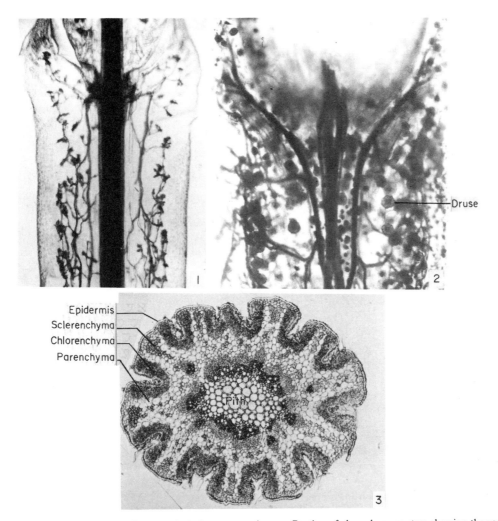

FIG. 116. Succulent and xeromorphic stems. 1, *Arthrocnemum glaucum*. Portion of cleared young stem showing the vascular system to be contracted to a narrow cylinder in the centre of the stem and the anastomosing system of vascular bundles in the fleshy cortex and reduced leaves. × 14. 2, Portion of a cleared fleshy young stem of *Anabasis articulata*, from which the epidermis has been removed; in addition to the details distinguishable in no. 1, numerous crystals in the form of druses can be seen. × 30. 3, Micrograph of a cross-section of a young stem of *Retama raetam* in which the ribbed nature of the cortex can be seen. × 35.

are situated in the furrows in which there are also numerous hairs. The central portion of the ribs consists of a sclerenchymatous strand which is accompanied laterally, on the sides facing the furrows, by one or two rows of large colourless parenchyma cells which serve as water-storing cells and usually contain crystals. Between these cells and the epidermis is the photosynthesizing tissue, which consists of small and dense parenchyma cells containing chloroplasts and crystals. The epidermal cells have very thick outer walls and cuticle.

The *succulent stems* of desert plants are characterized by a well-developed water-storing tissue in the cortex. As a result of this the ratio between the cortex and vascular cylinder in these plants is considerably larger than that of other dicotyledons (Fig. 116, nos. 1, 3). *Anabasis articulata* (Fig. 116, no. 2; Figs 117, 118) may be cited as an example of desert plants with a succulent stem (Volkens, 1887; Evenari, 1938; Fahn and Arzee, 1959; Lyshede, 1977a). In the young green internodes the epidermis consists of 3–4 layers of thick-walled cells

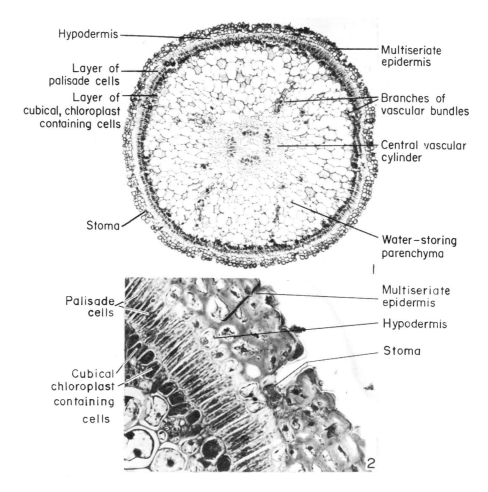

Hypodermis

Layer of palisade cells

Layer of cubical, chloroplast containing cells

Stoma

Multiseriate epidermis

Branches of vascular bundles

Central vascular cylinder

Water-storing parenchyma

Palisade cells

Cubical chloroplast containing cells

Multiseriate epidermis

Hypodermis

Stoma

FIG. 117. 1, Micrograph of a cross-section of the stem of *Anabasis articulata*. × 33. 2, Portion of no. 1, enlarged. × 400.

and it is covered by a thick cuticle and wax. The stomata are deeply sunken. In *Anabasis calcarea* where there are 8–11 layers of epidermis the stomata occur at the base of an obliquely oriented stomatal crypt (Bokhari and Wendelbo, 1978). Below the epidermis is a hypodermis of thinner-walled cells which, similarly to the epidermis, contain crystals. On the inside of this layer there is a layer of palisade cells with chloroplasts. Immediately inside the palisade tissue is a layer of more or less cubical cells which also contain chlorophyll. Still further inwards is the water-storing parenchyma which also contains, here and there, large druses. With the maturation of the stem, cork tissue develops in the outer phloem parenchyma.

Branches of the vascular bundles of the internodes pass through this cork tissue into the cortex. In still older branches the connection between the bundles and their branches is disrupted by the completion of the cork cylinder and, as a result, the outer layers dry out and are shed.

In *Salicornia fruticosa* (Fig. 119) which grows in salines, the structure of the cortex is simpler (Fahn and Arzee, 1959). The epidermis is single-layered and thin-walled. The photosynthesizing tissue consists of large palisade cells which store water, as do the parenchyma cells of the inner cortex.

In contrast to the succulent stems of desert plants mentioned above, in which the pith is very

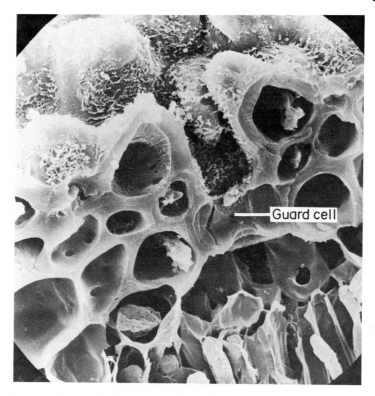

FIG. 118. Scanning electron micrograph of a transversely cut internode of *Anabasis articulata*. A guard cell, in its longitudinal view, is seen at the base of a crypt formed above the stoma. The outermost epidermal cells are papillate and covered by wax. × 730. (From Gedalovich and Fahn, 1983).

narrow, one of the evolutionary trends of the development of succulency in the Cactaceae has been the enlargement of the pith (Gibson, 1975; Gibson and Horak, 1978).

ADAPTATION TO AQUATIC HABITAT

The restricted penetration of light into water creates conditions similar to those in which densely shaded land plants grow. The submerged leaves (see Chapter 12, The Leaf) and stems have an increased distribution of chloroplasts and a reduced cuticle. There is a high surface area/volume ratio in the leaves, which are often dissected or fenestrated. The stems contain an extensive system of intercellular spaces through which diffusion of dissolved gases may freely occur. Absorption of gases is also facilitated by the thin walls of the epidermis and inner tissues. Cuticle is usually absent from submerged organs, but a very thin cuticle was reported to occur on the epidermis of stems and leaves of several species, e.g. *Elodea canadensis, Groenlandia densa, Hottonia palustris*, and *Myriophyllum spicatum* (Sculthorpe, 1967).

Chloroplasts generally occur throughout the mesophyll of leaves and outer cortex of many stems. In submerged leaves and stems of water plants chloroplasts occur in the epidermal cells, a feature common also to dicotyledons of deeply shaded habitats. In *Ceratophyllum, Myriophyllum, Potamogeton*, and some marine angiosperms, e.g. *Zostera*, the epidermis is richer in chloroplasts than the inner tissues. In these species the cortex and mesophyll function mainly as storage tissues for starch and lipids. In the Podostemaceae chloroplasts also occur in the epidermis of the flattened root thalli. In most submerged hydrophytes the epidermis lacks stomata.

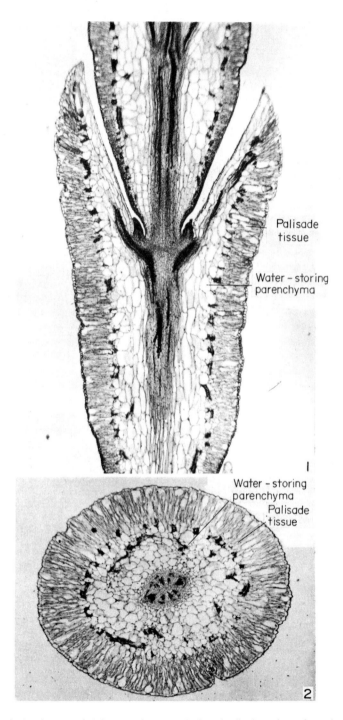

Palisade tissue

Water-storing parenchyma

Water-storing parenchyma
Palisade tissue

FIG. 119. Succulent articulated stem of *Salicornia fruticosa*. 1, Longitudinal section of portion of the stem. × 30. 2, Cross-section of an internode. × 33.

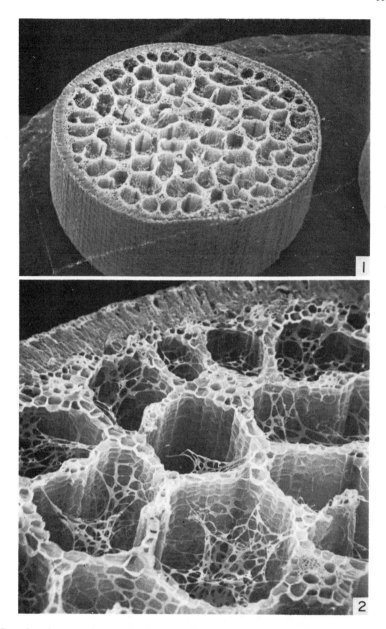

FIG. 120. Scanning electron micrographs of a stem of *Cyperus distachyos*, showing lacunae. 1, × 18. 2, × 90.

In the stem cortex and in the ground tissue of the petioles and mesophyll except for the very reduced leaves, schizogenous air passages or lacunae exist (Fig. 120, Fig. 123, no. 3; Fig. 132; Fig. 133, no. 2). The lacunae occur in the middle cortex of the stem. The outer cortex consists of a compact parenchyma or collenchyma. The inner cortex which surrounds the vascular cylinder also consists of a dense parenchyma. The number of lacunae, as seen in cross-section, is often characteristic to the species. The lacunae may be arranged in one ring, e.g. in *Ceratophyllum* and

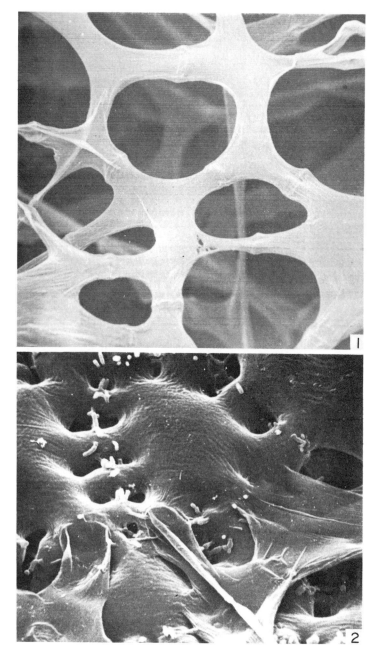

FIG. 121. Scanning electron micrographs of cells occurring in the lacunae of *Cyperus distachyos* stem. 1, Stellate parenchyma filling the spaces between the diaphragms. × 2000. 2, Cells of the diaphragm. × 1100.

Myriophyllum or in several rings, e.g. in *Hippuris.* In the petiole of *Nuphar* the lacunae are arranged in a reticulate pattern.

In some genera specialized cells occur in the tissue surrounding the lacunae. In *Myriophyllum,* cells with clustered crystals of calcium oxalate are found. In *Brasenia* mucilage glands project into the lacunae. In *Nuphar, Nymphaea,* and *Victoria* sclerenchymatous idioblasts of bizarre shape occur (Fig. 132). The lacunae are transversed at intervals by plates, or diaphragms, which strengthen the organs and may also eliminate the danger of waterlogging through wounds. Diaphragms in stems are most commonly found at the nodes. The diaphragms vary in structure and development. In *Nuphar* the lacunae are at first continuous, then groups of cells surrounding the lacunae at certain places swell, branch in the cavity, and divide to form a spongy mass which forms a diaphragm. In *Pontederia, Sagittaria, Potamogeton* and some other genera, the diaphragms are one or sometimes two to three cells thick. Between the cells characteristic intercellular spaces develop especially at the cell angles (Fig. 121, no. 2). In *Hippuris* the spaces enlarge considerably and the cells become stellate (Arber, 1920). Lacunae seldom occur in the Podostemaceae (Metcalfe and Chalk, 1950).

In non-submerged aquatic plants, e.g. in *Cyperus* (Fig. 121, no. 1) and *Scirpus*, the spaces between the diaphragms are filled with a stellate parenchyma (cf. Metcalfe, 1971; Kaul, 1971).

The most specialized tissue found in the stem and respiratory roots of many water plants is the *aerenchyma* (Fig. 163, no. 2). Aerenchyma is, strictly speaking, a phellem tissue derived from a typical phellogen of either cortical or epidermal origin (Haberlandt, 1918). During the development of this phellem, single cells from each cell layer elongate in a radial direction (in relation to the organ) at constant intervals, while the cells between them remain short. This lengthening of the cells pushes the previously formed layer of cells outwards so that numerous elongated air chambers, parallel to the axis of the plant, are formed. The elongated cells form the radial walls of the air chambers, while the non-elongated cells form the tangential walls. From the physiological point of view, however, any tissue that contains large intercellular spaces is termed aerenchyma.

Aerenchyma, which serves as an air transport system in waterlogged plants, may develop in crop plants growing in wet soil. The intercellular spaces of such plants develop mostly by disintegration of cells. It was suggested that in some plants (e.g. *Helianthus annuus* and *Lycopersicum esculentum*), when waterlogged the deficiency of oxygen triggers anaerobic stimulation of ethylene production, which causes an increase in cellulase activity (Kawase, 1979; Kawase and Whitmoyer, 1980). This activity brings about disintegration of cells and thus the development of aerenchyma. As only part of the cells in the tissue disintegrate, Kawase (1979) suggests that there is competition for water between neighbouring cells after the cell walls have been softened by the action of cellulase. Stronger cells survive and become larger whereas weaker ones loose water, become plasmolysed, and die.

The root system of hydrophytes is usually much reduced; it principally provides an anchorage in the soil since the uptake of water and salts is carried out by the stems and leaves. For the same reason the vascular system is also much reduced. The reduction is especially obvious in the xylem tissue. The roots of *Thalassia*, for instance, have no tracheary elements except close to the region of their insertion (Tomlinson, 1969). An outstanding character of the submerged organs of water plants is the reduction in number and degree of lignification of the tracheary elements. The tracheary elements usually have annular or spiral wall thickenings and in many species are reduced to tracheids. Often some or even all tracheary elements of a vascular bundle may become disorganized, leaving a xylem lacuna accompanied by a phloem strand. Cambium is absent or much reduced.

In most elongated submerged stems the vascular strands are arranged in a narrow condensed central cylinder, thus the *cortex/vascular* cylinder ratio is similar to that characteristic of roots. This rope-like core may be an adaptation of the submerged stems against pulling strain similar to that of roots. The phloem in most submerged organs is also reduced in comparison with that of land plants but is relatively more developed than the xylem of submerged water plants. For more details see Sculthorpe (1967).

REFERENCES

ALONI, R. (1987) Differentiation of vascular tissues. *Ann. Rev. Plant Physiol.* **38:** 179–204.

ALONI, R. (1988) Vascular differentiation within the plant. In: *Vascular Differentiation and Plant Growth Regulations* (eds. L. W. Roberts, P. B. Gahan and R. Aloni). Springer, Berlin Heidelberg New York. pp. 39–62.

ALONI, R. and SACHS, T. (1973) The three-dimensional structure of primary phloem systems. *Planta* **113**: 345–53.

ANCIBOR, E. (1979) Systematic anatomy of vegetative organs of the Hydrocharitaceae. *J. Linn. Soc. Bot.* **78**: 237–66.

ARBER, A. (1920) *Water Plants: A Study of Aquatic Angiosperms.* Cambridge Univ. Press.

ARBER, A. (1950) *The Natural Philosophy of Plant Form.* Cambridge Univ. Press.

AYENSU, E. S. (1969) Aspects of the complex nodal anatomy of the Dioscoreaceae. *J. Arnold Arb.* **50**: 124–37.

AYENSU, E. S. (1970) Analysis of the complex vascularity in stems of *Dioscorea composita. J. Arnold Arb.* **51**: 228–40.

AYENSU, E. S. (1972) Dioscorales. In: *Anatomy of the Monocotyledons* (ed. C. R. Metcalfe), Vol. VI. Clarendon Press, Oxford.

BAILEY, I. W. (1956) Nodal anatomy in retrospect. *J. Arnold Arb.* **37**: 269–87.

BALFOUR, E. E. and PHILIPSON, W. R. (1962) The development of the primary vascular system of certain dicotyledons. *Phytomorphology* **12**: 110–43.

BALL, E. (1948) Differentiation of primary shoots of *Lupinus albus* and *Tropaeolum majus. Symp. Soc. Exp. Biol.* **2**: 246–62.

BECK, C. B., SCHMID, R. and ROTHWELL, G. W. (1982) Stelar morphology and the primary vascular system of seed plants. *Bot. Rev.* **48**: 691–815.

BISALPUTRA, T. (1962) Anatomical and morphological studies in the Chenopodiaceae: III, The primary vascular system and nodal anatomy. *Aust. J. Bot.* **10**: 13–24.

BLYTH, A. (1958) Origin of primary extraxylary stem fibers in dicotyledons. *Univ. Calif. Publ. Bot.* **30**: 145–232.

BÖCHER, T. W. (1972) Comparative anatomy of three species of the apophyllous genus *Gymnophyton. Am. J. Bot.* **59**: 494–503.

BÖCHER, T. W. (1975) Structure of the multinodal photosynthetic thorns in *Prosopis cuntzei* Harms. *Biol. Skr.* **20** (8): 1–43.

BÖCHER, T. W. and LYSHEDE, O. B. (1968) Anatomical studies in xerophytic apophyllous plants. I. *Monttea aphylla, Bulnesia retama* and *Bredemeyera colletioides. Biol. Skr.* **16** (2): 1–44.

BÖCHER, T. W. and LYSHEDE, O. B. (1972) Anatomical studies in xerophytic apophyllous plants. II. Additional species from South American shrub steppes. *Biol. Skr.* **18** (4): 1–137.

BOKHARI, M. H. and WENDELBO, P. (1978) On anatomy, adaptations to xerophytism and taxonomy of *Anabasis* inclusive *Esfandiaria* (Chenopodiaceae). *Bot. Notiser* **131**: 279–92.

BOTHA, C. E. J. and EVERT, R. F. (1978) Observations of preferential feeding by the aphid *Rhopalosiphum maidis* on abaxial phloem of *Cucurbita maxima. Protoplasma* **96**: 75–80.

BOTHA, C. E. J., MALCOLM, S. B., and EVERT, R. F. (1977) An investigation of preferential feeding habit in four Asclepiadaceae by the aphid *Aphis nerii* B. de F. *Protoplasma* **92**: 1–19.

CARR, D. J., CARR, S. G. M. and JAHNKE, R. (1982) The eucalypt lignotuber: a position-dependent organ. *Ann. Bot.* **50**: 481–489.

CLOWES, F. A. L. (1961) *Apical Meristems.* Blackwell, Oxford.

CRAFTS, A. S. (1943) Vascular differentiation in the shoot apex of *Sequoia sempervirens. Am. J. Bot.* **30**: 110–21.

CUTTER, E. G. (1956) Experimental and analytical studies of pteridophytes: XXXIII, The experimental induction of

buds from leaf primordia in *Dryopteris aristata. Ann. Bot.* **20**: 143–68.

DE BARY, A. (1877) *Vergleichende Anatomie der Vegetationsorgane der Phanerogamen und Farne.* W. Engelman, Leipzig.

DENGLER, N. G. and DONNELLY, P. M. (1987) Comparison of shoot vascular organization in isophyllous (*Pillea cadieri*) and anisophyllous (*Pellionia daveauana*) species of the Urticaceae. *Bot. Gaz.* **148**: 188–197.

DEVADAS, C. and BECK, C. B. (1972) Comparative morphology of the primary vascular systems in some species of Rosaceae and Leguminosae. *Am. J. Bot.* **59**: 557–67.

DICKINSON, T. A. (1978) Epiphylly in angiosperms. *Bot. Rev.* **44**: 181–232.

DICKISON, W. C. and ENDRESS, P. K. (1983) Ontogeny of the stem-nodal-leaf vascular continuum of *Austrobaileya. Am. J. Bot.* **70**: 906–911.

EAMES, A. J. and MACDANIELS, L. H. (1947) *An Introduction to Plant Anatomy,* 2nd edn. McGraw-Hill, New York.

ESAU, K. (1950) Development and structure of the phloem tissue: II, *Bot. Rev.* **16**: 67–114.

ESAU, K. (1965a) *Plant Anatomy,* 2nd edn. Wiley, New York.

ESAU, K. (1965b) *Vascular Differentiation in Plants.* Holt, Rinehart and Winston, New York.

EVENARI, M. (1938) The physiological anatomy of the transpiratory organs and the conducting systems of certain plants typical of the wilderness of Judaea. *J. Linn. Soc. Bot.* **51**: 389–407.

FAHN, A. (1963) The fleshy cortex of articulated Chenopodiaceae. Maheshwari Comm. Vol. *J. Indian Bot. Soc.* **42**A: 39–45.

FAHN, A. (1964) Some anatomical adaptations of desert plants. *Phytomorphology* **14**: 93–102.

FAHN, A. and ARZEE, T. (1959) Vascularization of articulated Chenopodiaceae and the nature of their fleshy cortex. *Am. J. Bot.* **46**: 330–8.

FAHN, A. and BAILEY, I. W. (1957) The nodal anatomy and the primary vascular cylinder of the Calycanthaceae. *J. Arnold Arb.* **38**: 107–17.

FAHN, A. and BROIDO, S. (1963) The primary vascularization of the stems and leaves of the genera *Salsola* and *Suaeda* (Chenopodiaceae). *Phytomorphology* **13**: 156–65.

FAHN, A. and DEMBO, N. (1964) Structure and development of the epidermis in articulated Chenopodiaceae. *Israel J. Bot.* **13**: 172–92.

FISCHER, A. (1884) *Untersuchungen über das Siebröhren System der Cucurbitaceen.* Gebr. Borntraeger, Berlin.

FOSTER, A. S. and GIFFORD, E. M., JR. (1959) *Comparative Morphology of Vascular Plants.* W. H. Freeman, San Francisco.

FRENCH, J. C. and TOMLINSON, P. B. (1981a) Vascular patterns in stems of Araceae: Subfamilies Calloideae and Lasioideae. *Bot. Gaz.* **142**: 366–381.

FRENCH, J. C. and TOMLINSON, P. B. (1981b) Vascular patterns in stems of Araceae: Subfamily Pothoideae. *Am. J. Bot.* **68**: 713–729.

FRENCH, J. C. and TOMLINSON, P. B. (1981c) Vascular patterns in the stem of Araceae: Subfamily Monsteroideae. *Am. J. Bot.* **68**: 1115–1129.

FRENCH, J. C. and TOMLINSON, P. B. (1981d) Vascular patterns in stems of Araceae: Subfamily Philodendroideae. *Bot. Gaz.* **142**: 550–563.

FUKUDA, Y. (1967) Anatomical study of the internal phloem in stems of dicotyledons, with special reference to its histogenesis. *J. Fac. Sci. Univ. Tokyo,* sec. III, vol. 9, parts 10–11, pp. 313–75.

GEDALOVICH, E. and FAHN, A. (1983) Ultrastructure and development of the inactive stomata of *Anabasis articulata* (Forsk.) Moq. *Am. J. Bot.* **70**: 88–96.

GEWIRTZ, M. and FAHN, A. (1960) The anatomy of sporophyte and gametophyte of *Ophioglossum lusitanicum* L. spp. *lusitanicum*. *Phytomorphology* **10**: 342–51.

GIBSON, A. C. (1975) Another look at the cactus research of Irving Widmer Bailey. Suppl. vol. of the *Cactus & Succulent J.* (*US*): 76–85.

GIBSON, A. C. (1983) Anatomy of photosynthetic old stems of nonsucculent dicotyledons from North American deserts. *Bot. Gaz.* **144**: 347–362.

GIBSON, A. C. and HORAK, K. E. (1978) Systematic anatomy and phylogeny of Mexican columnar cacti. *Ann. Missouri Bot. Gard.* **65**: 999–1057.

GINZBURG, C. (1963) Some anatomic features of splitting of desert shrubs. *Phytomorphology* **13**: 92–97.

GUNCKEL, J. E. and WETMORE, R. H. (1946a) Studies of development in long shoots of *Ginkgo biloba* L.: I, The origin and pattern of development of the cortex, pith and procambium. *Am. J. Bot.* **33**: 285–95.

GUNCKEL, J. E. and WETMORE, R. H. (1946b) Studies on the development of long and short shoots of *Ginkgo biloba* L.: II, Phyllotaxis and organization of the primary vascular system, primary phloem and primary xylem. *Am. J. Bot.* **33**: 532–43.

HABERLANDT, G. (1918) *Physiologische Pflanzenanatomie*, 5th edn. W. Engelmann, Leipzig.

HASKINS, M. L. and HAYDEN, W. J. (1987) Anatomy and affinities of *Penthorum*. *Am. J. Bot.* **74**: 164–177.

HEIDE-JØRGENSEN, H. S. (1978) The xeromorphic leaves of *Hakea suaveolens* R. Br. I. Structure of photosynthetic tissue with intercellular pectic strands and tylosoids. *Bot. Tidsskr.* **72**: 87–103.

HICKS, G. S. and STEEVES, T. A. (1969) *In vitro* morphogensis in *Osmunda cinnamomea*. The role of the shoot apex in early leaf development. *Can. J. Bot.* **47**: 575–80.

JACOBS, W. P. (1952) The role of auxin in the differentiation of xylem around a wound. *Am. J. Bot.* **39**: 301–9.

JAMES, S. (1984) Lignotubers and burls—their structure, function and ecological significance in Mediterranean ecosystems. *Bot. Rev.* **50**: 225–266.

JEFFREY, E. C. (1910) The Pteropsida. *Bot. Gaz.* **50**: 401–14.

JONES, C. S. and LORD, E. M. (1982) The development of split axes in *Ambrosia dumosa* (Gray) Payne (Asteraceae). *Bot. Gaz.* **143**: 446–453.

KAUL, R. B. (1971) Diaphragms and aerenchyma in *Scirpus validus*. *Am. J. Bot.* **58**: 808–16.

KAWASE, M. (1979) Role of cellulase in aerenchyma development in sunflower. *Am. J. Bot.* **66**: 183–90.

KAWASE, M. and WHITMOYER, R. E. (1980) Aerenchyma development in waterlogged plants. *Am. J. Bot.* **67**: 18–22.

KIRCHOFF, B. K. (1984) On the relationship between phyllotaxy and vasculature: a synthesis. *Bot. J. Linn. Soc.* **89**: 37–51.

KIRCHOFF, B. K. and FAHN, A. (1984) The primary vascular system and medullary bundle structure of *Phytolacca dioica* (Phytolaccaceae). *Can. J. Bot.* **62**: 2432–2440.

KOCH, A. (1884) Ueber den Verlauf und die Endigungen der Siebröhren in den Blättern. *Bot. Z.* **42**: 401–11, 417–27.

KUMAZAWA, M. (1961) Studies on the vascular course in maize plant. *Phytomorphology* **11**: 128–39.

LACEY, C. J. (1983) Development of large plate-like lignotubers in *Eucalyptus botryoides* Sm. in relation to environmental factors. *Aust. J. Bot.* **31**: 105–118.

LARSON, P. R. (1975) Development and organization of the primary vascular system in *Populus deltoides* according to phyllotaxy. *Am. J. Bot.* **62**: 1084–99.

LESHEM, B. (1971) Bulb histogenesis in *Hordeum bulbosum* L. *Ann. Bot.* **35**: 57–62.

LYSHEDE, O. B. (1977a) Structure of the epidermal and subepidermal cells of some desert plants of Israel. *Anabasis articulata* and *Calligonum comosum*. *Israel J. Bot.* **26**: 1–10.

LYSHEDE, O. B. (1977b) Anatomical features of some stem assimilating desert plants of Israel. *Bot. Tidsskr.* **71**: 225–30.

LYSHEDE, O. B. (1979) Xeromorphic features of three stem assimilants in relation to their ecology. *J. Linn. Soc. Bot.* **78**: 85–98.

McARTHUR, I. C. S. and STEEVES, T. A. (1972) An experimental study of vascular differentiation in genus *Chiloense balbis*. *Bot. Gaz.* **133**: 276–87.

MARSDEN, M. P. F. and BAILEY, I. W. (1955) A fourth type of nodal anatomy in dicotyledons illustrated by *Clerodendron trichotomum* Thunb. *J. Arnold Arb.* **36**: 1–51.

MARSDEN, M. P. F. and STEEVES, T. A. (1955) On the primary vascular system and nodal anatomy of *Ephedra*. *J. Arnold Arb.* **36**: 241–58.

METCALFE, C. R. (1960) Gramineae. In: *Anatomy of the Monocotyledons*, vol. I (ed. C. R. Metcalfe). Clarendon Press, Oxford.

METCALFE, C. R. (1971) Cyperaceae. In: *Anatomy of Monocotyledons*, vol. V (ed. C. R. Metcalfe). Clarendon Press, Oxford.

METCALFE, C. R. and CHALK, L. (1950) *Anatomy of the Dicotyledons*. Clarendon Press, Oxford.

MIKESELL, J. E. and SCHROEDER, A. C. (1980) Development of chambered pith in stems of *Phytolacca americana* L. (Phytolaccaceae). *Am. J. Bot.* **67**: 111–18.

MIKESELL, J. and SCHROEDER, A. C. (1984) Internal phloem development in *Pharbitis nill* Chois. (Convolvulaceae). *Bot. Gaz.* **145**: 196–203.

MONTENEGRO, G., AVILA, G. and SCHATTE, P. (1983) Presence and development of lignotubers in shrubs of the Chilean matorral. *Can. J. Bot.* **61**: 1804–1808.

MOSS, E. H. (1940) Interxylary cork in *Artemisia* with a reference to its taxonomic significance. *Am. J. Bot.* **27**: 762–8.

NAMBOODIRI, K. K. and BECK, C. B. (1968a) A comparative study of primary vascular system of conifers. I. Genera with helical phyllotaxis. *Am. J. Bot.* **55**: 447–57.

NAMBOODIRI, K. K. and BECK, C. B. (1968b) A comparative study of the primary vascular system of conifers: III, Stelar evolution in gymnosperms. *Am. J. Bot.* **55**: 461–72.

NAST, C. G. (1944) The comparative morphology of the Winteraceae: VI, Vascular anatomy of the flowering shoot. *J. Arnold Arb.* **25**: 454–66.

NAVEH, Z. (1975) The evolutionary significance of fire in the Mediterranean region. *Vegetatio* **29**: 199–208.

NEUBAUER, H. F. (1981) Über Knotenbau und Blattgrundvaskularisation bei Dikotylen. Übersicht und Zusammenfassung. *Beitr. Biol. Pflanzen* **56**: 357–366.

O'BRIEN, T. P. and ZEE, S. Y. (1971) Vascular transfer cells in the vegetative nodes of wheat. *Aust. J. Biol. Sci.* **24**: 207–17.

O'NEILL, T. B. (1961) Primary vascular organization of *Lupinus* shoot. *Bot. Gaz.* **123**: 1–9.

ORSHAN, G. (1953) Note on the application of Raunkiaer's system of life forms in arid regions. *Palest. J. Bot.*, *Jerusalem* ser. **6**: 120–2.

PALLADIN, W. I. (1914) *Pflanzenanatomie*. B. G. Teubner, Leipzig.

PHILIPSON, W. R. (1949) The ontogeny of the shoot apex in dicotyledons. *Biol. Rev.* **24:** 21–50.

PHILIPSON, W. R. and BALFOUR, E. E. (1963) Vascular patterns in dicotyledons. *Bot. Rev.* **29:** 382–404.

PHILIPSON, W. R. and PHILIPSON, M. N. (1968) Diverse nodal types in *Rhododendron. J. Arnold Arb.* **49:** 193–217.

PLANTEFOL, L. (1947) Hélices foliaires, point végétatif et stèle chez les Dicotylédones. La notion d'anneau initial. *Rev. gén. Bot.* **54:** 49–80.

PRIESTLEY, J. H. (1926) Light and growth: II, On the anatomy of etiolated plants. *New Phytol.* **25:** 145–70.

RODIN, R. J. and PALIWAL, G. S. (1970) Nodal anatomy of *Gnetum ula. Phytomorphology* **20:** 103–11.

RUTISHAUSER, R. (1982) Der Plastochronquotient als Teil einer quantitativen Blattsetllungsanalyse bei Samenpflanzen. *Beitr. Biol. Pflanzen* **57:** 323–357.

SACHS, T. (1968a) The role of the root in the induction of xylem differentiation in peas. *Ann. Bot.* **32:** 391–9.

SACHS, T. (1968b) On the determination of the pattern of vascular tissue in peas. *Ann. Bot.* **32:** 781–90.

SACHS, T. (1969) Polarity and the induction of organized vascular tissues. *Ann. Bot.* **33:** 263–75.

SACHS, T. (1981) The control of the patterned differentiation of vascular tissues. In: *Advances in Botanical Research* Vol. 9 (ed. H. W. Woolhouse). Academic Press, New York, pp. 151–262.

SATTLER, R. (1974) A new conception of shoot of higher plants. *J. Theor. Biol.* **47:** 367–82.

SCHMID, R. (1982) The terminology and classification of steles: historical perspective and the outline of a system. *Bot. Rev.* **48:** 817–931.

SCULTHORPE, C. D. (1967) *The Biology of Aquatic Vascular Plants.* E. Arnold, London.

SELMAN, I. W. and KULASEGARAM, S. (1967) Development of stem tuber in kohlrabi. *J. Exp. Bot.* **18:** 471–90.

SHARMAN, B. C. (1942) Developmental anatomy of the shoot of *Zea mays* L. *Ann. Bot.* **6:** 245–82.

SINNOTT, E. W. (1914) Investigations on the phylogeny of angiosperms: I, The anatomy of the node as an aid in the classification of angiosperms. *Am. J. Bot.* **1:** 303–22.

SLADE, B. F. (1971) Stelar evolution in vascular plants. *New Phytol.* **70:** 879–84.

SMITH, G. M. (1955) *Cryptogamic Botany,* vol. II, *Bryophytes and Pteridophytes,* 2nd edn. McGraw-Hill, New York.

SNOW, M. and SNOW, R. (1947) On the determination of leaves. *New Phytol.* **46:** 5–19.

SNOW, R. (1955) Problems of phyllotaxis and leaf determination. *Endeavour* **14:** 190–9.

SPORNE, K. R. (1962) *The Morphology of Pteridophytes.* Hutchinson Univ. Library, London.

SPORNE, K. R. (1965) *The Morphology of Gymnosperms.* Hutchinson Univ. Library, London.

STEEVES, T. A. (1961) The developmental potentialities of excised leaf primordia in sterile culture. *Phytomorphology* **11:** 346–59.

STEEVES, T. A. and SUSSEX, I. M. (1972) *Patterns in Plant Development.* Prentice-Hall, Inc., Englewood Cliffs, New Jersey.

SUGIYAMA, M. (1979) A comparative study of nodal anatomy in the Magnoliales based on the vascular system in the node-leaf continuum. *J. Fac. Sci. Univ. Tokyo III* **12:** 199–279.

SUSSEX, I. M. (1955) Morphogenesis in *Solanum tuberosum* L.: Experimental investigation of leaf dorsiventrality and orientation in the juvenile shoot. *Phytomorphology* **5:** 286–300.

THOMPSON, N. P. and JACOBS, W. P. (1966) Polarity of IAA effect on sieve tube and xylem regeneration in *Coleus* and tomato stems. *Pl. Physiol.* **41:** 673–82.

TOMLINSON, P. B. (1969) On the morphology and anatomy of turtle grass, *Thalassia testudinum* (Hydrocharitaceae): II, Anatomy and development of the root in relation to function. *Bull. Marine Sci.* **19:** 57–71.

TROLL, W. (1948) *Allgemeine Botanik.* F. Enke, Stuttgart.

VAN FLEET, D. S. (1942a) The development and distribution of the endodermis and an associated oxidase system in monocotyledonous plants. *Am. J. Bot.* **29:** 1–15.

VAN FLEET, D. S. (1942b) The significance of oxidation in the endodermis. *Am. J. Bot.* **29:** 747–55.

VAN FLEET, D. S. (1950a) The cell forms and their common substance reactions in the parenchyma-vascular boundary. *Bull. Torrey Bot. Club* **77:** 340–53.

VAN FLEET, D. S. (1950b) A comparison of histochemical and anatomical characteristics of the hypodermis with the endodermis in vascular plants. *Am. J. Bot.* **37:** 721–5.

VAN FLEET, D. S. (1961) Histochemistry and function of the endodermis. *Bot. Rev.* **27:** 165–220.

VAN TIEGHEM, P. and DOULIOT, H. (1886) Sur la polystélie. *Annls Sci. nat. Bot.* **3:** 275–322.

VOLKENS, G. (1887) *Die Flora der aegyptisch-arabischen Wüste auf Grundlage anatomisch physiologischer Forschungen.* Gebr. Borntraeger, Berlin.

WAHLEN, M. A. (1987) Wood anatomy of the American Frankenias (Frankeniaceae) systematic and evolutionary implication. *Am. J. Bot.* **74:** 1211–1223.

WANGERMANN, E. (1967) The effect of the leaf on differentiation of primary xylem in the internode of *Coleus blumei* Benth. *New Phytol.* **66:** 747–54.

WARDEN, W. M. (1935) On the structure, development and distribution of the endodermis and its associated ducts in *Senecio vulgaris. New Phytol.* **34:** 361–85.

WARDLAW, C. W. (1948) Experimental and analytical studies of pteridophytes: XI, Preliminary observations on tensile stress as a factor in fern phyllotaxis. *Ann. Bot.,* **12:** 97–109.

WARDLAW, C. W. (1950) Experimental and analytical studies of pteridophytes: XVI, The induction of leaves and buds in *Dryopteris aristata* Druce. *Ann. Bot.* **14:** 435–55.

WARDLAW, C. W. (1955) Evidence relating to the diffusion-reaction theory of morphogenesis. *New Phytol.* **54:** 39–48.

WARDLAW, C. W. (1960) The inception of shoot organization. *Phytomorphology* **10:** 107–10.

WARDLAW, C. W. (1968) *Morphogenesis in Plants.* Methuen, London.

WETMORE, R. H. (1943) Leaf-stem relationships in the vascular plants. *Torreya* **43:** 16–28.

WETMORE, R. H. and RIER, J. P. (1963) Experimental induction of vascular tissues in callus of angiosperms. *Am. J. Bot.* **50:** 418–30.

ZAMSKI, E. and TSIVION, Y. (1977) Translocation in plants possessing supernumerary phloem. *J. Exp. Bot.* **28:** 117–26.

ZIMMERMANN, A. (1922) Die Cucurbitaceen, Heft 1. *Beiträge zur Anatomie und Physiologie.* Gustav Fischer, Jena.

ZIMMERMANN, M. H. and SPERRY, J. S. (1983) Anatomy of the palm *Rhapis excelsa.* IX. Xylem structure of the leaf insertion. *J. Arnold Arbor.* **64:** 599–609.

ZIMMERMANN, M. H. and TOMLINSON, P. B. (1965) Anatomy of the palm *Rhapis excelsa*: I, Mature vegetative axis. *J. Arnold Arb.* **46:** 166–78.

ZIMMERMANN, M. H. and TOMLINSON, P. B. (1966) Anatomy of the palm *Rhapis excelsa.* II. Rhizome. *J. Arnold Arb.* **47:** 248–61.

ZIMMERMANN, M. H. and TOMLINSON, P. B. (1969) The vascular system in the axis of *Dracaena fragrans* (Agavaceae): I, Distribution and development of primary strands. *J. Arnold Arb.* **50**: 370–83.

ZIMMERMANN, M. H. and TOMLINSON, P. B. (1970) The vascular system in the axis of *Dracaena fragrans* (Agavaceae). 2, Distribution and development of secondary vascular tissue. *J. Arnold Arb.* **51**: 478–91.

ZOHARY, M. (1961) On hydro-ecological relations of the Near East desert vegetation. Plant–water relationships in arid and semi-arid conditions. *Proc. Madrid Symp. Unesco, Arid Zone Res.* **16**: 199–212.

CHAPTER 12

THE LEAF

As HAS already been mentioned (p. 185), it is diffi-cult, both from theoretical and practical view-points, to distinguish clearly between the leaf and the stem. This difficulty is clarified in the light of the accepted assumption that the leaves in the Pteropsida (the higher Pteridophyta and the Sper-matophyta) have, phylogenetically, developed from a certain system of branches.

The structure of the conducting tissues in the petiole and main vein of the leaf is usually similar to that of the stem. Sometimes the same photo-synthetic and non-photosynthetic parenchyma-tous tissues are found both in the leaves and in the cortex of the stem (Fahn and Arzee, 1959). The most important characteristic of the leaf is the early termination of apical growth. In certain ferns the apical meristem remains active for a relatively long period, while in other ferns, e.g. *Ophioglossum*, and in the Spermatophyta the true apical activity ceases very early in the develop-ment of the leaf and then the shape and size of the leaf is determined by intercalary and marginal growth.

Morphologically and anatomically the leaf is the most variable plant organ. The collective term for all the types of leaves appearing on plants is the *phyllome* (Arber, 1950). The different phyl-lomes of the Spermatophyta are extremely vari-able both in external and in internal structure and in function. Because of this variability the follow-ing types of phyllomes have been classified: *foliage leaves, cataphylls, hypsophylls, cotyledons,* and others. The foliage leaves are the principal photosynthetic organs. The cataphylls are the scales that appear on buds and on underground stems and their function is protection or the stor-age of reserve materials. The first, lowermost leaves of a side branch are termed *prophylls*; in monocotyledons only one prophyll is usually

present and in dicotyledons, two. The hypsophylls are the various types of bracts that accompany the flowers and their function is, apparently, pro-tection. Sometimes the hypsophylls are coloured and then their function is similar to that of petals. The cotyledons are the first leaves of the plant. The floral organs are also considered as leaves.

MORPHOLOGY OF THE LEAF

In this chapter we shall deal with foliage leaves which, among the Angiospermae in particular, themselves exhibit great variation in anatomical and morphological structure. The foliage leaves in certain angiosperm genera are sessile and consist almost entirely of lamina, but in most genera several distinct parts can be distinguished, i.e. the *leaf base, petiole,* and *lamina.* The shape, structure, and relative size of these parts differ and are used to classify the foliage leaves into different types. In many dicotyledonous genera two appendages, i.e. the *stipules,* develop at the leaf base. The stipules may be attached to the leaf base or they may be free appendages. Even when they are free appen-dages they develop as outgrowths of the leaf pri-mordia. The vascular supply of the stipules is de-rived from the leaf traces (Fig. 133, no. 3). In some plants the stipules are green and resemble leaflets and have a photosynthetic function, but their main function is the protection of young develop-ing leaves. In some woody dicotyledons (e.g. *Ficus* and *Vitis*) the outermost scales protecting the buds are stipules (Fig. 122). In most monocoty-ledons and some dicotyledons (the Umbelliferae and Polygonaceae) the leaf base is widened so as to form a *sheath* which surrounds the node. In grasses, the sheath at the junction with the

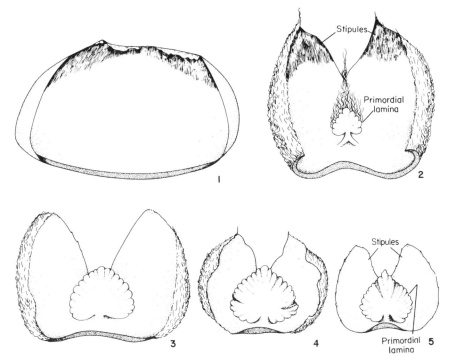

FIG. 122. Bud scales of *Vitis vinifera*. Numerals indicate the centripetal order of appearance of the scales and young foliage leaves. 1, Cataphyll which envelops the bud. 2–5, Young foliage leaves in order of their appearance in the bud. The stipules of the outer leaves are relatively large and they serve to protect the bud. (Adapted from drawings made by Z. Bernstein.)

lamina is usually provided with a hyaline or hairy projection known as *ligule*. At the corners of this junction, in some species, *auricles* are present. Usually a relationship exists between the anatomical structure of the node and the appearance of stipules and a sheath in the dicotyledons (Sinnott and Bailey, 1914). Most plants with trilacunar nodes (i.e. those in which there are three leaf gaps at each node) have stipules and those with multilacunar nodes have sheath-like leaf bases.

Leaves are divided into *simple* and *compound* depending on whether the stalk bears one or more leaflets. In the case of compound leaves the common stalk is termed the *rachis*. If the leaflets arise from the sides of the rachis, the leaf is said to be *pinnate*, and if all from one point, as rays, *palmate*. The margin of the leaf, and similarly the leaflets, may be entire or variously notched.

Although most leaves are dorsiventral appendages they may be flattened laterally as in *Iris*, almost cylindrical as in *Sansevieria cylindrica*, or

may have bizarre tubular forms as in some insectivorous plants.

In some plants, such as many species of *Acacia* of Australian origin, the lamina of the bipinnate leaves is reduced, and from the remaining parts a flat, leaf-like photosynthetic organ is developed; this type of organ is termed a *phyllode* (see p. 262). In some genera of the Umbelliferae *rachis leaves* occur which are radially symmetrical and resemble the axis of a compound leaf stripped of its pinnae. Kaplan (1970a) has shown that although pinnae are actually initiated during ontogeny they soon become arrested in their growth.

In certain plants (*Opuntia* spp., *Muehlenbeckia platyclados*) the stems are photosynthetic and have become flat; such organs are termed *platyclades*. When the platyclade appears more leaf-like, as in *Ruscus*, it is called a *phylloclade* or *cladode*.

A developmental anatomical study of phylloclades in four closely related species of the Asparagaceae revealed a continuum in phylloclade

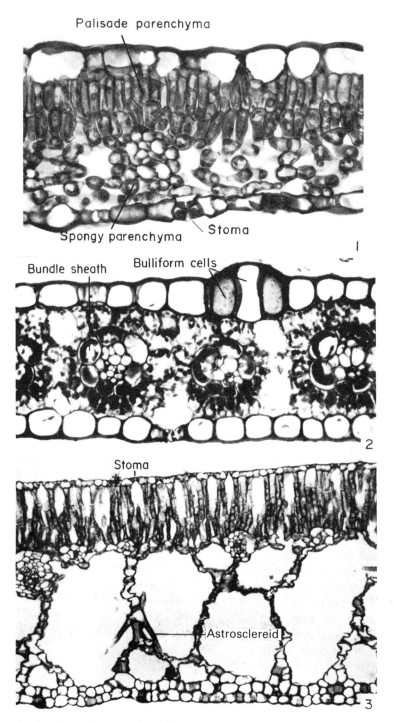

FIG. 123. Micrographs of portions of cross-sections of leaf laminae. 1, *Rosa* showing the dorsiventral arrangement of the mesophyll. ×175. 2, *Zea mays* showing a group of bulliform cells, the uniform structure of the mesophyll, and the bundle sheaths which consist of a single layer of thin-walled cells. ×240. 3, *Nymphaea alba* showing stomata in the adaxial epidermis, several layers of palisade cells with veins immediately below them and the wide region of tissue containing large air cavities, in the walls of which are astrosclereids. The abaxial epidermis is devoid of stomata. ×125.

development from very leaf-like forms (e.g. *Danae*), via more intermediate types (e.g. *Ruscus*), to a more shoot-like form (e.g. *Asparagus*). The fertile phylloclade of *Ruscus* represents an intermediate organ in which both stem and leaf features are combined (Cooney-Sovetts and Sattler, 1986).

HISTOLOGY OF THE FOLIAGE LEAF

Histologically the leaf is composed of three types of tissue systems: epidermis, mesophyll, and vascular tissues.

The epidermis

The epidermis of leaves of different plants varies in the number of layers, its shape, structure, arrangement of the stomata, appearance and arrangement of trichomes, and occurrence of specialized cells. (The above features are discussed in Chapter 10.) Because of the usually flat structure of the leaf, a distinction is made between the epidermal tissues of the two surfaces of the leaf; that surface of the leaf that is closer to the internode above it and which usually faces upwards is referred to as the *adaxial surface*, and the other surface as the *abaxial surface*.

The mesophyll

The mesophyll comprises the parenchymatous tissue internal to the epidermis. It usually undergoes differentiation to form the photosynthetic tissues and so contains chloroplasts. The various forms of mesophyll cells were reviewed in detail by Meyer (1962). In many plants, especially among the dicotyledons, two types of parenchyma can be distinguished in the mesophyll: *palisade parenchyma* and *spongy parenchyma* (Fig. 123, no. 1; Fig. 126, nos. 1, 2). The cells of typical palisade parenchyma are elongated, and in cross-section of the leaf they are rod-shaped and appear to be arranged in rows, while in a section parallel to the leaf surface these cells are seen to be rounded and

separated or only slightly attached to one another (Fig. 126, no. 1). In certain plants the palisade cells differ in shape from the typical ones. In certain species of the Xanthorrhoeaceae small papilla-like projections and constrictions which run around the cells are found (Fahn, 1954). In *Lilium* large lobes are present on the palisade cells which therefore appear branched (Fig. 124, no. 1).

The palisade cells are found immediately below the uni- or multiseriate epidermis, but sometimes a hypodermis may be present between the epidermis and the palisade tissue. The cells of palisade parenchyma may be arranged in one or more layers and in the latter case the length of the cells in the different rows may be equal, or they may become shorter towards the centre of the mesophyll. The palisade tissue is usually found on the adaxial surface of the leaf. In some plants, however, e.g. *Thymelaea*, *Tamarix* and species of *Frankenia*, in which the adaxial surface of the small leaves is adpressed to the stem, the palisade parenchyma is found only on the abaxial sides of the leaf (Fig. 124, nos. 2,3). In certain plants, including many xeromorphic species, e.g. *Dianthus caryophyllus*, *Atriplex portulacoides* (Fig. 125, no. 1), species of *Centaurea*, *Artemisia*, and *Myoporum*, palisade parenchyma is present on both sides of the leaf with the result that only a small strip of spongy parenchyma is present in the central portion of the lamina. A leaf in which the palisade parenchyma occurs on one side of the leaf and the spongy parenchyma on the other is termed *dorsiventral* or *bifacial*. When the palisade parenchyma is present on both sides of the leaf, the leaf is said to be *isolateral* or *isobilateral*.

The cells of the spongy parenchyma are variously shaped. They may resemble the palisade cells, or have equal diameters, or be elongated in a direction parallel to the leaf surface. However, a characteristic of all spongy parenchyma cells is the presence of lobes by which the neighbouring cells are connected.

The distinction between the palisade and spongy parenchyma is not always easy, especially when the palisade parenchyma consists of several layers. In the latter case the cells of the innermost layers greatly resemble those of adjacent spongy parenchyma.

In certain plants, such as *Zea* and many other

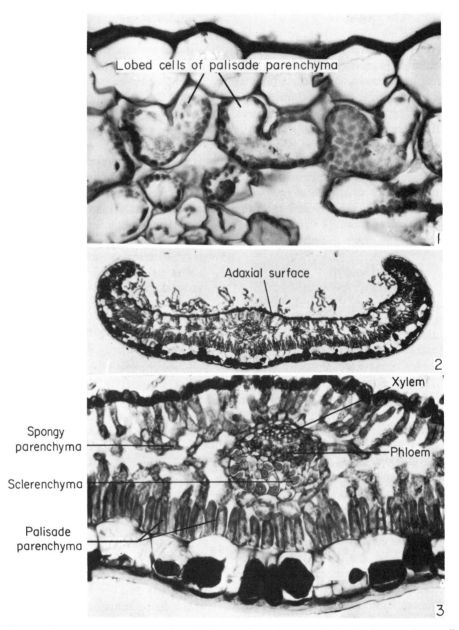

Fig. 124. 1, Portion of a cross-section of the lamina of *Lilium candidum* showing lobed palisade parenchyma cells immediately below the epidermis. × 500. 2, Cross-section of the leaf of *Thymelaea hirsuta* in which the palisade parenchyma is present on the abaxial side of the leaf and the spongy parenchyma on the adaxial side. × 63. 3, As in no. 2, but portion around mid-rib enlarged. × 230.

grasses, the mesophyll cells are more or less uniform in shape (Fig. 123, no. 2). In certain species of *Eucalyptus* (e.g. *E. hemiphloia*) and in *Atriplex* it is not possible to distinguish between the two types of parenchyma, and the mesophyll is entirely composed of palisade cells (Fig. 125, no. 2;

Fig. 128, no. 1).

The palisade tissue has become specialized in such a way that the efficiency of photosynthesis has been increased. In mesophyll that can be clearly divided into palisade and spongy parenchyma the large majority of the chloroplasts are

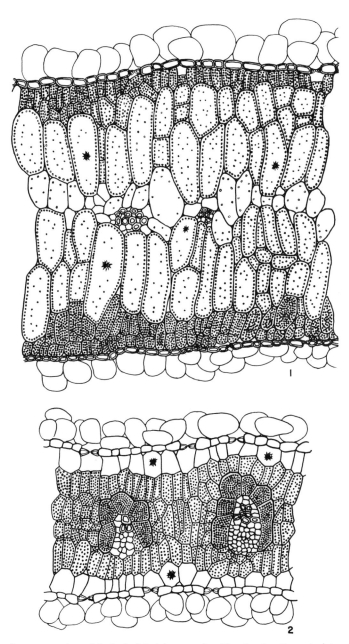

Fɪɢ. 125. 1, Portion of a cross-section of the leaf of *Atriplex portulacoides* showing the vesiculate salt trichomes, thick-walled epidermis, and isobilateral arrangement of the mesophyll tissues. The inner mesophyll cells are large, contain few chloroplasts, and store water; occasional druses occur in these cells. 2, As above, but of *Atriplex halimus* in which the epidermis is relatively thin-walled, the uniseriate hypodermis is devoid of chloroplasts, stores water, and contains occasional druses. The chlorenchyma, which consists of uniform elongated cells, is present between the abaxial and adaxial hypodermal layers. The bundle sheaths are open on the abaxial side of the veins and the cells of the sheaths are richer in chloroplasts than the neighbouring mesophyll cells.

found in the palisade cells. Because of the shape and arrangement of the palisade cells the chloroplasts can be placed so as to enable the maximum utilization of light. Under illumination the chloroplasts form a single layer lining the walls of the palisade cells.

Another important factor that increases photosynthetic efficiency is the presence of a well-developed system of intercellular spaces which is present in the mesophyll, and which facilitates rapid gas exchange. Because of the cell arrangement in the mesophyll large surface areas of the cells are exposed and so brought into contact with air. Together these surface areas have been termed the *internal surface area* of the leaf, as distinct from the *external surface area* of the leaf. The ratio of the volume of the intercellular spaces to the total volume of the leaf varies between 77:1000 and 713:1000 (Sifton, 1945). The ratio of the internal surface area to the external surface area is of ecological importance (Turrell, 1936, 1939, 1942, 1944). The internal surface area of

Styrax officinalis is eight times larger than the external surface area, and in *Olea* it is eighteen times larger.

The specialization of the palisade tissue that results in more efficient photosynthesis is brought about not only by the increased number of chloroplasts in the cells but also by the dimensions of its free surface area. Although the volume of the intercellular spaces in the spongy tissue is much larger than that in the palisade tissue, the free surface area is greater in the palisade tissue. This feature becomes obvious when sections parallel to the leaf surface are examined (Fig. 126, nos. 1, 2). In such sections it is seen that the palisade cells are round in cross-section and that the areas of contact between the cells are restricted to very narrow strips, while the areas of contact in the spongy tissue are flatter and wider. Thus, for example, in *Styrax officinalis* the free surface area of the palisade tissue is about twice as large as that of the spongy tissue.

The intercellular spaces of the mesophyll

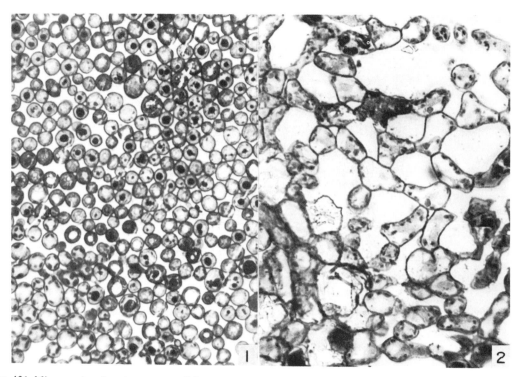

FIG. 126. Micrographs of sections cut parallel to the leaf surface of *Rosa*. 1, Section through the palisade parenchyma. × 660. 2, Section through spongy parenchyma. × 530.

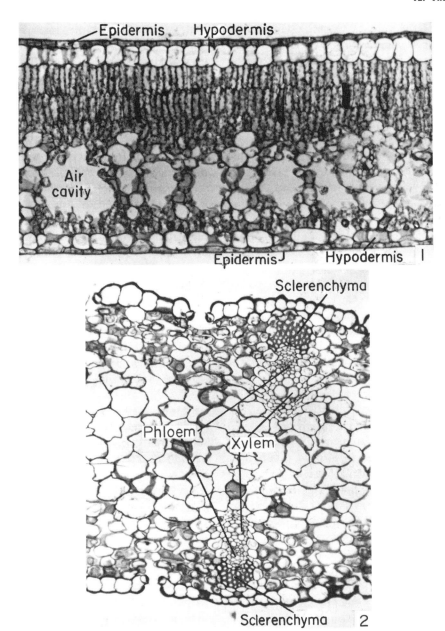

FIG. 127. Micrographs of portions of cross-sections of leaf laminae. 1, *Musa* (Dwarf Cavendish banana). × 190. 2, *Iris*, in which the leaf is unifacial. × 130.

usually develop schizogenously, but in certain plants the development may also be lysigenous by the disintegration of groups of cells. Examples of the latter type of development are seen in water and marshy plants, and also in the banana leaf (Skutch, 1927) (Fig. 127, no. 1).

Structural changes of epidermis and mesophyll in leaves of xerophytes

Xerophytes, according to the definition of Maximov (1931), are plants that grow in arid habitats and whose transpiration decreases to a minimum under conditions of water deficiency. Plants may develop structural characteristics that are adaptations to arid habitats. Such plants are termed *xeromorphic* plants. Xeromorphism, however, is not confined to xerophytes, and not all xerophytes exhibit xeromorphic characters.

One of the most obvious features of xeromorphic leaves is the small ratio of the external leaf surface to its volume. According to numerous workers (Schimper, 1898; Maximov, 1929; Weaver and Clements, 1929; Oppenheimer, 1960; and others) the reduction of the external surface is accompanied by certain changes in the internal structure of the leaf as, for instance, the reduction in cell size, the increase in the thickness of the cell walls, the greater density both of the vascular system (Wylie, 1949) and of the stomata, and the increased development of the palisade tissue at the expense of the spongy tissue. Xeromorphic leaves are often covered with trichomes. In xerophytes with succulent leaves water-storing tissue is developed.

The lack of nitrogenous compounds and/or water in the soil often results in the appearance of xeromorphic characters such as the formation of thick walls and cuticle and the additional development of sclerenchyma (Volkens, 1887; Kraus and Kraybill, 1918; Welton, 1928; Schneider, 1936; Böcher, 1979). A clear relationship can be established between the salinity of the soil and the appearance of succulent features in the plants growing in it (Mothes, 1932). Intense illumination and the retardation of the water flow due to water deficiency apparently results in the increased development of palisade tissue (Shields, 1950).

Some of the above changes, such as the increase in the number of stomata, allow a higher rate of gas exchange under conditions of favourable water supply. Furthermore, the increased development of palisade tissue probably results in an increase of photosynthetic activity.

The reduction in the size of the leaf is thought to be a feature correlated with the reduction of the rate of transpiration and in many places it is seen that plants with small leaves are commoner in dry habitats. In some cases the reduction of leaf size is connected with an increase in the total number of leaves on the plant. Thus, for example, the total external surface area of the entire foliage

of certain coniferous trees is usually greater than that of many dicotyledons (Groom, 1910).

Trichomes are very common on xerophytes. If a species exists in both xeromorphic and mesomorphic forms, the former has a denser indumentum (Coulter *et al.*, 1931). In many xeromorphic plants, such as, for instance, *Nerium* and *Xanthorrhoea*, the stomata are sunken in depressions or grooves which are covered by trichomes (Fig. 83, no. 4; Fig. 129).

Environmental factors can influence the development of the cuticle and wax (Hull *et al.*, 1975). In *Prosopis velutina*, Bleckmann *et al.* (1980) found that the cuticle of the leaflets of plants grown outdoors was more than ten times as thick as that of plants grown indoors. Insolation of photosynthesizing plant organs in deserts and semi-deserts is an important factor in reducing both the absorptance of photosynthetically active radiation and the temperature of the exposed organs. Hairs or scales, epicuticular wax and resin may play a role in insolation and light reflectance (Johnson, 1975; Ehleringer *et al.* 1976; Dell and McComb, 1978; Hartmann, 1979; Heide-Jørgensen, 1980). According to Shields (1950), living trichomes, which themselves lose water, do not protect the plant from excessive transpiration as do dead trichomes which form protective layers. As trichomes are more numerous on plants growing in dry habitats, and as they do not prevent excessive evaporation, Shields suggests that it is possible that the trichomes are symptoms of water loss rather than being structures that function to reduce evaporation (see p. 174). In some cases pubescence might play a role in deterring predators (Levin, 1973; Johnson, 1975). Mucilage wall layers of epidermal cells may be involved in water economy of leaves of xerophytes (Lyshede, 1977).

Volkens (1887) suggested that in some desert plants the stomata on the photosynthesizing plant organs become permanently closed during the summer season. This closure is caused by the additional thickening and cutinization of the guard-cell walls (e.g. *Aristida ciliata*, *Sporobolus spicatus*) or by the blocking of the sunken stomata from the exterior by resinous masses (e.g. *Pityranthus*) or by wax layers (e.g. *Capparis spinosa*). In *Anabasis articulata* and other related species prominent wall thickenings have been seen to de-

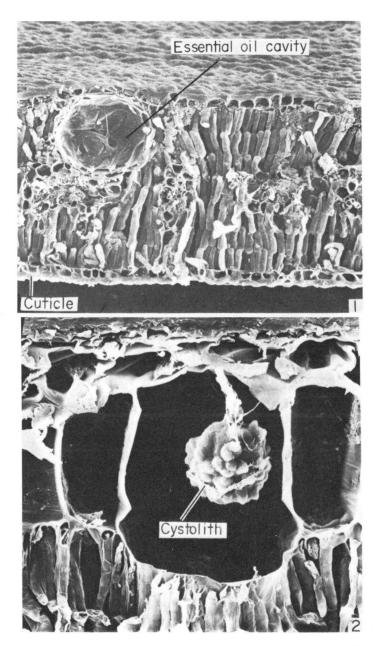

Fig. 128. Scanning electron micrographs of leaf portions. 1, *Eucalyptus camaldulensis*. ×150. 2, *Ficus elastica*, showing a lithocyst containing a cystolith.

velop in the guard cells in the hot summer months (Fahn and Dembo, 1964). The permanent closure of the stomata of desert plants in the dry season should be investigated experimentally, as this feature may explain how those green parts of desert plants that do not dry out manage to retain their water content (Fig. 92, nos. 1, 2).

In *Rumex acetosella* drops of resin or oil form in the epidermis and in the cells around the veins under drought conditions (Transeau, 1904); this feature apparently hinders the passage of water. It is possible that this is also the function of the tannins and resins found in species of *Quercus* and *Pistacia* of the Mediterranean maquis. Atay (1958)

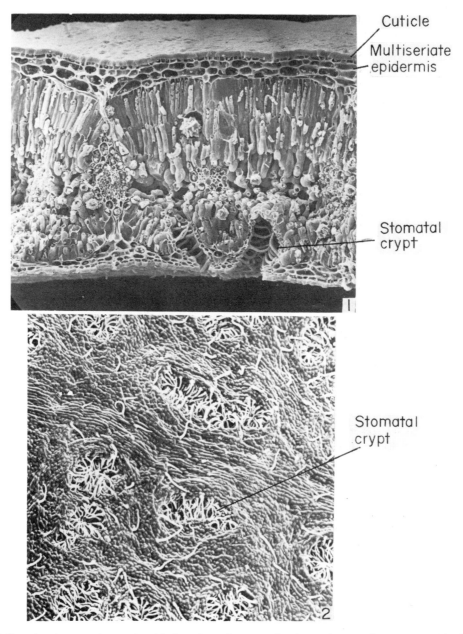

Fig. 129. Scanning electron micrographs of leaf portions of *Nerium oleander*. 1, Cross-section. × 110. 2, Abaxial surface. × 60.

and Heilbronn (1958) suggested that vapour of ethereal oils may lower the evaporation and transpiration rates.

Water in leaves is conducted not only by the veins and bundle-sheath extensions, but also by the mesophyll cells and epidermis (Shull, 1934; Wylie, 1943). Water transport towards the epi- dermis is much higher through the palisade tissue than through the spongy parenchyma. In centric xeromorphic leaves the palisade cells radiate around the central vascular bundles and there- fore, under favourable conditions of water supply, the transport of water from the bundles to the epidermis is enhanced (Thoday, 1931). The presence of intercellular spaces, especially between the palisade cells, however, limits water transport in the plane parallel to the leaf surface (Wylie, 1943).

The volume of intercellular spaces is smaller in xeromorphic leaves than in mesomorphic ones. However, the ratio between the internal free surface area of the leaf and its external surface is small in shade leaves (6.8 to 9.9) and large (17.2 to 31.3) in xeromorphic leaves (Turrell, 1936). Similar results have been obtained by us from plants of various ecological types; e.g. in *Styrax officinalis*, which is a mesophyte, the ratio is 8.91, while in *Olea europaea* and *Quercus calliprinos*, which are xerophytes, it is 17.95 and 18.52 respectively. The increase in the free internal surface area is due to the increased development of the palisade tissue. The latter is probably one of the reasons why, besides the increase in photosynthetic activity, the rate of transpiration of xerophytes is high under conditions of favourable water supply.

In some xerophytes, and generally in halophytes, well-developed water-storing tissues occur in the leaves (Fig. 131). Water-storage tissue in the leaf consists of large cells with large vacuoles containing a dilute and/or mucilaginous cell sap. These cells have a thin layer of cytoplasm lining the cell walls in which scattered chloroplasts may be found. The osmotic pressure in the photosynthesizing cells is higher than in the non-photosynthesizing ones and when water is lacking they obtain their water from the water-storing tissue. As a result of this, the thin-walled water-storing cells shrink, but under favourable conditions of water supply they rapidly return to their former state (Schimper, 1898). Hypertrophy of parenchyma cells of some coastal plants as compared with those of the same species grown in inland habitats is a well-known phenomenon (Boyce, 1954).

In the reduced leaves of *Salicornia*, broad and short tracheid-like cells are found among the palisade cells (Fahn and Arzee, 1959). The function of these cells has been variously interpreted by different workers. Duval-Jouve (1868) stated that they are filled with air. According to Holtermann (1907) these cells transport water to the peripheral layers. Other investigators (Heinricher, 1885; Volkens, 1887; Solereder, 1908; de Fraine, 1912) believe that these tracheid-like cells (tracheoids)

have a water-storing function (Fig. 130). *Tracheoid idioblasts* may be scattered throughout the mesophyll as, for instance, in *Pogonophora schomburgkiana* of the Euphorbiaceae (Foster, 1956) or in the central mesophyll, as in *Sansevieria* (Koller and Rost, 1988a, b).

Involution of the leaves, which is especially typical of grasses, is a character of xerophytes. This feature is brought about by the action of bulliform cells and/or other epidermal and mesophyll elements which may be parenchymatous or sclerenchymatous (Shields, 1951).

There are plants, such as *Nerium oleander*, which, although growing in favourable wet conditions, have xeromorphic leaves, as these are defined today. On the other hand, plants, such as *Prunus amygdalus* and *Anagyris*, which grow in dry habitats, have leaves of mesomorphic character. However, in the majority of cases there is a correlation between all the above-mentioned xeromorphic features, or some of them, and the dry conditions of the habitat. It is necessary to continue the investigation of additional anatomical and physiological features in order to understand better how desert plants withstand conditions of extreme drought.

Structural changes of leaves of water plants (hydrophytes)

In contrast with the different types of xerophytic habitats, water provides a uniform habitat and therefore the anatomical structure of water plants is less varied than that of the xerophytes. The factors that influence water plants are principally temperature, air, and the concentration and composition of salts in the water. The most striking structural features in the leaves of water plants are the reduction of supporting and protective tissues, the decrease in the amount of vascular tissue, especially xylem, and the presence of air chambers.

The epidermis of water plants does not have a protective function and it plays a part in the uptake of nutrient substances from the water and in gas exchange. The cuticle is very thin, as are the cell walls. The epidermal cells of many hydrophytes contain chloroplasts. Immersed leaves are

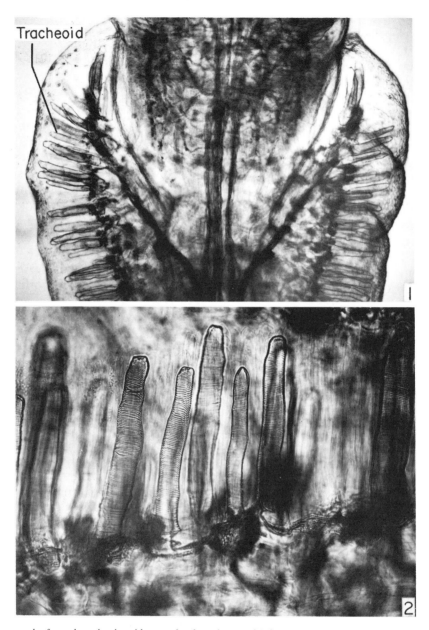

FIG. 130. Micrograph of a node and reduced leaves of a cleared stem of *Salicornia*, showing the tracheoids which are present in the region of the palisade parenchyma. These tracheoids are suggested to store water and they are connected to the vein endings present in the leaves and in the stem cortex. × 50. 2, Portion of the above in region of tracheoids, showing the helical thickening of the tracheoids. × 150.

usually devoid of stomata although sometimes vestigial stomata may be found. Many stomata are, however, present on the upper surface of floating leaves. In some submerged fresh water and marine plants epidermal cells with transfer cell characteristics have been reported (see Barnabas et al., 1977). In the epidermis of the immersed surface of leaves of some water plants small groups of cells appear which stain deeply with certain dyes. These cell groups, termed *hydropotes*, are thought to be structures which facilitate water and salt transport into and out of the plants (Lüttge and Krapf, 1969).

In many species of water plants, e.g. *Ceratophyllum*, *Myriophyllum*, and *Utricularia*, the leaves are divided into narrow cylindrical lobes which greatly increase the area of contact with the water. In *Ranunculus aquatilis* the floating leaves are entire, while the immersed leaves are divided into narrow lobes.

In the leaves and stems of hydrophytes *air chambers*, filled with gases, are found (Fig. 123, no. 3; Fig. 132, nos. 1, 2; Fig. 133, no. 2). These chambers are intercellular spaces which are usually regular in shape and which pass through the entire leaf. In many plants the air chambers

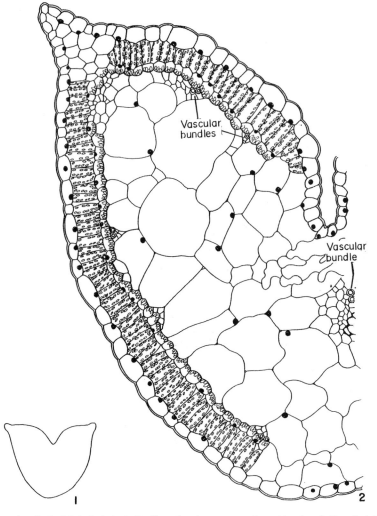

FIG. 131. 1 and 2. Succulent leaf of *Salsola kali*. 1, Outline of entire cross-section of lamina. 2, Detailed diagram of one half of the lamina showing palisade parenchyma on both sides of the leaf and the central portion of large, water-storing parenchyma cells. (Adapted from Shields, 1951.)

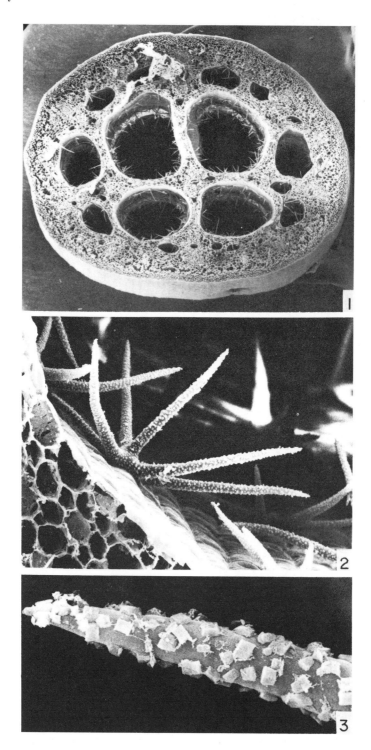

Fig. 132. Scanning electron micrographs of *Nymphaea* petiole. 1, Cross-section. × 12. 2, Portion of a lacuna with astrosclereids. × 160. 3, Portion of a branch of an astrosclereid with calcium oxalate crystals on its surface. × 1600.

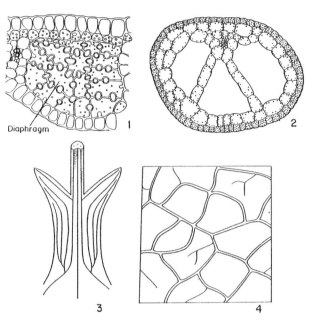

Diaphragm   1

2

3   4

FIG. 133. 1 and 2, Cross sections of leaves of water plants. 1, *Eriocaulon aquatile*. 2, *Ceratophyllum submersum*. 3, Diagram of the leaf base of *Trifolium* showing that the vascular supply of the stipules is derived from the lateral leaf traces. 4, Surface view of the venation of a mature portion of the leaf of *Quercus calliprinos*: the veins that are accompanied by bundle-sheath extensions are represented by double lines and those without such extensions by a single line. (No. 2 adapted from Troll, 1948; no. 3 adapted from Foster and Gifford, 1959; no. 1 adapted from Monteiro *et al.*, 1984.)

Structure of the petiole

A similarity exists between the tissues of the petiole and those of the stem. The epidermis of the petiole is continuous with that of the stem. The parenchyma cells of the petiole, like those of the cortex, contain only few chloroplasts especially as compared with the cells of the lamina. The supporting tissues of the petiole are collenchyma and/or sclerenchyma. The vascular bundles of the petiole may be collateral, e.g. *Ligustrum* (Fig. 134, no. 1) bicollateral, e.g. *Nerium*, or concentric, as in certain pteridophytes and many dicotyledons. The phloem is accompanied, in many species, by groups of fibres. The arrangement of the vascular tissues in the petiole differs in different plants. It may appear in a cross-section of the petiole as a continuous or an interrupted crescent, e.g. *Olea, Nicotiana, Nerium*, or as an entire or interrupted ring, e.g. *Ricinus, Quercus calliprinos, Q. boissieri* (Fig. 134, no. 2), *Citrus* (Fig. 140, no. 1), or as a ring with additional internal and external bundles, e.g. *Vitis, Platanus, Robinia, Pelargonium*. An arrangement of scattered bundles is seen in many monocotyledons and in *Rumex*. If there is a single collateral bundle in the petiole, the phloem is found on the abaxial side and if the bundles are arranged in a ring the phloem is external to the xylem on the periphery of the ring. Taking into account the variability of the vascular structure in the petiole Howard (1963, as cited in Howard 1974), proposed a classification relating the nodal structure at the level of the leaf gap to the vascular pattern in the leaf base and successive higher levels of the petiole or midrib. This approach to the study of the petiolar anatomy is important when using the latter as a taxonomic character (Howard, 1979; Dehgan, 1982).

In some plants, e.g. in *Mimosa* and *Albizzia* of the Leguminosae, the petioles or petiolules have pad-like swellings termed *pulvini (pulvinus)*. These are capable of bringing about movement of leaves or leaflets. The pulvinus contains a large amount of parenchyma and its surface is usually wrinkled. The closure and opening movements of the leaflets may be stimulated by internal or environmental factors, and are brought about by turgor changes in the motor cells of the pulvini. Toriyama

may penetrate deep into the tissues of the stem. This type of structure can be seen in the leaves of *Potamogeton* and *Eichhornia*. The air chambers are usually separated from one another by thin partitions of one or two layers of chloroplast-containing cells. *Diaphragms*—cross-partitions—occur in elongated air cavities. Such cavities are often termed air passages or lacunae. The diaphragms (Fig. 121, no. 2; Fig. 133, no. 1) consist of a single layer of cells with small intercellular spaces which appear as small pores and which apparently allow the passage of gases but not of water.

Immersed hydrophytes contain very little sclerenchyma and may even be devoid of this tissue. Strips of sclerenchyma are, however, sometimes found along the leaf margins. See also "Adaptation to aquatic habitat" in the previous chapter.

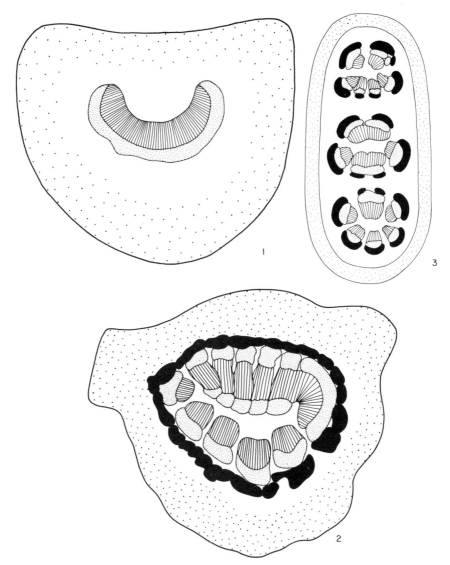

FIG. 134. Cross-sections of petioles. 1, *Ligustrum japonicum* in which the bundle is crescent-shaped. 2, *Quercus boissieri* in which the bundles are arranged in a ring. 3, *Populus angulata* in which the bundles are arranged in three rings. White areas—parenchyma; sparsely stippled areas—collenchyma; densely stippled areas—phloem; hatched areas—xylem; solid black areas—sclerenchyma.

(1960, 1962) observed thread-like structures in the petiolar parenchyma cells of *Mimosa*, which in the diurnal condition are thicker than in the nocturnal period. In *Mimosa pudica* there exists a symplastic continuity from the central vascular tissue of the pulvinus to the epidermis. It has been suggested that the numerous plasmodesmata may propagate the lateral conduction of the exitation (Fleurat-Lessard and Roblin, 1982). In *Albizzia julibrissin*

(Satter *et al.*, 1970) closure occurs when sub-epidermal cells of the dorsal side expand and those of the ventral side become compressed. Opening of the leaflets involves the reverse changes. The internal cells of the pulvinus surrounding the vascular tissue remain relatively unchanged during the movement of the leaflets. Electron micrographs show two features which may be connected with the movement: the presence of many vacuoles

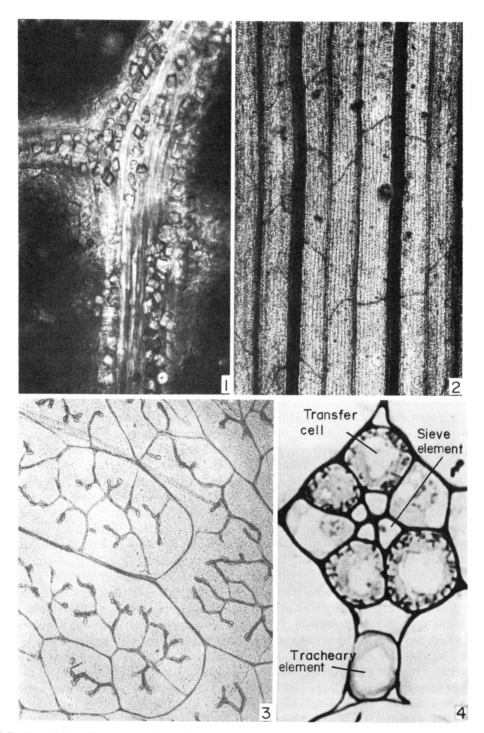

Fig. 135. Portions of cleared leaves as seen in surface view. 1, *Pistacia palaestina*, micrograph in polarized light in which the crystals in the cells of the bundle-sheath extensions can be seen. 2, *Lolium rigidum* showing minute veins which run transversely (commissural bundles) to the longitudinal axis of the leaf and which connect the parallel veins. × 40. 3, *Euphorbia milii* showing areoles with numerous blind vein endings. × 48. 4, Cross-section of a minor vein of a leaf of *Sherardia arvensis*, showing phloem transfer cells. × 2950. (No. 4 from Pate and Gunning, 1969).

which may contain a variety of enzymes (derived from spherical bodies), and the occurrence of thin fibrils oriented parallel to one another in the peripheral cytoplasm.

In the pulvinus of the solar-tracking leaf of *Lavatera cretica* the thickened walls of the epidermis and peripheral collenchyma consist of alternating thick and thin transverse strips.

This specialized architecture may enhance the flexibility of the collenchyma, while maintaining mechanical support (Werker and Koller, 1987). The vascular bundles of the veins are united in the pulvinus into a narrow flexible tube which constitutes the "hinge" of the pulvinus (Fisher *et al.*, 1987).

Vascular system of the leaf

As has already been noted in the chapter on the stem, one, two, three, or many leaf traces enter the leaf. The leaf traces may continue in the same number throughout the entire length of the leaf or they may divide, fuse and branch again later. Single or several closely associated vascular bundles form the *veins* (Isebrands and Larson, 1977, 1980; Larson, 1984a, b; Fisher, 1985). The term vein is sometimes used to include the vascular tissue together with the non-vascular tissue that surrounds it. There are plants, such as certain species of the Coniferales and *Equisetum*, in which the leaf has but a single vein. However, in most of the higher pteridophytes and the majority of the angiosperms the leaf contains numerous veins. The arrangement of the veins in the leaf is termed *venation*.

In the angiosperms two main types of venation are usually distinguished—reticulate and parallel venation. In leaves with reticulate venation, which is the commonest type among the dicotyledons, the veins are of different size depending on the degree of branching; i.e. central vein (or midrib), secondary veins, tertiary veins and so on. As a result of branching and fusing a network of veins is formed. In leaves with parallel venation, which is the commonest type among the monocotyledons, the main veins continue throughout the entire leaf and are almost parallel for most of their length but approach one another and fuse at the leaf tip or

both at the leaf tip and base. These "parallel" veins are interconnected by very thin commissural bundles which are scattered throughout the lamina (Fig. 135, no. 2). In certain monocotyledons, e.g. *Zantedeschia*, a special type of venation is found. Here the veins are parallel for a certain distance, after which they spread out in a feather-like pattern. In these leaves there are also small veins that connect the main veins. Parallel venation can also be found in certain dicotyledons, e.g. *Plantago, Geropogon,* and *Tragopogon,* and reticulate venation also occurs in certain monocotyledons, e.g. genera of the Orchidaceae, in *Smilax* and *Arum.*

When the venation is reticulate the largest vein passes through the median part of the leaf and forms the main or central vein (the midrib) from which smaller veins branch. In certain leaves numerous large veins can be seen spreading out, as rays, from the base of the leaf lamina towards its margins. Those parts of the lamina through which the larger veins, both main and secondary, pass are usually thicker and project as ribs on the abaxial side of the leaf. These ribs are formed of parenchymatous tissue which is poor in chloroplasts, and of supporting tissue which in the dicotyledons is usually collenchyma. Therefore the larger veins have no direct contact with the mesophyll, in the narrow sense of the word. With decreasing size of the veins, the direct contact between vascular and photosynthetic tissues increases (Fisher, 1985). The small veins that form a network between the larger veins, and which occur in the mesophyll proper, are usually situated in the outermost layer of the spongy mesophyll which borders the palisade cells (Fig. 127, no. 1; Fig. 129, no. 1).

The small veins usually form networks. These networks vary in size and shape, and they accordingly subdivide the area of the mesophyll. The smallest areas, which are bounded by the thinnest branches of the bundles, are called *areoles,* and they usually contain terminal vein-endings which end blindly in the mesophyll (Fig. 133, no. 4). The degree of branching of these vein-endings differs in the leaves of different plants. Thus, for instance, in the leaves of *Euphorbia* (Fig. 135, no. 3) or *Ricinus,* very many such blind ends may be found in a single areole, in *Morus* there are somewhat fewer, in *Quercus boissieri* very few, and in the leaves of *Q. calliprinos* (Fig. 133, no. 4) blind vein-endings are absent or almost so.

In monocotyledons with parallel venation, the veins that pass along the entire leaf may be almost of the same thickness or they may be of different thicknesses. In the latter case the thick and thinner veins are arranged alternately. The median vein is usually the thickest. As observed in *Zea mays*, the small longitudinal vascular bundles integrate into larger ones as they pass through the leaf from the tip of the blade towards the sheath. Only the large bundles and the intermediates, that arise midway between them, extend basipetally into the sheath and stem. While the number of the longitudinal bundles at the base of the blade decreases, their size and the cross-sectional areas of the sieve and tracheary elements increase (Russell and Evert, 1985).

In *Ginkgo* and many pteridophytes the veins do not form a close system since the adjacent branches do not anastomose (Arnott, 1959). In such leaves all the terminal branches terminate freely within the lamina or along its margins. In many leaves of this type the branching of the veins is dichotomous. Rodin (1967), who has studied the ontogeny of *Gnetum* leaves, found that the basic venation is one of dichotomies. The reticulate pattern develops with the increased size of the blade, by the anastomosis of the dichotomous vein branches in the submarginal region, and by the development of minor veins between the dichotomous ones. Open dichotomous venation has also been reported to occur in the leaves of the dicotyledonous species *Kingdonia uniflora* (Zheng-Hai *et al.*, 1964).

The control of the differentiation of vascular tissues by a polar flux of signals, from the shoot to the root, is mentioned in Chapter 11. This suggests that the growing tissues of the leaf produce signals in proportion to their growth, and that the vascular tissues differentiate along the channels of their drainage in the direction of the roots. In many ferns (Wagner, 1979) and most angiosperms there are, however, closed networks of veins. Such networks must include some cells which are not polar in terms of a preferred direction to the roots, and no flux of signals can be expected through such cells; yet they differentiate as part of a continuous vascular system. Sachs (1975)

presented evidence that the differentiation of closed networks depends on repeated changes in the direction of the flux, which occurs along a preferred axis but not necessarily in a preferred direction of polarity.

In most cases the arrangement of the vascular tissue in the main vein resembles that in the petiole.

The large veins in dicotyledonous leaves may consist of both primary and secondary tissues, while the smaller veins consist of primary tissues only. The large and medium-sized veins contain vessels and sieve tubes. In the smallest veins the tracheary elements are tracheids with annular and spiral wall thickenings. In the small or minor veins the parenchyma cells in contact with the sieve elements and tracheary elements constitute *transfer cells* (Gunning *et al.*, 1968; Gunning and Pate, 1969, 1974; Pate and Gunning, 1969, 1972). Some of these cells are believed to be concerned with short distance translocation between mesophyll and sieve elements (see Chapter 8), and others with an exchange of solutes between the phloem and xylem. The specialized transfer cells develop wall ingrowths which increase the internal cell surface (Fig. 135, no. 4). Some of the transfer cells (*A-type*) represent companion cells because of their ontogenetic relation to the sieve elements. These cells have labyrinthine wall ingrowths on all sides. Other transfer cells (*B-type*) are not companion cells, and their wall ingrowths are best developed on the walls contiguous with sieve elements or with A-type cells. Busby and O'Brien (1979) suggested that the xylem transfer cells, that are much developed in the wheat node, have an important role in redirecting solutes from the xylem of a mature leaf to developing leaves of the shoot apex. In mature plants these cells may play a role in the xylem transpiration pathway, especially in regions of xylem restrictions and discontinuities in the nodes. The phloem close to the vein-endings consists of parenchyma only, but in some dicotyledons, e.g. *Beta vulgaris*, the sieve elements of the minor veins accompany the xylem to the very ends of the veins (Esau, 1967, 1972).

The vein-endings in dicotyledons often contain only tracheids, which may be single or in pairs or in irregular groups. Sometimes terminal sclereids may be found as a continuation of the tracheids

(Foster, 1956; Fahn and Arzee, 1959) as, for instance, in *Mouriria, Boronia,* and *Arthrocnemum glaucum* (Fig. 53, no. 2) (see also Chapter 6, Sclereids). In the Magnoliaceae, Tucker (1964) reported, among others, the following types of cells: ordinary tracheids, clavate or irregularly shaped tracheids, thick-walled sclereids, and secretory cells. The thin veins that connect the parallel veins of grass leaves may contain a single row of tracheary elements and a single row of sieve elements.

Much importance is given to the problem of the density of veins in the leaf. The total length of the veins in a unit area of the leaf is usually great. Thus, for instance, we have found in *Quercus calliprinos* that the total length of the veins in a square millimetre is 11 mm and in *Q. boissieri* about 14 mm. According to Wylie (1939, 1946) the average distance between the veins of the dicotyledonous leaf is about 0.13 mm.

As has already been mentioned, the leaf contains tissues in which the cells have many lateral connections (i.e. the epidermis and the spongy parenchyma) and others in which the cells have few lateral connections (i.e. the palisade parenchyma). Wylie also came to the conclusion that a correlation exists between the density of the veins and the volume of the mesophyll tissues. On the one hand, with the increase in the volume of the palisade tissue, in which the conductivity in the direction parallel to the leaf surface is low, the distance between the veins becomes smaller and, on the other hand, with increase in the volume of the spongy tissue, in which conduction is efficient in the above-mentioned direction, the distance between the veins becomes larger. It should also be mentioned that leaf venation may, to some extent, be valuable in taxonomic and phylogenetic studies (Hickey and Wolfe, 1975; Rury and Dickison, 1977; Högermann, 1987).

Bundle sheaths

The large veins are surrounded by much parenchyma which is poor in chloroplasts. The smaller veins, also, are usually surrounded by a layer of tightly packed parenchyma cells; such a layer is termed the *bundle sheath*. In dicotyledons the cells of the bundle sheath are usually elongated in a direction parallel to the vein. Sometimes, however, for example in *Atriplex halimus*, the cells are more or less cubical (Fig. 136, no. 2; Fig. 125, no. 2). The cells of the bundle sheath are commonly thin-walled. They may contain as many chloroplasts as the mesophyll cells, or they may contain only a few chloroplasts, or they may be devoid of them. Sometimes crystals may occur in the bundle-sheath cells (Fig. 135, no. 1). The bundle sheaths may also surround the vein endings, e.g. in *Syringa vulgaris* (Morretes, 1967) and in *Atriplex halimus* (Fig. 136, no. 2). Although in most of the dicotyledons the bundle sheath consists of parenchyma cells, there are some families, such as the Winteraceae (Bailey and Nast, 1944), in which the sheath is sclerenchymatous.

In many dicotyledonous leaves, such as those of *Styrax officinalis* and *Quercus* spp. among others, the parenchyma of the bundle sheath extends to the epidermis on one side or on both sides of the leaf. These plates of parenchyma cells usually reach the epidermis itself, and they are termed *bundle-sheath extensions* (Fig. 136, no. 1). There is proof that the bundle-sheath extensions have a conduction function in the leaf (Wylie, 1943, 1947, 1949, 1951). They conduct from the bundles to the epidermal cells. The latter are very closely connected laterally and therefore function to conduct in a plane parallel to the leaf surface. Water may also move from the bundle sheath and its extensions through the mesophyll to the cells lining the substomatal chambers. Pizzolato *et al.* (1976) who used Wylie's (1943) method of precipitation of ferrocyanide by ferric ions in cotton leaves and examining sections in the electron microscope, found in the various cells precipitated crystals within the primary wall, in the space between the plasmalemma and cell wall, and within the protoplast. It thus seems that water flow from the tracheary elements to where it leaves the leaf takes place mainly in the apoplast. An apoplastic pathway of transpirational water from the xylem to the evaporating surface of the mesophyll and epidermis, has recently also been proposed by Evert *et al.* (1985) to occur in the leaves of *Zea mays*. In some plants the bundle-sheath extensions accompany the veins throughout almost their entire length, while there are other plants in which

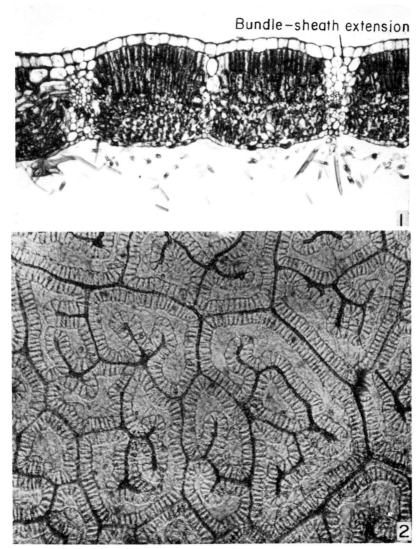

Bundle-sheath extension

FIG. 136. 1, Micrograph of portion of a cross-section of the leaf blade of *Styrax officinalis* in which stellate hairs on the abaxial surface and bundle-sheath extensions can be distinguished. × 140. 2, Surface view of cleared leaf of *Atriplex halimus* in which the bundle sheaths surrounding the veins can be seen. × 70.

bundle-sheath extensions are completely absent (*Olea, Pistacia lentiscus*). In *Quercus calliprinos* 94% of the total vein length is accompanied by bundle-sheath extensions; in *Q. boissieri*, 71%; in *Styrax officinalis*, 62%; and in *Pistacia palaestina*, 50%. According to Wylie, the density of the veins in mesomorphic leaves is inversely proportional to the total length of the bundle-sheath extensions.

Bundle sheaths also occur in the monocotyledons, especially in the grasses in which two types are distinguished, one- or two-layered. The outer sheath when two layers are present, or the single sheath where there is only one, generally consists of parenchymatous cells with thin walls. The cells of the parenchymatous sheath may store starch and form a "starch sheath". The cells of the inner sheath, the *mestome sheath*, are smaller in cross-section diameter, their walls are thickened and contain suberized lamellae. These lamellae are perforated by plasmodesmata (O'Brien and Carr,

1970; Carolin *et al.*, 1973; Eleftheriou and Tsekos, 1979). The mestome sheath is thus considered as analogous to an endodermis. Suberized lamellae were also observed to occur in the walls of the parenchyma sheath of maize (O'Brien and Carr, 1970; Evert *et al.*, 1977a). Based on histochemical and electron microscopical studies as well as on observations on the movement of a fluorescent apoplastic tracer, Eastman *et al.* (1988a,b) came to the conclusion that the bundle sheaths with suberized cell walls are permeable and do not function as endodermal layers. The plastids in the cells of the mestome sheath, if present, are few,

small and little differentiated. Single sheaths (Fig. 137, no. 2) are generally characteristic of panicoid grasses and double sheaths (Fig. 137, no. 1) of festucoid grasses. However, the distinction is not absolute (Metcalfe, 1960). On the small veins the inner layer may be present only on the side of the phloem.

Kranz type of leaf anatomy

Haberlandt (1918) distinguished in the leaf cross-section of certain plants a layer of radially oriented mesophyll cells which surrounds the vas-

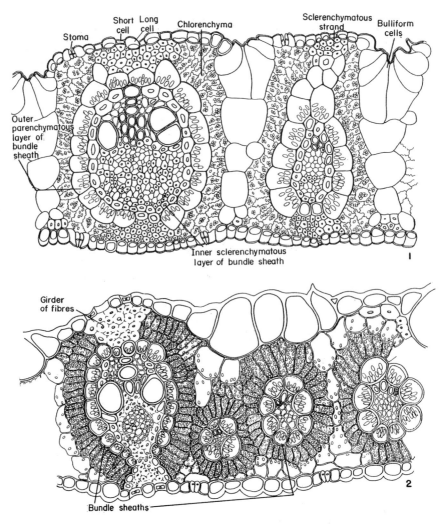

FIG. 137. Portions of cross-sections of grass leaves. 1, *Desmostachya bipinnata* in which the bundle sheath consists of two layers, the outer parenchymatous and the inner sclerenchymatous. × 260. 2, *Hyparrhenia hirta*, in which the bundle sheath consists of a single layer of chloroplast-containing cells. × 260.

FIG. 138. Extrafloral nectaries of *Sambucus nigra*. 1, Portion of a pinnate leaf. At the base of one leaflet there is a nectary, while at the base of the other leaflet there is a small leaf-like structure, which suggests that the nectaries represent modified stipules. × 2. 2, Nectaries at the base of two opposite leaves. × 6.

cular bundles (e.g. Fig. 123, no. 2; Fig. 125, no. 2; Fig. 136, no 2; Fig. 137, no. 2). He called this layer "Kranz" (wreath) and the type of leaf anatomy *Kranz type.*

In the literature dealing with C_4 plants (plants with the C_4 pathway of photosynthesis) the designation Kranz type of leaf anatomy includes both the mesophyll and bundle sheath (Dengler *et al.*, 1985). The C_4 pathway of photosynthesis (Björkman and Berry, 1973) has been recorded in a number of plants belonging to the following families: Aizoaceae, Amaranthaceae, Compositae, Chenopodiaceae, Cyperaceae, Euphorbiaceae, Gramineae, Nyctaginaceae, Portulacaceae, and Zygophyllaceae (Laetsch, 1974; Carolin *et al.*, 1975; Sanchez *et al.*, 1986). The C_4 plants are common in warm climates.

The chloroplasts of the bundle-sheath of these plants accumulate starch and may differ in ultrastructure from the mesophyll chloroplasts. An extreme case of this ultrastructural dimorphism is found in sugar-cane, where the bundle sheath chloroplasts lack grana. In other C_4 plants, grana are developed in all chloroplasts. Where differences in inner structure of chloroplasts of bundle sheath and mesophyll cells are lacking, there is considerable difference in their size. Numerous large mitochondria are present in the bundle

sheath cells (Chapman *et al.*, 1975) (Fig. 139). In the bundle sheath cells there usually are more microbodies than in the mesophyll cells.

A characteristic feature of C_4 chloroplasts is the occurrence of the peripheral reticulum (PR) (Carolin *et al.*, 1977). However, its presence in C_3 plants has also been reported (Gracen *et al.*, 1972). The PR appears in the peripheral stroma as a series of anastomosing tubules contiguous with the chloroplast envelope. The PR is generally more developed in mesophyll cell chloroplasts than in bundle-sheath cell chloroplasts (Laetsch, 1974).

It has been suggested that the PR may be involved in transport of materials between chloroplasts and cytoplasm. Evert *et al.* (1977a, b) who studied the mesophyll and bundle-sheath cells of the maize (C_4 plant) leaf, came to a conclusion that there is a direct connection between chloroplasts and plasmodesmatal desmotubules via ER in both types of cell.

The assimilates are transported from the mesophyll cells to the sieve tubes through the bundle sheath and vascular parenchyma. Whether the pathway is entirely symplastic or apoplastic or includes both ways is discussed in the literature. The frequency of plasmodesmata at the interfaces of the various cell types is used as an argument in support for the various views (Evert *et al.*, 1978; Fisher and Evert, 1982; Vassilyev *et al.*, 1982; Giaquinta, 1983; Russin and Evert, 1985; Evert and Mierzwa, 1986; Botha and Evert, 1988).

SUPPORTING TISSUES OF LEAF

The epidermis, itself, because of its compact structure and the strength of the cuticle, and the fact that the walls of its cells may sometimes be thick or impregnated with silica, gives support to the lamina. Collenchyma is usually found close to the larger veins immediately below the epidermis and also on the leaf margins in dicotyledonous leaves. The bundle-sheath extensions may also be collenchymatous. In addition to collenchyma, sclereids are present in the mesophyll of many dicotyledons. The large and medium-sized veins in many plants, e.g. *Pistacia palaestina*, are accompanied by groups of fibres. In monocotyledonous leaves the vascular bundles are accompanied by many fibres. In the Gramineae and in many other monocotyledons the fibres form girders on one or both sides of the bundles, and in many leaves they continue from the bundle sheaths to the epidermis, the cells of which, in such regions, may then also become fibre-like (Fig. 137, nos. 1, 2). In some monocotyledons, e.g. species of the Cyclanthaceae, strands of one or more fibres occur in the mesophyll, usually parallel to the longitudinal veins. The fibres are concentrated near the adaxial surfaces of the lamina (Wilder, 1985a).

In the mesophyll of many dicotyledons variously shaped sclereids are present (see Chapter 6).

Cork

Various structures with suberized cell walls have been reported to occur in leaves of some plants. Scattered areas of periderm occurring in various places of the leaf were observed in some species of the Cyclanthaceae (Wilder, 1985b). "Lenticell-like" structures situated below stomata, for instance, have been reported to occur in leaves of *Tripodanthus* (Morretes and Venturelli, 1985). Corkwart, structures similar to "lenticell-like" structures, occur in the leaves of some *Eucalyptus* species (Farooqui, 1982; Carr and Carr, 1987). It has been suggested that the above-mentioned suberized structures are artifacts that develop in response to insect punctures or mechanical injuries, or beneath micro-organisms.

Secretory structures

Secretory structures, which participate in the secretion of water or other substances, are a common feature of leaves. Many of these secretory structures are of epidermal nature and are discussed in Chapter 10.

The substances produced may be excreted from the cells or they may be retained only to be released upon the disintegration of the cells.

Glands are present on many foliage leaves and cataphylls. These structures consist of a mass of dense parenchyma cells in which there terminates a vascular bundle. The parenchyma is covered by a glandular epidermis. The cells of such an epidermis are mostly elongated in a direction at right angles to the surface of the gland (Fig. 97, no. 1) and have dense cytoplasm and large nuclei. The presence of these glands on the petioles and on the teeth of the laminar margins, for instance, are of great taxonomic importance because of their constant position in the species, and even varieties, in which they occur. The glands on different organs or parts of an organ of the same species may secrete different substances. In *Prunus persica* and other related species, for example, the glands on the laminar teeth secrete a bitter resinous substance, while those on the petiole secrete nectar.

Nectariferous glands are found on the petioles of many plants, e.g. *Passiflora*, *Ricinus*, and *Impatiens*. In *Vicia* nectariferous tissue is present in the central portion of the stipule where it is easily distinguished because of the anthocyanins present in its cells. Extrafloral stalk-like nectaries occur at the bases of the leaves and leaflets of *Sambucus nigra* (Fig. 138). These nectaries represent apparently both modified stipules and stipels (Fahn, 1979, 1987).

Another type of secretory structure is the essential oil cavity which occurs characteristically in the mesophyll of the leaves of *Citrus* (Heinrich, 1969; Bosabalidis and Tsekos, 1982) and the secretory cavity of *Gossypium*. These are examples of lysigenous cavities in which the secretions formed in the cells are released after the disintegration of

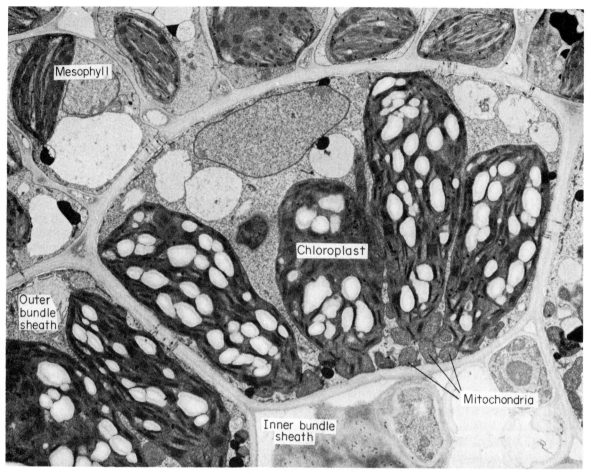

FIG. 139. Electron micrograph of a cross-section of a portion of a *Cynodon dactylon* leaf, a C$_4$ plant. The chloroplasts of the outer sheath (the "Kranz") are aggregated close to the inner wall. × 7500. (Courtesy of Maret Vesk of the University of Sydney.)

the cells, the lysis of which forms the cavity. Secretory cavities occurring in the mesophyll of *Eucalyptus* (Fig. 128, no. 1) develop lysigenously according to some authors (e.g. Fohn, 1935) and schizogenously according to others (Carr and Carr, 1970).

Examples of other types of secretory structures found in leaves are schizogenous resin ducts, which are characteristic of the Compositae, Anacardiaceae, and Coniferales; laticifers as found in *Euphorbia*; mucilage cavities found in the Sterculiaceae, different species of the Malvaceae, Moraceae, and other families.

In leaves, secretory substances may also be found in idioblasts. These secretory cells are classified according to the substances secreted, although cells that contain a mixture of different substances also exist. The secreting idioblasts, like other secretory structures, are of important taxonomic significance (Metcalfe and Chalk, 1950, 1979, 1983; Baas and Gregory, 1985). For instance, the secretory cells of the Lauraceae (Maron and Fahn, 1979), Simarubaceae, and Onagraceae contain oils. Cells containing the enzyme myrosin are found in the Cruciferae, Capparidaceae, Resedaceae, Tropaeolaceae, and Moringaceae. Cells with a resinous content are found in the Meliaceae and in many species of the Euphorbiaceae, Rutaceae, and Rubiaceae. Tannin-containing cells are found in the Anacardiaceae (especially in *Pistacia palaes-*

tina and *P. lentiscus*), Annonaceae, Crassulaceae (especially in *Sempervivum tectorum* and species of *Echeveria*), Ericaceae, Euphorbiaceae, Buxaceae, Polygonaceae, Rosaceae, Tamaricaceae, and Leguminosae. Cells with tannin compounds are also found in parenchyma of fruits, e.g. of *Ceratonia*. Cells with mucilaginous contents are found in the Rhamnaceae, many species of the Malvaceae, Chenopodiaceae, and Rubiaceae, and in many monocotyledonous plants. Cells containing secretory substances that have not been identified are found in many different families, such as the Anacardiaceae, Fagaceae, Buxaceae, Aristolochiaceae, Cruciferae, Platanaceae, Plumbaginaceae, Rutaceae, and Punicaceae.

Idioblasts with differing types of crystals and cystoliths are found in the leaves of different species and have already been described in previous chapters (especially 1 and 10).

Special secretory structures are the *hydathodes* (Fig. 140, no. 2; Fig. 141) which secrete water in the form of drops from within the leaf. This phenomenon is called *guttation*. Hydathodes secrete water which is brought to the surface by the terminal tracheids of the veins. This water passes through the intercellular spaces of the loosely packed parenchyma of the hydathode which is devoid of chloroplasts and which is called the *epithem*. The intercellular spaces open to the exterior by special pores which are of stomatal origin and which remain permanently open (Stevens, 1956). The epithem may be bounded by suberized cells or cells with Casparian strips. Some hydathodes lack a typical epithem. Haberlandt (1918) distinguished two different types of hydathodes—*active hydathodes* and *passive hydathodes*.

It is, however, more appropriate to use the term hydathode only for those organs through which water is secreted passively (Fahn, 1979). The active hydathodes of Haberlandt might better be considered as glands that secrete solutions of salt or other substances (see also p. 175). Typical hydathodes occur on the leaves of *Brassica* and plants belonging to the Gramineae, Crassulaceae, Rosaceae, and Saxifragaceae among others. The only known hydathode which develops from a shoot apex is that occurring at the tip of the grape tendril (Tucker and Hoefert, 1968).

Domatia

The term *domatia* refers to depressions, crypts, sacs, and tufts of hairs in the axils of main veins on the abaxial leaf surface, or even revolute basal margins of leaves of certain plants. (The anatomy and possible functions of domatia such as habitats for animals are discussed in Napp-Zinn, 1973–74 and Wilkinson, 1979; Pemberton and Turner, 1989.)

Galls

In many plants, leaves and other young organs may become swollen and deformed, sometimes to a great extent, producing *galls*. The latter are hypertrophic tissues which develop in response to stinging and deposition of eggs by insects or due to infection with micro-organisms or nemathodes. The form and anatomical structure of galls varies with plant species and the agent causing their formation (Kahl and Schell, 1982; Meyer and Maresquelle, 1983).

HISTOLOGY OF THE GYMNOSPERM LEAF

Most gymnosperms are evergreen and their leaves are usually xeromorphic. One of the peculiarities of gymnosperm leaves is the occurrence of *transfusion tissue*. This tissue accompanies the vascular bundles and is composed of tracheids, parenchyma, and albuminous cells. The transfusion tissue varies in amount and arrangement depending on the genera (Lederer, 1955; Ghouse, 1974; Ghouse and Yunus, 1974, 1975; Hu and Yao, 1981).

Two types of gymnosperm leaf will be described here—that of *Cycas* and that of conifers, such as *Pinus* and *Cedrus*.

The leaf of *Cycas* (Fig. 142, no. 1) is leathery and stiff, the epidermal cells are thick-walled and have a thick cuticle, and the stomata are sunken and occur on the abaxial surface of the leaf. The mesophyll consists of palisade and spongy parenchyma as in angiosperms. A uni- or biseriate hypodermis is present between the adaxial epi-

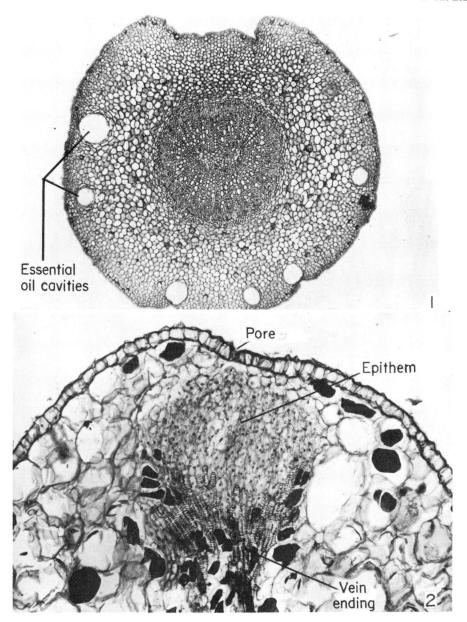

FIG. 140. 1, Cross-section of the petiole of *Citrus*; stem-like arrangement of vascular tissues and essential oil cavities can be distinguished. × 45. 2, Portion of section cut parallel to the surface of the leaf of *Sedum* sp. showing the structure of a hydathode. × 180.

dermis and the palisade parenchyma. The xylem of the median vein is of a special primitive type. The protoxylem accompanied by a small amount of parenchyma, is present on the abaxial side and the metaxylem on the adaxial side. Secondary xylem develops near the phloem from a cambium situated between the two types of vascular tissue. The median vein is surrounded by an endodermis.

A few transfusion tracheids are present on both sides of the metaxylem. Below the phloem there is a layer of transfusion-parenchyma cells. Outside the endodermis on both sides of the median vein there are plates of accessory transfusion tissue consisting of tracheids and parenchyma cells (Lederer, 1955).

The epidermis of the needle-like leaves of *Pinus*

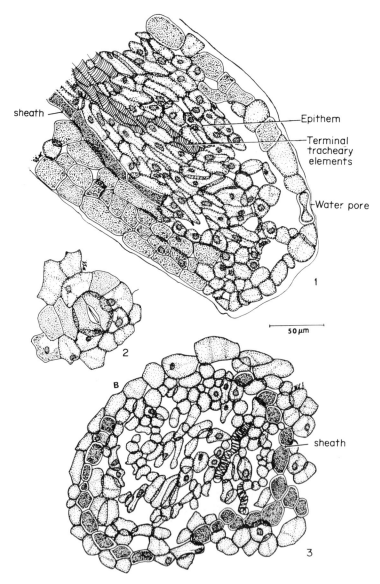

sheath

Epithem

Terminal
tracheary
elements

Water pore

1

50 μm

2

B

sheath

3

FIG. 141. Hydathode of *Ribes viburnifolium* leaf. 1, Longitudinal section. 2, A water pore (modified stoma) and surrounding epidermal cells. 3, Sheath surrounding epithem and tracheary elements with spiral thickenings. (From Stern *et al.*, 1970.)

(Fig. 142, no. 2) and *Cedrus* (Fig. 143, nos. 1, 2) consists of extremely thick-walled cells and is covered with a thick cuticle. The stomata are present on all sides of the leaf; they are sunken and are overarched by the subsidiary cells (Fig. 142, no. 2). A hypodermis of fibre-like sclerified parenchyma cells is present except in the areas below the stomata. The mesophyll is of a parenchymatous nature. The walls of the mesophyll cells have characteristic ridge-like invaginations into the cells. These cells contain chloroplasts. Resin ducts are also present in the mesophyll. In the centre of the leaf there is a single vascular bundle, or two, which are then close to one another. The arrangement of the proto- and metaxylem is as in angiosperms, i.e. the protoxylem is on the adaxial side and the metaxylem on the abaxial side close to the phloem. The bundle is surrounded by transfusion tissue consisting of tracheids and of living parenchyma cells. The parenchyma cells contain tan-

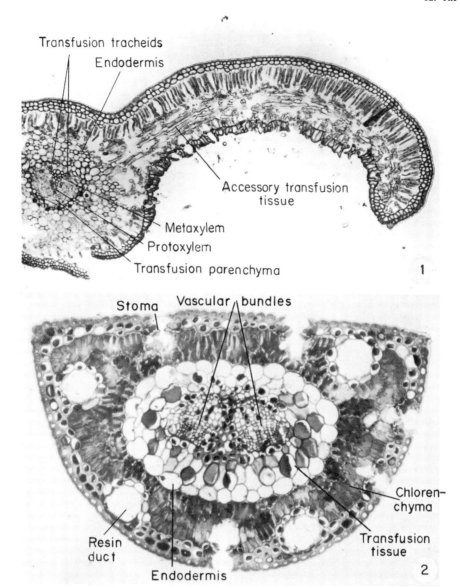

Fig. 142. Micrographs of cross-sections of gymnosperm leaves. 1, A leaflet of *Cycas revoluta.* × 50. 2, *Pinus halepensis.* × 155.

nins, resins, and also starch in certain seasons of the year. The tracheids closest to the bundles are long while those further away are more parenchyma-like in shape and have relatively thin, slightly lignified walls and bordered pits. Because of their thinner walls these tracheids are not able to withstand the pressure of the living cells around them in which the turgor is higher, so they become

somewhat crushed. In the transfusion tissue close to the phloem there are certain cells that have dense cytoplasm and which are similar to albuminous cells. The vascular bundles together with the transfusion tissue are surrounded by a sheath of relatively thick-walled cells—the endodermis (Fig. 142, no. 2; Fig. 143, no. 1). Electron microscopic studies of pine needles (Walles *et al.*,

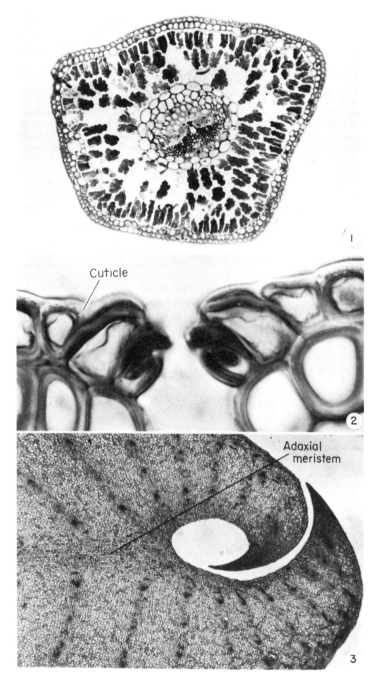

FIG. 143. 1, Micrograph of a cross-section of the leaf of *Cedrus deodara*. ×45. 2, Cross-section of a stoma of *C. deodara*. ×470. 3, Portion of a cross-section of a developing leaf of *Musa* showing the adaxial meristem. ×45.

1973; Carde, 1978) revealed that their endodermal cells do not possess true Casparian strips. All anticlinal walls contain a secondary encrusting substance devoid of suberin but heavily impregnated with lignin. The tonoplast of the central vacuole is thick. The vacuole contains dense spherical bodies. Numerous primary pit fields occur in the radial and tangential walls. The parietal cytoplasm contains smooth ER, vesicles, long slender chloroplasts, and lipid bodies.

DEVELOPMENT OF THE FOLIAGE LEAF

The development of the leaf can be divided, although artificially, into the following stages: initiation, early differentiation, development of the leaf axis, origin of the lamina, and histogenesis of the tissues of the lamina.

Initiation

The initiation of the leaf commences with periclinal divisions in a small group of cells on the sides of the apex. However, the number of cell layers that begin to divide thus and their position on the apex varies considerably in different plants. For example, in many grasses it was found that leaf initiation starts with periclinal division in the cells of the surface layer of the apex (i.e. in the outermost layer of the tunica) and in cells of the layer immediately below it (Sharman, 1942, 1945; Thielke, 1951). In this case the main portion of the leaf primordium originates from the outermost cell layer of the shoot apex.

Contrary to the situation in the grasses, in other monocotyledons, e.g. *Tulipa* (Sass, 1944), and apparently in most dicotyledons, the first periclinal divisions do not take place in the cells of the surface layer, but in the cells of one or more layers below it. In the apices of such plants, therefore, the surface layer does not take part in the initiation of the inner tissues of the leaf. This layer enlarges by numerous anticlinal divisions of its cells and so becomes adapted to the growth of the primordium. The surface layer gives rise to the protoderm of the young leaf.

Recently Cunninghame and Lyndon (1986) studied the correlation between the distribution of periclinal divisions in the apex and the development of the leaf primordia in *Pisum* and *Silene*. Using light and electron microscopy they determined the orientation of new cell walls on the basis of their thickness, and came to a conclusion that the formation of the leaf primordia is preceded by the occurrence of periclinal divisions in the tunica and corpus, apparently mainly in the latter.

In the gymnosperms there is the same amount of variation in the initiation of the leaf primordia as there is in the angiosperms. In *Taxodium distichum*, for example, the initiation of the leaf primordium originates with periclinal divisions in the cell layer immediately below the surface layer of the apex (Cross, 1940), while in most of the Coniferales and in *Zamia* (Korody, 1937; Johnson, 1943) the periclinal divisions take place in the surface layer of the apex as well as in the layer below it.

Most commonly the initiation of the leaf primordium commences in cell layers below the surface layer. In this case the degree to which the inner cell layers of the tunica and the neighbouring cell layers of the corpus participate in the initiation of the primordium differs, and it is difficult to determine accurately the part played by each of them. In order to clarify this problem periclinal cytochimeras, produced by the application of colchicine, have been used. With the aid of such cytochimeras it has been possible to show from which layers of the apex the various tissues of the leaf develop (Satina and Blakeslee, 1941; Dermen, 1947, 1951). Dermen thus was able to determine that in *Vaccinium* and *Pyrus malus*, for instance, three cell layers of the apical meristem take part in the formation of the leaf. The leaf epidermis develops from the outermost layer (i.e. from the outer tunica layer) by anticlinal divisions. The second and third layers (i.e. the inner layer of the tunica and the outermost layer of the corpus) give rise to the mesophyll and the vascular bundles.

Early differentiation

As a result of continued cell division the leaf primordium protrudes from the shoot apex as a buttress which has the form of a small papilla or crescent. This leaf buttress consists of a protoderm layer, an inner mass of ground meristem and a procambial strand which develops acropetally from the nearby procambium of the stem (see Chapter 11).

Development of the leaf axis

In many dicotyledons and gymnosperms the development of the leaf axis precedes that of the lamina or of the leaflets (in a compound leaf).

As a result of the rapid development the primordium becomes shaped like a gradually tapering cone the adaxial side of which is flattened (Fig. 144, no. 1). The tip of the cone functions for a while as an apical meristem, but in most spermatophytes the cells at the tip of the leaf exhibit histological signs of maturation relatively soon. In certain plants, from that early stage of development when the primordium is still less than 1 mm long, all further increase in length is due to the

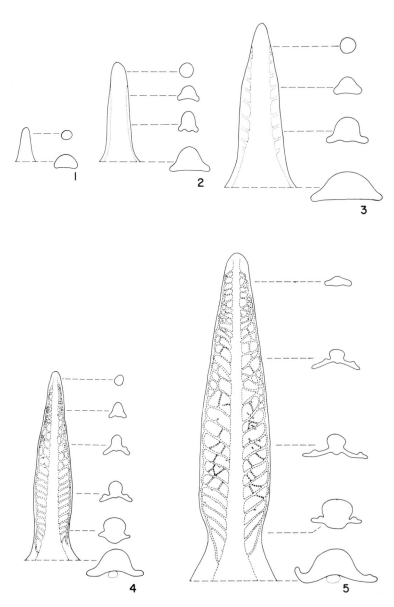

Fig. 144. Diagrams of longitudinal and cross-sections of leaf primordia of *Nicotiana tabacum* at different ontogenetic stages. 1, A young, more or less cone-shaped primordium. 2, Primordium in which the narrow margins, from which the lamina will develop, can be seen. 3, Primordium in which the beginning of development of the main lateral veins can be seen. 4, Primordium 5 mm long in which the early development of the provascular system can be seen. 5, A more advanced stage. Nos. 1–3, × 25; nos. 4 and 5, × 10. (Adapted from Avery, 1933.)

division and elongation of cells distant from the tip of the primordium, i.e. by intercalary growth. In leaves of ferns and in pinnate leaves of a few angiosperms (Steingraeber and Fisher, 1986), however, the apical growth continues for a long period together with the addition of cells by intercalary growth in an acropetal direction. The increase in length of the axis is usually accompanied by an obvious increase in width. This thickening is particularly striking in those numerous plants in which a cambium-like region, in which the cells divide tangentially, develops along the adaxial side of the primordium. This cambium-like region is termed the *adaxial meristem* or *ventral meristem* (Fig. 143, no. 3) and from it *accessory bundles* may develop (Foster, 1936; Kaufman, 1959).

Origin of the lamina

During the early elongation and thickening of the axis of the young leaf, the cells of the adaxial margins continue to divide very frequently— much more so than the inner cells of the ground meristem. In simple leaves two wing-like strips (Fig. 144, nos. 2, 3) usually develop on the margins as a result of the accelerated growth of these cells. In leaves with a petiole the marginal growth is depressed in the basal portion of the leaf axis which then develops into the petiole. In a cross-section both sides of the developing lamina can be seen to consist of protoderm enclosing a few layers of ground tissue. The new cells that are added to the different layers of the lamina originate from rows of *marginal initials* and *submarginal initials* (Fig. 145, nos. 1–2).

The marginal initials are the cells of the outermost layer on the margins of the young lamina. Generally, in the angiosperms these initials divide only anticlinally and so add new cells to the abaxial and adaxial protoderm. In certain monocotyledons and in the bud scales of *Rhododendron* spp. periclinal divisions also occur, resulting in the addition of new cells to the nearby ground meristem (Foster, 1937; Sharman, 1942, 1945). In the variegated leaves of certain plants the white margins may develop as a result of periclinal divisions in the protoderm (Renner and Voss, 1942). In *Daphne odora* (Hara, 1957) and in the hemiparasite *Eubrachion ambiguum* of the Loranthaceae (Bhandari, 1969) the whole lamina is produced by the marginal initials only.

The submarginal initials undergo divisions in various planes. These cells give rise to new cells which are added to the inner layers of the young lamina.

Some investigators have recently challenged the concept of the origin of the internal tissues of the leaf blade from a single row of submarginal initials (Dubuc-Lebreux and Sattler, 1980; Poething, 1984, 1987; Beardsell and Considine, 1987).

In pinnately and palmately compound leaves the lateral leaflets develop from the adaxial marginal meristem of the axis of the young leaf as two rows of papillae. In certain plants the order of development of the leaflets is acropetal, e.g. *Carya* (Foster, 1932, 1935), but in many other plants the order is basipetal. As is seen in *Carya*, the primordium of each lateral leaflet first develops an axis and the laminae of the leaflets later develop from the margins of these axes. The tip of the main axis develops into a terminal leaflet.

The development of a lobed leaf is a result of differential activity of the leaf margin which starts early in ontogenesis. In the regions of the future lobes meristematic activity is considerable, while in the regions of future sinuses there is little or no activity (Fuchs, 1975).

In fenestrated leaves, e.g. in *Monstera deliciosa*, holes develop at an early ontogenetic stage by necrosis of small patches of tissue (Melville and Wrigley, 1969).

The formation of the feather leaf of palms starts with the appearance of a series of folds within each half of the developing lamina. With progressing intercalary growth, the pleats (pleacations) resemble the foldings of a camera bellows. The pleats are ultimately cleaved by tissue separation into free pinna segments of the mature palm leaf frond (Kaplan, 1984).

Histogenesis of the tissues of the lamina

The marginal growth apparently continues longer than does the apical growth but it, too, ceases relatively early. In *Nicotiana tabacum*, for example, Avery (1933) observed that marginal growth continues, at least in the lower part of the lamina, until that stage where the leaf is several centimetres long. In *Cercis siliquastrum* (Slade, 1957) marginal growth of the lamina is completed

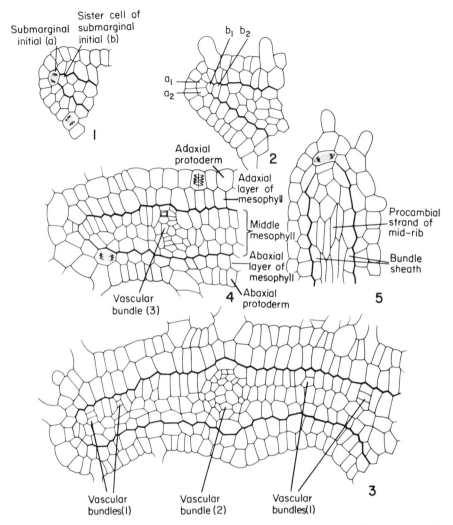

FIG. 145. 1–4, Cross-sections of the marginal portions of leaves of *Nicotiana tabacum*, at different stages of development, showing the divisions which, together with the enlargement of the cells, result in the growth of the young lamina. 1, Showing the direction of division in the protoderm and submarginal initial. The cell *b* is a sister cell of the submarginal initial (a) and both are derived from a periclinal division of a submarginal initial in that position. 2, Showing the two cells, a_1 and a_2, resulting from the anticlinal division of *a*, and the cells b_1 and b_2 which arose from the periclinal division of *b*. 3 and 4, Development of mesophyll and provascular strands; different stages of development of the vascular bundles are indicated by numerals. 5, Median longitudinal section of a primordium showing the provascular strand of the mid-rib. (Adapted from Avery, 1933.)

by the time the leaf measures 2–2.5 mm. In the leaf of the Dwarf Cavendish banana, in which a marginal vein is differentiated in the early stages of development of the leaf primordium, marginal growth in the laminar part of the primordium ceases as early as between the fourth and sixth plastochron. Marginal growth on the right side of the mid-rib ceases in the fourth plastochron when the primordium is about 5 mm long whereas, on

the left side, it ceases during the sixth plastochron when the primordium is about 20 mm long. After the cessation of marginal growth, further growth of the lamina is brought about by cell division in the various cell layers of the lamina. These divisions are mostly anticlinal and thus a *plate meristem* (Fig. 145, nos. 3, 4) is formed. A plate meristem is one in which the planes of cell divisions in each layer are perpendicular to the surface of the

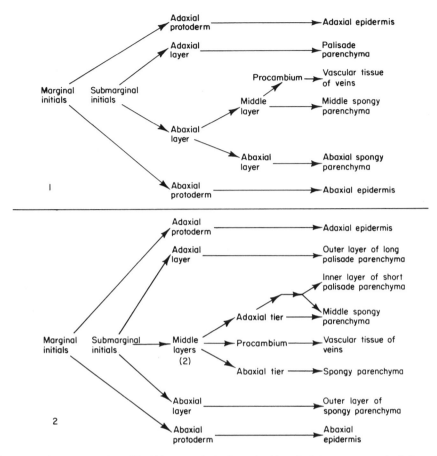

FIG. 146. Diagrammatic representation of leaf histogenesis. 1, *Carya buckleyi.* 2, *Pelargonium zonale.* (Adapted from Foster, 1936.)

organ in which the meristem occurs. The activity of such a meristem results in increase in surface area but not in thickness of the organ. In the lamina the cells of this meristem have a stratified arrangement and therefore it is possible to trace, with relative ease, the origin of the epidermis, palisade and spongy tissues, and the vascular bundles. In Figs 146 and 147, it can be seen how, in different plants, these tissues develop from the cell layers of the young lamina.

The regular arrangement of the cell layers is interrupted to differing extents by the development of the vascular bundles, their sheaths and supporting tissues. As a result of this, in the final stages of the expansion of the leaf surface the regular arrangement of the cell layers becomes restricted to those areas of the lamina between the large lateral veins. The palisade parenchyma is

one of the last tissues to cease growing and dividing. This tissue may continue to function as a meristem for some time after the cells of the spongy parenchyma and of the epidermis have ceased to divide. Most of the knowledge of leaf initiation and development is based on ordinary histological observations. However, several interesting attempts were made to study the pattern of contribution of the various cell layers of the shoot apex to the leaf buttress formation and to the leaf axis and laminar growth using chimeras (see Stewart and Dermen, 1975; Stewart, 1978; Poething, 1984).

The different parts of the leaf expand at different rates and in different directions (Avery, 1933). This type of growth has been termed anisotropic growth (Ashby, 1948a). The type of growth of a leaf is controlled by genetic factors, but it is also

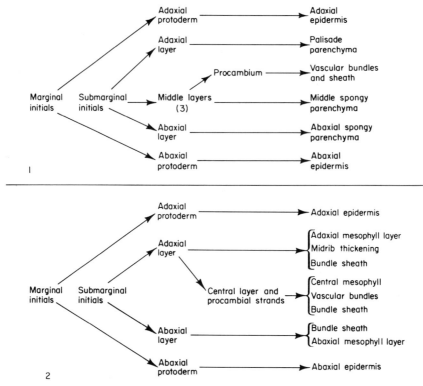

Fig. 147. Diagrammatic representation of leaf histogenesis. 1, *Nicotiana tabacum*. 2, *Oryza*. (No. 1 adapted from Foster, 1936; no. 2 adapted from Kaufman, 1959.)

influenced by internal and external environmental conditions (McCallum, 1902; Ashby, 1948b; Allsopp, 1955; Jones, 1956). Thus, the shape of leaves on different parts of the stem of the same plant is apparently influenced by internal factors. Among the external factors that influence leaf shape are water supply, nutrients, day length, amount of light, etc.

Development of the veins

The development of the vascular system in the leaf has, as yet, been studied only in a small number of plants. From what is known of the development of dicotyledonous leaves, it appears that development of the procambial strand of the mid-rib precedes that of the lamina and it proceeds in the acropetal direction (Fig. 145, no. 5). With the commencement of the development of the lamina the procambial strands of the large lateral veins and, later, of the smaller veins begin to form gradually. As was seen in *Nicotiana taba-*

cum (Avery, 1933), the procambial strands of the small veins, which form in a basipetal direction, develop mainly during the intercalary growth of the lamina (Fig. 144, nos. 4, 5). However, deviations from the above pattern of differentiation are also known to occur (Slade, 1957).

Recently Isebrands and Larson (1980) investigated the ontogeny of the major veins in the lamina of *Populus deltoides* and found that first a procambial strand arising from the axial leaf trace develops acropetally and becomes the precursor of the first vascular bundle of the midrib of the lamina. This is followed by acropetal differentiation of discrete subsidiary bundles in the meristematic regions. These regions precede the acropetally developing procambial bundles. Later subsidiary bundles diverge obliquely in the lamina margin giving rise to the secondary veins in a basipetal fashion. Subsequent differentiation and maturation of the secondary veins occurs within the lamina. There they initiate basipetally from the acropetally developing original procambial bundles, subsidiary bundles, and/or their derivatives. The term *subsidiary bundles* as used by Isebrands and Larson (1977) refers to "vascular bundles that differentiate both basipetally and acropetally from a point of origin near the leaf primordium base and on both sides of the original procambial bundles; these bundles maintain continuity between stem, node and petiole".

In the leaves of *Zea* (Sharman, 1942) the procambial strands of the median vein and of the principal lateral veins develop acropetally while those of the smaller lateral veins, which are arranged alternately with the larger ones, develop basipetally, i.e. from the tip to the base of the leaf. The latter development takes place only after the appearance of protophloem in the larger veins. In the leaf primordia of wheat the median and lateral procambial strands originate independently from the vascular system of the rest of the plant (Sharman and Hitch, 1967). They have their point of origin in the axis, in the disc of insertion of the primordium. The later developing laterals are initiated up in the primordium and differentiate both acropetally and basipetally. For four plastochrons the leaf primordium has no vascular connection with the rest of the plant. It would appear, therefore, that materials necessary for growth of the apex and of the first four leaf primordia are transported through meristematic and parenchymatous tissue. The cross veins which connect the parallel veins are the last to form and they develop basipetally. As seen in some grasses, the cross veins arise from single files of young mesophyll cells (Tiba and Johnson, 1987). The procambial strands of the marginal veins of the banana leaf also develop in a basipetal direction. The differentiation of the conducting elements begins before the completion of the procambial system. The protophloem and protoxylem both differentiate acropetally and the differentiation of the protoxylem follows that of the protophloem. After the final elongation of the veins, the development of the metaphloem and metaxylem commences in a more or less definite basipetal direction, first in the large strands in which the differentiation of the protophloem and protoxylem is completed, and then in basipetally developing procambial strands, in which protophloem and protoxylem do not develop.

The blind vein endings occurring in the areoles are, according to Slade (1957, 1959), caused by the rupture of the minor vascular network during that stage of development when the leaf expands as a result of mesophyll cell enlargement. According to Pray (1963), however, the vein endings do not result from rupture, but there is a progressive differentiation of procambium from the ground meristem during the expansion of the lamina. Lersten (1965) came to a similar conclusion.

DEVELOPMENT OF LEAVES DIFFERING FROM THE TYPICAL

Cataphyll formation in the shoot apices of the dicotyledons is indicated in the early stages of development of the primordium. Cataphylls are distinguished from foliage leaves by the following characteristics: the slightly developed mesophyll which is usually devoid of palisade parenchyma; the reduced system of vascular bundles which often form an open dichotomous type of venation; and the small number or even absence of stomata. In certain cataphylls there is very little and sometimes even no sclerenchyma, while in others there is an excessive development of sclerenchyma. In some plants, such as *Aesculus*, for example, a periderm develops below the abaxial epidermis of the outer scales.

The following are the differences that take place in the development of the cataphyll as compared with that of a foliage leaf: the adaxial meristem of the axis of the leaf primordium is only slightly active or not at all; the marginal growth is accelerated and is truly lateral and not lateroadaxial, as in the foliage leaf; as a result of rapid growth and the absence of thickening of the midrib the scale becomes typically sheath-like.

Glochidia. In the Cactaceae there are small outgrowths on the stem called *areoles*, which bear spines. The spines of some genera are barbed and called *glochidia*. The areole meristem is similar to that of the shoot. The glochidia differentiate from the areole meristem as do leaves at the shoot apex. The early ontogeny involves apical growth. When the glochidium is about 100 μm long, additional growth is intercalary. When it is 400 μm long, the retrorse barbs which develop from extensions of the epidermal cells start to initiate

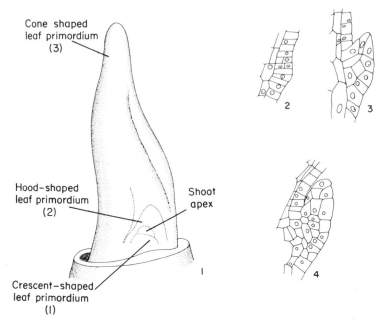

Cone shaped
leaf primordium
(3)

Hood-shaped
leaf primordium
(2)

Shoot
apex

Crescent-shaped
leaf primordium
(1)

FIG. 148. 1, Three-dimensional drawing of the shoot apex of *Oryza*, showing the apex and first three leaf primordia. Numerals indicate the relative plastochrons. 2–4, Median longitudinal sections of ligules at different stages of development. (Adapted from Kaufman, 1959.)

at the tip of the glochidium. As maturation continues glochidia and leaves differ significantly. No vascular tissue, chlorenchyma, or stomata differentiate in the glochidia. All the cells except those of the basal meristem become sclerified and the glochidia are easily detached at this place (Freeman, 1970).

Monocotyledonous leaves

Grass leaves have linear laminae and sheathing bases surrounding the stem. The development of the leaf of *Oryza sativa* as described by Kaufman (1959) will be used here as an example of the development of this type of leaf.

The leaf primordia are initiated in the tunica from which the ground meristem and protoderm of the leaf develop. At the shoot apex, in the early stage of initiation, a localized protuberance appears which later becomes crescent-shaped and then as a result of further marginal and apical growth eventually surrounds the apex. As the young primordium grows upwards it becomes hood-shaped (Fig. 148, no. 1). Apical growth of the leaf ceases during the third plastochron when

the primordium is about 0.9 mm long, but the margins continue to grow and the primordium elongates further. The continued marginal growth is brought about by the activity of the marginal meristem and the elongation of the primordium by a rib-meristem form of growth. A meristem of this kind is characterized by parallel series of cells in which transverse divisions take place. The rib meristem and adaxial meristem (Fig. 143, no. 3) become distinguishable during the second and third plastochrons.

As described above, in the primordia of monocotyledonous leaves the processes of apical and marginal growth are simultaneous during the early plastochrons. This is contrary to the development of dicotyledonous leaves where there are two distinct stages as have been previously described (Kaufman, 1959).

In *Oryza* the ligule is initiated during the third plastochron by periclinal divisions in the adaxial protoderm (Fig. 148, nos. 2–4). At first the ligule is a small adaxial protuberance which, as a result of continued periclinal divisions in the protoderm, expands laterally towards the margin of the sheath. The auricular primordia are apparently

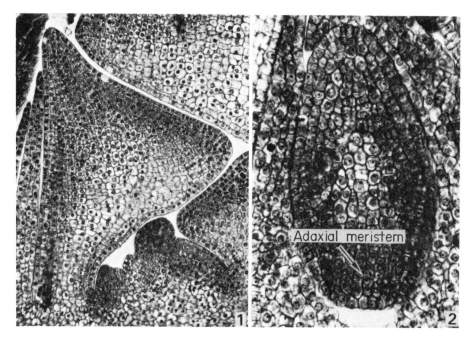

FIG. 149. Micrographs of sections of leaf primordia of *Acorus calamus*. 1, Median longitudinal section of a 385 μm long leaf, showing a marked adaxial growth. × 175. 2, Cross-section of a 224 μm long leaf, showing the adaxial meristem. × 450. (Courtesy of D. R. Kaplan.)

derived from both the sheath and ligule protoderm as well as from the ground meristem.

Continued elongation of the lamina and sheath, at this stage, still results from the activity of the rib meristem, whereas the extension of the lamina wings is caused by the activities of the marginal and of the distinct plate meristems.

The development of the sheath differs from that of the lamina in that no distinct plate meristem is seen in the wings of the sheath and the extension of them is accomplished, primarily, by the activity of the marginal meristem and by the enlargement of the cells derived from it.

As the differentiation in the lamina precedes that in the sheath, the meristematic activity becomes more and more restricted to the base of the sheath where the region of actively dividing and enlarging cells should, therefore, be regarded as an intercalary meristem. The direction of the cellular differentiation and maturation in the leaf of *Oryza* and *Musa*, for instance, is basipetal. The development of the laminar mesophyll in the former is depicted in Fig. 147, no. 2. Vein development in monocotyledons was discussed earlier in this chapter.

Unifacial leaves. Monocotyledonous leaves borne vertically, without distinction of adaxial and abaxial sides, are called *unifacial* leaves. This type of leaf was thought to develop on a sympodial pattern. The primary apical meristem of the primordium was considered to be responsible for initial radial growth. It ceases its activity very early and a new centre of growth arises on the abaxial side of the primary leaf apex. This secondary apex was held to be responsible for the longitudinal extension of the leaf (Knoll, 1948; Thielke, 1948). The unifacial leaf portion, above the sheathing base, which develops from the secondary apex may be cylindrical, as, for instance, in *Allium cepa* and *Juncus maritimus*, or it may be flattened laterally as in *Iris* (Fig. 127, no. 2).

Kaplan (1970b), who has studied the ontogeny of the leaf of *Acorus calamus*, has shown that the radial orientation of this ensiform or sword-shaped leaf is a result of emphasized adaxial meristematic activity and suppressed marginal meristematic activity. Adaxial meristematic activity is expressed very early in development of the leaf and causes the ventral surface of the primordium to protrude considerably (Fig. 149, nos. 1, 2). This protuberance was in the past misinterpreted, as stated by Kaplan, as a primary leaf apex. Apical growth in *Acorus* ceases early in ontogeny and subsequent growth in length is largely basal and intercalary.

The radial growth of the laterally flattened portion of the leaf is first a result of the emphasized activity of the adaxial meristem and subsequent intercalary meristematic activity on both sides (adaxial and abaxial) of a secondary midrib. This midrib develops early along the centre of the leaf axis, and procambium differentiates in it. Then procambial strands are initiated in pairs, in both sides of the primordium, from derivatives of intercalary meristem in the adaxial and abaxial wing of the leaf. Procambial differentiation is continu-

ous and acropetal.

The radial type of unifacial blade occurring in many monocotyledons has, according to Kaplan (1975), evolved by a divergent course of lamina morphogenesis in which growth in volume dominates over surface expansion.

Kaplan's (1975) view of the *Acacia* phyllode, is that this organ is homologous with the rachis of the bipinnate leaf plus a small sector of the petiole and is thus a unifacial leaf form of dicotyledons. The metamorphosis of the phyllode from a bipinnate-type leaf does not involve distal laminar suppression and compensatory petiolar elaboration, but rather takes place by an alternative mode of morphogenesis. Instead of the developing blade

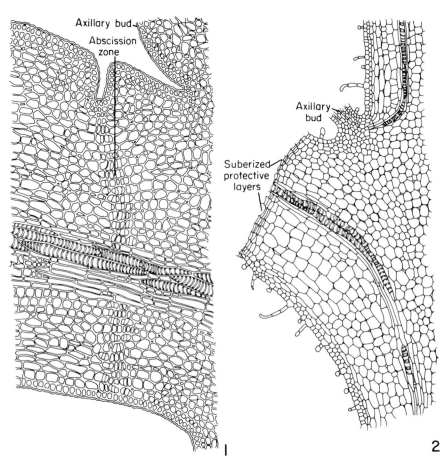

FIG. 150. Leaf abscission. 1, Longitudinal section of the leaf base of *Prunus* showing the cells that divide to form the separation layer. 2, *Coleus*, longitudinal section of portion of the stem together with the leaf base after the abscission of the leaf. (Adapted from Gibbs, 1950.)

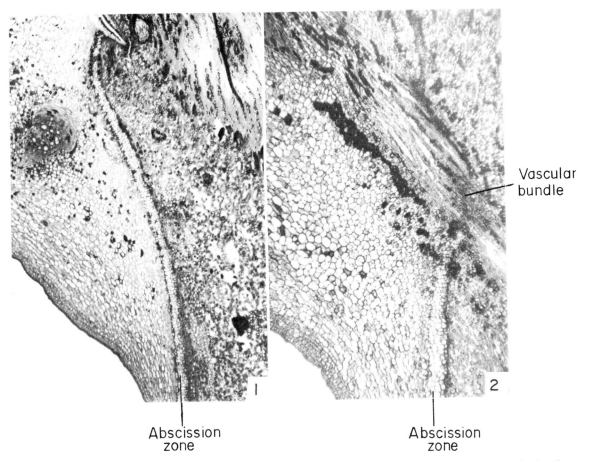

Vascular bundle

Abscission zone   Abscission zone

FIG. 151. Micrographs of longitudinal sections of the leaf base of *Populus*. In no. 2 the section includes a vascular bundle.
1, × 10; 2, × 42.

forming a series of adaxially situated pinna primordia, a single adaxial meristem is responsible for the vertical orientation of the phyllode-type of blade.

LEAF ABSCISSION

One of the well-known phenomena occurring in plants is seasonal loss of organs. This loss is brought about by the process of *abscission*. Leaves, floral parts, fruits, and often branches, e.g. *Taxodium*, *Populus*, *Naucleopsis*, and *Perebea* (Koek-Noorman and ter Welle, 1976) may be shed. In some species, e.g. in the tumbleweed *Kochia indica*, the whole aerial part of the plant separates from the root (Zeroni *et al.*, 1978).

Leaves of gymnosperms and woody dicotyledons are usually shed as a result of changes that take place in the tissues of the leaf base prior to leaf death. In the base of mature deciduous leaves a narrow zone, the so-called *abscission zone* or *region* can be seen. This zone can be distinguished histologically from the surrounding tissues (Figs. 150, 151) and externally its location is marked by the presence of a shallow groove or by a difference in the colour of the epidermis. The vascular system in this region is usually concentrated in the centre and the sclerenchyma and collenchyma are less well developed or absent. In the abscission zone two layers are discernible—an *abscission* or *separation layer* through which the detachment of the organ occurs and a *protective layer* which protects from desiccation and from the entry of parasites to the exposed surface. In many abscissing organs the preparation of the separation layer takes place during normal ontogeny, but may also occur a short time before shedding of the organ in response to conditions causing abscission.

Several histological processes were observed during abscission, not all of them in the same species (Hagemann, 1971). In *Phaseolus*, Webster (1970) described four intergrading, sometimes overlapping, histological stages: (1) breakdown of pith cells; (2) cell divisions in the "cortex"; (3) cellular differentiation involving cell enlargement; (4) breakdown of cortical and vascular cells.

In the abscission zone lignification of the parenchyma cells, tylose formation in the tracheary elements, and callose deposition in sieve elements was also observed (Poovaiah, 1974).

The final event in abscission involves the enzymatic degradation of cell walls. This starts with removal of calcium and pectin from the middle lamellae and continues with the hydrolysis of the cellulose walls (Rascio *et al.*, 1985). At the end, the sclerified tracheary elements rupture. According to Facey (1950), the middle lamellae between the cells of the vascular tissue also disintegrate. Xylem vessels and interstitial parenchyma of vascular bundles may also become weakened by hydrolysis of cell-wall material prior to separation (Moline and Bostrack, 1972). Golgi vesicles may be involved in secretion of abscission-specific enzymes (Sexton and Hall, 1974; Osborne and Sargent, 1976a, b). In some species no dissolution takes place and abscission is apparently effected by physical stresses. The latter is the case in most monocotyledons and herbaceous dicotyledons. In guayule (*Parthenium argentatum*), for example, the separation layer consists of suberized cells, but it is not directly involved in the separation of the leaf from the stem. After the leaf dies its base breaks away from the stem through the weakened region of the abscission zone (Addicott, 1945, 1982; Facey, 1950; Addicott and Lynch, 1955). The formation of an abscission zone in the leaves of a monocotyledonous plant (*Molinia caerulea*, Gramineae) has recently been reported (Salim *et al.*, 1988).

The *protective layer* may be of two types—primary or secondary, i.e. periderm. The primary protective layer is formed as a result of the lignification and suberization of the parenchyma cells in this region or of the cells arising from them by irregular cell divisions (Pfeiffer, 1928). An opinion exists that the substance appearing in the cell walls of the protective tissue, and which has been defined as lignin, is really wound gum which gives reactions similar to lignin (Hewitt, 1938).

The time at which the primary protective layer and periderm appear differs in different plants. This fact has resulted in the development of a complicated classification of abscission types (Pfeiffer, 1928).

It has been shown repeatedly that auxin applied to leaf blades inhibits the abscission of the leaf. Ethylene, on the other hand, is regarded as responsible for the induction of the enzymes active in cell-wall degradation (Gawadi and Avery, 1950; Bornman *et al.*, 1967; Horton and Osborne, 1967; Abeles, 1967; Osborne and Sargent, 1976a, b). According to Hagemann (1971) peroxidase and phenoloxidase activity increases during differentiation of the abscission zone. This author suggested that peroxidase brings about an increased synthesis of ethylene. The abscission process is usually regarded as a senescence phenomenon (Osborne and Moss, 1963; Leopold, 1967), but Webster (1970) opposes this view on the grounds that it involves cell division.

(For additional references on leaf anatomy see Napp-Zinn, 1984, 1988.)

REFERENCES

ABELES, F. B. (1967) Mechanism of action of abscission accelerators. *Physiol. Plant.* **20**: 442–54.

ADDICOTT, F. T. (1945) The anatomy of the leaf abscission and experimental defoliation in guayule. *Am. J. Bot.* **32**: 250–6.

ADDICOTT, F. T. (1982) *Abscission.* University of California Press, Berkeley.

ADDICOTT, F. T. and LYNCH, R. S. (1955) Physiology of abscission. *A. Rev. Pl. Physiol.* **6**: 211–38.

ALLSOPP, A. (1955) Experimental and analytical studies of pteridophytes: XXVII, Investigations on *Marsilea*: 5, Cultural conditions and morphogenesis with special reference to the origin of land and water forms. *Ann. Bot.* **19**: 247–64.

ARBER, A. (1950) *The Natural Philosophy of Plant Form.* Cambridge University Press, Cambridge.

ARNOTT, H. J. (1959) Anastomoses in the venation of *Ginkgo biloba. Am. J. Bot.* **46**: 405–11.

ASHBY, E. (1948a) Studies in the morphogenesis of leaves: I, An essay on the leaf shape. *New Phytol.* **47**: 153–76.

ASHBY, E. (1948b) Studies in the morphogenesis of leaves: II, The area, cell size and cell number of leaves of *Ipomoea* in relation to their position on the shoot. *New Phytol.* **47**: 177–95.

ATAY, S. (1958) Über die Einwirkung der ätherischen Öle auf die Evaporation und Transpiration. *Istanb. Üniv. Fen. Fak. Mecm.* ser. B, **23**: 143–70.

AVERY, G. S., JR (1933) Structure and development of tobacco leaf. *Am. J. Bot.* **20**: 565–92.

BAAS, P. and GREGORY, M, (1985) A survey of oil cells in the dicotyledons with comments on their replacement by and joint occurrence with mucilage cells. *Israel J. Bot.* **34**: 167–186.

BAILEY, I. W. and NAST, C. G. (1944) The comparative morphology of the Winteraceae: V, Foliar epidermis and sclerenchyma. *J. Arnold Arb.* **25**: 342–8.

BARNABAS, A. D., BUTLER, V., and STEINKE, T. D. (1977) *Zostera capensis* Setchell. I. Observations on the fine structure of the leaf epidermis. *Z. Pflanzenphysiol.* **85**: 417–27.

BEARDSELL, D. V. and CONSIDINE, J. A. (1987) Lineages, lineage stability and pattern formation in leaves of variegated chimeras of *Lophostemon confertus* (R. Br.) Wilson & Waterhouse and *Tristanopsis laurina* (Smith) Wilson & Waterhouse (Myrtaceae). *Aust. J. Bot.* **35**: 701–714.

BHANDARI, N. N. (1969) Ontogeny and marginal growth in the leaf of *Eubrachion ambiguum. Ann. Bot.* **33**: 537–40.

BJÖRKMAN, O. and BERRY, J. (1973) High-efficiency photosynthesis. *Scient. Am.* **229** (4): 80–93.

BLECKMANN, C. A., HULL, H. M., and HOSHAW, R. W. (1980) Cuticular ultrastructure of *Prosopis velutina* and *Acacia greggii* leaflets. *Bot. Gaz.* **141**: 1–8.

BÖCHER, T. W. (1979) Xeromorphic leaf types. Evolutionary strategies and tentative semophyletic sequences. *Biol. Skr.* **22** (8): 1–71.

BORNMAN, C. H., SPURR, A. R., and ADDICOTT, F. T. (1967) Abscisin, auxin and gibberellin effects on the developmental aspects of abscission in cotton (*Gossypium hirsutum*). *Am. J. Bot.* **54**: 125–35.

BOSABALIDIS, A. and TSEKOS, I. (1982) Ultrastructural studies on the secretory cavities of *Citrus deliciosa* Ten. II. Development of the oil-accumulating central space of the gland and process of active secretion. *Protoplasma* **112**: 63–70.

BOTHA, C. E. J. and EVERT, R. F. (1988) Plasmodesmatal distribution and frequency in vascular bundles and contiguous tissues of the leaf of *Themeda triandra*. *Planta* **173**: 433–441.

BOYCE, S. G. (1954) The salt spray community. *Ecol. Monogr.* **24**: 29–67.

BUSBY, C. H. and O'BRIEN, T. P. (1979). Aspects of vascular anatomy and differentiation of vascular tissues and transfer cells in vegetative nodes of wheat. *Aust. J. Bot.* **27**: 703–11.

CARDE, J. P. (1978) Ultrastructural studies of *Pinus pinaster* needles: The endodermis. *Am. J. Bot.* **65**: 1041–54.

CAROLIN, R. C., JACOBS, S. W. L., and VESK, M. (1973) The structure of the cells of the mesophyll and parenchymatous bundle sheath of the Gramineae. *J. Linn. Soc. Bot.* **66**: 259–75.

CAROLIN, R. C., JACOBS, S. W. L. and VESK, M. (1975) Leaf structure in Chenopodiaceae. *Bot. Jahrb. Syst.* **95**: 226–55.

CAROLIN, R. C., JACOBS, S. W. L., and VESK, M. (1977) The ultrastructure of Kranz cells in the family Cyperaceae. *Bot. Gaz.* **138**: 413–19.

CARR, D. J. and CARR, S. G. M. (1970) Oil glands and ducts in *Eucalyptus* L'Hérit.: II, Development and structure of oil glands in the embryo. *Aust. J. Bot.* **18**: 191–212.

CARR, D. J. and CARR, S. G. M. (1987) *Eucalyptus* II. The rubber cuticle, and other studies of the Corymbosae. Phy-

toglyph Press, Canberra.

CHAPMAN, E. A., BAIN, J. M., and GOVE, D. W. (1975) Mitochondria and chloroplast peripheral reticulum in C₄ plants *Amaranthus edulis* and *Atriplex spongiosa*. *Aust. J. Plant Physiol.* **2**: 207–23.

COONEY-SOVETTS, C. and SATTLER, R. (1986) Phylloclade development in the Asparagaceae: an example of homoeosis. *Bot. J. Linn. Soc.* **94**: 327–371.

COULTER, J. M., BARNES, C. R., and CLOWES, H. C. (1931) *A Textbook of Botany for Colleges and Universities*, Vol. III. *Ecology*. American Book Co., New York.

CROSS, G. L. (1940) Development of the foliage leaves of *Taxodium distichum*. *Am. J. Bot.* **27**: 471–82.

CUNNINGHAME, M. E. and LYNDON, R. F. (1986) The relationship between the distribution of periclinal cell divisions in the shoot apex and leaf initiation. *Ann. Bot.* **57**: 737–746.

DE FRAINE, E. (1912) The anatomy of the genus *Salicornia*. *J. Linn. Soc. Bot.* **41**: 317–348.

DEHGAN, B. (1982) Comparative anatomy of the petiole and infrageneric relationships in *Jatropha* (Euphorbiaceae). *Am. J. Bot.* **69**: 1283–1295.

DELL, B. and McCOMB, A. J. (1978) Plant resins—their formation, secretion and possible functions. In: *Advances in Botanical Research* (ed. E. W. Woolhouse). Academic Press, London, pp. 277–316.

DENGLER, N. G., DENGLER, R. E. and Hattersley, P. W. (1985) Differing ontogenetic origins of PCR ("Kranz") sheaths in leaf blades of C₄ grasses (Poaceae). *Am. J. Bot.* **72**: 284–302.

DERMEN, H. (1947) Periclinal cytochimeras and histogenesis in cranberry. *Am. J. Bot.* **34**: 32–43.

DERMEN, H. (1951) Ontogeny of tissues in stems and leaf of cytochimeral apples. *Am. J. Bot.* **38**: 753–60.

DUBUC-LEBREUX, M. A. and SATTLER, R. (1980) Dévelopement des organes foliacés chez *Nicotiana tabacum* et le problème des méristèmes marginaux. *Phytomorphology* **30**: 17–32.

DUVAL-JOUVE, J. (1868) Des *Salicornia* de l'Hérault. *Bull. Soc. bot. Fr.* **15**: 132–40.

EASTMAN, P. A. K., DENGLER, N. G. and PETERSON, C. A. (1988a) Suberized bundle sheaths in grasses (Poaceae). I. Anatomy, ultrastructure and histochemistry. *Protoplasma* **142**: 92–111.

EASTMAN, P. A. K., PETERSON, C. A. and DENGLER, N. G. (1988b) Suberized bundle sheaths in grasses (Poaceae) of different photosynthetic types. II. Apoplastic permeability. *Protoplasma* **142**: 112–126.

EHLERINGER, J., BJÖRKMAN, O., and MOONEY, H. A. (1976) Leaf pubescence: effects on absorptance and photosynthesis in a desert shrub. *Science* **192**: 376–7.

ELEFTHERIOU, E. P. and TSEKOS, I. (1979) Development of mestome sheath cells in leaves of *Aegilops comosa* var. *thessalica*. *Protoplasma* **100**: 139–53.

ESAU, K. (1967) Minor veins in *Beta* leaves: Structure related to function. *Proc. Am. Phil. Soc.* **111**: 219–33.

ESAU, K. (1972) Cytology of sieve elements in minor veins of sugar beet leaves. *New Phytol.* **71**: 161–168.

EVERT, R. F. and MIERZWA, R. J. (1986) Pathway(s) of assimilate movement from mesophyll cells to sieve tubes in the *Beta vulgaris* leaf. In: *Phloem Transport* (eds. J. Cronshaw, W. J. Lucas and R. T. Giaquinta). Alan R. Liss, Inc., New York, pp. 419–432.

EVERT, R. F., BOTHA, C. E. J. and MIERZWA, R. J. (1985) Free-space marker studies on the leaf of *Zea mays* L. *Protoplasma* **126**: 62–73.

EVERT, R. F., ESCHRICH, W., and HEYSER, W. (1977a) Distribution and structure of the plasmodesmata in mesophyll and bundle-sheath cells of *Zea mays* L. *Planta* **136**: 77–89.

EVERT, R. F., ESCHRICH, W. and HEYSER, W. (1978) Leaf structure in relation to solute transport and phloem loading in *Zea mays* L. *Planta* **138**: 279–294.

EVERT, R. F., ESCHRICH, W., NEUBERGER, D. S., and EICHHORN, S. E. (1977b) Tubular extensions of the plasmalemma in leaf cells of *Zea mays* L. *Planta* **135**: 203–5.

FACEY, V. (1950) Abscission of leaves in *Fraxinus americana* L. *New Phytol.* **49**: 103–16.

FAHN, A. (1954) The anatomical structure of the Xanthorrhoeaceae Dumort. *J. Linn. Soc., Bot.* **55**: 158–84.

FAHN, A. (1979) *Secretory Tissues in Plants.* Academic Press, London.

FAHN, A. (1987) Extrafloral nectaries of *Sambucus nigra* L. *Ann. Bot.* **60**: 299–308.

FAHN, A. and ARZEE, T. (1959) Vascularization of articulated Chenopodiaceae and the nature of their fleshy cortex. *Am. J. Bot.* **46**: 330–8.

FAHN, A. and DEMBO, N. (1964) Structure and development of the epidermis in articulated Chenopodiaceae. *Israel J. Bot.* **13**: 177–92.

FAROOQUI, P. (1982) Cork-warts in *Eucalyptus* species. *Proc. Indian Acad. Sci.* (Plant Sci.) **91**: 289–295.

FISHER, D. G. (1985) Morphology and anatomy of the leaf of *Coleus blumei* (Lamiaceae). *Am. J. Bot.* **72**: 392–402.

FISHER, D. G. and EVERT, R. F. (1982) Studies on the leaf of *Amaranthus retroflexus* (Amaranthaceae): Ultrastructure, plasmodesmatal frequency and solute concentration in relation to phloem loading. *Planta* **155**: 377–387.

FISHER, F. J. F., EHRET, D. L. and HOLLINGDALE, J. (1987) The pattern of vascular development near the pulvinus of the solar-tracking leaf of *Lavatera cretica* (Malvaceae). *Can. J. Bot.* **65**: 2109–2117.

FLEURAT-LESSARD, P. and ROBLIN, G. (1982) Comparative histo-cytology of the petiole and the main pulvinus in *Mimosa pudica* L. *Ann. Bot.* **50**: 83–92.

FOHN, M. (1935) Zur Entstehung und Weiterbildung der Exkreträume von *Citrus medica* L. und *Eucalyptus globulus* Labill. *Öst. bot. Z.* **84**: 198–209.

FOSTER, A. S. (1932) Investigations on the morphology and comparative history of development of foliar organs: III, Cataphyll and foliage-leaf ontogeny in the black hickory (*Carya buckleyi* var. *arkansana*). *Am. J. Bot.* **19**: 75–99.

FOSTER, A. S. (1935) A histogenetic study of foliar determination in *Carya buckleyi* var. *arkansana*. *Am. J. Bot.* **22**: 88–147.

FOSTER, A. S. (1936) Leaf differentiation in angiosperms. *Bot. Rev.* **2**: 349–72.

FOSTER, A. S. (1937) Structure and behavior of the marginal meristem in the bud scales of *Rhododendron*. *Am. J. Bot.* **24**: 304–16.

FOSTER, A. S. (1956) Plant idioblasts: remarkable examples of cell specialization. *Protoplasma* **46**: 184–93.

FOSTER, A. S. and GIFFORD, E. M., JR (1959) *Comparative Morphology of Vascular Plants.* W. H. Freeman, San Francisco.

FREEMAN, T. P. (1970) The developmental anatomy of *Opuntia basilaris*: II, Apical meristem, leaves, areoles, glochids. *Am. J. Bot.* **57**: 616–22.

FUCHS, C. (1975) Ontogenèse foliaire et acquisition de la forme chez le *Tropaeolum peregrinum* L. *Annls Sci. nat. Bot.* ser. 12, **16**: 321–89.

GAWADI, A. G. and AVERY, G. S., JR. (1950) Leaf abscission and the so-called "abscission layer". *Am. J. Bot.* **37**: 172–9.

GHOUSE, A. K. M. (1974) Transfusion tissue in the leaves of *Taxus baccata* L. *Cellule* **70**: 159–62.

GHOUSE, A. K. M. and YUNUS, M. (1974) Transfusion tissue in the leaves of *Cunninghamia lanceolata* (Lambert) Hooker (Taxodiaceae). *J. Linn. Soc. Bot.* **69**: 147–51.

GHOUSE, A. K. M. and YUNUS, M. (1975) Transfusion tissue in the leaves of *Thuja orientalis* L. *Ann. Bot.* **39**: 225–7.

GIAQUINTA, R. T. (1983) Phloem loading of sucrose. *Annu. Rev. Plant Physiol.* **34**: 347–387.

GIBBS, R. D. (1950) *Botany: an Evolutionary Approach.* Blackiston, Toronto.

GRACEN, V. E. JR., HILLIARD, J. H., BROWN, R. H., and WEST, S. H. (1972) Peripheral reticulum in chloroplasts of plants differing in CO_2 fixation pathways and photorespiration. *Planta* **107**: 189–204.

GROOM, P. (1910) Remarks on the ecology of Coniferae. *Ann. Bot.* **24**: 241–69.

GUNNING, B. E. S. and PATE, J. S. (1969) "Transfer Cells"—Plant cells with wall ingrowths, specialized in relation to short distance transport of solutes: their occurrence, structure and development. *Protoplasma* **68**: 107–33.

GUNNING, B. E. S. and PATE, J. S. (1974) Transfer cells. In: *Dynamic Aspects of Plant Ultrastructure* (ed. A. W. Robards). McGraw-Hill, London, pp. 441–80.

GUNNING, B. E. S., PATE, J. S., and BRIARTY, L. G. (1968) Specialized "Transfer Cells" in minor veins of leaves and their possible significance in phloem translocation. *J. Cell Biol.* **37**: C7–12.

HABERLANDT, G. (1918) *Physiologische Pflanzenanatomie*, 5th edn. W. Engelmann, Leipzig.

HAGEMANN, P. (1971) Histochemische Muster beim Blattfall. *Ber. schweiz. bot. Ges.* **81**: 97–138.

HARA, N. (1957) On the types of marginal growth in dicotyledonous foliage leaves. *Bot. Mag. Tokyo* **70**: 108–14.

HARTMANN, H. (1979) Surface structures of leaves: their ecological and taxonomical significance in members of the subfamily Ruschioideae SCHW. (Mesembryanthemaceae FENZL). In: Cutler, D. F. and Hartmann, H., *Scanning Electron Microscope Studies of the Leaf Epidermis in Some Succulents.* Mainz, Akad. d. Wiss. u. Lit. (Tropische u. Subtropische Pflanzenwelt, 28), pp. 31–55.

HEIDE-JØRGENSEN, H. S. (1980) The xeromorphic leaves of *Hakea suaveolens* R.Br. III. Ontogeny, structure and function of the T-shaped trichomes. *Bot. Tidsskrift* **75**: 181–98.

HEILBRONN, A. (1958) Über die Oberflächenaktivität ätherischer Öle und die biologische Bedeutung dieses Phänomens. *Istanb. Üniv. Fen. Fak.Mecm.*, ser. B, **23**: 131–41.

HEINRICH, G. (1969) Elektronenmikroskopische Beobachtungen zur Entstehungsweise der Exkret-behälter von *Ruta graveolens*, *Citrus limon* und *Poncirus trifoliata*. *Öst. bot. Z.* **117**: 397–403.

HEINRICHER, E. (1885) Über einige im Laube dikotyler Pflanzen trockenen Standortes auftretende Einrichtungen, welche mutmasslich eine ausreichende Wasserversorgung des Blattmesophylls bezwecken. *Bot. Zbl.* **23**: 25–31, 56–61.

HEWITT, W. B. (1938) Leaf-scar infection in relation to the olive knot disease. *Hilgardia* **12**: 41–71.

HICKEY, L. J. and WOLFE, J. A. (1975) The bases of angiosperm phylogeny: vegetative morphology. *Ann. Mo. bot. Gdn.* **68**: 538–589.

HOGERMANN, C. (1987) *Untersuchungen über die Blattnervatur einiger tropischer Baumarten von Venezolanisch Guayana im Hinblick auf taxonomische Typisierbarkeit und Waldschichtung.* Veroffentrichung der Naturforschenden Gesellschaft zu Emden von 1814. Vol. 9 Ser. 3–D4.

HOLTERMANN, K. (1907) *Der Einfluss des Klimas auf den Bau der Pflanzengewebe.* W. Engelmann, Leipzig.

HORTON, R. F. and OSBORNE, D. J. (1967) Senescence, abscission and cellulase in *Phaseolus vulgaris. Nature* **214:** 1086–8.

HOWARD, R. A. (1974) The stem-node-leaf continuum of the Dicotyledoneae. *J. Arnold Arb.* **55:** 125–81.

HOWARD, R. A. (1979) The petiole. In: C. R. Metcalfe and L. Chalk, *Anatomy of the Dicotyledons,* Vol. 1, 2nd edn. Clarendon Press, Oxford, pp. 88–96.

HU, Y. S. and YAO, B. J. (1981) Transfusion tissue in gymnosperm leaves. *Bot. J. Linn. Soc.* **83:** 263–272.

HULL, H. M., MORTON, H. L., and WHARRIE, J. R. (1975) Environmental influences on cuticle development and resultant foliar penetration. *Bot. Rev.* **41:** 421–52.

ISEBRANDS, J. G. and LARSON, P. R. (1977) Organization and ontogeny of the vascular system in the petiole of eastern cottonwood. *Am. J. Bot.* **64:** 65–77.

ISEBRANDS, J. G. and LARSON, P. R. (1980) Ontogeny of major veins in the lamina of *Populus deltoides* Bartr. *Am. J. Bot.* **67:** 23–33.

JOHNSON, H. B. (1975) Plant pubescence: an ecological perspective. *Bot. Rev.* **41:** 233–58.

JOHNSON, M. A. (1943) Foliar development in *Zamia. Am. J. Bot.* **30:** 366–78.

JONES, H. (1956) Morphological aspects of leaf expansion, especially in relation to changes in leaf form. In: *The Growth of Leaves* (ed. F. L. Milthorpe). Butterworths, London, pp. 93–106.

KAHL, G. and SCHELL, J. S. (1982) *The Molecular Biology of Plant Tumors.* Academic Press, New York.

KAPLAN, D. R. (1970a) Comparative development and morphological interpretation of "rachis-leaves" in Umbelliferae. *New Research in Plant Anatomy* (ed. N. K. B. Robson, D. F. Cutler and M. Gregory). Academic Press, London. *J. Linn. Soc., Bot.* **63,** suppl. 1: 101–25.

KAPLAN, D. R. (1970b) Comparative foliar histogenesis in *Acorus calamus* and its bearings on the phyllode theory of monocotyledonous leaves. *Am. J. Bot.* **57:** 331–61.

KAPLAN, D. R. (1975) Comparative developmental evaluation of the morphology of unifacial leaves in the monocotyledons. *Bot. Jahrb. Syst.* **95:** 1–105.

KAPLAN, D. R. (1984) Alternative modes of organogenesis in higher plants. In: *Contemporary Problems in Plant Anatomy* (eds. R. A. White and W. C. Dickison). Academic Press, Orlando, pp. 261–300.

KAUFMAN, P. B. (1959) Development of the shoot of *Oryza sativa* L.: II, Leaf histogenesis. *Phytomorphology* **9:** 277–311.

KNOLL, F. (1948) Bau, Entwicklung und morphologische Bedeutung unifazialer Vorläuferspitzen an Monokotylenblättern. *Öst. bot. Z.* **95:** 163–93.

KOEK-NOORMAN, J. and TER WELLE, B. J. H. (1976) The anatomy of branch abscission layers in *Perebea mollis* and *Naucleopsis guianensis* (Castilleae, Moraceae), *Leiden Botanical Series* **3:** 196–203.

KOLLER, A. L. and ROST, T. L. (1988a) Leaf anatomy in *Sansevieria* (Agavaceae). *Am. J. Bot.* **75:** 615-633.

KOLLER, A. L. and ROST, T. L. (1988b) Structural analysis of water-storage tissue in leaves of *Sansevieria* (Agavaceae). *Bot. Gaz.* **149:** 260–274.

KORODY, E. (1937) Studien am Spross-Vegetationspunkt von *Abies concolor, Picea excelsa* und *Pinus montana. Beitr. Biol. Pfl.* **25:** 23–59.

KRAUS, E. J. and KRAYBILL, H. R. (1918) Vegetation and

reproduction with special reference to tomato. *Bull. Oregon. Agric. Exp. Stn,* no. 149.

LAETSCH, W. M. (1974) The C_4 syndrome: a structural analysis. *Ann. Rev. Plant Physiol.* **25:** 27–52.

LARSON, P. R. (1984a) Vascularization of developing leaves of *Gleditsia triacanthos* L. I. The node, rachis, and rachillae. *Am. J. Bot.* **71:** 1201–1210.

LARSON, P. R. (1984b) Vascularisation of developing leaves of *Gledistsia triacanthos* L. II. Leaflet initiation and early vascularization. *Am. J. Bot.* **71:** 1211–1220.

LEDERER, B. (1955) Vergleichende Untersuchungen über das Transfusionsgewebe einiger rezente Gymnospermen. *Bot. Stud.* no. 4: 1–42.

LEOPOLD, A. C. (1967) The mechanism of foliar abscission. *Symp. Soc. Exp. Biol.* **21:** 507–16.

LERSTEN, N. (1965) Histogenesis of leaf venation in *Trifolium wormskioldii* (Leguminosae). *Am. J. Bot.* **52:** 767–74.

LEVIN, D. A. (1973) The role of trichomes in plant defence. *Q. Rev. of Biol.* **48:** 3–15.

LÜTTGE, U. and KRAPF, G. (1969) Die Ultrastruktur der *Nymphaea*-Hydropoten in Zusammenhang mit ihrer Funktion als Salztransportierende Drüsen. *Cytobiologie* **1:** 121–31.

LYSHEDE, O. B. (1977) Studies on the mucilaginous cells in the leaf of *Spartocytisus filipes* W. B. *Planta* **133:** 255–60.

MARON, R. and FAHN, A. (1979) Ultrastructure and development of oil cells in *Laurus nobilis* L. leaves. *Bot. J. Linn. Soc.* **78:** 31–40.

MAXIMOV, N. A. (1929) *The Plant in Relation to Water. A Study of the Physiological Basis of Drought Resistance.* Allen & Unwin, London.

MAXIMOV, N. A. (1931) The physiological significance of the xeromorphic structure of plants. *J. Ecol.* **19:** 272–82.

McCALLUM, W. B. (1902) On the nature of the stimulus causing the changes of form and structure in *Proserpinaca palustris. Bot. Gaz.* **34:** 93–108.

MELVILLE, R. and WRIGLEY, F. A. (1969) Fenestration in the leaves of *Monstera* and its bearing on the morphogenesis and colour patterns of leaves. *J. Linn. Soc., Bot.* **62:** 1–16.

METCALFE, C. R. (1960) Gramineae. In: *Anatomy of the Monocotyledons* (ed. C. R. Metcalfe), Vol. I, Clarendon Press, Oxford.

METCALFE, C. R. and CHALK, L. (1950) *Anatomy of the Dicotyledons.* Clarendon Press, Oxford.

METCALFE, C. R. and CHALK, L. (1979) *Anatomy of the Dicotyledons,* Vol. I, 2nd edn. Clarendon Press, Oxford.

METCALFE, C. R. and CHALK, L. (1983) *Anatomy of the Dicotyledons* 2nd edn. Vol. 2. Clarendon Press, Oxford.

MEYER, F. J. (1962) Das trophische Parenchym: A, Assimilationsgewebe. In: *Handbuch der Pflanzenanatomie,* Spez. Teil, Bd. 4, T. 7A. Gebr. Borntraeger, Berlin.

MEYER, J. and MARESQUELLE, K. J. (1983) Anatomie des Galles. In: K. Linsbauer, *Handbuch der Pflanzenanatomie,* Spez. Teil. Bd. 13, T. 1. Gebr. Borntraeger, Berlin.

MOLINE, H. E. and BOSTRACK, J. M. (1972) Abscission of leaves and leaflets in *Acer negundo* and *Fraxinus americana. Am. J. Bot.* **59:** 83–88.

MONTEIRO, W. R., GIULETTI, A. M. and CASTRO, M. M. (1984) Aspects of leaf structure of some species of *Eriocaulon* L. (Eriocaulaceae) from Serra do Cipo (Minas Gerais, Brazil) *Revta brasil. Bot.* **7:** 137–147.

MORRETES, B. L. DE (1967) Floema terminal em feixes vasculares do mesofilo de *Syringa vulgaris* e *Boerhaavia coccinea. Bolm. Fac. Filos. Ciênc. Univ. S. Paulo* 305, *Botânica* **22:** 291–312.

MORRETES, B. L. DE and VENTURELLI, M. (1985) Ocorrencia de "lesiticelas" em folhas de *Tripodanthus acutifolius* (R. & P.) Tiegh. (Loranthaceae). *Revta brasil. Bot.* **8**: 157–162.

MOTHES, K. (1932) Ernährung, Struktur und Transpiration. Ein Beitrag zur kausalen Analyse der Xeromorphosen. *Biol. Zbl.* **52**: 193–223.

NAPP-ZINN, K. (1973–4) Anatomie des Blattes. II. Blattanatomie der Angiospermen. A. Entwicklungsgeschichtliche und topograpische Anatomie des Angiospermenblattes. In: *Handbuch der Pflanzenanatomie*, Spez. Teil, Bd. 8 T. 2A, 2 vols. Gebr. Borntraeger, Berlin and Stuttgart.

NAPP-ZINN, K. (1984, 1988) Anatomie des Blattes. II **Blattanatomie der Angiospermen. In: *Handbuch der Pflanzenanatomie*, Spez. Teil, Bd. 8 T. 2B, 2 vols. Gebr. Borntraeger, Berlin and Stuttgart.**

O'BRIEN, T. P. and CARR, D. J. (1970) A suberized layer in the cell walls of the bundle sheath of grasses. *Aust. J. Biol. Sci.* **23**: 275–87.

OPPENHEIMER, H. R. (1960) Adaptation to drought: Xerophytism. Plant–water relationships in arid and semiarid conditions. Reviews of research, Unesco. *Arid Zone Res.* **15**: 105–38.

OSBORNE, D. J. and MOSS, S. E. (1963) Effect of kinetin on senescence and abscission in explants of *Phaseolus vulgaris*. *Nature* **200**: 1299–301.

OSBORNE, D. J. and SARGENT, J. A. (1976a) The positional differentiation of ethylene-responsive cells in rachis abscission zones in leaves of *Sambucus nigra* and their growth and ultrastructural changes at senescence and separation. *Planta* **130**: 203–10.

OSBORNE, D. J. and SARGENT, J. A. (1976b) The positional differentiation of abscission zones during development of leaves of *Sambucus nigra* and the response of cells to auxin and ethylene. *Planta* **132**: 197–204.

PATE, J. S. and GUNNING, B. E. S. (1969) Vascular transfer cells in angiosperm leaves. A taxonomic and morphological survey. *Protoplasma* **68**: 135–56.

PATE, J. S. and GUNNING, B. E. S. (1972) Transfer cells. *Ann. Rev. Pl. Physiol.* **23**: 173–96.

PEMBERTON, R. W. and TURNER, C. E. (1989) Occurrence of predatory and fungivorous mites in leaf domatia. *Am. J. Bot.* **76**: 105–112.

PFEIFFER, H. (1928) Die pflanzlichen Trennungsgewebe. In: K. Linsbauer, *Handbuch der Pflanzen-anatomie*, Bd. 5, Lief. 22. Gebr. Borntraeger, Berlin.

PIZZOLATO, T. D., BURBANO, J. L., BERLIN, J. D., MOREY, P. R., and PEASE, R. W. (1976) An electron microscope study of the path of water movement in transpiring leaves or cotton (*Gossypium hirsutum* L.). *J. Exp. Bot.* **27**: 145–61.

POETHIG, R. S. (1984) Cellular parameters of leaf morphogenesis in Maize and tabacco. In: *Contemporary Problems in Plant Anatomy* (eds. R. A. White and W. C. Dickison). Academic Press, Orlando, pp. 235–259.

POETHING, R. S. (1987) Clonal analysis of cell lineage patterns in plant development. *Am. J. Bot.* **74**: 581–594.

POOVAIAH, B. W. (1974) Formation of callose and lignin during leaf abscission. *Am. J. Bot.* **61**: 829–34.

PRAY, T. R. (1963) Origin of the vein endings in angiosperm leaves. *Phytomorphology* **13**: 60–81.

RASCIO, N., CASADORO, G., RAMINA, A. and MASIA, A. (1985) Structural and biochemical aspects of peach fruit abscission (*Prunus persica* L. Batsch). *Planta* **164**: 1–11.

RENNER, O. and VOSS, M. (1942) Zur Entwicklungsgeschichte randpanaschierter Formen von *Prunus, Pelargonium, Veronica, Dracaena. Flora* **35**: 356–76.

RODIN, R. J. (1967) Ontogeny of foliage leaves of *Gnetum. Phytomorphology* **17**: 118–28.

RURY, P. M. and DICKISON, W. C. (1977) Leaf venation patterns of the genus *Hibbertia* (Dilleniaceae). *J. Arnold Arb.* **58**: 209–41.

RUSSELL, S. H. and EVERT, R. F. (1985) Leaf vasculature in *Zea mays* L. *Planta* **164**: 448–458.

RUSSIN, W. A. and EVERT, R. F. (1985) Studies on the leaf of *Populus deltoides* (Salicaceae): ultrastructure, plasmodesmatal frequency and solute concentrations. *Am. J. Bot.* **72**: 1232–1247.

SACHS, T. (1975) The control of the differentiation of vascular networks. *Ann. Bot.* **39**: 197–204.

SALIM, K. A., CARTER, P. L., SHAW, S. and SMITH, C. A. (1988) Leaf abscission zone in *Molinia caerulea* (L.) Moench, the purple moor grass. *Ann. Bot.* **62**: 429–434.

SANCHEZ, E., ARRIAGA, M. O. and PANARELLO, H. O. (1986) El sindrome de "Kranz" en Asteraceae de la flora Argentina. *Bol. Soc. Argent. Bot.* **24**: 249–259.

SASS, J. E. (1944) The initiation and development of foliar and floral organs in the tulip. *Iowa St. Coll. J. Sci.* **18**: 447–56.

SATINA, S. and BLAKESLEE, A. F. (1941) Periclinal chimeras in *Datura stramonium* in relation to development of leaf and flower. *Am. J. Bot.* **28**: 862–71.

SATTER, R. L., SABNIS, D. D. and GALSTON, A. W. (1970) Phytochrome controlled nyctinasty in *Albizzia julibrissin*: I, Anatomy and fine structure of the pulvinule. *Am. J. Bot.* **57**: 374–81.

SCHIMPER, A. F. W. (1898) *Pflanzengeographie auf physiologischer Grundlage.* G. Fischer, Jena.

SCHNEIDER, K. (1936) Beeinflussung von N-Stoffwechsel und Stengelanatomie durch Ernährung. *Z. Bot.* **29**: 545–69.

SEXTON, R. and HALL, J. L. (1974) Fine structure and cytochemistry of the abscission zone cells of *Phaseolus* leaves. I. Ultrastructural changes occurring during abscission. *Ann. Bot.* **38**: 849–54.

SHARMAN, B. C. (1942) Developmental anatomy of the shoot of *Zea mays* L. *Ann. Bot.*, **6**: 245–82.

SHARMAN, B. C. (1945) Leaf and bud initiation in the Gramineae. *Bot. Gaz.* **106**: 269–89.

SHARMAN, B. C. and HITCH, P. A. (1967) Initiation of procambial strands in leaf primordia of bread wheat, *Triticum aestivum* L. *Ann Bot.* **31**: 229–43.

SHIELDS, L. M. (1950) Leaf xeromorphy as related to physiological and structural influences. *Bot. Rev.* **16**: 399–447.

SHIELDS, L. M. (1951) The involution mechanism in leaves of certain xeric grasses. *Phytomorphology* **1**: 225–41.

SHULL, C. A. (1934) Lateral water transfer in leaves of *Ginkgo biloba. Pl. Physiol.* **9**: 387–9.

SIFTON, H. B. (1945) Air-space tissue in plants. *Bot. Rev.* **11**: 108–43.

SINNOTT, E. W. and BAILEY, I. W. (1914) Investigations on the phylogeny of the angiosperms: 3, Nodal anatomy and the morphology of stipules. *Am. J. Bot.* **1**: 441–53.

SKUTCH, A. F. (1927) Anatomy of leaf of banana, *Musa sapientum* L. var. Hort. Gros Michel. *Bot. Gaz.* **84**: 337–91.

SLADE, B. F. (1957) Leaf development in relation to venation, as shown in *Cercis siliquastrum* L. *Prunus serratula* Lindl. and *Acer pseudoplatanus* L. *New Phytol.* **56**: 281–300.

SLADE, B. F. (1959) The mode of the origin of vein-endings in the leaf of *Liriodendron tulipifera* L. *New Phytol.* **58**: 299–305.

SOLEREDER, H. (1908) *Systematic Anatomy of the Dicotyledons.*

Clarendon Press, Oxford.

STEINGRAEBER, D. A. and FISHER, J. B. (1986) Indeterminate growth of leaves in *Gaurea* (Meliaceae): a twig analogue. *Am. J. Bot.* **73:** 852–862.

STERN, W. L., SWEITZER, E. M. and PHIPPS, R. E. (1970) Comparative anatomy and systematics of woody Saxifragaceae, *Ribes.* In: *New Research in Plant Anatomy* (eds. N. K. B. Robson, D. F. Cutler and M. Gregory). Academic Press, London and New York, and *J. Linn. Soc. Bot.* **63,** suppl. 1: 215–37.

STEVENS, A. B. P. (1956) The structure and development of hydathodes of *Caltha palustris* L. *New Phytol.* **55:** 339–45.

STEWART, R. N. (1978) Ontogeny of the primary body in chimeral forms of higher plants. In: *The Clonal Basis of Development* (eds. S. Subtelny and I. M. Sussex), *36th Symp. Soc. Develop. Biol.* Academic Press, New York, pp. 131–60.

STEWART, R. N. and DERMEN, H. (1975) Flexibility in ontogeny as shown by the contribution of the shoot apical layers to leaves of periclinal chimeras. *Am. J. Bot.* **62:** 935–48.

THIELKE, C. (1948) Beiträge zur Entwicklungsgeschichte unifazialer Blätter. *Planta* **36:** 154–77.

THIELKE, C. (1951) Über die Möglichkeiten der Periklinalchimärenbildung bei Gräsern. *Planta* **39:** 402–30.

THODAY, D. (1931) The significance of reduction in the size of leaves. *J. Ecol.* **19:** 297–303.

TIBA, S. D. and JOHNSON, C. T. (1987) A note on the ontogeny of cross veins in *Digitaria eriantha. Ann. Bot.* **60:** 603–605.

TORIYAMA, H. (1960) Observational and experimental studies of sensitive plants. XI. On the thread-like apparatus and the chloroplasts in the parenchymatous cells of the petiole of *Mimosa pudica. Cytologia* **25:** 267–279.

TORIYAMA, H. (1962) Observational and experimental studies of sensitive plants. XIV. On the changes of a new cellular element of *Mimosa pudica* in diurnal and nocturnal conditions. *Cytologia* **27:** 276–284.

TRANSEAU, E. N. (1904) On the development of palisade tissue and resinous deposits in leaves. *Science* **19:** 866–7.

TROLL, W. (1948) *Allgemeine Botanik.* F. Enke, Stuttgart.

TUCKER, S. C. (1964) The terminal idioblasts in magnoliaceous leaves. *Am. J. Bot.* **51:** 1051–62.

TUCKER, S. C. and HOEFFERT, L. L. (1968) Ontogeny of the tendril in *Vitis vinifera. Am. J. Bot.* **55:** 1110–19.

TURRELL, F. M. (1936) The area of the internal exposed surface of dicotyledon leaves. *Am. J. Bot.* **23:** 255–64.

TURRELL, F. M. (1939) The relation between chlorophyll concentration and the internal surface of mesomorphic and xeromorphic leaves grown under artificial light. *Proc. Iowa Acad. Sci.* **46:** 107–17.

TURRELL, F. M. (1942) A quantitative morphological analysis of large and small leaves of alfalfa with special reference to internal surface. *Am. J. Bot.* **29:** 400–15.

TURRELL, F. M. (1944) Correlation between internal surface and transpiration rate in mesomorphic and xeromorphic leaves grown under artificial light. *Bot. Gaz.* **105:** 413–25.

VASSILYEV, A. E., CRANG, R. E., MIROSLAVOV. E. A. and YOON, J. S. (1982) Cellular ultrastructure of minor veins in *Populus trichocarpa* (Salicaceae) leaves. *Bot. Zh. SSSR* **67:** 278–284. (In Russian with an English summary.)

VOLKENS, G. (1887) *Die Flora der aegyptisch-arabischen Wüste auf Grundlage anatomisch-physiologischer Forschungen.* Gebr. Borntraeger, Berlin.

WAGNER, W. H., JR. (1979) Reticulate veins in the systematics of modern ferns. *Taxon* **28:** 87–95.

WALLES, B., NYMAN, B., and ALDEN, T. (1973) On the ultrastructure of needles of *Pinus silvestris* L. *Stud. Forest. Suec.* **106:** 1–26.

WEAVER, J. E. and CLEMENTS, F. E. (1929) *Plant Ecology.* McGraw-Hill, New York.

WEBSTER, B. D. (1970) A morphogenetic study of leaf abscission in *Phaseolus. Am. J. Bot.* **57:** 443–51.

WELTON, F. A. (1928) Lodging in oats and wheat. *Bot. Gaz.* **85:** 121–51.

WERKER, E. and KOLLER, D. (1987) Structural specialization of the site of response to vectorial photo-excitation in the solar-tracking leaf of *Lavatera cretica. Am. J. Bot.* **74:** 1339–1349.

WILDER, G. J. (1985a) Anatomy of noncostal portions of lamina in the Cyclanthaceae (Monocotyledonae). II. Regions of mesophyll, monomorphic and dimorphic ordinary parenchyma cells, mesophyll fibers, and parenchyma-like dead cells. *Bot. Gaz.* **146:** 213–231.

WILDER, G. J. (1985b) Anatomy of noncostal portions of lamina in the Cyclanthaceae (Monocotyledonae). III. Crystal sacs, periderm, and boundary layers of the mesophyll. *Bot. Gaz.* **146:** 375–394.

WILKINSON, H. P. (1979) The plant surface (mainly leaf). In: C. R. Metcalfe and L. Chalk, *Anatomy of Dicotyledons,* Vol. I, 2nd edn. Clarendon Press, Oxford.

WYLIE, R. B. (1939) Relations between tissue organization and vein distribution in dicotyledon leaves. *Am. J. Bot.* **26:** 219–25.

WYLIE, R. B. (1943) The role of the epidermis in foliar organization and its relations to the minor venation. *Am. J. Bot.* **30:** 273–80.

WYLIE, R. B. (1946) Relations between tissue organization and vascularization in leaves of certain tropical and subtropical dicotyledons. *Am. J. Bot.* **33:** 721–6.

WYLIE, R. B. (1947) Conduction in dicotyledon leaves. *Proc. Iowa Acad. Sci.* **53:** 195–202.

WYLIE, R. B. (1949) Differences in foliar organization among leaves from four locations in the crown of an isolated tree (*Acer platanoides*). *Proc. Iowa Acad. Sci.* **56:** 189–98.

WYLIE, R. B. (1951) Principles of foliar organization shown by sun-shade leaves from ten species of deciduous dicotyledonous trees. *Am. J. Bot.* **38:** 355–61.

ZERONI, M., HOLLANDER, E., and ARZEE, T. (1978) Abscission in the tumbleweed *Kochia indica:* ethylene, cellulase, and anatomical structure. *Bot. Gaz.* **139:** 299–305.

ZHENG-HAI, H., KUANG-MIN, L. and LEE, C. L. (1964) Distribution and general morphology in *Kingdonia uniflora. Acta Bot. Sinica.* **12:** 351–358 and 4 plates. (In Chinese with an English summary.)

CHAPTER 13

THE ROOT

THE root constitutes the lower portion of the plant axis and it usually develops below the soil surface, although there are roots that grow in the air as there are stems that develop below soil surface. However, basic differences in the development and arrangement of the primary tissues in these two organs are always distinguishable. The histogenesis of the epidermis of the root differs from that of the stem (see Chapter 3). In spermatophytes the primary xylem in the root is exarch and that in the stem is endarch. The xylem and phloem strands in the root do not form common bundles but are arranged alternately, while in the stem the vascular bundles are collateral, bicollateral, or amphivasal. Roots bear no appendages that are comparable to the leaves of the stem; roots are devoid of stomata and their branches originate in the relatively mature tissue of the pericycle in contrast to the stem where the branches originate from the apical meristem. Roots also possess a root cap which has no parallel in stems.

Much variability exists in the shape and structure of roots. This variability, in many cases, is related to the function of the roots, i.e. whether they are storage roots, succulent roots, aerial roots, pneumatophores, climbing roots, prop roots, or whether they contain symbiotic fungi (mycorrhiza). Environmental conditions often influence the root system. Plants growing in dry soils usually have better developed root systems. Many plants growing in sandy soils develop shallow, horizontal, lateral roots which spread out, close below the soil surface, over a distance of tens of metres (e.g. *Tamarix* and *Retama*).

On the basis of origin two types of roots—primary roots and adventitious roots—are distinguished. Primary roots develop from the apex of the embryo that is destined, from its origin, to give rise to a root, and from the pericycle of relatively mature parts of roots, while adventitious roots develop from other tissues of mature roots or from other parts of the plant body, such as stems and leaves. Special importance has been given to those adventitious roots that develop from the callus of cuttings.

In most dicotyledons and gymnosperms the root system consists of a tap root from which side branches arise. The order of appearance of the lateral roots is from the root neck (that part where the root joins the stem) towards the root tip, but in some cases the primordia of some of the lateral roots remain dormant. The mature portions of the root, which usually undergo secondary thickening, function only as a holdfast in the soil and to store reserve materials. The uptake of water and salts is carried out mainly by the extremities of the root system which are still in the process of primary growth.

The roots of mature monocotyledons are usually adventitious and develop from the stem (Fig. 161, no. 1). They may branch several times, as do the roots of dicotyledons, or they may be unbranched. Generally such roots do not develop secondary thickening. The most common type of root system among monocotyledons is the fibrous root system. In grasses the adventitious roots begin to develop from the hypocotyl when the latter is still in the embryonic state, i.e. they are seminal roots.

The radicle present in the seed consists of the root meristem and it gives rise to the first root on the germination of the seed. In the gymnosperms and dicotyledons this root develops into the tap root with its branches. In monocotyledons this root usually dries out early in the growth of the plant and the root system of the mature plant consists of numerous adventitious roots.

The apical meristems of lateral roots develop deep within the inner tissues in contrast to the buds of the shoot which develop from outer tissues. Therefore the branching of roots is *endogenous* and that of stems, *exogenous*.

Cortical roots are present in several monocotyledons (Pant, 1943; McLean and Ivimey-Cook, 1951). In *Tillandsia*, for example, these roots, which originate in the pericycle of the stem, grow directly downward within the cortex and emerge near the base of the stem. In *Asphodelus tenuifolius*, contrary to most other herbaceous monocotyledons with fibrous root systems, the root developing from the radicle of the embryo persists, and a large number of cortical roots arise from the base of the flattened, condensed stem. The cortex of this primary root is penetrated by the adventitious cortical roots which pass vertically downwards through it for some distance before they emerge into the soil.

ARRANGEMENT OF THE PRIMARY TISSUES IN THE ROOT

At a certain distance from the apical initials (see Chapter 3) of the root the following zones can be distinguished: *root cap, epidermis, root cortex,* and *vascular* or *central cylinder*.

The root cap

The root cap is situated at the tip of roots (Fig. 40, no. 2; Fig. 41), it protects the root promeristem and aids the penetration of the growing root into the soil. The root cap consists of living parenchyma cells which often contain starch. These cells may have no special arrangement or they may be arranged in radiating rows which originate from the initials. In many plants the central cells of the root cap form a more distinct and constant structure which is termed the *columella* (see Chapter 3).

Root cap cells secrete a polysaccharide slime. It was found that the process of secretion is accompanied by hypertrophy of the dictyosome cisternae forming large vesicles. The vesicle contents are subsequently released from the protoplast by fusion of the membranes of the vesicles with the plasmalemma. The secretion moves outwards through the wall (Northcote and Pickett-Heaps, 1966; Morré *et al.*, 1967; Juniper and Pask, 1973; Paul and Jones, 1976).

It has long been known that the root cap controls the geotropic growth of the root. Scientists of the last century had already demonstrated that removal of the extreme tips of roots prevents their geotropic response. Later it was suggested that certain solid cell inclusions, termed *statoliths*, principally starch-containing plastids, occurring in the root cap cells transmit gravitational stimuli to the plasmalemma of the statolith-containing cells (*statocysts* or *statocytes*). Starch grains were, indeed, seen to collect at the lower side of the root cap cells (cf. Haberlandt, 1918). Electron microscope studies of root cap cells of *Vicia faba* roots subjected to geotropic stimulation (Griffiths and Audus, 1964) showed that the amyloplasts sedimented to the lower side of the cells, thus displacing the ER and other organelles to the upper halves of the cells (Fig. 152, no. 1).

The existence of two highly mobile components, i.e. amyloplasts and ER, in statocytes is also supported by experiments carried out by Juniper and French (1973). However, Sievers and Volkmann (1972) and Volkmann (1974), who worked on roots of *Lepidium sativum* and some other plants, suggested that geoperception in roots may be a function of pressure extended differentially by amyloplasts on static ER complexes (Fig. 152, no. 2).

The site of geoperception, the root tip, is separated from the site of gravity reaction, the elongating zone of the root. The nature of the stimulus transmitted from the root tip to the zone of cell elongation is as yet not known but there is some evidence that it may be an unequal distribution of abscissic acid (Wilkins, 1978). Moore and Evans (1986) and Moore *et al.* (1987) have recently suggested that the response to gravity by roots results from a gravity-induced and electrical-driven downward movement of Ca. The Ca, some of which moves through the root-cap mucilage, accumulates along the lower side of the root tip. This Ca asymmetry induces a migration of auxin to the lower side of the elongating zone of the root.

The root cap develops continuously. The outer-

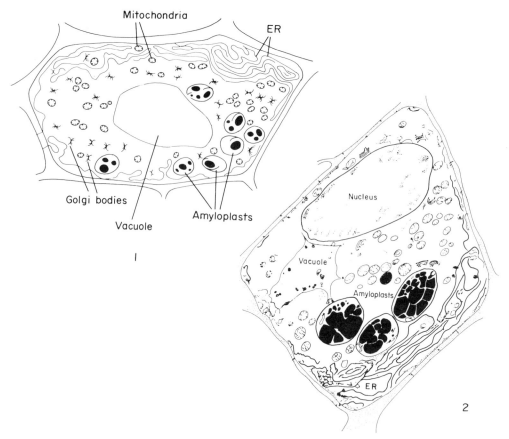

FIG. 152. Drawings of statocytes of root tips showing sedimentation of the amyloplast statoliths to the lower side of the cell resulting from geotropic stimulation. 1, *Vicia faba*, the amyloplasts sedimented to the lower side of the cell and the ER is displaced to its upper side. (Adapted from Griffiths and Audus, 1964.) 2, *Lepidium sativum*; amyloplasts on top of static ER complex. (Drawing based on electron micrograph of Sievers and Volkmann, 1972.)

most cells die, become separated from one another, and disintegrate, and they are replaced by new cells which are produced by the initials. The rate of production and sloughing of the cap cells is high. Clowes and Woolston (1978) found that the number of cap cells sloughed per root of *Zea mays* growing in water for 24 h at 23°C was between 3000 to 7000. Root caps are apparently found on roots of all plants except for the roots of some parasites and some mycorrhizal roots. External factors influence the structure of the root cap (Richardson, 1955). Root caps develop in true water plants but they degenerate early.

The epidermis

The epidermal cells of roots are thin-walled and are usually devoid of cuticle, although sometimes the outermost cell walls, including those of the root hairs, undergo cutinization (Guttenberg, 1940; Scott *et al.*, 1963). On those parts of roots that are exposed to the air and on those parts in the soil on which the epidermis persists for a long time, the outermost cell walls become thick, and may sometimes contain lignin or dark-coloured substances which have not been fully identified. The epidermis of roots is usually uniseriate, but exceptions do exist. On the aerial roots of plants belonging to the Orchidaceae and in the epiphytic, tropical genera of the Araceae the epidermis is multiseriate and it is specialized to form a *velamen* (Fig. 96, nos. 4, 5) (see Aerial Roots in this chapter).

The most characteristic feature of the root epidermis is the production of root hairs which are organs well adapted to the efficient uptake of water and salts. The region of root hairs is usually restricted to one or a few centimetres from the root apex. Root hairs are absent close to the apical meristem and they usually die and dry out on the more mature portions of the root. Certain herbaceous plants, and especially water plants, e.g. *Pistia* and *Eichhornia*, lack root hairs. Plants that usually grow in soil and which produce root hairs fail to do so when they are grown in water. Calcium is one of the factors controlling the normal development of root hairs (Cormack *et al.*, 1963). In some plants the root hairs remain on the root for a long time. In *Gleditschia triacanthos*, for example, the root hairs remain viable for some months and their walls become thickened. Long-lived root hairs have been found in certain species of the Compositae and in some plants of other families (Cormack, 1949; Scott *et al.*, 1963) but it is doubtful whether these root hairs take part in the uptake of water from the soil. In many cases the presence of such long-lived root hairs is connected with a small amount of secondary thickening and absence of periderm.

In certain plants all the epidermal cells may give rise to root hairs while in others only certain cells, *trichoblasts*, may do so. Some workers have found that root hairs develop from a subepidermal layer in the Commelinaceae and related families, and also in *Citrus* (Hayward and Long, 1942). (For more details see Chapter 10.)

Epidermal cells of young roots and root hairs of some genera secrete fibrillar mucilage forming a mucilaginous layer (Greaves and Darbyshire, 1972; Leppard, 1974; Chaboud and Rougier, 1986). In plants grown on soil this layer was found to be colonized by bacteria (Foster and Rovira, 1976). In *Sorghum*, in addition to fibrillar mucilage, drops of mucilaginous character are secreted near the tip of the root hairs (Werker and Kislev, 1978).

In the epidermis of seedling roots of some plants, e.g. *Ceratonia siliqua* and *Pisum arvense*, stomata have been observed to occur (Christodoulakis and Psaras, 1987; Tarkowska and Wacowska, 1988).

The root cortex

In most of the dicotyledons and gymnosperms the cortex of the root consists mainly of parenchyma cells. In many monocotyledons in which the root cortex is not shed while the root remains viable, much sclerenchyma develops in addition to the parenchyma. The root cortex is usually wider than the stem cortex (Fig. 154, no. 1) and therefore it plays a larger role in storage. The innermost layer of the cortex constitutes the *endodermis* (Fig. 153, nos. 1–3; Fig. 154, no. 2). In certain plants such as *Smilax, Iris* (Fig. 154, no. 1), *Citrus* (Cossmann, 1940), *Oryza* (Clark and Harris, 1981), and *Phoenix*, for example, there is a special layer below the epidermis termed the *exodermis*. In the roots of *Zea mays* and *Allium cepa*, the layer of cortical cells adjacent to the epidermis (the hypodermis) possess Casparian strips in the anticlinal cell walls. These strips develop about 20 mm further from the root tip than the Casparian strips of the endodermis. On the basis of experiments with fluorescent tracers it has been concluded that in the roots of the above-mentioned species a continuous symplastic pathway, from the epidermis to the stele, is present (Peterson *et al.*, 1982; Peterson and Emanuel, 1983; Peterson and Perumalla, 1984).

The arrangement of the cells of the cortex, as seen in cross-section of the root, may be in radial rows, at least in the inner layers, or the cells of two adjacent concentric layers may be arranged alternately. The radial arrangement is the result of the way in which the cells divide during the formation of the cortex (Guttenberg, 1940; Heimsch, 1960). Repeated periclinal divisions increase the number of cell layers in a radial direction, while anticlinal divisions add to the periphery and length of the cortex. The cells that undergo periclinal divisions are the inner cortical cells and, after the periclinal divisions are completed, the innermost layer of the cortex differentiates to form the endodermis.

Schizogenous intercellular spaces, which appear in the early ontogenetic stages, are very common in the root cortex. In certain plants, such as the Gramineae and Cyperaceae, large lysigenous intercellular spaces often develop in addition to the schizogenous ones. Large air canals are common in the root cortex of the Palmae (Tomlinson, 1961). A lacunate cortex may also be found in roots of some dicotyledonous plants, e.g. *Penthorum* (Haskins and Hayden, 1987). In hydro-

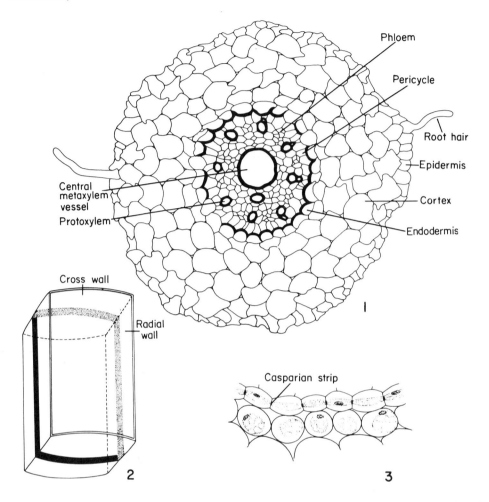

FIG. 153. 1, Cross-section of a root of a seedling of *Triticum.* 2, Three-dimensional diagram of a single endodermal cell with Casparian strip. 3, Portion of a cross-section of a root showing part of the endodermis and a row of cortical parenchyma cells in a state of plasmolysis. It can be seen that the protoplasts of the endodermal cells remain attached to the Casparian strips. (No. 1 adapted from Avery, 1930; no. 3 adapted from Esau, 1953.)

phytes, e.g. in the Hydrocharitaceae, the air canals occurring in the middle part of the cortex are traversed by perforated diaphragms (Ancibor, 1979).

Aerenchyma may develop in the cortex of roots of plants growing in flooded soils. The lacunae, which are formed both schizogenously and lysigenously, develop preferentially in roots in which the cortical cells, as seen in cross-section, are arranged in radial rows (Justin and Armstrong, 1987).

The parenchyma cells of the root cortex usually lack chlorophyll. Chloroplasts are found in the roots only of certain water plants and in the aerial roots of many epiphytes. Secretory cells, resin ducts, and laticifers are found in the root cortex of different plants. If sclerenchyma is developed it is usually in the form of a cylinder within the epidermis, within the exodermis or adjacent to the endodermis.

In many gymnosperms and certain dicotyledons, such as plants belonging to the Cruciferae,

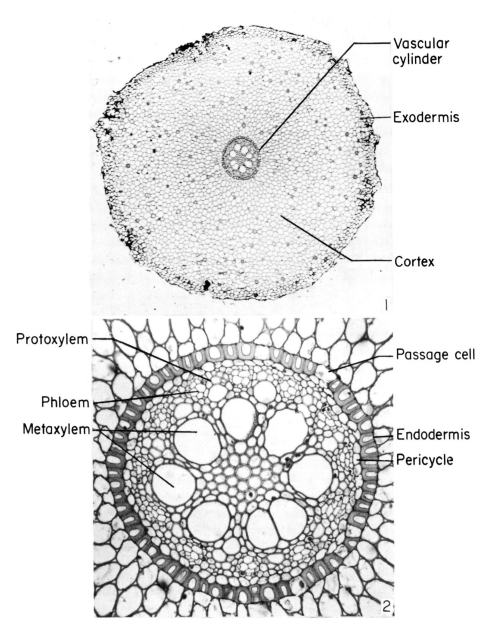

FIG. 154. Micrographs of cross-sections of the root of *Iris*. 1, Entire cross-section. ×13. 2, Vascular cylinder, enlarged. ×140.

Caprifoliaceae, and Rosaceae, reticulate or annular wall thickenings can be found in the cells outside the endodermis. Collenchyma is sometimes also found in the root cortex (Guttenberg, 1940).

The exodermis

In many plants the walls of the cells of the outer subepidermal layers of the cortex become

suberized (Fig. 154,no. 1). In this way the exodermis, a protective tisue, is formed (Guttenberg,
1943). The exodermis is similar in structure and
cytochemical characteristics to the endodermis
(van Fleet, 1950). An almost continuous suberin
lamella lines the primary cell wall internally, and
it in turn is usually lined with layers of cellulose
which develop centripetally. Also lignin is often
deposited in the walls of these cells and, in certain
cases, Casparian strips have also been distinguished (van Fleet, 1950). The cells of the exodermis contain viable protoplasts even when
mature. In the pteridophytes no exodermis is developed. Sometimes, however, for instance in
Ophioglossum, certain fatty substances are deposited in the walls of the subepidermal cells, but
no special suberin lamella is formed.

The thickness of the exodermis varies from a
single cell layer to many layers. The exodermis
may sometimes be accompanied, on its inner side,
by sclerenchyma as, for example, in the root
of *Ananas* (Krauss, 1949), of some grasses and
Cyperaceae. In *Phoenix* the exodermis is fibrous
(Tomlinson, 1961).

The endodermis

The endodermis consists of a uniseriate
cylinder of cells and develops, apart from a few
exceptions, in all vascular plants. This layer of
cells represents the inner boundary of the root
cortex (Fig. 153, nos. 1–3; Fig. 154, no. 2). Because
of its physiological and phylogenetic importance,
the endodermis has been very thoroughly investigated. In that part of the root where the primary
vascular system is starting to mature, *Casparian
strips* appear in the radial and cross walls of the
endodermal cells (Fig. 153, no. 2). These strips are
not merely wall thickenings but integral parts of
the primary wall and the middle lamella in which
suberin and lignin are deposited (van Fleet,
1942a). In the electron microscope the Casparian
strip appears as a slightly thickened, homogeneous electron opaque wall region. If plasmolysis is brought about in endodermal cells the
protoplast withdraws from the tangential walls
but remains attached to the Casparian strips
(Guttenberg, 1943). This type of plasmolysis has

been termed *band plasmolysis*. Plasmodesmata
occur in all wall portions of the endodermal cells
except for the Casparian strips to which the plasmalemma adheres (Bonnett, 1968). The Casparian
strips prevent inward flow of water and nutrients
through the apoplast. (For this and other functions of the endodermis see Clarkson and
Robards, 1975.) During the primary development
of the root the endodermal cells are capable of
much additional growth. This feature is striking
during the endogenous initiation of lateral roots
from the pericycle which is accompanied by the
division and stretching of the neighbouring endodermal cells. In addition, the endodermal cells in
certain roots continue to divide anticlinally during the early stages of secondary thickening. Casparian strips develop in many of the cells thus
formed.

In many of the angiosperms, pteridophytes, and
some gymnosperms, the endodermis remains in
the primary form and is shed together with the
cortex with the development of secondary thickening and periderm. However, in other angiosperms in which there is no secondary thickening,
an almost continuous lamella of suberin develops
on the inner side of the entire primary wall, including the Casparian strips. This lamella charac-
terizes the second stage of development of the
wall and in the third stage a layer of cellulose is
laid down centripetally on the inside of the
suberin lamella. This layer may reach a very considerable thickness on the radial walls, the walls
parallel to a cross-section of the root, and on the
inner tangential walls of the endodermal cells
(Fig. 154, no. 2). This type of endodermal cell is
common in the roots of most monocotyledons
(**Guttenberg, 1943; Clark and Harris, 1981). These**
thickened endodermal cell walls may becomme
lignified. The endodermis of the conifers is characterized by the second stage of wall development
only, i.e. by the development of a suberin lamella
on the inner side of the walls (Guttenberg, 1941;
Wilcox, 1962a).

The additional wall layers of the endodermal
cells do not develop simultaneously in all the
endodermal cells as seen in a single cross-section.
Casparian strips and the successive stages of the
development of the typical wall first appear opposite the phloem strands from where the develop-

ment spreads towards those endodermal cells that are opposite the xylem strands (van Fleet, 1942a, b; Guttenberg, 1943; Clowes, 1951). Because of the delay in wall differentiation of the endodermal cells opposite xylem these cells often have Casparian strips only. These cells are termed *passage cells* as they are thought to provide passage for substances between the cortex and vascular cylinder. The passage cells may remain unaltered throughout the entire life of the root or they, too, may develop thick secondary walls as do the other endodermal cells. Clarkson *et al.* (1971) concluded that in barley roots the passage cells make only a minor contribution to the current of water and phosphate passing into the stele, whereas the plasmodesmata which occur in great numbers in the inner tangential wall of the endodermal cells provide the principal channel of transport.

The production of the suberin lamellae on the endodermal cell walls results from the polymerization of unsaturated fatty compounds which is brought about by oxidases and peroxidases. The peroxidases are brought to the endodermal cells via the sieve elements. This has led van Fleet (1942b) to suggest that this is the reason why the greatest amount of suberin is laid down on the inner walls of the endodermal cells and why the passage cells, which lack suberin, appear mostly opposite the xylem, and not the phloem, strands. In some plants, e.g. *Pyrus* and *Malus*, the cells of the innermost layer of the cortex, in the differentiating region of the root, divide periclinally to form on the inside the future endodermis and outside it the so called *phi* layer. In the anticlinal walls of the cells of the phi layer lignin is deposited causing formation of wall thickenings similar to the Casparian strips (see MacKenzie, 1979).

In roots of some Gramineae, e.g. in *Sprobolus rigens*, the cells of one or sometimes two cortical layers next to the endodermis are similar to this endodermis because they develop suberin lamellae and become thick-walled (Böcher and Olesen, 1978).

The vascular cylinder

The vascular cylinder occupies the central por-

tion of the root. It is more clearly delimited from the cortex in roots than in the stem, because of the presence of the endodermis which is characteristically better developed in roots.

The primary vascular tissue is surrounded by a region of cells which is termed the *pericycle* (Fig. 154, no. 2). The pericycle generally consists of one or more layers of thin-walled parenchyma cells. In some plants, e.g. *Hordeum vulgare*, the cell walls become thickened (Warmbrodt, 1985). The pericycle is in direct contact with the protophloem and protoxylem and can already be distinguished prior to the lignification of the protoxylem elements. The pericycle retains its meristematic characteristics. The primordia of the lateral roots in all spermatophytes and the phellogen and portions of the vascular cambium in the dicotyledons develop from the pericycle. In monocotyledons the phellogen usually develops in the outer parts of the cortex. In roots of many Gramineae and Cyperaceae the outermost elements of the protoxylem may develop in the regions of the pericycle and, in the Potamogetonaceae, even the phloem elements may do so. In such cases the pericycle is not continuous (Guttenberg, 1943).

In monocotyledons, where there is usually no secondary thickening, sclerification takes place in part or all of the pericycle. In most angiosperms the pericycle is uniseriate but in many monocotyledons (such as the Gramineae, Palmae, *Agave*, and *Smilax*) and in a few dicotyledons (such as *Celtis*, *Morus*, and *Salix*) the pericycle consists of several layers of cells. Sometimes the pericycle is uniseriate opposite the phloem and is wider opposite the xylem. In gymnosperms the pericycle is usually multiseriate. In the roots of certain waterplants and parasites the pericycle is absent. Laticifers and secretory ducts may be present in the pericycle.

One of the principal features by which roots and stems can be differentiated is the arrangement of the primary vascular tissues. In the primary body of the root the pericycle is bordered directly on its inner surface by the phloem and xylem strands (Fig. 154, no. 2; Fig. 156, no. 1). The phloem strands are always separate and they are concentrated on the periphery of the vascular cylinder. The xylem strands may be in separate units on the periphery of the vascular cylinder or

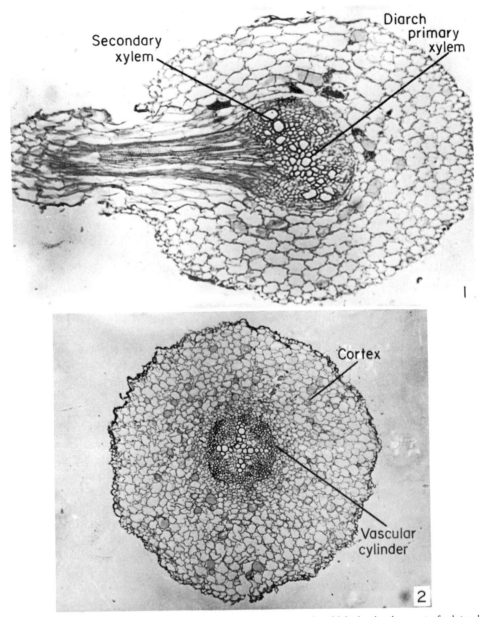

Secondary xylem

Diarch primary xylem

Cortex

Vascular cylinder

1

2

FIG. 155. Micrographs of cross-sections of roots. 1, *Lycopersicon esculentum* in which the development of a lateral root can be seen. × 90. 2, A young root of *Vicia faba*. × 50.

they may extend into the centre and then, as in many plants, the xylem appears star-shaped in cross-section. This structure led many workers to regard the vascular cylinder of the root as being a protostele. In many plants, and especially in the monocotyledons, the xylem strands do not reach the centre of the vascular cylinder which is then occupied by a pith.

The tracheary elements in the root mature centripetally and therefore the xylem is exarch, i.e. the protoxylem is situated on the outer side of the metaxylem. The differentiation of the phloem is also centripetal so that the protophloem is closest to the pericycle, while the metaphloem is closest to the axis of the root.

The number of protoxylem groups in the root, i.e. whether one, two, three, etc., is expressed by the terms *monarch*, *diarch*, *triarch* respectively, and a root in which there are many protoxylem groups is said to be *polyarch*. Diarch roots (Fig. 157, no. 1) are found, for example, in *Lycopersicon*, *Nicotiana*, *Beta*, *Raphanus*, *Daucus*, and *Linum*. In *Pisum* the root is triarch, while in *Vicia* (Fig. 155, no. 2), *Ranunculus* and *Gossypium*, it is tetrarch. Polyarch arrangement is characteristic of the adventitious roots of monocotyledons (Fig. 154, no. 2). A correlation exists between the diameter of the vascular cylinder and the number of protoxylem groups and the presence or absence of a pith. When the diameter of the vascular cylinder is large, a pith is usually present and the number of protoxylem groups is large. Variations in these features may be found even within the same plant. For example, in one plant of *Libocedrus decurrens* di-, tri-, tetra-, penta-, and hexarch roots have been found (Wilcox, 1962b). Similar variation has been observed in certain dicotyledonous species (Jost, 1932; Torrey, 1957).

In gymnosperms and dicotyledons the number of the xylem strands in the root is generally small. In these groups of plants the roots are usually di-, tri-, or tetrarch but there are some dicotyledonous species in which there are more xylem strands. The water-plant, *Trapa natans*, has a thin root which is monarch. In monocotyledons the number of xylem strands in the seminal roots is small, as in dicotyledonous roots, but the adventitious roots are polyarch and the number of strands in the Palmae and the Pandanaceae may be 100 or more. In the roots of the Filicinae different numbers of xylem strands may be found— from one, as in *Ophioglossum lusitanicum*, to many as in *Marattia fraxinea*.

In certain monocotyledons, such as *Triticum* (Fig. 153, no. 1), one large vessel is found in the centre of the vascular cylinder. Between this metaxylem vessel and the peripheral strands there is usually parenchyma. In other plants, e.g. *Zea* and *Iris* (Fig. 154), the large metaxylem vessels form a circle around the pith. The number of these large vessels is not always equal to that of the peripheral strands. In certain plants two peripheral xylem strands are associated with a single large inner vessel.

In woody monocotyledons the inner vessels may be arranged in two or three circles (e.g. *Latania* of the Palmae) or they may be scattered in the centre of the cylinder (e.g. *Raphia hookeri*). In a few other monocotyledons, e.g. *Musa*, *Cordyline*, and *Pandanus*, phloem strands are scattered in the centre of the root.

In most plants there are interconnections between the different tracheary elements which appear separate in cross-section of the root. In roots with a pith it is also possible to find, here and there, lateral connections between the phloem groups. However, there are plants that have no lateral connections between the phloem strands or the different tracheary elements.

The primary phloem of the roots of most plants does not contain fibres but in certain plants, such as those of the Papilionaceae, Malvaceae, and Annonaceae (Guttenberg, 1943), fibres are found in the primary phloem (Fig. 156).

In many mature roots that do not have secondary thickening the parenchyma associated with the primary vascular tissues becomes sclerified. In many Coniferales, except for the Taxodiaceae, Cupressaceae, and Taxaceae, resin ducts are present in the region of the primary conducting tissues.

Some roots are polystelic. In cross-sections of such roots a number of vascular cylinders each of which is surrounded by an endodermis can be seen. Examples of such roots are the root tubers of some species of *Orchis* (Arber, 1925) and members of the Palmae (Tomlinson, 1961).

MYCORRHIZAE

The epidermis and cortex of roots of many plants are often associated with soil fungi. This widespread association between fungal hyphae and the young roots of higher plants is known as *mycorrhiza* (Greek *mykes* mushroom, *rhiza* root). Usually this is a symbiosis: both the higher plant and the fungus derive benefit from the association.

Because of the very interesting problem of sym-

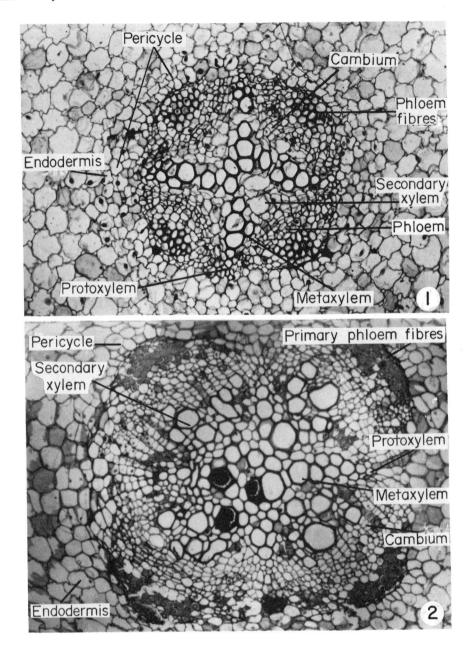

FIG. 156. 1, Micrograph of a cross-section of the tetrarch vascular cylinder of a young root of *Vicia faba* in which recently formed elements of secondary xylem, still thin-walled, can also be distinguished. × 150. 2, As above, but of an older root in which a relatively large amount of secondary xylem has been formed and in which the cambium is already almost cylindrical. × 150.

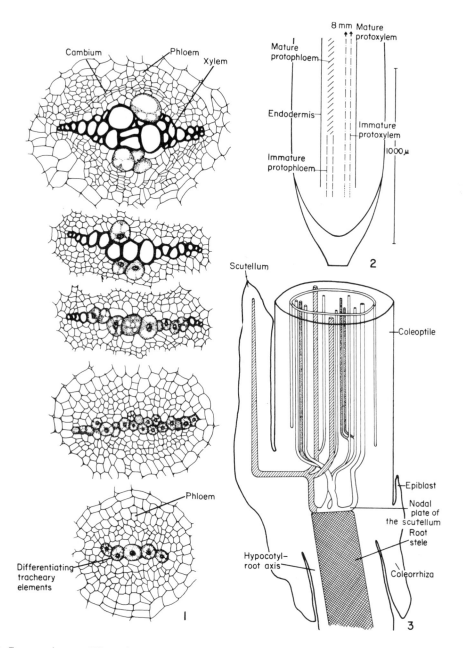

FIG. 157. 1, Cross-sections, at different levels, of a root of *Raphanus sativus* showing the differentiation of the primary xylem. 2, Diagram of a longitudinal section of the root tip of *Hordeum* showing the areas of differentiation and maturation of the vascular tissues. 3, Three-dimensional diagram of a portion of a seedling of *Triticum* showing the connection of the vascular systems of scutellum, hypocotyl-root axis, coleoptile and the first foliage leaves. Coleoptile and scutellum bundles— diagonally stripped; mid-vein and six lateral bundles of first foliage leaf—uncoloured; mid-vein and two lateral bundles of second foliage leaf—dotted. (No. 1 adapted from Stover, 1951; no. 2 adapted from Heimsch, 1951; no. 3 adapted from Boyd and Avery, 1936.)

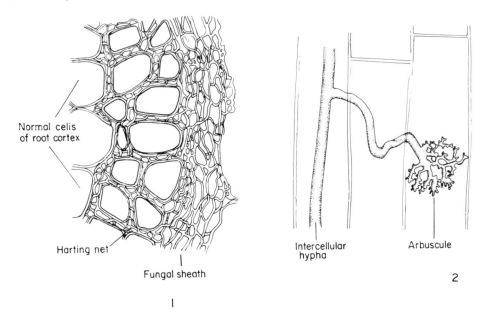

Normal cells
of root cortex

Harting net

Fungal sheath

1

Intercellular
hypha

Arbuscule

2

FIG. 158. Diagrams of mycorrhizae. 1, Ectomycorrhiza based on a micrograph of Meyer (1973). 2, Endomycorrhiza based on a micrograph of Cox and Sanders (1974).

biosis and the great economic importance of the mycorrhizae, a great amount of research has been devoted to their biology, histology and recently also to their ultrastructure (Harley, 1969; Clowes, 1951, 1981; Scannerini and Bonfante-Fasolo, 1983).

Two main types of mycorrhizae are recognized: *ectomycorrhizae* and *endomycorrhizae*. In the ectomycorrhizae the fungus produces a dense thick mycelium on the root surface. Hyphal strands penetrate the root between the cortical cells and form there a net (Harting net) (Fig. 158, no. 1). In the endomycorrhizae the fungus forms an inconspicuous mycelium on the root surface but invades the interior of the root cells and produce characteristic vesicles and arbuscules (Fig. 149, no. 2). These mycorrhizae are thus called *vesicular-arbuscular mycorrhizae* or *VA mycorrhizae* (Powell and Bagyaraj, 1984; Brundrett *et al.*, 1985) (Fig. 158, no. 2).

Ectomycorrhizal roots are short, branched, often dichotomous (Mexal *et al.*, 1979) and appear thicker than the normal roots. Ectomycorrhizae occur in many plants, e.g. *Pinus, Picea, Abies, Cedrus, Larix, Quercus, Castanea, Fagus, Betula,* *Populus, Carya, Salix,* and *Eucalyptus.* Endomycorrhizae occur in the Orchidaceae, Ericaceae, *Acer, Liriodendron, Ornithogalum umbellatum,* and others. Certain mycorrhizae have been called ecto-endomycorrhizae because of the occurrence of penetration hyphae of the ectomycorrhizae type into the host cells.

Electron microscopical studies of endomycorrhiza roots showed that the hyphae after, passing the wall of the host cells, do not penetrate into their protoplasts. The developing individual branches of the fungal arbuscule remain surrounded by the plasmalemma of the host cell (Cox and Sanders, 1974; Bonfante-Fasolo and Scannerini, 1977a).

Mycorrhizae occur on all types of plants: ferns, wheat, orchids, and forest trees (Harley, 1969). They are very often essential for successful plant growth (Bowen, 1973). The role of the mycorrhizal fungus appears to be the absorption of nutrients from the soil. In return the chlorophyllous host supplies the fungus with carbohydrates, amino acids, vitamins, and other organic substances. Starch is absent from the plastids of cortical cells of ectomycorrhiza roots. This may indicate that soluble sugars, translocated to the root, are rapidly transferred to the mycorrhizal fungus and compete with starch storage in the cortical root cells (Marks and Foster, 1973; Bonfante-Fasolo and Scannerini, 1977b). Some mycorrhizae increase also resistance of the host plants to pathogenic infections (Marx, 1973). Mycorrhizae may also make the host plant less susceptible to drought (Reid, 1979, as cited by Mexal *et al.*, 1979).

ROOT NODULES

Another type of association between roots of higher plants and lower organisms occurs in the Leguminosae. On the roots of these plants there are swellings, called *nodules* (Fig. 159). They develop as a result of penetration of nitrogen-fixing bacteria (species of *Rhizobium*) into the root cortex (Bond, 1948; Dart, 1975; Newcomb, 1976; Bergersen, 1982; Baird *et al.*, 1985). The bacteria invade the root primarily through the root hairs. While multiplying the bacteria form *infection threads* by becoming embedded in a matrix at the outside of which there is a layer similar to the host primary cell-wall, and continuous with it. The threads penetrate deeply into the root cortex in-

ducing a proliferation of its cells. The number of cells in the nodule increases at first by division throughout a spherical mass of cells and later by the activity of an apically localized meristematic region which is not penetrated by bacteria. The differentiated cells of the inner region, the bacterioid zone, contains bacteria released from the infection threads. The nodule at this stage resembles superficially a primordium of a lateral root.

The root epidermis is broken as the nodule enlarges but the nodule does not emerge from the cortex. The cells of the root cortex divide and stretch considerably so that they remain as the outermost layer of the nodule.

Branches developing from the root vascular tissue surround the bacterioid zone. Each bundle has a parenchymatic sheath and an endodermis. In some plants, the sheath cells of these bundles develop wall protuberances characteristic of transfer cells serving in short distance transport (Pate *et al.*, 1969; Libbenga and Harkes, 1973).

According to Libbenga *et al.* (1973) "nodule initiation may be controlled by stimulative and inhibitory host factors and their opposite gradient systems in combination with secretion of auxin and cytokinins by the *Rhizobium*".

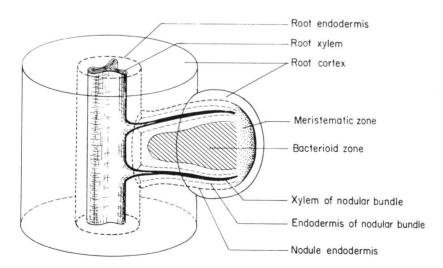

Root endodermis
Root xylem
Root cortex

Meristematic zone

Bacterioid zone

Xylem of nodular bundle

Endodermis of nodular bundle

Nodule endodermis

FIG. 159. Diagramatic representation of leguminous root and nodule. (Adapted from Bond, 1948.)

Recently a number of non-legume type of nitrogen-fixing nodules were described (Becking, 1975; Torrey, 1976, 1978; Heisey *et al.*, 1980).

TISSUE DIFFERENTIATION IN THE ROOT

Some distance from the apical promeristem of the root the epidermis, cortex, and vascular cylinder can be distinguished. The pericycle can also be identified close to the apical meristem. As it is not possible to distinguish clearly between the meristems of the vascular and non-vascular tissues in the vascular cylinder, it is not yet clear whether the pericycle develops from the procambium or from the ground meristem. The cells of the procambium that differentiate into the tracheary elements soon become distinguishable from those cells from which the phloem elements will develop. The former cells enlarge and they have large vacuoles, while the latter undergo numerous divisions without enlarging so that they become very small.

The order of appearance of the different tracheary elements, in comparison to the order in which they undergo maturation, is of interest. The cells that develop into metaxylem elements enlarge, together with the vacuoles in them, prior to those cells that differentiate into the protoxylem elements while the order of maturation is, of course, the contrary. Therefore the final dimensions of the metaxylem elements are far larger than are those of the protoxylem. This is especially obvious in the monocotyledons (Heimsch, 1951).

The ontogenetic development of the primary vascular system of the root is simpler than that of the stem because the differentiation of the vascular system of the latter is connected with the development of the leaves. The vascular system of the root develops independently of the lateral organs and the procambium develops acropetally as an uninterrupted continuation of the vascular tissues in the more mature parts of the root. The differentiation and maturation of the xylem and phloem are also acropetal (Popham, 1955) and follow that of the procambium. From the accurate investigations that have been carried out it appears that the protophloem elements mature

closer to the apical meristem than do the earliest tracheary elements (Fig. 157, no. 2). From this it is seen that the process of maturation of the protoxylem and protophloem elements is also simpler in the root than in the stem where the early differentiation of the xylem close to a leaf primordium is in two directions.

Generally the differentiation of the root tissues behind the apical promeristem can be summarized as follows: periclinal divisions in the cortex cease near the level where the sieve elements mature; beyond this region the root undergoes rapid elongation, and the maturation of the protoxylem usually takes place only when the process of elongation is almost completed; Casparian strips develop in the endodermal cells before the maturation of the protoxylem elements and generally also before the appearance of root hairs.

The proximity of the mature conducting elements to the root apex is dependent on the rate of growth and both are dependent on the external conditions, the type of root, and the stage of its development (Wilcox, 1962a). Heimsch (1951) found the following distances between the root apex and the first mature vascular elements in different roots of *Hordeum*: protophloem elements, 0.25–0.75 mm; protoxylem elements, 0.40–8.5 mm; elements of the early metaxylem, 0.55–21.6 mm or more, while the large central vessels mature at even greater distances (Fig. 157, no. 2). The earliest appearance of Casparian strips is at a distance of about 0.75 mm from the apex.

Rost and Baum (1988) found in pea seedlings a correlation between the length of the primary root, meristem height and position of the protoxylem. In the early stages of root growth, the meristem height increased and the protoxylem could be distinguished correspondingly further from the root tip. In relatively longer roots, as they continue to elongate, the meristem became shorter and the protoxylem position became closer to the tip.

Cambium in roots

There is great variation in the secondary growth in different roots. The tap-root and main

Lateral root primordia

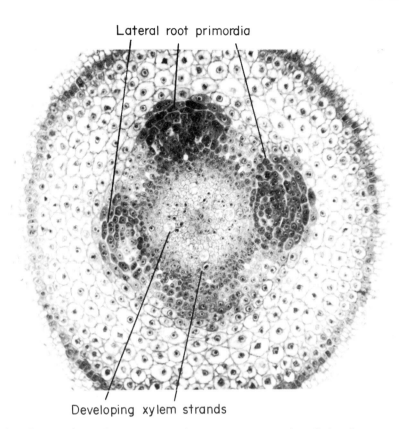

Developing xylem strands

FIG. 160. Cross-section of a root of *Cucurbita maxima* showing three lateral root primordia in adjacent sectors at the same level on the parent root. × 200. (Courtesy of T. E. Mallory.)

lateral roots of gymnosperms and woody dicotyledons usually have secondary thickening, but the smallest branches do not. In the roots of some herbaceous dicotyledons secondary thickening may be completely absent, vestigial (e.g. *Ranunculus*) or it may be well developed (e.g. *Medicago*).

Roots of most monocotyledons are devoid of secondary thickening. However, in some, e.g. *Dracaena*, such thickening does occur.

In the roots of gymnosperms and dicotyledons that do exhibit secondary thickening, the cambium first appears on the inner side of the phloem (Fig. 156, no. 1). After these cambial cells have produced a few secondary elements the pericycle cells on the outer sides of the protoxylem groups begin to divide, and the inner cells resulting from

these divisions form cambial cells. These strands of cambium unite with those on the inner sides of the primary phloem strands. At first the cambium has an undulating shape, as seen in cross-section of the root; but as the development of the secondary xylem on the inner side of the phloem strands precedes that of the secondary xylem external to the protoxylem groups, the cambium soon becomes circular in cross-section (Fig. 156, no. 2).

The secondary tissues of the root differ to some extent from those of the stem. The root may have a higher bark to wood ratio; relatively less fibres in the bark; less fibres in the wood; larger and more uniform vessels; less distinct growth increments; a larger volume of ray tissue and wider and longer tracheids in the gymnosperms, often with multiseriate arrangement of pits; a relatively large volume of living cells; less tannic substances (Esau, 1965; Cutler *et al.*, 1987).

Development of lateral roots

Lateral root primordia seem to occur in many plants in a more or less regular sequence (Riopel, 1966; Mallory *et al.*, 1970). It was found that the smaller the number of protoxylem poles (potential sites of lateral root formation) the greater the degree of regularity in the arrangement of the lateral roots. It seems that the spacing of lateral root primordia in the horizontal plane is determined by their relationship to the developing vascular system, and that little or no inhibition or competition exists between primordia situated opposite adjacent protoxylem poles (Mallory *et al.*, 1970; Pulgarin *et al.*, 1988) (Fig. 160).

In longitudinal plane, several factors seem to control the spacing of lateral root primordia. These are the effect of older primordia on the future sites of lateral root initiation in the same sector, and the effect of parent root apex (Torrey, 1965; Mallory *et al.*, 1970). In some angiosperms and gymnosperms it has been observed that the growth of a number of lateral root primordia may be arrested while at the same time vigorous laterals emerge both above and below them.

As has already been mentioned, one of the most prominent characteristics by which roots and stems can be distinguished is the manner in which the lateral appendages develop from the axis. In the stem, the primordia of the branches and leaves are initiated in the apical meristem from the outer cell layers. Unlike this, lateral roots develop endogenously from inner cell layers in regions of relatively mature tissues. The initiation of lateral roots usually commences immediately behind the region of root hairs; but in certain plants, especially water plants, e.g. *Eichhornia*, and *Pistia* the lateral roots begin to form within the apical meristem of the mother root (Clowes, 1985). In gymnosperms and angiosperms the lateral roots are commonly initiated in the pericycle whence they pass through the cortex of the parent root to the exterior. In pteridophytes the lateral roots mostly are initiated in the endodermis (Ogura, 1938).

A definite relationship exists between the position of the protoxylem groups and phloem strands and the position of the initiation of lateral roots. In triarch roots or those with more than three protoxylem groups the lateral roots usually develop opposite the protoxylem groups. But in certain plants, for instance species of the Gramineae, Cyperaceae, and Juncaceae, the lateral roots develop opposite the phloem strands. In diarch roots the primordia of the lateral roots appear opposite the phloem strands or close to the protoxylem groups (Fig. 155, no. 1), and so two rows of lateral roots develop. In some diarch roots, lateral roots may develop on both sides of the protoxylem groups so that, in such cases, four rows of lateral roots develop. In the fleshy carrot root additional lateral roots develop at the base of earlier formed lateral roots that have dried out (Esau, 1940, 1953).

In angiosperms the primordia of the lateral roots are formed by the periclinal and anticlinal divisions of a group of pericycle cells. The initiating divisions are periclinal. Bell and McCully (1970) have observed that in *Zea mays* the early stages of lateral root initiation, adjacent to a protoxylem pole, involve not only pericycle cells but also a considerable number of parenchyma cells of the adjacent portion of the stele. As a result of further growth the primordium penetrates through the cortex of the parent root. It is possible to distinguish the zones of primary tissues, apical meristem and root cap of the lateral root primordium even before it appears on the surface of the parent root (Fig. 161, nos. 2–4). Different opinions exist as to how the passage of the growing lateral root is effected through the cortex of the original root. According to one view the lateral roots partially digest the cortical tissue during penetration, while according to another view the process of penetration is purely a mechanical one. However, it is generally agreed that the developing lateral roots do not form any connection with the tissues through which they penetrate.

In many plants, as, for instance, *Daucus carota*, *Pistia*, *Hydrocharis* and *Zea mays*, the endodermis of the parent root takes part in the formation of the primordium of the lateral roots (Esau, 1940; Bell and McCully, 1970; Clowes, 1985). In such cases

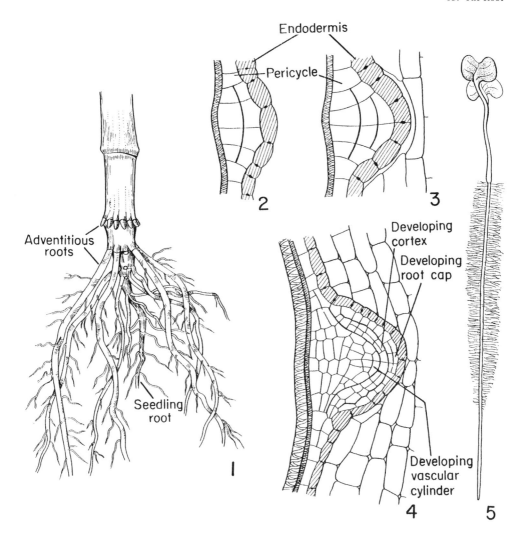

Fig. 161. 1, Basal portion of a plant of *Zea mays* in which the remains of the seedling root and the adventitious roots, arising at the base of the internodes, can be seen. 2–4, Portions of radial longitudinal sections of roots of *Hypericum* showing early stages in the development of lateral roots. 5, A seedling of *Sinapis alba* in which the region of root hairs can be seen. (Adapted from Troll, 1948.)

the endodermis may divide only anticlinally, but sometimes it may divide periclinally as well and thus forms more than one layer. With the eruption of the lateral root on the surface of the parent root, or even prior to it, the tissue that developed from the endodermis dies and it is eventually shed. According to Clowes (1978), in *Zea mays* the endodermis forms the root cap of the young primordium.

In certain plants the derivatives of the endodermis, together with those of other cortical layers, form a root-cap-like structure called pocket (Tasche) (Guttenberg, 1960). The pocket is particularly well developed in water plants in which the root cap may be lacking (e.g. *Eichhornia*, *Hydrocharis*, *Lemna*).

The connection between the vascular systems of the lateral and parent roots is brought about by intervening cells. As the lateral roots originate in the pericycle the distance between the two vascular systems is small. Of the intervening cells, which also develop from the pericycle, some differentiate into sieve elements and some into tracheary elements.

The xylem of the lateral roots of many monocotyledons is connected with two or more xylem strands of the original root. This can be seen in *Monstera*, for example, where the connections are not only with the peripheral xylem strands but also with the innermost large vessels of the metaxylem. This is brought about by the modification, into tracheary elements, of parenchyma cells between the xylem and phloem strands (Rywosch, 1909).

ADVENTITIOUS ROOTS

Adventitious roots, as defined above, may develop from large roots, from the hypocotyl of young plants, from the primary and secondary body of stems, and from leaves. In the roots and stems of most plants adventitious roots develop endogenously, but there are examples in which the development is exogenous. Primordia of adventitious roots may be formed by the following tissues: the epidermis, together with cortical tissue, of buds and hypocotyls (e.g. *Cardamine pratensis, Rorippa austriaca*); stem pericycle (e.g. *Coleus, Zea mays*); ray parenchyma between pericycle and cambium (e.g. *Tropaeolum majus, Lonicera japonica, Tamarix*); phloem ray parenchyma (e.g. *Hedera helix*); non-differentiated secondary phloem and cambium between the vascular bundles (e.g. *Rosa*); interfascicular cambium and pericycle (e.g. *Portulaca oleracea*); interfascicular cambium, pericycle, and phloem (e.g. *Begonia*); the pith of the stem (e.g. *Portulaca oleracea*); parenchymatous interruptions in the secondary xylem which are formed due to the presence of leaf gaps (e.g. *Ribes nigrum*) or buds (e.g. *Cotoneaster dammeri*); tissues of leaf margins and petioles (e.g. *Begonia, Kalanchoë*) (Hayward, 1938; Boureau, 1954; Ginzburg, 1967; Girouard, 1967a, b).

Auxins are often used to promote adventitious root formation. Gramberg (1971) has observed, in leaf petioles of *Phaseolus vulgaris* treated with IAA, that the first symptom of adventitious root induction is the formation of new protein and nucleic acids in the nucleus.

The development of adventitious roots has been described in some species of *Salix* (Carlson, 1938, 1950; Fjel, 1985a, b). In these species the adventitious roots develop from primordia which appear in the stem prior to its removal as a cutting. These primordia are formed from secondary parenchymatous tissue in the leaf or branch gaps. Several layers of cells external to the cambium take part in the formation of a primordium, to whose inner side cells are also added by the cambium. The primordium becomes dome-shaped as a result of the intensified growth of the secondary xylem immediately internal to it. These primordia remain dormant within the inner bark as long as the branch is not removed from the tree. The differentiation of these primordia is extremely slow so that even in 9-year-old branches, the typical root tip structure is not discernible. After the first year of growth, additional primordia may develop vertically above and below the first-formed primordia on both branches left on the tree and on cuttings. On cuttings most of the primordia develop rapidly into roots. Similar adventitious root primordia have been observed on woody roots of *Zygophyllum dumosum*.

Plant species differ from one another in their ability to produce roots on cuttings. Cuttings of plants with dormant or preformed adventitious root primordia (e.g. *Salix*) root easily as do many plants with broad vascular rays but without such primordia (e.g. *Vitis vinifera, Tamarix* spp.). Cuttings of *Ceratonia, Pyrus*, and *Carya*, for example, which have no dormant primordia and in which the rays are narrow, root with difficulty. In the gymnosperm species, *Agathis australis*, it may take weeks or months from taking a cutting to root emergence (White and Lovell, 1984a, b). MacKenzie *et al.* (1986) states that successful development of adventitious roots in cuttings depends on the cambium participating in callus formation and cambium subsequently regenerating within this callus in the form of an outward-pointing salient.

In the palm *Rhapis excelsa* (Tomlinson and Zimmermann, 1968) adventitious roots originate endogenously, rupturing the surface of the stem. The vascular tissues of the root are attached directly to the stem vascular bundles, often penetrating deeply into the stem. An extensive peripheral attachment via girdling vascular bundles is also present. The root vessels break up into aggregates or clusters of narrow elements which run parallel to the stem bundles. The distal elements of these clusters are applied to the meta-xylem vessels of the stem bundles. The terminal clusters of the root xylem frequently spread around the stem vessels, thus making very wide contacts. The metaxylem of the stem bundles is not usually modified.

In some palms, e.g. *Cryosophila*, adventitious roots are modified and act as spines (McArthur and Steeves, 1969).

The ability to produce adventitious roots varies with age—generally, they develop more easily on younger plants and plant organs.

ROOT STRUCTURE IN RELATION TO FUNCTION

Relation between structure and uptake of water and solutes

The uptake of water and solutes is accomplished mainly by the young parts of the roots.

The accepted opinion is that only a small amount of water penetrates through the root cap and through the apical meristem (Kramer, 1945; Brouwer, 1959). The main uptake of water is apparently in that region where the primary xylem is almost completely mature. In this region the root hairs, whose important role in the uptake of water is not doubted, are produced. In mature regions in which a periderm is present, water uptake also takes place and is probably effected through the lenticels (Kramer, 1946). The amount of water uptake in different places along the root depends on various factors. To a certain extent these differences are dependent on differences in structure. In many trees the region of uptake may become inactive with the onset of unfavourable ecological conditions as a result of the production of impermeable layers. The endodermis and sometimes the exodermis, both of which contain

suberin lamellae, develop a short distance from the apical meristem. In certain plants fatty substances have been found in the walls of the cells of the outer layers of the root cap, and in the epidermal cells between the root cap and that region where the exodermis begins to form. This process has been termed *metacutization* (Plaut, 1920; Guttenberg, 1943; Leshem, 1965). As observed by Wilcox (1968) in *Pinus resinosa*, the process of metacutization in a root apex which enters dormancy proceeds gradually. A metacutized cell layer develops first in the root cap, then the endodermis becomes suberized, and, finally, a bridge of metacutized cells is formed in the cortex joining the outer metacutized cells of the root cap to the suberized endodermis.

Ions are selectively transported and accumulated by roots. In the various processes involved the different tissues of the root participate in different ways. The undifferentiated non-vacuolated cells of the root apex (up to 0.5 mm in the root of *Zea mays*, for example) do not accumulate ions, and ions enter and leave the cells passively (Handley and Overstreet, 1963). The vacuolated and differentiated cells of the cortex have a marked ability to accumulate solutes. However, in different plants the main sites of entry of the various ions along the roots differ (Scott and Martin, 1962). The vascular cylinder, however, which serves as the main transport system for water and ions, exhibits a low rate of metabolism and has almost no capacity for accumulation.

The main barrier to transport across the root is believed to be the endodermis. It is assumed that the fatty substances, deposited in the Casparian strips of the endodermal cells, prevent the free passage of water and solutes through the cell walls and restrict all transport to the protoplast. It is well known that the cytoplasm of the endodermal cells is strongly attached to the Casparian strips. Therefore the passage of solutes from the cortex into the xylem through the free space is prevented (Priestley, 1920, 1922; Priestley and North, 1922; Brouwer, 1959). The endodermis not only takes part in selective transport of solutes from the outer solutions into the vessels, but also constitutes a barrier to water movement, thus making possible the existence of a hydrostatic pressure—usually termed the root pressure. The endodermal barrier appears to be very effective in ion selectivity except in cases where the ionic strength of the outer solution is very high or when the roots are deprived of energy sources.

Environmental conditions may influence root anatomy. So for instance it was observed that Casparian strips are much wider in plants growing in extremely dry or saline habitats than in those growing in ordinary mesophytic conditions.

In the root epidermis and root hairs of *Atriplex hastata* wall protuberances developed in response to salt treatment on the inner side of the outer walls. The epidermal cells thus become transfer cells (Kramer *et al.*, 1978). In response to iron deficiency peripheral cells of the zone near the root apex of *Helianthus annuus* were also found to differentiate into transfer cells (Kramer *et al.*, 1980).

Structure of storage roots

In all primary roots reserve substances (mostly starch) are stored in the cortex, which in most plants is relatively thick. In ordinary roots with secondary thickening reserve substances are stored as in stems, i.e. in the parenchyma and sometimes also in the sclerenchyma of the secondary xylem and phloem. Usually roots contain more parenchyma than do stems.

There are plants in which certain parts of the root system develop into thick fleshy organs which function especially as storage organs. In many plants the tap root and hypocotyl undergo such modification.

The origin of the storage tissue may differ. In the carrot, for example (Esau, 1940), the hypocotyl and tap root become thickened and, with the development of the periderm, the narrow cortex is shed. The organ becomes fleshy as a result of the excessive development of parenchyma in the secondary xylem and especially in the secondary phloem.

In the sugar beet, according to Artschwager (1926), the hypocotyl and root become fleshy as a result of an anomalous secondary thickening which is characteristic of the Chenopodiaceae and which is discussed in more detail in a later chapter. Here it will only be mentioned that, as a result of the activity of numerous cambia, layers of secondary tissue are formed consisting of parenchyma in which groups of conducting elements are scattered. The sugar is found as a reserve substance in the cells of this secondary parenchyma.

In *Ipomoea batatas* (Fig. 162) the fleshiness of the root is due to the following development (Hayward, 1938). Both the primary and secondary xylem develop normally and contain a large amount of parenchyma. However, with further development many anomolous secondary cambia are formed around single vessels or groups of them. These cambia, which are annular in cross-section of the root, produce some phloem but mainly parenchyma. Some distance from the vessels, laticifers are also formed. Tertiary tracheary elements develop close to and around the vessels that are encircled by these special secondary cambia. Still later, secondary cambia may be formed in the parenchyma not associated with vascular elements.

In the radish the fleshiness of the root and hypocotyl is due to the excessive development of parenchyma in the secondary xylem (which is produced by the normal cambium), as well as secondary parenchyma produced by additional cambia (which also produce tertiary conducting elements) (Hayward, 1938).

Roots as anchorage organs

The anchorage function of the root in the soil is aided by the following structural features: the branching of many lateral roots from a tap root and the development of many adventitious roots in fibrous root systems; the growth of root hairs which are of great importance in young roots; the development of sclerified tissues (principally xylem) in the centre of young roots and the development of sclerenchyma in old roots.

Contractile roots

The renewal buds of certain plants occupy a definite position within the soil or on its surface. This position is mostly obtained by the pull of special roots, which have been termed *contractile*

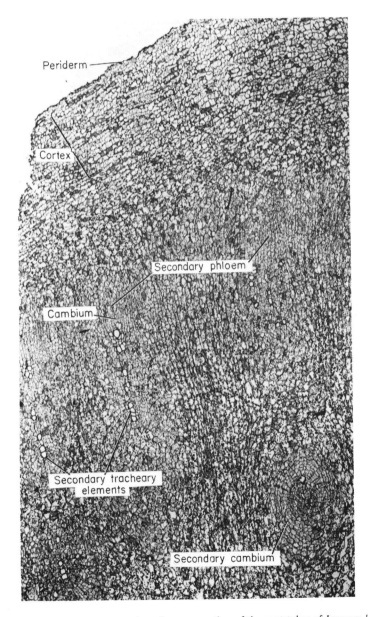

Fig. 162. Micrograph of the outer portion of a cross-section of the root tuber of *Ipomoea batatas*. × 25.

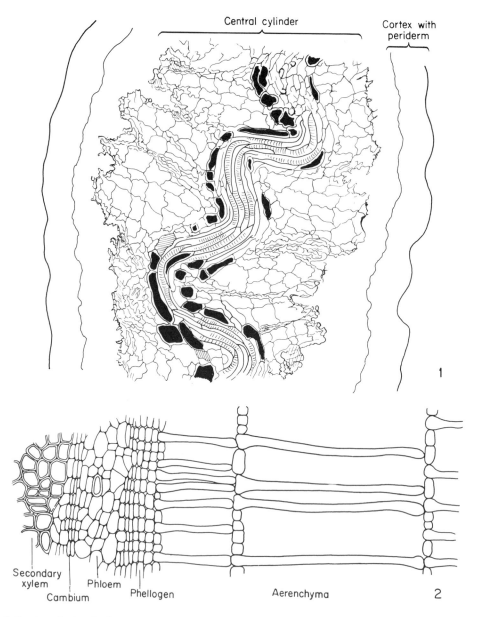

FIG. 163. 1, Portion of a longitudinal section of a contractile root of *Oxalis hirta* showing the arrangement of turgid and collapsed cells in the contracted central cylinder, and the contorted xylem. Solid black cells represent secretory cells. 2, Portion of a cross-section of a pneumatophore of *Jussiaea peruviana* in which secondary vascular tissue, phellogen and aerenchyma can be distinguished. (No. 1 adapted from Davey, 1946; no. 2 adapted from Palladin, 1914.)

roots (Rimbach, 1895, 1899, 1929, 1932; Arber, 1925; Bottum, 1941; Davey, 1946; Dittmer, 1948; Galil, 1958, 1968). Such roots are known to exist in many herbaceous dicotyledons (e.g. *Taraxacum, Medicago sativa, Daucus, Trifolium, Oxalis,* *Gymnarrhena micrantha,* sugar beet) and in many bulbous and cormous monocotyledons (e.g. *Phaedranassa chloracra, Hypoxis setosa, Bellevalia flexuosa, Gladiolus segetum, Colchicum steveni, Ixiolirion montanum, Muscari parviflorum, Allium neapolitanum*). Contractile roots or parts of roots are distinguishable from normal roots by their outer, wrinkled appearance.

Contraction of roots is a result of cellular growth and collapse. According to Arber (1925), who studied *Hypoxis setosa*, only the outer cortex is wrinkled, whereas the central cylinder and the inner cortex are unaffected. Rimbach (1899) and later investigators explain the shortening of the inner core as being due to the change in form of the inner cortical cells. These cells, according to them, increase in radial and tangential diameter and decrease in length.

Davey (1946) described the histological changes that are involved in the root contraction of *Oxalis hirta* seedlings. According to him a small amount of the contraction is due to the active growth of the phloem parenchyma cells in a transverse direction and their shortening in a longitudinal direction. The main contraction mechanism, however, is as follows: horizontal zones of cells of the uniform secondary phloem parenchyma lose their protoplasts and sap and so collapse. Alternating with these zones of collapsing cells there remain narrow zones of living turgid cells. The vertical walls of the collapsing cells fold so that the horizontal walls are brought together. Each collapsing zone becomes inclined upwards so that the diameter of the root core becomes reduced, and it tears away from the periderm and the remnants of the cortex. The cortical tissues then exhibit wrinkling. The phloem strands become contorted but remain alive. The xylem strands and their associated cambium become spirally contorted (Fig. 163, no. 1).

According to Wilson and Honey (1966) root contraction in *Hyacinthus orientalis* is associated with the growth of inner cortical cells. After these cells become fully elongated during the normal growth of the root, they expand radially and contract longitudinally. According to the above authors the contraction is a growth process since it occurs in turgid cells and is partly reversible by plasmolysis. In addition they found that the radial walls of these cells increase in area and the cells themselves increase in volume. The changes in cell shape were found to involve changes in the fine structure of the walls.

Variations of the above-mentioned mechanisms occurring in other plants were recorded by several authors (e.g. Galil, 1969; Chen, 1969; Sterling, 1972; Montenegro, 1974; Reyneke and van der Schijff, 1974; Ruzin, 1979; Deloire, 1980; Zamski *et al.*, 1983).

Aerial roots

Many tropical plants, e.g. some *Ficus* species, *Rhizophora*, epiphytic tropical Araceae, and Orchidaceae, produce roots from the stem or branches which remain free in the air. If these roots grow downwards into the soil they serve as *prop roots*. If they attach themselves to solid objects, they represent climbing or adhesive roots. Tropical orchids possess tufts of roots protruding free into the humid air. These aerial roots exhibit special adaptive anatomical characters. The cortical cells frequently contain chloroplasts, and in some cases (e.g. *Taeniophyllum*) the roots take over the main photosynthetic activity. It was suggested that absorption of water is accomplished with the aid of a specialized multiseriate epidermis, the *velamen* (Fig. 96, nos. 4, 5). The velamen is a sheath of compactly arranged dead cells, the walls of which are strengthened by band-like or reticulate thickenings and which contain many primary pit-fields. When the air is dry, these cells are filled with air, but when rain falls they become filled with water. The exodermis situated at the inner edge of the velamen is interrupted by thin-walled *passage cells* (Fig. 96, no. 5) through which the absorbed water may be transferred. In the innermost velamen cells of some orchid species occur variously shaped outgrowths of the inner tangential and radial walls ("Stabkorper" or "tilosome") (Pridgeon, 1987; Porembski and Barthlott, 1988). Special structures termed *pneumathodes* are present in the velamen. The function of these structures is to enable gas exchange during the periods when the root is saturated with moisture. The pneumathodes consist of groups of cells with usually very dense spiral wall thickenings. These groups extend, in a ray-like fashion, from the periphery of the epidermis to the endodermis. Oil droplets can be discerned in these cells. The endodermal cells that are continuous with the pneumathodes are filled with air (Gessner, 1956).

The function of water absorption by the velamen was questioned by Dycus and Knudson (1957). Based on experimental work on orchids these authors stated that the intact mature velamen and exodermis seem to be nearly impermeable to water and certain solutes. If the velamen is modified by injury when the aerial root affixes itself to a solid substrate or enters a potting medium, the root may serve as an absorbing organ. They thus concluded that the principal roles of the velamen are mechanical protection and prevention of excess loss of water from the cortex. (See also Gill and Tomlinson, 1975.)

Climbing roots occur, for instance, in many tropical species of the family Araceae and in ivy (*Hedera helix*). The climbing stems of the juvenile ivy become attached to their support by dense brush-like clusters of adventitious roots. It appears that light and contact mechanisms operate together in the process of attachment of the stem. Light was suggested to inhibit root formation on the illuminated side of the stem and to enhance it on the shade side. Light causes also growth of the branches towards the support. Contact of the stems with the support enables a thigmomorphogenetic potential in root formation to be realized (Negbi *et al.*, 1982).

Roots as organs of aeration

The root systems of trees growing in littoral swamps, in which the soil is periodically inundated and lacking in oxygen, exhibit various adaptations to their habitat. These involve features that ensure sufficient aeration and additional support.

The Rhizophoraceae are characterized by stilt roots which descend from the stems and whose lower portions only are subterranean. As the root enters into the soil, its apex undergoes a pronounced morphogenetic change. The surface layers lose the chlorophyll which is present in the aerial part of the root, and the cortical parenchyma becomes lacunated. Secondary thickening proceeds preferentially along the aerial arching portions of the roots. In the "columns", the root branches leading to the underground system of roots, there is relatively little secondary thickening. Just above

the soil the aerenchyma is much developed and is associated with abundant lenticels (Gill and Tomlinson, 1971, 1975, 1977).

In *Phoenix paludosa* there are, at the base of the stem, special roots which descend into the mud and which contain lenticels and aerenchyma produced by the phellogen.

Aerial, negatively geotropic root projections, which are termed *pneumatophores*, are commonly produced in swampy habitats. These roots serve for gas exchange. In *Jussiaea* and *Ludwigia* and some other emerged hydrophytes, they grow erect from the lower nodes. In cross-section the narrow stele of the pneumatophore is surrounded by a very wide aerenchyma which is produced by a phellogen (Fig. 163, no. 2). In *Avicennia* the pneumatophores are erect, peg-like aerial projections of the lateral subterranean roots. Here the whole pneumatophore, including its apex, is covered by a cork tissue with numerous lenticels. The cortex is penetrated by large air spaces (Chapman, 1939). In *Bruguiera eriopetala* knee-like aerial projections, which are part of the horizontal roots, are produced. The morphology and anatomy of the pneumatophores of *Amoora*, *Carapa*, and *Heritiera* were studied in detail by Groom and Wilson (1925). In these genera the aerial projections are peg-like protuberances which are produced on the upper surface of the horizontal roots by intensified cambial activity in these regions. In the three last-mentioned genera it was seen that the pneumatophores all possess lenticels, that they contain only few xylem vessels, and that the fibres are prevailingly thin-walled. The bulk of the tissue of the root consists of thin-walled fibres and parenchyma tissue (axial and ray parenchyma). At certain times this parenchyma was seen to contain much starch, and so it may sometimes act as a storage tissue. However, most of the cells have no solid contents, and therefore it is possible that all the cells, including the vessels, may act as air reservoirs. In *Amoora* and *Carapa* the intercellular spaces in the wood are no larger than usual.

Buttress roots

Many tropical trees possess remarkable *buttress roots* (Fig. 155) around the base of the trunk.

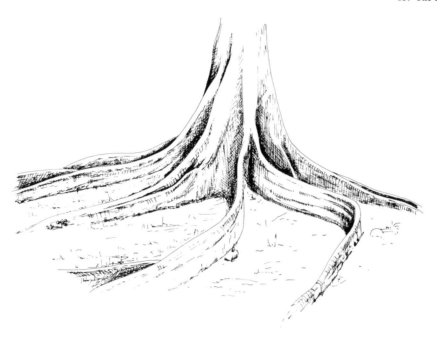

FIG. 164. Buttress roots of *Eriodendron aufractuosum*. (Adapted from a photograph in McLean and Ivimey-Cook, 1951.)

They are produced by the bases of the main roots in which the secondary thickening is asymmetrical (being mainly on the upper side), thus forming board- or plank-like structures. The anatomy of the wood of the buttresses stresses the function of supporting rather than the function of water conducting (Stahel, 1971). The percentage of fibres in the buttresses of *Khaya ivorensis* was found to be considerably higher and the percentage of vessels and parenchyma lower than in the stem. Fisher (1982) who examined the anatomy of buttresses of large numbers of plant species, found that only very few of them contained reaction wood (see Chapter 15). He therefore concluded that reaction wood is not essential for production or function of buttresses. Reaction wood was found to be very common in aerial roots.

Haustoria

Parasitic angiosperms have specialized structures, *haustoria*, which connect to the host plant and serve as channels for nutrient flow. Musselman and Dickison (1975), who worked on parasitic Scrophulariaceae, distinguish between primary haustoria, which develop by direct transformation of a root apex into a haustorium, and secondary haustoria which arise as such laterally along the root. In *Orobanche* the primary root of the seedling penetrates the host root, later secondary haustoria are also formed in most *Orobanche* species (Dörr and Kollmann, 1974). *Cuscuta* has only a temporary radicle in the seedling stage. The seedling has no leaves, its radicle withers, and the thread-like stem twines around the stem of the host and develops numerous haustoria (Fig. 165, no. 1) (Lyshede, 1985).

The stem of *Cuscuta* has four cortical layers. A flat pad develops from the outer two layers and attaches itself firmly to the epidermis of the host plant. In the centre of this pad a growth involving all four cortical layers forms the penetrating organ, the haustorium. The tip of the haustorium, which penetrates between the epidermal cells into the host cortex, consists of elongated cells—the hyphae. In the cortex the hyphae grow independently, spread through the tissues and form contacts with the vascular elements (Fig. 165, nos. 2–4). A strand of phloem and xylem then differentiates in the haustorium and establishes a connection between the vascular tissues of the host and the *Cuscuta* plant (Fig. 165, no. 1). The various forms of the hyphae are: (1) intracellular growing "searching" hyphae, (2) intercellular "searching" hyphae, and (3) "contact" hyphae which are attached to the sieve tubes of the host plant (Schumacher, 1934; Dörr, 1968).

Description of development and mature structure of haustoria of parasitic plants belonging to various angiosperm families were reviewed and described by Kuijt (1969, 1977), Musselman and Dickison (1975), Weber (1976a, b), and Renaudin and Cheguillaume (1977). Studies of the ultrastructure and development of the haustorium of *Orobanche* (Dörr and Kollmann, 1974, 1975, 1976) have shown that the haustorial cells have a cytoplasm rich in ribosomes, highly developed rough ER, and numerous organelles. No plasmodesmata interconnecting haustorial cells and host cells could be detected. In the developed haustorial sieve elements there are sieve pores and a smooth-surfaced ER occurs adjacent to the wall. No direct contact was observed, however, between the haustorial sieve elements and the host sieve elements. The "contact cells" next to the sieve elements of the host plant contain a nucleus and a dense cytoplasm (Dörr and Kollmann, 1975).

The haustorial tracheary elements are connected to those of the host by pits. These elements develop in a peculiar manner (Fig. 166). First, those parts of haustorial cell-wall oriented towards the tracheary elements of the host become thickened and develop wall protuberances, typical of transfer cells. During further differentiation the cell walls change, the protoplasts degenerate, and typical tracheary elements are formed.

Riopel and Musselman (1979) were able to induce *in vitro* the initiation of haustoria of *Agalinis* (Scrophulariaceae) by root exudate of a host plant. In hemi-parasites, e.g. in species of the Santalaceae, the vascular cambium of the haustorium is unusual in producing only a few parenchyma cells and no phloem (Fineran, 1979).

ADAPTATIONS OF ROOTS TO XERIC CONDITIONS

The following adaptations of roots to xeric conditions are well known: the form of the root system, the succulence of the roots, the development of a thick bark, the sclerification of the cortical cells, and the isolation of the vascular cylinder by periderm formation or by the necrosis of the cortical parenchyma (Hayden, 1919; Weaver, 1920; Evenari, 1938; Zohary and Orshan, 1954; Killian and Lemée, 1956). Secondary wall thickening in endodermal cells of ferns and monocotyledons was observed by van Fleet (1957) to be more prominent in plants growing in unfavourable conditions. In addition to the above, there are other characters that may be of adaptive value (Fahn, 1964).

One of the commonly accepted characters of xerophytes is the presence of well-developed xylem tissues which help rapid conduction when water is available. In this connection it is of interest to mention the roots of *Retama raetam*. In this plant, which grows in sandy soils and wadi beds, there are in addition to the usual vertical and diageotropic roots, horizontal roots which have been observed to reach a length of up to 10 m. The vessel members of the vertical roots are narrower and shorter than those of the diageotropic roots and even more so than those of the horizontal roots. Moreover, in the latter type of root a gradient in the length and width of vessel members exists from the distal ends of the roots to the part where they are attached to the vertical roots. The advantage of this feature may be in its compensating effect on what would otherwise be a much more pronounced gradient in suction pressure from the stem toward the distal parts of the roots. Both the greater width and length of the vessel members in the distal parts of the horizontal roots presumably ensure a more efficient flow of water in these very long, horizontal roots. This feature is doubtless of great importance as the horizontal roots take up water from the upper layers, which in sandy soils, drain rapidly.

Two other interesting features have been observed in the primary roots of desert plants. Firstly, in the roots of plants growing under extreme desert conditions the number of cortical

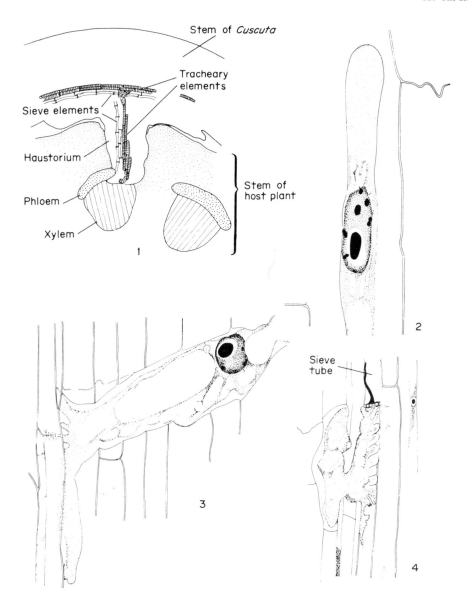

FIG. 165. *Cuscuta*. 1, Schematic drawing showing relation between parasite and host plants. 2, A growing end portion of a haustorial cell seen in a parenchyma cell. 3, A haustorial cell with a "foot" which started to develop while reaching a sieve tube. 4, A "foot" of a haustorial cell attached to the wall of a sieve tube. 2–4, × 600. (Nos. 2–4 adapted from Schumacher, 1934.)

layers is reduced. The advantage of this feature may be that it shortens the distance between the soil and the stele. Secondly, it has been observed that the Casparian strips are much wider in plants growing in extremely dry habitats and in salt marshes as compared with those growing under mesophytic conditions. In extreme cases the Casparian strips were seen to occupy the entire radial and transverse walls of the endodermal cells. As has already been mentioned, it appears that the endodermis represents a semi-permeable barrier that controls the movement of solutes into the

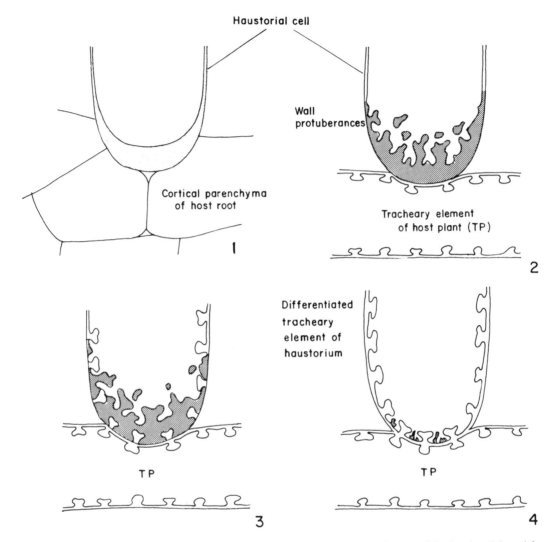

Haustorial cell

Wall
protuberances

Cortical parenchyma
of host root

Tracheary element
of host plant (TP)

1

2

Differentiated
tracheary
element of
haustorium

TP

TP

3

4

FIG. 166. Schematic drawing of four development stages of a haustorial tracheary element of *Orobanche*. (Adapted from Dörr and Kollmann, 1976.)

central cylinder. If this is correct, the endodermis would function more efficiently when the protoplasts are attached to larger portions of the radial and transverse walls of the endodermal cells. Such a feature seems, therefore, to be an adaptive character in plants growing under saline conditions.

Roots of many xeric grasses growing on moving sand (e.g. *Andropogon foveolatus, Aristida* spp. *Elionurus hirsutus, Panicum turgidum* and *Sporobolus spicata*) possess sand grain sheaths (*rhizosheaths*) (Volkens, 1887; Price, 1911; Wullstein *et*

al., 1979). The sand grains become attached to the root hairs by secreted mucilage. As the root hairs persist the sheath covers the whole root. Volkens suggested that the function of the rhizosheaths in the xeric grasses is similar to that of the peripheral cork layers in roots of dicotyledons. Both structures prevent the loss of water from the inner root tissue. Because of the high absorption power of mucilage for water, Price expressed the view that the strong connection between the mucilage and the root hairs must also promote water absorption. According to Wullstein *et al.*, nitrogen fixation (acytylene reduction) is associated with the occurrence of rhizosheaths in the xeric grasses examined by them.

In some xerophytes, small proturberances consisting of very small rootlets, stunted in growth (in *Zygophyllum dumosum*), or small clusters of appendages with rootlets (in *Opuntia arenaria*; termed "spurs"), occur on the ordinary roots. Near the base of the stunted rootlets, primordia of new rootlets are present. From these primordia variously sized adventitious roots may develop (Ginzburg, 1964; Boke, 1979). The development of this type of root seems to be adaptive by the possibility of rapid production of roots when water is available.

In many taxa of the Proteaceae, very dense clusters of rootlets occur at certain sites of the ordinary long roots. The rootlets, plus the portion of the parent root bearing them comprise the structures known as the *proteoid root*. In *Hakea*, for instance, the rootlets occur in 3–7 longitudinal rows. It has been suggested that concentration of nutrients in the soil, especially nitrogen availability largely determines the relative contribution of proteoid roots to the root system of *Hakea* species (Lamont, 1972a, b, 1973; Dell *et al.*, 1980).

CONNECTION BETWEEN THE VASCULAR SYSTEMS OF THE ROOT AND STEM

The primary vascular systems of the root and stem are distinguished from one another, as has already been described, by structure and by the direction of the radial development. The protoxylem in the root is exarch while that of the stem is endarch. The xylem and phloem are arranged alternately in the root, while the arrangement in the stem is usually collateral. Because of the differences in the functional design of the purely axial structure of the root and that of the appendage-bearing stem, there are necessarily basic differences in the pattern of the vascular systems of these two organs. The pattern of the vascularization of the stem, unlike that of the root, is determined by the presence of the leaves. At the level where the vascular systems of the root and stem

meet, they must necessarily become adapted to one another. This region of the plant axis where one system gradually passes into the other has been termed the *transition region*. As has already been explained in an earlier chapter, in the embryo the shoot apex is found on one side of the hypocotyl and the root apex on the other. Therefore it is in the hypocotyl, and sometimes also in the lowermost internodes, that the one type of vascular system must change into the other.

The structure of the transition region is complex and differs in different plant species. Only general explanations are usually given of the transition between the conducting systems of the root and stem, and not of the transition between the conducting system of the root and the conducting system serving the cotyledons and the shoot above them.

The explanations given by many research workers and quoted in a large number of textbooks are based on the study only of serial sections of seedlings in which primary vascular tissue is fully developed. This method of investigation caused the workers involved to conclude that the separate strands of phloem twist and that their orientation is inverted during their passage through the hypocotyl and into the part of the stem above the cotyledons. These conclusions have not been confirmed by more recent ontogenetic studies, which are based on tracing the connection between the simple, axial conducting system of the root on the one hand, and the complex vascular system of the shoot, on the other.

In *Daucus carota* (Esau, 1940), for example, three bundles enter into each cotyledon. The median bundle of these three traces consists of a strand of exarch xylem which is continuous with the protoxylem of the root and is accompanied laterally by two phloem strands. In this strand centripetal differentiation can be followed for some distance into the cotyledon. In contrast with this, each of the lateral cotyledonary traces is collateral with external phloem and endarch xylem on the inside. Throughout the entire length of these bundles the differentiation of the xylem proceeds in a centrifugal direction. These bundles originate from the central portion of the diarch xylem of the root. Therefore, in the case of

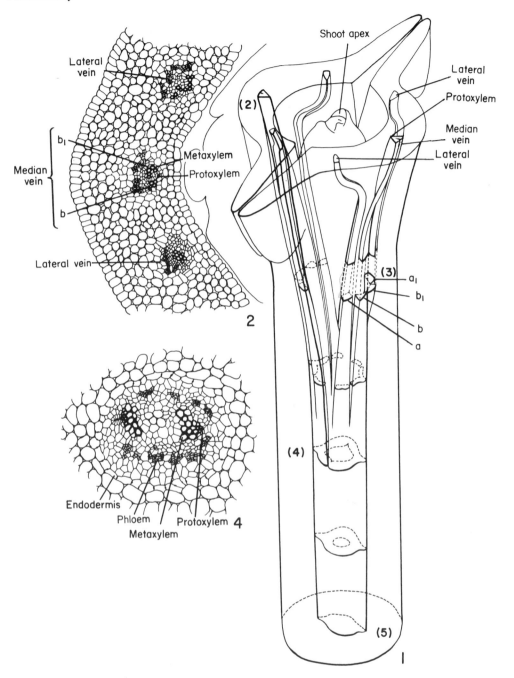

FIG. 167 (continued on opposite page).

Daucus, there is a definite continuation, without any inversion, between the primary vascular system of the cotyledons and that of the common axis of the hypocotyl–radicle. In *Daucus* the apical

meristem of the epicotyl commences to produce leaf primordia after secondary thickening is initiated in the region of the hypocotyl–radicle. The collateral traces of these foliage leaves join with

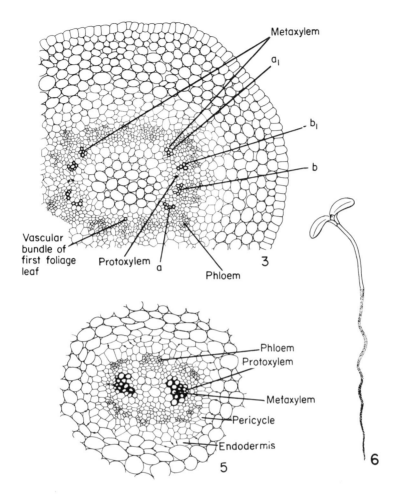

FIG. 167. Transition region in a seedling of *Linum usitatissimum*. 1, Three-dimensional diagram of the hypocotyl and cotyledonary node. 2–5, Portions of cross-sections at various levels of the hypocotyl and of one cotyledon. The numerals in no. 1 indicate the levels at which the cross-sections were made. 6, An entire seedling at the stage of development corresponding to that depicted in the other drawings. For further explanations, see text. (Adapted from Crooks, 1933.)

the secondary vascular tissues of the hypocotyl and root.

According to Crooks (1933) the transition between the vascular systems of the stem and root in *Linum* (Fig. 167), which also has a diarch root system, is as follows.

In the lower part of the hypocotyl the stele is similar to that of the root. A little higher a parenchymatous pith develops in the centre of the stele and at the same time each of the phloem strands divides, resulting in four strands (Fig. 167, no. 4). The development of the metaxylem elements gradually passes to the sides of the protoxylem

and so four groups of metaxylem are obtained, each of which is in contact with the inner side of a phloem strand. About half-way along the hypocotyl each of the four metaxylem strands, together with its external accompanying phloem strand, divides into two, so that eight collateral bundles, which constitute cotyledonary traces, are obtained. Still higher, just below the cotyledonary node, the eight traces become arranged in two opposing groups of four (Fig. 167, no. 3, a, a_1 and b, b_1). At the cotyledonary node the two lateral traces of each group (a and a_1) become greatly separated from the inner two and they enter into the cotyledons where they form the two lateral veins. Still higher, the middle two metaxylem strands (b and b$_1$) again approach one another on the outer side of the protoxylem, so that at the base of the cotyledons the primary xylem is endarch. The two metaxylem strands, which had approached one another, eventually fuse and form the median vein of each of the cotyledons. The phloem groups of the median vein fuse at a higher level within the cotyledon (Fig. 167, no. 2).

The phloem, which is associated with the two leaves of the epicotyl, is already differentiated in the lower part of the hypocotyl at the level where the two strands of the root phloem diverge to form four strands. However, the xylem of the epicotyl develops somewhat later than that of the cotyledons and hypocotyl. The development of the xylem of the collateral bundles of the first foliage leaves above the cotyledonary node is endarch. These bundles develop basipetally into the hypocotyl where they may join with the strands of metaxylem and phloem or where they may terminate blindly within the parenchyma. In the axes of older seedlings (more than 1 week old), the protoxylem is the first to be stretched and crushed. Later, the earliest-formed metaxylem is obliterated in the middle and upper portions of the hypocotyl. Finally, nearly all the primary xylem disappears, but fragments of it may be distinguished in the parenchyma of the transition region, especially in longitudinal sections, for some time or even throughout the life of the plant.

In certain plants, e.g. *Beta*, in which the root is also diarch, one trace of double nature enters each of the cotyledons. Each of these cotyledonary traces consists of two bundles which are partially fused along the protoxylem. In this case the protoxylem is also brought closer to the centre by the change in the position of differentiation of the metaxylem in the upper part of the hypocotyl and the bundles become completely collateral and endarch only in the cotyledons (Artschwager, 1926; Hayward, 1938). The double nature of the cotyledonary trace has phylogenetic importance (Bailey, 1956).

In *Medicago sativa* the root is usually triarch. In transition to the hypocotyl one of the xylem strands becomes smaller than the other two. The two large strands become situated one opposite the other as in a diarch arrangement. The small third bundle is at right-angles to the two large ones. Higher up a fourth strand of protoxylem develops opposite the small third one and so the stele becomes tetrarch at the base of the hypocotyl. During this process a fourth strand of phloem is also developed and the phloem strands alternate with the xylem strands. Higher up in the hypocotyl pith is developed. The phloem strands divide so that they number eight; these strands become orientated so as to take up a position almost collateral to the xylem strands. The differentiation of the metaxylem elements, a short distance below the cotyledonary node, takes place in such a position that two V-shaped groups, as seen in cross-section, are formed. The protoxylem is located in the angle between the two arms of the V. The two triads, i.e. the xylem groups together with their associated phloem, constitute the cotyledonary bundles (Hayward, 1938).

The length of the transition region in dicotyledons differs—in some plants it is short while in others it is long. In some plants the changes in orientation are gradual and continue throughout the entire length of the hypocotyl, while in others these changes are restricted to the upper portion of the hypocotyl alone. In the latter case the hypocotyl is referred to as being of root structure. The transition region is longest in seedlings with subterranean cotyledons (hypogeal germination) as it extends for one or more nodes above the cotyledons. Vascular development in the transition region in *Populus deltoides* has recently been described (Sundberg, 1983).

In monocotyledons the transition between the vascular tissues of the root and stem is affected by the presence of a single cotyledon and by the shortness of the lowermost internodes (Esau, 1953). In many monocotyledons part of the vascular system of the root is connected with the vascular system of the cotyledon, while part is connected with the vascular tissues of the first foliage leaf. In both these cases the connecting vascular strands exhibit features typical of transition. However, in a small number of monocotyledons the transition takes place between the root and cotyledon alone as is common in the dicotyledons (Arber, 1925).

As an example of a transition region in monocotyledons we shall cite that of the seedling of *Triticum* (Fig. 157, no. 3) as described by Boyd and Avery (1936). The polyarch vascular cylinder of the root is connected to that of the leaves by the presence of plate-like vascular tissue which is present below the insertion of the scutellum. This plate is termed the *nodal plate*. Separate bundles arise acropetally from the nodal plate. These bundles are irregularly arranged in the basal portions where they exhibit transitional features. Higher up these bundles continue to branch and produce a cylinder of bundles with endarch xylem and with collateral arrangement of xylem and phloem. This system consists of bundles that enter the scutellum, coleoptile and the first two foliage leaves.

The transition region of the gymnosperms is generally similar to that common in the dicotyledons in which the connection occurs primarily between the root and the cotyledons, but it is more complex than in the dicotyledons because of the increased number of cotyledons (Boureau, 1954).

The complex structure of the transition region between the root and shoot is apparently, as already suggested by Esau (1961), the result of the meeting of influences of two morphogenetic centres—the shoot apex and the root apex. These influences are especially strong in the embryonic stage of the plant where the two centres (apices) are close to each other. The differentiation in that region where the two opposite trends meet should be intermediate between the two. During the growth of the seedling the two poles move further apart and the influence of each of them on the region near the other is weakened. Accordingly, the extension of the transition region into one or more internodes above the cotyledons in seedlings in which the germination is hypogeal may be explained by the extended influence of the root apex on the basal internodes of the stem as a result of the retarded growth of the hypocotyl in such seedlings.

REFERENCES

ANCIBOR, E. (1979) Systematic anatomy of vegetative organs of the Hydrocharitaceae. *J. Linn. Soc. Bot.* **78**: 237–266.

ARBER, A. (1925) *Monocotyledons. A morphological study.* Cambridge University Press, Cambridge.

ARTSCHWAGER, E. (1926) Anatomy of the vegetative organs of the sugar beet. *J. Agric. Res.* **33**: 143–76.

AVERY, G. S., JR. (1930) Comparative anatomy and morphology of embryos and seedlings of maize, oats, and wheat. *Bot. Gaz.* **89**: 1–39.

BAILEY, I. W. (1956) Nodal anatomy in retrospect. *J. Arnold Arab.* **37**: 269–87.

BAIRD, M. L. VIRGINIA, R. A. and WEBSTER, B. B. (1985) Development of root nodules in a woody legume, *Prosopis glandulosa* Torr. *Bot. Gaz.* **146**: 39–43.

BECKING, J. H. (1975) Root nodules in non-legumes. In: *The Development and Function of Roots* (eds. J. G. Torrey and D. T. Clarkson). Academic Press, London, pp. 507–66.

BELL, J. K. and MCCULLY, M. E. (1970) A histological study of lateral root initiation and development in *Zea mays*. *Protoplasma* **70**: 179–205.

BERGERSEN, F. J. (1982) *Root Nodules of Leguminosae: Structure and Functions.* Research Studies Press, New York.

BÖCHER, T. W. and OLESEN, P. (1978) Structural and ecophysiological pattern in the xero-halophytic C_4 grass *Sporobolus rigens* (Tr.) Desv. *Biol. Skr.* **22** (3): 1–48.

BOKE, N. H. (1979) Root glochids and root spurs of Opuntia arenaria (Cactaceae). *Am. J. Bot.* **66**: 1085–1092.

BOND, L. (1948) Origin and developmental morphology of root nodules of *Pisum sativum*. *Bot. Gaz.* **109**: 411–34.

BONFANTE-FASOLO, P. and SCANNERINI, S. (1977a) A cytological study of the vesicular-arbuscular mycorrhiza in *Ornithogalum umbellatum* L. *Allionia* **22**: 5–21.

BONFANTE-FASOLO, P. and SCANNERINI, S. (1977b) Cytological observations on the mycorrhiza *Endogone flammicorona-Pinus strobus*. *Allionia* **22**: 23–34.

BONNETT, H. T., JR. (1968) The root endodermis: fine structure and function. *J. Cell Biol.* **37**: 199–205.

BOTTUM, F. R. (1941) Histological studies on the root of *Melilotus alba*. *Bot. Gaz.* **103**: 132–45.

BOUREAU, E. (1954) *Anatomie végétale*, Vol. I. Presses Universitaires de France, Paris.

BOWEN, G. D. (1973) Mineral nutrition of ectomycorrhizae. In: *Ectomycorrhizae: Their Ecology and Physiology* (eds. G. C. Marks and T. T. Kozlowski). Academic Press, New York, pp. 151–205.

BOYD, L. and AVERY, G. S., JR. (1936) Grass seedling anatomy: the first internode of *Avena* and *Triticum*. *Bot. Gaz.* **97**: 765–79.

BROUWER, R. (1959) Diffusible and exchangeable Rb ions in pea roots. *Acta bot. neerl.* **8**: 68–76.

BRUNDRETT, M. C., PICHE. Y. and PETERSON, R. L. (1985) A developmental study of the early stages in vesicules—arbuscular mycorrhiza formation. *Can. J. Bot.* **63**: 184–194.

CARLSON, M. C. (1938) The formation of nodal adventitious roots in *Salix cordata*. *Am. J. Bot.* **25**: 721–25.

CARLSON, M. C. (1950) Nodal adventitious roots in willow stems of different ages. *Am. J. Bot.* **37**: 555–61.

CHABOUD, A. and ROUGIER, M. (1986) Ultrastructural study of the maize epidermal root surface. I. Preservation and extent of the mucilage layer. *Protoplasma* **130**: 73–79.

CHAPMAN, V. J. (1939–45) 1939 Cambridge University Expedition to Jamaica: Part 3, The morphology of *Avicennia nitida* Jacq. and the function of its pneumatophores. *J. Linn. Soc., Bot.* **52**: 487–533.

CHEN, S. (1969) The contractile roots of *Narcissus*. *Ann. Bot.* **33**: 421–6.

CHRISTODOULAKIS, N. S. and PSARAS, G. K. (1987) Stomata on the primary root of *Ceratonia siliqua*. *Ann. Bot.* **60**: 295–297.

CLARK, L. H. and HARRIS, W. H. (1981) Observations on the root anatomy of rice (*Oryza sativa* L.). *Am. J. Bot.* **68:** 154–161.

CLARKSON, D. T. and ROBARDS, A. W. (1975) The endodermis, its structural development and physiological role. In: *The Development and Function of Roots* (eds. J. G. Torrey and D. T. Clarkson). Academic Press, London, pp. 415–36.

CLARKSON, D. T., ROBARDS, A. W. and SANDERSON, J. (1971) The tertiary endodermis in barley roots: fine structure in relation to radial transport of ions and water. *Planta* **96:** 292–305.

CLOWES, F. A. L. (1951) The structure of mycorrhizal roots of *Fagus sylvatica*. *New Phytol.* **50:** 1–16.

CLOWES, F. A. L. (1978) Chimeras and the origin of lateral root primordia in *Zea mays*. *Ann. Bot.* **42:** 801–7.

CLOWES, F. A. L. (1981) Cell proliferation in ectotrophic mycorrhizas of *Fagus sylvatica* L. *New Phytol.* **87:** 547–555.

CLOWES, F. A. L. (1985) Origin of epidermis and development of root primordia in *Pistia*, *Hydrocharis* and *Eichhornia*. *Ann. Bot.* **55:** 849–857.

CLOWES, F. A. L. and JUNIPER, B. E. (1968) *Plant Cells*. Blackwell, Oxford.

CLOWES, F. A. L. and WOOLSTON, R. E. (1978) Sloughing of root cap cells. *Ann. Bot.* **42:** 83–89.

CORMACK, R. G. H. (1949) The development of root hairs in angiosperms. *Bot. Rev.* **15:** 583–612.

CORMACK, R. G. H., LEMAY, P., and MACLACHLAN, G. A. (1963) Calcium in root-hair wall. *J. Exp. Bot.* **14:** 311–15.

COSSMANN, K. F. (1940) Citrus roots: their anatomy, osmotic pressure and periodicity of growth. *Palest. J. Bot., Rehovot ser.* **3:** 65–104.

COX, G. and SANDERS, F. (1974) Ultrastructure of the host-fungus interface in a vesicular-arbuscular mycorrhiza. *New Phytol.* **73:** 901–12.

CROOKS, D. M. (1933) Histological and regenerative studies on the flax seedling. *Bot. Gaz.* **95:** 209–39.

CUTLER, D. F., RUDALL, P. J., GASSON, P. E. and GALE, R. M. O. (1987) *Root Identification Manual of Trees and Shrubs*. Chapman and Hall, London.

DART, P. J. (1975) Legume root nodule initiation and development. In: *The Development of Roots* (eds. J. G. Torrey and D. T. Clarkson). Academic Press, London, pp. 467–506.

DAVEY, A. J. (1946) On the seedling of *Oxalis hirta* L. *Ann. Bot.* **10:** 237–56.

DELL, B., KUO, J. and THOMSON, G. J. (1980) Development of proteoid roots in *Hakea obliqua* R.Br. (Proteaceae) grown in water culture. *Aust. J. Bot.* **28:** 27–37.

DELOIRE, A. (1980) Les racines tractrices de l'*Allium polyanthum* Roem. et Schult: une étude morphologique, anatomique et hitoenzymologique. *Rev. Cytol. Biol. végét.-Bot.* **3:** 383–90.

DITTMER, H. J. (1948) A comparative study of the number and length of roots produced in nineteen angiosperm species. *Bot. Gaz.* **109:** 354–8.

DÖRR, J. (1968) Zur lokalisierung von Zellkontakten zwischen *Cuscuta odorata* und verschiedenen höheren Wirtspflanzen. *Protoplasma* **65:** 435–48.

DÖRR, I. and KOLLMANN, R. (1974) Strukturelle Grundlage des Parasitismus bei *Orobanche*. I. Wachstum der Haustorialzelln im Wirtsgewebe. *Protoplasma* **80:** 245–59.

DÖRR, I. and KOLLMANN, R. (1975) Strukturelle Grundlagen des Parasitismus bei *Orobanche*. II. Die Differenzierung der Assimilat-Leitungsbahn im Haustorialgewebe. *Protoplasma* **83:** 185–99.

DÖRR, I. and KOLLMANN, R. (1976) Strukturelle Grundlagen des Parasitismus bei *Orobanche*. III. Die Differenzierung des xylemanschlusses bei *O. crenata*. *Protoplasma* **89:** 235–49.

DYCUS, A. M. and KNUDSON, L. (1957) The role of the velamen of the aerial roots of orchids. *Bot. Gaz.* **119:** 78–87.

ESAU, K. (1940) Developmental anatomy of the fleshy storage organ of *Daucus carota*. *Hilgardia* **13:** 175–226.

ESAU, K. (1953) *Plant Anatomy*. Wiley, New York.

ESAU, K. (1961) *Anatomy of Seed Plants*. Wiley, New York.

ESAU, K. (1965) *Plant Anatomy*, 2nd edn. Wiley, New York.

EVENARI, M. (1938) The physiological anatomy of the transpiratory organs and the conducting system of certain plants typical of the Wilderness of Judaea. *J. Linn. Soc., Bot.* **51:** 389–407.

FAHN, A. (1964) Some anatomical adaptations of desert plants. *Phytomorphology* **14:** 93–102.

FINERAN, B. A. (1979) Ultrastructure of differentiating graniferous tracheary elements in the haustorium of *Exocarpus bidwillii* (Santalaceae). *Protoplasma* **98:** 199–221.

FISHER, J. B. (1982) A survey of buttresses and aerial roots of tropical trees for presence of reaction wood. *Biotropica* **14:** 56–61.

FJELL, I. (1985a) Preformation of root primordia in shoots and root morphogenesis in *Salix viminalis*. *Nord. J. Bot.* **5:** 357–376.

FJELL, I. (1985b) Morphogenesis of root cap in adventitious roots of *Salix viminalis*. *Nord. J. Bot.* **5:** 555–573.

FOSTER, R. C. and ROVIRA, A. D. (1976) Ultrastructure of wheat rhizosphere. *New Phytol.* **76:** 343–52.

GALIL, J. (1958) Physiological studies on the development of contractile roots in geophytes. *Bull. Res. Counc. Israel* **6D:** 221–36.

GALIL, J. (1968) Vegetative dispersal in *Oxalis cernua*. *Am. J. Bot.* **55:** 68–73.

GALIL, J. (1969) On the lateral-contracting root of *Colchicum steveni*. *Beitr. Biol. Pfl.* **46:** 315–22.

GESSNER, F. (1956) Der Wasserhaushalt der Epiphyten und Lianen. In: W. Ruhland, *Handbuch der Pflanzenphysiologie* **3:** 915–50.

GILL, A. M. and TOMLINSON, P. B. (1975) Aerial roots: an array of forms and functions. In: *The Development and Function of Roots* (eds. J. G. Torrey and D. T. Clarkson). Academic Press, London, pp. 237–60.

GILL, A. M. and TOMLINSON, P. B. (1971) Studies on the growth of red mangrove (*Rhizophora mangle* L.). 2. Growth and differentiation of aerial roots. *Biotropica* **3:** 63–77.

GILL, A. M. and TOMLINSON, P. B. (1977) Studies on the growth of red mangrove (*Rhizophora mangle* L.). 4. The adult root system. *Biotropica* **9:** 145–155.

GINZBURG, C. (1964) Ecological anatomy of roots. Ph.D. thesis of The Hebrew University of Jerusalem (In Hebrew with an English summary).

GINZBURG, C. (1967) Organization of the adventitious root apex in *Tamarix aphylla*. *Am. J. Bot.* **54:** 4–8.

GIROUARD, R. M. (1967a) Initiation and development of adventitious roots in stem cuttings of *Hedera helix*: anatomical studies of juvenile growth phase. *Can. J. Bot.* **45:** 1877–81.

GIROUARD, R. M. (1967b) Initiation and development of adventitious roots in stem cuttings of *Hedera helix*: anatomical studies of mature growth phase. *Can. J. Bot.* **45:** 1883–6.

GRAMBERG, J. J. (1971) The first stages of the formation of adventitious roots in petioles of *Phaseolus vulgaris*. *Proc.*

Nederl. Akad. Wet., ser. C, **74**: 42–45.

GREAVES, M. P. and DARBYSHIRE, J. F. (1972) The ultra-structure of the mucilaginous layer on plant roots. *Soil Biol. Biochem.* **4**: 443–9.

GRIFFITHS, H. J. and AUDUS, L. J. (1964) Organelle distribution in the statocyte cells of the root-tip of *Vicia faba* in relation to geotropic stimulation. *New Phytol.* **63**: 319–33.

GROOM, P. and WILSON, S. E. (1925) On the pneumatophores of paludal species of *Amoora, Carapa* and *Heritiera. Ann. Bot.* **39**: 9–24.

GUTTENBERG, H. V. (1940) Der primäre Bau der Angiospermenwurzel. In: K. Linsbauer, *Handbuch der Pflanzenanatomie*, Bd. 8, Lief. 39. Gebr. Borntraeger, Berlin.

GUTTENBERG, H. V. (1941) Der primäre Bau der Gymnospermenwurzel. In: K. Linsbauer, *Handbuch der Pflanzenanatomie*, Bd. 8, Lief. 41. Gebr. Borntraeger, Berlin.

GUTTENBERG, H. V. (1943) Die physiologischen Scheiden. In: K. Linsbauer, *Handbuch der Pflanzenanatomie*, Bd. 5, Lief. 42. Gebr. Borntraeger, Berlin.

GUTTENBERG, H. V. (1960) Grundzüge der Histogenese höheren Pflanzen: I. Die Angiospermen. In: K. Linsbauer, *Handbuch der Pflanzenanatomie*. Spez. Teil, Bd. 8, T. 3. Gebr. Borntraeger, Berlin.

HABERLANDT, G. (1918) *Physiologische Pflanzenanatomie*, 5th edn. W. Engelmann, Leipzig.

HANDLEY, R. and OVERSTREET, R. (1963) Uptake of strontium by roots of *Zea mays. Pl. Physiol.* **38**: 180–4.

HARLEY, J. L. (1969) *The Biology of Mycorrhiza.* Leonard Hill, London.

HASKINS, M. L. and HAYDEN, W. J. (1987) Anatomy and affinities of *Penthorum. Am. J. Bot.* **74**: 164–177.

HAYDEN, A. (1919) The ecologic subterranean anatomy of some plants of a prairie province in Central Iowa. *Am. J. Bot.* **6**: 87–105.

HAYWARD, H. E. (1938) *The Structure of Economic Plants.* Macmillan, New York.

HAYWARD, H. E. and LONG, E. M. (1942) The anatomy of the seedling and roots of the Valencia orange. *USDA Tech. Bull.* 786.

HEIMSCH, C. (1951) Development of vascular tissues in barley roots. *Am. J. Bot.* **38**: 523–37.

HEIMSCH, C. (1960) A new aspect of cortical development in roots. *Am. J. Bot.* **47**: 195–201.

HEISEY, R. M., DELWICHE, C. C., VIRGINIA, R. A., WRONA, A. F., and BRYAN, B. A. (1980) A new nitrogen-fixing non-legume *Chamaebatia foliolosa* (Rosaceae). *Am. J. Bot.* **67**: 429–31.

JOST, L. (1932) Die Determinierung der Wurzelstruktur. *Z. Bot.* **25**: 481–522.

JUNIPER, B. E. and FRENCH, A. (1973) The distribution and redistribution of endoplasmic reticulum (ER) in geoperceptive cells. *Planta* **109**: 211–24.

JUNIPER, B. E. and PASK, G. (1973) Directional secretion by Golgi bodies in maize root cells. *Planta* **109**: 225–31.

JUSTIN, S. H. F. W. and ARMSTRONG, W. (1987) The anatomical characteristics of roots and plant response to soil flooding. *New Phytol.* **106**: 465–495.

KILLIAN, T. and LEMÉE, G. (1956) Les xérophytes: leur économie d'eau. In: W. Ruhland, *Handbuch der Pflanzenphysiologie* **3**: 787–824.

KRAMER, D., ANDERSON, W. P. and Preston, J. (1978) Transfer cells in the root epidermis of *Atriplex hastata* L. as a response to salinity: a comparative cytological and X-ray microprobe investigation. *Aust. J. Plant Physiol.* **5**:

739–47.

KRAMER, D., RÖMHELD, V., LANDSBERG, E. and MARSCHNER, H. (1980) Induction of transfer-cell formation by iron deficiency in the root epidermis of *Helianthus annuus* L. *Planta* **147**: 335–9.

KRAMER, P. J. (1945) Absorption of water by plants. *Bot. Rev.* **11**: 310–55.

KRAMER, P. J. (1946) Absorption of water through suberized roots of trees. *Pl. Physiol.* **21**: 37–41.

KRAUSS, B. H. (1949) Anatomy of the vegetative organs of the pineapple *Ananas comosus* (L.) Merr.: III, The root and the cork. *Bot. Gaz.* **110**: 550–87.

KUIJT, J. (1969) *The Biology of Parasitic Flowering Plants.* University of California Press, Berkeley.

KUIJT, J. (1977) Haustoria of phanerogamic parasites. *Ann. Rev. Phytopathol.* **17**: 91–118.

LAMONT, B. (1972a) The effect of soil nutrients on the production of proteoid roots by *Hakea* species. *Aust. J. Bot.* **20**: 27–40.

LAMONT, B. (1972b) The morphology and anatomy of proteoid roots in the genus *Hakea. Aust. J. Bot.* **20**: 155–174.

LAMONT, B. (1973) Factors affecting the distribution of proteoid roots within the root system of two *Hakea* species. *Aust. J. Bot.* **21**: 165–187.

LEPPARD, G. G. (1974) Rhizoplane fibrils in wheat: demonstration and derivation. *Science, NY* **185**: 1066–7.

LESHEM, B. (1965) The annual activity of intermediary roots of the aleppo pine. *Forest Sci.* **11**: 291–8.

LIBBENGA, K. R. and HARKES, P. A. A. (1973) Initial proliferation of cortical cells in the formation of root nodules in *Pisum sativum* L. *Planta* **114**: 17–28.

LIBBENGA, K. R., VAN IREN, F., BOGERS, R. J. and SCHRAAG-LAMERS, M. F. (1973) The role of hormones and gradients in the initiation of cortex proliferation and nodule formation in *Pisum sativum* L. *Planta* **114**: 29–39.

LYSHEDE, O. B. (1985) Morphological and anatomical features of *Cuscuta pedicellata* and *C. campestris. Nord. J. Bot.* **5**: 65–77.

MACKENZIE, K. A. D. (1979) The development of the endodermis and *phi* layer of apple roots. *Protoplasma* **100**: 21–32.

MACKENZIE, K. A. D., HOWARD, B. H. and HARRISON-MURRAY, R. S. (1986) The anatomical relationship between cambial regeneration and root initiation in wounded winter cuttings of the apple rootstock M. 26. *Ann. Bot* **58**: 649–661.

MCARTHUR, I. C. S. and STEEVES, T. A. (1969) On the occurrence of root thorns on a Central American palm. *Can. J. Bot.* **47**: 1377–82.

MCLEAN, R. C. and IVIMEY-COOK, W. R. (1951) *Textbook of Theoretical Botany*, Vol. 1. Longmans, London.

MALLORY, T. E., CHIANG, S., CUTTER, E. G., and GIFFORD, E. M., JR. (1970) Sequence and pattern of lateral root formation in five selected species. *Am. J. Bot.* **57**: 800–9.

MARKS, G. C. and FOSTER, R. C. (1973) Structure, morphogenesis and ultrastructure of ectomycorrhizae. In: *Ectomycorrhizae: Their Ecology and Physiology* (eds. G. C. Marks and T. T. Kozlowski). Academic Press, New York, pp. 1–41.

MARX, D. H. (1973) Mycorrhizae and feeder root diseases. In: *Ectomycorrhizae: Their Ecology and Physiology* (eds. G. C. Marks and T. T. Kozlowski). Academic Press, New York, pp. 351–82.

MEXAL, J. G., REID, C. P. P., and BURKE, E. J. (1979) Scanning electron microscopy of lodgepole pine roots. *Bot. Gaz.* **140:** 318–23.

MEYER, F. H. (1973) Distribution of ectomycorrhizae in native and man-made forests. In: *Ectomycorrhizae: Their Ecology and Physiology* (eds. G. C. Marks and T. T. Kozlowski). Academic Press, New York, pp. 79–105.

MONTENEGRO, G. R. (1974) Desarrollo de ráices contráctiles en *Hippeastrum chilense*, geofita del matarral chileno. *Acta Cientif. Venezolana* **25:** 82–86.

MOORE, R. and McCLELEN, C. E. (1983) A morphometric analysis of cellular differentiation in the root cap of *Zea mays. Am. J. Bot.* **70:** 611–617.

MOORE, R. and EVANS, M. L. (1986) How roots perceive and respond to gravity. *Am. J. Bot.* **73:** 574–587.

MOORE, R., FONDREN, W. M. and MARCUM, H. (1987) Characterization of root agravitropism induced by genetic, chemical and developmental constraints. *Am. J. Bot.* **74:** 329–336.

MORRÉ, D. J., JONES, D. D., and MOLLENHAUER, H. H. (1967) Golgi apparatus mediated polysaccharide secretion by outer root cap cells of *Zea mays.* I. Kinetics and secretory pathway. *Planta* **74:** 286–301.

MUSSELMAN, L. J. and DICKISON, W. C. (1975) The structure and development of the haustorium in parasitic Scrophulariaceae. *J. Linn. Soc. Bot.* **70:** 183–212.

NEGBI, M., ZAMSKI, E. and ZE'EVI, O. (1982) Photo- and thigmomorphogenetic control of the attachment of the ivy (Hedera helix L.) to its support. *Z. Pflanzenphysiol.* **108:** 9–15.

NEWCOMB, W. (1976) A correlated light and electron microscopic study of symbiotic growth and differentiation in *Pisum sativum* root nodules. *Can. J. Bot.* **54:** 2163–2186.

NORTHCOTE, D. H. and PICKETT-HEAPS, J. D. (1966) A function of the Golgi apparatus in polysaccharide synthesis and transport in root-cap cells of wheat. *Biochem. J.* **98:** 159–67.

OGURA, Y. (1938) Anatomie der Vegetationsorgane der Pteridophyten. In: K. Linsbauer, *Handbuch der Pflanzenanatomie*, Bd. 7, Lief. 36. Gebr. Borntraeger. Berlin.

PANT, D. D. (1943) On the morphology and anatomy of the root system in *Asphodelus tenuifolius* Cavan. *J. Ind. Bot. Soc.* **22:** 1–26.

PATE, J. S., GUNNING, B. E. S., and BRIARTY, L. G. (1969) Ultrastructure and functioning of the transport system of the leguminous root nodule. *Planta* **85:** 11–34.

PAUL, R. E. and JONES, R. L. (1976) Studies on the secretion of maize root cap slime. IV. Evidence for the involvement of dictyosomes. *Plant Physiol.* **57:** 249–56.

PETERSON, C. A. and EMANUEL, M. E. (1983) Casparian strips occur in onion root hypodermal cells: evidence from band plasmolysis. *Ann. Bot.* **51:** 135–137.

PETERSON, C. A. and PERUMALLA, C. J. (1984) Development of the hypodermal Casparian band in corn and onion roots. *J. Exp. Bot.* **35:** 51–57.

PETERSON, C. A., EMANUEL, M. E. and WILSON, C. (1982) Identification of a Casparian band in the hypodermis of onion and corn roots. *Can. J. Bot.* **60:** 1529–1535.

PLAUT, M. (1920) Über die morphologischen und mikroskopischen Merkmale der Periodizität der Wurzel, sowie über die Verbreitung der Metakutisierung der Wurzelhaube im Pflanzenreich. *Festschrift zur Feier des 100-jährigen Bestehens der Kgl. Württ. Landwirtschaftlichen Hochschule Hohenheim.* E. Ulmer, Stuttgart, 129–51.

POPHAM, R. A. (1955) Levels of tissue differentiation in primary roots of *Pisum sativum. Am. J. Bot.* **42:** 529–40.

POREMBSKI, S. and BARTHLOTT, W. (1988) Velamen radicum micromorphology and classification of Orchidaceae. *Nord. J. Bot.* **8:** 117–137.

POWELL, C. L. and BAGYARAJ, D. J. (1984) *VA Mycorrhiza.* CRC Press: Boca Raton, Florida.

PRICE, R. S. (1911) The roots of some North African desert grasses. *New Phytol.* **10:** 328–339.

PRIDGEON, A. M. (1987) The velamen and exodermis of orchid roots. In: *Orchid Biology IV* (ed. J. Arditti), pp. 139–192.

PRIESTLEY, J. H. (1920) The mechanism of root pressure. *New Phytol.* **19:** 189–200.

PRIESTLEY, J. H. (1922) The mechanism of root pressure. *New Phytol.* **21:** 41–47.

PRIESTLEY, J. H. and NORTH, E. K. (1922) The structure of the endodermis in relation to its function. *New Phytol.* **21:** 111–39.

PULGARIN, A., NAVASCUES, J., CASERO, P. J. and LLORET, P. G. (1988) Branching pattern in onion adventitious roots. *Am. J. Bot.* **75:** 425–432.

RENAUDIN, S. and CHEGUILLAUME, N. (1977) Sur quelques aspects de l'ultrastructure des sucoirs de *Thesium humifusum* L. *Protoplasma* **91:** 55–69.

REYNEKE, W. F. and VAN DER SCHIJFF, H. P. (1974) The anatomy of contractile roots in *Eucomis* L'Hérit. *Ann. Bot.* **38:** 977–82.

RICHARDSON, S. D. (1955) The influence of rooting medium on the structure and development of the root cap in seedlings of *Acer saccharinum* L. *New Phytol.* **54:** 336–7.

RIMBACH, A. (1895) Zur Biologie der Pflanzen mit unterirdischem Spross. *Ber. dt. bot. Ges.* **13:** 141–55.

RIMBACH, A. (1899) Beiträge zur Physiologie der Wurzeln. *Ber. dt. bot. Ges.* **17:** 18–35.

RIMBACH, A. (1929) Die Verbreitung der Wurzelverkürzung im Pflanzenreich. *Ber. dt. bot. Ges.* **47:** 22–31.

RIMBACH, A. (1932) Nachträgliche Dickenzunahme kontraktiler Monokotylen-Wurzeln. *Ber. dt. bot. Ges.* **50:** 215–19.

RIOPEL, J. L. (1966) The distribution of lateral roots in *Musa acuminata* "Gros Michel". *Am. J. Bot.* **53:** 403–6.

RIOPEL, J. L. and MUSSELMAN, L. J. (1979) Experimental initiation of haustoria in *Agalinis purpurea* (Scrophulariaceae). *Am. J. Bot.* **66:** 570–5.

ROST, T. L. and BAUM, S. (1988) On the correlation of primary root length, meristem size and protoxylem tracheary element position in pea seedlings. *Am. J. Bot.* **75:** 414–424.

RUZIN, S. E. (1979) Root contraction in *Freesia* (Iridaceae). *Am. J. Bot.* **66:** 522–31.

RYWOSCH, S. (1909) Untersuchungen über die Entwicklungsgeschichte der Seitenwurzeln der Monocotylen. *Z. Bot.* **1:** 253–83.

SCANNERINI, S. and BONFANTE-FASOLO, P. (1983) Comparative ultrastructural analysis of mycorrhizal associations. *Can. J. Bot.* **61:** 917–943.

SCHUMACHER, W. (1934) Die Absorptionsorgane von *Cuscuta odorata* und der Stoffübertritt aus Siebröhren der Wirtspflanze. *Jb. wiss. Bot.* **80:** 74–91.

SCOTT, B. I. H. and MARTIN, D. W. (1962) Bioelectric fields of bean roots and their relation to salt accumulation. *Austr. J. Biol. Sci.* **15:** 83–100.

SCOTT, F. M., BYSTROM, B. G., and BOWLER, E. (1963) Root hairs, cuticle and pits. *Science* **140:** 63–64.

SIEVERS, A. and VOLKMANN, D. (1972) Verursacht differentieller Druck der Amyloplasten auf ein komplexes Endomembransystem die Geoperzeption in Wurzeln? *Planta* **102:** 160–72.

STAHEL, J. (1971) Anatomische Untersuchungen an Brettwurzeln von *Khaya ivorensis* A. Chev. und *Piptadeniastrum africanum* (Hook. f.) Brenan. *Holz Roh- u. Werkstoff* **29**: 314–18.

STERLING, C. (1972) Mechanism of root contraction in *Gladiolus*. *Ann. Bot.* **36**: 589–98.

STOVER, B. L. (1951) *An Introduction to the Anatomy of Seed Plants*. D. C. Heath, Boston.

SUNDBERG, M. D. (1983) Vascular development in the transition region of *Populus deltoides* Bartr. ex Marsh. seedlings. *Am. J. Bot.* **70**: 735–743.

TARKOWSKA, J. A. and WACOWSKA, M. (1988) The significance of the presence of stomata on seedling roots. *Ann. Bot.* **61**: 305–310.

TOMLINSON, P. B. (1961) Palmae. In: *Anatomy of the Monocotyledons* (ed. C. R. Metcalfe), Vol. II. Clarendon Press, Oxford.

TOMLINSON, P. B. and ZIMMERMANN, M. H. (1968) Anatomy of the palm *Rhapis excelsa*: VI, Root and branch insertion. *J. Arnold Arb.* **49**: 307–16.

TORREY, J. G. (1957) Auxin control of vascular pattern formation in regenerating pea root meristems grown *in vitro*. *Am. J. Bot.* **44**: 859–70.

TORREY, J. G. (1965) Physiological bases of organization and development in the root. In: W. Ruhland, *Handbuch der Pflanzenphysiologie* **15**: 1256–327.

TORREY, J. G. (1976) Initiation and development of root nodules of *Casuarina* (Casuarinaceae). *Am. J. Bot.* **63**: 335–44.

TORREY, J. G. (1978) Nitrogen fixation by actinomycete-nodulated angiosperms. *BioScience* **28**: 586–592.

TROLL, W. (1948) *Allgemeine Botanik*. F. Enke, Stuttgart.

VAN FLEET, D. S. (1942a) The development and distribution of the endodermis and an associated oxidase system in monocotyledonous plants. *Am. J. Bot.* **29**: 1–15.

VAN FLEET, D. S. (1942b) The significance of oxidation in the endodermis. *Am. J. Bot.* **29**: 747–55.

VAN FLEET, D. S. (1950) A comparison of histochemical and anatomical characteristics of the hypodermis with the endodermis in vascular plants. *Am. J. Bot.* **37**: 721–5.

VAN FLEET, D. S. (1957) Histochemical studies on phenolase and polyphenols in the development of the endodermis in the genus *Smilax*. *Bull. Torrey Bot. Club* **84**: 9–28.

VOLKENS, G. (1887) *Die Flora der aegyptisch-arabischen Wüste auf Grundlage anatomisch physiologischer Forschungen*. Gebr. Borntraeger, Berlin.

VOLKMANN, D. (1974) Amyloplasten und Endomembranen: Das Geoperzeptionssystem der Primärwurzel. *Protoplasma* **79**: 159–83.

WARMBRODT, R. D. (1985) Studies on the root of *Hordeum*

vulgare L.—ultrastructure of the seminal root with special reference to the phloem. *Am. J. Bot.* **72**: 414–432.

WEAVER, J. E. (1920) *Root Development of the Grassland Formation. A Correlation of the Root Systems of Native Vegetation and Crop Plants*. Carnegie Inst., Washington Publ. No. 292.

WEBER, H. C. (1976a) Anatomische Studien an den Haustorien einiger parasitischer Scrophulariaceen Mitteleuropas. *Ber. dt. bot. Ges.* **89**: 57–84.

WEBER, H. C. (1976b) Studies on new types of haustoria in some Central European Rhinanthoideae (Scrophulariaceae). *Plant Syst. Evol.* **125**: 223–32.

WERKER, E. and KISLEV, M. (1978) Mucilage on the root surface and root hairs of *Sorghum*: heterogeneity in structure, manner of production and site of accumulation. *Ann. Bot.* **42**: 809–16.

WHITE, J. and LOVELL, P. H. (1984a) Anatomical changes which occur in cuttings of *Agathis australis* (D.Don) Lindl. 1. Wounding responses. *Ann. Bot.* **54**: 621–632.

WHITE, J. and LOVELL, P. H. (1984b) Anatomical changes which occur in cuttings of Agathis australis (D.Don) Lindl. 2. The initiation of root primordia and early root development. Ann. Bot. **54**: 633–645.

WILCOX, H. (1962a) Growth studies of the root of incense cedar *Libocedrus decurrens*: I, The origin and development of primary tissues. *Am. J. Bot.* **49**: 221–36.

WILCOX, H. (1962b) Growth studies of the root of incense cedar *Libocedrus decurrens*: II, Morphological features of the root system and growth behaviour. *Am. J. Bot.* **49**: 237–11.

WILCOX, H. E. (1968) Morphological studies of the root of red pine, *Pinus resinosa*: I, Growth characteristics and patterns of branching. *Am. J. Bot.* **55**: 247–54.

WILKINS, M. B. (1978) Gravity-sensing guidance mechanisms in root and shoot. *Bot. Mag. Tokyo*, special issue **1**: 255–77.

WILSON, K. and HONEY, J. N. (1966) Root contraction in *Hyacinthus orientalis*. *Ann. Bot.* **30**: 47–61.

WULLSTEIN, L. H., BRUENING, M. L. and BOLLEN, W. B. (1979) Nitrogen fixation associated with sand grain root sheaths (rhizosheaths) of certain xeric grasses. *Physiol. Plant.* **46**: 1–4.

ZAMSKI, E., UCKO, O. and KOLLER, D. (1983) The mechanism of root contraction in *Gymnarrhena micrantha*, a desert plant. *New Phytol.* **95**: 29–35.

ZOHARY, M. and ORSHAN, G. (1954) Ecological studies in the vegetation of the Near East desert: V, The *Zygophylletum dumosi* and its hydro-ecology in the Negev of Israel. *Vegetatio* **5–6**: 341–50.

SECONDARY BODY OF THE PLANT

GROWTH in thickness that occurs distant from the apices is called *secondary growth*, and the tissues thus produced are termed *secondary tissues*. These tissues constitute the secondary body of the plant. Secondary tissues develop from secondary meristems, i.e. from the *vascular cambium* and the *phellogen* or *cork cambium*. Commonly, the main stem, which in certain plants may reach a diameter of several metres, the branches, roots, and often even petioles and the main veins of leaves, contain secondary tissues (Fig. 168, no. 1).

The development of secondary vascular tissues from the cambium is characteristic of the dicotyledons and the gymnosperms. In certain monocotyledons the vascular tissues are also increased after the primary growth is completed, but the cambium of these plants is of a different nature. In the pteridophytes secondary thickening was more common among those species that have become extinct. In the living pteridophytes this feature is rare but occurs, for example, in *Isoetes* and *Botrychium*. Certain monocotyledons, as, for instance, some Palmae, exhibit considerable thickening that is the result of a *primary thickening meristem* only, but these plants never reach the diameter of old dicotyledonous trees.

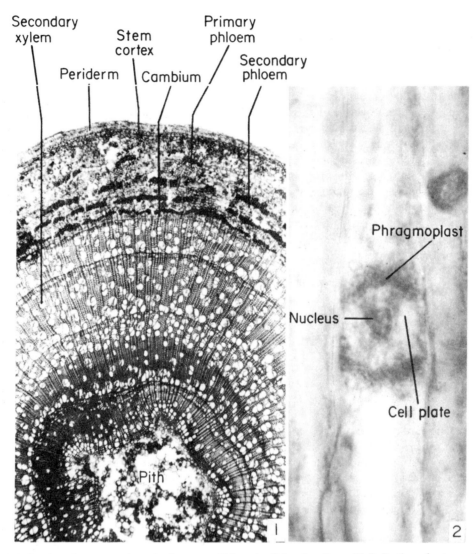

Secondary xylem

Periderm

Stem cortex

Cambium

Primary phloem

Secondary phloem

Pith

Phragmoplast

Nucleus

Cell plate

1

2

FIG. 168. 1, Portion of a cross-section of a four-year-old branch of *Populus alba*. × 50. 2, Portion of a tangential section through the cambium of the stem of *Nicotiana tabacum*, showing the middle portion of a dividing fusiform initial. × 1500.

CHAPTER 14

VASCULAR CAMBIUM

THE vascular cambium is a lateral meristem that develops either as longitudinal strands or as a hollow cylinder. In the woody dicotyledons and gymnosperms the primary vascular tissues of the stem and root exist for only a relatively short period, and their function is taken over by the secondary vascular tissues which are produced by the cambium. In many herbaceous angiosperms, and also in most of the recent lower vascular plants, cambium is absent or vestigial. The vascular cambium of the woody carboniferous pteridophytes differed, apparently fundamentally, from that of modern seed plants (Cichan, 1985a, b).

GENERAL DEVELOPMENT AND STRUCTURE OF THE VASCULAR CAMBIUM

In certain plants, including monocotyledons, all the cells of the procambium undergo differentiation into primary vascular tissues. In almost all the dicotyledons and gymnosperms a portion of the procambium remains meristematic even after the completion of primary growth and develops into the cambium of the secondary body. The cambium that arises within the bundles of primary vascular tissue of the stem is called *fascicular cambium* (Fig. 99, nos. 2, 3). The strips of fascicular cambium usually become joined by additional strips of cambium which constitute the *interfascicular cambium* (Fig. 100). The interfascicular cambium is not a continuation of the procambium but develops from the interfascicular parenchyma. Therefore this part of the cambium constitutes a secondary meristem also from the point of view of its origin. Thus a complete

hollow cylinder of cambium is developed which is present throughout the length of the main plant axis and to which narrower cylinders of cambium, belonging to the stem and root branches, are connected. Sometimes the cambium extends into the leaves. In most dicotyledons and gymnosperms the cambial cylinder develops between the primary xylem and phloem, a position that is retained throughout the life of the plant. From this position the cambium produces the secondary xylem centripetally and the secondary phloem centrifugally. In certain dicotyledonous plants, e.g. Chenopodiaceae, the secondary thickening of the axis is anomalous and it is brought about in a manner that differs from that described above.

The cambium usually consists of two types of cells.

1. *Fusiform initials* (Fig. 173) which are elongated cells with tapered ends. These cells are very long and in old trunks of *Sequoia sempervirens*, for example, they reach a maximum length of 8.7 mm (Bailey, 1923).

2. *Ray initials* (Fig. 173) which are much smaller cells than those of the above type and which are almost isodiametric.

Both these types of initials are larger in older trunks than in very young ones. The longitudinally orientated elements in an organ, such as the tracheary elements, fibres, xylem and phloem parenchyma, and the sieve elements, develop from the fusiform initials. The cells of the vascular rays, which are orientated horizontally in the organ, develop from the ray initials.

One of the interesting features of cambial cells is their intense vacuolation. The walls of these cells possess primary pit-fields with plasmodesmata (Fig. 174, no. 1). The radial walls, especially of the xylem and phloem mother cells, are thicker than the tangential ones; this feature is a result of the predominantly periclinal divisions in the cambial cells during which the thickening of the radial walls is continuous.

Few attempts were made to determine whether a definite distinction can be made between the procambium and cambium (Sterling, 1946; Catesson, 1964; Esau, 1965; Cumbie, 1967b; Philipson *et al.*, 1971; Fahn *et al.*, 1972; Butterfield, 1976; Larson, 1976, 1982; Soh *et al.*, 1988). The main criteria proposed for distinguishing between the two types of vascular meristem were as follows: the procambial cells have gabled ends (Fig. 172, no. 3), as seen in radial view, and they stain deeply, whereas the cells of the cambium have flat endings in radial view and their protoplasts do not stain strongly. The procambium is not differentiated into long and short cells, while in the cambium long fusiform and short ray cells can be distinguished (Fig. 172, no. 4). The last-mentioned difference appears to be the most characteristic one. However, as observed in *Ricinus* (Fahn *et al.*, 1972) it is not possible to define the exact place of ray initiation because the initiation of the rays is very gradual. It appears, therefore, that the procambium and cambium represent the same meristem, but at different stages of development of the region in which it is found. This view is also supported by wounding experiments. When stems were wounded close to their apices, regions of cambium which were continuous with procambium both above and below it appeared near the wounds (Fahn *et al.*, 1972). While accepting the view that the procambium and the cambium are only two stages of the same meristem, treating them separately may be justified nevertheless; firstly because there are groups of plants in which the second stage does not occur, secondly the cambium develops also in regions where a procambium was not in existence before, and, thirdly, because of the usual uniformity of the cambium when fully developed.

Another question is the exact location on the stem of the first periclinal divisions characteristic to the cambium. In mature cambium the cells are arranged in radial rows. In cross-section of *Ricinus* a radial row of cambial-like cells in which repeated periclinal divisions take place can already be distinguished in each procambial strand (Fig. 172, no. 2). The divisions found in a mature cambium, however, are ordered in the manner seen in cross-sections, where there is usually a close correspondence in the location of divisions between neighbouring cells. The divisions of the mature cambium, furthermore, proceed in a characteristic order, and there is a region of initial cells. Ordered periclinal divisions were found in *Ricinus* to occur first on both sides of the radial row and not within it. They were seen to extend into the radial row only later, shortly before the interfascicular cambium was completed (Fahn *et al.*, 1972).

Among the important investigations of the structure of the cambium and the manner of its cell division those of Bailey (1920a, b, 1923, 1930), Bannan (1950, 1951a, b, 1955; Bannan and Whal-

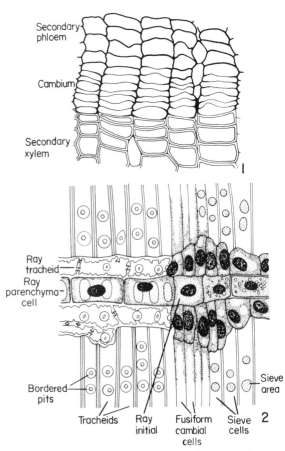

FIG. 169. 1, Portion of a cross-section of a stem of *Pinus* showing the cambial zone and neighbouring tissues. 2, As above but a radial section and showing a vascular ray. (No. 2 adapted from Haberlandt, 1918.)

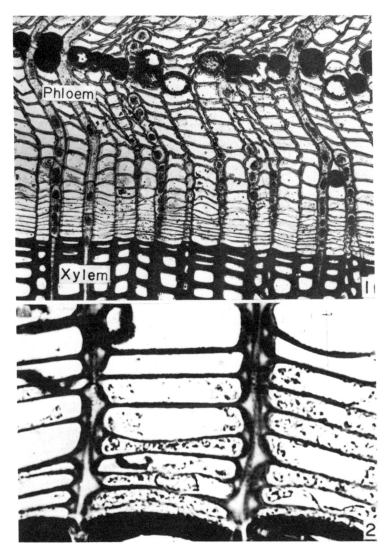

FIG. 170. Cross-section of cambium of *Pinus strobus*. 1, Cambial zone and adjacent portions of xylem and phloem. 2, Enlarged portion of cambial zone showing thick radial walls in which layering is observed. Groups of two and four cells are seen to have common wall layers. (Courtesy of I. W. Bailey.)

ley (1950), Newman (1956), Catesson (1964), and Wilson (1964), should be mentioned.

During the growing period the *cambial initials,* together with their immediate derivatives, form the *cambial zone.* In a cross-section the cells of the cambial zone are seen, as mentioned before, to be **arranged in radial rows** (Fig. 169, no. 1; Fig. 170, no. 1). The cells on either side of this zone gradually widen until they acquire the shape and features of mature phloem and xylem elements. In the narrow sense of the word the cambium con-

sists only of the single layer of initials, but it is customary to refer to the entire cambial zone by the term cambium as it is difficult to distinguish between the initials and the neighbouring cells that are derived from them. Some authors (Catesson, 1964) doubt the existence of one layer of cambial initials and consider the cells of the whole homogeneous part of the cambial zone to have the properties of initials. Bannan (1955, 1968) and Newman (1956) succeeded in distinguishing single initial cells between the *phloem* and *xylem mother*

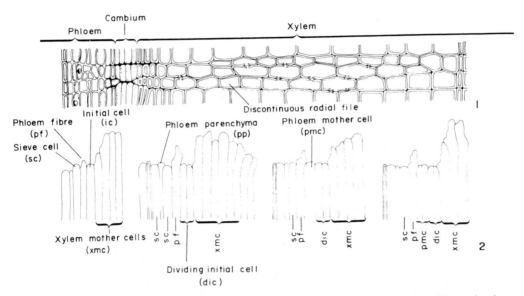

Fɪɢ. 171. Vascular cambium of *Thuja occidentalis*. 1, Cross-section through xylem, cambium and phloem, showing a radial file of xylem and phloem cells discontinuous in the cambial zone. 2, Drawings from radial sections of the cambial zone, showing differences in length of the various types of cells. (Adapted from Bannan, 1955.)

cells in conifers. Bannan (1955), in radial sections of the cambium of *Thuja occidentalis,* distinguished between the initials and phloem and xylem mother cells by their length. In each radial file of elements there are 1–4 cells adjoining the phloem which are about the same length as the last differentiated sieve cell or phloem parenchyma. The xylem mother cells, however, are somewhat longer. One of the shorter cells, which abuts on the xylem mother cells, represents the initial cell (Fig. 171, no. 2). Bannan finds support for the concept of a single layer of initials in the cambium in the radial continuity of cells across the cambium and in the simultaneous origin of new radial files and the cessation of old ones in the phloem and xylem (Fig. 171, no. 1). The initials produce, by periclinal divisions, alternately xylem and phloem mother cells. Subsequently each phloem mother cell forms groups of two cells by further division, and each xylem mother cell divides usually twice to form groups of four cells. Due to the fact that around each daughter protoplast a new primary wall is deposited, walls become thicker with successive divisions (Fig. 170, no. 2). Using the criterion of wall thickness Newman (1956) and Mahmood (1968) found it possible to distinguish in *Pinus* the groups of cells which include the initial. In the

same way Murmanis (1977) was able to distinguish initials in branches of *Quercus rubra*. As discussed in Chapter 3 on the primary meristems, it should also be mentioned here that the permanent initials, in the case of the cambium, are determined by their location. When a permanent initial of the cambium divides, one of the daughter cells which is situated at the proper relative distance from the phloem and xylem will inherit the function of the permanent cambial initial.

It is of interest to mention that in the needle leaves of several conifers (e.g. *Pinus longaeva* and *P. flexilis*), the cambium is unidirectional and produces only phloem (Ewers, 1982a, b).

When the cambium is active the cambial zone is wide and consists of many cell layers, but when it is dormant the zone is usually reduced to one or a few cell layers only. In conifers, according to Bannan (1962), the cambial zone in the resting state may consist of five layers, but it is usually two- or three-layered. In the three-layered condition the layer nearest the more or less immature phloem is recognizable as that of the cambial initials and the inner layers constitute the xylem mother cells. In these plants the first divisions, on the renewal of cambial activity, may occur in any of the three layers of the cambial zone, but the

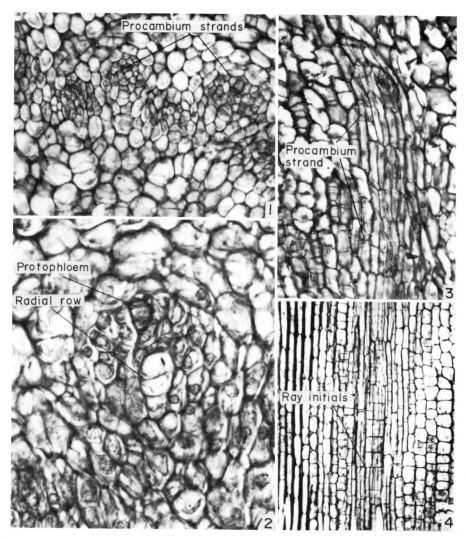

Fig. 172. Procambium and early stage of cambium in the stem of *Ricinus communis*. 1, Portion of a cross-section at the base of the second leaf primordium. ×95. 2, Portion of a cross-section below the second leaf primordium showing protophloem and a radial row of highly vacuolated cells below it, formed by tangential divisions. ×800. 3, Radial longitudinal section of a young portion of a procambium strand, showing gabled ends of its cells. ×400. 4, Radial-longitudinal section at the level at which the cambium has already differentiated. ×95.

usual site of the first divisions is among the xylem mother cells closest to the already differentiated xylem and not, as might be expected, in the cambial initials. The initiation of the divisions closest to the xylem is of interest and may be connected with the supply of water as well as with the presence of growth hormones (Bannan, 1962). According to Evert (1963), in *Pyrus malus* final differentiation of the phloem elements from cells produced in the previous season precedes xylem

differentiation by about 6 weeks. In some tropical trees (e.g. *Delonix regia* and *Polyalthia longifolia*) xylogenesis has been reported to begin about two months before phloem formation (Ghouse and Hashmi, 1982).

Types of cambium

Two types of cambium can be distinguished on

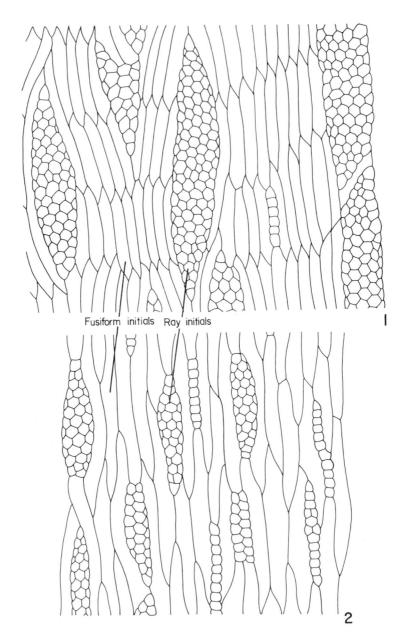

Fusiform initials Ray initials

1

2

FIG. 173. Tangential sections of different cambial types. 1, Storied cambium of *Robinia*. 2, Non-storied cambium of *Fraxinus*.

the basis of the arrangement of the fusiform cells as seen in tangential section.

1. *Storied or stratified cambium* (Fig. 173, no. 1) in which the fusiform initials are arranged in horizontal rows so that their ends are approximately at the same level. The length of these initials is found to vary between 140 μm and 520 μm (Bailey, 1923). This type of cambium is found in *Tamarix* and *Robinia*, for example.

2. *Non-storied* or *non-stratified cambium* (Fig. 173, no. 2) in which the fusiform initials partially overlap one another. The length of the initials of this type of cambium is found to vary between 320 and 2300 μm in many dicotyledons. In vessel-less

dicotyledons the fusiform initials may reach a maximum length of 6200 μm. In the gymnosperms initials 1000–8700 μm long were reported (Bailey, 1923).

The fusiform initials in the second type are longer and this type is the more common. Bailey (1923) saw in the initials of storied cambium a higher phylogenetic stage, and he suggested that they have developed by gradual reduction of the size of the cells and of longitudinal sliding growth.

The fusiform initials vary in size over the span of a year (Sharma *et al.*, 1979, 1980).

Structure of protoplast

The pioneering works of Bailey (1920a, b, c, 1930) stated that in spite of their large size the fusiform initials are, as a rule, uninucleate. However, more recent researchers observed, that in species of a number of families, polynucleate fusiform cambial cells may occur in addition to uninucleate ones (Ghouse and Khan, 1977; Iqbal, 1981). Bailey developed a technique which allowed him to examine sections of the living cambium under the microscope. He described changes occurring in the protoplast during transition from low to high temperatures and vice versa. In high summer temperatures protoplasmic streaming is active and there is a very large vacuole penetrated by a system of anastomosing threads. When the temperature is low, the vacuome tends to divide into a varying number of small independent vacuoles. Later some electron microscope work has been carried out on cambial cells (Buvat, 1956; Srivastava, 1966; Srivastava and O'Brien, 1966; Robards and Kidwai, 1969; Murmanis, 1971). These studies showed that the fusiform and ray initials are essentially alike, and both have the basic content of organelles and membranes typical of parenchyma cells. Active growing cambial cells were found to have one or two large vacuoles and contain rough endoplasmic reticulum (RER) and polyribosomes. The dictyosomes are active in the production of vesicles. According to Rao and Catesson (1987), their activity increases after the cessation of the meristematic activity. Mitochondria occur singly. Microtubules were found in the peripheral cytoplasm. In *Fraxinus excelsior*, Goosen-de Roo *et al.* (1983) have observed

in the fusiform cells of the active cambium, in addition to cisternae of RER, tubular ER which always appeared associated with microfilament (see Chapter 2) bundles. The two associated organelles were oriented parallel to the longitudinal axis of the cells. It was suggested that the association of the specialized structures is related to the strong longitudinal cytoplasmic streaming in the active cytoplasmic cells. The association of the tubular ER and microfilament bundles has not been seen in the cambial ray cells. In the winter the vacuoles were small and numerous, the plasmalemma was thrown into folds and the endoplasmic reticulum occurred mostly in the form of smooth vesicles. Mitochondria were sometimes found to occur in a chain. In the cells of the resting cambium lipid droplets and protein bodies were found. Robards and Kidwai suggest that these inclusions are storage materials which are required during the first stages of differentiation at the beginning of the growing period.

Ultrastructural differences between dormant and active state were also observed in procambial cells of *Salix* (Berggren, 1987).

Fluctuations in starch content in the secondary xylem and phloem correlate with seasonal changes in cambial activity and bud burst (Fahn, 1959a, b; Lawton, 1972; Boscagli, 1982; Essiamah and Eschrich, 1985).

Cell divisions and growth

The cambial initials and the cells that are derived from them but which have not yet undergone differentiation divide periclinally and anticlinally in a longitudinal plane. As a result of the periclinal divisions, which are the more numerous, new cells are added to the secondary phloem and xylem. The derivatives of each initial therefore form radial rows, which can sometimes also be distinguished in the xylem and phloem. Usually, however, this order is lost in the vascular tissues because of the changes in shape that take place during the differentiation and maturation of their cells.

The early stage of differentiation of the cambial cells is characterized by their pronounced radial expansion. On the basis of ultrastructural cytochemical examinations Roland (1978) suggested that at this stage large portions of the cell walls, particularly of the radial walls, are devoid of cellulose.

As a result of the secondary thickening, the circumference of the xylem cylinder increases. Together with this the cambium also increases in circumference by the addition of new cells. In storied cambium the addition of new fusiform initials is brought about by longitudinal anticlinal divisions (Fig. 174, no. 3) of the existing initials.

In non-storied cambium, on the other hand, the fusiform initials undergo oblique, pseudo-transverse, anticlinal divisions, followed by intrusive growth, and each of the new cells becomes as long as, or even longer than, the cell from which it was derived (Fig. 174, nos. 2, 4–7). In *Fraxinus excelsior* a transition from oblique divisions to longitudinal anticlinal divisions takes place. Short cells tend to divide longitudinally and longer ones tend to divide obliquely (Krawczyszyn, 1977).

In the oblique division of the fusiform cells the

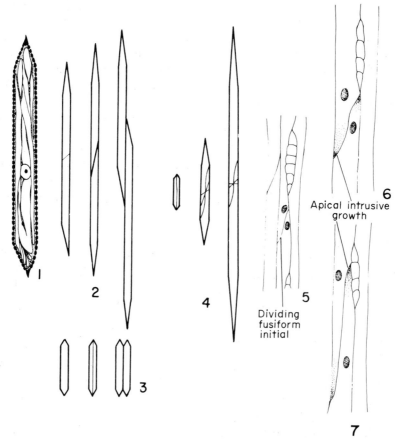

Fig. 174. 1, Diagram of a fusiform initial of the cambium of *Robinia pseudacacia* showing its highly vacuolated protoplast and the presence of numerous primary pit-fields. 2–4, Diagrammatic representation of the anticlinal division of fusiform initials that results in the increase in girth of the cambium. 2, Division of a fusiform initial in non-storied cambium showing the relative position of the daughter cells during their elongation. The plane of the division is diagonal, and the cells come to lie one next to the other as a result of intrusive growth. 3, Three stages in the anticlinal division of a fusiform initial in a storied cambium. 4, Three possible orientations of the new wall formed during the anticlinal division of a fusiform initial. From right to left: in conifers, in non-storied dicotyledonous cambium, in storied dicotyledonous cambium. 5–7, Portions of successive tangential sections of cambial zone, starting in the initial layer. 5, Showing pseudo-transverse division of a fusiform initial. 6 and 7, Apical intrusive growth of the two daughter fusiform cells. (Nos. 1–4 adapted from Bailey, 1923, 1930; nos. 5–7 adapted from Zimmermann and Brown, 1971.)

pseudo-transverse new wall has usually the same orientation as the neighbouring cells of the cambium (Bannan, 1966; Hejnowicz, 1964; Hejnowicz and Krawczszyn, 1969). From time to time reversals occur in the direction of the tilt of the pseudo-transverse division, from a Z-type to an S-type and vice versa. Both the orientation of pseudo-transverse division and direction of cell elongation after division seem to be under general polar control. Local areas in the cambium in which all the pseudotransverse divisions are oriented predominately in the same direction are called domains (Hejnowicz and Romberger, 1973). Changes of S to Z orientation of fusiform initials occur in cycles, the length of which apparently has a closer relation to calender time than to the amount of radial growth of the stem (Krawczyszyn and Romberger, 1980). The changes in the inclinations of the fusiform cambial initials determine the type of the wood grain (see Chapter 15).

Changes in fusiform cells orientation may occur not only in non-storied cambium, but also in storied cambium (Hejnowicz and Zagorska-Marek, 1974; Wloch, 1987; Wloch and Zagorska-Marek, 1982).

When trunks are girdled in such a manner that narrow bridges with a transverse orientation are left between the upper and lower parts of the stem, reorientation of the cambial initials in the bridges occurs. It has been shown that in the first stage of reorientation the fusiform cells divide transversely, then the short cells achieved by the divisions grow intrusively in the direction of the bridge. The change of direction occurs gradually (Kirschner et al., 1971).

In other experiments, however, the polarity of the cambium can be very stable or determinate. This is most clearly seen in grafts in which one of the two members is reoriented, since the new vascular tissues express the original shoot-root polarity for indefinite periods (Vöchtig, 1892; Thair and Steeves, 1976). This was one of the reasons for the claim that plant cells have a completely determined polarity (Vöchtig, 1892). The difference between the different degrees of cambial polarity may be connected with the determinate polarity of auxin transport; polarity is maintained whenever possible, but if transport is stopped, a diffusion-dependent flux induces a new polarity (Sachs, 1981). Zagorska-Marek and Little (1986)

studied the reorientation of fusiform cambial cells in helical bridges produced by girdling of *Abies balsamea* branches and by indol-3-ylacetic acid (IAA). Their results support the view that the orientation of the fusiform cambial initials is parallel to the direction of IAA transport.

In untreated normally growing trees the daughter cells resulting from anticlinal division of the fusiform cells may undergo various types of transformation. Some grow and become new fusiform initials, whereas others may decline and either lose their generative capacity and eventually mature into abnormal xylem or phloem elements or develop into ray initials. Bannan regards the fusiform initials as competitive units in an overcrowded environment. In the competition for survival those persist which have the greatest length and the maximal contacts with rays. The survival of fusiform initials may also be under polar influence. Bannan (1968) found that when the upper cell persists it elongates predominantly acropetally and when the lower one persists basipetal elongation predominates. According to Bannan the correlation between the survival of the daughter cells after the pseudo-transverse division and the direction of major cell elongation indicates linkage to the same polar factor. Bannan

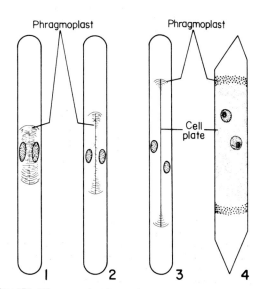

FIG. 175. Diagrams showing various stages in the division of a fusiform initial of *Robinia pseudacacia*. 1–3, Dividing cells as seen in radial section. 4, As seen in tangential section. (Adapted from Bailey, 1920c.)

suggests that certain substances concerned with viability and growth may be unequally distributed in the fusiform initials, greater concentration occurring towards the upper cell tip in some cases and towards the lower tip in others, with capacity for reversal.

Because of the great length of the fusiform initials the formation of the cell plate during the process of longitudinal division is peculiar to these cells. The cell plate begins to form between the two new nuclei and it spreads slowly. A relatively long period passes before it reaches the end walls. While the cell plate is not complete its free margins are surrounded by the phragmoplast (Fig. 168, no. 2; Fig. 175).

In the conifers (Bannan, 1962) intensively dividing cambial cells divide once every 4–6 days whereas apical meristematic cells may divide every 8–18 hours. Possibly, the slower division of the cambial cells is due to the time required for the phragmoplasts to reach the ends of the elongated cells, which may be up to a few millimetres long. Wilson (1964) estimated that the length of the cell cycle (from mitosis to mitosis including division, synthesis, and enlargement) is about 10 days in *Pinus strobus*: the process of division one day, during which mitosis occupies 5 hours and cell plate growth 19 hours. Later Wilson (1966) observed that the length of the cell cycle in *P. strobus*, based on the mitotic index (observed divisions expressed as a percentage of the total number of cambial cells in an investigated area), changes during the growth season. The mitotic index dropped from 4.2 on 6 May, when the number of cells in the cambial zone has stabilized, to 1.9 on 21 July. Assuming that the length of time for mitosis stayed the same during this period, then the length of the cell cycle approximately doubled.

Variation in the number of cells in the cambial zone during the season seems to reflect a balance between the rate of differentiation and the rate of cell production. During the first 2 weeks after mitotic activity starts in spring, there is little differentiation and thus a rapid build up in the width of the cambial zone. After commencement of differentiation of new derivatives a balance is established between the rate of differentiation and cell production. The decrease in number is due to

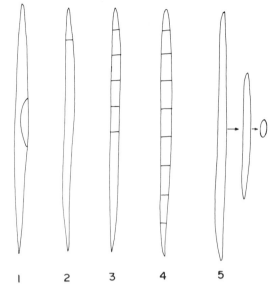

Fig. 176. Schematic drawings showing the origin of rays from fusiform initials. 1, Lateral division. 2, Division at the end; formation of one ray initial. 3, Several divisions occurring in half of the fusiform initial. 4, Division of an entire fusiform initial into a row of ray initials. 5, Reduction in height of a fusiform initial from which a single ray initial is formed.

a faster rate of differentiation as compared with production of cells. Most divisions, as seen in radial longitudinal sections, take place among the xylem mother cells. The rate of division in the cambial initials and the phloem mother cells is lower than in the xylem mother cells. The relative rates of xylem and phloem formation have been found to change during the growing season only in some plants.

Ray initiation

During the enlargement of the cambial cylinder new ray initials develop and single fusiform initials are continually lost from the cambium and are replaced by new ones (Bannan, 1950, 1951a, b). Following are summarized the main ways by which ray initials are formed in the cambium. (1) A single cell may be cut off the side of a fusiform initial–lateral division (Fig. 176, no. 1). (2) A single cell may be cut off the end of a fusiform initial (Fig. 176, no. 2; Fig. 202, nos. 1–4). (3) A declining fusiform initial may be reduced to a single ray initial (Fig. 176, no. 5). (4) The whole or part of a fusiform initial may be segmented by transverse divisions to form a tier of ray initials (Fig. 176, nos. 3, 4). Various degrees of transition may occur between these types. (See Bannan, 1951a; Barghoorn, 1940; Braun, 1955; Cumbie, 1967a; Srivastava, 1963; Evert, 1961.)

Bünning (1965) suggested that new rays are initiated when growth in circumference of the cambium produces areas which are sufficiently far from all rays surrounding them. Carmi *et al.* (1972) have shown that when growth in circumference of the cambium was prevented artificially, rapid radial growth occurred which was associated with a rapid increase in the number and volume of rays. They thus suggested that the rays differentiate along channels of a stimulus moving between the phloem and the differentiating xylem and their spacing is controlled not in the cambium but in the differentiating vascular tissues. The excess stimulus which is not drained by existing rays may form new channels through the cambium, leading to an increase in the size of rays or the formation of new ones.

CAMBIAL ACTIVITY

The pattern of radial growth is correlated to the rate of the cambial activity. As mentioned before (Wilson, 1966), the variation in the number of cells across the cambial zone seems to express the balance between the rate of cell division and the rate of differentiation of the new derivatives. The process of division is faster than differentiation when the cambium becomes active. As a result, the cambial zone becomes wide. With the beginning of differentiation a balance is established, and the width of the zone remains more or less constant. When the rate of division becomes reduced and the differentiation proceeds faster than division, the cambial zone becomes narrow.

The vascular cambium shows great variation in the period and intensity of activity. These variations are the result of many internal and external factors (see Reinders-Gouwentak, 1965; Philipson *et al.*, 1971).

Seasonal activity of cambium

There are plants whose cambium is active throughout the entire life of the plant, i.e., the cambial cells divide continuously and the resulting cells undergo gradual differentiation to form the xylem and phloem elements. This type of activity is usually found in plants growing in tropical regions. However, not all tropical trees exhibit a continuous cambial activity (Coster, 1927–8; Mariaux, 1967; Fahn *et al.*, 1981; Ash, 1983). So, for instance, the percentage of ringless trees in the rain forest of India was reported to be 75 (Chowdhury, 1961), in the rain forest of the Amazon basin 43 (Alvim, 1964), and in Malaysia only 15% of the species show continuous radial growth (Koriba, 1958). In warm temperate climates the percentage of ringless trees is still lower. In regions with definite seasonal climates the cambium ceases its activity with the onset of unfavourable conditions, usually the autumn, and it enters a dormant state which may last from the end of summer till the following spring. In spring the cambium again becomes active. From an anatomical point of view the commencement of the cambial activity consists usually of two stages: (1) the cambial cells expand radially; and (2) the cells begin to divide as described above. With the enlargement of the cambial cells their radial walls usually become weakened, so that in this stage the bark of the stems and roots may easily be peeled. In later stages this easy separation of the bark from the xylem is also possible because of the increase in number of cells in the cambial zone as a result of the cell divisions. The separation usually occurs in the region of the young xylem cells which have already reached their maximum diameter, but which still have thin primary walls.

Various methods are used to determine cambial activity. The ease by which the bark may be peeled is often used as an indication of cambial activity. A second method is based on the anatomical examination of cross-sections cut in the region of the cambium and the tissues adjacent to it. In this method the number of cell layers in the cambial zone and of the not yet fully differentiated xylem elements is taken as an indication of the rate of cambial activity. A more accurate method has been developed by which both the rates

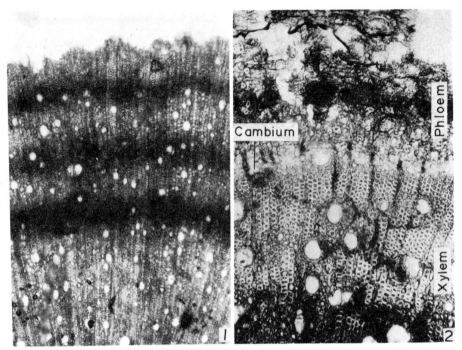

FIG. 177. Micro-autoradiograms of cross-sections of outer portions of a stem of *Eucalyptus camaldulensis* labelled with $^{14}CO_2$ in order to measure the amount of tissue produced by the cambium during definite periods. 1, Plant labelled three times at about monthly intervals, showing three labelled zones in the xylem (phloem and cambium removed). × 70. 2, Plant labelled once showing two labelled zones, one in the xylem and one in the phloem. × 135.

of cell division and of cell differentiation can be precisely determined (Waisel and Fahn, 1965a). This method involves the application of radioactive carbon to photosynthesizing plants. If the cambium is active the radioactive carbon is incorporated in the newly formed cell walls where it can easily be detected by autoradiographic techniques (Fig. 177).

As has already been mentioned, there are plants in which the cambium is active throughout the year and those in which there is a break, which may be as long as 8 months in duration, in the cambial activity. In the Mediterranean region and in hot desert regions it is possible to find both the above two as well as intermediate types (Oppenheimer, 1945; Messeri, 1948; Minervini, 1948; Fahn, 1953, 1955, 1958, 1959a, b, c, 1962; Fahn and Sarnat, 1963; Liphshitz and Lev-Yadun, 1986). In these regions the range of temperature is such that the cambium may remain active throughout the year if such activity is in accordance with the hereditary characteristics of the plant. In arid regions, such as these, however, the amount of available water in the soil is an important factor in the control of cambial activity, and, of course, the general ability to grow depends on this factor. It is of interest that in certain plants whose roots reach those moist levels, cambial activity occurs throughout the year. Such an uninterrupted cambial activity occurs, e.g. in *Tamarix aphylla, Acacia raddiana,* and *A. tortilis*, which grow in the desert regions of Israel, and *Acacia cavena*, which grows in the Mediterranean semiarid region of Central Chile (Fahn, 1958b; Aljaro *et al.*, 1972). From the point of view of cambial activity, trees and shrubs in Israel may be divided into the following types (Fig. 178) (Fahn, 1962).

1. Woody plants exhibiting more or less distinct growth rings, the development of which commences in the early winter months, i.e. between November and January. The cambium in these species is dormant for a fairly long period of drought (e.g. *Retama raetam, Artemisia monosperma, Zygophyllum dumosum,* and *Reaumuria*

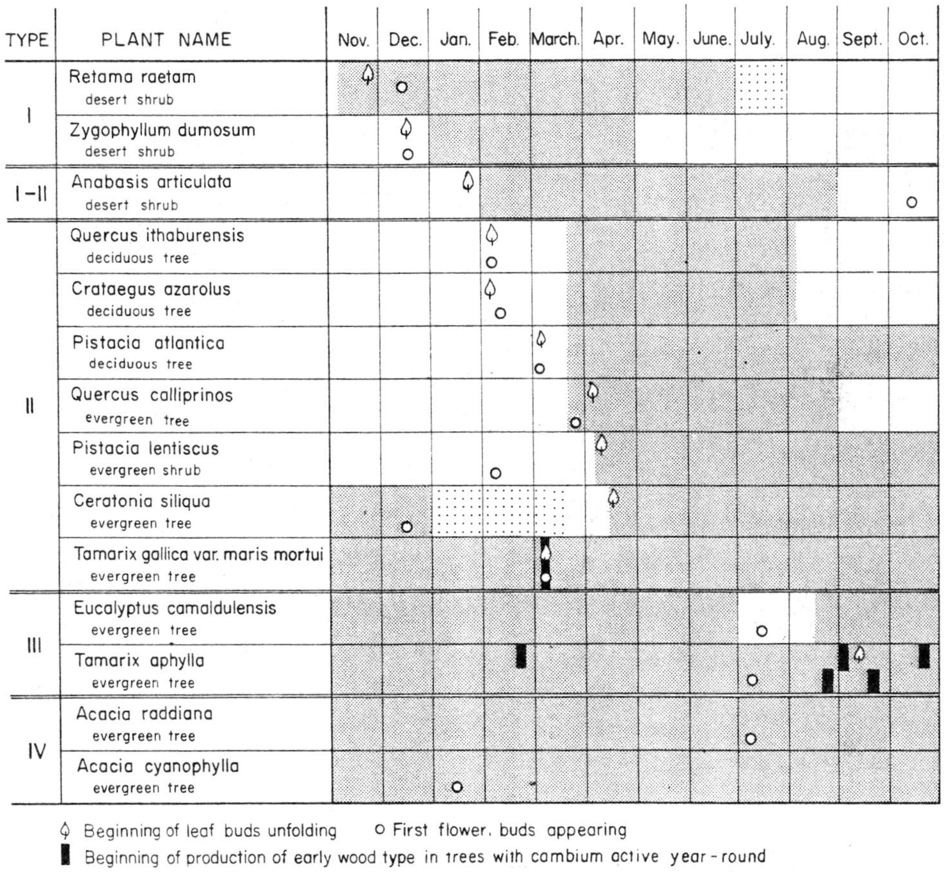

TYPE	PLANT NAME	Nov.	Dec.	Jan.	Feb.	March	Apr.	May.	June.	July.	Aug.	Sept.	Oct.
I	Retama raetam (desert shrub)	φ	o										
I	Zygophyllum dumosum (desert shrub)		φ / o										
I–II	Anabasis articulata (desert shrub)			φ									o
II	Quercus ithaburensis (deciduous tree)				φ / o								
II	Crataegus azarolus (deciduous tree)				φ / o								
II	Pistacia atlantica (deciduous tree)					φ / o							
II	Quercus calliprinos (evergreen tree)					o	φ						
II	Pistacia lentiscus (evergreen shrub)				o		φ						
II	Ceratonia siliqua (evergreen tree)		o				φ						
II	Tamarix gallica var. maris mortui (evergreen tree)				■ / o								
III	Eucalyptus camaldulensis (evergreen tree)										o		
III	Tamarix aphylla (evergreen tree)			■							o / ■	φ / ■	
IV	Acacia raddiana (evergreen tree)										o		
IV	Acacia cyanophylla (evergreen tree)		o										

φ Beginning of leaf buds unfolding   o First flower, buds appearing
■ Beginning of production of early wood type in trees with cambium active year-round

FIG. 178. Diagram showing the duration of cambial activity in various plants during the year. Active cambium at least in part of the plants examined—hatched; dormant cambium—blank.

palaestina). *Proustia cuneifolia* of Central Chile was also found to be active only during the rainy season (Aljaro *et al.*, 1972).

2. Trees and shrubs which exhibit more or less distinct growth rings, the development of which commences in the period March to May, i.e. in the spring. To this type belong many species, e.g. *Quercus* spp., *Pistacia* spp., *Calligonum commosum*, *Ceratonia siliqua*, and *Tamarix* spp. Some of the plants of this group have a marked dormant period, while in others, such as *Ceratonia* and two of the *Tamarix* species, the cambium is inactive for a very short period only, and may even be active throughout the year. In the latter case only the seasons of early and late wood production can be determined.

The length of the period of radial growth is generally 2–4 months in north temperate dicotyle-donous trees. It tends to shorten in plants of high altitudes and higher latitudes (as little as 4–6 weeks) and lengthen in plants from regions closer to the equator. The radial growth period is generally longer in conifers than in dicotyledons (Studhalter *et al.*, 1963).

3. Plants which are intermediate between the first two groups in that the commencement of the growth-ring production is in February (e.g. *Anabasis articulata* and *Salsola rosmarinus*).

4. Trees, such as *Eucalyptus camaldulensis* Dehn. and *Tamarix aphylla* Karst., in which the formation of the early wood starts in September (August), i.e. towards the end of the dry summer season. In *Eucalyptus* the late wood, which consists of one or two bands of flattened fibres two or three layers thick, is produced during the spring or in early summer, and the cambium is

inactive or almost so during July–August. In some specimens of *Tamarix aphylla* commencement of growth-ring production was found to be in August–September, while in other specimens two such periods were seen—one in the late summer and one at the end of February—resulting in the production of two growth rings annually.

5. Trees and shrubs in which there are no growth rings and in which the same type of wood is produced throughout the year.

In a number of tropical trees growing in India, interruption in cambial activity may also take place when extension growth is not continuous but intermittent, occurring in discrete flushes (Iqbal, 1981).

The South Californian pine species, *Pinus radiata*, grown in the North Island of New Zealand, was found not to enter full dormancy even in the winter (Barnett, 1971, 1973).

In trees of *Eucalyptus camaldulensis*, which either have no dormant period or have a short dormancy of a month or so during July–August, growth rings may sometimes be hard to distinguish, but usually one or two bands of late wood fibres are produced between April and June. Early wood formation starts in autumn (September). Amos *et al.* (1950) reported that the growth-ring formation in *Eucalyptus gigantea* in South Australia begins at the same calendar time, in September. Thus the course of wood formation of *E. camaldulensis*, in Israel, may suggest preservation of a rhythm adapted to the Australian origin in eucalypts. This feature of preservation of an endogenous growth rhythm is, however, confined to evergreens, as in deciduous plants the endogenous rhythm of cambial activity may become suppressed under the influence of sudden changes in climate that bring about leaf fall and bud burst. In the grapevine a second bud burst, which was accompanied by the formation of a second growth ring, could be artificially induced by defoliation (Bernstein and Fahn, 1960).

From the behaviour of tropical woody species and of *Eucalyptus* introduced into an area with a mild climate, it appears that the annual rhythm of growth-ring production, at least in evergreens, may be considered as a hereditary character. Within the limits set by the genotype, cambial activity is dependent on internal and external factors such as growth substances, temperature, photoperiod, and soil moisture.

Factors influencing cambial activity

The relationship between the cambial activity and the activity of the vegetative buds differs in different species. The activity of the cambium usually starts below the sprouting buds from where it spreads downwards. In deciduous diffuse-porous dicotyledons it was found to be rather slow, and divisions in the cambium at the base of the trunk may not begin until several weeks after they begin in the twigs (Cockerham, 1930; Wareing, 1958a). In decideous ring-porous dicotyledons (Priestley, 1930; Wareing, 1951; Wilcox, 1962) the downward spread of cambial activity is very rapid, and usually too rapid to detect a time lapse. In conifers the condition is intermediate and the spread of divisions in the cambium takes about a week (Wilcox, 1962). A correlation between extension growth and cambial activity could not always be defined in *Acacia raddiana* and *Eucalyptus camaldulensis* in which no clear dormant season of the cambium exists (Waisel *et al.*, 1966a; Fahn *et al.*, 1968).

Jost (1891) and later Coster (1927–1928) and Snow (1933, 1935) suggested that stimuli which activate the cambium are hormonal in nature. Up to the discovery and isolation of natural growth substances in the early 1930s, the evidence of the hormonal nature of the stimulus was based only on experiments of girdling and debudding of shoots. These experiments confirmed the role of buds and leaves in producing certain substances which are translocated downwards. Avery *et al.* (1937) concluded that these substances, which are produced in the buds, flow downwards from them along the axis where activity is induced. The transport of the signals from the developing buds downward can be understood on the basis of the polar properties of the tissues involved in auxin transport (Sachs, 1981). The identification of IAA (indole-3 acetic acid) led to the conclusion that this naturally-occurring and basipetally-transported substance, produced in the buds and developing shoots, is the stimulus reactivating the cambium.

Although auxin was regarded as the primary

stimulus causing resumption of cambial activity, further research suggested that substances other than auxin are also involved. Gouwentak and Mass (1940) and Gouwentak (1941) tried to induce activity of the cambium in shoots of some trees in the autumn, winter and early spring months by applying auxin, and concluded that an additional factor breaking the winter rest period is needed. It is possible that inhibitory substances accumulated during the previous growing season must be removed during the rest period before auxin can exert its effect. It is also possible that certain amounts of other growth substances are needed for the auxin to act efficiently (see also Reinders-Gouwentak, 1965; Brown, 1971).

Wareing (1958b) and Wareing et al. (1964) showed the interacting role of IAA and GA (gibberellic acid) in activating the cambium and in controlling the developmental pattern of the secondary xylem. When completely disbudded shoots of Acer pseudoplatanus were treated with IAA alone, a narrow zone of new xylem, in which only the vessels were lignified, was formed. The shoots treated with GA alone produced a zone of new xylem consisting of thin walled cells without vessels. In the shoots treated with both IAA and GA a wide zone of normal wood was formed. The importance of IAA and GA for cambial reactivation and normal xylem differentiation was also demonstrated by Lachuad (1983).

The application of ethrel to seedlings of Pinus halepensis and Ulmus americana greatly increases the growth of secondary xylem and phloem (Yamamoto and Kozlowski, 1987; Yamamoto et al., 1987). It should be mentioned that many effects previously considered to be induced directly by auxin are now known to be a result of auxin-induced ethylene formation.

Very little is as yet known about the factors that cause the cessation of cambial activity. Wareing (1951) and Wareing and Roberts (1956) have stressed the role of photoperiodism in the activity of the cambium. These investigators showed that in juvenile plants of Robinia pseudacacia and Pinus sylvestris cambial dormancy can be induced by the application of short day (SD) conditions and that the cambium can be reactivated by long day (LD) conditions. However, other external factors, such as temperature (Waisel and Fahn,

1965b), as well as internal ones, especially in adult plants, seem to play a major role in the control of the rhythm of cambial activity.

Waisel and Fahn (1965b) found that under SD conditions and high temperature (HT) seedling of Robinia exhibited a continuous cambial activity for long periods, although the terminal buds became inactive and abscised. In the tropical tree Polyalthia longifolia, Paliwal et al. (1975) and Ghouse and Hashmi (1979) found a correlation between the commencement of cambial activity and the high summer temperature.

Under LD conditions the presence of active buds is accompanied by the production of wood of the early wood type (see Chapter 15). The transfer of plants to SD conditions inactivates the buds, causes their abscission and is followed by the formation of wood of the late wood type (Wareing et al., 1964).

The fact that SD conditions (in cool temperatures) cause trees to stop growing and form resting buds, led Wareing and his co-workers to support the idea previously formulated by other authors, that dormancy may be a result of the action of growth inhibitors. Such a growth-inhibiting substance was identified in the late 1960s as abscisic acid (ABA). The application of synthetic ABA to seedlings of some woody plants, stopped growth and active apices became transformed into resting buds (El Antably et al., 1967; Wareing and Ryback, 1970). More recent research raises some uncertainty about the role of ABA in controlling cambial activity (Little and Wareing, 1981).

The relative production of xylem and phloem by the cambium was reported by Wilson (1964) to vary considerably in conifers. According to him the ratio of the number of xylem layers produced to the number of phloem layers produced may be 10:1 in vigorous conifers while only 1:1 in slow-growing ones. In Eucalyptus Waisel et al. (1966b) have found a constant ratio of about 4:1 (Fig. 177, no. 2) which was practically independent of such factors as photoperiod, temperature, NAA, kinetin, and triiodobenzoic acid. Only gibberellic acid seemed to accelerate the production of xylem while the formation of phloem remained quite normal or just a little suppressed, thus causing an increase in the xylem–phloem ratio to 10:1.

PART PLAYED BY CAMBIUM IN HEALING OF WOUNDS

One of the important functions of the cambium is to form *callus* or *wound tissue* over wounds. This tissue consists of masses of soft parenchyma tissue which is rapidly formed on or below the damaged surface of stems and roots. Callus may be formed by the division of parenchyma cells of the phloem, cortex, or vascular rays, but it is mostly formed by the cambium. The outer cells of the parenchymatous masses become suberized or a periderm develops in them (Chapter 17). Below this protective layer a reorganized cambium produces new vascular tissues.

Callus is also developed, at the beginning of the growth season, on the circumference of wounds caused by pruning. With the continued production, by the cambium, of secondary xylem in the undamaged area around the scar, and because of the eventual fusion of the cambial layers, the wound becomes completely covered. The production of the secondary wood continues so that the layer covering the wound is continually thickened.

If the cambium is damaged during the growing season it may be re-formed from the immature xylem below it if it is protected from drying out immediately after the wound is formed. The majority of the callus cells in such a case are derived from the proliferating ray cells. However, some of the differentiating fusiform cells of the xylem may divide transversely and also contribute cells to the callus. Sometimes in ringing experiments it is difficult to prevent the formation of new cambium even if the ringed surface is scratched with a knife, as the living immature xylem cells below it produce callus tissue in which the new cambium is developed (Li and Cui, 1988).

Several hypotheses have been suggested as to the physiology of cambial regeneration (see Wilson and Wilson, 1961). Many workers attributed it to hormonal stimuli coming from the existing cambium. A new cambium, however, may also develop inside the callus when, in a region around the periphery of the wound, the normal cambium and young xylem are removed. In this case the stimulus for differentiation of the vascular and cork cambia, if hormonal in nature, does not have

to be supplied by the existing intact cambium. It has been suggested that some kind of gradient, whether nutritional or hormonal, is established across the callus and determines the exact depth at which the cell layers will become the appropriate meristems.

EFFECT OF PRESSURE ON DIFFERENTIATION OF SECONDARY XYLEM AND PHLOEM

Brown and Sax (1962) have separated longitudinal strips or bark from *Populus trichocarpa* and *Pinus strobus* trees and left them attached at their upper ends. They have encased these strips in polythene bags to prevent desiccation and either left them free or replaced them in their original position under a membrane pressure of 0.5–1.0 atmosphere. In the first case, callus developed along the inner bark surface, new cork and vascular cambia began to differentiate at the outer margins of the strip in continuation with the existing cambium. The new cambia extended through the callus until they restored a continuous cambial ring. Within 3–4 weeks a normal pattern of xylem and phloem formation was established both by the original and newly differentiated cambia. Finally, a new "stem" with a radial symmetry was developed. In the second case where the bark strips were under pressure, a thin layer of callus was formed on the inner surface of the strip, which filled the free space between the strip and the stem portion from which it has been separated. As soon as the callus cells came under presure, cell division in the callus ceased and the cell walls became thickened and lignified. The original vascular cambium of the strip continued to produce xylem and phloem in the usual manner.

GRAFTING

Natural grafting of stems has often been observed in forest trees. Root grafts are probably even more common (La Rue, 1934). Natural root grafts are formed as a result of a pressure point of contact which develops by continuous growth in diameter of parallel or intersecting contiguous roots. Near this pressure point the characteristic vascular union becomes established (Graham and Bormann, 1966; Rao, 1966). Grafts are successful between plants of the same species and variety as well as between plants of different species belonging to the same family. In some families grafts are possible only between closely related species. Grafting is commonly used in agriculture (Mendel, 1936; Roberts, 1949; Mosse, 1953; Buck, 1954; Mosse and Labern, 1960). Several types of grafting are employed in plants with secondary thickening. In *bud grafting* a bud and some of the surrounding bark from one stem are transferred to another. The bark of the stem to which the bud is transferred is cut and either raised or removed, and the cambium of the bud portion has to join closely the cambium of the stem.

In *stem grafts* a portion of one stem (scion) is transferred to another stem (stock) in such a way that the cambia of the two are placed in contact. The union of stock and scion is not only through the cambia. The wood rays also proliferate and take part in the establishment of the union. Scions may also be grafted to roots with secondary xylem.

The process of grafting in dicotyledons and conifers is based: 1, on the ability of the cambium or of another appropriate tissue to form callus at the junction of the two grafted organs; 2, on the differentiation (in the callus) of a new cambium and vascular tissue, which are in continuation to those of the two grafts (Fig. 179) (Deloire, 1981; Barnett and Weatherhead, 1988).

Establishment of symplastic connections (development of continuous plasmodesmata) between callus cells of stock and scion is of major significance in graft union (Jeffree and Yeoman, 1983; Kollmann and Glockmann, 1985; Kollmann et al., 1985). In cases where the stock and scion are incompatible, weak unions occur between them. These are due to abnormal arrangement of the xylem elements, particularly the fibres which, instead of interlocking across the union, either curve in a horizontal direction or are separated by a layer of parenchymatous tissue. The discontinuity of the vascular tissues is due to some interruption in normal cambial activity at the point of union. In the phloem many necrotic cells occur singly or in groups; the rays may show abnormal proliferation (Roberts, 1949; Herrero, 1951; Mosse and Herrero, 1951; Mosse, 1955; Mosse and Scaramuzzi, 1956). Several physiological factors causing incompatibility of grafting have been suggested, e.g. differences in season of growth, in vigour, or in rate of callus production and metabolic differences between stock and scion. Virus diseases may also be responsible for some

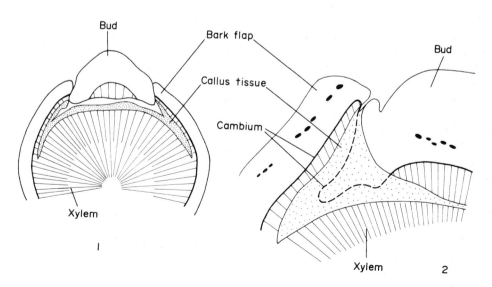

Fig. 179. 1, Diagram of a cross-section of apple stem showing distribution of tissues in a 3-week-old bud union. 2, As in no. 1, but in an 8-week-old bud union and showing the location of the connecting cambium.

union failures (see Rogers and Beakbane, 1957).

Closely related herbaceous plants can also be grafted, e.g. annual *Medicago* species (Simon, 1967). La Rue and Reissig (1946) found that dicotyledonous leaves grafted readily where callus formation could supply meristematic tissue for union.

Successful grafting of monocotyledons has also been reported (La Rue, 1944; Muzik and La Rue, 1954; Muzik, 1958). It was shown that, contrary to the accepted practice, a cambium is not essential for graft union but that any meristematic tissue is suitable for this purpose. In grafted monocotyledons a regular process of union occurs. Firstly, a darkly coloured contact layer appears, then an enlargement of parenchyma cells occurs next to this layer. These parenchyma cells divide and the contact layer disappears between the vascular bundles. Some of the new cells then differentiate into tracheary elements which unite the vascular bundles.

Chimeras may develop as a result of grafting. Graft chimeras are formed in cases where the apex develops by common growth of mixed meristematic tissues of different plant components. For details on chimeras and problems of grafting in general, see Brabec (1965).

REFERENCES

ALJARO, M. E., AVILA, G., HOFFMANN, A., and KUMMEROW, J. (1972) The annual rhythm of cambial activity in two woody species of the Chilean "Matorral". *Am. J. Bot.* **59**: 879–85.

ALVIM, P. DE T. (1964) Tree growth and periodicity in tropical climates. In: *The Formation of Wood in Forest Trees* (ed. M. H. Zimmermann). Academic Press, New York, pp. 479–95.

AMOS, G. L., BISSET, I. J. W., and DADSWELL, H. E. (1950) Wood structure in relation to growth in *Eucalyptus gigantea* Hook. f. *Aust. J. Sci. Res.* B, **3**: 393–413.

ASH, J. (1983) Growth rings in *Agathis robusta* and *Araucaria cunninghamii* from Tropical Australia. *Aust. J. Bot.* **31**: 269–275.

AVERY, G. S., JR., BURKHOLDER, P. R., and CREIGHTON, H. B. (1937) Production and distribution of growth hormone in shoots of *Aesculus* and *Malus*, and its probable role in stimulating cambial activity. *Am. J. Bot.* **24**: 51–58.

BAILEY, I. W. (1920a) The cambium and its derivative tissues: II, Size variations of cambial initials in gymnosperms and angiosperms. *Am. J. Bot.* **7**: 355–67.

BAILEY, I. W. (1920b) The cambium and its derivative tissues: III, A reconnaissance of cytological phenomena in the cambium. *Am. J. Bot.* **7**: 417–34.

BAILEY, I. W. (1920c) The formation of cell plate in the cambium of the higher plants. *Proc. Natn. Acad. Sci.* **6**: 197–200.

BAILEY, I. W. (1923) The cambium and its derivative tissues: IV, The increase in girth of the cambium. *Am. J. Bot.* **10**: 499–509.

BAILEY, I. W. (1930) The cambium and its derivative tissues: V, A reconnaissance of the vacuome in living cells. *Z. Zellforsch. mikrosk. Anat.* **10**: 651–82.

BANNAN, M. W. (1950) The frequency of anticlinal divisions in fusiform cambial cells of *Chamaecyparis*. *Am. J. Bot.* **37**: 511–19.

BANNAN, M. W. (1951a) The reduction of fusiform cambial cells in *Chamaecyparis* and *Thuja*. *Can. J. Bot.* **29**: 57–67.

BANNAN, M. W. (1951b) The annual cycle of size changes in the fusiform cambial cells of *Chamaecyparis* and *Thuja*. *Can. J. Bot.* **29**: 421–37.

BANNAN, M. W. (1955) The vascular cambium and radial growth in *Thuja occidentalis* L. *Can. J. Bot.* **33**: 113–38.

BANNAN, M. W. (1962) The vascular cambium and tree-ring development. In: *Tree Growth* (ed. T. T. Kozlowski). Ronald Press, New York, pp. 3–21.

BANNAN, M. W. (1966) Spiral grain and anticlinal divisions in the cambium of conifers. *Can. J. Bot.* **44**: 1515–38.

BANNAN, M. W. (1968) Polarity in the survival and elongation of fusiform initials in conifer cambium. *Can. J. Bot.* **46**: 1005–8.

BANNAN, M. W. and WHALLEY, B. E. (1950) The elongation of fusiform cambial cells in *Chamaecyparis*. *Can. J. Res.*, Sect. C, *Bot. Sci.* **28**: 341–55.

BARGHOORN, E. S., JR. (1940) Origin and development of the uniseriate ray in the Coniferae. *Bull. Torrey Bot. Club* **67**: 303–28.

BARNETT, J. R. (1971) Winter activity in the cambium of *Pinus radiata*. *N.Z. J. For. Sci.* **1**: 208–22.

BARNETT, J. R. (1973) Seasonal variation in ultrastructure of the cambium in New Zealand grown *Pinus radiata* D. Don. *Ann. Bot.* **37**: 1005–11.

BARNETT, J. R. and WEATHERHEAD, J. (1988) Graft formation in sitca spruce: a scanning electron microscope study. *Ann. Bot.* **61**: 581–587.

BERGGREN, B. (1987) Structure and cytochemistry of the procambium in *Salix* buds during dormancy and dormancy breaking. *Nord. J. Bot.* **7**: 153–167.

BERNSTEIN, Z. and FAHN, A. (1960) The effect of annual and bi-annual pruning on the seasonal changes in xylem formation in the grapevine. *Ann. Bot.* **24**: 159–71.

BOSCAGLI, A. (1982) The starch content in *Fraxinus ornus* L. during the year cycle. Histological observations. *G. bot. ital.* **116**: 41–49.

BRABEC, F. (1965) Pfropfung und Chimären, unter besonderer Berücksichtigung der entwicklungsphysiologischen Problematic. In: W. Ruhland, *Handbuch der Pflanzenphysiologie* **15** (2): 388–498.

BRAUN, H. J. (1955) Beiträge zur Entwicklungsgeschichte der Markstrahlen. *Bot. Stud. Jena* **4**: 73–131.

BROWN, C. L. (1971) Secondary growth. In: *Trees, Structure and Function* (eds. M. H. Zimmermann and C. L. Brown). Springer, Berlin, pp. 67–123.

BROWN, C. L. and SAX, K. (1962) The influence of pressure on the differentiation of secondary tissues. *Am. J. Bot.* **49**: 683–91.

BUCK, G. J. (1954) The histology of the bud–graft union in roses. *Iowa State Coll. J. Sci.* **28**: 587–602.

BÜNNING, E. (1965) Die Entstehung von Mustern in der

Entwicklung von Pflanzen. In: W. Ruhland, *Handbuch der Pflanzenphysiologie* **15** (1): 383–408.

BUTTERFIELD, B. G. (1976) The ontogeny of the vascular cambium in *Hoheria angustifolia* Raoul. *New Phytol.* **77**: 409–20.

BUVAT, R. (1956) Variations saisonnières du chondriome dans le cambium de *Robinia pseudoacacia*. *CR Acad. Sci. (Paris)* **243**: 1908–11.

CARMI, A., SACHS, T. and FAHN, A. (1972) The relation of ray spacing to cambial growth. *New Phytol.* **71**: 349–53.

CATESSON, A. M. (1964) Origine, fonctionnement et variations cytologiques saisonnières du cambium de l'*Acer pseudoplatanus* L. (Aceracées). *Annls Sci. nat., Bot. ser.* 12, **5**: 229–498.

CHOWDHURY, K. A. (1961) Growth rings in tropical trees and taxonomy. *10th Pac. Sci. Congr.* Abstracts 280.

CICHAN, M. A. (1985a) Vascular cambium and wood development in carboniferous plants. I. Lepidodendrales. *Am. J. Bot.* **72**: 1163–1176.

CICHAN, M. A. (1985b) Vascular cambium and wood development in carboniferous plants. II. *Sphenophyllum pluri foliatum* Williamson and Scott (Sphenophyllales). *Bot. Gaz.* **146**: 395–403.

COCKERHAM, G. (1930) Some observations on cambial activity and seasonal starch content in sycamore (*Acer pseudoplatanus*). *Proc. Leeds Phil. Lit. Soc.* **2**: 64–80.

COSTER, CH. (1927–8) Zur Anatomie und Physiologie der Zuwachszonen und Jahresringbildung in den Tropen. *Annls Jard. bot. Buitenz.* **37**: 49–160; **38**: 1–114.

CUMBIE, B. G. (1967a) Developmental changes in the vascular cambium of *Leitneria floridana*. *Am. J. Bot.* **54**: 414–24.

CUMBIE, B. G. (1967b) Development and structure of the cambium of *Canavalia*. *Bull. Torrey Bot. Club* **94**: 162–75.

DELOIRE, A. (1981) Etude histogénétique du greffage herbacé de combinaisons compatibles du genre *Vitis*. *Vitis* **20**: 85–92.

EL ANTABLY, H. M. M., WAREING, P. F. and HILLMAN, J. (1967) Some physiological responses to D,L abscisin (dormin). *Planta* **73**: 74–90.

ESAU, K. (1965) *Vascular Differentiation in Plants.* Holt, Rinehart & Winston, New York.

ESSIAMAH, S. and ESCHRICH, W. (1985) Changes of starch content in the storage tissue of deciduous trees during winter and spring. *IAWA Bull.* **6**: 97–106.

EVERT, R. F. (1961) Some aspects of cambial development in *Pyrus communis*. *Am. J. Bot.* **48**: 479–88.

EVERT, R. F. (1963) The cambium and seasonal development of the phloem in *Pyrus malus*. *Am. J. Bot.* **50**: 149–59.

EWERS, F. W. (1982a) Secondary growth in needle leaves of *Pinus longaeva* (Bristlecone pine) and other conifers: quantitative data. *Am. J. Bot.* **69**: 1552–1559.

EWERS, F. W. (1982b) Development and cytological evidence for mode of origin of secondary phloem in needle leaves of *Pinus longaeva* (bristlecone pine) and *P. flexilis*. *Bot. Jahrb. Syst.* **103**: 59–88.

FAHN, A. (1953) Annual wood ring development in maquis trees of Israel. *Palest. J. Bot. Jerusalem ser.* **6**: 1–26.

FAHN, A. (1955) The development of the growth ring in wood of *Quercus infectoria* and *Pistacia lentiscus* in the hill region of Israel. *Trop. Woods* **101**: 52–59.

FAHN, A. (1958) Xylem structure and annual rhythm of development in trees and shrubs of the desert: I, *Tamarix aphylla, T. jordanis* var. *negevensis, T. gallica* var. *maris-mortui. Trop. Woods* **109**: 81–94.

FAHN, A. (1959a) Xylem structure and annual rhythm of development in trees and shrubs of the desert: II. *Acacia tortilis* and *A. raddiana. Bull. Res. Counc. Israel* **7D**: 23–28.

FAHN, A. (1959b) Xylem structure and annual rhythm of development in trees and shrubs of the desert: III, *Eucalyptus camaldulensis* and *Acacia cyanophylla. Bull. Res. Counc. Israel* **7D**: 122–9.

FAHN, A. (1959c) Annual rhythm of xylem development in trees and shrubs in Israel. *Proc. IX. Int. Bot. Congr. Montreal,* **110**.

FAHN, A. (1962) Xylem structure and the annual rhythm of cambial activity in woody species of the East Mediterranean regions. *News Bull. Int. Ass. Wood Anat.* **1962** (1): 2–6.

FAHN, A. and SARNAT, C. (1963) Xylem structure and annual rhythm of development in trees and shrubs of the desert: IV, Shrubs. *Bull. Res. Counc. Israel* **11D**: 198–209.

FAHN, A., WAISEL, Y., and BENJAMINI, L. (1968) Cambial activity in *Acacia raddiana* Savi. *Ann. Bot.* **32**: 677–86.

FAHN, A., BEN-SASSON, R., and SACHS, T. (1972) The relation between the procambium and the cambium. In: *Research Trends in Plant Anatomy* (eds. A. K. M. Ghouse and Mohd. Yunus). Tata McGraw-Hill, New Delhi, pp. 161–70.

FAHN, A., BURLEY, J., LONGMAN, K. A., MARIAUX, A., and TOMLINSON, P. B. (1981) Possible contributions of wood anatomy to the determination of the age of tropical trees. In: *Age and Growth of Tropical Trees* (eds. F. H. Bormann and G. Berlyn). School of Forestry & Environmental Studies, Bull **94**. Yale University, New Haven, pp. 31–54.

GHOUSE, A. K. M. and HASHMI, S. (1979) Cambium periodicity in *Polyalthia longifolia. Phytomorphology* **29**: 64–67.

GHOUSE, A. K. M. and HASHMI, S. (1982) Impact of extension growth and flowering on the cambial activity of *Delonix regia* Rafin. *Proc. Indian Acad. Sci. (Plant Sci.)* **91**: 201–209.

GHOUSE, A. K. M. and KHAN, M. I. H. (1977) Seasonal variation in the nuclear number of fusiform cambial initials in *Psidium guajava* L. *Caryologia* **30**: 441–444.

GOOSEN-DE ROO, L., BURGGRAAF, P. D. and LIBBENGA, K. R. (1983) Microfilament bundles associated with tubular endoplasmic reticulum in fusiform cells in the active cambial zone of *Fraxinus excelsior* L. *Protoplasma* **116**: 204–208.

GOUWENTAK, C. A. (1941) Cambial activity as dependent on the presence of growth hormone and the non-resting conditions of stems. *Proc. Acad. Sci. Amst.* **44**: 654–63.

GOUWENTAK, C. A. and MASS, A. L. (1940) Kambiumtätigkeit und Wuchsstoff: II, *Meded. Landbouwhoogesch. Wageningen* **44**: 3–16.

GRAHAM, B. F., JR., and BORMANN, F. H. (1966) Natural root grafts. *Bot. Rev.* **32**: 255–92.

HABERLANDT, G. (1918) *Physiologische Pflanzenanatomie*, 5th edn. W. Engelmann, Leipzig.

HEJNOWICZ, Z. (1964) Orientation of the partition in pseudotransverse division in cambia of some conifers. *Can. J. Bot.* **42**: 1685–91.

HEJNOWICZ, Z. and KRAWCZYSZYN, J. (1969) Oriented morphogenetic phenomena in cambium of broadleaved trees. *Acta Soc. Bot. Pol.* **38**: 547–60.

HEJNOWICZ, Z. and ROMBERGER, J. A. (1973) Migrating cambial domains and the origin of wavy grain in xylem of broadleaved trees. *Am. J. Bot.* **60**: 209–22.

HEINOWICZ, Z. and ZAGORSKA-MAREK, B. (1974) Mechanism of changes in grain inclination in wood produced by storied cambium. *Acta Soc. Bot. Pol.* **43**: 381–398.

HERRERO, J. (1951) Studies of compatible and incompatible graft combinations with special reference to hardy fruit trees. *J. Hort. Sci.* **26**: 186–237.

IQBAL, M. (1981) *A decade of research in plant anatomy (1971–80)*. Aligarh Muslim Univ. India.

JEFFREE, C. E. and YEOMAN, M. M. (1983) Development of intercellular connections between opposing cells in graft union. *New Phytol.* **93**: 491–509.

JOST, L. (1891) Über Dickenwachstum und Jahresringbildung. *Bot. Ztg.* **49**: 482–499.

KIRSCHNER, H., SACHS, T., and FAHN, A. (1971) Secondary xylem reorientation as a special case of vascular tissue differentiation. *Israel J. Bot.* **20**: 184–98.

KOLLMANN, R. and GLOCKMANN, C. (1985) Studies on graft unions. I. Plasmodesmata between cells of plants belonging to different unrelated taxa. *Protoplasma* **124**: 224–235.

KOLLMANN, R., YANG, S. and GLOCKMANN, C. (1985) Studies on graft unions. II. Continuous and half plasmodesmata in different regions of the graft interface. *Protoplasma* **126**: 19–29.

KORIBA, K. (1958) On the periodicity of tree-growth in the tropics, with reference to the mode of branching, the leaf fall, and the formation of the resting bud. *Gard. Bull. Straits Settlements* **17**: 11–81.

KRAWCZYSZYN, J. (1977) The transition from nonstoried to storied cambium in *Fraxinus excelsior*. I. The occurrence of radial anticlinal divisions. *Can. J. Bot.* **55**: 3034–41.

KRAWCZYSZYN, J. and ROMBERGER, J. A. (1980) Interlocked grain, cambial domains, endogenous rhythms, and time relations, with emphasis on *Nyssa sylvatica*. *Am. J. Bot.* **67**: 228–36.

LACHAUD, S. (1983) Xylogénèse chez les Dictotylédones arborescentes. IV. Influence des bourgeons, de l'acide β-indolyl acétique et l'acide gibberellique sur la réactivation cambiale et la xylogénèse dans les jeunes tiges de Hêtre. *Can. J. Bot.* **61**: 1768–1774.

LARSON, P. R. (1976) Procambium vs. cambium and protoxylem vs. metaxylem in *Populus deltoides* seedlings. *Am. J. Bot.* **63**: 1332–48.

LARSON, P. R. (1982) The concept of cambium. In: *New Perspectives in Wood Anatomy* (ed. P. Baas). Martinus Nijhoff/Dr. W. Junk Publishers, The Hague, pp. 85–121.

LA RUE, C. D. (1934) Root grafting in trees. *Am. J. Bot.* **21**: 121–6.

LA RUE, C. D. (1944) Grafts of monocotyledons secured by the use of intercalary meristem (abstract). *Am. J. Bot.* **31**: 3–4.

LA RUE, C. D. and REISSIG, F. (1946) Grafting in leaves (abstract). *Am. J. Bot.* **33**: 220.

LAWTON, J. R. (1972) Seasonal variations in the secondary phloem of some forest trees from Nigeria. II. Structure of the phloem. *New Phytol.* **71**: 335–348.

LI ZHENGLI (LEE CHENGLEE) and CUI KEMING. (1988) Differentiation of secondary xylem after girdling. *IAWA Bull.* **9**: 375–383.

LIPHSHITZ, N. and LEV-YADUN, S. (1986) Cambial activity of evergreen seasonal dimorphics around the Mediterranean. *IAWA Bull.* **7**: 145–153.

LITTLE, C. H. A. and WAREING, P. F. (1981) Control of cambial activity and dormancy in *Picea sitchensis* by indol-3-ylacetic and abscisic acid. *Can. J. Bot.* **59**: 1480–1493.

MAHMOOD, A. (1968) Cell grouping and primary wall generations in the cambial zone, xylem and phloem in *Pinus*. *Aust. J. Bot.* **16**: 177–95.

MARIAUX, A. (1967) Les cornes dans les bois tropicaux africains, nature et périodicité. *Bois et Forêts des Tropiques* **113**: 3–14; **114**: 23–37.

MENDEL, K. (1936) The anatomy and histology of the bud-union in citrus. *Palest. J. Bot. Hort. Sci.* **1**: 3–46.

MESSERI, A. (1948) L'evoluzione della cerchia legnosa in *Pinus halepensis* Mill. in Bari dal lugio 1946 al lugio 1947. *Nuovo G. bot. ital.*, NS **55**: 111–32.

MINERVINI, J. (1948) Ciclo di accrescimento e differenziazione della gemme in piante perenni nel territorio di Bari: IV, L'evoluzione della cerchia legnosa in *Viburnum tinus* L. dal dicembre 1946 al novembre 1947 a Bari. *Nuovo G. bot. ital.*, NS **55**: 433–45.

MOSSE, B. (1953) The origin and structure of bridge tissues in ring-grafted apple stems. *J. Hort. Sci.* **28**: 41–48.

MOSSE, B. (1955) Symptoms of incompatibility induced in a peach by ring grafting with an incompatible rootstock variety. *A. Rep. E. Malling Res. Sta.* 1954, pp. 76–77.

MOSSE, B. and HERRERO, J. (1951) Studies on incompatibility between some pear and quince grafts. *J. Hort. Sci.* **26**: 238–45.

MOSSE, B. and LABERN, M. V. (1960) The structure and development of vascular nodules in apple bud-unions. *Ann. Bot.* **24**: 500–7.

MOSSE, B. and SCARAMUZZI, F. (1956) Observations on the nature and development of structural defects in the unions between pear and quince. *J. Hort. Sci.* **31**: 47–54.

MURMANIS, L. (1971) Structural changes in vascular cambium of *Pinus strobus* L. during an annual cycle. *Ann. Bot.* **35**: 133–41.

MURMANIS, L. (1977) Development of vascular cambium into secondary tissue of *Quercus rubra* L. *Ann. Bot.* **41**: 617–20.

MUZIK, T. J. (1958) Role of parenchyma cells in graft union in vanilla orchid. *Science* **127**: 82.

MUZIK, T. J. and LA RUE, C. D. (1954) Further studies on the grafting of monocotyledonous plants. *Am. J. Bot.* **41**: 448–55.

NEWMAN, I. V. (1956) Pattern in the meristems of vascular plants: I, Cell partition in living apices and in the cambial zone in relation to the concepts of initial cells and apical cells. *Phytomorphology* **6**: 1–19.

OPPENHEIMER, H. R. (1945) Cambial wood production in stems of *Pinus halepensis*. *Palest. J. Bot., Rehovot ser.* **5**: 22–51.

PALIWAL, G. S., PRASAD, N. V. K., SAJWAN, V. S. and AGGARWAL, S. K. (1975) Seasonal activity of cambium in some tropical trees. II. *Polyaltia longifolia*. *Phytomorphology* **25**: 478–484.

PHILIPSON, W. R., WARD, J. M., and BUTTERFIELD, B. G. (1971) *The Vascular Cambium*. Chapman & Hall, London.

PRIESTLEY, J. H. (1930) Studies in the physiology of cambial activity: III, The seasonal activity of the cambium. *New Phytol.* **29**: 316–54.

RAO, A. N. (1966) Developmental anatomy of natural root grafts in *Ficus globosa*. *Aust. J. Bot.* **14**: 269–76.

RAO, K. S. and CATESSON, A.-M. (1987) Changes in the membrane components of nondividing cambial cells. *Can. J. Bot.* **65**: 246–254.

REINDERS-GOUWENTAK, C. A. (1965) Physiology of the cambium and other secondary meristems of the shoot. In: W. Ruhland, *Handbuch der Pflanzenphysiologie* **15** (1): 1077–1105.

ROBARDS, A. W. and KIDWAI, P. (1969) A comparative study of the ultrastructure of resting and active cambium of *Salix fragilis* L. *Planta* **84**: 239–49.

ROBERTS, R. H. (1949) Theoretical aspects of graftage. *Bot. Rev.* **15**: 423–63.

ROGERS, W. S. and BEAKBANE, A. B. (1957) Stock and scion relations. *A. Rev. Pl. Physiol.* **8**: 217–36.

ROLAND, J.-C. (1978) Early differences between radial walls and tangential walls of actively growing cambial zone. *IAWA Bull. 1978* (1): 7–10.

SACHS, T. (1981) The control of the patterns of vascular differentiation in plants. In: *Advances in Botanical Research*, Vol. 9 (ed. H. W. Woolhouse). Academic Press, London, pp. 151–262.

SHARMA, H. K., SHARMA, D. D., and PALIWAL, G. S. (1979) Annual rhythm of size variations in cambial initials of *Azadirachta indica* A. Juss. *Geobios* 6: 127–9.

SHARMA, D. D., SHARMA, H. K., and PALIWAL, G. S. (1980) Annual rhythm of size variations in the cambial initials of *Chorisia speciosa* St. Hil. *Science and Culture* 45: 96–97.

SIMON, J. P. (1967) Relationship in annual species of *Medicago*: IV, Interspecific graft affinities between selected species. *Aust. J. Bot.* 15: 75–82.

SNOW, R. (1933) The nature of the cambial stimulus. *New Phytol.* 32: 288–296.

SNOW, R. (1935) Activation of cambial growth by pure hormones. *New Phytol.* 34: 347–360.

SOH, W. Y., HONG, S. S. and CHO, D. Y. (1988) The ontogeny of the vascular cambium in *Ginkgo biloba* roots. *Bot. Mag. Tokyo* 101: 39–53.

SRIVASTAVA, L. M. (1963) Cambium and vascular derivatives of *Ginkgo biloba*. *J. Arnold Arb.* 44: 165–92.

SRIVASTAVA, L. M. (1966) On the fine structure of the cambium of *Fraxinus americana* L. *J. Cell Biol.* 31: 79–93.

SRIVASTAVA, L. M. and O'BRIEN, T. P. (1966) On the ultrastructure of cambium and its vascular derivatives: I, Cambium of *Pinus strobus* L. *Protoplasma* 61: 257–76.

STERLING, C. (1946) Cytological aspects of vascularization in *Sequoia*. *Am. J. Bot.* 33: 35–45.

STUDHALTER, R. A., GLOCK, W. S., and AGERTER, S. R. (1963) Tree growth. *Bot. Rev.* 29: 245–365.

THAIR, B. W. and STEEVES, T. A. (1976) Response of the vascular cambium to reorientation in patch grafts. *Can. J. Bot.* 54: 361–73.

VÖCHTIG, H. (1892) *Über Transplantation am Pflanzenkörper.* Verlag H. Laupp'schen Buchhandlung, Tübingen.

WAISEL, Y. and FAHN, A. (1965a) A radiological method for the determination of cambial activity. *Physiol. Plant.* 18: 44–46.

WAISEL, Y. and FAHN, A. (1965b) The effects of environment on wood formation and cambial activity in *Robinia pseudacacia* L. *New Phytol.* 64: 436–42.

WAISEL, Y., NOAH, I., and FAHN, A. (1966a) Cambial activity in *Eucalyptus camaldulensis* Dehn.: I, The relation to extension growth in young saplings. *La-Yaaran* 16: 59–73. (English summary 103–8.)

WAISEL, Y., NOAH, I., and FAHN, A. (1966b) Cambial activity in *Eucalyptus camaldulensis* Dehn.: II, The production of phloem and xylem elements. *New Phytol.* 65: 319–24.

WAREING, P. F. (1951) Growth studies in woody species: IV, The initiation of cambial activity in ring-porous species. *Physiol. Plant.* 4: 546–62.

WAREING, P. F. (1958a) The physiology of cambial activity. *J. Inst. Wood Sci.* 1: 34–42.

WAREING, P. F. (1958b) Interaction between IAA and GA in cambial activity. *Nature* 181: 1744–5.

WAREING, P. F. and ROBERTS, D. L. (1956) Photoperiodic control of cambial activity in *Robinia pseudacacia*. *New Phytol.* 55: 356–66.

WAREING, P. F. and RYBACK, G. (1970) Abscisic acid: a newly discovered growth-regulating substance in plants. *Endeavour* 29: No. 107, pp. 84–88.

WAREING, P. F., HANNEY, C. E. A. and DIGBY, J. (1964) The role of endogenous hormones in cambial activity and xylem differentiation. In: *The Formation of Wood in Forest Trees* (ed. M. H. Zimmermann). Academic Press, New York, pp. 323–344.

WILCOX, H. (1962) Cambial growth characteristics. In: *Tree Growth* (ed. T. T. Kozlowski). Ronald Press, New York, pp. 57–88.

WILSON, B. F. (1964) A model for cell production by the cambium of conifers. In: *The Formation of Wood in Forest Trees* (ed. M. H. Zimmermann). Academic Press, New York, pp. 19–36.

WILSON, B. F. (1966) Mitotic activity in the cambial zone of *Pinus strobus*. *Am. J. Bot.* 53: 364–72.

WILSON, J. W. and WILSON, P. M. W. (1961) The position of regenerating cambia: a new hypothesis. *New Phytol.* 60: 63–73.

WLOCH, W. (1987) Transition areas in the domain patterns of storeyed cambium of *Tilia cordata* Mill. *Acta Soc. Bot. Pol.* 56: 645–665.

WLOCH, W. and ZAGORSKA-MAREK, B. (1982) Reconstruction of storeyed cambium in the linden. *Acta Soc. Bot. Pol.* 51: 215–228.

YAMAMOTO, F. and KOZLOWSKI, T. T. (1987) Effect of ethrel on growth and stem anatomy of *Pinus halepensis* seedlings. *IAWA Bull.* 8: 11–19.

YAMAMOTO, F., ANGELES, G. and KOZLOWSKI, T. T. (1987) Effect of ethrel on stem anatomy of *Ulmus americana* seedlings. *IAWA Bull.* 8: 3–9.

ZAGORSKA-MAREK, B. and LITTLE, C. H. A. (1986) Control of fusiform initial orientation in the vascular cambium of *Abies balsamea* stem by indol-3-ylacetic acid. *Can. J. Bot.* 64: 1120–1128.

ZIMMERMANN, M. H. and BROWN, C. L. (1971) *Trees—Structure and Function.* Springer-Verlag, Berlin.

CHAPTER 15

SECONDARY XYLEM

THE cambium, which is described in the preceding chapter, produces, towards the centre of the stem and root, secondary xylem which comprises various elements—tracheids, vessel members, different types of fibres, parenchyma cells, xylem ray cells, and sometimes secretory cells. The occurrence and the arrangement of these elements vary in different groups of plants. The quantitative differences in the number of cells, as well as in the size of the elements that exist between the species of a single genus, make it possible to identify the plant by its secondary xylem alone.

Usually it is difficult to distinguish clearly between the primary and secondary xylem. The best distinguishing feature between these two tissues is the length of the tracheary elements (Sanio, 1872; Bailey and Tupper, 1918; Bailey, 1944). The first-formed tracheary elements of secondary xylem are generally much shorter than the tracheary elements of primary xylem; they are even shorter than the pitted tracheary elements of primary xylem which are themselves usually shorter than the spirally thickened elements. This feature may be the result of stretching, which takes place during the development of the primary xylem elements, but not during that of elements formed by the cambium. Also it is possible, that, prior to the differentiation of the cambium from the procambium, the cells of the latter divide transversely.

In some growth forms such as rosette trees, however, it has been observed that a rapid decrease in length of tracheary elements of the secondary xylem as related to the primary xylem does not occur. In these plants the decrease in length of elements is either gradual or almost imperceptible. Woody plants of this type are suggested by Carlquist (1962) to be permanently juvenile. Carlquist also mentions some other primitive characters which occur in the secondary xylem of such plants and he ascribes these also to prolonged juvenility. Carlquist's theory of "paedomorphosis" was challenged by Mabberley (1974) who studied giant *Senecio* plants.

A great amount of research has been done on the relationship between the anatomy of the xylem and water movement in the plant. This aspect of wood anatomy has been thoroughly dealt with in the book *Xylem Structure and the Ascent of Sap* by Zimmerman (1983).

Taxonomic, phylogenetic and ecological wood anatomy is recently going through a phase of renewed interest. A short history and discussion on the systematic wood anatomy was presented by Baas (1982). A great number of articles on taxonomic wood anatomy is now published in the Bulletin of the International Association of Wood Anatomists (IAWA).

The structure, ontogeny, and phylogeny of the various xylem elements, both primary and secondary, have already been dealt with in Chapters 6 and 7. This chapter will mainly discuss the arrangement of the elements in the secondary xylem.

BASIC STRUCTURE OF SECONDARY XYLEM

The most distinctive feature characterizing the secondary xylem is the existence of two systems of elements which differ in the orientation of their longitudinal axes—one system is vertical and the other horizontal. The horizontal system comprises the xylem rays (Figs. 181, 192; Fig. 187, no. 2), and the vertical or axial system, the tracheary elements, fibres, and wood parenchyma. The living cells of the rays and of the axial system are

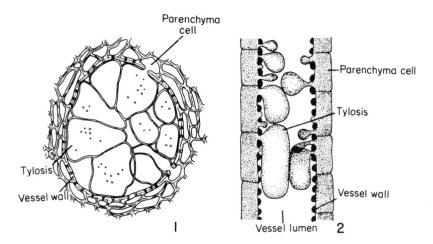

FIG. 180. 1, Cross-section of a vessel of *Robinia pseudacacia* showing tyloses. 2, Longitudinal section of a vessel of *Vitis vinifera* showing the development of tyloses from the neighbouring parenchyma cells. (No. 1 adapted from Strasburger.)

usually interconnected, so that it is possible to speak of a continuous system of living cells. This system is generally connected with the living cells of the pith, phloem, and cortex.

WOOD PARENCHYMA

Two types of parenchyma are found in the secondary xylem—the *axial parenchyma* and the *ray parenchyma* (Fig. 192). The ray parenchyma cells originate from special, relatively short cambial initials, while the cells of the axial parenchyma develop from fusiform initials. The cells of the axial parenchyma may be as long as the fusiform initials from which they are derived or much shorter as a result of transverse division prior to differentiation. The shorter axial parenchyma cells are the more common.

The ray-parenchyma cells may be variously shaped, but the following two forms are the most common: that in which the longest axis of the cells is radial, and that in which it is vertical. All ray parenchyma cells may have secondary walls, or only primary walls may be developed. Where secondary walls are developed the pit-pairs may be simple, half-bordered, and sometimes even bordered.

A distinction was made between two categories of xylem parenchyma cells (Czaninski, 1977): 1. *storage parenchyma cells; 2. specialised-vessel associated cells* or *contact cells*, which are always contiguous with vessels.

The cells of the storage parenchyma (of vertical and horizontal system) store reserve materials, such as starch and fats. Tannins, crystals, silica bodies (ter Welle, 1976; Richter, 1980), and other substances are also frequently found in many of these cells. Sometimes the parenchyma cells containing crystals divide so that chambers, each containing one crystal, are formed (Chattaway, 1955, 1956). In some species cystoliths occur in ray or axial parenchyma cells (ter Welle, 1980). Because of the variability in form, crystals may sometimes be of taxonomic significance (Baas *et al.*, 1988).

The contact cells have numerous large pits between them and the adjacent vessels (Braun, 1970; Sauter, 1972). They contain numerous vacuoles and mitochondria, well-developed endoplasmic reticulum and little or no storage material (Czaninski, 1977). Like transfer cells (see Chapters 4 and 11), contact cells seem to be involved especially, in carbohydrate metabolism and short-distance transport (Sauter, 1972; Sauter *et al.*, 1973; Czaninski, 1977).

Braun (1983) suggested that the vessel associated tissues, consisting of the paratracheal contact parenchyma and the parenchymatous contact cells of the xylem rays, may play a role in the mechanism of water ascent in trees by releasing osmotically active substances into the vessels.

In some trees investigated, radial intercellular spaces were found to constitute a complex network of cavities. This network of spaces is important for radial gas exchange in the tree axis (Kučera, 1985).

In many plants the cells of both axial and ray parenchyma form protuberances which penetrate through the pits into the vessels after they become inactive, or into the vessels of xylem tissue that has been injured. These outgrowths are termed *tyloses* (singular: *tylosis*) (Figs: 180, 188; Fig. 187 no. 4). The nucleus and part of the cytoplasm of the parenchyma cells, from which the tylosis is formed, enter the tylosis. Tyloses may divide. Although tylosis formation is considered to be a normal phenomenon, in many species it has also been found to be induced by mechanical injury and by diseases (see Zimmermann, 1979).

RAYS, CELL ARRANGEMENT, HEARTWOOD, AND SAPWOOD

The number of xylem rays in a trunk increases with the increase in its girth. The length, width, and height of each ray can be measured. The length of the ray is determined in cross-sections of the wood. The width of the ray is measured in tangential sections and it is usually expressed as the maximal number of cells in a horizontal direction. The height of the ray, parallel to the longitudinal axis of the stem or root, is also measured from tangential sections and it is usually expressed in one of two ways—if it is not very large, in the number of cells, and if it is very large, in microns or millimetres. The dimensions of the rays vary in the different plants and sometimes even in the same plant (Fig. 195, no. 2; Fig. 199, no. 2). When the ray is one cell wide, it is said to be a *uniserate ray* (Fig. 183, no. 3; Fig. 193, no. 1; Fig. 194, no. 2); when two cells wide, *biseriate*, and when more than two cells wide, *multiseriate* (Fig. 190, no. 1; Fig. 198, no. 2). In a tangential section a multiseriate ray is seen to become narrow towards both its upper and lower edges, where it is usually uniseriate.

In species that have a storied cambium (see Chapter 14) a similar arrangement may exist in the xylem (Fig. 188, no. 2; Fig. 190, no. 1) Sometimes the storied arrangement becomes indistinct

because of the intrusive growth of the ends of the developing fibres and tracheids. The blurring of the arrangement occurs to different extents so that it is possible to distinguish different degrees of storied arrangement from that where the fibres, tracheary elements and axial parenchyma cells are equal in length and are arranged in horizontal rows, to that in which arrangement of the xylem elements is similar to that developed from a non-storied cambium. In storied xylem the vessel members are usually short. Phylogenetically the storied arrangement is thought to be the more advanced.

The outer part of the secondary xylem contains living cells and at least part of it is active in the transport of water. The outer part is termed the *sapwood* or *alburnum*. In most trees the inner portion of the secondary xylem completely ceases to conduct water and living cells in it die. This is accompanied by the disintegration of the protoplast, the loss of the cell sap and the removal of reserve materials from cells that stored them. In those species in which tyloses are a characteristic feature of the wood, the vessels in this inner portion become totally blocked, at this stage, by the formation of tyloses. The cell walls of the parenchyma cells which were little lignified may become more heavily lignified. Certain substances, such as oils, gums, resins, tannins, coloured substances, and aromatic compounds, develop in the cells or are accumulated in them. The development of coloured substances in the heartwood is a gradual process of oxidation and polymerization of phenols which in turn follows the disappearance of starch and an apparent breakdown in enzymatic control over the activities of living cells (Frey-Wyssling and Bosshard, 1959; Shah *et al.*, 1981; Nobuchi *et al.*, 1982). In the gymnospserms the flexible pit membrane becomes rigid and fixed in such a position that the torus closes the pit aperture. In this condition the pit is not functioning in water conduction and is called *aspirated*. The tori, which in sapwood consist mainly of pectin and a limited amount of cellulose and hemicellulose, become lignified (Bauch *et al.*, 1968). Secondary xylem that has undergone such changes is termed *heartwood* or *duramen* (Fig. 191, no. 1). The above-mentioned changes make the heartwood more resistant to decay. The accumulation of coloured substances in this part of the xylem makes it easily recognizable from the sapwood. Heartwood may sometimes be developed as the result of pathological conditions.

Many authors consider heartwood formation as the end result of changes in the living cells of the sapwood connected with their ageing. Bamber (1976) suggested that heartwood formation is a developmental process of the tree analogous to other phases of plant development, such as to abscission for instance. He considers heartwood formation as a regulatory process which keeps the amount of sapwood at an optimum level.

The quantitative relation between the amount of heart- and sapwood, and the degree of difference between them varies greatly in the different species, and the differences are influenced by the conditions under which the plants are grown. Bosshard (1966, 1967, 1968) distinguished four types of sapwood–heartwood relationship in trees: (1) *Trees having light heartwood* (e.g. *Abies alba*). Here necrobiosis of the storage cells occurs without the production of large quantities of pigmented heartwood substances, although precursors of these are apparently formed in the cambium. (2) *Trees with retarded formation of heartwood* (e.g. *Carpinus betulus*). (3) *Trees with obligatorily coloured heartwood* (e.g. *Quercus robur*). In this type coloured heartwood substances are invariably formed in the storage tissue and are generally capable of penetrating into the cell walls of all elements. (4) *Trees with facultatively coloured heartwood* (e.g. *Fraxinus excelsior, Beilschmiedia tawa*). In these trees the coloured substances need not be present in all samples and need not affect the entire heartwood. In addition, the coloured heartwood substances are commonly retained as wall coating or drop-like inclusions in

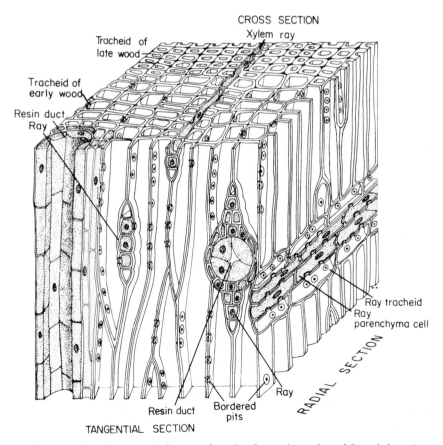

FIG. 181. Three-dimensional diagram of a cube of secondary xylem of *Pinus halepensis*.

the storage cells. In this case the cell walls of any of the xylem elements are not impregnated by the pigmented substances.

There are fundamental differences in the histological structure of the wood of dicotyledons and that of gymnosperms, and especially of that of the conifers. In the timber trade the wood of dicotyledons is known as *hardwood* and that of gymnosperms as *softwood*. These terms do not accurately express the degree of hardness, as in both groups wood with both hard and soft structure can be found.

SECONDARY XYLEM OF GYMNOSPERMAE

The structure of the secondary xylem of the gymnosperm (Fig. 181) is simpler and more homogeneous than that of the angiosperms. The principal differences are the absence of vessels in the wood of gymnosperms (with the exception of the Gnetales) and their presence in almost all angiosperms, and the relatively small amount of wood parenchyma, especially axial parenchyma, in the gymnosperms.

The vertical system

In most gymnosperms, the tracheary elements of the vertical system are, as has already been mentioned, tracheids. However, the tracheids of the late wood (i.e. those formed at the end of the growing season) develop relatively thick walls and their pits have small pit chambers and long canals. Because of this structure these tracheids may be termed fibre-tracheids. Libriform fibres are not found in the secondary xylem of gymnosperms. The xylem formed at the end of the growing season appears darker because of the special structure of its tracheids, and so growth rings in conifers are easily distinguishable.

The tracheids (and the fibre-tracheids) are from 0.5 mm to 11 mm long. Because of this great length each tracheid comes into contact with one or more rays. The tracheids overlap one another with flat, chisel-shaped ends. Neighbouring tracheids are joined by bordered pits which may be arranged in a single longitudinal row or in a few rows. In the latter case the pitting may be opposite or alternate. From various investigations it has been shown that the number of pits per tracheid may be from 50 to 300.

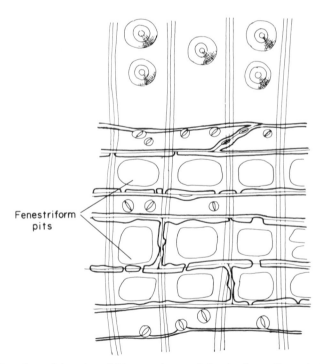

Fenestriform pits

FIG. 182. Portion of a radial section of *Pinus excelsa* wood showing fenestriform pits in the ray cells.

The size of the pits, the shape of the pit border and of the pit aperture, vary very greatly and therefore these features are important in the identification of gymnosperm wood. The pits are more numerous at the ends of the tracheids where they overlap one another. Usually the pits are present only on the radial walls of the tracheids, but in the tracheids of the late wood pits are also present on the tangential walls. Pits with tori are found in *Ginkgo*, the Gnetales and most of the Coniferales. In a microscopic radial longitudinal section of the wood of many gymnosperm species transversely orientated thickenings of the wall can be observed above and below the pits (Fig. 183, no. 4). These are thickenings of the middle lamella and the primary walls, and they are termed *crassulae* (see also Chapter 2).

Another feature of gymnosperm wood is the occurrence of *trabeculae* in the tracheids. Trabeculae are rod-shaped outgrowths of the tangential cell walls which grow across the cell lumen so as to connect the tangential walls. Tracheids containing trabeculae are usually arranged in long radial rows.

In some conifers (e.g. in *Taxus*) spiral thickenings have been observed in the internal wall surface of the pitted tracheids.

When axial parenchyma is present in the wood of conifers it is usually arranged in strips that are equally distributed throughout the secondary xylem. In certain conifers, e.g. *Araucaria* and *Taxus*, axial parenchyma is completely absent. In *Pinus*, axial parenchyma is present only in connection with the resin ducts (Fig. 181).

Cupressaceae) or secondary as well (e.g. in most species of the Pinaceae).

The ray tracheids all have lignified secondary walls. In certain conifers these cells have very thick walls with tooth-like or band-like thickenings which project into the cell lumen (Fig. 169, no. 2). The ray tracheids occur singly or in rows. They may be at the upper or lower edges of the ray, or they may be scattered among the ray parenchyma cells.

In the large majority of gymnosperms the rays are uniseriate and they are usually from one to twenty cells high, but they may be as high as sixty cells. If a resin duct passes through a ray it passes through the centre of the ray, which becomes more than one cell wide in that position.

Where the vertical tracheids come in contact with ray parenchyma cells the pit-pairs are usually half-bordered, i.e. the bordered pit is situated on the side of the tracheid and the simple pit on the side of the parenchyma cell. This area of contact between a ray parenchyma cell and a single vertical tracheid is termed a *cross-field*. The type of pits, their number and distribution in the cross-field are important features in the identification of gymnosperms wood (Fig. 184, no. 4).

In ray cells of some *Pinus* species very large window-like pits with extremely narrow borders, which extend almost throughout the width of the cell, occur. These pits are termed *fenestriform pits* (Fig. 182).

Rays

The rays in gymnosperms may comprise parenchyma cells only, i.e. *homocellular rays*, or parenchyma cells and tracheids, i.e. *heterocellular rays* (Fig. 181). The ray tracheids are distinguished from the ray parenchyma mainly by the presence of bordered pits and by the absence of a protoplast. The ray parenchyma cells contain living protoplasts in the sapwood, and generally in the heartwood, darkly coloured resins. The walls of the parenchyma cells may be primary only (e.g. in the Taxodiaceae, Araucariaceae, Taxaceae, and

Resin ducts

Resin ducts are developed in the vertical or both vertical and horizontal systems of a large number of gymnosperms (Bannan, 1936; Werker and Fahn, 1969; Bosshard and Hug, 1980; Sato and Ishida, 1982, 1983). The ducts develop schizogenously between resin-producing parenchyma cells which then form the epithelium of the duct. Sometimes a resin duct may become blocked by the enlargement of the epithelial cells; such structures are termed *tylosoids*. Differences exist in the thickness and lignification of the cell wall of the

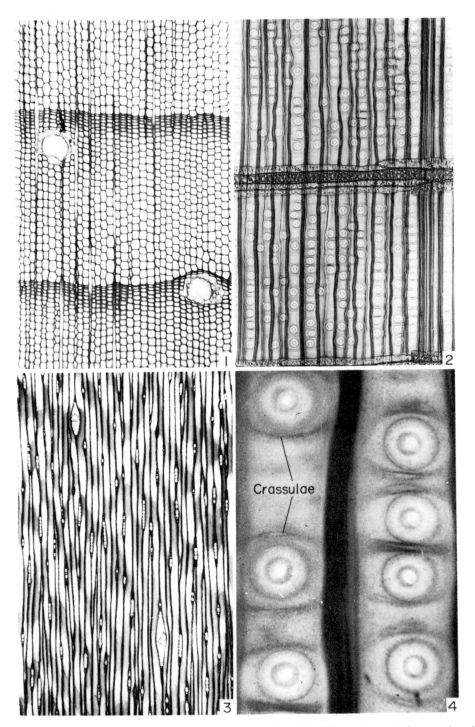

FIG. 183. Micrographs of the secondary xylem of *Pinus ponderosa* sectioned in different planes. 1, Cross-section showing growth rings and two resin ducts. ×55. 2, Radial section. ×55. 3, Tangential section. ×55. 4, Portion of a radial section showing tracheids with bordered pits and crassulae. ×400.

FIG. 184. 1, Cross-section of the secondary xylem of *Cupressus sempervirens.* × 45. 2, Cross-section of the secondary xylem of *Cedrus libani* in which resin ducts, developed as a result of wounding, can be seen. × 40. 3, Dark-field micrograph of bordered pits of *Cedrus libani* showing fringed torus. × 2400. 4, Portion of a radial section of *Picea glauca* wood. × 280.
(No. 3 from *Huber* in *Handbuch der Mikroskopie in der Technik,* Umschau Verlag, 1951.)

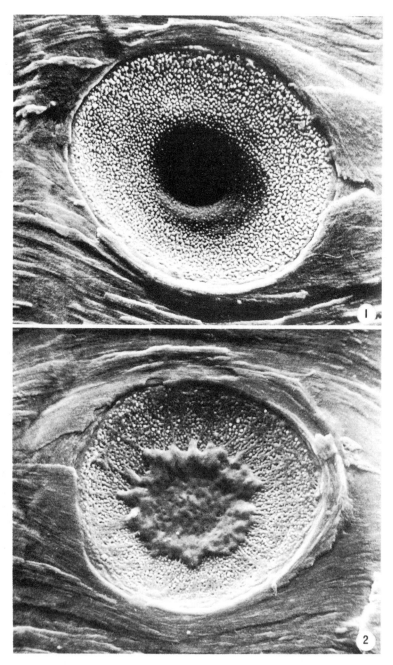

FIG. 185. Scanning electron micrographs of a pit of an untreated *Cedrus libani* tracheid. × 7500. 1, Interior surface of pit chamber densely covered with warts. 2, Pit membrane showing a scalloped torus which is depressed in the centre; the margo is heavily incrusted with warts which obscure the network of fibril bundles.

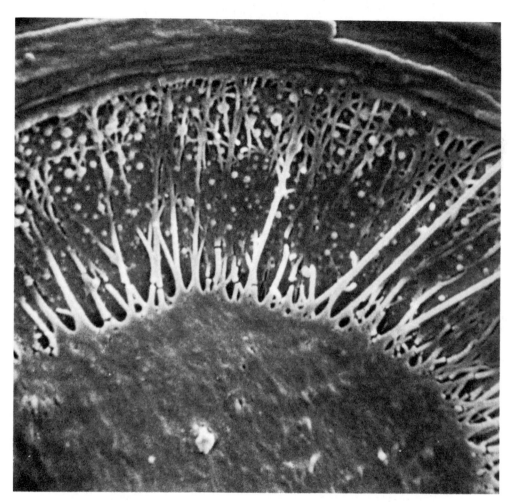

FIG. 186. Scanning electron micrograph of a portion of a pit membrane of a *Pinus halepensis* tracheid. × 15000.

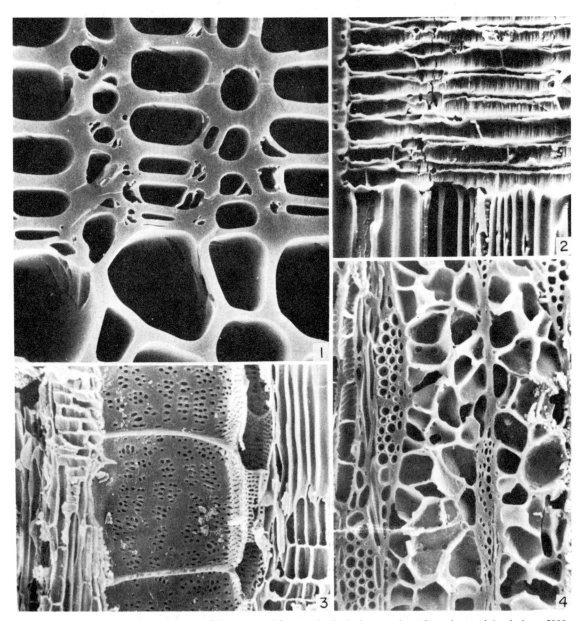

FIG. 187. Scanning electron micrographs of burned wood from archeological excavations (Jerusalem and Arad about 2000 years ago). 1 and 2, *Cupressus sempervirens.* 1, Cross-section. × 2000. 2, Radial-longitudinal section showing a ray. × 390. 3, Radial section of *Ficus sycomorus*, showing a vessel. × 200. 4, Tangential section of *Pistacia atlantica* showing vessels with tyloses. × 240.

FIG. 188. Secondary xylem of *Robinia pseudacacia* showing vessels filled with tyloses. × 45. 1, Cross-section. 2, Longitudinal–tangential section.

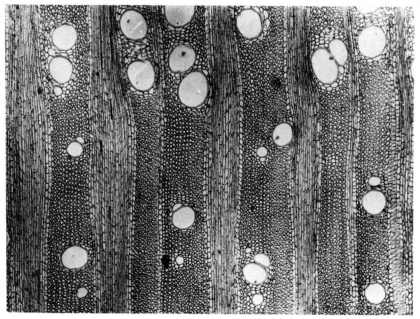

FIG. 189. Cross-section of the secondary xylem of the stem of *Tamarix aphylla* showing the expansion of the rays between adjacent growth rings and the vasicentric parenchyma. × 45.

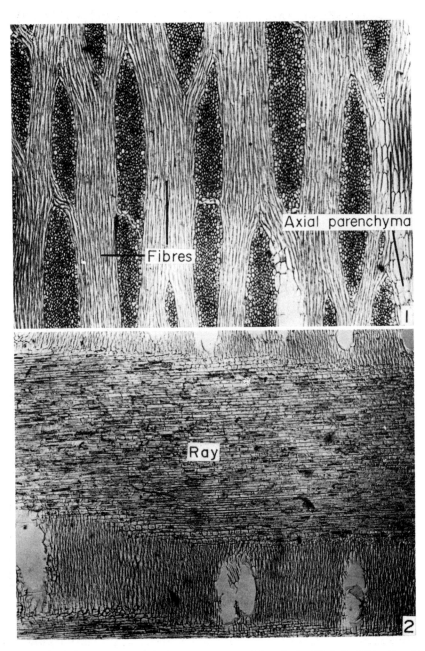

Fig. 190. Secondary xylem of the stem of *Tamarix aphylla*. 1, Tangential section. ×45. 2, Radial section showing heterogeneous rays. ×45.

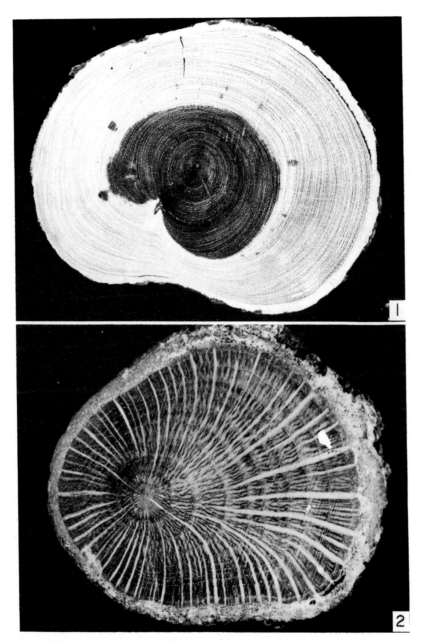

FIG. 191. 1, An entire cross-section of a branch of *Acacia raddiana* in which, from the outside inwards, the bark, sapwood and dark heartwood can be distinguished. ×0.6. 2, As above, but of *Quercus boissieri* in which the very broad rays and eccentric growth rings are readily distinguished. ×1.2.

epithelial cells in the various conifer genera. In those genera, where the resin-secreting cells have thick, lignified walls and in which these cells die after one season, relatively little resin is produced. In those plants where the epithelial cells are thin walled and function during several seasons, a large amount of resin is produced. Resin ducts, the epithelial cells of which have lignified walls, occur in *Abies* and *Cedrus,* for example while in *Pinus* the secretory cells are thin walled and not lignified (Fig. 181; Fig. 183, no. 1: Fig. 184, no. 2).

In the secondary xylem of conifers resin ducts are produced as a result of injuries, such as wounding, pressure and frost, among others. In *Pinus, Picea, Larix,* and *Pseudotsuga* resin ducts also appear as a normal feature of the wood. In certain conifers such as *Cupressus,* for example (Fig. 184, no. 1), resin ducts are never developed in the secondary xylem. The location of the resin ducts, when formed in the xylem, depends on the type of injury and on the plant species. For instance, an open wound results in the formation of dense or scanty tangential groups of resin ducts around the wound. Injuries resulting from pressure or any other factor that acts on a relatively large area results in the formation of scattered ducts. The extent of this scattering depends on the genus: in *Pinus,* for example, the ducts are more scattered than in *Abies* or *Cedrus.* In the last two the ducts are short and branched. In *Pinus* the ducts which develop a great distance from the centre of the injury are very long and are not arranged in groups, but are scattered to a great extent. From the results of experiments it has been shown that the largest number of resin ducts is produced when the cambium of the injured branches is intensively active (Thomson and Sifton, 1925; Bannan, 1933, 1934, 1936; Messeri, 1959; Fahn and Zamski, 1970) (see also Chapter 9).

SECONDARY XYLEM OF DICOTYLEDONAE

The secondary xylem of dicotyledons (Fig. 192) is more complex than that of the gymnosperms. Dicotyledonous wood comprises elements that vary in size, shape, type, and arrangement. In the secondary xylem of *Quercus,* for example, vessel members, tracheids, fibre-tracheids, libriform fibres, gelatinous fibres (Chapter 6), wood-parenchyma, and rays of different sizes, are found. However, there are some dicotyledonous trees in which the wood is comprised of a smaller number of element types. For instance, in many species of the Juglandaceae, apart from vessel members and parenchyma cells, only fibre-tracheids are found in the wood. The secondary xylem of small plant species of Cactaceae often lacks libriform fibres and consists only of vessel elements with helical to reticulate secondary thickenings, vascular tracheids with only annular thickenings, and parenchyma cells (Gibson, 1973, 1977).

In trees of tropical origin, growth rings are not usually distinguishable, while in trees of temperate origin, growth rings can usually be distinguished in the secondary xylem. These growth rings are generally annual, that is, one such ring is produced each year; but there are trees, for instance, certain specimens of *Tamarix aphylla* (Fahn, 1958), in which two rings are produced in a single year. The growth rings are obvious because of the seasonal differences in amount and shape of certain of the cells developing from the cambium. The cells produced at the end of the growth season when the cambial activity is slow are narrower, especially in a radial direction, and mostly have thicker walls. The early-wood consists of relatively broad cells with large lumina. The differences between these cells and the thick-walled smaller cells, which terminate the growth of the preceding year, emphasize the growth rings. To the unaided eye these rings are distinguishable because of the differences in colour between the early and late-wood. In certain plants, such as *Tilia, Ceratonia, Spartium junceum,* and *Zygophyllum dumosum* (Fig. 193, no. 4), one to three cell layers between the rings consist of wood-parenchyma cells. The walls of these cells are thinner than those of the neighbouring fibres. In the stems of many plants, such as *Zygophyllum,* for example, this border parenchyma forms a pale line which clearly marks the borders of the growth rings.

Trabeculae, rod-like extensions of cell wall material, which traverse the cell lumen from one tangential wall to another, and which often occur in tracheids of gymnosperms, have also been

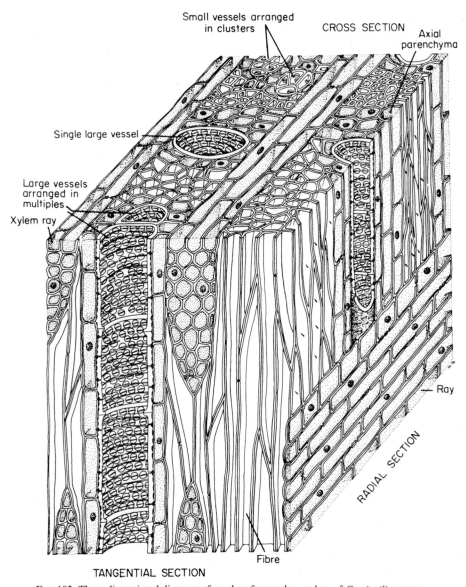

Small vessels arranged
in clusters

CROSS SECTION

Axial
parenchyma

Single large vessel

Large vessels
arranged in
multiples

Xylem ray

Ray

RADIAL SECTION

TANGENTIAL SECTION

Fibre

FIG. 192. Three-dimensional diagram of a cube of secondary xylem of *Cercis siliquastrum*.

found in vessels of some dicotyledons, e.g. in *Knightia excelsa* of the Proteaceae (Butterfield and Meylan, 1972). In some species of other families trabeculae were found also in fibres and axial and ray parenchyma cells (Butterfield and Meylan, 1979; Werker and Baas, 1981).

Arrangement of vessels

The arrangement of the vessels in the secondary xylem of dicotyledons is a characteristic feature

and is used in the identification of species. When the vessels are more or less equal in diameter and uniformly distributed throughout the wood, or when there is only a gradual change in size and distribution throughout the growth ring, the wood is termed *diffuse-porous wood* (Fig. 193, no. 3; Fig. 194, nos. 1, 3). Examples of species with such wood are *Acer* spp., *Populus alba*, *Acacia cyanophylla*, *Olea europaea*, and *Eucalyptus* spp. When the wood contains vessels of different diameters and in which those produced at the beginning

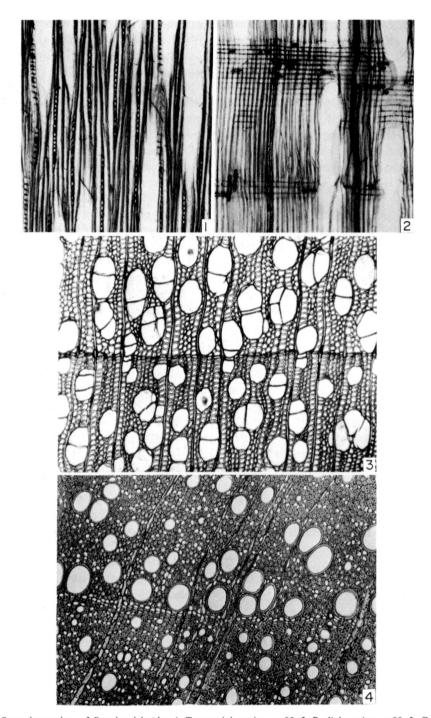

FIG. 193. 1–3, Secondary xylem of *Populus deltoides*. 1, Tangential section. ×80. 2, Radial section. ×80. 3, Cross-section. ×65. 4, Cross-section of the secondary xylem of *Zygophyllum dumosum* in which the initial parenchyma on the border between the adjacent growth rings can be distinguished. ×110.

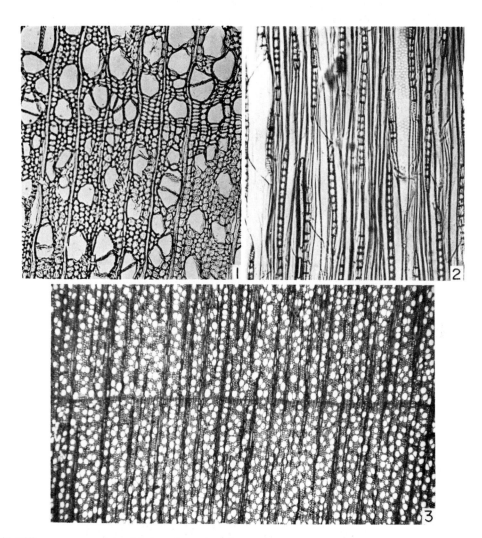

FIG. 194. Diffuse-porous wood. 1 and 2, Secondary xylem of *Salix babylonica*. 1. Cross-section showing the border between two growth rings. × 115. 2, Tangential section. × 115. 3, Cross-section of the secondary xylem of Crataegus azarolus showing the border between two growth rings. × 40.

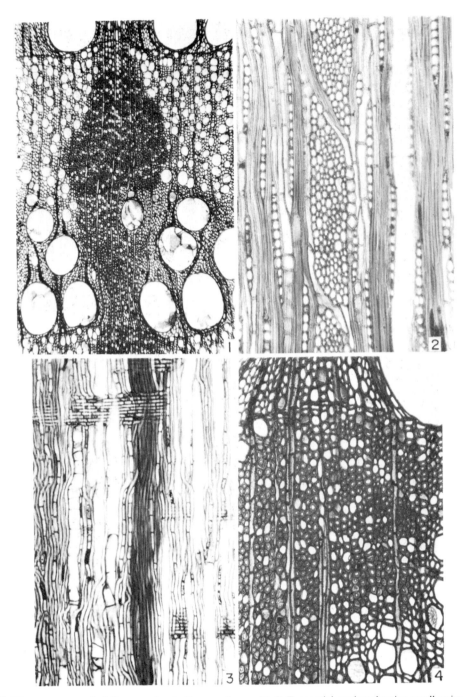

FIG. 195. Ring-porous wood of *Quercus alba*. 1, Cross-section. × 35. 2, Tangential section showing small uniseriate and a very large multiseriate aggregate ray. × 120. 3, Radial section. × 48. 4, Cross-section showing diffuse apotracheal paren-chyma. × 115.

FIG. 196. 1, Cross-section of the secondary xylem of *Eucalyptus camaldulensis* showing the border between two growth rings. ×65. 2, Radial section of the secondary xylem of *Olea europaea* showing a heterogeneous ray. ×200.

of the season are distinctly larger than those of the late-wood, the wood is said to be *ring-porous wood* (Fig. 195, no. 1). The wood of *Fraxinus* spp., *Quercus robur*, *Q. ithaburensis*, *Robinia pseud-acacia*, and *Pistacia atlantica* may be given as examples of this type. Many intermediate forms occur between the above two extreme types. Environmental conditions and the age of the plant also influence, to a certain extent, the arrangement of the vessels.

Ring-porous wood is thought to be more advanced than diffuse-porous from the phylogenetic point of view. The former is found only in relatively few species and mainly in plants from the northern hemisphere (Gilbert, 1940). It has also been recorded that ring-porous wood is relatively common in plants growing in arid habitats (Huber, 1935). This observation is not supported by the results of the investigations made on woody plants growing in the Negev desert (Israel). Bailey (1924) suggests that typical ring porosity developed in plants already adapted to tropical environments that became subjected to climates with cold winters or with alternating very dry and wet seasons. This development is, in other words, connected with seasonal activity.

The pattern of distribution of the vessels is studied in cross-sections of the wood. Here the vessels can be seen to be single as, for example, in *Eucalyptus* (Fig. 196, no. 1) and *Quercus*, or in groups of different size and shape. For instance, the groups may consist of radial, oblique or tangential rows of two to many vessels, the walls of which are in contact with one another and which

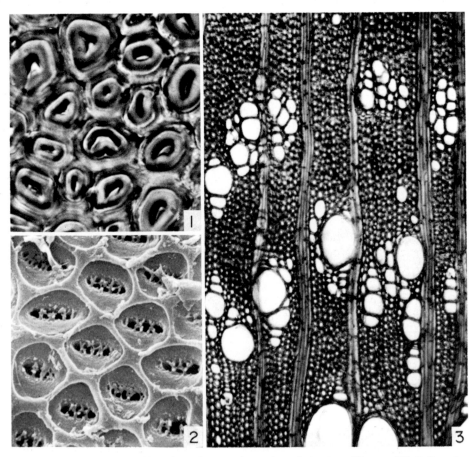

FIG. 197. 1, Cross-section of gelatinous fibres from the secondary xylem of *Acacia raddiana*. × 900. 2, Scanning electron micrograph of a vessel of the secondary xylem of *A. tortilis* in radial view, showing vestured pits. Pit membrane removed. × 1700. 3, Cross-section of the secondary xylem of *Pistacia atlantica*. × 125.

are called *multiples* (Fig. 194, no. 1). Such distribution may be seen, for example, in the wood of *Ceratonia*, *Acacia* and *Populus*. The groups may be in the form of *clusters* (Fig. 197, no. 3; Fig. 199, no. 1), i.e. irregular groups which consist of varying numbers of vessels in both radial and tangential directions. *Pistacia* may be cited as an example of such distribution. The single vessels may be circular or elliptical as seen in cross-section. Vessels arranged in groups are usually flattened where they are in contact with one another.

According to Priestley and Scott (1936), the development of the vessels in early ring-porous wood is very rapid and sudden, while that in diffuse-porous wood is slow. Handley (1936) measured the length of vessels and concluded that the vessels in ring-porous wood are longer than those in diffuse-porous wood. According to Huber (1935) and Kozlowski and Winget (1963), in ring-porous wood the transport of water is almost entirely restricted to the outermost ring, and the flow of water in plants with such wood is ten times faster than that in plants with diffuse-porous wood.

In some plants, e.g. *Acer rubra*, a continuous increase in vessel diameter and vessel length from twigs to branches, down along the stem and into the roots, was observed (Zimmermann and Potter, 1982). Vessel diameter and vessel-member length also increased in the long conducting roots of *Retama raetam* with increasing distance from the stem (Fahn, 1964). Application of auxin to decapitated bean stems caused a very sharp basipetal gradient of decreasing vessel number and increasing vessel diameter (Aloni and Zimmermann, 1983). On the basis of these experiments it was suggested that the lower auxin concentration further down the stem results in slower differentiation and therefore the formation of fewer and larger vessels.

It is worth mentioning that, in the light of recent research, it appears that all the vessels of one ring, or apparently even of the entire secondary xylem, are interconnected so that a network is formed (Braun, 1961).

Arrangement of the axial wood parenchyma

The amount of the axial parenchyma varies in the different dicotyledonous species. In some species there is very little axial parenchyma, or it is entirely absent, while in others it constitutes a very large portion of the wood. Apart from the differences in the amount of axial parenchyma there are also differences in its distribution among the other elements of the secondary xylem.

Much taxonomic importance is attached to the type of distribution of the axial parenchyma. There are two basic types: *apotracheal* (Fig. 195, no. 4) in which the parenchyma is typically independent of the vessels though it may come in contact with them here and there; and *paratracheal* (Fig. 198, no. 1) in which the parenchyma is distinctly associated with the vessels. Both these types are subdivided into the following variations. When the apotracheal parenchyma is in the form of small uniseriate bands or single cells scattered irregularly among the fibres, it is said to be *diffuse parenchyma* (Fig. 195, no. 4). When, in a cross-section of the wood, the axial parenchyma is seen to form concentric bands, it is said to be *banded* or *metatracheal parenchyma*. Single apotracheal parenchyma cells or those arranged in more or less continuous layers, which may be of variable width, at the border between two growth rings, are termed *marginal parenchyma*. If this parenchyma is produced at the end of the growing season it is termed *terminal parenchyma*. Similar parenchyma formed at the beginning of the growth ring is termed *initial parenchyma* (Fig. 193, no. 4). Initial parenchyma occurs in *Ceratonia*, *Zygophyllum*, and *Spartium*, for example.

The paratracheal parenchyma also may be variously distributed. If the parenchyma does not form a continuous sheath around the vessels, as for example in *Acer*, it is said to be *scanty paratracheal parenchyma*. When the paratracheal parenchyma occurs on one side, either external (abaxial) or internal (adaxial), of the vessels it is said to be *unilaterally paratracheal parenchyma*. Parenchyma which forms entire sheaths, of different width, around the vessels, e.g. *Tamarix* (Fig. 189), is termed *vasicentric parenchyma*. The shapes of such sheaths as seen in cross-section of the wood may be circular or somewhat elliptical. In some plants, e.g. *Acacia cyanophylla*, *Cercis siliquastrum*, the sheaths in cross-section can be seen to have lateral wing-like extensions (Fig. 198, no. 1); such parenchyma is called *aliform parenchyma*.

In the wood of certain species, such as *Acacia raddiana* and *A. albida*, the aliform parenchyma is seen, in cross-section, to form diagonal or tangential bands; this type of parenchyma is termed *confluent parenchyma* (Fig. 200, no. 1).

The distribution of septate wood fibres, if present, is similar to that of the axial parenchyma. In those species that contain a large number of septate fibres there is little axial parenchyma (Spackman and Swamy, 1949).

In certain plants, e.g. *Eucalyptus*, short irregularly shaped tracheids are present in the immediate proximity of the vessels. These tracheids do not, however, form a separate continuous vertical system and they are termed *vasicentric tracheids*.

Kribs (1937) describes the phylogenetic evolution of the axial parenchyma as having taken the following course: (1) from the diffuse apotracheal type, through various intermediate forms and apotracheal banded parenchyma types to paratracheal distribution, and, finally, in the most advanced form, to the development of vasicentric parenchyma with numerous cell layers in the sheath; (2) the individual parenchyma cells became shorter and broader with the increased specialization of the secondary xylem, just as in the vessel members. According to Kribs the absence of axial parenchyma is a primitive feature and the presence of terminal parenchyma is an advanced feature which has resulted from reduction.

Structure of the rays

In dicotyledons the rays usually consist only of parenchyma cells. On the basis of the orientation of the longest axis of the cells, as seen in radial longitudinal section, parenchyma cells that form the ray may be of one type only or of two types. If the ray cells are all elongated in a radial direction, i.e. if all the cells are procumbent, the ray is *homogeneous* (Fig. 193, no. 2). This type of ray is included, by some workers, with the similar type found in the conifers in the term *homocellular ray*. When the ray in dicotyledonous wood consists of the two types of cells, i.e. procumbent and square or vertically elongated cells, it is said to be *heterogeneous* (Fig. 190, no. 2; Fig. 196, no. 2). It has

been proposed by some workers to include this type of ray with the coniferous rays composed of different cell types in the term *heterocellular ray*.

Heterogeneous rays may be uni- or multiseriate. The most common type of heterogeneous ray is that in which the central portion of the ray is multiseriate and consists of the radially elongated cells, while the upper and lower edges contain the square or vertically elongated cells. Sometimes the radially elongated cells are surrounded by square or vertically elongated cells. There are also a few plants in which the square or vertically elongated cells are mingled with the radially elongated cells.

The above nomenclature concerning the structure and arrangement of the elements of the secondary xylem is based on that published by the Committee on Nomenclature of the International Association of Wood Anatomists (Int. Ass. Wood Anatomists, 1947; IAWA Committee, 1989).

In addition to the above-mentioned types of rays, Braun (1967) and Braun et al. (1967, 1968a, b) have suggested a classification of wood rays based on the presence or absence of pits between ray cells and vessels.

It is thought that the presence of two types of rays—uniseriate comprising vertically elongated cells and multiseriate heterogeneous—is a primitive feature from the phylogenetic viewpoint. From this type the many other different types of rays evolved. The evolution was apparently in various directions. One trend has led to the enlargement of the multiseriate rays, while another has been to their reduction in size and number. The uniseriate rays became reduced in height and number. In many plants one type of ray was lost and in a few examples both. In *Quercus* the specialization has resulted in the development of very large multiseriate rays and small uniseriate ones (Fig. 195, no. 2). In *Salix* and *Populus* the phylogenetic specialization is expressed in the development of only one type of ray—the uniseriate (Fig. 193, no. 1; Fig. 194, no. 2). In *Hebe* (Scrophulariaceae), *Aeonium arboreum* (Crassulaceae), and *Halophytum* (Chenopodiaceae), for example, the secondary xylem, or at least its inner portion that develops from the cambium in the first years of its activity, is completely devoid of rays. In plants that have such secondary xylem a few uniseriate, non-

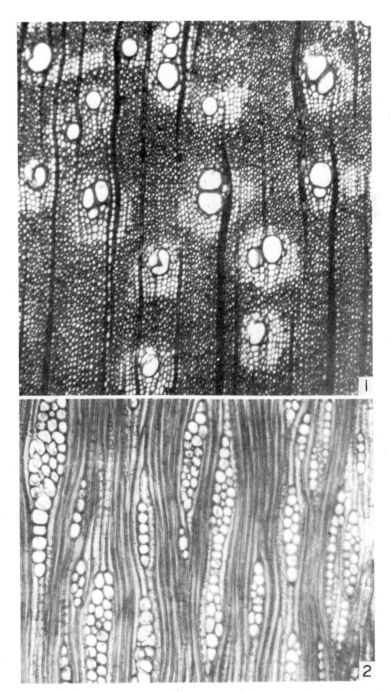

FIG. 198. Secondary xylem of *Acacia cyanophylla*. 1, Cross-section. ×60. 2, Tangential section. ×130.

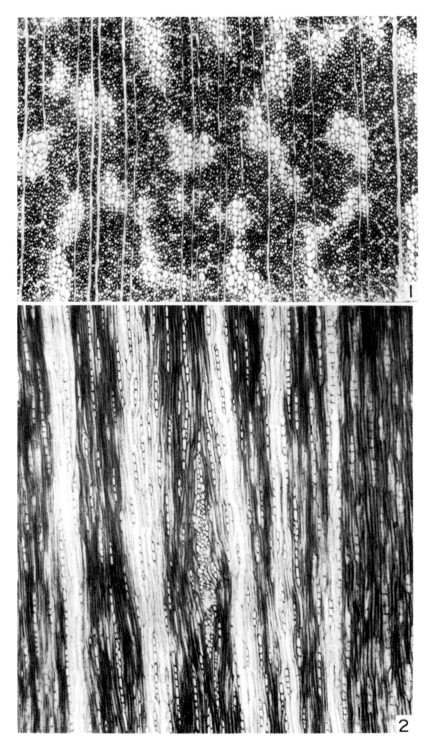

FIG. 199. Secondary xylem of *Thymelaea hirsuta*. 1, Cross-section, showing the small vessels to be arranged in dendritic, diagonal or radial patterns. × 75. 2, Tangential section. × 90.

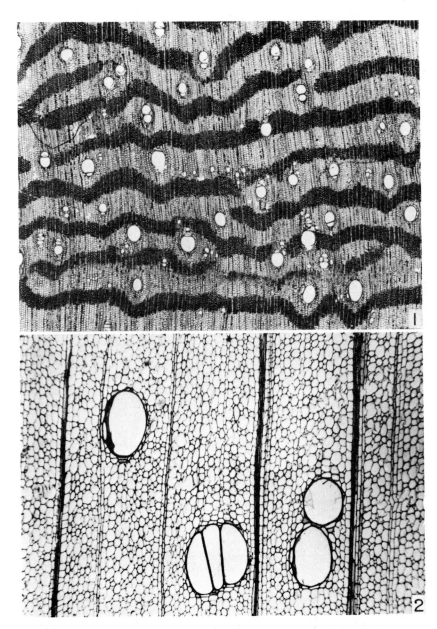

FIG. 200. 1, Cross-section of the secondary xylem of *Acacia albida*, showing broad bands of parenchyma. ×15. 2, Cross-section of the secondary xylem of *Ochroma*. ×50.

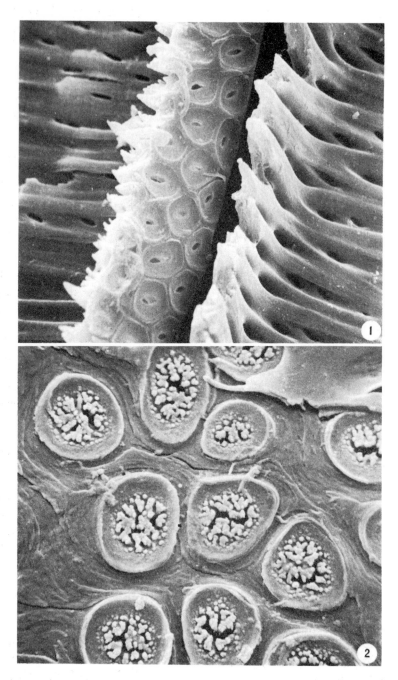

Fig. 201. Scanning electron micrographs of vessel elements. 1, Portions of two vessels of *Cercis siliquastrum*, showing alternately arranged inter-vessel pits (left vessel) and helical thickenings on the inner surface of the vessels. × 2500. 2, Vestured inter-vessel pits of *Ceratonia siliqua*. The wall layer with the pit membranes was removed. × 4000.

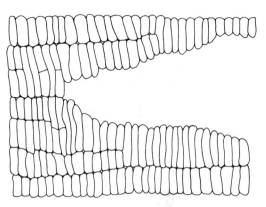

FIG. 203. Radial section of the secondary xylem of *Casearia nitida* illustrating the fusion of two rays by the vertical elongation of the marginal initials and their derivatives. Cambium on left. (Adapted from Barghoorn, 1940.)

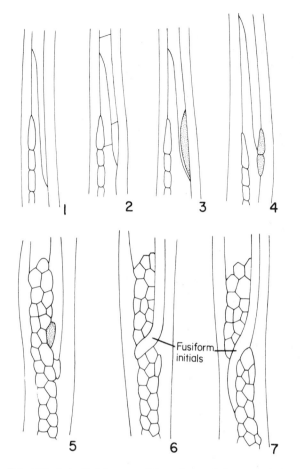

FIG. 202. 1–4. Serial tangential sections of the secondary xylem of *Viburnum odoratissimum* showing the origin of a ray initial at the end of a fusiform initial. Newly formed ray initial stippled. 5–7, Serial tangential sections of the secondary xylem of *Trochodendron aralioides* showing the splitting of a ray by the apical elongation of a fusiform initial. The entry of the fusiform initial into the ray is made possible by the loss of the ray initial that is stippled in no. 5. (Adapted from Barghoorn, 1940.)

continuous rays can be seen in the outer portion of the wood that develops later (Barghoorn, 1941b; Gibson, 1978). From the point of view of cell composition the homogeneous rays are considered to be more advanced than the heterogeneous ones.

The decrease in the size of rays with phylogenetic advancement may have been brought about by changes that took place in the cambium (Barghoorn, 1941a) (Fig. 202, nos. 5–7). Some of the ray initials may become lost and their place be taken up by fusiform initials of the cambium. If this

occurs on the ray margins the ray simply becomes narrower, but if it occurs within the ray it results in the splitting of the ray into two or more parts. A ray may also become split as a result of the intrusive growth of the tips of the fusiform initials into a group of ray initials. Further, a ray may become split by the changing of some ray initials into fusiform initials. In this manner *aggregate rays* (Fig. 195, no. 2) are often formed from large rays. An aggregate ray is defined as a ray comprising a group of small and narrow rays which appears to the unaided eye or under low magnification to be a single large ray.

As was mentioned above, the rays may also have increased in size during evolution. This enlargement may be brought about by the merging of rays (Fig. 203), by the anticlinal division of the ray initials, or by the horizontal subdivision of the fusiform initials adjacent to the ray so as to form additional ray initials. The merging of rays is brought about by the loss of the fusiform initials that separate the two rays.

In some plant species there are perforated ray cells which are a part of an axial vessel passing obliquely through a ray. Radially oriented vessels were found in the rays of two genera of the Combretaceae (van Vliet, 1976).

Secretory structures

Various types of secretory structures may occur in the secondary xylem (Metcalfe and Chalk, 1983).

Secretory ducts and cavities in the wood may occur normally or as a result of injury (traumatic ducts or cavities). The ducts are either axially aligned in the wood ot radially, in the rays. The two types very rarely occur together. The epithelial cells of the ducts and cavities usually produce resin, gum or gum-resins. A special type of duct, kino ducts, are characteristic of *Eucalyptus* species. Laticifers are found in the wood of relatively few plant species. They occur for instance, in *Euphorbia hierosolymitana*, some *Ficus* species and in *Calotropis procera*. (For more information on secretory tissues see Chapter 9 and Fahn, 1979).

EFFECT OF SECONDARY THICKENING ON LEAF TRACES

As the cambium is situated between the xylem and the phloem, the newly formed xylem continually pushes the phloem and the cambium itself outwards. This results in the separation of the xylem of the inner portion of the leaf trace, which is situated in the primary body, from its associated primary phloem. The xylem portion of the leaf trace, therefore, becomes buried in the secondary xylem of the stem (Fig. 204, nos. 1, 2).

With the continued secondary growth and further pushing out of the cortex and the phloem, in which the outer portion of the leaf trace is situated, the xylem of the leaf trace becomes torn into two parts (Fig. 204, no. 3). The tearing of the trace takes place only after the leaf is shed— usually in the first or second season after shedding.

In evergreen plants the leaf traces elongate as a result of a special type of secondary growth which is brought about by the addition of new xylem to the middle of the trace. The primary xylem of the leaf base gradually ruptures obliquely (because the leaf traces pass out, towards the leaves, obliquely through the cortex) and the cambium in that region then adds new cells which replace those destroyed. In *Araucaria* and *Ceratonia*, for example, leaf traces may reach a considerable length and may pass through several growth rings (Fig. 204, no. 4; Fig. 205). In evergreens the trace is ruptured after the leaf is shed as is the case in deciduous plants (Eames and MacDaniels, 1947).

VARIATION IN SIZE OF XYLEM ELEMENTS

The length of fibres, tracheids, and vessel members has been found to increase from the centre of the trunk towards its periphery through at least a certain number of annual rings (Gerry, 1915; Hejnowicz and Hejnowicz, 1958; Stern-Cohen and Fahn, 1964). Fibre length also varies along the trunk: in *Eucalyptus gomphocephala*, for example, the longest fibres were found to occur at a height of about 2 m above ground level. In a growth ring the late-wood fibres in dicotyledons and tracheids in gymnosperms have been found to be longer than those of the early-wood (Kribs, 1928; Bisset *et al.*, 1950; Stern-Cohen and Fahn, 1964). In dicotyledons the variations in fibre length within the same growth ring are probably due to changes which take place during the differentiation of the fibres and are not due to an alteration in size of the fusiform initials in the cambium. Such a view has also been expressed by Chalk *et al.* (1955) as a result of investigations carried out on trees with storied wood. The increase in wall thickness, and length of the fibres of the late-wood is probably the result of the difference in the rate of growth, i.e. the period of differentiation of an element produced towards the end of the growth season is longer than that of one produced at the beginning of the season and, therefore, the elements of the late-wood may become longer and have thicker walls (Bisset *et al.*, 1950). The variation in fibre length along the tree as well as from the centre of the trunk toward the cambium in trees with non-storied wood is generally accepted to be connected with the change in size of the fusiform initials.

Comparison of trees growing in mesophytic and arid habitats showed that the dimensions of width, wall thickness and length of the fibres were all larger in the trees growing in the mesophytic habitat (Stern-Cohen and Fahn, 1964).

DENDROCHRONOLOGY

The variation in width of the annual increment has suggested to scientists a correlation between the width of tree-rings and environmental factors.

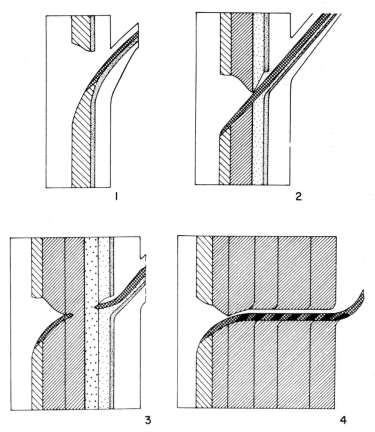

FIG. 204. Diagrams illustrating the relationship of the leaf trace to the expanding secondary plant body in woody dicotyledons, as seen in radial section. 1, Showing the condition of the leaf trace in a stem devoid of secondary thickening; the leaf-gap is still open. 2, A stem after 1 year of secondary growth showing that the gap is partly closed by the secondary xylem and phloem. 3, Showing the stem of a deciduous plant after 2 years of secondary growth; the ruptured ends of the trace are separated by secondary vascular tissues. 4, Showing the extension of the leaf trace in an evergreen plant after 5 years of secondary growth (parts outside the cambium not represented in this diagram). In this case the trace ruptures from time to time and the gaps thus formed are filled by secondary tissue produced by that part of the cambium on the under side of the trace where it is in contact with the secondary xylem. In all the figures primary xylem is represented by widely spaced hatching, secondary xylem by more dense hatching, primary phloem by fine stippling, secondary phloem by coarse stippling, the xylem of the diverging part of the leaf trace by cross-hatching. (Adapted from Eames and MacDaniels, 1947.)

Douglas began his investigation of tree-rings in 1901 while searching for a method to be used in studying sunspot cycles. Later, growth-ring analysis, or dendrochronology, became a tool for the study of climatic history and archaeological dating (Douglas, 1936; Glock, 1955; Bannister, 1963; Fritts, 1965; Eckstein and Bauch, 1969). Dendochronology was most widely applied in semi-arid and arid conditions where low moisture limits growth (Schulman, 1956; Fahn *et al.*, 1963; Waisel and Liphshitz, 1968; Liphshitz *et al.*, 1979), and at northern latitudes where low tem-

perature during the growing season usually limits growth (Holmsgaard, 1955). In sites where soil moisture is adequate, the annual rings may be wide and exhibit little variation from year to year except for gradual decrease in width with increase in age of the tree. In arid and semi-arid sites they are narrow and show marked variations in width. Trees from such sites exhibit *sensitive* ring series.

Irregular changes of environmental conditions may cause the formation of *false rings*. Such rings show more gradual changes in cell size along their outer margin than true rings. As a result of exceptionally unfavourable conditions in some years, rings may be locally or completely absent. In order to avoid misinterpretations and to verify the identification of the false and missing rings, the *cross-dating* technique was developed. This technique is based on the principle that, when the growth of several plants is influenced by a common fluctuating environment, similar and synchronous ring patterns will be produced. Cross-dating is carried out by comparing many ring series (Fig. 206) and, where a lack of synchrony is noted, re-examination of the samples will often show that a false ring was confused with a true ring or that a very narrow ring was overlooked, or that a ring is missing. Because environmental factors probably limit growth, mostly during drought years, particular attention is paid to the matching of narrow rings. Once a reliable master chronology has been prepared for a certain area, other samples may be dated by careful comparison with it.

Successively older and older specimens of unknown cutting date can be added to a modern established chronology by cross-dating and thus extend the sequence into old periods. Such a dating is of great help in archaeology.

RELATIONSHIP BETWEEN THE MICROSCOPIC STRUCTURE AND WOOD PROPERTIES

The woods of different species possess certain properties which make them suitable for different uses. These properties depend on the histological and chemical structure of the xylem tissue. This is discussed widely by Bosshard (1974–5). The histological features that affect the characteristics of the wood are the presence and distribution of vessels, the presence or absence of fibres and their relative number, the diameter of the fibres and thickness of their walls, the length of the fibres and the extent to which they overlap, the form of the fibres— whether straight or curved, the width and number of the rays, and the presence or absence of tyloses.

The chemical structure is of importance in connection with certain properties and especially those by which heartwood differs from sapwood.

The cell walls themselves differ in the relative amounts of cellulose, lignin, etc. Tannin compounds may be accumulated in large quantities in the cell walls and the cells may contain different amounts of gum, resins, and tannins.

Weight of wood

The specific gravity of the wall substance of the secondary xylem of all plants is more or less the same, and is about 1.53. Therefore the differences in the weight of woods depend on the proportion between the amount of wall substance and lumen. Plants such as *Diospyros*, for example, in which the cell walls are thick and the lumina small and which contain many fibres, have heavy wood. Plants in which the cell walls are thin and in which the lumina of parenchyma and fibres are large, have light wood. In some genera, such as *Populus* and *Tilia*, the specific gravity of the secondary xylem is low although the fibre walls are not particularly thin; in these woods the low specific gravity is due to the presence of numerous thin-walled vessels. An extremely light wood is that of *Ochroma* (balsa wood), which belongs to the type of wood known as "corkwood". Such wood contains a high proportion of large, thin-walled parenchyma cells (Fig. 200, no. 2).

The specific gravity of wood varies from 0.04 (*Aeschynomene* of the Leguminosae) to about 1.4 (*Krugiodendron* of the Rhamnaceae). The specific gravity of timber commonly used in trade is between 0.35–0.65 (Eames and MacDaniels, 1947). The specific gravity of the wood of some trees growing in Israel is as follows: *Eucalyptus camaldulensis*, 0.52–0.68; *Pinus halepensis*, 0.48; *Ficus sycomorus*, 0.40; *Quercus calliprinos*, 0.80; *Phillyrea media*, 0.79; *Tamarix aphylla*, 0.51.

Of the very light type of wood, balsa wood (*Ochroma*), the specific gravity of which is 0.1–0.16, is much utilized in industry, especially as insulating material in the aircraft and lifeboat industries. Balsa wood is strong in relation to its specific gravity. Histologically, two types of light wood can be distinguished: that in which there are bands of lignified, thick-walled

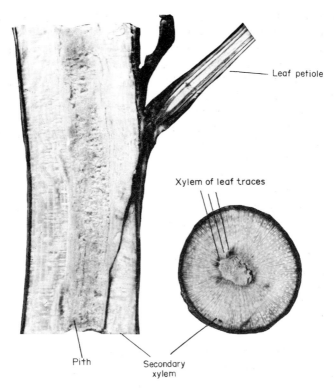

FIG. 205. A branch portion of *Ceratonia siliqua* sectioned: 1, longitudinally, and 2, transversely. Before sectioning a coloured solution was sucked into the branch through the cut leaf petiole. In no. 1, the dark stripe represents the coloured leaf trace embedded in the secondary xylem. In no. 2, the cross-sectioned leaf traces appear as dark spots near the pith. × 5. (Photograph by M. Grundwag.)

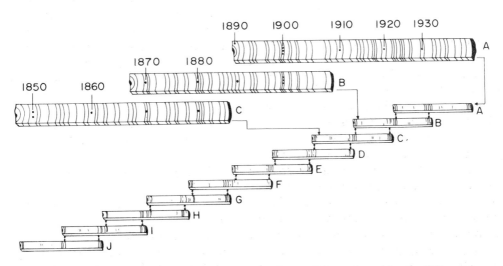

FIG. 206. Chronology building by cross-dating. A, Radial sample from a living tree cut after the 1939 growing season. B–J, Specimens taken from old houses and successively older ruins. (Adapted from Bannister, 1963.)

cells alternating with bands of non-lignified thin-walled cells, and that in which the above two types of cells are homogeneously arranged (Eames and MacDaniels, 1947). In the wood of certain plants, such as *Carica papaya* and *Phytolacca dioica*, the only lignified elements are the vessels or vessels and fibres associated with them.

Strength

Wood that contains many libriform fibres or fibre-tracheids is a strong wood and therefore it is seen that dense and heavy woods must also be strong woods. The length of the fibres and the degree to which their ends overlap are apparently of less importance to the strength of the wood. The wood-rays play an important role in the strength of the wood in radial direction (Futo, 1969). According to some workers (Pillow and Luxford, 1937) the orientation of the cellulosic microfibrils in the walls influences the strength of the wood. For example, it is thought that when the microfibrils are so orientated as to be almost parallel to the horizontal axis of the fibre, the wood is weakened.

Durability

The ability of wood to withstand rotting, as a result of bacterial or fungal action, mainly depends on the chemical composition of the wood. The degree of durability is determined by the presence of substances such as resins, tannins, and oils in the walls and lumina of the cells. Also, the appearance of tyloses is of some importance, as they block the passage of fungal hyphae as well as of oxygen and water through the vessels.

Wood of trees such as *Sequoia*, *Catalpa*, *Robinia*, *Maclura*, and *Castanea* are very durable, while that of trees such as *Populus*, *Acer*, *Tilia*, and *Carya* rot very rapidly (Eames and MacDaniels, 1947). The heartwood is usually more resistant to decay than is the sapwood. There is no definite correlation between the depth of colour of wood and its durability, but generally darker woods are more resistant to decay. The colour is usually an indicator of the amount of preservative substances in the wood.

The ability of wood to remain intact under mechanical strain depends on the hardness and density of the wood. For technical use wood durability can be raised by adequate treatment with preservatives.

Pliability

A flexible wood is one in which the wood structure is homogeneous with long, straight fibres that overlap for some distance, and with straight rays. Pliability is also influenced by the amount of water in the wood (Pillow and Luxford, 1937).

Many properties of wood are connected with hygroscopic moisture, a factor on which numerous technological problems, the main of which is the dimensional stability of wood, depend. For details of this and other problems of wood technology see Brown *et al.*, vol. 2 (1952) and Tiemann (1951).

Grain, texture, and figure

Grain is the term used when referring to the direction of the elements in the wood relative to the longitudinal axis of the trunk. *Texture* is a term applied to the relative size and amount of variation in size of the elements. *Figure* refers to the pattern of wood as seen when the wood is cut in the longitudinal direction. Figure depends on the grain and texture and their exposure by direction of sawing (Brown *et al.*, 1952; Desch, 1968; Harris, 1989).

Different types of grain may be distinguished. *Straight*—all the elements are oriented parallel to the longitudinal axis. *Irregular*—the elements are at varying and irregular inclinations to the longitudinal axis. The figure commonly known as *bird's eye* is produced, in a tangentially cut wood, as a result of numerous conical indentations of the growth rings. *Spiral*—the elements are spirally arranged, the inclination of which may vary at different heights.

The inclination of elements may change from

left (*S*) to right (*Z*) repeatedly in the radial and axial direction of the stem (see p. 319). When the length along the stem of each domain of uniform inclination is large the grain is *interlocked*, when the changes in the axial direction are short (a few millimeters to a few decimeters) the grain is *wavy* (Hejnowicz and Romberger, 1973; Krawczyszyn and Romberger, 1980).

A radial cut of wood with interlocked grain results in timber that exhibits a characteristic figure known in commerce as *ribbon* or *stripe* figure. A wood with wavy grain when cut tangentially exhibits undulated figures. When the undulations are close to one another and the changes in direction of the elements are abrupt, the pattern obtained is termed *fiddleback* figure, as such designs, chiefly from maple and mahogany stock, have long been used in the manufacture of violins.

Texture can be described as *coarse*—having large vessels and broad rays, and *fine*—when the vessels are small and the rays narrow. It may also be defined as *even*—if there is no perceptible difference between the early and late wood, or *uneven*—when distinct differences exist between the early and the late wood of a growth ring.

BRANCH AND STEM WOOD

The cambium which surrounds the secondary xylem of the stem is continuous with that of the branches. However, the cambial cells of the upper side of the branch turn downward, and the xylem elements developing from them surround the base of the branch forming a "collar", which becomes later covered by a "collar" produced by the stem xylem (Fig. 207). This manner of xylem development repeats itself in each growth season

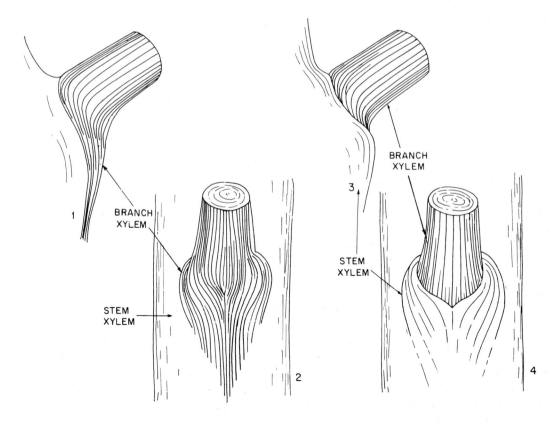

Fig. 207. Diagrams illustrating the relationship between the secondary xylem of branch and stem as described by Shigo, 1985. 1 and 2, Showing a "collar" formed by the branch xylem. 3 and 4, Showing a "collar" formed by the stem xylem, which covers the branch "collar".

and strengthens the branch attachment (Shigo, 1985). When a branch dies the cambium in the branch ceases to function. No new vascular tissues are formed in the branch, although they continue to develop in the stem. From this time on there is a break in the continuity of the new xylem formed. Most of the dead branch usually breaks off and the portion left on the tree becomes gradually embedded in the wood of the stem. The portion of the branch which becomes embedded forms a *knot* in the lumber. The branch wood involved in the production of the knot undergoes changes similar to those of heartwood.

MODIFICATIONS OF WOOD RESULTING FROM GROWTH STRESSES

Environmental factors may cause the production of wood with special structure and properties. One such type of wood is *reaction wood* which develops in leaning trunks and limbs (Wardrop and Dadswell, 1948; Scurfield and Wardrop, 1963; Wardrop and Davies, 1964; Wardrop, 1965; Westing, 1968; Philipson *et al.*, 1971). This wood brings about the recovery of the organs. There are considerable differences in the place of development, nature, and shape of the reaction wood of conifers and that of dicotyledons. In the conifers the reaction wood develops on the underside of the leaning trunk or branch, while in the dicotyledons it develops on the upper side. The reaction wood of the conifers is termed *compression wood*, and that of the dicotyledons *tension wood.*

Reaction xylem is usually paralleled by a similar eccentric development of *reaction phloem* on the same side of the stem. In the reaction phloem the fibres possess greatly thickened unlignified cell walls, both in dicotyledons and in many examined gymnosperms (Wardrop, 1964; Höster and Liese, 1966).

Several experiments have shown that the formation of reaction wood is usually a response to the stimulus of gravity. Reaction wood may develop in vertical stems under natural conditions and may be related to slight deflections possibly due to uneven growth of the stem. Its distribution around the stem at different heights is spiral or irregular.

There is some evidence to suggest that gravity affects the distribution of growth substances in inclined stems, which in turn causes eccentric radial growth. Necesany (1958) suggested that high concentration of IAA promotes the formation of compression wood in conifers, but that low concentration is required for the development of tension wood in dicotyledons. Cronshaw and Morey (1968) also stated that development of tension wood appears to be correlated with a reduced auxin level on the upper side of the horizontal stem. Leach and Wareing (1967) have found that the concentration of auxin in horizontal stems is greater on the lower than on the upper side. The lower activity of the cambium on the lower side of dicotyledonous stems is explained by these workers as resulting from the presence of inhibitory substances.

Boyd (1977) concluded on the basis of a review of available information that reaction wood is a response to imposed stress. Metzger (1908, cited by Boyd) was one of the first to suggest that tension wood generates a contractive force which tends to counteract and reverse the bending of branches and stems caused by gravity or other displacing force.

Compression wood

Compression wood is produced by the local increased activity of the cambium on the lower side of the stem. There is, however, little direct correlation between the amount of compression wood formed and the extent of recovery of the stem. According to some workers, a leaning stem can also recover by the production of normal wood in addition to the compression wood. Compression wood is recognizable by the presence of eccentric growth rings. Typical compression wood is 15–40% heavier than the normal wood of the same species. In compression as compared with normal wood there is a more gradual transition between early- and late-wood. The tracheids of compression wood are seen to be rounder in cross-section, and intercellular spaces can be found between them. The microfibrils of cellulose in the secondary wall of these tracheids are orientated in such a way as to form a large angle with the longitudinal axis of the cells. The inner layer of the secondary wall (S_3) is absent and the inner surface of the central layer (S_2) towards the lumen is helically grooved and ribbed. The grooves are deep (Scurfield and Silva, 1969). Helical grooves are lacking in *Ginkgo biloba* and in the Araucariaceae (Timell, 1983). The walls of the compression wood tracheids that are formed in the spring are slightly thicker than those of the normal wood. The compression wood tracheids are usually shorter than those of the normal wood. There is usually a somewhat higher lignin content in compression wood than in normal wood. More recently it has been shown that callose deposits are present in the walls of compression wood tracheids (Brodzki, 1972; Wloch, 1975; Waterkeyn *et al.*, 1982). Generally, it is possible to summarize the characteristics of compression wood, in comparison with those of normal wood, as being heavier, more brittle and capable of unusually high and irregular longitudinal shrinkage but of less transverse shrinkage. Because of these differences in shrinkage, lumber containing compression wood twists as it dries out.

Tension wood

Tension wood develops on the upper side of leaning dicotyledonous stems and, similarly to compression wood, the presence of tension wood is recognized in a stem by the formation of eccentric growth rings (Fig. 191, no. 2). This assymetry. is the result of the more rapid or more continuous cambial activity on one side of the stem (Wardrop and Dadswell, 1955). Histologically the distinguishing feature of tension wood is the presence of gelatinous fibres. The orientation of the cellulosic microfibrils in the outer, non-gelatinous wall layers of the gelatinous fibres is about 45° in relation to the longitudinal axis of the fibres. In the inner, gelatinous layers, the orientation of the microfibrils is almost parallel to the longitudinal axis of the fibre (Münch, 1938; Wardrop and Dadswell, 1948, 1955). From these facts some workers have concluded that it is the gelatinous wall layer (*G*-layer) that is mainly responsible for the contraction of the tension wood. There is less lignin in tension wood than in normal wood, but the amount of cellulose is higher. The *G*-layer normally replaces the S_3 layer but can also replace both S_2 and S_3 or be added inside S_3 (Fig. 208). It has been suggested that the absence of hemicelluloses and lignin is responsible for the looser connection between the cellulose fibrils of the *G*-layer (Côté *et al.*, 1969) (see also Chapter 6 and Fig. 197, no. 1).

The facts gathered from different investigations and cited by Wardrop (1964) and Philipson *et al.* (1971) show that eccentric growth in some dicotyledons (e.g. *Buxus, Gardenia*) may be on the lower side, as in conifers.

Accentuated growth and anatomical modification do not always occur together. In some plants (e.g. *Sassafras officinale*) much more growth is

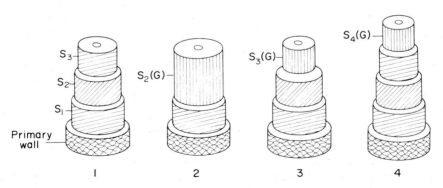

FIG. 208. Diagrammatic representation of variation in cell-wall layers of wood fibres associated with the formation of tension wood. 1, Normal wood fibre. 2–4, Different types of tension wood fibres. The orientation of microfibrils is indicated by striation. (Adapted from Wardrop, 1964.)

found on the lower side of a branch whereas the gelatinous fibres appear on the upper side. In others reaction anatomy may occur with no asymmetry of growth rings, and many woody plants do not produce reaction wood at all. The reaction wood of *Entelea* and *Aristotelia* lacks gelatinous fibres but it differs in several other respects from normal wood. The tension wood of *Ochroma pyramidale* and *Carica papaya* also lacks gelatinous fibres. Eccentric growth in these species occurs both, in the xylem and in the phloem. In the latter the rays become dilated (Fisher and Mueller, 1983).

Tension wood also occurs in roots of many species, and is located there around the entire circumference of the organ (Patel, 1964; Höster and Liese, 1966).

Two types of tension wood are distinguishable: *compact tension wood* in which the gelatinous fibres form continuous regions (e.g. *Acer*) and *diffuse tension wood* in which single or groups of gelatinous fibres are scattered among the normal fibres (e.g. *Acacia*).

The strength of the tension wood of the different species varies, but in most cases this wood tends to form horizontal breaks. Some research workers found that such breaks occur in single fibres. When tension wood is sawn the cut surface appears woolly due to the tearing of the groups of gelatinous fibres.

Tension usually exists between the peripheral secondary xylem and that near the centre of the stem. When a piece of stem is cut longitudinally the peripheral parts contract more than those portions nearer the centre. In asymmetric stems containing tension wood the difference in bending is more obvious. In branches that were bent into loops (Fig. 209, nos. 1–3) tension wood was seen to develop on those sides of the loop which were directed upwards. When such a loop was cut, after the formation of the tension wood, contraction again occurred in those places where the tension wood had formed. Similar experiments were made with conifer branches in order to establish the properties of the compression wood (Jaccard, 1938).

As a result of bending in different ways and directions, it was possible to obtain tension wood on different sides of the same branch. As it became clear from these experiments that a relationship exists between the production of tension wood and the tendency of the stem or branch to shrink longitudinally (Wardrop, 1956), it was possible to understand the fact that tension wood develops in those places where it enables the stem or branches to return to the normal position.

There are several theories that attempt to explain the mechanism by which a bent stem or branch recovers due to the development of the tension wood. According to Frey-Wyssling (1952)

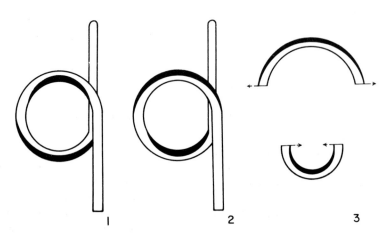

FIG. 209. 1 and 2, Diagrams showing the development of reaction wood (represented by solid black areas) in looped branches. 1, Compression wood. 2, Tension wood. 3, Diagrams illustrating the results obtained when a loop in which tension wood has been formed is cut in half. The arrows indicate the direction of the movement of the cut ends. (Adapted from Jaccard, 1938.)

it is accomplished by forces in the cambium which far exceed the normal osmotic pressure. According to Münch (1938) the cells shrink because of the characteristic orientation of the cellulose units in the different wall layers. According to Wardrop (1956) the acting force is the volume contraction of the cell wall during the crystallization of the cellulose. Hejnowicz (1967) showed that the mechanism of the movement, both in conifers and dicotyledons, is due to stresses related to imbibition of water by the cell walls of the reaction fibres and tracheids. This, however, may not apply to those dicotyledons which lack gelatinous fibres in the tension wood. None of the theories suggested as yet have been sufficiently proved to explain the mechanism by which the bent stems straighten.

AN EXAMPLE OF A KEY BY WHICH TREES AND SHRUBS MAY BE IDENTIFIED ACCORDING TO THE STRUCTURE OF THE SECONDARY XYLEM

The chief trends of evolutionary specialization in the xylem have been established by statistical analysis. This method, however, minimizes the minor variations and localized divergent specializations. While dealing with the taxonomy of particular plant groups such variations and localized deviations become increasingly important (Bailey, 1957; Stern, 1978). Attention has recently been paid to this and the wood anatomy of particular plant groups has been thoroughly studied (see bibliography compiled by Gregory, 1980).

As has already been discussed, the secondary xylem of the different species varies in the shape, size, arrangement, and relative amounts of its constituent elements. Usually there is a connection between the similarity of the internal structure of the wood and the genetic relationship of the plants (cf. Metcalfe and Chalk, 1950). This variability and similarity make it possible to devise a key by which the plant species can be identified according to the structure of their secondary xylem only. This possibility of identifying the botanical species from a piece of timber is of great taxonomic as well as commercial and industrial importance. This method is used to identify the wood from which objects found in archaeological excavations were made and for forensic purposes.

There are different ways in which such a key may be made, including the use of a punched-card system (Clarke, 1938a, b; Phillips, 1948; Greguss, 1955, 1959) and the computor. Following is an example of a dichotomous key (e.g. Fahn *et al.*, 1986) in which a small number of the trees can be determined. In such a key two alternatives are given at each stage. Only one of the alternatives will be entirely applicable to the wood in question. This alternative leads to another stage in the key from which the wood may be identified immediately, or which leads to further alternatives from which the wood can eventually be identified.

1. Wood lacking vessels (gymnosperms) . . .   2
–. Wood with vessels (dicotyledons)   4
2. Resin ducts present in each growth ring . . . . .
    Pinus
–. Resin ducts not present in all growth rings (only present in case of injury), or completely absent   3
3. Resin ducts sometimes present; tori with fringed edges . . . . *Cedrus*
–. Resin ducts never present; tori not as above . .
    Cupressus
4. (1) Included phloem present   5
–. Included phloem absent   6
5. Rays contain cells that are radially elongated .
    Salvadora
–. Rays lacking; sometimes the wood parenchyma may be arranged in short radial rows similar to rays . . . .
    Bougainvillea and genera of the Chenopodiaceae
6. (4) Growth rings not distinct   7
–. Growth rings distinct   8
7. Rays clearly heterogeneous   *Ficus sycomorus*
–. Rays almost homogeneous . . . .
    Acacia raddiana, A. tortilis
8. (6) Wood more or less distinctly ring-porous— the vessels formed at the beginning of the season are conspicuously larger than those formed later and this may be seen without the aid of a microscope   9
–. Wood diffuse-porous—the diameter of the vessels almost equal throughout the growth

ring, or gradually smaller towards the termination of the ring 13

9. Vessels solitary or arranged in multiples of 2–4.
Rays 2–3 cells wide. Paratracheal wood parenchyma around the narrow vessels is aliform-
confluent . . . . *Fraxinus*

–. Vessels and wood parenchyma not arranged
as above 10

10. Wood rays 1–4 cells wide 11

–. Large wood rays wider than 4 cells 12

11. Vasicentric parenchyma scanty . . . *Pistacia*

–. Vasicentric parenchyma aliform to aliform-
confluent . . . . *Cercis*

12. (10) Large, aggregate, multiseriate rays (more
than 20 cells wide) present; vessels devoid of
gum deposits . . . . *Quercus*

–. Rays not aggregate, less than 10 cells wide;
vessels filled with gums
Prunus amygdalus

13. (8) Rays not wider than one cell 14

–. Rays wider than one cell 15

14. Rays homogeneous . . . . *Populus*

–. Rays heterogeneous . . . . *Salix*

15. (13) Maximum width of the rays exceeds 6
cells 16

–. Maximum width of rays less than 4 cells   18

16. Wood parenchyma storied, paratracheal and
vasicentric . . . . *Tamarix*

–. Wood parenchyma not as above 17

17. Height of large rays exceeds 70 cells; in cross-
section of wood the rays widen perceptibly
between growth rings. Vessels are not filled
with gum . . . . *Platanus*

–. Height of large rays less than 60 cells; rays do
not widen perceptibly. Most vessels filled with
gum . . . . *Prunus amygdalus*

18. (15) Apotracheal terminal and paratracheal
vasicentric wood parenchyma present. Vessels
mostly arranged in multiples of 2–6 vessels . . .
19

Wood not as above 20

19. Spiral thickening clearly visible on inner pitted walls of the vessels. Diameter of the early-
formed vessels greater than that of the later-
formed vessels . . . . *Cercis*

–. Spiral thickening not present on inner walls of
vessels. Early and late-formed vessels of same
diameter . . . . *Ceratonia*

20. (18)   Rays   conspicuously   heterogeneous;

radially elongated cells in the middle and wings
of several rows of vertically elongated cells
. . . . *Olea*

–. Rays not as above 21

21. Vessels single . . . . *Eucalyptus*

–. Most of the vessels arranged in multiples . . . .
Acer

REFERENCES

ALONI, R. and ZIMMERMAN, M. H. (1983) The control of vessel
size and density along the plant axis. *Differentiation* **24**:
203–208.

BAAS, P. (1982) Systematic, phylogenetic and ecological wood
anatomy—History and perspectives. In: *New perspectives in
Wood Anatomy* (ed. P. Baas). Martinus Nijhoff/Dr. W.
Junk Publishers, The Hague, pp. 23–58.

BAAS, P., ESSER, P. M., VAN DER WESTEN, M. E. T. and ZANDEE,
M. (1988) Wood anatomy of Oleaceae. *IAWA Bull.* **9**:
103–182.

BAILEY, I. W. (1924) The problem of identifying the wood of
cretaceous and later dicotyledons: *Paraphyllanthoxylon
arizonense. Ann. Bot.* **38**: 439–451.

BAILEY, I. W. (1944) The development of vessels in angiosperms and its significance in morphological research.
Am. J. Bot. **31**: 421–8.

BAILEY, I. W. (1957) The potentialities and limitations of wood
anatomy in the study of the phylogeny and classification
of angiosperms. *J. Arnold Arb.* **38**: 243–54.

BAILEY, I. W. and TUPPER, W. W. (1918) Size variation in
tracheary cells. I. A. comparison between the secondary
xylem of vascular cryptogams, gymnosperms and angiosperms. *Proc. Am. Acad. Arts Sci.* **54**: 149–204.

BAMBER, R. K. (1976) Heartwood, its function and formation.
Wood Sci. Tech. **10**: 1–8.

BANNAN, M. W. (1933) Factors influencing the distribution of
vertical resin canals in the wood of the larch—*Larix laricina* (Du Roi) Koch. *Trans. R. Soc. Can.*, sect. 5, **27**:
203–18.

BANNAN, M. W. (1934) Seasonal wounding and resin cyst production in the hemlock *Tsuga canadensis* (L.) Carr. *Ann.
Bot.* **48**: 857–68.

BANNAN, M. W. (1936) Vertical resin ducts in the secondary
wood of the Abietineae. *New Phytol.* **35**: 11–46.

BANNISTER, B. (1963) Dendrochronology. In: *Science in Archaeology* (ed. D. Brothwell and E. Higgs). Basic Books,
New York, pp. 162–76.

BARGHOORN, E. S., JR. (1940) The ontogenetic development
and phylogenetic specialization of rays in the xylem of
dicotyledons: I, The primitive ray structure. *Am. J. Bot.*
27: 918–28.

BARGHOORN, E. S., JR. (1941a) The ontogenetic development
and phylogenetic specialization of rays in the xylem of
dicotyledons: II, Modifications of the multiseriate and
uniseriate rays. *Am. J. Bot.* **28**: 273–82.

BARGHOORN, E. S., JR. (1941b) The ontogenetic development
and phylogenetic specialization of rays in the xylem of
dicotyledons: III, The elimination of rays. *Bull. Torrey
Bot. Club* **68**: 317–25.

BAUCH, J., LIESE, W., and SCHOLZ, F. (1968) Über die Entwicklung und stoffliche Zusammensetzung der Hoftüpfelmembranen von Längstracheiden in Coniferen. *Holzforschung* **22**: 144–53.

BISSET, I. J. W., DADSWELL, H. E., and AMOS, G. L. (1950) Changes in fibre length within one growth ring of certain angiosperms. *Nature* **165**: 348–9.

BOSSHARD, H. H. (1966) Notes on the biology of heartwood formation. *News Bull. Int. Ass. Wood Anat.* 1966 (1): 11–14.

BOSSHARD, H. H. (1967) Über die fakultative Farbkernbildung. *Holz Roh- u. Werkstoff* **25**: 409–16.

BOSSHARD, H. H. (1968) On the formation of facultatively colored heartwood in *Beilschmiedia tawa*. *Wood Sci. Tech.* **2**: 1–12.

BOSSHARD, H. H. (1974–5) *Holzkunde*. Vols. I–III. Birkhauser Verlag, Basel und Stuttgart.

BOSSHARD, H. H. and HUG, U. E. (1980) The anastomoses of the resin canal system in *Picea abies* (L.) Karst. *Larix decidua* Mill. and *Pinus sylvestris* L. *Holz als Roh- und Werkstoff* **38**: 325–328.

BOYD, J. D. (1977) Basic cause of differentiation of tension wood and compression wood. *Aust. For. Res.* **7**: 121–43.

BRAUN, H. J. (1961) The organization of the hydrosystem in the stemwood of trees and shrubs. *News Bull. Int. Ass. Wood Anat.* **1961** (2): 2–10.

BRAUN, H. J. (1967) Entwicklung und Bau der Holzstrahlen unter dem Aspekt der Kontakt–Isolations–Differenzierung gegenüber dem Hydrosystem: I, Das Prinzip der Kontakt–Isolations–Differenzierung. *Holzforschung* **21**: 33–37.

BRAUN, H. J. (1970) Funktionelle Histologie der Sekundären Sprossachse. I. Das. Holz. In. K. Linsbauer. *Handbuch der Pflanzenanatomie*, 9, 1. Gebr. Borntraeger, Berlin.

BRAUN, H. J. (1983) Zur Dynamik des Wassertransportes in Bäumen. *Beitr. Deutsch. Bot. Ges.* **96**: 29–47.

BRAUN, H. J., WOLKINGER, F., and BÖHME, H. (1967) Entwicklung und Bau der Holzstrahlen unter dem Aspect der Kontakt–Isolations–Differenzierung gegenüber dem Hydrosystem: II, Die Typen der Kontakt–Holzstrahlen. *Holzforschung* **21**: 145–53.

BRAUN, H. J., WOLKINGER, F., and BÖHME, H. (1968a) Entwicklung und Bau der Holzstrahlen unter dem Aspect der Kontakt–Isolations–Differenzierung gegenüber dem Hydrosystem: III, Die Typen der Kontakt–Isolations–Holzstrahlen und der Isolations–Holzstrahlen. *Holzforschung* **22**: 53–60.

BRAUN, H. J., WOLKINGER, F., and BÖHME, H. (1968b) Entwicklung und Bau der Holzstrahlen unter dem aspect der Kontakt–Isolations–Differenzierung gegenüber dem Hydrosystem: IV, Die Organisation der Holzstrahlen. *Holzforschung* **22**: 153–7.

BRODZKI, P. (1972) Callose in compression wood tracheids. *Acta Soc. Bot. Pol.* **41**: 321–327.

BROWN, H. P., PANSHIN, A. J., and FORSAITH, C. C. (1949, 1952) *Textbook of Wood Technology*, Vols. 1 and 2. McGraw-Hill, New York.

BUTTERFIELD, B. G. and MEYLAN, B. A. (1972) Trabeculae in a hardwood. *Bull. Int. Ass. Wood Anat.* **1972** (1): 3–9.

BUTTERFIELD, B. G. and MEYLAN, B. A. (1979) Observations of trabeculae in New Zealand hardwoods. *Wood Sci. Tech.* **13**: 59–65.

CARLQUIST, S. (1962) A theory of paedomorphosis in dicotyledonous woods. *Phytomorphology* **12**: 30–45.

CHALK, L., MARSTAND, E. B., and WALSH, J. P. C. DE (1955) Fibre length in storied hardwoods. *Acta bot. neerl.* **4**: 339–47.

CHATTAWAY, M. M. (1955) Crystals in woody tissue: I, *Trop. Woods* **102**: 55–74.

CHATTAWAY, M. M. (1956) Crystals in woody tissue: II, *Trop. Woods* **104**: 100–24.

CLARKE, S. H. (1938a) A multiple-entry perforated card-key with special reference to the identification of hardwoods. *New Phytol.* **37**: 369–74.

CLARKE, S. H. (1938b) The use of perforated cards in multiple-entry identification keys and in the study of the inter-relation of variable properties. *Chron. Bot.* **4**: 517–18.

CÔTÉ, W. A., JR., DAY, A. C., and TIMELL, T. E. (1969) A contribution to the ultrastructure of tension wood fibres. *Wood Sci. Tech.* **3**: 257–71.

CRONSHAW, J. and MOREY, P. R. (1968) The effect of plant growth substances on the development of tension wood in horizontally inclined stems of *Acer rubrum* seedling. *Protoplasma* **65**: 379–91.

CZANINSKI, Y. (1977) Vessel-associated cells. *IAWA Bull.* 1977/3: 51–55.

DESCH, H. E. (1968) *Timber—its Structure and Properties*. Macmillan, London, Melbourne, and Toronto.

DOUGLAS, A. E. (1936) *Climatic Cycles and Tree Growth*, Vol. 3, *A Study of Cycles*. Carnegie Inst., Washington.

EAMES, A. J. and MACDANIELS, L. H. (1947) *An Introduction to Plant Anatomy*. McGraw-Hill, New York.

ECKSTEIN, D. and BAUCH, J. (1969) Beitrag zur Rationalisierung eines dendrochronologischen Verfahrens und zur Analyse seines Aussagesicherheit. *Forstwiss. Zbl.* **88**: 230–50.

FAHN, A. (1958) Xylem structure and rhythm of development in trees and shrubs of the desert: I, *Tamarix aphylla, T. jordanis* var. *negevensis, T. gallica* var. *maris-mortui. Trop. Woods* **109**: 81–94.

FAHN, A. (1964) Some anatomical adaptations of desert plants. *Phytomorphology* **14**: 93–102.

FAHN, A. (1979) *Secretory Tissues in Plants*. Academic Press, London.

FAHN, A. and ZAMSKI, E. (1970) The influence of pressure wind, wounding and growth substances on the rate of resin duct formation in *Pinus halepensis* wood. *Israel J. Bot.* **19**: 429–46.

FAHN, A., WACHS, N., and GINZBURG, C. (1963) Dendrochronological studies in the Negev. *Israel Explor. J.* **13**: 291–9.

FAHN, A., WERKER, E. and BAAS, P. (1986) *Wood Anatomy and Identification of Trees and Shrubs from Israel and Adjacent Regions*. The Israel Academy of Sciences and Humanities. Jerusalem.

FISHER, J. B. and MUELLER, R. J. (1983) Reaction anatomy and reorientation in leaning stems of balsa *Ochroma* and papaya (Carica). *Can. J. Bot.* **61**: 880–887.

FREY-WYSSLING, A. (1952) Wachstumsleistung des pflanzlichen Zytoplasmas. *Ber. schweiz. bot. Ges.* **62**: 583–91.

FREY-WYSSLING, A. and BOSSHARD, H. H. (1959) Cytology of ray cells in sapwood and heartwood. *Holzforschung* **13**: 129–37.

FRITTS, H. C. (1965) Dendrochronology. In: *The Quaternary of the United States* (ed. H. E. Wright, Jr., and D. G. Frey). Princeton University Press, Princeton.

FUTO, L. P. (1969) Qualitative und quantitative Ermittlung der Mikrozugeigenschaften von Holz. *Holz Roh- u. Werkstoff* **27**: 192–201.

GERRY, E. (1915) Fiber measurement studies. Length variations, where they occur, and their relation to the strength and uses of wood. *Science* **41**: 179.

GIBSON, A. C. (1973) Comparative anatomy of secondary xylem in Cactoideae (Cactaceae). *Biotropica* **5**: 29–65.

GIBSON, A. C. (1977) Wood anatomy of opuntias with cylindrical to globular stems. *Bot. Gaz.* **138**: 334–51.

GIBSON, A. C. (1978) Rayless secondary xylem of *Halophytum. Bull. Torrey Bot. Club* **105**: 39–44.

GILBERT, S. G. (1940) Evolutionary significance of ring porosity in woody angiosperms. *Bot. Gaz.* **102**: 105–20.

GLOCK, W. S. (1955) Growth rings and climate. *Bot. Rev.* **21**: 73–188.

GREGORY, M. (1980) Wood identification: an annotated bibliography. *IAWA Bull. NS* **1**: 3–41.

GREGUSS, P. (1955) *Xylotomische Bestimmung der heute lebenden Gymnospermen.* Akadémiai Kiadó, Budapest.

GREGUSS, P. (1959) *Holzanatomie der europäischen Laubhölzer und Sträucher.* Akadémiai Kiadó, Budapest.

HANDLEY, W. R. C. (1936) Some observations on the problem of vessel length determination in woody dicotyledons. *New Phytol.* **35**: 456–71.

HARRIS, J. M. (1989) *Spiral Grain and Wave Phenomena in Wood Formation.* Springer Verlag, Berlin.

HEJNOWICZ, Z. (1967) Some observations on the mechanism of orientation movement of woody stems. *Am. J. Bot.* **54**: 684–9.

HEJNOWICZ, A. and HEJNOWICZ, Z. (1958) Variation of length of vessel members and fibres in the trunk of *Populus tremula* L. *Acta Soc. Bot. Pol.* **27**: 131–59.

HEJNOWICZ, Z. and ROMBERGER, J. A. (1973) Migrating cambial domains and the origin of wavy grain in xylem of broadleaved trees. *Am. J. Bot.* **60**: 209–22.

HOLMSGAARD, E. (1955) Tree-ring analysis of Danish forest trees. *Forst. Forsoksv. Danm.* **22**: 1–246 (with English summary).

HÖSTER, H. R. and LIESE, W. (1966) Über das Vorkommen von Reaktionsgewebe in Wurzeln und Ästen der Dicotyledonen. *Holzforschung* **20**: 80–90.

HUBER, B. (1935) Die physiologische Bedeutung der Ring- und Zerstreutporigkeit. *Ber. dt. bot. Ges.* **53**: 711–19.

IAWA Committee (1989) IAWA List of microscopic features for hardwood identification. *IAWA Bull.* **10**: 219–332.

INTERNATIONAL ASSOCIATION OF WOOD ANATOMISTS (1947) International glossary of terms used in wood anatomy. *Trop. Woods* **107**: 1–36.

JACCARD, P. (1938) Exzentrisches Dickenwachstum und anatomisch-histologische Differenzierung des Holzes. *Ber. schweiz. bot. Ges.* **48**: 491–537.

KOZLOWSKI, T. T. and WINGET, C. H. (1963) Patterns of water movement in forest trees. *Bot. Gaz.* **124**: 301–11.

KRAWCZYSZYN, J. and ROMBERGER, J. A. (1980) Interlocked grain, cambial domains, endogenous rhythms, and time relations, with emphasis on *Nyssa sylvatica. Am. J. Bot.* **67**: 228–36.

KRIBS, D. A. (1928) Length of tracheids in Jack pine in relation to their position in the vertical and horizontal axes of the tree. *Bull. Minn. Agric. Exp. Sta.* **54**: 1–14.

KRIBS, D. A. (1937) Salient lines of structural specialization in the wood parenchyma of dicotyledons. *Bull. Torrey Bot. Club* **64**: 177–88.

KUČERA, L. J. (1985) Zur Morphologie der Intercellularen in den Markstrahlen. Teil 4: Deltamikroskopische Untersuchungen und Gesamatschau. *Vierteljahrsschr. Natur-*

forsch. Ges. Zurich **130**: 374–397.

LEACH, R. W. A. and WAREING, P. F. (1967) Distribution of auxin in horizontal woody stems in relation to gravimorphism. *Nature* **214**: 1025–7.

LIPHSCHITZ, N., WAISEL, Y. and LEV-YADUN, S. (1979) Dendrochronological investigations in Iran. *Tree-Ring Bull.* **39**: 39–45.

MABBERLEY, D. J. (1974) Pachycauly, vessel-elements, islands and the evolution of arborescence in "herbaceous" families. *New Phytol.* **73**: 977–84.

MESSERI, A. (1959) Contributo alla conoscenza die meccanismi anatomici e fisiologici della resinazione. I, Alterazioni anatomiche in legno resinato e tamponato di *Pinus pinea. Ann. Acc. Sci. Forest.* **8**: 203–25.

METCALFE, C. R. and CHALK, L. (1950) *Anatomy of the Dicotyledons,* Vols. 1 and 2. Clarendon Press, Oxford.

METCALFE, C. R. and CHALK, L. (1983) *Anatomy of the Dicotyledons.* 2nd ed. Vol. II. Clarendon Press, Oxford.

MÜNCH, E. (1938) Statik und Dynamik des schraubigen Baues der Zellwand, besonders des Druck-und Zugholzes. *Flora* (Jena), NS, **32**: 357–424.

NEČESANÝ, V. (1958) Effect of β-indoleacetic acid in the formation of reaction wood. *Phyton* **11**: 117–27.

NOBUCHI, T., KURODA, K., IWATA, R. and HARADA, H. (1982) Cytological study of the seasonal features of heartwood formation of sugi (*Cryptomeria japonica* D. Don). *Mokuzai Gakkaishi* **28**: 669–676.

PATEL, R. N. (1964) On the occurrence of gelatinous fibres with special reference to root wood. *J. Inst. Wood Sci.* **12**: 67–80.

PHILIPSON, W. R., WARD, J. M., and BUTTERFIELD, B. G. (1971) *The Vascular Cambium.* Chapman & Hall, London.

PHILLIPS, E. W. J. (1948) Identification of softwoods by their microscopic structure. London, Dept. Sci. and Ind. Res., *Forest Prod. Res. Bull.,* no. 22.

PILLOW, M. Y. and LUXFORD, R. F. (1937) Structure, occurrence and properties of compression wood. *USDA Tech. Publ.,* no. 546.

PRIESTLEY, J. H. and SCOTT, L. I. (1936) A note upon summer wood production in the tree. *Proc. Leeds Phil. Soc.* **3**: 235–48.

RICHTER, H. G. (1980) Occurrence, morphology and taxonomic implications of crystalline and siliceous inclusions in the secondary xylem of the Lauraceae and related families. *Wood Sci. Tech.* **14**: 35–44.

SANIO, C. (1872) Über die Grösse der Holzzellen bei der gemeiner Kiefer (*Pinus sylvestris*). *Jb. wiss. Bot.* **8**: 401–20.

SATO, K. and ISHIDA, S. (1982) Resin in wood of *Larix leptolepis* Gord. (II) Morphology of vertical resin canals. *Res. Bull. College. Exp. For., Hokkaido Univ.* **39**: 297–316.

SATO, K. and ISHIDA, S. (1983) Resin canals in the wood of *Larix leptolepis* Gord. (V) Formation of vertical resin canals. *Res. Bull. College Exp. For., Hokkaido Univ.* **40**: 723–740.

SAUTER, J. J. (1972) Respiratory and phosphatase activities in contact cells of wood rays and their possible role in sugar secretion. *Zeit. Pflanzenphysiol.* **67**: 135–145.

SAUTER, J. J., ITEN, W. and ZIMMERMANN, M. H. (1973) Studies on the release of sugar into the vessels of sugar maple (*Acer saccharum*). *Can. J. Bot.* **51**: 1–8.

SCHULMAN, E. (1956) *Dendroclimatic Changes in Semiarid America.* University Arizona Press, Tucson.

SCURFIELD, G. and SILVA, S. (1969) The structure of reaction wood as indicated by scanning electron microscopy. *Aust. J. Bot.* **17**: 391–402.

SCURFIELD, G. and WARDROP, A. B. (1963) The nature of reaction wood: VII, Lignification in reaction wood. *Aust. J. Bot.* **11**: 107–16.

SHAH, J. J., BAQUI, S., PANDALAI, R. C. and PATEL, K. R. (1981) Histochemical changes in *Acacia nilotica* L. during transition from sapwood to heartwood. *IAWA Bull.* **2**: 31–36.

SHIGO, A. L. (1985) How tree branches are attached to trunks. *Can. J. Bot.* **63**: 1391–1401.

SPACKMAN, W. and SWAMY, B. G. L. (1949) The nature and occurrence of septate fibers in dicotyledons. (abstract). *Am. J. Bot.* **36**: 804.

STERN, W. L. (1978) A retrospective view of comparative anatomy, phylogeny, and plant taxonomy. *IAWA Bull. 1978* (2, 3): 33–39.

STERN-COHEN, S. and FAHN, A. (1964) Structure and variation of the wood fibres of *Eucalyptus gomphocephala* DC along and across the stem. *La-Yaaran* **14**: 106–17. (In Hebrew with English summary, 132–3).

TER WELLE, B. J. H. (1976) Silica grains in woody plants of the neotropics, especially Surinam. *Leiden bot. Ser.* **3**: 107–42.

TER WELLE, B. J. H. (1980) Cystoliths in the secondary xylem of *Sparattanthelium* (Hernandiaceae). *IAWA Bull.* NS, **1**: 43–48.

THOMSON, R. B. and SIFTON, H. B. (1925) Resin canals in the Canadian spruce (*Picea canadensis*). *Phil. Trans. R. Soc.* **B214**: 63–111.

TIEMANN, D. H. (1951) *Wood Technology.* Pitman, New York.

TIMELL, T. E. (1983) Origin and evolution of compression wood. *Holzforschung* **37**: 1–10.

VAN VLIET, G. J. C. M. (1976) Radial vessels in rays. *IAWA Bull. 1976* (3): 35–37.

WAISEL, Y. and LIPHSCHITZ, N. (1968) Dendrochronological studies in Israel. II. *Juniperus phoenicea* of north and central Sinai. *La Yaaran* **18**: 1–22, 63–67.

WARDROP, A. B. (1956) The nature of reaction wood: V, The distribution and formation of tension wood in some species of *Eucalyptus. Aust. J. Bot.* **4**: 152–66.

WARDROP, A. B. (1964) The reaction anatomy of arborescent angiosperms. In: *The Formation of Wood in Forest Trees* (ed. M. H. Zimmermann). Academic Press, New York.

WARDROP, A. B. (1965) The formation and function of reaction wood. In: *Cellular Ultrastructure of Woody Plants* (ed. W. A. Côté, Jr.). Syracuse Univ. Press, New York, pp. 371–90.

WARDROP, A. B. and DADSWELL, H. E. (1948) The nature of reaction wood: I, The structure and properties of tension wood fibres. *Aust. J. Sci. Res.* **B1**: 3–16.

WARDROP, A. B. and DADSWELL, H. E. (1955) The nature of reaction wood: IV, Variations in cell wall organization of tension wood fibres. *Aust. J. Bot.* **3**: 177–89.

WARDROP, A. B. and DAVIES, G. W. (1964) The nature of reaction wood: VIII, The structure and differentiation of compression wood. *Aust. J. Bot.* **12**: 24–36.

WATERKEYN, L., CAEYMAEX, S. and DECAMPS, E. (1982) La callose des trachéides du bois de compression chez *Pinus silvestris* et *Larix decidua. Bull. Soc. Roy. Bot. Belg.* **115**: 149–155.

WERKER, E. and BAAS, P. (1981) Trabeculae of Sanio in secondary tissues of *Inula viscosa* (L.) Desf. and *Salvia fruticosa* Mill. *IAWA Bull.* **2**: 69–76.

WERKER, E. and FAHN, A. (1969) Resin ducts in *Pinus halepensis* Mill.: their structure, development and pattern of arrangement. *J. Linn Soc. Bot.* **62**: 379–411.

WESTING, A. H. (1968) Formation and function of compression wood in gymnosperms: II, *Bot. Rev.* **34**: 51–78.

WLOCH, W. (1975) Longitudinal shrinkage of compression wood in dependence on water content and cell wall structure. *Acta. Soc. Bot. Pol.* **44**: 217–229.

ZIMMERMAN, M. H. (1979) The discovery of tylose formation by a Viennese Lady in 1845. *IAWA Bull. 1979* (2, 3): 51–56.

ZIMMERMANN, M. H. (1983) *Xylem Structure and Ascent of Sap.* Springer-Verlag, Berlin.

ZIMMERMANN, M. H. and POTTER, D. (1982) Vessel-length distribution in branches, stem and roots of *Acer rubrum* L. *IAWA Bull.* **3**: 103–109.

CHAPTER 16

SECONDARY PHLOEM

THE arrangement of the elements of the secondary phloem is parallel to that of the secondary xylem. In the phloem there are also two systems, the vertical and horizontal, which are derived from those initials that give rise to the similar systems in the xylem.

The principal components of the vertical system of the phloem are the sieve elements, phloem parenchyma, and phloem fibres. The horizontal system comprises the parenchyma of the phloem rays.

Storied, non-storied, and intermediate arrangements of the elements can also be distinguished in the phloem. As in the xylem, the arrangement of the tissue is primarily determined by the nature of the cambium, i.e. whether it is storied or not. Secondly, the arrangement is determined by the extent of elongation of the various elements of the vertical system during the differentiation of the cells.

In many species of dicotyledonous trees growth rings may also be observed in the phloem, but they are less distinct than those seen in the xylem. The growth rings, as seen in the phloem, are due to the differences in the cells produced at the beginning and at the end of the season—at the beginning of the growth season the cells are conspicuously extended radially, while those produced at the end of the season are flattened. The arrangement of growth rings becomes obscured after some growth seasons as a result of the obliteration of the sieve elements, which cease to function, and because of the changes that take place in the other cells. These changes involve the enlargement of the parenchyma cells, for instance. In many gymnosperms and angiosperms, tangential bands of fibres are developed in the secondary phloem (Fig. 211, no. 2). The number of these bands is not constant in each season and there-

fore they cannot be used as an indication of the age of the secondary phloem (Esau, 1948; Huber, 1949; Artschwager, 1950).

The ray initials in the cambium produce cells towards both the xylem and the phloem (Fig. 169, no. 2), so that the xylem and phloem rays are continuous. In the vicinity of the cambium the xylem and phloem rays are equal in size, but in many plants the mature outer portions of the phloem rays are of increased width (Fig. 211, no. 1). This feature is, of course, connected with the increase in circumference of the trunk as a result of secondary thickening. The widening of the phloem rays may be accomplished solely by the lateral expansion of the existing cells or, as is more common, by the increase, as a result of radial cell division, of the number of cells on the periphery. These dilated parts of the rays constitute the expansion tissue which is discussed more fully in the following chapter. Sometimes only some rays become enlarged, while others do not change. The outer parts of the phloem rays become cut off from the inner portions by the development of cork tissue. This tissue is produced by the phellogen, and its formation results in the interruption of the connection between the inner living layers of the phloem and the outer layers which dry out.

SECONDARY PHLOEM OF CONIFERS

As in the secondary xylem, the secondary phloem of the conifers is of relatively simple structure (Fig. 210, nos. 1–3). The vertical system is comprised of sieve cells and parenchyma cells, including albuminous cells (see Chapter 8), and in many plants fibres are also present. The phloem rays of the conifers are usually uniseriate and they

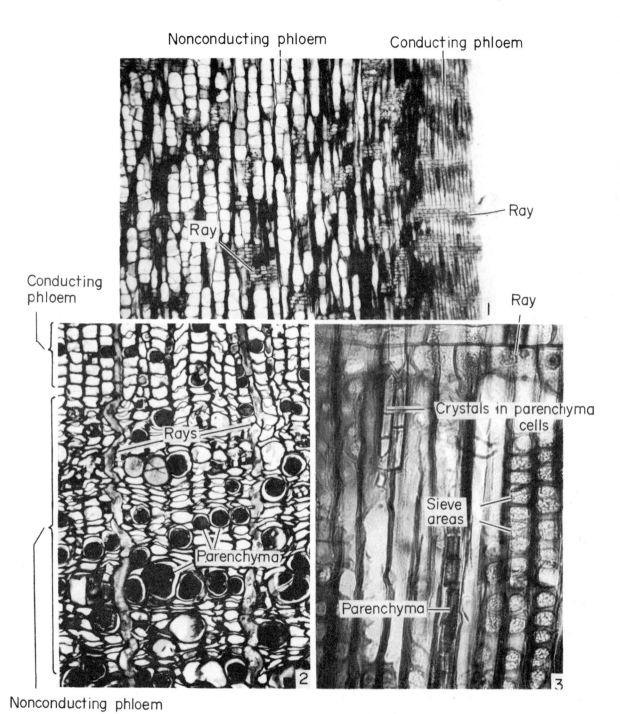

FIG. 210. Secondary phloem of *Pinus halepensis*. 1 and 2, Micrographs showing conducting and nonconducting phloem. 1, Radial section. ×42. 2, Cross-section. ×126. 3, Radial section of conducting phloem. ×300.

Secondary xylem Secondary phloem Phloem ray Periderm

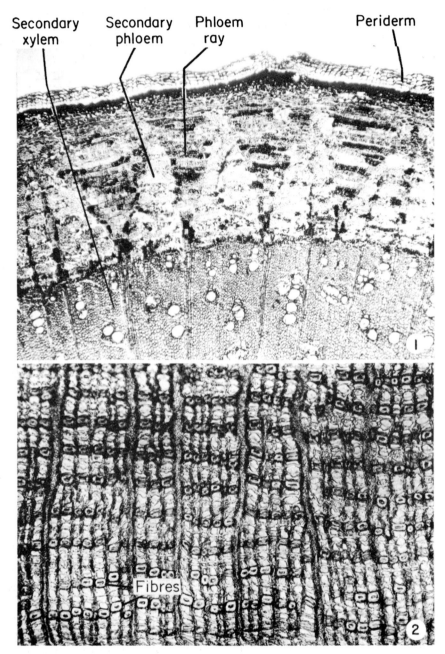

Fibres

FIG. 211. 1, Micrograph of the outer portion of the stem of *Gossypium* in which the dilatation of the vascular ray towards the periderm can be distinguished. × 35. 2, Cross-section of the secondary phloem of *Cupressus sempervirens* in which bands of phloem fibres can be seen to alternate with bands of phloem parenchyma and sieve cells. × 150.

usually consist of parenchyma cells only, but albuminous cells may also be present. The radial arrangement of the secondary phloem elements of the conifers is retained, even in the mature regions, because the elements do not change their shape much during differentiation. The ends of the sieve cells overlap one another and lengthwise each sieve cell comes into contact with a few rays. As in the distribution of the pits on the tracheids, there are more sieve areas on the overlapping end portions of the sieve cells than there are on the other portions of these cells. Usually the sieve areas develop on the radial walls of the sieve cells (Fig. 210, no. 3). The parenchyma cells, with the exception of the albuminous cells (Srivastava, 1963), of the vertical system store starch at certain seasons of the year and many of them contain resins, fats, tannins and also crystals. Those cells containing calcium oxalate crystals are lined, as observed in *Larix*, with a suberin layer (Wattendorf, 1969). Murmanis and Evert (1967) have observed in *Pinus strobus* the development of perforations in transverse walls between axial parenchyma cells of the secondary phloem, and the resulting union of the cellular content of two such connected cells. In the Pinaceae the cells of the phloem parenchyma develop in tangential rows or bands. In the Cupressaceae and Taxodiaceae there are alternating bands of phloem parenchyma, sieve cells and fibres (Fig. 211, no. 2). The secondary phloem of the Pinaceae contains no fibres but the sieve cells develop thick secondary walls which are not lignified (Abbe and Crafts, 1939; Don, 1965). Abundant trabeculae (see Chapters 2 and 15) were found in thin-walled fibres of the secondary phloem of *Libocedrus bidwillii* (Chan, 1985). Resin ducts may be present in the secondary phloem of the conifers. In some conifers, e.g. *Abies balsamea*, these schizogenous ducts form blisters in the secondary phloem which can be seen externally on the bark of the trunk. This resin, known as Canada balsam, is used as a mounting medium in microscopy because its refractive index is the same as that of glass. In conifers only a narrow zone of the phloem may be active. In species lacking fibres, e.g. in *Pinus*, the collapse of the sieve cells of nonconducting phloem results in a distorted appearance, and the rays become wavy (Fig. 170, no. 1; Fig. 210, no. 2). The axial parenchyma cells enlarge in the nonconducting phloem and remain alive until they are cut off by periderm. In this region the ray parenchyma cells also remain active and as seen in *Larix decidua*, increase also in volume (Ott, 1982). However, the albuminous cells collapse in nonconducting phloem.

SECONDARY PHLOEM OF DICOTYLEDONS

As in the secondary xylem of the dicotyledons, the secondary phloem has a relatively complicated structure (Fig. 212). The vertical system contains sieve-tube members, companion cells, parenchyma cells, and fibres. In *Trochodendron*, although the xylem is vesselless, there are sieve-tube members in the phloem (Jørgensen *et al.*, 1975). The horizontal system consists of variously sized rays, from uni- to multiseriate, which consist of parenchyma cells only. In both systems, sclereids, lysigenous or schizogenous secretory structures (Fig. 212), laticifers, and other cells that contain special substances may be present (Fahn and Evert, 1974; Gibson, 1981; Gómez-Vazquez and Engleman, 1984). Many of the parenchyma cells contain crystals, and sometimes such cells become subdivided to form chambers, each of which contains a single crystal. Crystals may also be formed in the rays and in the sclerenchyma cells of the phloem.

The parenchyma plays a major role in storage of carbohydrates, lipids and nitrogen compounds. A special type of axial parenchyma cells storing protein, has been found in *Hevea brasiliensis*. These cells are characterized by the presence of a large amount of proteinaceous fibrils in the central vacuole (Wu Jilin and Hao Binzhog, 1987a, b).

The secondary phloem fibres of dicotyledons are variously arranged. In *Carya*, which has a hard bark, the fibres constitute the largest portion of the phloem, and they include scattered groups of the other phloem elements. In *Nothofagus* the fibres are arranged in numerous scattered bundles, encased in a sheath of chambered crystaliferous cells (Patel and Shand, 1985). In *Vitis*, for example, the fibres, a great proportion of which are septate, are arranged in tangential bands which alternate with bands of the sieve tubes, companion cells, and

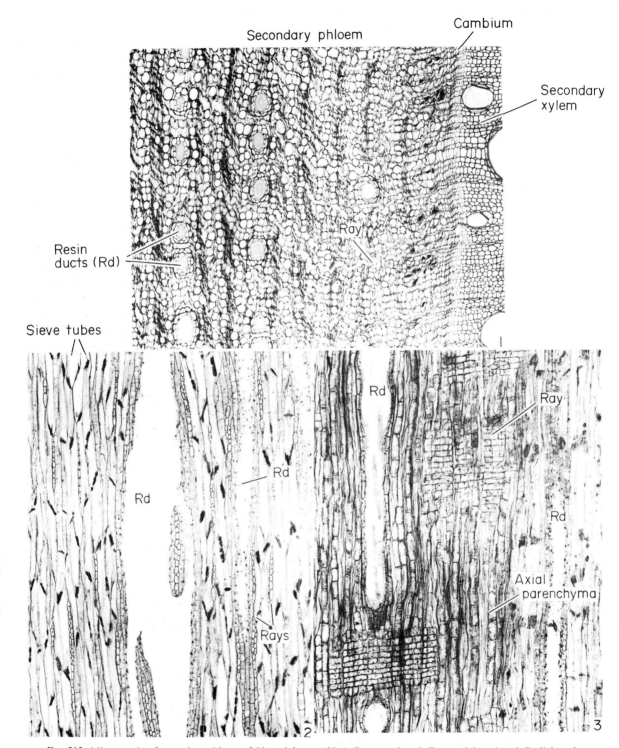

FIG. 212. Micrographs of secondary phloem of *Rhus glabra*. × 120. 1, Cross-section. 2, Tangential section. 3, Radial section.

phloem parenchyma. In *Laurus, Nicotiana,* and *Stenolobium* there are few fibres, and they are scattered among the rest of the elements of the vertical system. In *Aristolochia* no fibres are present.

An increase in length as the cambium is approached occurs in phloem fibres and sieve-tube members as it is in xylem fibre and tracheary elements (Parameswaran and Liese, 1974).

Phloem fibres that do not develop directly from fusiform cambial initials but from parenchyma cells of non-functioning phloem have been termed *fibre-sclereids* (Esau et al., 1953). Fibre-sclereids occur, for example, in the secondary phloem of *Pyrus malus* (Evert, 1963).

Sclereids occur both in functioning and nonfunctioning phloem where they arise from parenchyma cells. In secondary phloem the sclereids may occur separately from, or together with, the fibres. In *Platanus* and *Fagus,* for example, the sclereids are the only sclerified elements present in the secondary phloem. In functioning phloem, sclereids are usually not as abundant as are the fibres, but in many species the parenchyma cells of the nonfunctioning phloem differentiate into sclereids. In certain plants, such as *Prunus,* for example, there is no sclerenchyma in the functioning phloem, but, in the phloem which has ceased to conduct, both fibres and sclereids differentiate. A special type of sclereid (crystalliferous sclereids), the asymmetrically thickened walls of which are non-pitted and non-lignified, occurs singly or in groups of 2–5 in the secondary phloem of *Schisandra.* Each such sclereid is filled with resinous material and its outer wall is lined on its inner surface by a layer of crystals (Jalan, 1968). Sclereids with protrusions of the lignified cell walls into the lumen, and lignified sclereids with membrane-like cross walls of cellulosic nature, are also found in the secondary phloem of some trees (Parameswaran, 1968).

The arrangement of the sieve tubes and the parenchyma cells differs in various plants. The sieve tubes and parenchyma cells may form separate alternating bands as in *Robinia* and *Aristolochia,* for example, or sieve tubes may be arranged in radial rows as in *Prunus.*

In the sieve-tube members, the sieve areas are distinctly better developed on the sieve plates than on the lateral walls. However, in some plants, e.g. the subfamily Pomoideae of the Rosaceae, this difference is relatively slight. In genera such as *Quercus, Juglans, Vitis,* and *Populus,* the secondary phloem is not storied and the sieve-tube members are elongated and bear mostly compound sieve plates on the oblique end walls. In *Acer,* for example, the sieve-tube members are shorter than the above and the end walls are only slightly oblique and bear simple sieve plates. In *Robinia, Tamarix, Ulmus,* and *Fraxinus* the simple sieve plates are horizontal. In *Robinia* and *Tamarix* the phloem is storied and the sieve-tube members are short. The oblique end walls are usually so orientated that, in a radial longitudinal section of the stem, the surface of the sieve plate is seen, and in a tangential section, the sieve plates are sectioned longitudinally. This arrangement is well demonstrated in *Vitis,* for example.

Comparative studies of the anatomical features of the bark, especially of its phloem component, were made in a large number of plant species, mostly tropical (Zahur, 1959; Roth, 1981; Chavan and Shah, 1983; Den Outer, 1983).

DURATION OF THE ACTIVITY OF SECONDARY PHLOEM

In conifers, e.g. *Abies balsamea* and *Picea mariana,* some of the phloem formed in late summer overwinter and in spring constitute the first functional sieve cells (Alfieri and Evert, 1973).

In most dicotyledons the functioning part of the phloem is restricted to that secondary phloem that is produced in the last growth season. Sometimes, before the cambium begins to produce new phloem, all or most of the sieve elements produced in the previous season cease to function. However, in some plants, e.g. *Tilia,* the sieve tubes are active throughout a number of years, and no changes have been observed to take place during the winter. In *Grewia tiliaefolia,* a deciduous tropical tree of West India, the sieve elements also function for more than one season, but most of the elements develop callose plugs in the leafless period. The different phloem increments, in this species, are separated by a zone of very narrow sieve elements, which mature just before and immediately after the period of dormancy (Deshpande and Rajendrabaubu, 1985). In *Vitis* the phloem was observed to be active for two seasons, but, unlike *Tilia*, *Vitis* lays down thick layers of callose with the onset of winter. These layers are subsequently resorbed in the spring before the renewal of cambial activity (Esau, 1948; Bernstein and Fahn, 1960). In *Fraxinus americana* the last sieve tubes of the previous year are reactivated in spring and remain functional during the period when the buds develop and the young leaves grow (Zamski and Zimmermann, 1979). It should be mentioned that in plants with included phloem, e.g. *Bougainvillea* and the woody species of the Chenopodiaceae, the phloem strands remain active for many years (Fahn and Shchori, 1967). The duration of the activity of the phloem in the secondary bundles of the long-lived monocotyledons has yet to be investigated. In most conifers at least a portion of a seasonal phloem increment appears to function for two growing seasons (Esau, 1969).

That part of the secondary phloem in which the sieve tubes no longer serve as a conducting system are termed *nonfunctioning* or *nonactive phloem*. In many plants the parenchyma cells in this phloem remain viable and continue to store starch until such time when the region is cut off from the more central region by the formation of the periderm. The characteristic features of the non-functioning phloem are the presence of thick layers of callose, termed *definitive callose*, which cover the sieve areas, the disorganization and disintegration of the protoplast, and the collapse and crushing of the elements, especially in the older regions of the phloem. The definitive callus may not always be seen as it tends to peel away. The most definite indication of non-functioning phloem is the presence of crushed elements. The companion cells cease to function together with the sieve tubes to which they are attached.

In certain plants, e.g. *Robinia pseudacacia* and some species of *Pinus*, the nonfunctioning phloem contains only a narrow band of intact phloem elements. In *Populus* and *Tilia* crushed sieve tubes are found only a great distance from the cambium and, although only the sieve tubes close to the cambium function, the shape of the nonfunctioning sieve tubes is retained. In *Salix* the sieve tubes are never obliterated, and their shape and size is retained even after the formation of the periderm which separates them from the inner tissues.

In *Vitis vinifera*, *Antiaris*, *Bombax*, and *Ricinodendron* the non-functioning sieve tubes become filled with tylose-like proliferations from the neighbouring parenchyma cells. These proliferations or tylosoids differ from tyloses in not growing through pits (Esau, 1948; Lawton and Lawton, 1971).

The amount of non-functioning phloem that accumulates depends on the manner in which the phellogen is formed. If the phellogen is formed close to the stem surface, and when it is not replaced for many years by deeper-formed phellogens, the stem may be encircled by a thick layer of non-functioning phloem (e.g. *Prunus*). If new, more deeply formed phellogens are formed annually within the phloem, e.g. *Vitis*, there is no accumulation of nonfunctioning secondary phloem.

The secondary phloem, in addition to transporting the organic substances, plays also a major role in storage of material during the dormant periods (Kramer and Kozlowski, 1979).

REFERENCES

ABBE, L. B. and CRAFTS, A. S. (1939) Phloem of white pine and other coniferous species. *Bot. Gaz.* **100**: 695–722.

ALFIERI, F. J. and EVERT, R. F. (1973) Structure and seasonal development of the secondary phloem in the Pinaceae. *Bot. Gaz.* **134**: 17–25.

ARTSCHWAGER, E. (1950) The time factor in the differentiation of the secondary xylem and phloem in pecan. *Am. J. Bot.* **37**: 15–24.

BERNSTEIN, Z. and FAHN, A. (1960) The effect of annual and bi-annual pruning on the seasonal changes in xylem formation in the grapevine. *Ann. Bot.* **24**: 159–71.

BOSSHARD, H. H. and STAHEL, J. (1969) Modifikationen in der sekundären Rinde von *Populus robusta*. *Holzforsch. Holzverwert.* **5**: 1–5.

CHAN, L. L. (1985) The anatomy of the bark of *Libocedrus* in New Zealand. *IAWA Bull.* **6**: 23–34.

CHAVAN, R. R. and SHAH, J. J. (1983) Statistical approach for the understanding of secondary phloem in 125 tropical dicotyledons. *Proc. Indian Nat. Sci. Acad. Part B, Biol. Sci.* **49**: 28–36.

DEN OUTER, R. W. (1983) Comparative study of the secondary phloem of some woody dicotyledons. *Aca Bot. Neerl.* **32**: 29–38.

DESHPANDE, B. P. and RAJENDRABABU, T. (1985) Seasonal changes in the structure of the secondary phloem of *Grewia tiliaefolia*, a deciduous tree from India. *Ann. Bot.* **56**: 61–75.

DON, D. (1965) Secondary walls in phloem of *Pinus radiata*. *Nature* **207**: 657–8.

ESAU, K. (1948) Phloem structure in the grapevine, and its seasonal changes. *Hilgardia* **18**: 217–96.

ESAU, K. (1969) The Phloem. In: K. Linsbauer, *Handbuch der Pflanzenanatomie*, Bd. V, T. 2. Gebr. Borntraeger, Berlin.

ESAU, K., CHEADLE, V. I., and GIFFORD, E. M., JR. (1953) Comparative structure and possible trends of specialization of the phloem. *Am. J. Bot.* **40**: 9–19.

EVERT, R. F. (1963) Ontogeny and structure of the secondary phloem in *Pyrus malus*. *Am. J. Bot.* **50**: 8–37.

FAHN, A. and EVERT, R. F. (1974) Ultrastructure of the secretory ducts of *Rhus glabra* L. *Am. J. Bot.* **61**: 1–14.

FAHN, A. and SHCHORI, Y. (1967) The organization of the secondary conducting tissues in some species of the Chenopodiaceae. *Phytomorphology* **17**: 147–54.

GIBSON, A. C. (1981) Vegetative anatomy of *Pachycormus* (Anacardiaceae). *Bot. J. Linn. Soc.* **83**: 273–84.

GOMEZ-VAZQUEZ, B. G. and ENGLEMAN, E. M. (1984) Bark anatomy of *Bursera longipes* (Rose) Stanley and *Bursera copallifera* (Sessé & Moc.) Bullock. *IAWA Bull.* **5**: 335–40.

HUBER, B. (1949) Zur Phylogenie des Jahrringbaues der Rinde. *Svensk bot. Tidskr.* **43**: 376–82.

JALAN, S. (1968) Observations on the crystalliferous sclereids of some Schisandraceae. *Beitr. Biol. Pfl.* **44**: 277–88.

JØRGENSEN, L. B., MØLLER, J. D., and WAGNER, P. (1975) Secondary phloem of *Trochodendron aralioides*. *Bot. Tidsskr.* **69**: 217–38.

KRAMER, P. J. and KOZLOWSKI, T. T. (1979) *Physiology of Woody Plants*. Academic Press, New York.

LAWTON, J. R. and LAWTON, J. R. S. (1971) Seasonal variations in the secondary phloem of some forest trees from Nigeria. *New Phytol.* **70**: 187–96.

MURMANIS, L. and EVERT, R. F. (1967) Parenchyma cells of secondary phloem in *Pinus strobus*. *Planta* **73**: 301–18.

OTT, E. (1982) Morphologische und physiologische Alterung von sekundaren Rindengewebe in *Larix decidua* Mill. *Vierteljahrsschr. Naturforsch. Ges. Zurich* **127**: 89–166.

PARAMESWARAN, N. (1968) Strukturbesonderheiten von Sklereiden im sekundären Phloem der Bäume. *Ber. dt. bot. Ges.* **81**: 199–202.

PARAMESWARAN, N. and LIESE, W. (1974) Variation in cell length in bark and wood of tropical trees. *Wood Sci. Tech.* **8**: 81–90.

PATEL, R. N. and SHAND, J. E. (1985) Bark anatomy of *Nothofagus* species indigenous to New Zealand. *NZ. J. Bot.* **23**: 511–32.

ROTH, I. (1981) Structural patterns of tropical barks. In: *Handbuch der Pflanzenanatomie*, Band 9, Teil 3. Gebruder Borntraeger, Berlin.

SRIVASTAVA, L. M. (1963) Secondary phloem in the Pinaceae. *Univ. Calif. Publ. Bot.* **36**: 1–142.

WATTENDORF, J. (1969) Feinbau und Entwicklung der verkorkten Calciumoxalat-Kristallzellen in der Rinde von *Larix decidua* Mill. *Z. Pflanzenphysiol.* **60**: 307–47.

WU JILIN and HAO BINGZHONG (1978a) Protein-storing cells in secondary phloem of *Hevea brasiliensis*. *Xexue Tongbao* **32**: 118–121.

WU JILIN and HAO BINGZHONG (1987b) Ultrastructure and differentiation of protein-storing cells in secondary phloem of *Hevea brasiliensis* stem. *Ann. Bot.* **60**: 505–512.

ZAHUR, M. S. (1959) Comparative study of secondary phloem of 423 species of woody dicotyledons belonging to 85 families. *Cornell Univ. Agr. Exp. Sta. Mem.* **358**: 1–160.

ZAMSKI, E. and ZIMMERMANN, M. H. (1979) Sieve tube longevity in white ash (*Fraxinus americana*) studied with a new histochemical test for the identification of sugar. *Can. J. Bot.* **57**: 650–6.

CHAPTER 17

PERIDERM

THE development of the secondary vascular tissue is usually accompanied by the formation of cork. Functionally, this cork tissue forms a protective layer that replaces the epidermis which dies and is shed. Cork usually forms in the roots and stems of dicotyledons that have continuous and pronounced secondary thickening, but it is usually not formed on leaves with the exception of the scales of winter buds of certain plants. In the rhizomes of some pteridophytes, e.g. *Ophioglossum*, the epidermis and underlying outer cortical cells become suberized.

Cork is part of the compound secondary tissue which is termed the *periderm* (Fig. 211, no. 1; Fig. 213, no. 1). The periderm usually consists of three parts: the *phellogen*, which is the cork cambium; the *phellem*, which is the cork and which is produced centrifugally by the phellogen; and the *phelloderm*, which is the parenchymatous tissue, in some but not all species, produced centripetally by the phellogen.

The development of the periderm sometimes commences only after the production of the secondary vascular tissues has reached considerable dimensions. In such cases the circumference of the epidermis increases together with the secondary and other tissues on the outer side of the cambium. Examples of trees with the above type of development are *Laurus*, *Citrus*, *Cornus*, some species of *Eucalyptus*, *Acer*, and *Acacia*. In *Viscum*, for example, cork tissue is never formed, and the epidermis, the cell walls of which thicken, increases in circumference and persists on the stem throughout the life of the plant.

In addition to the above-mentioned places, cork tissue also develops at the site of leaf abscission, around those areas of the plant which are damaged by disease, and below wounds.

STRUCTURE OF THE PERIDERM COMPONENTS

Phellogen

The phellogen is a secondary meristematic tissue from all points of view—it originates from cells that have undergone differentiation and it produces tissues that comprise part of the secondary plant body. By virtue of its position, the phellogen is a lateral meristem as, like the cambium, it results in an increase of the diameter of the axis by periclinal divisions in its cells. Histologically the phellogen is simpler than the vascular cambium as it consists of only one type of initial cells. These cells appear rectangular in cross-section with their shorter axis in the radial direction and, in longitudinal tangential section, they are seen to be regular polygons. The protoplasts of the phellogen cells contain variously sized vacuoles and they may contain chloroplasts and tannins. There are no intercellular spaces in the phellogen except in those regions where lenticels develop.

The phellogen, similar to the vascular cambium, exhibits periods of activity and inactivity. In some plants, e.g. in *Quercus ithaburensis* and *Q. infectoria*, the periods of activity of the phellogen and of the vascular cambium coincide (Arzee *et al.*, 1978). This is not true, however, of other plants. In *Robinia pseudacacia*, for instance, two periods of phellogen activity occur during the single annual period of cambial activity. In *Acacia raddiana* three alternating periods of phellogen activity and inactivity were reported to occur during the year (Waisel *et al.*, 1967; Arzee *et al.*, 1970). Several factors seem to affect phellogen activity and initiation (Waisel *et al.*, 1967; Arzee *et al.*, 1968; Liphschitz and Waisel, 1970; Borger and Kozlowski, 1972a, b, c). In *Robinia* the phellogen was found to be active mainly under a combination of short day and high temperature or long day and low temperature. Gibberellic acid and naphthalene acetic acid were reported to have a distinct retarding effect on initiation of phellogen in this plant. In *Eucalyptus camaldulensis* high humidity and/or a continuous flush of oxygen were found to cause early initiation of phellogen (see also "Wound cork" in this chapter).

Phellem

Like the cells of the phellogen, the cells of the phellem (the cork cells) are usually polygonal as seen in tangential section and, in cross-section, they are flattened radially. In cross-section (Fig. 213, nos. 1, 2), the cork cells are usually seen to be arranged in compact radial rows which are devoid of intercellular spaces. This radial arrangement indicates that the phellogen cells divide tangentially.

Cork cells are dead cells. Various types of cork cells can be distinguished and in a few plants crystal-containing cells and sclereids may be found among the cork cells. Sometimes non-suberized cells, which are termed *phelloids*, occur in the phellem. Two common types of cork cells are those which are hollow, thin-walled, and somewhat widened radially, and those which are thick-walled and radially flattened. The cells of the latter type may often be filled with dark resiniferous or tanniniferous substances as, for example, can be seen in *Eucalyptus*. These two types of phellem cells may occur together in the same plant as, for example, in *Arbutus* and *Betula* where they occur in alternating layers. In *Betula* this feature causes the cork to peel like sheets of paper. The exfoliation of the bark in some species of *Bursera* and *Pachycormus*, allows the passage of sun-light to the chlorenchymatous cells of the phelloderm, improving photosynthesis (Gibson, 1981; Gomez-Vazquez and Engleman, 1984).

The primary wall of the phellem cells consists of cellulose and may sometimes also contain lignin or suberin. Internally the primary wall is lined by a relatively thick layer of suberin, the *suberin layer*, which consists of fine alternating lamellae

of suberin and wax. In the process of suberin production the endoplasmic reticulum apparently plays a major role (Wattendorff, 1974). A thin cellulose layer, which in certain plants may be lignified, may be present on the inside of this lamella. This layer is mostly absent in the bottle cork (of *Quercus suber*). Ultra-thin pores are found in the wall of the bottle-cork cells. These develop from plasmodesmata and become blocked by a dense material (Sitte, 1962). The suberin layer is impermeable to water and gases, and it withstands the action of acids. The protoplast of the phellem cells is lost after the various wall layers have been formed and the cell lumen becomes filled with air or the pigmented substances mentioned above. In *Betula* Schönherr and Ziegler (1980) showed that the middle lamellae and primary walls of the phellem are at most only partly suberized. According to these authors this feature explains their finding that the *Betula* periderm is quite permeable to water.

In the phellem of some plants, e.g. species of *Haloxylon* and *Anabasis* (Fig. 213, no. 7), bands or large groups of hollow, thick-walled cells occur among the usual thin-walled cork cells. These cells have a lignified primary wall and a thick outer layer of secondary wall on the inside of which is a relatively thin suberin lamella. This suberin lamella, in turn, is lined by a thinnish cellulose layer which may sometimes be lignified.

The cork tissue of certain plants, such as that of *Quercus suber*, for example, is very elastic, but the cork of most plants lacks this quality.

Phelloderm

The phelloderm cells are living cells with non-suberized walls. They are similar to the parenchyma cells of the cortex but, if the phelloderm is multiseriate, they are usually arranged in radial rows. In certain plants the phelloderm cells contain chloroplasts and they are photosynthetic (e.g. in **Bursera** and **Pachycormus**; Gibson, 1981; Gomez-Vazquez and Engleman, 1984). These cells may also store starch. Sclereids and other special cells are sometimes present among the phelloderm cells.

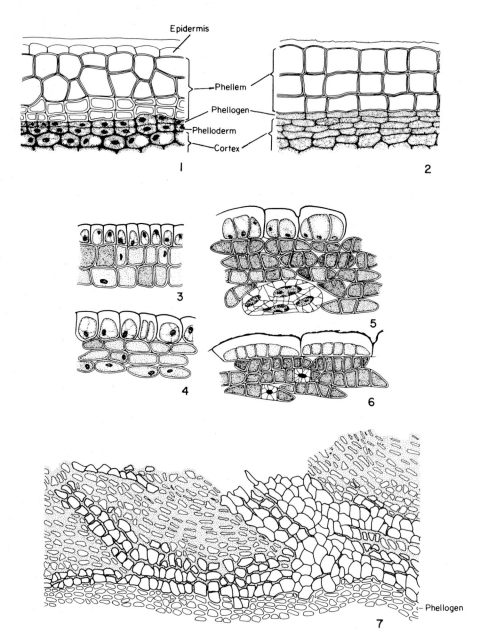

FIG. 213. 1, Outermost portion of a cross-section of a branch of *Populus deltoides* showing the development of the phellogen in the outermost layer of the cortex. 2, As above but of *Solanum dulcamara* in which the phellogen develops in the epidermis. The phellogen is formed from the inner cells resulting from a periclinal division of the epidermal cells; the outer cells form the outermost layer of the phellem and are covered by cuticle. No phelloderm is formed. 3–6, Portions of the outermost layers as seen in cross-section of the stem of *Eucalyptus gigantea* showing various stages in the anticlinal division of the epidermal and cortical cells; the divisions bring about the increase in the circumference of the organ. In the early stages groups of cells related to a single mother cell can be distinguished. 7, Portion of a cross-section of the periderm of *Anabasis articulata*, in which two types of phellem cells—thin-walled and thick-walled—can be seen. Thickened walls stippled. (Nos. 1 and 2 adapted from Eames and MacDaniels, 1947; nos. 3–6 adapted from Chattaway, 1953.)

DEVELOPMENT OF PERIDERM

The phellogen may develop in living epidermal (Fig. 213, no. 2), parenchyma, or collenchyma cells (Fig. 213, no. 1). The cells become meristematic, lose their central vacuoles, the volume of the cytoplasm increases, and they undergo periclinal division. With the commencement of these divisions, starch and tannins are gradually lost from those cells that contained them. As a result of the first periclinal division two cells, which are similar in appearance, are formed. The inner cell is capable of further division, but often it does not do so. In both cases, however, this cell is regarded as a phelloderm cell. The outer cell undergoes a periclinal division resulting in the formation of two cells. The outer of these two cells differentiates into a cork cell and the inner cell constitutes the phellogen initial and continues to divide. Sometimes the cork and phellogen cells are formed after the first division and then no phelloderm cell is formed. In addition to periclinal divisions the initials of the phellogen undergo occasional anticlinal divisions, so that the circumference of the cork cylinder is continuously increased.

The number of phellem layers is usually greater than the number of phelloderm layers. In certain plants the phelloderm is completely absent but in many plants it consists of one to three layers of cells, while in a few other plants it may be up to six layers thick. The number of layers in the phelloderm may also alter with the age of the plant. The number of layers of phellem cells produced in a single season varies, in different species, and may be very large. If the first-formed periderm remains on the axial organ for many years, the outer layers of cork become cracked and are shed so that the layer of cork remaining on a plant is of more or less constant thickness.

In certain plants, such as *Quercus suber* and *Aristolochia*, thick layers of cork are added on the surface of the stems. The first-formed periderms, which are replaced by more internal periderms, are relatively thin and contain only a few layers of cork cells.

In most dicotyledons and gymnosperms the first periderm is usually developed in the first year of growth of the axial organ, on those portions which have ceased to elongate. The first periderm in *Ceratonia siliqua* develops in vertical rows along the stem and then extends laterally until it forms a complete cylinder (Arzee *et al.*, 1977). In some *Quercus* species the periderm develops first below tufted hairs which apparently serve as centres of phellogen initiation. Here, too, the periderm first expands in a vertical direction and only later spreads laterally (Arzee *et al.*, 1978).

LOCATION OF PHELLOGEN FORMATION

As has already been mentioned above, the periderm replaces the primary protective tissues (epidermis and cortex) of the axial organs. With the continuation of the process of secondary thickening the periderms themselves are replaced, from time to time, by new periderms which are formed each time deeper in the living tissues of the axis. Therefore, it is necessary to distinguish between the first periderm and those that are formed later.

The development of the first phellogen may take place in different cell layers external to the vascular cambium. In many stems, e.g. *Solanum dulcamara*, *Quercus suber*, *Malus pumila*, *Pyrus communis*, and *Nerium oleander*, the first phellogen is formed in the epidermis itself (Fig. 213, no. 2). More commonly the first phellogen develops in the layer of cells immediately below the epidermis. Such development can be seen in *Populus* (Fig. 213, no. 1), *Juglans* and *Ulmus*, among others. In the potato tuber, the phellogen develops in the epidermis as well as in the subepidermal cell layer, but the phellogen formed in the epidermis does not continue to function after its formation. In the stems of certain plants, e.g. *Robinia pseudacacia*, species of *Aristolochia* and *Pinus*, the first phellogen forms in the second or third cortical layer. In *Thuja*, *Punica*, *Arbutus*, *Vitis* and *Anabasis* the cambium of the first-formed periderm develops near the phloem or in the phloem parenchyma itself. In roots of gymnosperms and dicotyledons, the first phellogen is characteristically formed in inner layers, usually in the pericycle. In the roots of monocotyledons the first phellogen usually develops in the outer layers of the cortex.

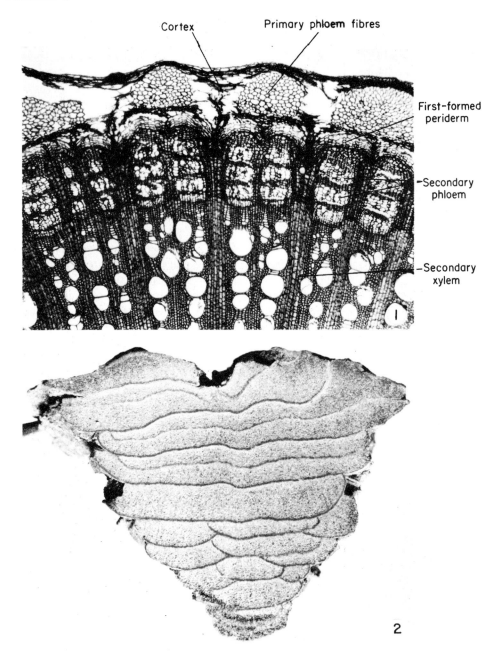

FIG. 214. 1, Micrograph of the outer portion of a cross-section of a branch of *Vitis vinifera* showing the first-formed periderm to develop in the phloem parenchyma. × 40. 2, Photograph of a cross-section of rhytidome of *Pinus* showing scale-shaped additional periderms. The periderms are distinguishable as dark lines. × 7.

If subsequent periderms form one inside the other, one additional phellogen may be produced in each growth season. The later-formed periderms each develop deeper in the cortex or primary phloem and, with continued secondary thickening, deeper and deeper within the secondary phloem. Two types of formation of subsequent periderms may be distinguished. In those plants in which the first-formed periderm develops in an inner layer, e.g. *Vitis*, the additional periderms usually form entire cylinders similar to the first-formed periderm, while in plants in which the first periderm is formed in the epidermis or the outer layers of the cortex, e.g. *Pinus*, the ad-ditional periderms develop in the form of scales or shells, the concave side of which is directed outwards (Fig. 214, no. 2; Fig. 215, nos. 1–5).

In certain genera the subsequent periderm

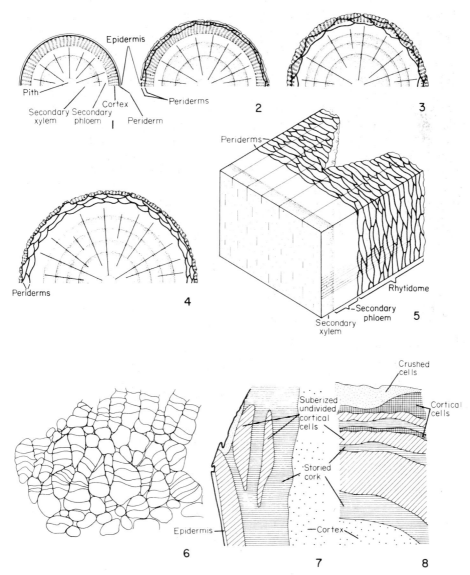

FIG. 215. 1–5, Diagrams showing the position, shape, and extent of the additional periderms formed in a woody stem on which the first-formed periderm develops as an entire cylinder close to the epidermis. 1–4, As seen in cross-section of branches aged from 1 to 4 years. In nos. 3 and 4 the first-formed periderm together with the tissues external to it have sloughed away. 5, Three-dimensional diagram of an outer portion of a stem showing the peripheral tissues, which here include a narrow zone of functioning phloem and a wide rhytidome with deep grooves; a considerable amount of the latter tissue has been weathered away. 6–8, Protective layers of monocotyledons. 6, Cross-section of the outer part of the cortex of *Curcuma longa* showing storied cork. 7, Radial section of the stem of *Cordyline australis* showing the position of the layers of storied cork, which enclose patches of suberized undivided cortical cells. 8, Diagram of portion of a cross-section of the stem of *Cordyline indivisa* showing a superficial layer of crushed cells and alternating tangential bands of suberized, undivided cortical cells, and storied cork. (Nos. 1–5 adapted from Eames and MacDaniels, 1947; nos. 6–8 adapted from Philipp, 1923.)

already begins to develop in the first year of growth of the stem or branch. In apple and pear trees subsequent periderms begin to develop in the sixth or eighth year of growth and, according to Evert (1963), the first phellogen may even remain active for about 20 years. In *Punica* and in a few species of *Populus* and *Prunus*, the first-formed phellogen may remain active for 20 or 30 years and in *Ceratonia* for about 40 years (Arzee et al., 1977), while in *Quercus suber*, some species of *Fagus, Anabasis*, and *Haloxylon*, and in a few other genera and species, normally no subsequent phellogens are formed during the life of the plant. In *Quercus suber* and other species in which the first phellogen is active throughout the entire life of the plant, or for many years, there are seasonal differences in the types of phellem cells produced. As a result of this, bands, which can apparently be regarded as annual growth rings of phellem are developed. In bottle corks such annual rings can be well demonstrated (Fig. 217, no. 1).

In roots the first-formed periderm may persist as a continuous layer over the entire length of the organ, with the exception of the tips. The increase in diameter of the cork cylinder is brought about by anticlinal divisions in the phellogen cells and in the living cells below them. Usually the cork of roots is thin and smooth. The conditions in the soil apparently hasten the rotting and sloughing of the cracked outer portions of the cork. In the roots of many herbaceous plants no periderm is formed, but the outer layers of cells become suberized (i.e. deposits of suberin are laid down in the cell walls).

In certain plants, as, for example, many species of *Artemisia* and especially those from arid regions (Moss, 1940; Moss and Gorham, 1953) and in *Achillea fragrantissima*, layers of cork cells occur in the secondary xylem, on the borders of the annual growth rings.

With the formation of each subsequent periderm the tissues exterior to it become cut off from the nutrient and water supply and so die. As a result of this a hard outer crust develops on the periphery of the axis. This crust increases in thickness due to the addition of further cork layers which enclose pockets of cortical tissue and dry phloem. All the cork layers, together with the cortical and phloem tissues, external to the inner-most phellogen, are termed *rhytidome* or *outer bark*, while all the tissues external to the vascular cambium are included in the term *bark*. The rhytidomes of some plants may contain many periderms and be very thick (Fig. 214, no. 2, Fig. 215, no. 5). In other plants, the rhytidomes may consist of only one or two poorly defined periderms e.g. in *Notofagus* (Patel and Shand, 1985). The living part of the bark inside the rhytidome is often termed the *inner bark*. With the increase in diameter of the secondary xylem the circumference of the cambial cylinder enlarges. As a result of this, the new-formed layers of secondary xylem are larger in circumference than are the outer layers of the inner bark which are, therefore, brought under strain. This strain is accommodated by the production of *expansion tissue* and *proliferation tissue* (Whitmore, 1962a). Expansion tissue is an intercalary tissue formed mainly by the phloem rays (Fig. 211, no. 1), and proliferation tissue develops as a result of the proliferation of the axial phloem parenchyma.

MORPHOLOGY OF BARK

The outer appearance of stems differs in different species of plants and the type of bark is used in many cases as a taxonomic character. These differences result from the manner of growth of the periderm, the structure of the phellem and the nature and amount of tissues that are separated by the periderm from the stem. Therefore it is possible to conclude that the outer appearance of the stem is determined by the type of rhytidome.

In plants in which the first periderm forms close to the epidermis a small amount of primary tissue is cut off from the stem and is eventually shed. In this case the phellem becomes exposed and then no rhytidome is considered to be present on the stem. When such a phellem is thin its surface is usually smooth, while if it is thick the surface is cracked and ridged. In plants where the first-formed periderm develops deep within the axis, thicker layers of tissue, which are usually connected to the cork, remain on the stem surface, and therefore these plants exhibit a rhytidome.

Certain rhytidomes, e.g. those of *Ulmus americana, Magnolia acuminata* and *Calotropis* (Fig.

FIG. 216. 1, Photograph of a branch of *Calotropis procera* showing the deeply grooved rhytidome. ×0.7. 2, Branch of *Tamarix* sp. in which transverse lenticels can be seen. ×0.9.

216, no. 1), consist mainly of parenchyma tissue and soft phellem (Pereira, 1988), while in others, e.g. species of *Quercus* and *Carya*, the rhytidome contains large quantities of fibres (mostly phloem fibres) which are associated with hard cork cells. The manner of formation of the periderm influences the shape of the bark in general and of the rhytidome in particular. When the subsequent periderms develop in the form of overlapping scales or shells the outer layers are sloughed accordingly, and so a *scaly bark* is formed. This type of bark occurs on relatively young stems of *Pinus, Pyrus communis*, and others. In *Vitis, Lonicera, Clematis*, and *Cupressus*, for example, the subsequent periderms are formed as entire cylinders and so the dead outer tissues are sloughed as hollow cylinders. This type of bark is termed *ring bark*. The bark of *Platanus, Arbutus*, and species of

Eucalyptus, for example, is intermediate between the above two types. In these plants the outer layers of the bark peel off in the form of relatively large sheets.

The sloughing of the outer layers of the bark is brought about in various ways. In *Arbutus* and *Platanus* the large plates of dead outer tissue separate from the inner portions of the bark through a layer of thin-walled cork cells, and the thick-walled cork cells below them remain attached to the stem which, therefore, has a smooth surface. In some *Eucalyptus* species the sheets of dead outer tissues of the bark exfoliate through layers of parenchyma cells with unthickened walls, which occur on the periphery of the phellem (Chattaway, 1953). In *Eucalyptus* species with furrowed bark the cracks extend through the rhytidome to the outermost phloem. The cracks originate in wedges of parenchyma that develop towards the periphery of the living phloem as the stem increases in girth. The cells in the middle of the wedge at first divide and later separate forming intercellular spaces which enlarge till a conspicuous crack is formed (Chattaway, 1955). In some trees, e.g. *Fagus*, the inner bark grows slowly and therefore much expansion tissue is formed. In this case the subsequent periderms cut off small amounts of secondary phloem and the sloughing of the outer bark is slow, resulting in the fall of minute scales and even powder (Whitmore, 1962b).

In many plants the different layers of the rhytidome adhere to one another and remain on the stem for many years and the outer bark becomes very thick and is deeply grooved. Such bark occurs in species of *Pinus*, *Quercus*, and many other trees.

COMMERCIAL CORK

Commercial cork is made from the bark of trees and, in particular, from that of *Quercus suber* (Eames and MacDaniels, 1947). In the stem of this plant the first phellogen forms in the epidermis. This phellogen may remain on the plant indefinitely but in order to obtain commercial cork this first-formed periderm is removed when the tree is about 20 years old and about 40 cm in diameter. After the first periderm is removed the exposed cells of the phelloderm and cortex dry out and die, and a new phellogen is formed a few millimetres within the cortex, below the first-formed phellogen. This subsequent phellogen produces cork more rapidly and in about 10 years a cork layer thick enough to be of commercial value is obtained. This cork is of better value than the virgin cork, which has almost no commercial value. However, it is of poorer quality than that of cork obtained from periderms that are formed as a result of subsequent strippings made at about 10-year intervals till the tree is about 150 or more years old. After a few strippings the phellogens are formed in the secondary phloem. The pieces of cork that are stripped from the tree exhibit surfaces with different structure—the outer surface is rough because of weathering and the

presence of remnants of dead tissues outside of the periderm, while the inner surface is smooth. On the radial surfaces and in cross-sections of such pieces of cork, bands, which apparently represent annual increments, can be distinguished (Fig. 217, no. 1)

The dark brown spots that can be seen on tangential surfaces of cork and the similar stripes seen on radial surfaces and in cross-sections (Fig. 217, nos. 1, 2) are lenticels (see below).

The features that give commercial value to cork are its imperviousness to gases and liquids and its strength, elasticity and lightness.

PROTECTIVE TISSUES OF MONOCOTYLEDONS

In herbaceous monocotyledons the epidermis, which has a cuticle, is the only means of external protection on the plant axis. When the epidermis is ruptured the cortical cells below become suberized. The suberin lamellae in these cells are laid down as in typical cork cells. This feature is common in the Gramineae, Juncaceae, Typhaceae, and other families.

Different types of protective tissues are found on the stems of perennial monocotyledons (Floresta, 1905; Philipp, 1923; Eames and MacDaniels, 1947; Tomlinson, 1961). For instance, in the palm *Roystonea*, in which the trunk is white and smooth, a hard periderm, which remains on the stem throughout the life of the plant, and which is similar to that of dicotyledons, is developed. According to the above authors, in thickened stems of the monocotyledons such as *Curcuma*, *Cordyline*, and many palms, a special kind of protective tissue is developed. This tissue is formed by the secondary activity of a storied meristem which appears in the outer cortex. The initials of this meristem undergo three to eight periclinal divisions and so radially arranged layers of cells, which become suberized, are produced (Fig. 215, no. 6). Cork of this type is called *storied cork*. The initials here, unlike those of a typical phellogen, do not form regular uninterrupted layers.

The radial rows of cells that divide to form the storied cork are arranged in irregular tangential bands which enclose between them large cells

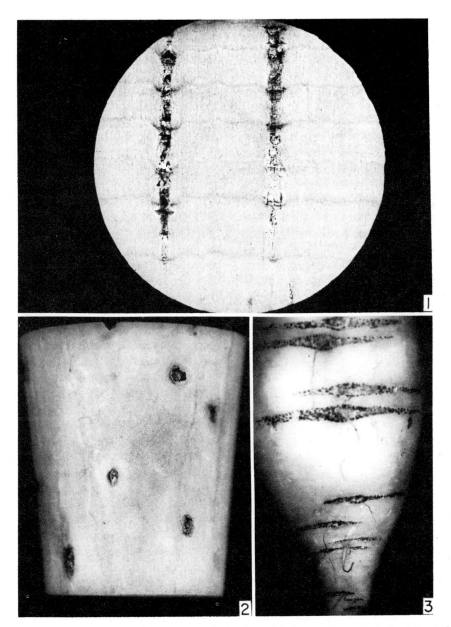

FIG. 217. 1 and 2, A bottle-stopper made of cork. 1, As seen from above (i.e. cross-section of the cork relative to its position on the stem) in which wide growth rings and two lenticels can be seen. × 3. 2, As seen from the side (i.e. in tangential section of the cork relative to its position on the stem) in which the lenticels can be seen as dark patches. × 3. 3, *Raphanus sativus* root showing each lateral root to be accompanied by a pair of lenticels.

which do not divide but which, however, are also suberized. The individual bands of storied cork may fuse both radially and tangentially (Fig. 215, nos. 7, 8). The layers of storied cork may be formed further in towards the centre of the trunk, and between these layers alternating layers of undivided suberized cells and non-suberized destroyed cells occur (Fig. 215, no. 8). In this way layers that are analogous to but less organized than the rhytidome of dicotyledons are formed.

From observations made on *Dracaena* sp., *Aloë arborescens* and *Yucca*, it was found difficult to detect differences between the secondary protective tissues developed in them and those of dicotyledons which are produced by a typical phellogen.

WOUND CORK

Generally, in such places where living plant tissue is exposed to the air as a result of wounding, *wound cork* is developed. Usually the outer dead tissues are separated from the inner intact ones by a layer of cells which become lignified and suberized. Apart from this impervious boundary layer, a phellogen may be developed in the living undamaged layers. This phellogen produces phelloderm and phellem in the usual manner (Biggs and Stobbs, 1986; Biggs *et al.*, 1984). The layer of cork thus formed prevents the loss of water through the wound and it protects the plants against the entry of fungi and bacteria. Apparently wound cork may develop on all parts of the plant, including even fruits and leaves. However, there are differences in the type and amount of cork developed in different species, organs, and tissues, and under different environmental conditions. Usually wound cork is more easily developed on woody plants than on herbaceous or monocotyledonous ones. Low temperatures and low humidity may delay the development of wound cork even in those places where it develops readily as, for example, on potato tubers (Küster, 1925; Artschwager and Starrett, 1933; Bloch, 1941). (See also "Phellogen" earlier in this chapter.)

POLYDERM

In certain species of the Rosaceae, Myrtaceae, Hypericaceae, and Onagraceae, a special phellogen is formed in the pericycle of the root or underground stem. This phellogen produces, centrifugally, a few layers of thin-walled non-suberized cells which alternate with a layer of endodermal-like cells. At the start of the differentiation of the latter into cork cells, Casparian strips appear on the walls, which, with further development, become entirely lined by a suberin layer. This type of complex tissue is termed *polyderm*. Its inner layers, including the cork cells, are living and may serve as a storage tissue (Mylius, 1913; Luhan, 1955; Nelson and Wilhelm, 1957).

LENTICELS

In the large majority of plants there are restricted areas of relatively loosely arranged cells, suberized or non-suberized, in the periderm. These areas are termed *lenticels*. Lenticels protrude above the surrounding periderm because of the bigger size and loose arrangement of the cells which, themselves, are usually more numerous in these regions. Because of the continuity of the intercellular spaces of the lenticels and those of the inner tissues of the axial organ, it is assumed that the function of the lenticels is connected with gas exchange, similar to that of the stomata on organs covered by an epidermis only.

Distribution of lenticels

Lenticels usually occur on stems and roots, and they appear on young branches and other smooth organs as rough, dark patches which are somewhat raised above the epidermis through which they erupt (Fig. 216, no. 2). Lenticels are also sometimes present on fruits. In apples and pears, for example, they appear as small dots on the surface of the fruit. Only a few plants, e.g. *Philadelphus*, *Anabasis*, *Haloxylon*, *Campsis radicans*, *Vitis*, and some other species, many of which are climbers, do not possess lenticels although they form a periderm.

The number of lenticels occurring on a unit of surface area of a stem differs in the various species (Eames and MacDaniels, 1947). In certain species a lenticel develops under each stoma, or under a group of stomata. In others, lenticels may also develop between stomata if the latter are sparsely distributed, and, if the stomata are very numerous, lenticels may develop only under some of them. The arrangement of the lenticels on the stem also varies—sometimes they appear in longitudinal or horizontal rows, but generally they are

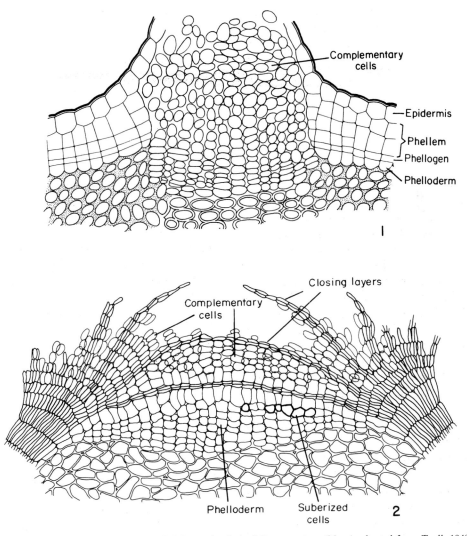

Fig. 218. 1, Young lenticel of *Sambucus nigra*. 2, Mature lenticel of *Prunus avium*. (No. 1 adapted from Troll, 1948; no. 2 adapted from Boureau, 1954.)

irregularly scattered over the entire surface. In young roots the lenticels usually appear in pairs—one lenticel on each side of a lateral root (Fig. 217, no. 3). In storage roots, such as those of *Daucus*, for example, the pairs of lenticels appear in vertical rows alongside the rows of lateral roots. In older roots the arrangement of the lenticels is irregular.

Externally mature lenticels are usually lens-shaped and they are convex both towards the exterior and the interior (Fig. 218, no. 2). According to the orientation of the rupture of the epidermis the lenticels are described as being longi-

tudinal or transverse (Fig. 216, no. 2). The lenticels which are situated close to the outer ends of the phloem rays enable relatively free passage of gases between the inner tissues of the axis and the atmosphere. In stems with larger vascular rays it may be seen that the lenticels usually appear directly opposite the rays.

In the roots of *Phoenix dactylifera* lenticel-like structures occur which take part in aeration of the root but which differ from the above-described ordinary lenticels. Here the lenticels form collar-like structures around the thinner roots. The complementary tissue is, however, of typical structure.

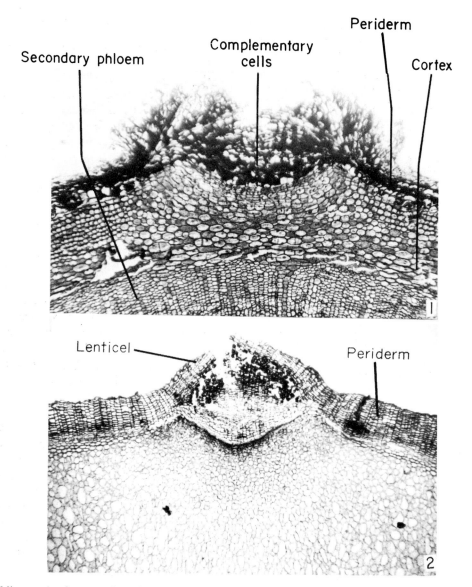

Secondary phloem

Complementary
cells

Periderm

Cortex

Lenticel

Periderm

FIG. 219. Micrographs of cross-sections of lenticels. 1, In stem of *Sambucus nigra.* × 60. 2, In aerial root of *Avicennia marina.* × 50. (No. 2 courtesy of Allen Witztum.)

Development and structure of lenticels

Lenticels begin to form together with the first periderm or shortly before. The time of the formation of the lenticels differs in various species and it is dependent on the persistence of the epidermis on the stem. In most species the development of the lenticels is already commenced during the first season of growth of the organ and sometimes even before the elongation of the organ is completed.

The first-formed lenticels generally appear below a stoma or group of stomata. The cells in these regions begin to divide in different directions and the chlorophyll in them disappears so that a loose colourless tissue is formed. The division of the cells progresses in the cortex inwards and the orientation of the divisions becomes more and more periclinal until the phellogen of the lenticel is formed. The cells that are derived from the divisions of the substomatal cells, as well as those produced towards the exterior by the phellogen of the lenticel, are termed *complementary cells*. The increase in number of these cells causes the rupture of the epidermis so that masses of complementary cells are pushed out and rise above the surface of the organ. The exposed cells die and wither away and they are replaced by new cells produced by the phellogen (Fig. 218, no. 1). Complementary cells may be suberized or non-suberized, and they are usually more or less spherical and thin-walled. Centripetally the phellogen of the lenticel may produce phelloderm.

In some species, apart from the complementary cells, the phellogen of the lenticel also produces *closing layers* centrifugally (Fig. 218, no. 2). These layers, which consist of compact tissue, alternate with the complementary tissue.

Two types of complementary tissue are distinguished: that in which the connection between the cells is relatively strong, as, for example, in *Sambucus nigra*, *Salix*, and *Ginkgo*, and that in which the cells have almost no attachment between them, giving the tissue a powder-like appearance, as, for example, in the stems of *Pyrus*, *Prunus*, and *Robinia* and in the roots of *Morus* (Eames and MacDaniels, 1947). In the latter type the cells of the complementary tissue are held in place by the closing layers (Fig. 218, no. 2). In spite of their compactness, the closing layers contain intercellular spaces which allow the passage of gases. Similar gaps are found in the phellogen itself. The closing layers are ruptured as a result of the continual production of new complementary cells. In the temperate zones the lenticels become closed at the end of the growing season by a closing layer. This layer is ruptured only with the renewal of the activity of the plant which is accompanied by the rapid and excessive production of new complementary cells.

In the central region of the lenticels of *Picea* there are layers of unlignified sclerotic cells that resemble the sclerotic phelloids of the periderm of this plant. Between the sclerotic cells in the lenticel there are intercellular spaces and they thus correspond to the complementary cells of angiosperm lenticels (Parameswaran *et al.*, 1975).

Duration of lenticels

In plants in which additional inner periderms are formed relatively early in the life of the plant, the lenticels are cut off early from the inner tissues and are shed together with the outer tissues of the bark. In plants in which the first-formed outer periderm remains for a long time on the axial organ, the lenticels may remain active for many years. In these cases the lenticels elongate transversely as a result of secondary growth. The elongation of the phellogen of the lenticel in a tangential direction, with the increase of the circumference of the organ, is brought about by anticlinal divisions of the initials as occurs in the phellogen surrounding the lenticel. On the stems of some plants, e.g. *Acacia raddiana*, *Tamarix gallica*, and *Betula*, and on the roots of *Morus*, for example, the transversely elongated lenticels form conspicuous marks on the surface of the smooth bark. In some plants the lenticels do not increase in size with age, but they become split into several lenticels. In certain plants, e.g. *Quercus suber* and *Ailanthus*, there is no appreciable increase in size of the lenticels with age.

In the subsequent periderms new lenticels are produced as a result of the activity of special areas of the new phellogens. In plants with rough barks the lenticels are not easily seen as they occur in the additional inner periderms at the base of the cracks in the outer bark. These cracks are a result of the continued secondary growth.

In *Quercus suber*, where the cork layer is several centimetres thick, the lenticels remain active for a long time and result in the formation of cylinders of complementary tissue (Fig. 217, nos. 1, 2) which extend from the phellogen to the surface of the phellem. This complementary tissue forms the patches of dark brown crumbling tissue found in commercial cork. Because of the radial orientation of these cylindrical masses of complementary tissue, bottle corks are cut from the cork tissue in a direction parallel to the surface of the trunk so that the cylindrical lenticels extend transversely through them.

REFERENCES

ARTSCHWAGER, E. F. and STARRETT, R. C. (1933) Suberization and woundcork formation in the sugar beet as affected by temperature and relative humidity. *J. Agric. Res.* **47**: 669–74.

ARZEE, T., LIPHSCHITZ, N., and WAISEL, Y. (1968) The origin and development of the phellogen in *Robinia pseudacacia* L. *New Phytol.* **67**: 87–93.

ARZEE, T., WAISEL, Y., and LIPHSCHITZ, N. (1970) Periderm development and phellogen activity in shoots of *Acacia raddiana* Savi. *New Phytol.* **69**: 395–8.

ARZEE, T., ARBEL, E., and COHEN, L. (1977) Ontogeny of periderm and phellogen activity in *Ceratonia siliqua* L. *Bot. Gaz.* **138**: 329–33.

ARZEE, T., KAMIR, D., and COHEN, L. (1978) On the relationship of hairs to periderm development in *Quercus ithaburensis* and *Q. infectoria. Bot. Gaz.* **139**: 95–101.

BIGGS, A. R. and STOBBS, L. W. (1986) Fine structure of the suberized cell walls in the boundry zone and nectrophylactic periderm in wounded peach bark. *Can. J. Bot.* **64**: 1606–1610.

BIGGS, A. R., MERRILL, W. and DAVIS, D. D. (1984) Discussion: Response of bark tissues to injury and infection. *Can. J. For. Res.* **14**: 351–356.

BLOCH, R. (1941) Wound healing in higher plants. *Bot. Rev.* **7**: 110–46.

BORGER, G. A. and KOZLOWSKI, T. T. (1972a) Effect of growth regulators and herbicides on normal and wound periderm ontogeny in *Fraxinus pennsylvanica* seedlings. *Weed Res.* **12**: 190–4.

BORGER, G. A. and KOZLOWSKI, T. T. (1972b) Effects of temperature on first periderm and xylem development in *Fraxinus pennsylvanica*, *Robinia pseudacacia* and *Ailanthus altissima. Can. Jour. Forest Res.* **2**: 198–205.

BORGER, G. A. and KOZLOWSKI, T. T. (1972c) Effects of cotyledons, leaves and stem apex on early periderm development in *Fraxinus pennsylvanica* seedlings. *New Phytol.* **71**: 691–702.

BOUREAU, E. (1954) *Anatomie végétale*, Vol. 1. Presses Universitaires de France, Paris.

CHATTAWAY, M. M. (1953) The anatomy of bark: I, The genus *Eucalyptus. Aust. J. Bot.* **1**: 402–33.

CHATTAWAY, M. M. (1955) The anatomy of bark. VI. Peppermints, boxes, ironbarks and other eucalypts with cracked and furrowed barks. *Aust. J. Bot.* **3**: 170–6.

EAMES, A. J. and MACDANIELS, L. H. (1947) *An Introduction to Plant Anatomy.* McGraw-Hill, New York.

EVERT, R. F. (1963) Ontogeny and structure of the secondary phloem in *Pyrus malus. Am. J. Bot.* **50**: 8–37.

FLORESTA, P. LA (1905) Ricerche sul periderma delle Palme. *Contr. Biol. Veg. Palermo* **3**: 333–54, tav. 18–19.

GIBSON, A. C. (1981) Vegetative anatomy of *Pachycormus*

(Anacardiaceae). *Bot. J. Linn. Soc.* **83**: 273–284.

GOMEZ-VAZQUEZ, B. G. and ENGLEMAN, E. M. (1984) Bark anatomy of *Bursera longipes* (Rose) Standley and *Bursera copallifera* (Sesse & Moc.) Bullock. *IAWA Bull.* **5**: 335–340.

KÜSTER, E. (1925) *Pathologische Pflanzenanatomie*, 3rd edn. G. Fischer, Jena.

LIPHSCHITZ, N. and WAISEL, Y. (1970) Phellogen initiation in the stem of *Eucalyptus camaldulensis* Dehn. *Aust. J. Bot.* **18**: 185–9.

LUHAN, M. (1955) Das Abschlussgewebe der Wurzeln unserer Alpenpflanzen. *Ber. dt. bot. Ges.* **68**: 87–92.

MOSS, E. H. (1940) Interxylary cork in *Artemisia* with a reference to its taxonomic significance. *Am. J. Bot.* **27**: 762–8.

MOSS, E. H. and GORHAM, A. I. (1953) Interxylary cork and fission of stems and roots. *Phytomorphology* **3**: 285–94.

MYLIUS, G. (1913) Das Polyderm. Eine vergleichende Untersuchung über die physiologischen Scheiden: Polyderm, Periderm und Endodermis. *Bibliotheca bot.* **18**: 1–119.

NELSON, P. E. and WILHELM, S. (1957) Some aspects of the strawberry root. *Hilgardia* **26**: 631–42.

PARAMESWARAN, N., KRUSE, J., and LIESE, W. (1975) Aufbau und Feinstruktur von Periderm und Lenticellen der Fichtenrinde. *Z. Pflanzenphysiol.* **77**: 212–21.

PATEL, R. N. and SHAND, J. E. (1985) Bark anatomy of *Nothofagus* species indigenous to New Zealand. *New Zealand J. Bot.* **23**: 511–532.

PEREIRA, H. (1988) Structure and chemical composition of cork from *Calotropis procera* (Ait.) R. BR. *IAWA Bull.* **9**: 53–58.

PHILIPP, M. (1923) Über die verkorkten Abschlussgewebe der Monokotylen. *Bibliotheca bot.* **92**: 1–28.

SCHÖNHERR, J. and ZIEGLER, H. (1980) Water permeability of *Betula* periderm. *Planta* **147**: 345–54.

SITTE, P. (1962) Zum Feinbau der Suberinschichten im Flaschenkork. *Protoplasma* **54**: 555–9.

TOMLINSON, P. B. (1961) Palmae. In: *Anatomy of the Monocotyledons* (ed. C. R. Metcalfe), Vol. II. Clarendon Press, Oxford.

TROLL, W. (1948) *Allgemeine Botanik*. F. Enke, Stuttgart.

WAISEL, Y., LIPHSCHITZ, N., and ARZEE, T. (1967) Phellogen activity in *Robinia pseudacacia* L. *New Phytol.* **66**: 331–5.

WATTENDORFF, J. (1974) The formation of cork cells in the periderm of *Acacia senegal* Willd. and their ultrastructure during suberin deposition. *Z. Pflanzenphysiol.* **72**: 119–34.

WHITMORE, T. C. (1962a) Studies in systematic bark morphology: I, Bark morphology in Dipterocarpaceae; II, General features of bark construction in Dipterocarpaceae. *New Phytol.* **61**: 191–220.

WHITMORE, T. C. (1962b) Why do trees have different sorts of bark? *New Scient.* **16**: 330–1.

CHAPTER 18

UNUSUAL SECONDARY GROWTH

THE usual structure of the secondary conducting tissues of most spermatophytes is as described in the previous chapters. The processes of secondary growth that give rise to this type of structure may be termed the *common type of secondary growth*. However, in many plants there are deviations from this type of secondary growth. Such deviations may include such features as the unequal activity of different portions of the cambium on the circumference of the axis, the alteration of the relative amounts and position of the xylem and phloem, and the appearance of additional cambia. The secondary growth that results in the development of a secondary body differing from the common has been termed *anomalous secondary growth*.

If, in a cross-section of a stem, the cambium produces more xylem than phloem in certain places and more phloem than xylem in others, the xylem cylinder becomes ridged and sometimes even more complex structures may develop. Such a stem configuration occurs, for instance, in *Passiflora glandulosa* (Fig. 220, no. 3) (Ayensu and Stern, 1964). In certain plants, such as *Aristolochia*, there are strands of cambium that produce only ray-like parenchyma; these strands increase in number with the increase of the circumference of the cambium (Fig. 220, no. 1). The reduction of cambial activity to certain restricted areas results in the formation of ridged stems, which often split. In certain lianes of the Sapindaceae, e.g. *Serjania*, the cambium first appears in separate strands each of which surrounds a group of vascular bundles or even a single primary vascular bundle (Fig. 220, no. 2). Stems that develop in this way appear as if having originated by fusion of a number of stems. With the ageing of such stems and with the production of periderm layers, the stems split into numerous parts (splits). A similar

structure can result from the excessive development of the xylem and phloem parenchyma which results in the splitting of both the conducting tissues and the cambium cylinder. The above-mentioned stem structures are adaptive for lianes which need a pliant axis capable of torsion movements without damaging the conducting elements.

In desert plants such as *Peganum harmala* and *Zygophyllum dumosum* (Fig. 221, no. 1) the stem becomes lobed and finally splits as a result of the death of strips of the cambium. In *Achillea fragrantissima* and some species of *Artemisia*, a layer of cork is produced each year on the border between two growth rings of xylem, i.e. *interxylary cork*. This feature, when accompanied by the suberization of the rays, as in *Artemisia herba-alba*, for example (Ginzburg, 1963), or by the cessation of activity in certain portions of the cambium, also results in the splitting of the stem (Fig. 221, no. 2).

INCLUDED PHLOEM

In certain plants there are strands of secondary phloem within the secondary xylem, e.g. in *Avicennia* (Avicenniaceae), *Strychnos* (Loganiaceae), *Leptadenia* (Asclepiadaceae), *Thunbergia* (Acanthaceae), *Bougainvillea* (Nyctaginaceae), *Salvadora* (Salvadoraceae), *Dicranostyles* (Convolvulaceae), *Simmondsia* (Simmondsiaceae), *Phytolacca* (Phytolaccaceae), and the families Amaranthaceae and Chenopodiaceae. In these plants a cambium differentiates outside the primary vascular bundles, in the pericycle or in the inner cortical layers. Later series of vascular cambia arise successively further outward. Each successive cambium produces xylem toward the inside and phloem toward the outside until a new cambium develops

FIG. 220. Diagrams of cross-sections of stems with unusual secondary thickening. 1, *Aristolochia triangularis*, portion of a cross-section. 2, *Serjania clematidifolia*. 3, *Passiflora* sp. Periderm—cross-hatched; secondary phloem—stippled; secondary xylem—hatched; parenchyma colourless. (No. 1 adapted from Schenk, 1892; no. 2 adapted from Pfeiffer, 1926; no. 3 adapted from a micrograph of Ayensu and Stern, 1964.)

from parenchyma cells on the outside of the phloem (Fig. 224, no. 2) (Artschwager, 1926; Pfeiffer, 1926; Iljin, 1950; Studholme and Philipson, 1966; Esau and Cheadle, 1969; Mennega, 1969; Philipson *et al.*, 1971; Stevenson and Popham, 1973; Wheat, 1977; Zamski, 1979; Mikesell, 1979; Bailey, 1980; Fahn, 1985; Kirchoff and Fahn, 1984). Fahn and Zimmermann (1982) showed that in *Atriplex halimus* (Chenopodiaceae) the first successive cambial strands initiate from cambial bands which develop in continuation with the still active intrafascicular cambia of the primary vascular bundles. The additional successive cambia also appear to start developing continuous with preceding still-active cambia. (Fig. 223, no. 2; Fig. 224, no. 3). On the basis of this and other

researches (Fahn and Shchori, 1967; Zamski and Azenkot, 1981; Kirchoff and Fahn, 1984; Stieber and Beringer, 1984) it can be concluded: 1. Secondary xylem with included phloem derives from numerous successive cambia. 2. The first supernumerary cambial strands develop outside the primary vascular bundles, either directly in connection with the primary intrafascicular cambia or the two become connected shortly after the initiation of secondary cambium. The additional supernumerary cambia appear to develop similarly in the outer parenchyma of the phloem strands formed previously. 3. The primary and all successive supernumerary "vascular bundles" form one continuous system. This permits the movement of stimuli for cambial development and activity, from

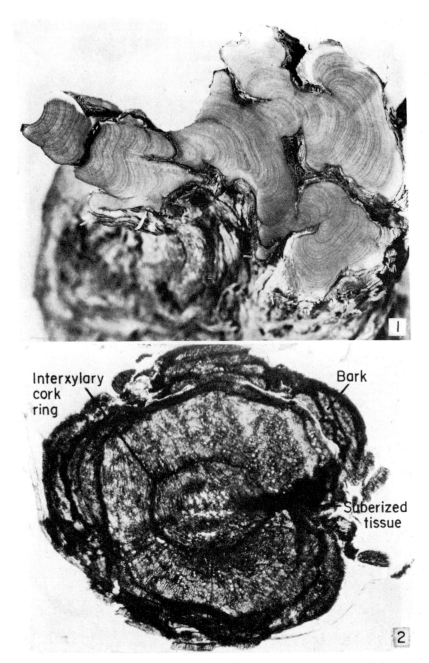

FIG. 221. 1, Surface view of a cross-section of the stem of *Zygophyllum dumosum* showing the splitting that results from the cessation of cambial activity in localized areas. × 1. 2, As above but of a 2-year-old stem of *Artemisia herba-alba* showing an interxylary cork ring and suberized tissue which splits the last growth ring. × 20.

the leaves to the most distant part of the plant.

In the Chenopodiaceae the successive cambia are, as seen in cross-section, in the form of long or short arches. They produce irregularly or spirally arranged phloem strands (Fig. 222, no. 1). Frequently the additional cambia in this family form more or less entire rings (Fig. 222, no. 2).

FIG. 222. Surface view of cross-sectioned stems of plants belonging to the Chenopodiaceae showing different arrangements of the included phloem and its associated tracheary elements. 1, *Hammada scoparia*. × 10. 2, *Anabasis articulata*. × 7.

Some authors (Balfour, 1965; Studholme and Philipson, 1966) interpreted the mode of development of the successive cambia of the Nyctaginaceae, Amaranthceae, and Chenopodiaceae differently from the way described above. According to these workers there is a cambial zone without the layer of permanent initials characteristic of a normal cambium. Divisions producing xylem and phloem may be concentrated temporarily in a narrow region. The site of maximum division, however, soon passes to other cells of the cambial zone which lie further out. In this way the cambial zone continuously moves outwards leaving, behind it xylem with included phloem. This view on the manner of development of the secondary xylem with included phloem did not gain support from most of the other researchers.

A special type of development of additional cambia around vessels or groups of vessels is found in the root of the sweet potato (*Ipomoea batatas*) which has been described in Chapter 13.

In many species of the Chenopodiaceae, in *Bougainvillea*, and in other plants with similar secondary thickening, each strand of secondary phloem is seen in cross-section to be accompanied on its inner surface by a group of xylem vessels (Fig. 223). Thus in cross-section such stems appear to consist of a ground tissue of fibres and parenchyma in which scattered vascular strands occur. Serial sections, however, reveal that the vessel groups and phloem strands behave independently in their longitudinal course (Fahn and Shchori, 1967). The phloem strands and vessel groups of the secondary body may anastomose both tangentially and radially, the tangential connections being the more common. The parenchyma associated with the included phloem is termed *conjunctive parenchyma*. It is variously arranged and developed, and may be ray-like or in bands that connect the phloem strands, or it may surround and intermingle with the vessel groups.

In the perennial species of the Chenopodiaceae growing in the desert the anastomosing system of included phloem may have important adaptive value, especially as the phloem in this family may remain active for many years. Even in cases where most of the outer tissues of the stem dry out during the long summer, the included phloem strands remain viable and so can supply nutrients to the buds, which, at the onset of the growing season, can then commence to develop.

Included phloem has also been suggested to be of advantage for lianas (Dobbins and Fisher, 1986). The included phloem reduces the chances of disturbing translocation of materials to the roots, and is of advantage to rapid vigorous regeneration of tissues following wounding.

THICKENING OF THE AXIS IN MONOCOTYLEDONS

Stem thickening during primary growth

In most monocotyledons, secondary thickening is absent, but in those plants that have thick axial organs, e.g. the stems of palms, the rhizomes of *Musa* spp. (Fig. 225, no. 3) and *Veratrum album* and the bulbs of *Galanthus nivalis*, *Tulipa*, and *Allium cepa* (DeMason, 1979a, b, 1980) considerable and rapid thickening takes place below the apical meristem (Skutch, 1932; Ball, 1941; Clowes, 1961; Zimmermann and Tomlinson, 1967, 1968, 1969). During early ontogeny the stem diameter increases progressively, so that each successive internode becomes wider than the preceding one until the stem obtains its mature width. Because of this type of early growth the base of the stem has an obconical shape (Fig. 224, no. 1). This gradual increase in the width of the stem has been called *establishment growth* (Zimmermann and Tomlinson, 1970; Fisher and Tomlinson, 1972; Tomlinson and Esler, 1973). In these plants the apex proper is not large, but immediately behind it cell division is intensive. This causes an increase in diameter of the stem before extension growth ceases. This activity, therefore, is regarded as part of the primary growth of the shoot. Additional increase in diameter, mainly due to the increase in size of cells, may take place below this meristematic region. The meristem, which is responsible for the abrupt dilation of the apical region to a wide crown, lies below the young leaf bases (Fig. 224, no. 4). Within the meristematic. region localized mitotic activity occurs, resulting in the formation of procambial strands running

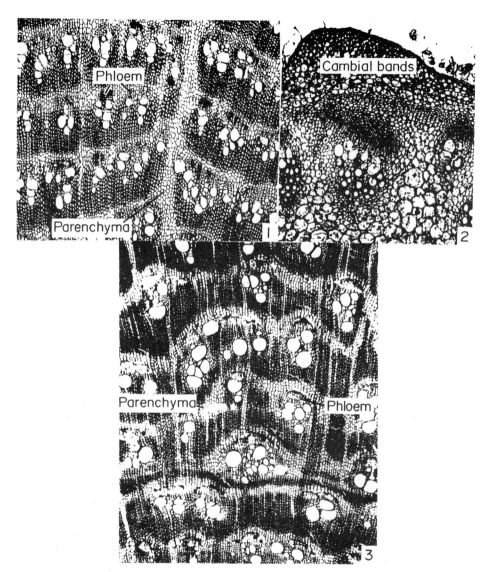

FIG. 223. Micrographs of cross-sections of secondary xylem with included phloem. 1. *Suaeda fruticosa,* × 40. 2. Cross-section of a stem of *Atriplex halimus* 3 mm below the apex, showing cambial bands extending from the intrafascicular cambia of primary vascular bundles into the pericycle outside neighbouring bundles, × 180. 3. *Bougainvillea glabra.* × 40.

almost horizontally and more or less parallel to the surface of the extended apex (Fig. 225, no. 3). This zone of procambium formation is termed by Zimmermann and Tomlinson the *meristematic cap.* The bulk of the meristem causing the stem thickening is situated below the cap, although the ground tissue between the procambial strands of the cap also contributes to it. Because the meristematic cap is easily observed in longitudinal sections, many authors (e.g. Eckardt, 1941; Ball,

1941; Clowes, 1961; De Mason, 1983) attributed to it the main role in stem growth and coined it *primary thickening meristem* (PTM). Zimmermann and Tomlinson (1968) suggested that this term be discarded, because the tissue which really merits the name, according to them, is diffusely located below the tissue for which the term was originally coined. The cap or PTM may, therefore, be regarded as a specialized zone of a massive thickening region which functions in a somewhat

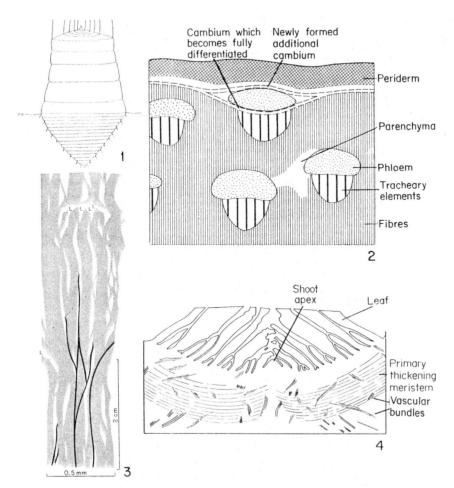

Fig. 224. 1, Diagram showing a type of establishment growth occurring in most palms and many other monocotyledons. The first internodes are underground. These are short and, starting from the base, are progressively wider until their diameter has approximately reached that of the adult stem, when growth in height begins with longer internodes (adapted from Zimmermann and Tomlinson, 1970). 2, Diagram of portion of a cross-section of a stem with anomalous secondary thickening illustrating how the strands of included phloem are formed. 3. Vascular system of a sector of the stem *Atriplex halimus* (Chenopodiaceae) reconstructed from a film, in tangential view. The grey vertical bands show the anastomosing primary bundles of the inner ring. The first strands which branch off from the primary bundles to form the first ring of secondary bundles are represented by black lines. (Note that the axial scale is foreshortened about 20 times.) *L,* leaf trace. Diagram of a median longitudinal section of the shoot apex of *Washingtonia filifera.* (No. 3 from Fahn and Zimmermann, 1982; no. 4 adapted from a micrograph of Ball, 1941.)

similar way to the special cambium of some mono-cotyledons described later in this section (cf. Philipson *et al.,* 1971). Leaf traces are always continuous with vertical bundles of the stem, and this connection must take place within the PTM.

If the PTM and the monocotyledonous vascular cambium occur in one and the same plant then the latter develops from the former.

Ball (1941) described the primary thickening of the stem in the Palmae in the following way. Dur-ing the process of growth of the stem, the activity of the primary thickening meristem is largely independent of the apical meristem (Fig. 224, no. 4). In the embryo the primary thickening meristem constitutes a flat zone below the leaf and sheath primordia and, in the seedling, a steep cone. In later stages of development this meristem again appears as a flat zone and eventually it becomes concave. Most of the stem tissues de-velop from it. During the maturation of the palm

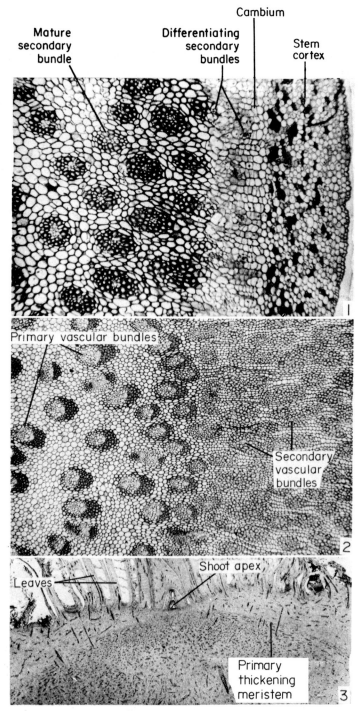

Cambium

Mature
secondary
bundle

Differentiating
secondary
bundles

Stem
cortex

1

Primary vascular bundles

Secondary
vascular
bundles

2

Shoot apex

Leaves

Primary
thickening
meristem

3

FIG. 225. 1 and 2, Portions of cross-sections of monocotyledonous stems showing the type of secondary growth characteristic in this group. 1, *Aloë arborescens*. × 52. 2, *Dracaena*. × 35. 3, Upper portion of a median longitudinal section of the vegetative shoot apex of *Musa* (Dwarf Cavendish banana) in which the zone of the PTM can be distinguished. × 3.

the primary thickening meristem at first contributes mainly to the thickening of the stem, but later it is also responsible for the increase in stem height. Below the concave meristematic zone of the mature palm longitudinally orientated rows of cells can be distinguished; the activity of this zone results in the increased length of the trunk.

In the palms and in the banana corm the provascular strands (the procambium) are derived from two sources—to a smaller extent from the shoot apex and to a larger extent from the PTM.

Diffuse secondary thickening

In addition to the above-mentioned meristem which brings about primary thickening, in palm stems the ground tissue close to the apex expands and thus also results in additional thickening of the stem. It has been reported (Zodda, 1904; Schoute, 1912; Tomlinson, 1961) that in some palms the expansion of the ground tissue continues for a long period and that it is very obvious in the older parts of the stem which are considerably distant from the shoot apex (e.g. in *Roystonea* and *Actinophloeus*). Here the central parenchyma cells and the not yet fully differentiated outer fibres of the bundle sheaths continue to undergo divisions, which are followed by cell expansion, for a long period. The intercellular spaces also increase in size proportionally to the increase in size of the parenchyma cells. This type of secondary thickening has been termed *diffuse secondary thickening* (Tomlinson, 1961). The ventricose shape of some palm trees is, according to Tomlinson, probably due to the increased vigour of the leafy crown. Such stems are mainly of primary origin, but some secondary thickening may also take place.

The cambium of monocotyledons

Secondary thickening proper, from a lateral meristem distant from the stem apex, takes place in different monocotyledonous species such as *Aloë arborescens* of the Liliaceae; *Alpinia* of the Zingiberaceae; species of *Cordyline, Dasylirion,* *Dracaena, Sansevieria,* and *Yucca* of the Agavaceae; *Tamus, Dioscorea,* and *Testudinaria* of the Dioscoreaceae; and *Xanthorrhoea, Kingia,* and *Lamandra* of the Australian family the Xanthorrhoeaceae (Chouard, 1937; Cheadle, 1937; Fahn, 1954; Philipson *et al.,* 1971; Bell, 1980a, b). Such secondary thickening in the monocotyledons is brought about by a special cambium, also termed *secondary thickening meristem* (STM). This meristem is continuous with the PTM (Diggle and DeMason, 1983a, b; DeMason and Diggle, 1984). The two meristems were reported to be discontinuous in adult shoots of *Beaucarnea recurvata* and *Cordyline terminalis* (Stevenson, 1980; Stevenson and Fisher, 1980). This cambium develops in the parenchyma of the stem external to the entire mass of primary vascular bundles. The cells of this secondary meristem are, as seen in tangential view, of different shapes—they may be fusiform (that is, long with tapered ends), rectangular, or with one tapered and one truncate end.

The monocotyledonous cambium develops and functions in the manner described below. According to Schoute (1912) it originates as a tiered meristem (*Etagenmeristem*), which involves a succession of tangential divisions in individual parenchyma cells resulting in radial series of derivatives. In the outer derivatives such divisions are repeated. Some of the outermost derivatives may not divide; they form a secondary cortex. This type of activity may only be a temporary stage in the development of the cambium. In a later stage a true cambium with a single layer of initials develops. These initials add now on the outside more consistently derivatives to the parenchymatous secondary cortex. On the inside they produce cells which differentiate in part into parenchyma and in part into the vascular bundles (Fig. 225, nos. 1, 2). The inner parenchymatous ground tissue is termed *conjunctive tissue* and the walls of its cells may sometimes become thickened. The bundles develop from longitudinal files of single cells that are cut off from the cambial initials. As seen in cross-section each of these single cells represents the centre of a future vascular strand. These cells divide anticlinally to produce two or three rows of cells which then divide periclinally; later the direction of the divisions becomes haphazard but still longitudinal. In this manner the secondary vascular bundles are formed. The cells of the bundle undergo intrusive growth during their development. The xylem elements elongate fifteen to forty times their original length, while the xylem parenchyma and the phloem elements undergo little or no elongation. The thickening of the walls of those tracheids closest to the centre of the axis commences, in most cases, prior to the completion of the cell division in the developing bundle (Cheadle, 1937).

The secondary bundles may be amphivasal, i.e. the xylem surrounds the phloem as is seen in *Xanthorrhoea, Lomandra, Dracaena* (Fig. 115, no. 3) and *Aloë arborescens* (Fig. 225, no. 1) for example, or the xylem may surround the phloem on three sides and then the bundle appears U-shaped in cross-section as is seen, for example, in *Kingia* (Fig. 101, no. 1). The tracheary elements of these bundles in all the species that have as yet been studied are all tracheids. The walls of the parenchyma cells in which the vascular bundles are scattered may be thin or thick and lignified. The parenchyma that develops externally to the cambium usually remains thin-walled and in many plants many of the cells of this tissue contain crystals. In *Xanthorrhoea* resin is secreted by cells of the outer parenchyma so that a resin sheath is formed around the stem. (It is reported that the aborigines of Australia have almost completely destroyed the trees of *Xanthorrhoea* by burning them in order to enjoy the beautiful flames that result from the burning of this resin sheath.)

The connection between the primary and secondary bodies in monocotyledons with secondary thickening is strong, as it is in the dicotyledons. The union, in monocotyledons, is even more obvious because there are connections between the secondary bundles and the peripheral primary ones (Zimmermann and Tomlinson, 1970).

REFERENCES

ARTSCHWAGER, E. (1926) Anatomy of the vegetative organs of the sugar beet. *J. Agric. Res.* **33**: 143–76.

AYENSU, E. S. and STERN, W. L. (1964) Systematic anatomy and ontogeny of the stem in Passifloraceae. *Contr. US Natn. Herb.* **34**: 45–73.

BAILEY, D. C. (1980) Anomalous growth and vegetative anatomy of *Simmondsia chinensis. Am. J. Bot.* **67**: 147–61.

BALFOUR, E. (1965) Anomalous secondary thickening in Chenopodiaceae, Nyctaginaceae and Amaranthaceae. *Phytomorphology* **15**: 111–22.

BALL, E. (1941) The development of the shoot apex and the primary thickening meristem in *Phoenix canariensis* Chaub., with comparisons to *Washingtonia filifera* Wats. and *Trachycarpus excelsa* Wendl. *Am. J. Bot.* **28**: 820–32.

BELL, A. (1980a) The vascular pattern of a rhizomatous ginger (*Alpina speciosa* L. Zingiberaceae). 1. The aerial axis and its development. *Ann. Bot.* **46**: 203–212.

BELL, A. (1980b) The vascular pattern of a rhizomatous ginger (*Alpina speciosa* L. Zingiberaceae). 2. The Rhizome. *Ann. Bot.* **46**: 213–220.

CHEADLE, V. I. (1937) Secondary growth by means of a thickening ring in certain monocotyledons. *Bot. Gaz.* **98**: 535–55.

CHOUARD, P. (1937) La nature et le rôle des formations dites "secondaires" dans l'édification de la tige des monocotylédones. *Bull. Soc. Bot. Fr.* **83**: 819–36.

CLOWES, F. A. L. (1961) *Apical Meristems.* Blackwell Scientific Publ., Oxford.

DeMASON, D. A. (1979a) Function and development of the primary thickening meristem in the monocotyledon, *Allium cepa* L. *Bot. Gaz.* **140**: 51–66.

DeMASON, D. A. (1979b) Histochemistry of the primary thickening meristem in the vegetative stem of *Allium cepa* L. *Am. J. Bot.* **66**: 347–50.

DeMASON, D. A. (1980) Localization of cell division activity in the primary thickening meristem in *Allium cepa* L. *Am. J. Bot.* **67**: 393–9.

DeMASON, D. A. (1983) The primary thickening meristem: definition and function in monocotyledons. *Am. J. Bot.* **70**: 955–962.

DeMASON, D. A. and DIGGLE, P. K. (1984) The relationship between the primary thickening meristem and the secondary thickening meristem in *Yucca whipplei* Torr. III. Observations from histochemistry and autoradiography. *Am. J. Bot.* **71**: 1260–1267.

DIGGLE, P. K. and DeMASON, D. A. (1983a) The relationship between the primary thickening meristem and the secondary thickening meristem in *Yucca whipplei* Torr. I. Histology of the maure vegetative stem. *Am. J. Bot.* **70**: 1195–1204.

DIGGLE, P. K. and DeMASON, D. A. (1983b) The relationship between the primary thickening meristem and the secondary thickening meristem in *Yucca whipplei* Torr. II. Ontogenetic relationship within the vegetative stem. *Am. J. Bot.* **70**: 1205–1216.

DOBBINS, D. R. and FISHER, J. B. (1986) Wound responses in girdled stems of lianas. *Bot. Gaz.* **147**: 278–289.

ECKARDT, TH. (1941) Kritische Untersuchungen über das primäre Dickenwachstum bei Monocotylen mit Verdickung. *Bot. Arch.* **42**: 289–334.

ESAU, K. and CHEADLE, V. I. (1969) Secondary growth in *Bougainvillea. Ann. Bot.* **33**: 807–19.

FAHN, A. (1954) The anatomical structure of the Xanthorrhoeaceae Dumort. *J. Linn. Soc. Bot.* **55**; 158–84.

FAHN, A. (1985) The development of the secondary body in plants with interxylary phloem. In: *Xylorama. Trends in Wood Research* (ed. L. J. Kučera). Birkhauser Verlag, Basel. pp. 58–67.

FAHN, A. and SHCHORI, Y. (1967) The organization of the secondary conducting tissues in some species of the Chenopodiaceae. *Phytomorphology* **17**: 147–54.

FAHN, A and ZIMMERMANN, M. H. (1982) Development of the

successive cambia in *Atriplex halimus* (Chenopodiaceae). *Bot. Gaz.* **143:** 353–357.

FISHER, J. B. and TOMLINSON, P. B. (1972) Morphological studies in *Cordyline* (Agavaceae). II. Vegetative morphology of *Cordyline terminalis*. *J. Arnold Arb.* **53:** 113–27.

GINZBURG, C. (1963) Some anatomic features of splitting of desert shrubs. *Phytomorphology* **13:** 92–97.

ILJIN, M. M. (1950) Polykambialnost' i evoliutsya. In: *Problemy Bot.* **1:** 232–49.

KIRCHOFF, B. K. and FAHN, A. (1984) Initiation and structure of the secondary vascular system in *Phytolacca dioica* (Phytolaccaceae). *Can. J. Bot.* **62:** 2580–2586.

MENNEGA, A. M. W. (1969) The wood structure of *Dicranostyles* (Convolvulaceae). *Acta bot. neerl.* **18:** 173–9.

MIKESELL, J. E. (1979) Anomalous secondary thickening in *Phytolacca americana* L. (Phytolaccaceae). *Am. J. Bot.* **66:** 997–1005.

PFEIFFER, H. (1926) Das abnorme Dickenwachstum. In: K. Linsbauer, *Handbuch der Pflanzenanatomie*, Bd. 9, Lief. 15. Gebr. Borntraeger, Berlin.

PHILIPSON, W. R., WARD, J. M., and BUTTERFIELD, B. G. (1971) *The Vascular Cambium*. Chapman & Hall, London.

SCHENK, H. (1892) *Beiträge zur Biologie und Anatomie der Lianen*. G. Fischer, Jena.

SCHOUTE, J. C. (1912) Über das Dickenwachstum der Palmen. *Annls Jard. bot. Buitenz.* **26:** 1–209.

SKUTCH, A. F. (1932) Anatomy of the axis of the banana. *Bot. Gaz.* **93:** 233–58.

STEVENSON, D. W. (1980) Radial growth in *Beaucarnea recurvata*. *Am. J. Bot.* **67:** 476–89.

STEVENSON, D. W. and FISHER, J. B. (1980) The developmental relationship between primary and secondary thickening growth in *Cordylinae* (Agavaceae). *Bot. Gaz.* **141:** 264–268.

STEVENSON, D. W. and POPHAM, R. A. (1973) Ontogeny of the primary thickening meristem in seedlings of *Bougainvillea spectabilis*. *Am. J. Bot.* **60:** 1–9.

STIEBER, J. and BERINGER, H. (1984) Dynamic and structural relationships among leaves, roots, and storage tissue in the sugar beet. *Bot. Gaz.* **145:** 465–473.

STUDHOLME, W. P. and PHILIPSON, W. R. (1966) A comparison of the cambium in the woods with included phloem: *Heimerliodendron brunonianum* and *Avicennia resinifera*. *NZ J. Bot.* **4:** 355–65.

TOMLINSON, P. B. (1961) Palmae. In: *Anatomy of the Monocotyledons* (ed. C. R. Metcalfe), Vol. II. Clarendon Press, Oxford.

TOMLINSON, P. B. and ESLER, A. E. (1973) Establishment growth in woody monocotyledons native to New Zealand. *NZ J. Bot.* **11:** 627–44.

WHEAT, D. (1977) Successive cambia in the stem of *Phytolacca dioica*. *Am. J. Bot.* **64:** 1209–17.

ZAMSKI, E. (1979) The mode of secondary growth and the three-dimensional structure of the phloem in *Avicennia*. *Bot. Gaz.* **140:** 67–76.

ZAMSKI, E. and AZENKOT, A. (1981) Sugarbeet vasculature. I. Cambial development and the three-dimensional structure of the vascular system. *Bot. Gaz.* **142:** 334–343.

ZIMMERMANN, M. H. and TOMLINSON, P. B. (1967) Anatomy of the palm *Rhapis excelsa*: IV, Vascular development in apex of vegetative aerial axis and rhizome. *J. Arnold Arb.* **48:** 122–42.

ZIMMERMANN, M. H. and TOMLINSON, P. B. (1968) Vascular construction and development in the aerial stem of *Prionium* (Juncaceae). *Am. J. Bot.* **55:** 1100–9.

ZIMMERMANN, M. H. and TOMLINSON, P. B. (1969) The vascular system in the axis of *Dracaena fragrans* (Agavaceae): 1, Distribution and development of primary strands. *J. Arnold Arb.* **50:** 370–83.

ZIMMERMANN, M. H. and TOMLINSON, P. B. (1970) The vascular system in the axis of *Dracaena fragrans* (Agavaceae): 2, Distribution and development of secondary vascular tissue. *J. Arnold Arb.* **51:** 478–91.

ZODDA, G. (1904) Sull'ispessimento dello stipite di alcune palme. *Malpighia* **18:** 512–45.

REPRODUCTIVE ORGANS

CHAPTER 19

THE FLOWER

FLORAL ORGANS

THE problem of homology and morphological evolution of the flower has occupied research workers for a long time. Investigators such as Wolff and Goethe in the eighteenth century, de Candolle at the beginning of the nineteenth century, and many others since then, were interested in this problem (Arber, 1937, 1950). Opinions were expressed that floral organs are derived directly from foliage leaves. However, in the light of the view generally accepted today that the leaves and stem constitute a single unit which is termed the shoot, we can visualize the development of the flower as being parallel to that of a vegetative branch and not as being derived from it.

The flower consists of an axis on which the rest of the floral organs are borne. That part of the axis that represents the internode terminated by the flower is termed the *pedicel*. The distal end of the pedicel is swollen to various extents and this portion is termed the *floral receptacle* or *thalamus*. The floral organs are attached to the receptacle. A typical flower has four types of organs. The outermost organs of the flower are the *sepals* which together constitute the *calyx* which is usually green and is found lowest on the receptacle. On the inside of the sepals is the *corolla*, consisting of the *petals* which are generally coloured. These two types of organs together form the *perianth*; however, sometimes one of them may be lacking. When all the organs of the perianth are similar they are termed *tepals*. Within the perianth two kinds of reproductive organs are found: externally the *stamens* which together form the *androecium*, and internally the *carpels* which form the *gynoecium* (Fig. 226, no. 1).

In the following the reproductive organs only of the Angiospermae will be discussed. For inform-

ation on the reproductive organs of the Gymnospermae and lower vascular plants see Foster and Gifford (1974).

The arrangement of the floral organs on the receptacle may be spiral or whorled, and both types of arrangement may occur in the same flower. In most flowers in which the arrangement is whorled, the organs of each whorl alternate with those of the neighbouring whorl. The floral organs may be free or fused. Fusion of organs of the same type is termed *cohesion*, and that of different types of organs, *adnation*.

The term *pistil* is no longer used as it is not sufficiently well defined. This term was used both for each of the free carpels of a flower as well as for the unit which is formed by the fusion of a few carpels. Therefore, in this book we shall refer to all the carpels of a flower, whether they are free or fused, as the gynoecium.

The stamen consists of a *filament* which distally bears the *anther*. Two lobes can be distinguished in the anther and they are attached to a continuation of the filament which is termed the *connective*. Each of these lobes contains two *pollen sacs* in which *pollen grains* are found. The pollen grains in their early stage of development are the microspores of the angiosperms.

The free carpel or fused gynoecium usually consists of the following three parts: the *ovary*, a hollow body which contains one or more *ovules*; the *style* which results from the elongation of the ovary wall; and the *stigma* which is that part at the top of the style which has a surface structure enabling pollination. The ovules are attached to a special thickened region of the carpel wall which is termed the *placenta*.

When the carpels are found on the highest level of the floral axis the ovary is termed *superior* and the flower *hypogynous*. In certain plants the per-

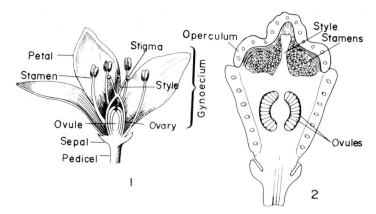

FIG. 226. 1, A flower of *Fagopyrum* sectioned longitudinally. The gynoecium consists of a unilocular ovary and three styles formed by three carpels. 2, A flower bud of *Eucalyptus globulus* sectioned longitudinally. (No. 2 adapted from McLean and Ivimey-Cook, 1956.)

ianth and stamens are located on the edge of a laterally expanded disc which raises them above the ovary; such a flower is termed *perigynous*, and its ovary is said to be *intermediate* or *pseudo-inferior*. The concave disc may completely enclose the ovary so that it is then found below the other floral organs; in such a flower the ovary is said to be *inferior*, and the flower *epigynous*.

In some plants, e.g. *Capparis*, only that part of the receptacle that bears carpels is elongated; such a structure is termed a *gynophore*. An elongation of that part of the receptacle that bears the carpels and stamens, e.g. *Passiflora*, is termed *androgynophore*.

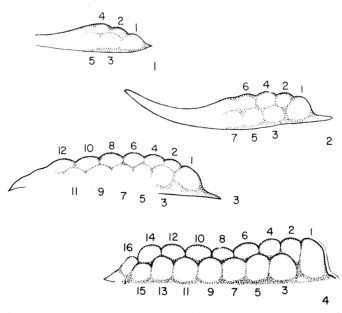

FIG. 227. Four stages in the ontogeny of a flower cluster of *Musa*; the numerals indicate the order of the appearance of the floral primordia.

ONTOGENY OF THE FLOWER

The transition of the shoot apex from the vegetative phase to the reproductive phase was discussed in Chapter 3. The morphological changes that take place are usually rapid and obvious. In many plants the axis elongates abruptly during this time or immediately afterwards and the apex widens and often becomes slightly flattened (Popham and Chan, 1952; Fahn et al., 1963). The sudden elongation of the flower-bearing axis prior to flowering is especially noticeable in those plants in which the internodes are very short and in which the foliage leaves form basal rosettes, e.g. in many species of the Compositae, in the Gramineae, and in the banana (Skutch, 1932; Barnard, 1957; and others). The elongated axis may bear a single flower but usually it bears an inflorescence (Fig. 231). The terminal meristem and the meristems in the axils of the bracts may develop flowers in the order (Fig. 227) that reflects the type of inflorescence (Fahn, 1953a). In a vegetative apex the apical meristem usually continues to function above the developing leaf primordia, but in the case of a developing flower the meristematic zone in the apex diminishes until it is lost completely, or until a small, non-active residue alone remains (Tepfer, 1953; Leroy, 1955).

In the apical meristem of the inflorescences and of the individual flowers of most plants that have been examined, a central zone of cells with large vacuoles has been observed to be covered by a zone of relatively small cells which are rich in protoplasm and which mostly divide anticlinally. With this development the activity of the rib meristem diminishes. It will not be discussed here whether in the reproductive apex a tunica-corpus arrangement is present, but it should be mentioned that the number of cell layers in the outer zone is usually larger than that found in the vegetative apex. The initiation and the first stages of histological differentiation of the different floral organs are, in principle, similar to the early ontogenetic stages of bracts and foliage leaves (Tepfer, 1953).

Many characteristics of flowers are related to the phyllotaxis of floral organs. In primitive angiosperms floral phyllotaxis, spiral and whorled patterns coexist together with unordered arrangements. In more evolved angiosperm flowers, as suggested by Endress (1987), such elaborations as sympetaly and syncarpy restricted the plasticity of phyllotaxis and strongly favoured whorled pattern. Whorled phyllotaxis was, according to Endress, a prerequisite for low and fixed organ number, for sympetaly and syncarpy and for synorganization of the adroecium with other parts of the flower.

The morphological and functional differences between the different floral organs are apparently related to a series of physiological processes which take place during the different stages of floral differentiation. This assumption is supported by the results of experiments which involved the incising and dissecting of the primordia of floral organs at different developmental stages (Cusick, 1956).

The developing floral organs usually appear in a distinct acropetal order (Fig. 228, nos. 1, 2), i.e. the youngest organs are closest to the apex. However, in certain genera and families, e.g. *Paeonia*, the Bixaceae, Dilleniaceae, Tiliaceae and in phytelephantoid palm group, the floral apex stops growing and then basipetal or centrifugal development of the staminal primordia can be distinguished in the zone of the stamens (Corner, 1946; Sporne, 1958; Tucker, 1972; Uhl and Moore, 1977). Other deviations from the usual ontogenetic development of the floral organs are also known (Cheung and Sattler, 1967; Gemmeke, 1982). Sometimes there may be a delay in the development of the petals although their primordia appear prior to those of the outermost stamens. In zygomorphic flowers the bilateral symmetry may be expressed at various stages, from prior to organ initiation up to the stage of enlargement of petals or stamens. In zygomorphic flowers of the Leguminosae, the whorls of flower organs are initiated undirectionally, from the abaxial to the axial side of the floral apex (Tucker, 1984a,b). In *Reseda*, for example (Fig. 228, nos. 4, 5), the adaxial side of the receptacle is already better developed than the abaxial side in the earliest stage of development, and the abaxial petal appears only after the adaxial stamens are almost fully developed.

In those flowers in which certain organs are partially reduced, as, for instance, in the case of the sterility of some of the stamens, the development of these organs is retarded in relation to

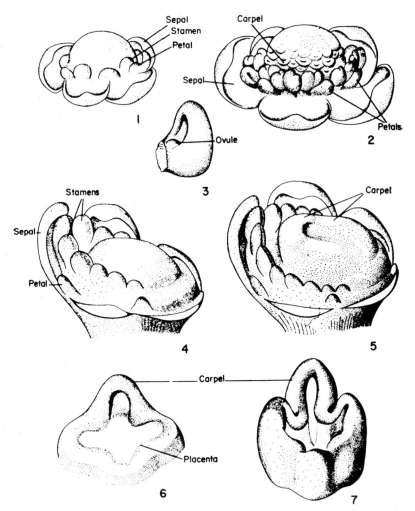

FIG. 228. Ontogeny of flowers. 1–3, *Ranunculus trilobus*. 1 and 2, Two stages in the development of the entire flower. 3, Developing carpel. 4–7, *Reseda odorata*. 4 and 5, Two stages in the development of the entire flower. 6 and 7, Developing gynoecium at stages later than those depicted in nos. 4 and 5. (Adapted from Payer, 1857.)

that of the other normal organs. The development of the reduced organs ceases completely before mature shape is obtained. The female flower of *Musa* may be given as an example of such development. In certain species the appearance of primordia of reduced organs may be retarded in relation to that of the similar normal organs. Examples also exist in which the carpels mature before the stamens. This feature is apparently brought about by a local concentration of growth substances (McLean and Ivimey-Cook, 1956).

In flowers in which the gynoecium is *apocarpous*, i.e. in which the carpels are free, each carpel primordium appears in its earliest stage of development as a rounded buttress which is similar to the primordia of the other floral organs and leaves. At a later stage of development the carpel primordium resembles that of a peltate leaf. Still later a depression appears at the tip of the primordium and as a result of unequal development, which then commences, an abaxial lip, from which the dorsal side of the carpel develops, is formed. The adaxial lip of the primordium develops more slowly and forms the "sill" of the carpel (McLean and Ivimey-Cook, 1956). This sill may enlarge and form the margins of the carpel

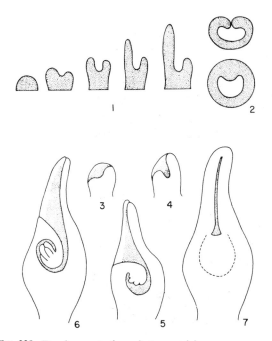

FIG. 229. Development of a peltate carpel in an apocarpous gynoecium. 1 and 2, Diagrams of sections of the carpel at different ontogenetic stages. 1, Longitudinal sections. 2, Cross-sections. 3–7, Different stages in the development of the carpel in *Thalictrum* showing the development of the sill, the ovule and the portion above the sill which folds over and closes the locule. 3–6, Developing carpel sectioned longitudinally. 7, Surface view of the side of the carpel on which the closure took place. (Nos. 1 and 2 adapted from Goebel, 1928–33; nos. 3–7 adapted from Troll, 1948.)

(Fig. 229). It is assumed that the sill originally consisted of the two basal laminar lobes which, as in the ontogeny of the peltate leaf, have fused. The area where the two margins fuse is termed the *cross-zone*. The ovule or ovules develop from this zone (Fig. 229, nos. 5, 6). In most cases the dorsal side of the carpel folds and closes over the sill. In a few genera of the monocotyledons, e.g. plants belonging to the Butomaceae, the primordia of the carpels are horseshoe-shaped, in cross-section, and remain so almost until the maturation of the carpels, and so no sill is developed. In this case the fusion of the margins of the carpels is incomplete even in the mature carpel (Fig. 230, nos. 1–4).

In flowers with *syncarpous* gynoecia, i.e. those in which the carpels are fused, the ovaries may develop in two ways. In the first type, carpel primordia first appear separate and fuse later as a result of lateral growth (*ontogenetic fusion*). Sub-

sequently, the carpels grow upwards as a unit on which the primordial apices are borne (Fig. 230, nos. 5–10). In the other type of development the carpels are already joined in the earliest stages of primordial development (*congenitally fused*) so that the ovary wall rises, from its initiation, as a ring. The regions of fusion between the carpels are distinguishable by inwardly directed folds (Fig. 228, nos. 5–7). Fusion of flower organs, whether postgenital (ontogenetic) or congenital (which is considered as phylogenetic), presents many developmental and evolutionary problems (Cusick, 1966; Sattler, 1978; Endress, 1983; Nishino, 1983; Ramp, 1987).

During the very early stages of development of an inferior ovary, a depression can be distinguished in the centre of the developing flower (Fig. 232, no. 1). This depression gradually deepens with the development and growth of the flower organs. In the flower of *Musa*, three separate rounded carpel primordia can be distinguished before the formation of this depression. In *Downingia* (Campanulaceae), as described by Kaplan (1967), this depression is a result of the growth of the common bases of the developing floral organs. Firstly, the common bases of the sepal lobes grow to form a floral cup. The initiation of the following floral whorls occurs along the inner margins of the cup. Continued basal growth of the cup-shaped bud results in the formation of the elongated inferior ovary.

The introduction of epi-illumination light microscopy (Poslushny *et al.*, 1980) and the use of the scanning electron microscope greatly advanced the study of the ontogeny of flowers. The latter appears to be of great value in dealing with taxonomic and evolutionary problems.

VASCULARIZATION OF THE FLOWER

The anatomical structure of the axis of the inflorescence and of the pedicel is similar to that of a typical stem; the vascular cylinder may be entire or split. Within the receptacle the shape of the stele becomes similar to that of the receptacle itself, i.e. the stele is usually widened at the base and narrows towards the upper part of the recep-

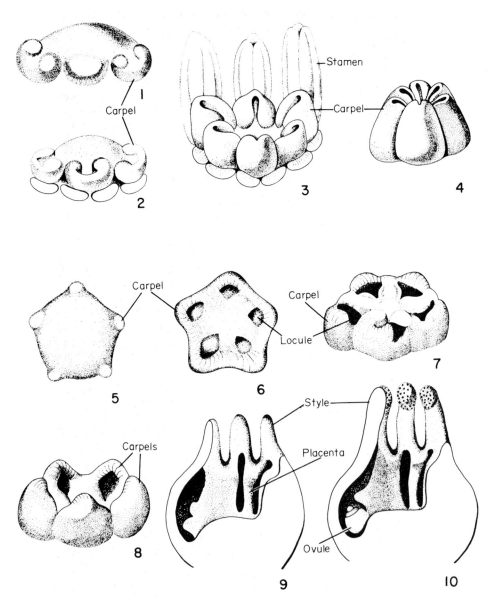

FIG. 230. Ontogeny of flowers. 1–4, Development of the apocarpous gynoecium of *Butomus umbellatus*. 5–10, *Linum perenne*. Stages in the development of the syncarpous gynoecium from separate primordia which expand laterally and fuse to form a single ring. (Adapted from Payer, 1857.)

tacle. From the stele of the receptacle the traces pass out to the various floral organs. The traces and the gaps split the stele into a characteristic network of vascular bundles (Fig. 235, no. 1; Fig. 237). The stele may be further split as a result of the reduction of vascular tissues, as occurs in a typical stem (Eames, 1931; Eames and Mac-Daniels, 1947; Tepfer, 1953). The traces of the various floral organs leave the stele in whorls or along a spiral line, in accordance with the arrangement of the organs on the receptacle. The traces of the uppermost carpels may, in certain plants, terminate the stele of the receptacle (Fig. 238, no. 1). In other plants, however, the stele continues above the level of the traces to the uppermost carpels. In this case the terminal

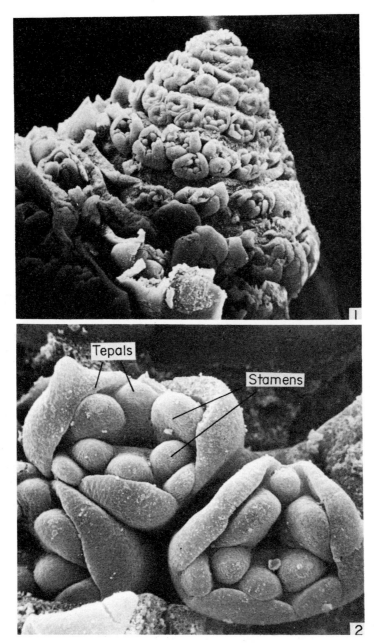

FIG. 231. Scanning electron micrograph of developing banana flowers. 1, Apical region of an inflorescence. The bracts of the flower groups were removed so that the individual primordia can be seen. × 20. 2, Young flowers showing tepal and stamen primordia. × 120.

bundles of the stele of the receptacle, which consist only of phloem or procambium, gradually fade out within the top of the receptacle (Fig. 238, no. 2).

The number of traces to the various floral organs varies. The number of traces to a sepal is usually equal to that supplying the foliage leaves of the same plant. Petals generally have one trace, but in some families three or more traces may be present. Stamens usually have a single trace but in some ranalian families and certain species of a few other families, e.g. the Lauraceae and Musaceae,

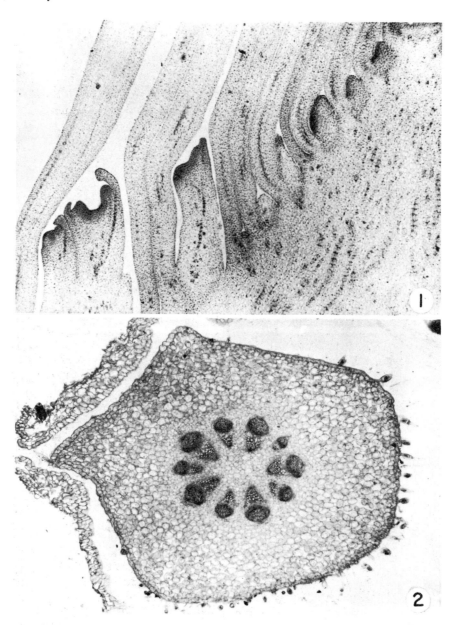

FIG. 232. 1, Portion of a longitudinal section of a developing inflorescence of *Musa* (Dwarf Cavendish banana) showing floral primordia in different stages of development. × 43. In the second primordium from the left, the depression that represents an early stage in the development of the inferior ovary can be seen. 2, Cross-section of the pedicel of *Aquilegia caerulea* in which ten bundles, five large alternating with five small, can be distinguished. × 43.

three traces enter each stamen. Carpels may have one, three, five, or more traces; three traces is the most common condition, but five also occur frequently. The median carpel trace leaves the stele at a lower level than do the other traces. The continuation of this median trace constitutes the dorsal bundle of the carpel and the trace itself is termed the *dorsal trace*. The dorsal bundle is actually homologous to the leaf mid-rib. The two outermost traces on each side are termed the *marginal* or *ventral traces*, as the bundles that pass along the carpel margins are continuous with

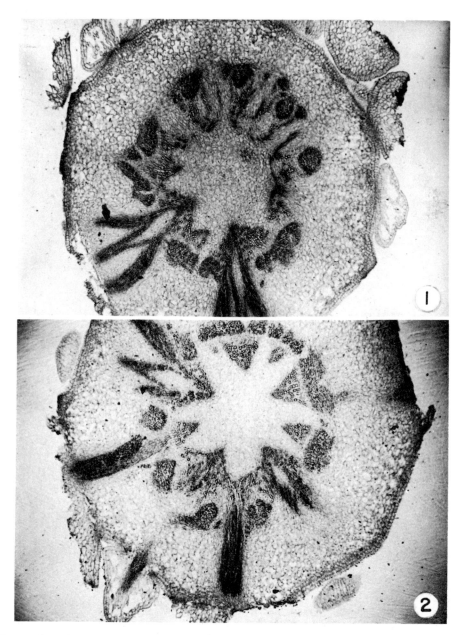

FIG. 233. Cross-section of the receptacle of *Aquilegia caerulea*. 1, At the level of the exit of the sepal traces, showing the traces to two sepals. Each sepal is supplied by three traces. × 46. 2, At a level a little higher than in no. 1 showing two single petal traces and the vascular supply, consisting of three traces, to one sepal. × 48.

them. If the carpels are fused to form a syncarpous gynoecium, these marginal bundles are found lateral to the dorsal bundle, and if the carpels are folded inwards they are found ventrally relative to the dorsal bundle (Fig. 236, nos. 1, 2). As a result of the inward-folding of the carpel margins the ventral bundles are inverted—their phloem is directed towards the locules of the ovary, and their xylem outwards. If more than three traces are present in a carpel the additional traces are found between the ventral and dorsal traces, and they are termed *lateral traces*. The vas-

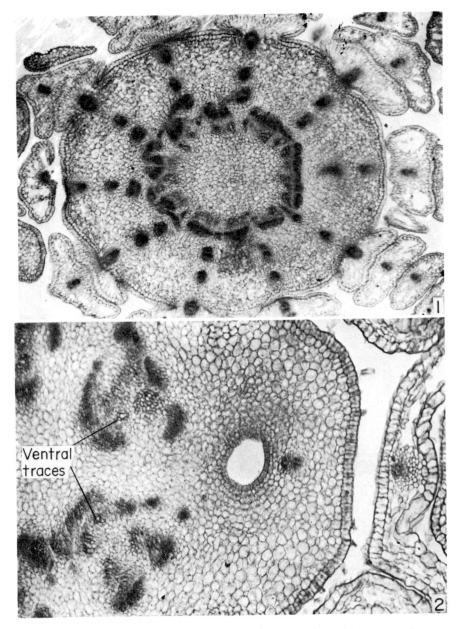

Ventral
traces

FIG. 234. *Aquilegia caerulea*. 1, Cross-section in the receptacle at the level of the exit of the traces to the numerous whorls of stamens. Within the receptacle the staminal traces are seen to be arranged in radial rows. × 45. 2, Cross-section in the basal portion of a single carpel in which the single dorsal trace and two ventral traces can be distinguished. In the ventral traces (but not in the branches of them that appear close to them in the micrograph) it can be seen that the xylem is directed outwards and the phloem inwards. × 140.

cular bundles of the carpels may branch and may even continue to do so during the development of the fruit.

The vascular bundles that supply the ovules usually originate from the ventral carpel bundles, or from branches of them that are present in the placenta (Fig. 236, no. 1). The ovular bundle is single and thin, and reaches the zone of the chalaza; it does not penetrate into the nucellus but in some genera, branches of it enter into the

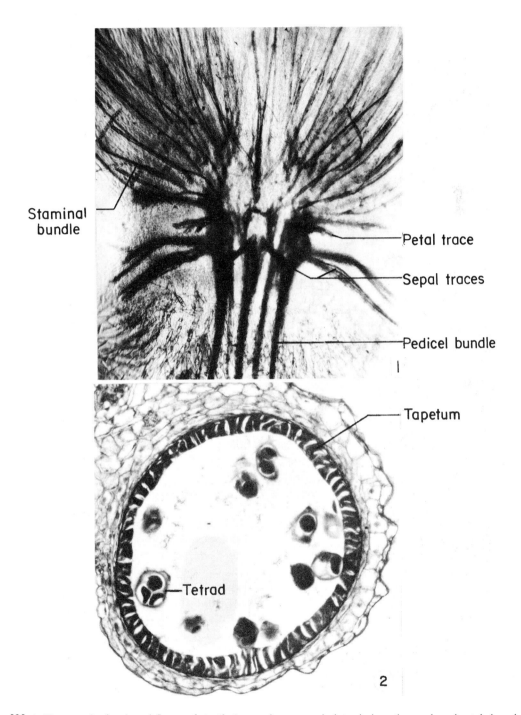

FIG. 235. 1, Photograph of a cleared flower of *Aquilegia caerulea* as seen in lateral view; the sepals and petals have been removed. × 27. 2, Micrograph of a cross-section of a young pollen sac of *Passiflora caerulea* showing a glandular tapetum. × 180.

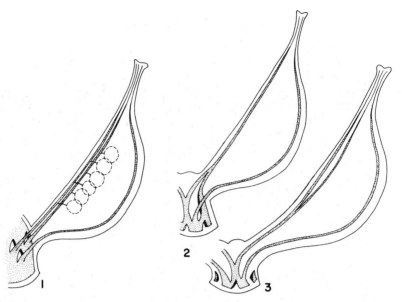

FIG. 236. Diagrams illustrating the carpellary vascular supply. 1, A carpel supplied by three traces each of which is accompanied by a separate gap; bundles remain unfused within the carpel to the stigma. 2, A carpel in which the two ventral traces are fused from the base of the carpel. 3, A carpel in which the ventral traces arise fused, but in which they separate in the basal portion of the carpel. (Adapted from Eames and MacDaniels, 1947.)

integuments.

The structure of the receptacular stele and the system of vascular bundles within the floral organs is relatively complex even in the simpler types of flowers (i.e. those that are phylogenetically relatively primitive). The interpretation of floral vascularization becomes even more difficult in flowers in which fusion of the traces has taken place during the process of evolution. Eames (Eames, 1931; Eames and MacDaniels, 1947) gives *Aquilegia* of the Ranunculaceae and *Pyrola* of the Ericaceae as examples of flowers with simple vascular systems. In species of *Aquilegia* the pedicel contains five thick bundles which alternate with five thin bundles (Fig. 232, no. 2). These bundles fuse at the base of the flower to form an uninterrupted ring. Above this level five groups of sepal traces, each group consisting of three traces per single gap, depart from the stele (Fig. 233, no. 1). A little higher, alternating with the sepal traces, a single trace passes out to each of the five petals (Fig. 233, no. 2). Above the petal gaps the traces to the numerous stamens are found. Each stamen has a single trace (Fig. 234, no. 1). Above the uppermost whorl of stamens the.

stele once again becomes an uninterrupted ring. Shortly above this level five compound gaps of the carpels are formed. From the base of each such gap, the dorsal trace is given off and, from the sides, the pair of ventral traces (Fig. 237; Fig. 238, no. 2). The ventral traces invert immediately on their exit from the stele and so enter the carpel with the xylem directed outwards (Fig. 234, no. 2). Above this level the stele consists of five bundles which mainly consist of phloem. This vascular tissue gradually fades towards the rounded tip of the receptacle.

The type of vascular system that is exemplified by *Pyrola* (Fig. 238, no. 1) differs principally from the type of *Aquilegia* in the following features: (a) each sepal has only one trace; (b) above that level at which the dorsal carpel traces depart from the stele the remaining stelar bundles fuse in pairs to form five strands, each of which represents the two ventral traces; (c) no vestigial vascular tissue is found at the top of the receptacle.

An unusual phenomenon of xylem discontinuity occurs in the tribe Psoraleae of the Leguminosae. The ovary bundles merge at the base of the ovary where they end in a horizontal

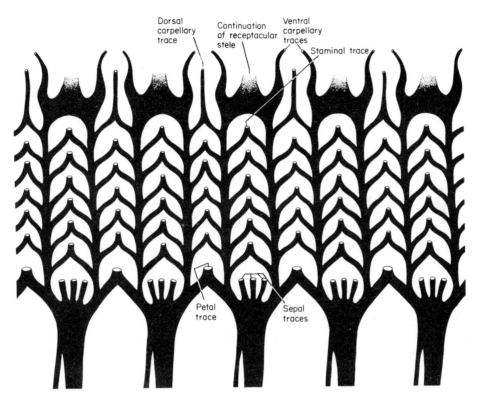

FIG. 237. Diagrammatic representation of the stele of *Aquilegia* spread out in a single plane. (Adapted from Tepfer, 1953.)

proliferation of tracheary elements. This plate-like xylem expansion is not connected to the xylem of the pedicel. The xylem of the petal and stamen bundles in the receptacle is also commonly discontinuous. The phloem, however, is continuous throughout the flower (Lersten and Wemple, 1966).

Variations in the vascularization of the flower

During phylogenetic development of the flower, processes of cohesion, adnation and organ abortion have taken place. It is commonly accepted that during the process of evolution the external fusion of organs precedes that of the inner tissues, and that the fusion of the vascular tissue represents the last stage in this process. Fusion of vascular bundles involves those bundles that were close to one another. The fusion may involve the traces alone or may include part of the bundles within the organs. Infrequently it continues throughout the entire length of the bundles. Usually there is ontogenetic and histological evidence that fusion has taken place. Evidence of aborted organs is assumed from the presence of rudimentary organs and persisting traces as compared with related flowers.

Bundle fusion

As a result of the cohesion of organs, in many cases the lateral and marginal vascular bundles and traces fuse. In the calyces of plants of the Labiatae different stages of vein fusion can be found. Thus, according to Eames (1931), in *Nepeta* and *Monarda* it is possible to discern that each sepal has a main vein and two lateral veins (Fig. 239, nos. 1, 3); in the calyces of *Ajuga* and *Physostegia* the neighbouring lateral veins are seen to be fused almost up to the base of the sinuses between the calyx lobes (Fig. 239, nos. 2, 4); in *Salvia* (Fig. 239, no. 5) two pairs of lateral veins are fused while the others remain free.

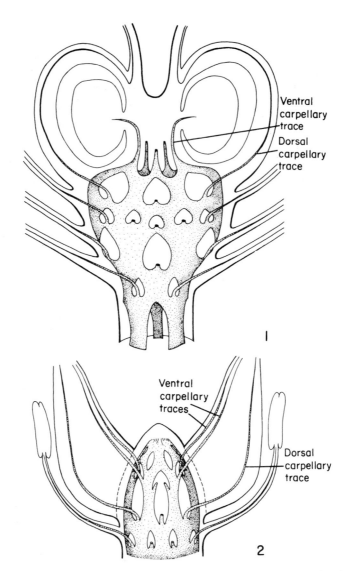

FIG. 238. Diagrammatic representation of different types of floral vascularization. 1, *Pyrola*, showing an entire flower. 2, *Aquilegia*, showing the upper part of a flower. Vascular tissue stippled. (Adapted from Eames and MacDaniels, 1947.)

FIG. 239. 1–5, Stages in the fusion of lateral bundles in gamosepalous calyces. 1, Calyx of *Nepeta veronica* cut and spread out; the lateral bundles are unfused. 2, Calyx of *Ajuga reptans* in which the lateral bundles are fused. 3, Diagram of the cross-section of the calyx of *Monarda didyma* in which the lateral bundles are free. 4, Cross-section of the calyx of *Physostegia virginiana* in which the lateral bundles are fused. 5, Cross-section of the calyx of *Salvia patens* in which two pairs of lateral bundles are fused. 6–10, Stages in the fusion of lateral bundles in gamopetalous corollas. 6, *Hamelia patens*, in which the lateral bundles are unfused. 7, *Senecio fremontii*. 8, *Anastraphia ilicifolia*. 9, *Chrysanthemum leucanthemum*. 10, *Xanthium orientale*. 11–18, Diagrams of cross-sections of carpels showing different stages of cohesion. 11, *Reseda odorata* in which the placentation is parietal. 12–15, Different stages in the fusion of the ventral bundles in follicles. 16–18, Different stages in the phylogenetic development of the syncarpous ovary from the stage where the tissues and vascular bundles of each carpel are unfused through the stage where the carpellary tissues but not the vascular tissues are fused to the final stage where the carpellary tissues and lateral and ventral bundles are fused. In nos. 11–18 the xylem is represented by solid black and the phloem is white. (Nos. 1–5 and 11–18 adapted from Eames and MacDaniels, 1947; nos. 6–10 adapted from Koch, 1930.)

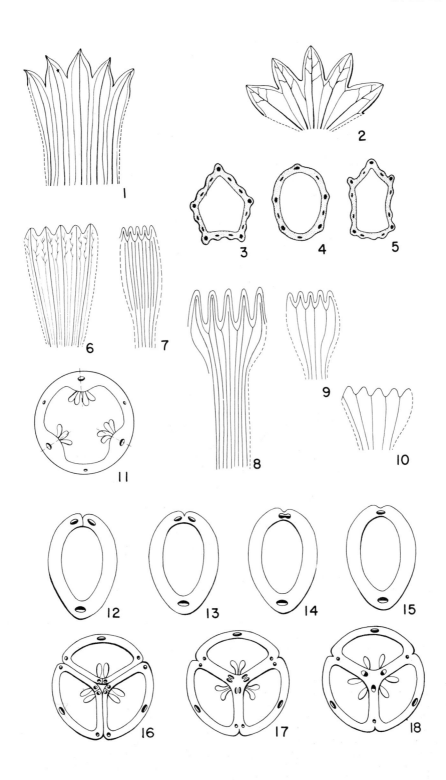

In sympetalous corollas vein fusion may also be found (Fig. 239, nos. 6–10). In the sympetalous flowers of the Rubiaceae, for example, it is still possible to discern fifteen veins, three in each petal. In the Compositae it is possible to find evidence of the fusion of the lateral bundles and of the loss of the median bundles (Koch, 1930).

In the gynoecium the fusion of the bundles takes place in the following ways. If the carpels are free the two ventral bundles approach one another on the ventral side of the primitive type of carpel that develops into a follicle (Fig. 236, no. 1). The ventral bundles may fuse from the base and the fusion may continue along the entire length of the carpel or along only part of it (Fig. 236, nos. 2, 3). A carpel with a single dorsal and a single ventral bundle may have two or three traces. In the former case one trace is related to the dorsal bundle and one to the ventral. In very reduced carpels, as, for instance, those in cypselae, there is sometimes only one trace, and it splits at the base of the locule of the carpel. In such a trace the three original traces are fused. In *Filipendula*, for instance, which appears to be a highly evolved genus of the Spiraeoideae (Rosaceae), fusion of the carpellary bundles has occurred in such a way (Sterling, 1966). If the carpels are fused, two types of fusion are distinguished. On the one hand, when a large portion of the carpel is folded inwards and the placenta is axile, the ventral bundles are brought to the centre of the ovary where a pair of inverted ventral bundles of a single carpel or those of two adjacent carpels fuse (Fig. 239, nos. 16–18). The lateral bundles (i.e. those between the dorsal and ventral bundles) of two neighbouring carpels may also fuse in such an ovary (Fig. 239, nos. 16–18). On the other hand, if the carpels become fused while they are still open and so form a single common locule with parietal placentation, the ventral bundles are not inverted. They occur in pairs along those lines where the cohesion of the carpel margins takes place, or they may fuse to form common bundles (Fig. 239, no. 11).

In addition to the fusion of bundles as a result of cohesion of organs, bundle fusion also results from adnation, i.e. the fusion of organs of different whorls. Bundles that are radially or tangentially close to one another may fuse, and this process may involve different numbers of bundles which may belong to two or more whorls. Different stages in the fusion of the staminal bundles with those of the petals (epipetalous stamens) may be seen, for example, in the Crassulaceae (Fig. 240, no. 7). In flowers of plants belonging to the Rosaceae (Fig. 240, nos. 3–6) it is possible to discern the fusion of bundles from more than two whorls (Jackson, 1934).

The inferior ovary

The problem of the phylogenetic development of the inferior ovary still challenges flower morphologists. The problem is whether the gynoecium has sunk into a concave receptacle which then surrounded the carpels and fused with them (*axial theory*), or whether the inferior ovary arose as a result of the cohesion and adnation of the lower portions of all the floral organs to the gynoecium, which they surround (*appendicular theory*). In order to solve this problem different methods within the field of comparative morphology, involving ontogeny, histology, teratology and palaeontology have been used. Much attention has been paid to the floral vascular anatomy (Douglas, 1944). The latter line of investigation in connection with the inferior ovary has revealed the following features. In the wall of the inferior ovary of certain plants, e.g. *Hedera* (Fig. 240, no. 1), separate traces which are related to the different floral organs are found. In many plants, e.g. *Juglans* (Fig. 240, no. 2), Lecythidaceae and a few species of the Ericaceae, different stages of bundle fusion have been observed in the wall of the inferior ovary (Manning, 1940; Douglas, 1944; Eames and MacDaniels, 1947; Eames, 1961; Monteiro-Scanavacca, 1974). In relation to these plants, therefore, it is possible to conclude from the vascular anatomy that the wall of the inferior ovary consists of the ovary wall proper together with the other floral organs that are fused with it.

In species belonging to the Cactaceae and Santalaceae (Fig. 241, no. 1), inverted vascular bundles, i.e. those with inwardly directed phloem and outwardly directed xylem, can be found along the entire length of the inferior ovary wall (Smith and Smith, 1942; Tiagi, 1955; Boke, 1966). This feature suggests that in these species the inferior

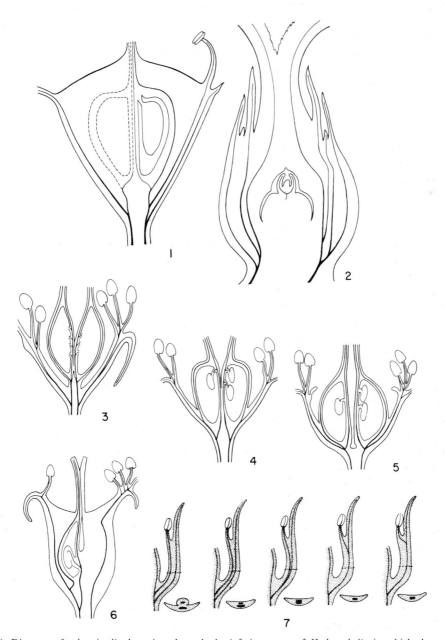

Fig. 240. 1, Diagram of a longitudinal section through the inferior ovary of *Hedera helix* in which there is no fusion between the sepal, petal and staminal bundles, or between these bundles and the carpellary bundles. 2, Longitudinal section of the flower, including bracts, of *Juglans nigra* showing the fusion, in their basal portions, of the bundles of the different organs. 3–6, Longitudinal sections of flowers of related genera of the Rosaceae showing different degrees of adnation of the hypanthium to the carpels. 3, *Physocarpus opulifolius* in which the ovary is superior and the sepals, petals and stamens are fused, but not to the carpels. 4, *Sorbus sorbifolia* in which the hypanthium is adnated to the base of the carpels. 5, *Spiraea vanhouttei* in which the hypanthium is adnated to the carpels to a level half way along the ovary. 6, *Pyrus malus* var. *paradisiaca* in which the hypanthium is adnated to the ovary throughout its length. 7, Diagrams showing different degrees of vascular fusion during the adnation of a stamen and petal, as seen in both longitudinal and cross-sections. (Nos. 1, 6 and 7 adapted from Eames and MacDaniels, 1947; no. 2 adapted from Manning, 1940; nos. 3–5 adapted from Jackson, 1934.)

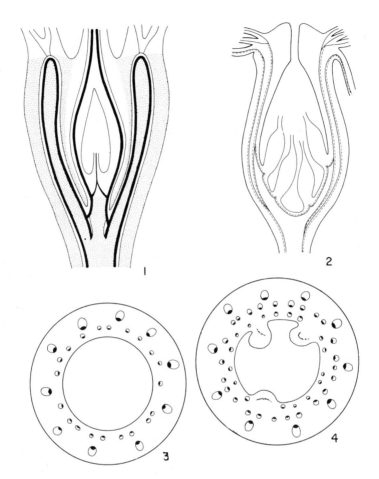

FIG. 241. 1, Diagram of a longitudinal section of the inferior ovary of *Darbya* (Santalaceae) showing the vascularization; xylem—black, phloem—white. 2–4, *Rosa helenae*. 2, Diagram of a longitudinal section of the flower; broken line, phloem; solid line, xylem. 4, Diagram of a cross-section in the lower quarter of the hypanthium showing the vascular bundles (xylem—black, phloem—white). The arrangement of the vascular tissues in the innermost row is seen to be inverted, a fact which is evidence of invagination at this level. 3, A cross-section at a level higher than in no. 4 in which there is no inversion of the vascular tissues, indicating that at this level there is no invagination. (No. 1 adapted from Douglas, 1944; nos. 2–4 adapted from Jackson, 1934.)

ovary has developed as a result of the involution of the receptacle, as the sinking of the gynoecium into the receptacle necessitates an inward folding of the upper portion of the receptacular stele. In *Rosa*, Jackson (1934) found, according to the vascular anatomy, that the lower portion of the fleshy fruit consists of receptacular tissue, while the upper portion of it consists of fused floral organs (*hypanthium*) (Fig. 241, nos. 2–4). Jackson also showed that various intermediate stages, which demonstrate the evolution of the rose fruit, can be seen in related genera. The fruit of *Pyrus malus* var. *paradisiaca* and related genera apparently consist mainly of true ovary wall and of other floral organs that are fused to it, and the receptacle constitutes a very small portion, which is not easily discernible, at the base of the fruit (Fig. 289, no. 6) (MacDaniels, 1940). This adnation of the bundles in an inferior ovary that is developed from the fusion of the ovary wall proper to other floral organs is not equal in different radii, neither in the number of traces involved nor to the extent to which they are fused.

On the basis of floral vascular anatomy and floral ontogeny and histogenesis the inferior ovary of *Downingia* (Campanulaceae) was also interpreted as wholly appendicular, representing adnation of the outer floral organs to the gynoecium (Kaplan, 1967).

In addition to fusion between the floral organs of a single flower fusion may occur between flowers and the bracts accompanying them, e.g. *Lonicera* and *Juglans*, and even between several flowers, e.g. in certain species of *Lonicera*, *Cornus* and *Maclura*. These features are also often accompanied by the fusion of bundles (Manning, 1940; Eames and MacDaniels, 1947).

VESTIGIAL VASCULAR BUNDLES

The presence of vestigial vascular bundles, in certain positions, is frequently taken as proof of the degeneration of the organs that were originally present in the same position in the ancestral floral type. Vestigial petal traces may be found in the receptacle of apetalous flowers, e.g. *Aristolochia*, species of *Rhamnus*, *Salix* and certain species of *Quercus*. Vestigial staminal traces can be found in the Scrophulariaceae, Labiatae, Cucurbitaceae, and in certain genera of other families. In the Caprifoliaceae, Ericaceae, Rutaceae, and Valerianaceae vestigial carpel traces are found. Such vestigial traces may also be found in unisexual flowers. Degenerate ovules are commonly associated with vestigial bundles.

The reduction that has taken place during the evolutionary specialization of the floral organs is often accompanied by the reduction of the vascular supply to the organs. This is exemplified by the development of the achenium or cypsela, which contains a single ovule, from a legume or follicle, which contain several ovules, and by the reduction of the corolla in the capitulum of the Compositae. Such reduction is recognizable by the shortening of the bundles.

In some carpels, in which much reduction has taken place, only one bundle, which enters the ovule directly from the receptacular stele, is found. In stamens the bundles may sometimes terminate at the base of the filaments instead of reaching the anther. In the sympetalous corollas of many of the Compositae the median bundle is lost and lateral bundles may also become reduced, and they even disappear in the corollas of certain plants (Chute, 1930; Eames, 1931).

The importance of the structure of the floral vascular system in the interpretation of the morphological structure and the clarification of the phylogenetic relationships between different plants is still debatable (Henslow, 1891; Sporne, 1958; Carlquist, 1969; Schmid, 1972). Puri (1951), who reviewed the relevant literature on floral anatomy, came to the conclusion that this method of research has contributed much to the better understanding of the angiosperm flower. However, he stresses the necessity of regarding the structure of the vascular system of the flower as only one important aspect of the problem of floral morphology. According to Puri the study of the external structure of organs and their ontogenetic development should not be neglected. In spite of the above-mentioned reservations, the vascularization of the flower represents an additional feature which enriches our ability to tackle evolutionary and taxonomic problems.

HISTOLOGY OF SEPALS AND PETALS

The primordia of all floral organs usually initiate by anticlinal cell divisions in the outermost cell layer of the floral apex and by periclinal plus anticlinal or oblique cell divisions in the second layer (Tepfer, 1953; Cheung and Sattler, 1967). In some plants sepals and petals were reported to initiate by periclinal divisions in the second and third layers and primordia of stamens in the third and fourth layers (Richardson, 1969). Sepals and petals undergo apical, marginal, and intercalary growth, similar to foliage leaves (Kaplan, 1968). Dubuc-Lebreux and Sattler (1985) reject the view that the laminar growth of petals is initiated by a marginal meristem (see Chapter 12).

The external structure of sepals and petals may be leaf-like. However, the internal structure only of green sepals resembles that of foliage leaves and the internal structure of coloured sepals and petals is distinctly different.

In the perianth the vascular system is usually only poorly developed and the veins usually lack sclerenchyma. The mesophyll commonly consists of spongy parenchyma only and its cells contain chromoplasts or pigments in the cell sap, or both.

Usually the epidermal cell walls are thin. The anticlinal walls of these cells are, in many flowers, folded or undulated and are arranged so that the projections of the cells dove-tail one into the other, a feature that strengthens the epidermis. At the base of the petals, as well as along the length of the veins, the anticlinal walls are usually straight even when they are wavy in the other epidermal cells. The outer walls of the epidermal cells commonly have papillae, which make the petals glisten. More papillae are present on the adaxial than the abaxial epidermis and they are not developed at the base of the petals. Stomata, if present, are scanty and non-functioning. Trichomes may sometimes be present on sepals and petals. Sometimes intercellular spaces, covered by cuticle (Hiller, 1884), are formed in the epidermis of petals. The thickness of the cuticle varies in different plants and it may be variously sculptured (Martens, 1934) (Fig. 267, nos. 2, 3).

In the upper epidermis and in mesophyll cells variously shaped chromoplasts may occur. Both the carotenoids of the chromoplasts and betacyanins, occurring in the central vacuoles, or each of them separately contribute to the colours of the flowers. Many petals have UV-absorbing flavonoid pigments in their epidermal cells (Horovitz and Cohen, 1972; Kay and Daoud, 1981; Rieseberg and Schilling, 1985). These pigments are invisible to us, but bees can discern them. Fragrance also guides pollinators to flowers. In some plants, e.g. *Ceropegia* species, *Aristolochia* and species of the Orchidaceae and Araceae, restricted portions of flower organs, *osmophores*, bear a tissue specialized for synthesis and secretion of fragrant substances, mainly volatile terpenes (Vogel, 1962). In *Spartium junceum*, for instance, the osmophores are located on the "wings" of the Papilionoideae flower. In some Araceae species, which have a repugnant odour of rotting flesh, the osmophores produce amines and ammonia in addition to terpenes (Smith and Meeuse, 1966). The ultrastructure of the osmophores in some orchids has recently been studied by Pridgeon and Stern (1983, 1985).

The less specialized the petals are, the more their internal structure resembles that of foliage leaves, that is, the veins and the mesophyll are better developed, a palisade-like tissue is present, the epidermis is devoid of papillae and there are many stomata.

Sepals and petals may fuse to form a lid or *operculum* which dehisces circumscissily as in a certain group of *Eucalyptus* (Fig. 226, no. 2). The sepals and petals may form two separate opercula or may be fused and form one common operculum. In another group of *Eucalyptus* the operculum throughout its development is seen to be a single structure formed by the petals only (Carr and Carr, 1968).

THE STAMEN

Phylogeny

The stamens and carpels of highly specialized flowers differ greatly in structure and general shape from foliage leaves. Different opinions still exist as to the morphological interpretation and evolutionary development of the stamens and carpels (Parkin, 1951). The most commonly accepted theory is that these organs are homologous to leaves. The opinion also exists that the development of the stamen can be explained on the basis of the *telome theory*, that is, from primitive dichotomously branched axes. According to this theory the stamen has developed as the result of the reduction and fusion of a system of axes that bore sporangia at their tips (Wilson, 1937, 1942). New light was thrown on the phylogeny of stamens and carpels by Bailey and his co-workers from their morphological and comparative anatomical research in many families of the Ranales. From these investigations it has become apparent that in the woody species of the Ranales that exist today not only primitive stages of xylem development have been preserved, but also primitive types of stamens and carpels (Bailey and Smith, 1942; Bailey and Nast, 1943a, b; Bailey *et al.*, 1943; Bailey and Swamy, 1949, 1951; Canright, 1952; Bailey, 1956; Endress, 1980, 1983). A very primitive type of stamen is found in the genus *Degeneria*; here no filament, anther, or connective can be distinguished as the stamen is broad and leaf-like, and has three vascular bundles. The four pollen sacs (the microsporangia) are deeply sunk into the abaxial side of the stamen (Fig. 243, nos. 1, 2). The pollen sacs are found between the lateral and median bundles. Similar stamens are found in other ranalian genera, e.g. *Austrobaileya*, *Himantandra* and certain genera of the Magnoliaceae. In the Magnoliaceae intermediate stages may be found from broad stamens with three bundles and laminal pollen sacs, i.e. pollen sacs distant from the margins, as in *Degeneria*, to stamens with marginal pollen sacs and distinct filaments and anthers (Canright, 1952). Kunze (1978), on the basis of morphological and morphogenetic study of stamens of 56 species belonging to 32 families of angiosperms, came to the following conclusion: "probably all angiospermous stamens follow basically the same developmental pattern and can therefore be described as primary bifacial organs of phyllomic nature with a marginal growth zone developing into pollen sacs."

Structure and tissue differentiation

The epidermis of the filament possesses a cuticle and, in certain species, trichomes. The filament consists of parenchyma with well-developed vacuoles and small intercellular spaces. Often pigments are present in the cell sap. The size and external shape of angiosperm stamens vary greatly. In the Compositae, for instance, the distal part of the filament, that adjacent to the anther and termed the *anther collar*, differs considerably from the proximal part (Meiri and Dulberger, 1986). The anther generally contains four pollen sacs (microsporangia) which are paired in two lobes. The two lobes are separated by a zone of sterile tissue, termed the *connective,* through which a vascular bundle passes (Fig. 242). Complete and incomplete septa occur in the pollen sacs of some genera (Lersten, 1971).

According to Boke (1949), who worked on *Vinca rosea*, even some time after the region of the developing anther becomes discernible in the staminal primordium, the anther still consists only of protoderm and a mass of ground meristem. All of the subepidermal layer of the young anther may be sporogenous, but actually the sporogenous tissue is developed from four cell regions which are located in the four angles of the developing anther. In each of these regions there is a row of hypodermal initials which divide periclinally to form two layers (Fig. 243, nos. 4–8). The inner layer of these initials constitutes the *primary sporogenous cells*, which by further division form the *pollen mother cells*. The outer layer of the above initials constitute the *primary parietal cells*, from which the wall of the pollen sacs and a large portion of the *tapetum* develop, as a result of periclinal and anticlinal cell division. The tapetum apparently serves for the nourishment of the developing pollen mother cells and microspores (pollen grains). The outermost layer of parietal cells is located immediately below the epidermis of the anther. Prior to the liberation of the pollen grains several wall thickenings develop in each of the cells of this layer. However, no thickening is developed in the outer wall closest to the epidermis. Each thickening is U-shaped with the gap directed towards the epidermis (Fig. 242, no. 2). This cell layer is usually termed the *endothecium* and the opening of the pollen sacs is brought about by this layer. The opening mechanism is described as follows. During the dehydration of the anther the endothecium loses water. As the water content of these cells decreases the walls of each cell are drawn towards its centre as a result of the cohesion forces between the water molecules and the adhesion forces between the water and the cell walls. Because of the absence of thickenings in the outer periclinal wall, it becomes more folded than the anticlinal and the inner periclinal walls which have thickenings; thus, as a result of water loss, the cell appears trapezium-shaped in cross-section. As all the endothecial cells lose water almost at the same time and all the external walls fold and wrinkle, the endothecium shrinks in a manner that results in the opening of the anther. The cells in the region along which anthers dehisce are thin-walled. In most plants the regions of dehiscence are in the form of longitudinal slits between the two pollen sacs of each lobe. Each such slit is termed a *stomium*. Prior to dehiscence part of the partition between the two sacs of a lobe usually disintegrates. After this the region of dehiscence is covered only by the epidermis. The epidermal cells, in this region, are extremely small, especially in comparison with the neighbouring epidermal cells, and they are easily ruptured when the anther ripens (Fig. 242; Fig. 243, no. 3). Another type of stomium, common in

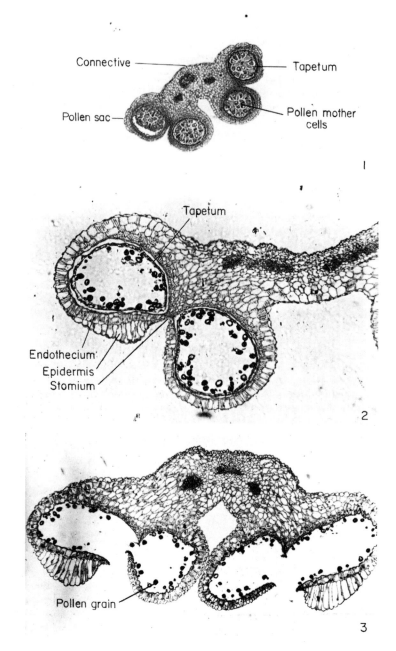

FIG. 242. Micrographs of cross-sections of the anthers of *Passiflora caerulea* at different stages of development. 3, Mature anther after dehiscence. 1 and 2, ×34; 3, ×30.

the Ericaceae and in *Solanum*, is that occurring at the top of the anther only (Fig. 244, no. 1; Fig. 245, nos. 1–3). Openings may also develop on the sides of the anther as, for example, in the Lauraceae (Fig. 244, nos. 2, 3; Fig. 245, no. 4).

The cell layer or layers below the endothecium both stretch and become compressed during the development of the pollen sac and in many plants

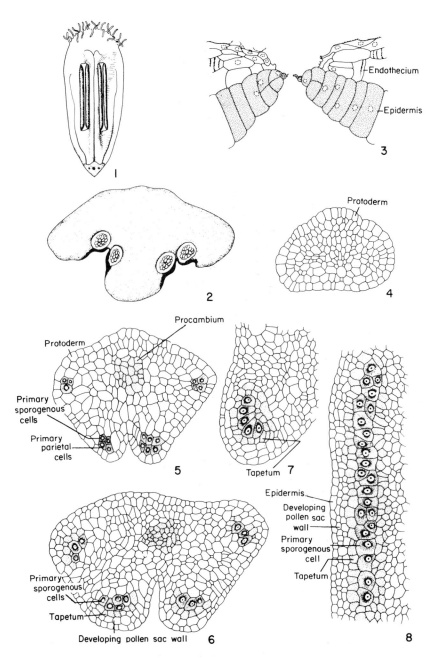

FIG. 243. 1 and 2, Stamen of *Degeneria vitiensis*. 1, Drawing of an entire stamen, showing a pair of sporangia between each of the lateral and median veins. 2, Diagram of a cross-section of the stamen showing the four deeply embedded sporangia and endothecium (shaded black). 3, Portion of a cross-section of the anther of *Lilium* showing the structure of the stomium. 4–8, Development of the pollen sacs of *Vinca rosea*. 4–7, In cross-section. 8, In longitudinal section. (Nos. 1 and 2 adapted from Bailey and Smith, 1942; no. 3 adapted from Esau, 1953; nos. 4–8 adapted from Boke, 1949.)

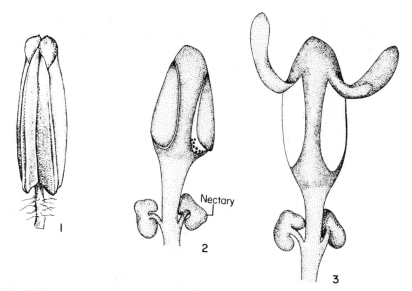

FIG. 244. Types of anther dehiscence differing from the ordinary. 1, Anther of *Solanum villosum* showing the apical pores. 2 and 3, Anther of *Laurus nobilis* showing different stages in dehiscence by lateral valves.

they are obliterated so that it is difficult to distinguish them in mature anthers immediately prior to dehiscence (Fig. 242, no. 2).

The formation of the tapetum takes place as a result of gradual differentiation in the anther wall. In those cases where additional tapetal layers develop, they arise from cell division in other cells of the anther, especially those on the inner side of the sporogenous tissue (Fig. 243, nos. 7, 8). In some cases the sporogenous tissue itself may take part in tapetum formation (Eames, 1961). The tapetal cells are distinctly enlarged, rich in protoplasm and they may be multinucleate or polyploid. Various irregular mitotic divisions and nuclear fusion take place in these cells (Maheshwari, 1950). Two types of tapetum are distinguished —*glandular* or *secretory tapetum* (Fig. 235, no. 2) in which the cells remain in their original position where they later disintegrate and their contents are absorbed by the pollen mother cells and the developing pollen grains, and *amoeboid tapetum* (Fig. 246, no. 1) in which the protoplasts of the tapetal cells penetrate between the pollen mother cells and the developing pollen grains where they fuse among themselves to form a *tapetal periplasmodium*.

Pollen grain development

The primary sporogenous cells commence to divide mitotically, in different planes, together with the development of the pollen sac wall. The derivatives of these divisions are the *pollen mother cells*, which are also known as *microsporocytes*. Each mother cell undergoes a meiotic division to form a *tetrad* of pollen grains, i.e. four haploid microspores. Shortly before the meiotic division, the primary walls of the microsporocyte are replaced by thick layers of callose, but massive protoplasmic strands develop between the cells. Thus the whole mass of pollen mother cells of a pollen sac forms a syncytium. At the conclusion of the meiotic divisions the haploid microspores round up within the tetrad, each wholly enveloped by callose wall and without plasmodesmatal connections with sibs or parent.

According to the orientation of the microspore callose walls in the tetrad, different types of arrangement of pollen grains in the tetrad are distinguished. The arrangement differs in different species and sometimes even within the same species (Fig. 246, no. 2). Apparently the two most common arrangements are the tetrahedral and the

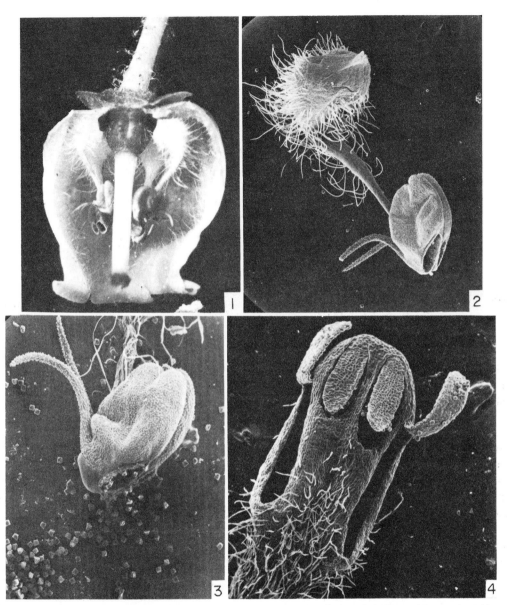

FIG. 245. Stamens with different types of anther dehiscence. 1–3, *Arbutus andrachne*. 1, A flower in natural position cut longitudinally. × 14. 2, A scanning electron micrograph of a stamen, showing an apical pore in each of the two anther lobes. Each lobe is equipped with a horn-like projection. × 44. 3, As in no. 2, but pollen was emitted from the pollen sacs as a result of stirring the horn-like projections (in nature this is due to pollinating insects). × 65. 4, Scanning electron micrograph of an anther of *Persea americana* showing each of the four pollen sacs opened by a raised valve. × 54.

isobilateral ones.

On the basis of the manner of the wall formation accompanying the meiotic division of the pollen mother cell, two types are distinguished: (1) the *successive type* in which each of the nuclear divisions is accompanied by wall formation (Fig. 247, nos. 1–5); (2) the *simultaneous type* in which peripheral constrictions commence to develop only after the four nuclei have been formed, and the wall formation proceeds from these constric-

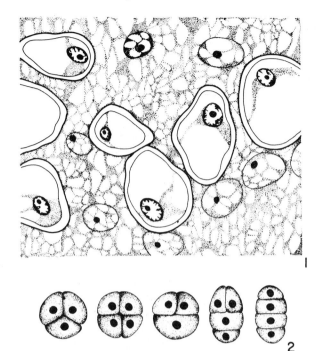

FIG. 246. 1, Diagram of portion of a cross-section of the pollen sac of *Symphoricarpus racemosus* showing an amoeboid tapetum. 2, Different types of arrangement of the pollen grains in the tetrad. From left to right—tetrahedral, isobilateral, decussate, T-shaped, linear. (Adapted from Maheshwari, 1950.)

tions inwards (Fig. 247, nos. 6–9). The simultaneous type is more typical of dicotyledons, while the successive type is typical of many monocotyledons, but this systematic division is not absolute and many exceptions exist. Also there is no correlation between these types of wall formation and the arrangement of the grains in the tetrad.

In most cases the pollen grains of each tetrad separate from one another and they lie freely in the pollen sac. Pollen grains are connected by viscin threads in many Onagraceae species and some Leguminosae (Cruden and Jensen, 1979). In some plants, e.g. the Ericaceae, the pollen grains remain in tetrads even when mature (Fig. 252, no. 4). In certain plants, e.g. *Acacia* and other species of the Mimosoideae the tetrads stuck together in groups, *polyads* (Niezgoda *et al.*, 1983; Feuer *et al.*, 1985). The polyads may contain as many as 64 pollen grains (Fig. 252, nos. 1, 2); these groups are found in separate chambers formed by the development of tranverse partitions in the pollen sacs. In some

plants, e.g. the Asclepiadaceae (Galil and Zeroni, 1969), all the pollen grains of a sac are united in a single compact mass; such a mass is termed a *pollinium* (Fig. 247, no. 10). In Orchidaceae pollinia are also formed, but in certain genera of this family the pollinium is less compact as it comprises smaller groups of pollen grains, i.e. *massulae*, which are loosely jointed among themselves, usually by means of elastoviscin strands (Fig. 247, no. 11) (Cocucci and Jensen, 1969; Schill and Wolter, 1986; Wolter and Schill, 1986; Yeung, 1987; Wolter *et al.*, 1988). In the Asclepiadaceae, the *translator,* which consists of the *adhesive body* and the *caudicles* (Fig. 247, no. 10), is formed by an emulsion-like mixture of lipophilic material and polysaccharides. These substances are secreted by the glandular epithelium of the gynostegium (Schnepf *et al.*, 1979). The caudicles of the Orchidaceae differentiate from the tapetum (Yeung and Law, 1987).

A young pollen grain has a large central vacuole, but with maturation the nucleus enlarges and the cytoplasm becomes denser and increases in amount, so that when the grain is mature the cytoplasm obliterates the vacuole. Mature pollen grains contain large amounts of starch or, in certain species, fatty substances which are apparently absorbed from the tapetum. In many plants the starch disappears from the pollen grains during the ripening of the anther, while in others the starch disintegrates only in the pollen tube. It is assumed that there is a connection between the disintegration of the starch and the high osmotic pressure of the pollen tubes which is higher than that of the cells of the style through which the tube passes.

The chemical analysis of mature pollen grains shows the following composition (McLean and Ivimey-Cook, 1956):

proteins	7.0–26.0%	ash	0.9–5.4%
carbohydrates	24.0–48.0%	water	7.0–16.0%
fats	0.9–14.5%		

A mature pollen grain is surrounded by a thin pectocellulosic wall, the *intine*. Outside the intine is another layer, termed *exine*. The principal component of the exine is *sporopollenin*, a tough substance which gives the remarkable durability to the pollen grain wall. The chemical nature of sporopollenins is somewhat obscure. It has been sug-

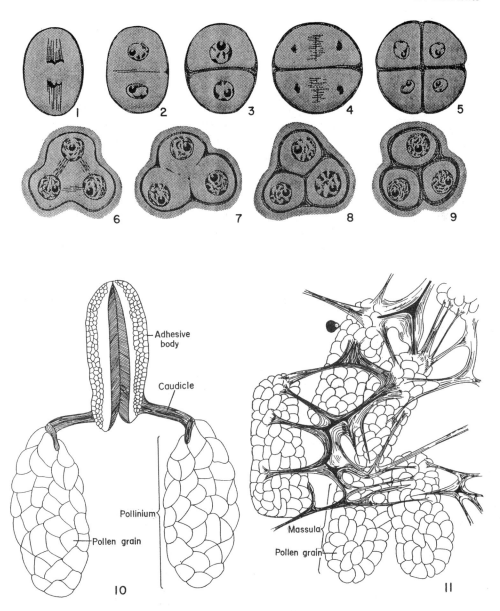

Fig. 247. 1–9, Different types of wall formation in pollen mother cells. 1–5, Successive type in *Zea*. 6–9, Simultaneous type in *Melilotus alba*. 10, Diagram of the paired pollinia of *Asclepias* which are joined by the translator, consisting of two caudicles and the adhesive body. 11, Diagram of a group of massulae from a pollinium of a plant belonging to the Orchidaceae showing the connecting elastoviscin strands. (Nos. 1–9 adapted from Foster and Gifford, 1959; nos. 10 and 11 adapted from Schoenichen, 1922.)

gested that they are oxidative polymers of carotenoids and/or carotenoid esters (Shaw, 1971).

The intine is permeated by proteinaceous material. Heslop-Harrison (1975) suggested that at the time of fertilization enzymes and enzyme precursors are released from the intine of the pol-len-grain apertures. These enzymes degrade the cuticle of the stigma papillae after appropriate activation and enable the pollen tube to enter the style (see also pp. 448–456).

The exine generally consists of an outer sculptured portion, the *sexine*, and an inner portion, the *nexine*. The nexine, which completely covers the intine, usually forms a smooth layer. The sculpturing of the sexine results from radially directed rods, the *bacula*, with enlarged heads. The bacula differ in size and may be either isolated or clustered in groups. In many genera the heads of the bacula are fused to form a roof or *tectum*, which may again be perforated or sculptured in characteristic ways (Fig. 250, nos. 1–8). The formation of air sacs, e.g. in *Pinus*, is brought about by the separation of the sexine from the nexine.

In *Heliconia* of the Musaceae and a few other monocotyledonous families, the exine of the pollen grains is greatly reduced, while the intine is elaborate (Skvarla and Rowley, 1970; Kress *et al.*, 1978).

For details on pollen structure, biology, and physiology see: Heslop-Harrison (1971, 1978), Mascarenhas (1975), Stanley and Linskens (1974), and Frankel and Galun (1977).

Development of the pollen grain wall. While still in the tetrad a new cell wall, the *primexine*, is deposited around the microspore protoplast within the wall of callose. The primexine appears to contain cellulose microfibrils (Heslop-Harrison and Heslop-Harrison, 1982c). In the regions immediately below the future apertures it is often possible to discern a plate of endoplasmic reticulum applied closely to the plasmalemma. It has been suggested that it is possible that this plate prevents the deposition of cellulose at these places by blocking the access of Golgi vesicles carrying cellulose precursors. The transition from primexine to exine is as follows (Fig. 248). Elements of the primexine give rise to the precursors of the rod-like bacula. The bacula, which form the sexine, enlarge and increase in electron density due to rapid deposition of sporopollenin, and their heads expand laterally to form the tectum. In some species it was observed that at this stage a lateral expansion on the bases of the bacula occurs which forms a foot-layer. Further deposition of sporopollenin is continued and the whole pollen grain wall expands radially and laterally as the pollen grain enlarges. The callose wall disappears and the pollen grains lie free in the pollen sac. During the early phases of expansion, the cellulosic substance of the primexine is thinned out, so that only a loosely fibrillar residue remains between the bacula.

The nexine is deposited below the sexine. Initially this process involves a number of very thin electron-transparent lamellae, which appear to arise from the cytoplasm and provide a locus around which sporopollenin is deposited. As the deposition proceeds, the lamellae thicken and merge with each other to form the nexine. No lamellae can be seen in the nexine of mature pollen grains.

In the aperture region the sexine is rather sparsely sculptured and shows very short bacula. The nexine becomes much thicker in the region of the aperture than in the rest of the pollen grain wall but it is discontinuous there. The final stage in development is the formation of the cellulosic intine inside the nexine. This process is associated with an increase in the activity of dictyosomes and randomly oriented microtubules in the peripheral cytoplasm.

Processes occurring in the tapetum. During maturation of the pollen grains structural changes also occur in the cells of the tapetum. At the same time that the tetrad is formed the tapetal cells of the anther enlarge and begin to form specific spherical bodies (*pro-orbicules*). In the late tetrad stage they appear between the plasmalemma and the disappearing wall of the tapetum. Later, at the same time that sporopollenin is deposited on the sexine bacula, it also coats the pro-orbicules forming *orbicules*, which are also called *Ubisch bodies*. In *Helleborus* the orbicules break free into the pollen sacs. In certain plants, e.g. in *Sorghum bicolor*, the orbicules and a reticulum produced by sporopollenin form an "orbicular wall" which coats the inner tangential surface of the tapetum. This wall, together with a fibrillar layer which is formed below it, remains intact even after disintegration of the protoplasts and their dispersion in the pollen sac. The function of the orbicules is not known.

It is not yet clear whether and how the tapetum contributes to the exine formation. Possibly the material from the autolysing tapetum is able to pass into the microspore cytoplasm for subsequent secretion or be deposited and subsequently polymerized *in situ* in the pollen grain wall. Figure 249 shows possible sites of sporopollenin precursors synthesis, polymerization, and

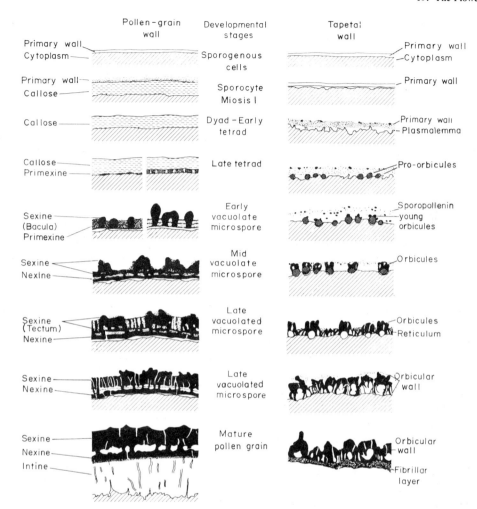

FIG. 248. Diagram showing the formation of pollen-grain wall and tapetal orbicular wall, in *Sorghum bicolor*. (Adapted from Christensen *et al.*, 1972.)

deposition of sporopollenin suggested by Heslop-Harrison and Dickinson (1969).

Pacini and Juniper (1983), who studied the ultrastructure of the ameboid tapetum in *Arum italicum*, did not find orbicule formation in this species. Instead, they observed that when the tapetal plasma membrane begins to retract from the exine, it leaves a roughly "cone-shaped zone" (of spines) which becomes filled with fibrillar material. This material begins to be deposited on the exine surface. The "cone-shaped zone" was found to contain polysaccharides, proteins and sporopollenin.

The deposits of the sticky lipoid material (*pollenkitt* or *pollen-coat*), often yellow or orange in colour, which covers the mature pollen grains of most insect-pollinated plants, apparently originates from the tapetum and not from the developing microspore (Skvarla and Larson, 1966; Rowley and Southworth, 1967; Echlin, 1968; Heslop-Harrison, 1968a, b; Echlin and Godwin, 1969; Mepham and Lane, 1970; Shaw, 1971; Christensen *et al.*, 1972; Knox, 1984; Fernandez and Rodriguez-Garcia, 1988; Chen *et al.*, 1988).

In the Apocynaceae, a glandular epithel, located on the cylindrical part of the stigma, secretes an adhesive which glues the pollen grains to the mouth parts of the pollinating insect (Schick, 1980, 1982).

Characteristics of mature pollen grains. The re-

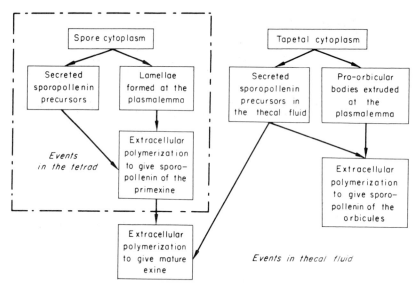

FIG. 249. Scheme for deposition of sporopollenin in the orbicules and exine, based upon the evidence from *Lilium* and other genera with a tapetum of the parietal, secretory type. (From Heslop-Harrison and Dickinson, 1969.)

ticulate, striped, or other patterns of sculpturing, which are visible on the surface of many pollen grains, result from the particular arrangement of the bacula (Fig. 250, nos. 1–8). The lipoid material (pollenkitt) occurring on the exine of many plants, apart from the spines and other appendages of the grain wall, aids the adhesion of the pollen to pollinating insects.

In addition to the different sculpturings on the surface of pollen grains, which result from the structure of the sexine, there are also many other morphological characteristics that are used in classification of the pollen (Figs. 251–254). From pollen analysis it is possible to determine the species of plants occurring in a certain area as well as the existence of plant species, of which no other relic has been preserved, in early geological times. Quantitative pollen analysis made on recent geological layers has greatly helped in the under-standing of the history of floras in many parts of the world. Because pollen grains are one of the most important causes of allergic diseases, especially those of the respiratory tract, much attention has been paid to the analysis of pollen present in the atmosphere in the different seasons. Much importance is also attached to the determination of the pollen collected by bees in research on honey plants. As a result of this wide range of

research, based on the variability of pollen grains, a very detailed nomenclature of the structural characteristics of pollen has been developed which enables exact morphological descriptions to be made of different pollen grains (Wodehouse, 1935; Erdtman, 1952).

Oval pollen grains are commoner among monocotyledons than among dicotyledons, but this is not a feature distinguishing between these two groups. In the monocotyledons the pollen grains of a single tetrad are usually arranged in one plane, while in dicotyledons the arrangement is usually tetrahedral. Pollen grains that are arranged in one plane are somewhat boat-shaped, and are bilaterally symmetrical. Monocotyledonous pollen grains usually have one aperture, and those of dicotyledons usually have three. In many types of pollen the pattern of pores and furrows bears a close relation to the contact geometry of the microspores during their association in the tetrad (Wodehouse, 1935).

Although these characteristics are generally reliable for distinguishing between pollen of mono- and dicotyledons, there are, however, exceptions. In dicotyledons, pollen grains with a single aperture are found in the Piperaceae and in woody species of different ranalian families. In the Nymphaeaceae there are both genera whose pollen

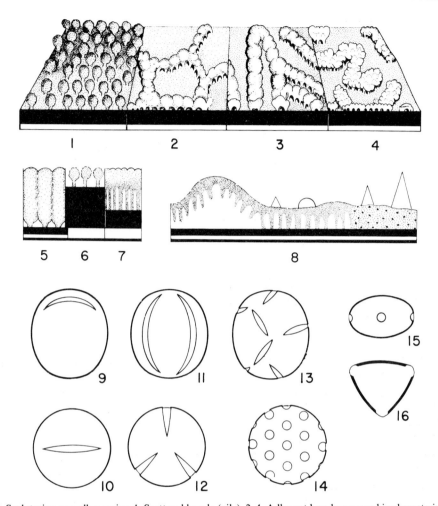

FIG. 250. 1–8, Sculpturing on pollen grains. 1, Scattered bacula (pila). 2–4, Adherent bacula arranged in characteristic reticulate patterns or in rows. 5, Cross-section of a pollen grain wall with a thick sexine. 6, As in no. 5, but with thin sexine. 7, Cross-section of pollen grain wall on which the heads of the bacula are fused to form a tectum. 8, Different types of bacula fusion, and the resulting types of structure. On the left, disappearance of the bacula bases; on the right, development of spines, etc. 9–16, Arrangement and types of apertures. 9 and 10, Unisulcate pollen grain. 9, Lateral view. 10, Polar view. 11 and 12, Tricolpate pollen grain. 11, Lateral view. 12, Polar view. 13, Rugate pollen grain. 14, Porate pollen grain. 15 and 16, Pollen grain with three pores. 15, Lateral view. 16, Polar view. (Nos. 1–8 adapted from Erdtman, 1952.)

grains possess one aperture and those that have three (Bailey and Nast, 1943a, b). Pollen grains with more than three apertures also exist.

Erdtman (1952) distinguishes between different types of aperture, four of which are given below.

1. *Sulcus*: an elongated furrow perpendicular to the longitudinal axis, at the pole of the grain (Fig. 250, nos. 9, 10).
2. *Colpa*: an elongated furrow at right angles to the equatorial plane; the ends of the furrow

are directed towards the poles of the grain (Fig. 250, nos. 11, 12).
3. *Ruga*: an elongated furrow the direction of which differs from both of the above types (Fig. 250, no. 13).
4. *Porus*: a circular aperture. When the number of pori is small they occur only in the equatorial region, but if they are numerous they may occur all over the surface of the grain (Fig. 250, nos. 14–16).

A circular pore may sometimes be present in

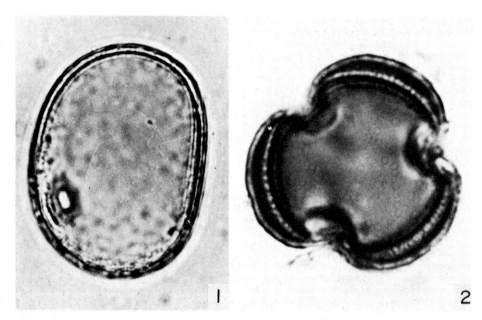

FIG. 251. Micrographs of pollen grains. 1, *Hordeum spontaneum*, showing smooth surface of the exine and a single aperture. × 1200. 2, *Artemisia* sp. polar view of grain with three apertures. × 920.

the middle of an elongated furrow (e.g. in *Nicotiana tabacum* and in *Centaurea*).

In *Zostera*, which is marine, the pollen grains are thread-like and lack sporopollenin. This feature is apparently connected with the hydrophilous mode of pollination. Pollen grains of most plants that are of typically wind-pollinated families are smooth and dry. If most of the genera of a family are insect pollinated and only certain genera of it are wind pollinated, as, for example, the genera *Artemisia* and *Ambrosia* of the Compositae, the wind-pollinated genera retain the sculptured structure typical of the entire family, but it may be developed to a lesser extent.

The size of pollen grains also varies very greatly. Erdtman classifies them, according to size, into the following groups: *perminuta*, in which the diameter is less than 10 μm; *minuta*, in which the diameter is 10–25 μm; *media*, 25–50 μm; *magna*, 50–100 μm; *permagna*, 100–200 μm; *giganta*, the diameter of which is greater than 200 μm. Very small grains may be found in *Myosotis alpestris* (2.5–3.5 μm) and *Echium vulgare* (10–14 μm); very large grains occur in *Cucurbita pepo* (230 μm) and *Mirabilis jalapa* (250 μm).

THE CARPEL

Phylogeny of the carpel

Several views have been expressed concerning the homology of the carpels. It has been suggested that the carpel is of axial nature, that is, the flower may be interpreted as a system of branches. According to Wilson (1942) the carpel, like the stamen, has developed from fertile telomes. In this case the sporangia-bearing telomes fused to form a leaf-like organ which bore ovules on its margins. The involution of the margins of this hypothetical organ represents the final stage in the development of the present-day ovary which encloses the ovules.

There are some theories which are based on the independence of origin of the placenta and carpel. The *gonophyll theory*, as proposed by Melville (1961, 1962), is an example. According to him the ovary consists of sterile leaves and ovule-bearing branches that are usually epiphyllous to the leaves. Each leaf, together with the fertile branch, is considered as a unit and is termed a *gonophyll* instead of carpel. (This theory is also applied to the stamen where the basic unit is termed an

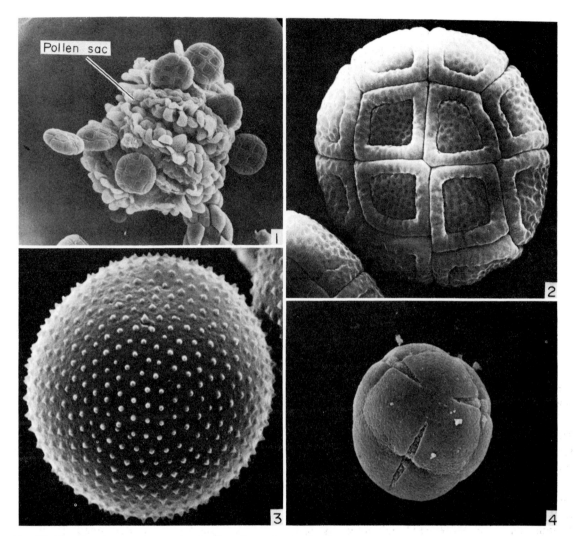

FIG. 252. Scanning electron micrographs of various dispersal units of pollen grains. 1 and 2, Dispersal units (polyads) of *Acacia cultriformis* consisting of many pollen grains. 1, Each polyad protruding from a separate chamber of the subdivided pollen sacs. × 360. 2, An enlarged polyad. × 1800. 3, A single pollen grain of *Persea americana*. × 3400. 4, A tetrad of *Arbutus andrachne* which forms a single dispersal unit. × 2000.

androphyll.) In a number of families it is supposed that the ovule-bearing branches have become ebracteate and then the ovary consists of sterile gonophylls, i.e. *tegophylls*, which alternate with the ebracteate ovuliferous branches. However, the anatomical arguments used by Melville against the classical theory of the foliar origin of the carpel are not convincing (Corner, 1963; Tucker, 1966).

Another concept is the *sui generis concept* by which it is claimed that the stamens and the car-

pels are neither homologous with leaves nor leaf-like.

According to Meeuse (1966) the angiospermous flower may be interpreted on the basis of one general unifying concept which assumes that the ovules are borne on axes or their homologues and not on leaf homologues.

The most generally accepted view is that the carpel is homologous with the leaf, and that the ovules develop on it.

Sattler and Lacroix (1988) who studied the onto-

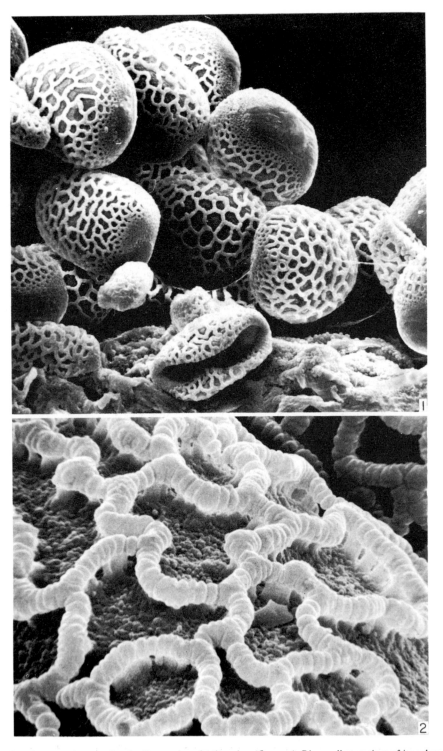

FIG. 253. Scanning electron micrographs of pollen grains of *Lilium longiflorum*. 1, Ripe pollen at time of its release from the anther. ×660. 2, Reticulately patterned wall of a pollen grain. Some of the outer amorphous coating has been removed to reveal the intricate structure. ×4500. (Courtesy of J. Heslop-Harrison.)

FIG. 254. Scanning electron micrographs. 1, *Cosmos bipinnatus*; mature pollen grain ethanol and ether washed. ×3100. 2, *Hibiscus rosa-sinensis*; stigmatic surface with pollen grains. ×160. (Courtesy of J. Heslop-Harrison.)

geny of the gynoecium of *Basella rubra*, one of the species with basal placentation, suggested three possible ways for the evolution of this type of placentation in the Angiospermae: (1) It is derived from the carpellate condition. (2) It is primitive and the carpellate condition is derived. (3) Both, carpellate and noncarpellate organization have co-existed during the evolution.

Several interpretations of the foliar origin of the carpels have been given. Troll and his co-workers, on the basis of the structure and ontogeny of carpels, such as those seen in *Thalictrum* (described earlier in this chapter), have put forward the *peltate carpel theory*. According to this theory the carpel is basically a peltate organ (Puri, 1960; Eames, 1961).

According to the classical interpretation, the carpel is derived from a fertile leaf, the margins of which bore ovules. The margins became involuted and fused between themselves or with the margins of other carpels. In this way the ovules became enclosed within the locule (de Candolle, 1819; Brown, 1826; Joshi, 1947; Puri, 1960; Lorch, 1963).

A different view concerning the original position of the ovules on the carpel was suggested by Bailey and Swamy who investigated some primitive woody genera of the Ranales. The carpel of *Degeneria* (Degeneriaceae) and *Drimys* (Winteraceae), for instance, differs in many respects from the typical carpel of the angiosperms as no closed ovary, style or stigma can be distinguished. In the early ontogenetic stages the carpel of *Degeneria* can be seen, in cross-section, to be a folded organ of which the margins flare outwards and remain unfused for a long time. This type of carpel is termed *conduplicate* (Fig. 255, nos. 1–3). Swamy (1949b) observed on the flared margins and extending into the locule, even beyond the attachment of the ovules, an excessive development of hairs. The cleft between the closely appressed margins is filled with interlocking hairs. On the inner side of the carpel the trichomes are small and papilla-like. According to Swamy, all the hairs together form the surface area of the stigma. When pollination takes place the pollen grains become attached to the hairy, outwardly flared margins of the open carpel. Swamy also observed that the germination tubes of the pollen grains grow into the carpel locule between these hairs

and that they never penetrate the carpel tissue. In the early stage of carpel development the two rows of ovules are very remote from the true margins. The vascular supply to the ovules may be derived from the ventral or dorsal bundles, or from both. After pollination and fertilization, the inner, appressed surfaces of the carpel, which are on the outside of the ovules, fuse, and the outwardly flared margins remain as suberized ribs on the mature fruit.

In *Drimys* (Fig. 255, nos. 1–3), like *Degeneria*, the carpels are also folded in half and they have widened margins with stigma-like characteristics. Bailey and Swamy (1951) found that the young carpel has a stalk, and the blade-like portion of the carpel encloses two rows of ovules which are situated on the inner adaxial surface a great distance from the margins (Fig. 255, nos. 2, 3). When such a mature carpel is cleared and unfolded, it can be seen that the ovules, which are situated between the dorsal and ventral bundles, are supplied with traces from both these bundles (Fig. 255, no. 3). According to Tucker (1975) at least in some species of *Drimys* the ventral bundles provide the major supply to the ovules. Here also the pollen grains reach the hairy margins of the carpels and the pollen tubes penetrate to the ovules by growing between the hairs on the inner surface of the carpels (Fig. 255, no. 2). A structure somewhat similar to this has been found in the genus *Cananga* of the Annonaceae (Periasamy and Swamy, 1956).

From this type of conduplicate carpel with flared margins, the present-day carpel has developed, according to Bailey, by closure and the concentration of the stigmatic margins to the upper part of the carpel only. According to Bailey, the loss of the marginal portions outside of the ovules has resulted in the impression that the carpel margins are involuted and that the placentae are marginal (Fig. 255, nos. 4–7). During the course of evolution the number of ovules was reduced, and they remained only on the lower portion of the carpel while the upper part underwent differentiation to form a style and stigma. Van Heel (1983), on the basis of his studies of the carpel ontogeny in species of the Winteraceae and some other families, suggested tha the angiosperm carpels may have evolved by allometric development of an oblique

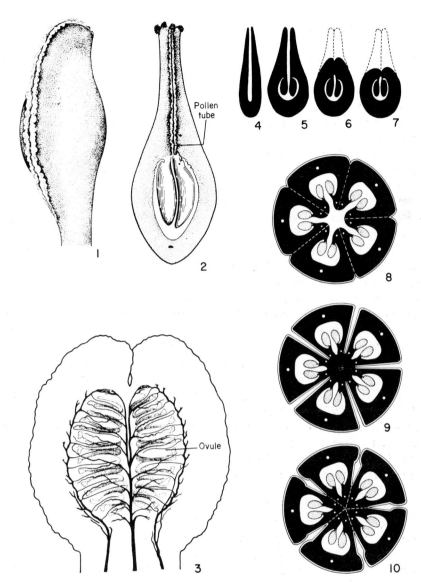

FIG. 255. 1–3, Conduplicate carpel of *Drimys piperita*. 1, Lateral view showing the stigmatic margins. 2, Cross-section showing the position of the ovules and the penetration of a pollen tube. 3, Cleared, unfolded carpel showing the general pattern of the vascularization and the source of the ovular supply. 4–7, Diagrammatic representation of the development of the present-day carpel from the conduplicate carpel. 8–10, Diagrams of cross-sections of syncarpous gynoecia showing the different ways in which fusion may take place. 8, Lateral cohesion in a whorl of open conduplicate carpels. 9, Adnation of the free margins of the conduplicate carpel to the torus. 10, Cohesion of ventral surfaces of the carpels. (Adapted from Bailey and Swamy, 1951.)

asciform structure, which led to the formation of a large ovary at the base, leaving the stigmatic part apical. Puri (1960, 1962) disputes the assertion that *Drimys* ovules are laminar rather than marginal in position and defends the classical concept of ventral involution of the carpels and the fusion of the ovule-bearing margins. However, ontogenetic-anatomical studies of Tucker and Gifford (1966) show that the margin of developing carpels of *Drimys* can be defined precisely by the location of the marginal and submarginal meristems. The coherent or free carpels of the Nymphaeceae are also interpreted as being conduplicately closed (Moseley, 1965; Richardson, 1969).

The evolution of the angiosperm gynoecium has also involved the fusion between two or more carpels of a single flower. This fusion has taken place in various ways (Murray, 1945; Baum, 1948a, b, c; Leinfellner, 1950; Bailey and Swamy, 1951). The margins of the carpels may have fused to the receptacle (Fig. 255, no. 9), or they may be fused to one another along their ventral parts (Fig. 255, no. 10), or along their lateral parts. In the latter case the carpels may remain open to form a unilocular ovary (Fig. 255, no. 8). In the case of the fusion of the margins of the carpels in the centre of the ovary, the number of locules formed is equal to the number of carpels (Fig. 255, nos. 9, 10). The carpels may fuse during their ontogeny or they may be fused already at their conception (see above "Ontogeny of the flower").

There are angiosperm flowers in which the structure of the ovary differs from those described above. In the Cruciferae, for instance, the ovary is divided into locules other than by the folding of the carpels. The placenta may develop on a central column which is not joined to the wall of the ovary throughout its length (*free-central placentation* as is seen, for example, in the Primulaceae), or in a unilocular ovary the placenta may develop only at the base of the ovary (*basal placentation* as is seen, for example, in the Compositae). The fusion of the carpels may not always take place along their entire length.

In an apocarpous gynoecium each carpel has a single style. In a syncarpous gynoecium the styles may be fused to different extents (Baum, 1948d). In certain plants, e.g. in the Hypericaceae, the carpels are fused only at the base and the styles are free or almost so. In highly specialized flowers, e.g. in the Solanaceae and Oleaceae, the styles and stigmas are completely fused.

Carr and Carr (1961) distinguished three types of gynoecia: apocarpous, pseudo-syncarpous, and eu-syncarpous. The distinction between the two syncarpous types is based on the paths of pollen tubes from the stigma to the ovules. In the pseudo-syncarpous gynoecium the carpels are fused to form a single structure, but in respect of the path of the pollen tube they are functionally apocarpous. In the eu-syncarpous gynoecium pollen tubes from any part of the stigma can reach ovules of any carpel, even in multilocular ovaries.

Histology of the carpel

At the time of *anthesis*, i.e. the maturation of the anthers and ovules, or prior to it, only slight histological differentiation is observable in the ovary wall. It then consists mainly of parenchyma and vascular tissues and is covered by a cuticle-bearing epidermis. As the ovary develops into a fruit, striking histological changes take place in the ovary wall (see Chapter 20).

The stigma and style have special structures and physiological characteristics that enable the pollen grains to germinate on the stigma and the pollen tube to penetrate to the ovules. The protoderm of the stigma differentiates into a glandular epidermis, the cells of which are rich in protoplasm. This epidermis is usually papillate and covered with a cuticle (Schnarf, 1928). Sometimes, e.g. in *Olea europaea*, cell layers below the epidermis form a secretory zone (Ciampolini *et al.*, 1983).

At the the receptive state the stigma surface may be "dry", without free-flowing secretion fluid, but with a hydrated proteinaceous pellicle, or its surface may be "wet", i.e. with a free-flowing secretion fluid (Fig. 258). There are also intermediate categories between the two types mentioned above. Some stigmas bear small droplets of secreted material which are scattered over their surface. In others the secreted fluid is held in subcuticular spaces (Heslop-Harrison, 1981).

The stigmatic fluid, such as in *Petunia*, consists primarily of oil, sugars, and amino acids (Konar and Linskens, 1966a, b). In the stigmatic exudate of *Aptenia cordifolia* Kristen *et al.* (1979) found, in addition to protein and polysaccharides, a high proportion of mono- and oligo-saccharides. In *Nicotiana tabacum* fatty acids and phenols were also found to be present in the stigmatic exudate. The ultrastructure of the stigma and the involvement of various cell organelles in the production of the secreted materials has been studied in a number of plant species (Cresti *et al.*, 1982, 1986; Kandasamy and Kristen, 1987; Baird *et al.*, 1988). The protein component of the stigmatic exudate,

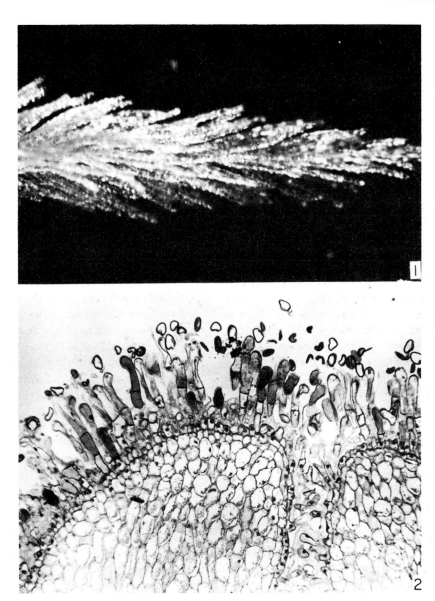

FIG. 256. 1, Photograph of portion of the feathery stigma of *Avena*. × 20. 2, Micrograph of a longitudinal section of the stigma of *Lilium* showing the multicellular hairs on its surface. × 60.

together with proteins released by the pollen grains (see p. 437, apparently play an important role in the process of pollen-stigma interaction (Shivanna *et al.*, 1978; Kristen *et al.*, 1979). An extracuticular protein layer occurring on the stigma papillae of many angiosperms has been suggested as the site of pollen recognition in incompatibility reactions (Mattison *et al.*, 1974). Dumas and Gaude (1982) suggested that the recognition macromolecules may be characterized by a dual structure, a constant proteic part and a variable (sugar) part.

The cuticles of the receptive parts of the stigma are adapted to enable the stigma to interact with the pollen grains. They should be permeable to a greater or lesser extent in order not to present a too great mechanical barrier. The cuticles on mature stigmas of different plants show considerable variations in structure. In some cases the cuticle is torn

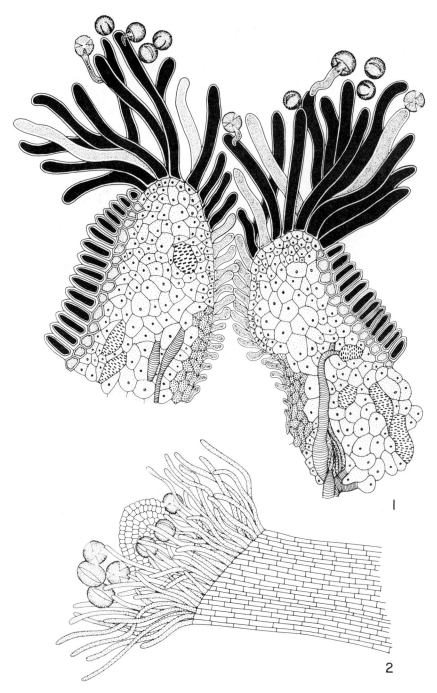

Fig. 257. 1, Diagram of a longitudinal section of the stigma of *Papaver rhoeas* in which the germination of pollen grains among the unicellular hairs on the surface of the stigma can be seen. 2, Tip of the style of *Lupinus luteus* showing pollen grains among the hairs on the stigma. (Adapted from Schoenichen, 1922.)

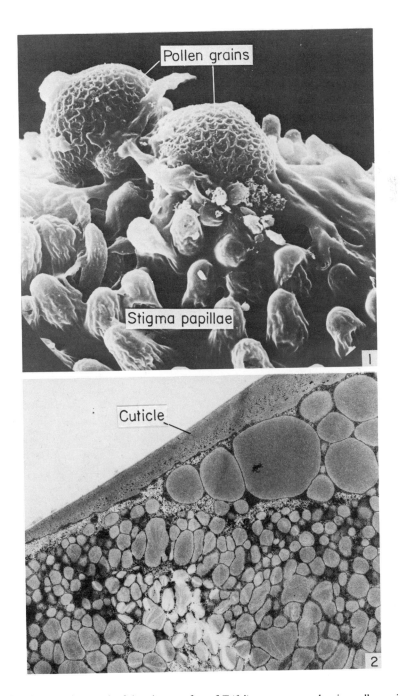

Fig. 258. 1, Scanning electron micrograph of the stigma surface of *Trifolium purpureum*, showing pollen grains immersed in the stigmatic exudate. × 1200. 2, Transmission electron micrograph of the top of a stigma papilla of *Vicia palaestina*, showing stigmatic exudate which has accumulated below the cuticle. The exudate occurs in the form of droplets as well as in the form of granular and fibrillar material. × 9000. (Courtesy of Gila Alon; from her PhD thesis, 1986.)

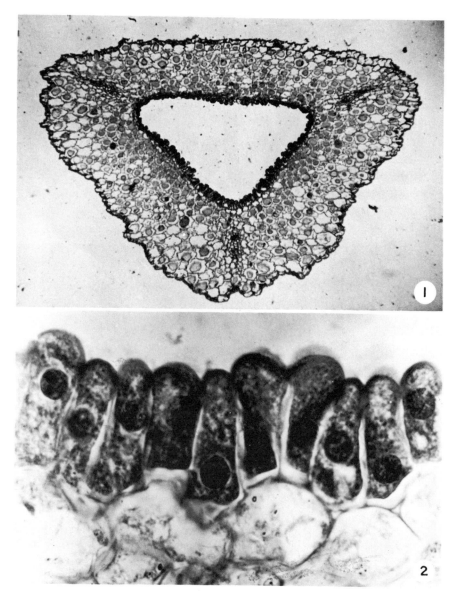

FIG. 259. 1, Micrograph of a cross-section of the style of *Lilium* in which the glandular epidermis lining the central canal can be seen. ×45. 2, Portion of the glandular epidermis, enlarged. ×640.

during the final expansion of the receptive cells (Heslop-Harrison and Heslop-Harrison, 1982a). In many plants (e.g. *Phaseolus, Hibiscus, Lilium, Papaver*, and *Lupinus*) the epidermal cells of the stigma develop into short dense hairs (Fig. 254, no. 2; Fig. 256, no. 2; Fig. 257), or they may develop into long, branched hairs, e.g. the Gramineae and other wind-pollinated plants (Fig. 256, no. 1).

Between the tissue of the stigma and the ovary there is a specialized tissue through which the germinating pollen tube penetrates. This tissue provides a nutrient substrate which aids the pollen tube to grow through the style into the ovary. This tissue was termed *transmitting tissue* by Arber (1937), and this is the term that will be used here; however, this tissue is also known by

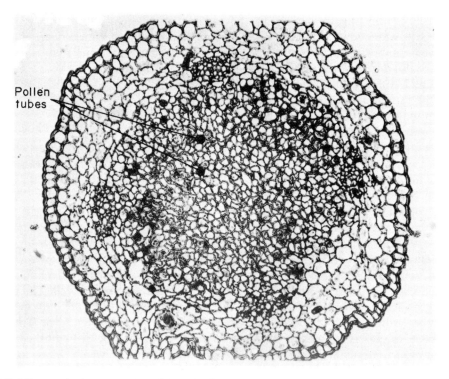

Pollen
tubes

FIG. 260. Micrograph of a cross-section of the style of *Oenothera drummondii* in which the pollen tubes can be seen between the cells of the central portion of the style (transmitting tissue). × 175.

various other terms. As has already been mentioned, in the most primitive dicotyledons (e.g. in ranalian families such as the Winteraceae and Degeneriaceae) the carpels do not develop styles and the pollen tubes reach the ovules by growing through the hairs present on the unfused margins of the carpels. In the phylogenetically more advanced forms the carpel margins fuse, a style is developed, and the stigmatic tissue is reduced to the upper portion of the style only. This tissue, however, remains in contact with the placenta by the transmitting tissue, which has features similar to those of the stigma. The style may be hollow or solid, depending on the degree of closure of the fused or free carpels. The hollow style of a syncarpous gynoecium may contain a single canal, or several canals, the number of which is equal to that of the carpels. The canals are lined entirely or in longitudinal strips by glandular transmitting tissue, which may be papillose (Fig. 259). In the canal cells of the transmitting tissue are covered by cuticle. The canal, for instance in the Liliaceae,

may be filled with a mucilaginous material rich in polysaccharides.

The study of the fine structure of the secretory cells which line the canal of *Lilium longiflorum* styles revealed that they are rich in organelles and contain abundant multivesicular bodies (paramural bodies). On the inner side of the wall facing the canal there is a very thick layer of elaborated wall protuberances (Rosen and Thomas, 1970; Dashek *et al.*, 1971; Gawlik, 1984). Similar protuberances occur in many other secretory cells (see nectaries). In many plants (e.g. *Cucurbita* and *Datura*) the transmitting tissue is several cell layers thick. It also covers the placenta and in certain species it is even present on the funiculus. In some plants the transmitting tissue is brought closer to the micropyle by the development of outgrowths of the funiculus, placenta or stylar canal (Tilton *et al.*, 1984b); these outgrowths have been termed *obturators* (Fig. 263, no. 1) (Schnarf, 1928). Ontogenetic research on the styles of *Cucurbita* and *Datura* has shown that the multiseriate transmitting tissue and the multiseriate glandular tissue of the stigma develop from the epidermal cells by periclinal division (Kirkwood, 1906; Satina, 1944). In most angiosperms the style is solid (Fig. 260), and then the transmitting tissue constitutes strands of elongated cells rich in cytoplasm. Electron microscope studies on the transmitting tissue of cotton styles show that its cells have few vacuoles, are rich in starch, ribosomes, endoplasmic reticulum, and dictyosomes. Druses of calcium salts may also occur. These cells exhibit thick lateral walls and relatively thin transverse walls. The thick lateral walls consist of several layers: (1) an innermost layer composed primarily of pectic substances and hemicellulose (2) outside it a thinner darker-appearing layer, richer in hemicellulose; (3) still further to the outside a thick loose-textured wall, showing concentric rings of fibrous material, relatively poor in hemicelluloses, but rich in pectic substances and also containing cellulose and non-cellulosic polysaccharides; (4) a thick middle lamella consisting primarily of pectic substances. The middle lamella and the outermost wall layer apparently contain protein. In the wall, especially in the last-mentioned layer, masses of small vesicles occur. The pollen tubes grow through the outermost wall layer (Jensen and Fisher, 1969). Cresti *et al.* (1976) and Ciampolini *et al.* (1978), who studied the ultrastructure of the transmitting tissue in *Lycopersicum peruvianum* and *Malus communis*, stated that the intercellular substance is not solid but rather a viscous fluid. In *Lycopersicum* at the early stage of style development only polysaccharides, apparently produced by dictyosomes, are secreted into the intercellular spaces, whereas at a later stage of development protein secretion occurs. At this later stage numerous rough ER cisternae and polyribosomes are present in the cells of the transmitting tissue.

The proteins present in the intercellular substance, besides being of nutritional value, may also play some role in incompatibility reactions (Linskens, 1975; Cresti *et al.,* 1976), which, as mentioned above, apparently involve the extra-cuticular protein of the stigma.

In a syncarpous gynoecium with a single, solid style, several strands of transmitting tissue develop and these are connected to the different placentae of the ovary.

Different opinions have been expressed as to the factors directing the growth of the germinating pollen tube. There are investigators who suggest that a chemotactic attraction exists between the pollen tube and the tissues of the stigma, transmitting tissue and ovule (Rosen, 1964, Welk *et al.*, 1965, Tilton and Horner, 1980). According to other workers the very structure and arrangement of the transmitting tissue in the style direct the growth of the pollen tube (Schnarf, 1928; Renner and Preuss-Herzog, 1943; Jensen and Fisher, 1969). Heslop-Harrison and Heslop-Harrison (1986) although supporting this view, do not exclude the possibility that at the early stage of growth, the pollen tube may respond chemitropically in locating the transmitting tissue and in locating the micropyle and nucellus.

The average rate of growth of the pollen tube differs in the various plant species; e.g. in apple 0.35 mm per hour, in *Lilium*-0.9 mm, in *Strelitzia* 1.8 mm, in *Secale* 5.4 mm, and in *Oenothera* 6.5 mm (Kronestedt *et al.*, 1986).

In hollow styles the tubes of the germinating pollen grains grow between the papillae of the transmitting tissue, and, if they are absent, on the outer surfaces of the epidermal cells. In many plants the cuticle on the transmitting tissue disappears before pollination and the walls of the glandular tissue soften and swell. The pollen tubes may sometimes penetrate deeper into the transmitting tissue and grow between the cells. The narrow stylar canal of *Citrus limon* is filled, already before anthesis, with a material consisting of polysaccharides, proteins and lipids which are secreted by the canal cells (Ciampolini *et al.*, 1981). In solid styles the pollen tubes grow between the cells of the transmitting tissue. In grasses the pollen tube may even grow between the cells already in the stigma. On the stigma of grasses there are large multicellular hairs, consisting of several longitudinal rows of cells. The pollen tube penetrates between the inner cell rows of these hairs and from there to the transmitting tissue of the style. In the ovary the pollen tube penetrates via the transmitting tissue, which lines the ovary wall and the placenta, and eventually it reaches the ovule (Pope, 1946; Kiesselbach, 1949). Before the pollen tubes penetrate the transmitting tissue the walls of its cells swell, so that the tissue appears collenchymatous

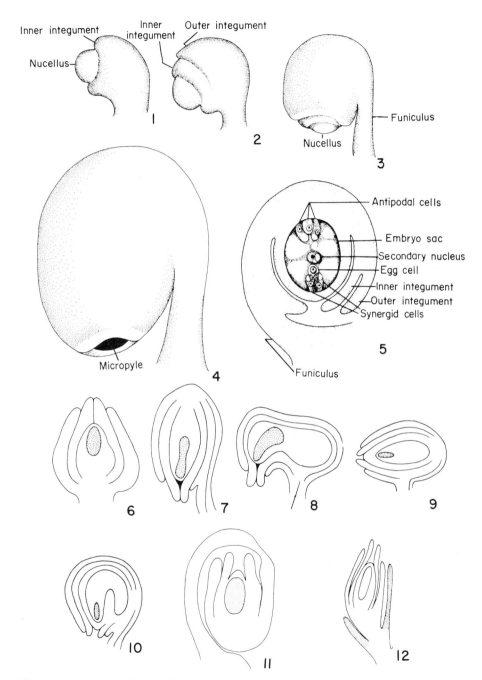

Fig. 261. 1–4, Stages in the ontogeny of an ovule. 5, Diagram of a longitudinal section of an ovule. 6–11, Different types of ovules. Embryo sacs stippled. 6, Atropous. 7, Anatropous. 8, Campylotropous. 9, Hemianatropous. 10, Amphitropous. 11, Circinotropous. 12, Longitudinal section of the ovule of *Asphodelus fistulosus* showing the development of the aril which is considered to be a third integument; aril stippled. (No. 5 adapted from Haupt, 1953; nos. 6–12 adapted from Maheshwari, 1950.)

with mucilaginous walls and the connections between the cells weaken. As a result of these changes it is easy to macerate the transmitting tissue at this stage of development. The pollen tubes pass through the swollen, mucilaginous parts of the walls which they apparently digest (Schoch-Bodmer and Huber, 1947). Proofs have also been given that pollen tubes contain enzymes that are capable of disintegrating pectic substances (Paton, 1921). The protoplasts of the transmitting tissue may also be utilized by the developing pollen tube, but in many plants they may contract and die. As a result of this the style does not increase in width even when it contains many pollen tubes. In some plants, e.g. *Trifolium pratense*, a lysigenously formed canal is formed in the maturing style (Heslop-Harrison and Heslop-Harrison, 1982b).

Apart from the transmitting tissue and vascular bundles, the style consists of thin-walled parenchyma and a typical cuticle-covered outer epidermis, in which stomata may sometimes be found.

THE OVULE

The ovule consists of the *nucellus* which is surrounded by one or two *integuments*, and it is attached to the placenta by a stalk, i.e. the funiculus. At the free end of the ovule a small gap is left by the integuments; this opening is termed the *micropyle*. The region where the integuments fuse with the funiculus is termed the *chalaza*. A nucellar cell, usually one of those below the outermost layer at the micropylar end, differentiates into the *macro-* or *megaspore mother cell*. The nucellus is, therefore, considered to be the *megasporangium*.

Ovules may be of different form. The following two main types may be distinguished: (1) *orthotropous* or *atropous* in which the nucellar apex is in a straight line with the funiculus and is continuous with it; (2) *anatropous* ovule in which the apex of the nucellus is directed backwards towards the base of the funiculus (Fig. 261, nos. 6, 7). Between these two extreme forms there are different intermediate forms in which the ovule axis is variously bent (Fig. 261, nos. 8–10). A detailed terminology has been developed for all these forms, i.e. *hemianatropous, campylotropous,* and *amphitropous* (Schnarf, 1927; Maheshwari,

1950). In the Plumbaginaceae, *Opuntia*, and some other genera of the Cactaceae, the funiculus is very long and it surrounds the ovule; this type of ovule is termed *circinotropous* (Fig. 261, no. 11).

The ovules develop from the placentae of the ovary. The ovule primordium originates by periclinal division of cells below the surface layer of the placenta. At first the primordium appears as a conical projection with a rounded tip. The first sporogenous cell is already distinguishable in the primordial nucellus in that it is larger than the neighbouring cells, and it has a larger nucleus and denser cytoplasm. The inner integument begins to develop some distance from the nucellar apex. The initiation of this integument takes place by periclinal divisions in the protoderm (subdermal initiation of the inner integument occurs very rarely). At first the integument appears as an annular ridge and later it grows towards the nucellar apex and so covers the nucellus except for the micropyle left at the free end of the ovule (Fig. 251, nos. 1–4). The initiation of the outer integument takes place by periclinal divisions of either the subdermal or dermal layer a little lower than the initiation of the inner integument. The development of the two integuments is similar. A subdermal initiation of the outer integument is rather common, while a dermal initiation is known, for example, from most families of the Centrospermae, Parietales, Capparidales, and Geraniales (Bouman and Calis, 1977). In many plants the outer integument does not reach the micropyle. In anatropous and bent ovules the growth of the integuments is asymmetric. In plants with sympetalous flowers the nucellus is usually enveloped by a single integument, while in more primitive dicotyledons and in many monocotyledons the ovule has two integuments. In some species the distal rim of one of the integuments is found to be lobed, e.g. the inner integument of *Hernandia peltata* (Heel, 1971).

There are two main possible ways in which the unitegmy (one integument) developed from the bitegmy (two integuments): (1) reduction of one of the integuments; (2) fusion of the integument primordia; (3) a process of integumentary shifting of one integument so it is formed over the other (Bouman and Calis, 1977; Bouman and Schier, 1979).

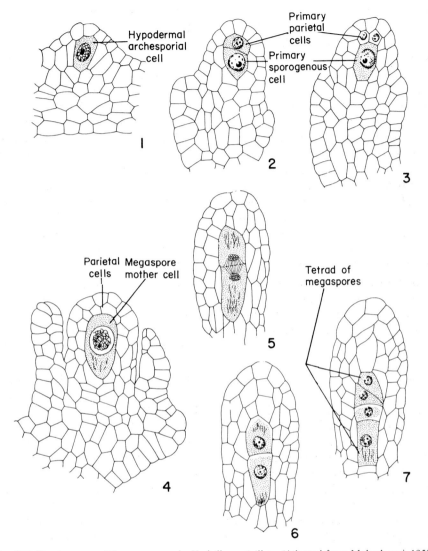

FIG. 262. Development of the megaspore in *Hydrilla verticillata*. (Adapted from Maheshwari, 1950.)

The nucellus is usually considered to be the megasporangium, but the homology of the integuments is still an unsolved problem. At the chalaza there is no differentiation between the tissues of the integuments and the funiculus.

In certain plants the structure of the ovules differs from that described above. There are ovules that lack integuments and those in which the number of integuments is greater than two. In certain plants, e.g. species of *Asphodelus*, a third integument develops from the base of the ovule; this structure is termed *aril* (Fig. 261, no. 12). The

nucellus may be fused entirely to the integuments. In some ovules the integuments grow more than usual and may even close the micropyle, while in others the integuments do not reach the nucellar tip. (See also Chapter 21.)

The thickness of the nucellus in a mature ovule differs in various plants. It may be very thin—one to two cell layers surrounding the embryo sac—e.g. in *Quinchamalium chilense* of the Santalaceae (Johri and Agarwal, 1965), or it may consist of numerous cell layers, e.g. in *Pistacia* of the Anacardiaceae (Fig. 263, no. 1) and *Dysphania* of

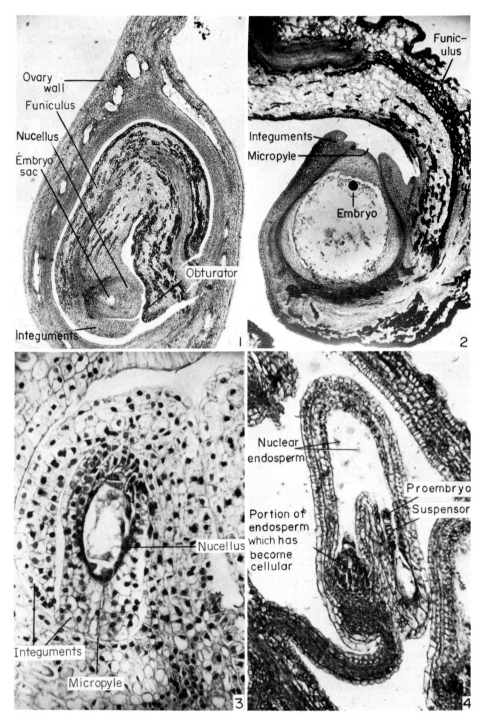

FIG. 263. Micrographs of longitudinal sections of ovules. 1, Longitudinal section of the ovary of *Pistacia vera* in which the large, folded funiculus can be seen. The ovule has a thick nucellus and the embryo sac lies close to the chalaza. × 125. 2, Longitudinal section of the ovule of *Pistacia vera* after fertilization in which an embryo can already be discerned. × 37. 3, Longitudinal section of the ovule of *Lilium candidum* before fertilization showing the very thin nucellus. Six nuclei can be distinguished in the embryo sac. × 143. 4, Longitudinal section of the ovule of *Capsella bursa-pastoris* with a young embryo. × 200.

the Centrospermae (Eckardt, 1967). Also the integuments may vary in thickness, and the thinnest may consist of the two epidermal layers only. However, in such integuments the part closest to the micropyle may be somewhat thicker.

The entire surfaces of all the ovular parts are covered with cuticle. Thus it is possible to distinguish an *outer cuticle* which covers the funiculus and the outer integument externally, a *middle cuticle* which is double and is present between the two integuments, and an *inner cuticle* which is also double and is present between the inner integument and the nucellus.

During the development of the embryo sac the vegetative tissue of the nucellus is completely or partly destroyed, and its content is absorbed by the other parts of the ovule. In certain plants, e.g. the Centrospermae, the nucellus may, in the seed, produce a nutritious tissue which is termed *perisperm*. With the maturation of the ovule the histological structure of the integuments alters. In many plants the inner epidermis of the integument develops into a nutritious layer which is termed the *integumental tapetum*, or *endothelium*. This layer consists of tall, dark-staining cells. This feature is characteristic of those families in which the nucellus is destroyed early so that the integument is brought into contact with the embryo sac. It is a common feature in the Sympetalae.

Megasporogenesis

There are plants in which several megaspore mother cells appear in a single ovule, but usually only a single mother cell develops in each nucellus. Generally the sporogenous cell develops directly from a hypodermal nucellar cell (Fig. 262, no. 1). This cell is distinguishable from the neighbouring cells by its size, the size of its nucleus and the density of its cytoplasm. In certain plants indirect development of the sporogenous cell has been observed; the hypodermal cell first divides into an outer parietal cell, which is usually

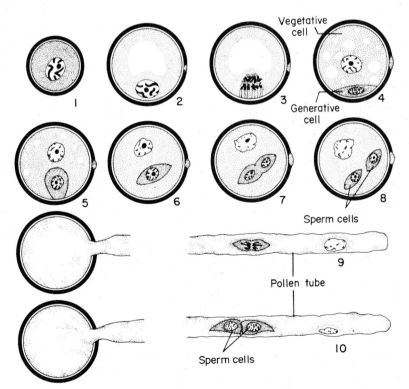

Fig. 264. Types of development of the male gametophyte in angiosperms. Explanation in text. (Adapted from Maheshwari, 1950.)

smaller, and a larger inner cell, which constitutes the primary sporogenous cell. The latter usually develops into the megaspore mother cell, and the parietal cell may divide in different planes to form numerous parietal cells. As a result of these divisions the megaspore mother cell is pushed deep into the nucellus in many plants (Fig. 262, nos. 2–7; Fig. 263, no. 1). Ovules with this type of nucellus are called *crassinucellate*. In ovules where no parietal cells develop the nucellus is thin (Fig. 263, no. 3), and the ovules are called *tenuinucellate*. The phylogenetic significance of the parietal cells is not clear, but it is thought that the trend, during the development of the angiosperms, has been towards their loss.

The megaspore mother cell undergoes a meiotic division which in most plants is accompanied by the formation of a separate wall on each of the four megaspores. The megaspores are arranged in one row, and generally the three closest to the micropyle degenerate, and the remaining one enlarges.

THE MALE GAMETOPHYTE

The mature male gametophyte (microgametophyte) consists of three cells resulting from two mitotic divisions which take place in the pollen grain. Prior to the first mitotic division the nucleus of the microspore (the young pollen grain) takes up a position close to the wall. The first division results in the formation of two cells, viz., the *vegetative cell* and the *generative cell* (Fig. 264, nos. 1–4). The generative cell initially has a callose or cellulose wall. Shortly after its formation, the generative cell separates from the pollen grain wall and loses its callose wall (Mepham and Lane, 1970). It becomes surrounded by the cytoplasm of the vegetative cell. Here it is seen to be oval or lens-shaped (Fig. 264, no. 6). It is at this stage that the pollen is shed from the anther although in many plants it has been found that the generative cell divides once to form two sperm cells (male gametes) before the opening of the anther (Fig. 264, nos. 7, 8). In other plants the generative cell is seen to divide only after it penetrates into the developing pollen tube (Fig. 264, nos. 9, 10).

The question how the generative cell and the vegetative nucleus (also termed *pollen-tube nucleus*) move in the growing pollen tube is still open. In a number of plant species microtubules occur in the pollen grain, tube, generative cell and sperms. The microtubules in the generative cell, which were seen to be orientated along the longitudinal axis of the pollen tube, may be involved in maintaining the elongated shape of the cell and its movement towards the pollen-tube tip (Cresti *et al.*, 1984). In certain plants the vegetative nucleus begins to degenerate a short while after its formation and, even when it is present for a long time, it does not always precede the male gametes and it may be found behind them.

The inner lamella of the wall of the pollen tube consists of callose in addition to cellulose (Tupy, 1959). The protoplast is present only in the distal part of the tube and it becomes separated from the proximal part by the formation of callose plugs which are formed from time to time by the protoplast. As a result it is possible to distinguish many such plugs in a long pollen tube.

In the growing pollen tube of *Lycopersicum peruvianum* four zones were distinguished: (1) apical growth zone; (2) subapical zone, (3) nuclear zone; and (4) callose plug-formation zone (Cresti *et al.*, 1977). The first zone is slightly swollen. It contains numerous small and large Golgi vesicles and it has only a pectocellulotic wall. In the subapical zone two wall layers are already evident; an outer pectocellulotic layer and an inner thin callose layer. The cytoplasm here is rich in organelles. The dictyosomes are very active, producing small and large vesicles. The small vesicles appear to carry polysaccharides and the large vesicles seem to contain precursors of callose. In the nuclear zone, which contains the vegetative nucleus and generative cell, the callose wall is clearly evident. The various organelles and small vacuoles are present in this zone. In the callose plug-formation zone the vacuoles are large and abundant and the callose wall is very thick. As observed in *Petunia* (Cresti and van Went, 1976), in this zone the ground cytoplasm is very electron-dense, there are mitochondria, rough ER, numerous ribosomes, some vacuoles, and lipid bodies. Other organelles appear absent. Plasma membrane is partially absent. In the cytoplasm there are numerous

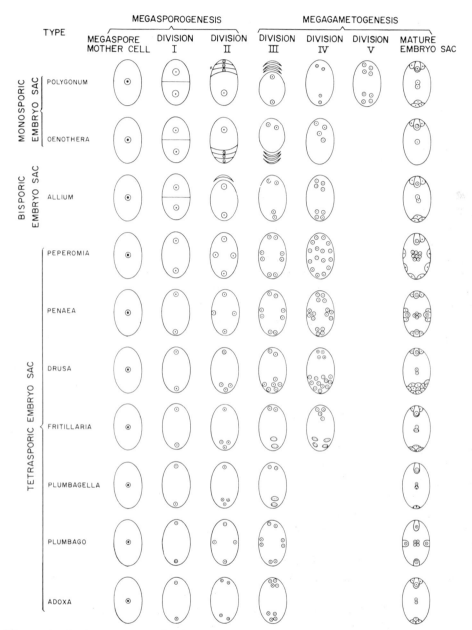

Fig. 265. Diagram showing important types of embryo sacs and their development in angiosperms. (Adapted from Maheshwari, 1950.)

spherical structures, apparently containing cellulose, which produce the callose plug.

The growth of the *Lilium* pollen tube is found to be restricted to a zone which extends back no more than 3–5 μm from the tip of the tube. Electron microscope studies revealed that the cytoplasm of the non-growing region of the tube contains an abundance of mitochondria, amyloplasts, dictyosomes, endoplasmic reticulum, lipid bodies, and vesicles. Approaching the growing tip there is a gradual increase in the amount of vesicles and disappearance of other cytoplasmic elements. At least some of the vesicles appear to have originated from dictyosomes and appear to contribute to tube wall and plasmalemma formation. Cyto-chemical analysis indicated that the tips of the pollen tubes are rich in RNA, protein and carbo-hydrates (Rosen *et al.*, 1964).

THE FEMALE GAMETOPHYTE

As previously described, the megaspore mother cell produces four megaspores by miotic division, only one of which remains and enlarges. This megaspore undergoes three successive mitotic divi-sions to give rise to the *embryo sac*, i.e. female angiosperm gametophyte (megagametophyte) with either nuclei (Fig. 261, no. 5). It should be men-tioned that in the Loranthaceae, e.g. in the semi-parasite *Struthanthus*, up to five embryo sacs may be formed, but only one completes its development (Venturelli, 1984). As a result of the extensive embryogenetic research that has been carried out in angiosperm families and genera, it has been shown that there are many deviations from the typical manner of development of the megaspores and the embryo sac. Maheshwari (1950) formulated a method of classification of the different types of development of the angiosperm embryo sac based on the following features: (1) the number of mega-spores or megaspore nuclei that participate in the formation of the embryo sac; (2) the total number of divisions that take place during the formation of the megaspore and the megagametophyte; (3) the number and arrangement of the nuclei and their chromosome number in the mature embryo sac.

A few of the most common types are described below and other types may be seen in Fig. 265.

ive divisions so that eight nuclei, four at each pole, are formed. Three of the four nuclei at the micropylar pole become organized in cells to form the *egg apparatus*. At the time of fertiliz-ation the middle cell of these three constitutes the female gamete, i.e. the *egg cell*. The two lateral cells are termed *synergids*. Three of the nuclei at the chalazal pole become organized into three cells which are termed the *antipodal cells*. The number of these cells increases, in certain plants, as a result of additional divisions. In the rest of the embryo sac, which becomes the *central cell*, the two remaining nuclei (*polar nuclei*) move into the centre where they may fuse to form a diploid nucleus, termed the *secondary nucleus* (Fig. 261, no. 5).

This type of embryo sac is the most common and was first clearly described by Strasburger in 1879.

As seen in *Nicotiana tabacum* (Mogensen and Suthar, 1979), a continuous wall around the cha-lazal end of the egg cell is formed only after ferti-lization and the formation of a zygote. The syner-gids possess a cell wall only around their micro-pylar half.

The end wall at the micropylar side of the synergids is thickened and proliferated. This por-tion of wall protuberances is termed the *filiform apparatus* (Fig. 274, no. 12). It has been suggested that the filiform apparatus, which greatly in-creases the surface area of the plasmalemma, plays a role in the transfer of substances from the nucellus and integuments into the embryo sac, es-pecially into the egg cell (Johri, 1962; Jensen, 1965; Schulz and Jensen, 1968a).

Monosporic embryo sac

Polygonum type—an eight-nucleate embryo sac

In this type four distinct megaspores develop and only one, usually that furthest from the mic-ropyle, develops into the embryo sac. This mega-spore enlarges and its nucleus divides into two nuclei, one of which moves to the micropylar pole and the other to the chalazal pole of the cell. Later, each of these nuclei undergoes two success-

Oenothera type—a four-nucleate embryo sac

In this type the megaspore closest to the micro-pyle is the one that usually remains viable and, as the result of two mitotic divisions, an embryo sac with four nuclei is formed. Three of the nuclei constitute the egg apparatus and the fourth forms a single polar nucleus. Fertilization in this type results in the formation of a diploid endosperm nucleus and not a triploid one as in the above type.

Bisporic embryo sac

Allium type

The sporogenous cell divides into only two cells—*dyad*—as a result of the first meiotic division. Of these two cells, either the cell directed towards the chalaza may remain viable (*Allium*), or that closest to the micropyle (*Scilla*). The nucleus of the viable cell divides to form two haploid nuclei which are regarded as megaspore nuclei. As the result of a further two divisions eight nuclei are obtained, which become organized and arranged as in the *Polygonum* type.

There are also plants in which intermediate stages between monosporic and bisporic embryo sacs can be found.

Tetrasporic embryo sac

Adoxa type

This type is found, among others, in *Sambucus*, *Ulmus*, and *Tulipa*. In this type, as a result of the meiotic division of the sporogenous cell, four nuclei, which lie free in the cytoplasm of the young embryo sac, are obtained. After an additional mitotic division a total of eight haploid nuclei, which organize in the typical way, are obtained.

Fritillaria type

This type is found in many genera among which are *Lilium* and *Fritillaria*. Here three of the four nuclei obtained from the meiotic division move to the chalazal pole of the young embryo sac, and the fourth is found at the micropylar pole. The latter nucleus divides in the usual manner, and the other three nuclei fuse to form a single triploid nucleus, which immediately divides into two. As a result of this, a second four-nucleate stage is obtained in which there are two haploid nuclei at the micropylar pole and two triploid ones at the chalazal pole. Later a third and last division takes place, which gives rise to four haploid nuclei at the micropylar pole and four triploid ones at the chalazal pole. The final arrangement in the mature embryo sac is a normal haploid egg apparatus, three triploid antipodal cells, and tetraploid secondary nucleus which results from the fusion of a haploid and a triploid polar nucleus.

NECTARIES

Nectar, a sugar-containing solution, is secreted by nectaries which most frequently occur on insect- and bird-pollinated plants. *Nectaries* may consist of specialized tissue which differs in structure from the neighbouring tissues, i.e. *structural nectaries*, or they may not differ in structure, i.e. *non-structural nectaries* (Zimmermann, 1932; Frey-Wyssling and Häusermann, 1960). Non-structural nectaries have been observed in many plants and on various organs, e.g. on the leaves of *Pteridium aquilinum* and *Dracaena reflexa*; on the floral bracts of *Sansevieria zeylanica*; on the sepals of *Paeonia albiflora*; and on the tepals of *Cattleya percivaliana*. Structural nectaries may form special outgrowths or they may occupy delimited regions of the surface layers of the various plant organs on which they are formed.

Location of nectaries

Nectaries may develop on all aerial parts of the plant. Nectaries that are connected with the floral organs are termed *floral nectaries*, and those developing on the vegetative parts of the plant, *extrafloral nectaries*. Extrafloral nectaries may be found on different organs such as petioles (*Passiflora*), stipules (*Vicia faba*), teeth of leaves (*Ailanthus altissima*, *Prunus*, and *Impatiens*), and on the margins of the cyathia of *Euphorbia*. Extrafloral nectaries are regarded, phylogenetically, as more primitive than floral nectaries (Frey-Wyssling, 1933). This chapter will deal only with floral nectaries.

Many workers used the form and location of the nectary as taxonomic characteristics by which they attempted to support their theories regarding the phylogenetic relationship between species, genera and families (Bonnier, 1879; Schniewind-Thies, 1897; Knuth, 1898–1905; Porsch, 1913; Daumann, 1928, 1930a, b, c, 1931a, b; Brown, (1938). It has been suggested (Fahn, 1953b, 1988) that a phylogenetic trend of development exists, expressed by the acrocentripetal change of position of the nectaries within the flower, i.e. from the sepals towards the ovary and up to the style. The following classification of the floral nectaries, according to their location, has been suggested (Fahn, 1952, 1953b).

1. *Perigonal nectaries*—those developing on the perianth:
   (a) close to the base of the perianth parts (*Ranunculus, Leontice, Althaea, Hibiscus, Fritillaria*);
   (b) in spurs formed by the perianth parts (*Garidella, Pelargonium, Tropaeolum, Centranthus*).

2. *Toral nectaries*—those developing on the receptacle:
   (a) *marginal*—between the base of the sepals and petals (*Capparis, Reseda*);
   (b) *annular*—the nectary forms a ring or part of one on the surface of the receptacle:
      (i) between the sepals and ovary (*Grevillea*);
      (ii) between the stamen bases (*Polygonum*, Cruciferae);
      (iii) a ring consisting of small swellings between the stamens and around the ovary (*Cistus*);
      (iv) a shallow or concave ring between the stamens and the ovary base (*Anagyris, Caesalpinia, Ceratonia, Cercis siliquastrum, Robinia, Prunus, Cydonia, Rubus, Punica*); or between the stamens and the styles when the ovary is inferior (*Eucalyptus*, Cucurbitaceae, Campanulaceae, Dipsaceae);
      (v) a prominent ring around the ovary base (Boraginaceae, most of the Labiatae, Bignoniaceae, *Citrus*);
   (c) *tubular*—the nectary lines a tube-like receptacle (*Bauhinia*).

3. *Staminal nectaries*—those related to stamens:
   (a) on the filaments (*Colchicum, Laurus, Dianthus, Silene*); or in the tube formed by the fusion of the filaments (many plants of the Papilionaceae);
   (b) on an appendage of the connective (*Viola*).

4. *Ovarial nectaries*—those developing on the ovary wall:
   (a) on all the free surface of the carpels (*Tofieldia palustris, Sarracenia*);
   (b) on the ovary base (*Gentiana*);
   (c) *septal*—on the partitions of monocotyledonous syncarpous ovaries (Liliaceae, Musaceae, Amaryllidaceae, Iridaceae).

5. *Stylar nectaries*—those developing at the base of the style:
   (a) common to the base of the style and stylopodium (Umbelliferae);
   (b) only on the base of the style (in most insect-pollinated species of the Compositae such as *Helianthus, Senecio* and *Calendula*).

It should be mentioned that studies of the early stages of flower development may in some case be important in determining the exact localization of the nectary in relation to the other floral parts (Brett and Poslushny, 1982).

Secretion of nectar and structure of nectaries

Sucrose, glucose, and fructose have been found to be among the most common constituents of nectar. Apart from these, other sugars, mucilage, amino acids, proteins, mineral ions, vitamins, enzymes hydrolyzing sucrose, and organic acids have also been recorded in nectar of different plants (Fahn, 1979a, b). The acids are responsible for the low pH found in the nectar of some plants (Beutler, 1930; Fahn, 1949). Zimmermann (1954)

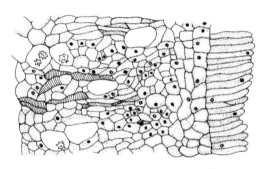

FIG. 266. Portion of a cross-section of the nectariferous gland on the base of the petiole of *Ricinus communis* showing the palisade-like epidermis and the vascular tissues which penetrate deeply into the nectariferous tissue. (Adapted from Agthe, 1951.)

found transglucosidases in the nectar of *Impatiens*. The concentration of nectar varies from 3 to 87%. It was found that the output of fresh nectar, and the amount of the dry matter in it, produced by different species in a 24 h period varies considerably. The variation per flower may be between 0.13 mg fresh nectar and 0.10 mg dry matter, and 268 mg fresh nectar and 47 mg dry matter (Fahn, 1949). In unisexual flowers there are striking differences in the amount of nectar secreted by the male and female flowers. Among the external factors that increase the amount of nectar secretion, temperature and soil moisture should be mentioned; these factors influence the general physiological activity of the plant (Fahn, 1949). The sugar content in the plant is the most important internal factor influencing nectar secretion (Helder, 1958).

The cells constituting the nectary form the nectariferous tissue. The nectariferous tissue usually contains branches of vascular bundles (Fig. 266). The vascular tissue is frequently very well developed and consists of a high proportion of phloem elements (Frei, 1955; Frey-Wyssling, 1955). This phenomenon is consistent with the hypothesis that the secreted sugar is brought to the nectary by the vascular bundles. According to Agthe (1951), the sugar concentration of the nectar is correlated with the type of vascular tissue that reaches the nectariferous tissue. Thus, he claims that, in *Euphorbia pulcherrima* and *Abutilon striatum*, in which the sugar concentration of the nectar is high, the endings of the terminal branches closest to the nectary consist only of phloem, while in *Ricinus communis*, *Fritillaria imperialis*, and *Ranunculus acer*, in which the sugar concentration of the nectar is low, the branch endings consist of equal amounts of xylem and phloem or of a larger proportion of xylem. One of the fundamental differences between nectaries and hydathodes is that the branch endings of the vascular bundles that come into contact with the hydathode tissue consist of tracheary elements only. Correlation between sugar concentration in the nectar and the type of vascular tissue was challenged by some other workers. Nectar secretion, i.e. the process of release of sugar solution from protoplast to the apoplast, may occur in all the cells constituting the nectary or in part of them only. Cells which

secrete the nectar are termed *secretory cells*. Secretory cells may be epidermal or parenchyma cells.

Exudation of nectar from the nectary is brought about in different ways related to the type of tissue from which the secretory cells have developed. When the secretory cells are parenchymatous, nectar is secreted into the intercellular spaces from which it is exuded via modified stomata (Fig. 267, no. 2). When secretion is brought about by epidermal cells which lack a discernible cuticle, nectar diffuses through the cells directly to the outer environment. When the secretory epidermis is covered by a cuticle the exudation of nectar is effected by the rupture of the cuticle, the occurrence of pores in it or by the cuticle being permeable (Frey-Wyssling, 1933).

The nectariferous tissue is generally composed of small cells with thin walls, relatively large nuclei, dense granular cytoplasm, and small vacuoles (Caspary, 1848; Bonnier, 1879; Behrens, 1879; Fahn, 1952). This description was generally confirmed by recent electron microscope observations (Schnepf, 1964a, b; Eymé, 1966; Fahn and Rachmilevitz, 1970; Fahn and Benouaiche, 1979; Fahn, 1979a, b). Under the electron microscope the nectariferous tissue, especially the secretory cells, shows characteristic ultrastructural features. Secretory cells show, at the stage of secretion, a dense cytoplasm rich in organelles, especially mitochondria, ER, dictyosomes, and ribosomes (Fig. 269). The vacuome is reduced. Such structural features are generally believed to be associated with high metabolic activity. The cytological properties of the secretory cells, the high rate of respiration in the nectariferous tissue, and the dependence of secretion on respiration indicate that nectar is secreted by an active mechanism located in the secretory cells. It has been suggested that sugar solution is secreted by vesicles derived from ER cisternae or dictyosomes which release their contents by fusion with the plasmalemma.

Some nectaries will be described here which demonstrate their variability as regards their location, form, and structure.

Lonicera japonica

The floral nectary of *Lonicera* forms a strip

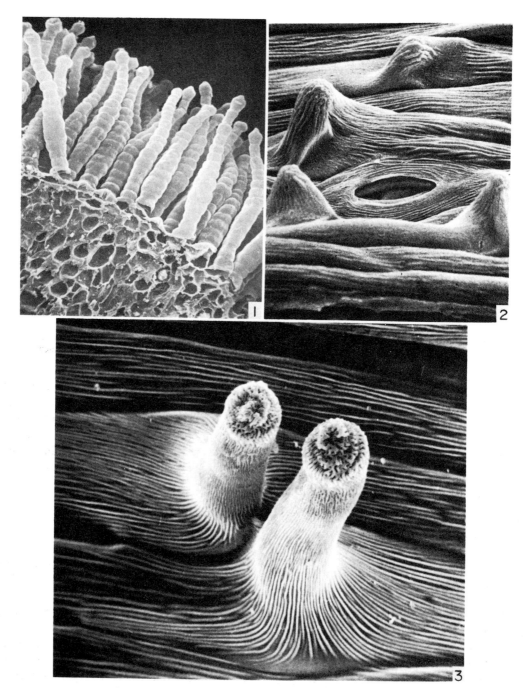

FIG. 267. Scanning electron micrographs of nectaries. 1, Multicellular secretory trichomes of a nectary occurring on the inner side of the calyx of *Abutilon pictum.* × 440. 2, Inner epidermis of the basal part of a sepal spur of *Tropaeolum majus* showing a stoma through which nectar is secreted from the nectariferous tissue. × 1000. 3, As in no. 2 but showing epidermal unicellular hairs which secrete polysaccharides together with small amounts of sugar. × 1550. (Nos. 2 and 3 from Rachmilevitz and Fahn, 1975.)

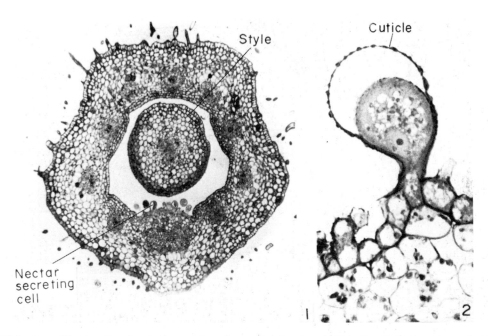

FIG. 268. Nectary of *Lonicera japonica*. 1, Cross-section of a corolla tube when still in bud. × 65. 2, A secretory cell at the stage of secretion. The cuticle is detached from the cell wall. × 800.

along the inner lower side of the basal part of the corolla tube. The nectariferous tissue consists of a secretory epidermis and a subtanding parenchyma below which a group of vascular bundles are evident. Nectar is secreted by specialized epidermal cells (Fig. 268) which comprise a large number of the epidermal cells of the nectary. The secretory cells differ from the non-secreting epidermal cells in size, shape, and ultrastructure. Each secretory cell consists of a narrow base situated between the neighbouring epidermal cells and an upper free part which consists of a spherical "head" and a narrow "neck". The epidermis is covered by a cuticle. The cuticle which covers the secretory cells has an elaborate structure. It is composed of thick and thin regions. A few days before the stage of secretion the cuticle of the "head" starts to detach from the cell wall (Fig. 268, no. 2; Fig. 269, no. 1). At this stage of secretion the space between the cuticle and the cell wall reaches a maximum volume. It is assumed that the extension of the cuticle occurs in the thin areas through which nectar may diffuse at the stage of secretion. In the upper part of the secretory cell an inner layer of wall protuberances develops which forms an

elaborate network at the stage of secretion. At this stage the cytoplasm is very dense and rich in mitochondria and ER elements. The ER forms stacks of parallel-arranged cisternae which give rise to vesicles that apparently fuse with the plasmalemma and release nectar (Fig. 269, no. 2).

Garidella unguicularis

The nectariferous tissue in this flower is found at the petal knee, i.e. in that place where the blade passes into the claw (Fig. 270, nos. 1–4). This knee forms a type of spur, the aperture of which is closed by a cover connected to the claw of the petal (Fig. 270, nos. 2, 3). Within the aperture of the spur, on the side of the petal blade and close to the cover margins, are brushes of unicellular hairs which thus block the spur aperture more tightly. The tongue of the bee is pushed into the spur, which contains the nectar, through the above-mentioned hairs. The nectariferous tissue consists of layers of extremely small cells which are strikingly different from the neighbouring cells. The outer thickened walls of the epidermal cells

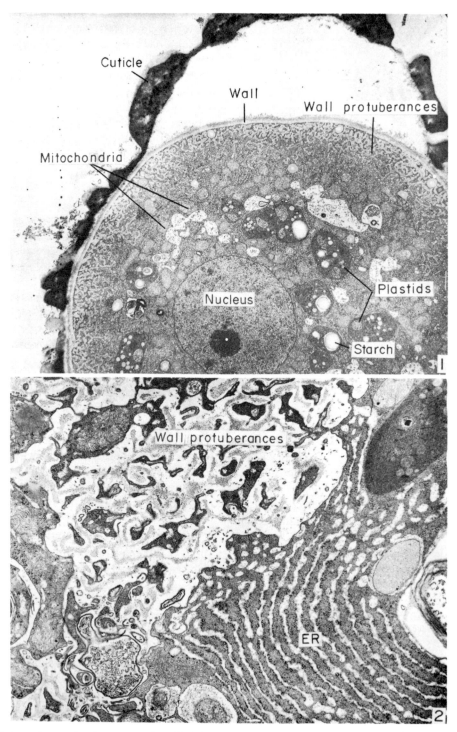

FIG. 269. Electron micrographs of nectar-secreting cells of *Lonicera japonica*. 1, Secretory cell of a flower bud. × 3200. 2, Portion of a secretory cell of a flower at the time of maximum secretion. × 18,000. (No. 2 from Fahn and Rachmilevitz, 1970.)

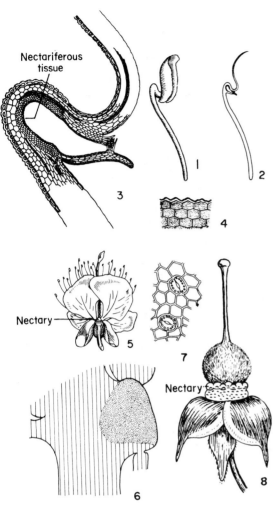

FIG. 270. Nectaries. 1–4, *Garidella unguicularis*. 1, An entire petal. 2, Median longitudinal section of a petal. 3, Knee portion, in longitudinal section, enlarged to show the position of the nectariferous tissue. 4, Outer portion of nectariferous tissue showing the thick cuticle on the epidermis. 5–7, *Capparis sicula*. 5, General aspect of the flower after the removal of the sepal, in which the nectar collects, to show the triangular nectary. 6, Median longitudinal section of the receptacle in the region of the nectary (stippled). 7, A portion of the epidermis over the nectary showing the modified stomata through which the nectar is secreted. 8, *Cistus villosus*. Flower from which the petals and stamens have been removed to expose the nectary.

are covered by cuticle and form low papillae. The contents of the epidermal cells do not differ from those of the other nectariferous cells, which are granular and yellow in colour. The vascular bundles come into contact with the nectariferous tissue (Fig. 270, no. 3).

Capparis sicula

Externally the nectary appears as a white triangle on the margin of the torus (Fig. 270, no. 5). The apex of the triangle is directed towards the stamens and the base is opposite that of the largest, boat-shaped sepal, in which large quantities of nectar accumulate. The other two sides of the triangle are bordered by two petals of which the margins of the basal portions, where they border the nectary, are somewhat thickened and folded downwards on themselves so as to form a narrow slit between them. The adjacent margins of the upper portions of the petals overlap one another. In order to suck the nectar from the boat-shaped sepal the bee stands on the two petals, separates them, and pushes its tongue through the slit formed by the lower margins of the petals (Fig. 270, no. 5). In a median section of the flower (Fig. 270, no. 6) it can be seen that the nectariferous tissue penetrates deeply into the receptacle. The epidermis of the nectariferous tissue is covered by a very thin cuticle and it contains many elliptical stomata (Fig. 270, no. 7).

Colchicum ritchii

The nectariferous tissue is found on the basal portion of the filaments (Fig. 271, no. 1), which are adnated to the tepals. Externally the nectariferous tissue appears as a yellow band surrounding the base of the filament. A swelling of the petal is present on each side of the stamen; the nectar accumulates between these swellings. The nectary consists of transparent epidermal cells which contain large central vacuoles surrounded by a thin layer of cytoplasm and rounded parenchyma cells with a granular content. The vacuome of the parenchyma cells is relatively small and the cytoplasm is rich in organelles. It is therefore assumed that only the parenchyma cells of the nectary function in nectar secretion. Nectar is secreted into the intercellular spaces from which it is exuded via large modified stomata. The epidermis is covered by a very thin cuticle (Fig. 271, nos. 2, 3).

Citrus

The nectary in *Citrus* forms a ring around the base of the ovary (Fig. 271, nos. 4–6). Stomata

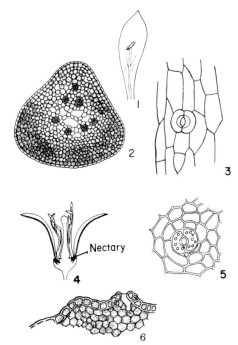

FIG. 271. Nectaries. 1–3, *Colchicum ritchii.* 1, An entire tepal with adnated stamen. Nectariferous tissue at the base of the filament, stippled. 2, Cross-section of the filament in the region of the nectary. 3, Portion of the epidermis of the nectary in surface view. 4–6, *Citrus limon.* 4, Median longitudinal section of an entire flower, showing the position of the nectary. 5, Portion of the epidermis of the nectary. 6, Portion of a cross-section of the nectary showing a modified stoma, and a chamber below it through which the nectar is secreted.

with wide apertures are present on raised portions of the ring. In tangential section of the nectary the stomata are seen to be strikingly rounded in shape (Fig. 271, no. 5), and in cross-section it is seen that the substomatal chambers are fairly deep and that the cells below the epidermis are small and compact. The epidermis itself consists of small cubical, thick-walled cells which are covered by a relatively thin cuticle. All the cells have a similar ultrastructure, and it is therefore assumed that all the cells constituting the nectary (epidermal and parenchyma cells) are capable of nectar secretion. Nectar is secreted to the intercellular spaces from which it is exuded via the stomata.

Cistus villosus

This plant is usually thought to be a pollen-producing plant only, but it is also frequently visited by bees that collect nectar. The upper portion of the receptacle is swollen around the bases of the innermost stamens, especially on the side closest to the ovary, and forms the nectariferous tissue (Fig. 270, no. 8). The epidermis of the nectary contains stomata and is apparently without cuticle.

Bauhinia purpurea

In this plant the nectariferous tissue lines a tubular cavity which has been formed either by the depression of the receptacle on one side of the gynophore, or by the fusion of the basal portions of the perianth and stamens (Fig. 272, no. 2). In a cross-section of the receptacle (Fig. 272, no. 1) two whorls of vascular bundles can be seen as well as the large central bundle of the gynophore. The epidermis lining the above cavity is thin-walled, fairly transparent, and consists of very small cells; there are no stomata in this epidermis. The nectariferous tissue, which consists of small cells with yellow, granular contents, is found below the entire epidermis with the exception of that part that overlies the gynophore.

Bupleurum subovatum

As in all the Umbelliferae, the nectariferous tissue in this plant is found on the upper portion of the inferior ovary, i.e. on the stylopodium. In a cross-section of the nectary (Fig. 272, no. 8) the epidermis is seen to be transparent and consists of thick-walled cells covered by a thick ridged cuticle (Fig. 272, no. 9). This epidermis contains very small, sunken stomata. Below the epidermis there is a thick region of compact cells with granular contents. These nectariferous cells are strikingly different from the neighbouring parenchyma cells below them.

Some species with septal nectaries

In *Muscari racemosum,* for example, the septal nectary consists of three very compressed cavities, one in each of the three partitions of the ovary, which are lined by nectariferous tissue (Fig. 272, no. 3). Parallel to each cavity, on the external

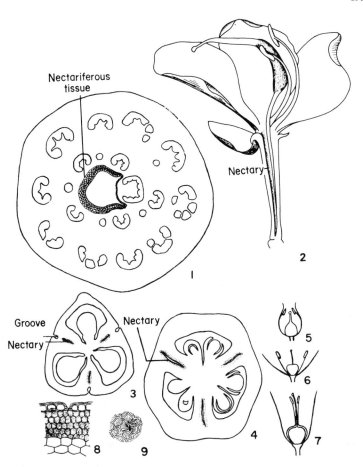

FIG. 272. Nectaries. 1 and 2, *Bauhinia purpurea*. 1, Cross-section of the flower in the region of the nectary. The vascular bundles of the various floral organs are shown in the tissues arround the nectary. 2, Median longitudinal section of the flower showing the position of the tube-like nectary. 3, Diagram of a cross-section of the ovary of *Muscari racemosum*. 4, Diagram of a cross-section of the ovary of *Asphodeline lutea*. 5–7, Diagrams of longitudinal sections of flowers. 5, *Muscari*. 6, *Allium*. 7, *Asphodelus*. 8 and 9, *Bupleurum subovatum*. 8, Portion of a cross-section of the epidermis and underlying nectariferous tissue. 9, Surface view of portion of the epidermis.

surface of the ovary, is a groove which is connected to the cavity at the upper portion of the ovary. In this species the nectar, secreted by the nectariferous tissue into the cavities, fills them, whence it overflows down the grooves and accumulates between the base of the ovary and the tepals. In *Allium*, the secretion of the nectar is as in *Muscari*, but it accumulates between the base of the ovary and the stamens (Fig. 272, nos. 5, 6). In *Asphodelus* and *Asphodeline* there are no grooves on the ovary wall (Fig. 272, no. 4) and the three nectariferous canals open at the top of the ovary. In these genera the filaments are bent at

their bases so that they cover the ovary (Fig. 272, no. 7). This results in the formation of a capillary cavity between the ovary and the filaments. The nectar, which is extruded at the top of the ovary, is drawn by capillary force to the ovary base. Without this structure the nectar would spill out of the flower because of its horizontal or pendulous position in these genera.

With the aid of the electron microscope a very thick inner layer of wall protuberances was found in the epithelial secretory cells facing the cavity of the septal nectary of *Gasteria* and other genera (Schnepf, 1964a).

ABSCISSION OF FLORAL PARTS

Petals, stamens or whole flowers, which do not produce fruit, may be shed as a result of the formation of a separation layer. The cells of this tissue are distinguishable from those of the neighbouring ones by their more circular or cubical shape (Pfeiffer, 1928; Lieberman *et al.*, 1982). In flowers the tissue of the separation layer is usually less well developed than in leaves of woody dicotyledons (see Chapter 12), and it appears only shortly before the floral parts are shed.

In plants with unisexual flowers, usually entire male flowers are shed after the release of the pollen (Yampolsky, 1934). Sometimes entire male inflorescences are shed, e.g. *Morus, Casuarina*, and *Ceratonia*. Female and bisexual flowers that fail to be fertilized may also be shed.

FORMATION OF ENDOSPERM AND EMBRYO

Fertilization

As already described above, the pollen grain germinates and produces a pollen tube on the stigma. The pollen tube, which carries within it the two sperm cells, passes through the style and reaches the ovule. In most plants the pollen tube penetrates into the ovule via the micropyle. In some plants the pollen tube penetrates through the chalazal region, i.e. *chalazogamy*. This feature occurs, for example, in *Casuarina* and species of *Pistacia*. After its entry into the ovule the pollen tube penetrates into the embryo sac where it may pass through a synergid, between the synergids and the embryo-sac wall or between the egg cell and the synergids. Usually one of the synergids is destroyed by the penetration of the pollen tube. Sometimes the sinergids have previously been degenerated (Went and Willemse, 1984). Later the tip of the pollen tube ruptures and the two sperm cells, sometimes together with remnants of the vegetative cell, enter into the cytoplasm of the embryo sac (the female gametophyte). One of the sperm nuclei fuses with the egg cell nucleus, and the second fuses with the two polar nuclei or with the secondary nucleus

if the latter two have fused previously. This process of fertilization is termed *double fertilization* (Fig. 273). As a result of the fusion of a sperm nucleus with the egg cell nucleus a *zygote*, which later develops into the *embryo,* is formed and, as a result of the fusion of the second sperm nucleus with the polar nuclei or with the secondary nucleus, the *endosperm* is formed.

Electron microscope studies on the embryo sac of *Euphorbia dulcis, Glycine max* and *Gossypium hirsutum*, in which endosperm and embryos have started to develop, has revealed a layer of wall protuberances on the inside of the wall of the micropylar part of the embryo sac. Such structures are usually related to the function of short distance transport of substances (Sarfatti and Gorri, 1969; Schulz and Jensen, 1977; Tilton *et al.*, 1984a).

Development of the endosperm

The endosperm is a storage tissue which provides nutrition for the embryo and the young seedling. In certain plants, e.g. *Pisum, Phaseolus*, and *Arachis*, the entire endosperm tissue is digested by the developing embryo. Generally, in the seeds of such plants the cotyledons thicken and store reserve substances which provide nutrients for the young seedlings. In other plants, e.g. *Ricinus* and species of the Gramineae, the endosperm tissue is still present at the time of germination.

The endosperm develops from mitotic division of the *endosperm nucleus*, which results from the fusion of one of the sperm nuclei with the two polar nuclei or with the secondary nucleus. This division usually precedes that of the zygote. The further development of the endosperm differs in the various groups of plants, and the following main types can be distinguished according to the manner of development.

Nuclear endosperm

In this type the first divisions are not followed by wall formation. The nuclei usually take up a parietal position and a large vacuole forms in the central cell of the embryo sac. These nuclei may remain free in the cytoplasm throughout the entire development, or later walls may develop in at least

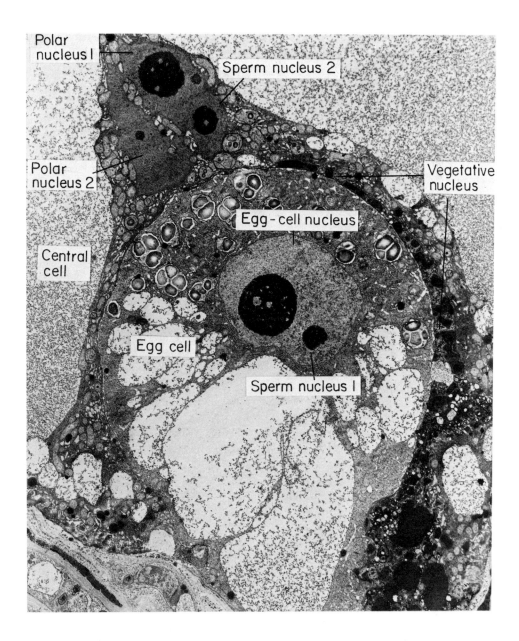

Fɪɢ. 273. Electron micrograph of part of an embryo sac of spinach at the stage of fertilization. × 2300. (Courtesy of H. J. Wilms; from Wilms, 1981.)

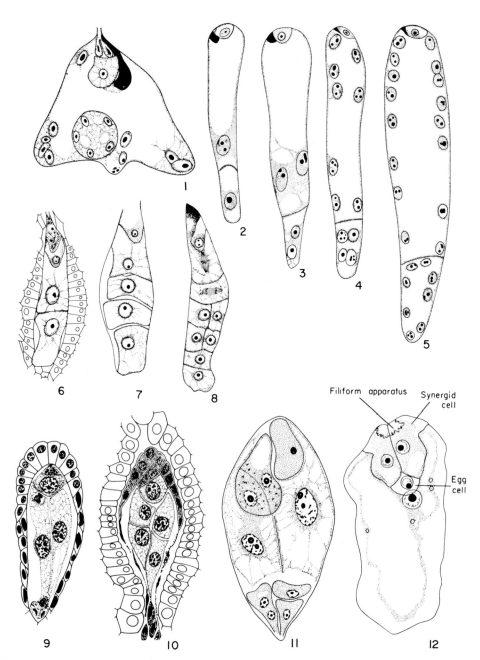

FIG. 274. 1, Embryo sac of *Musa errans* which has a nuclear endosperm. 2–5, Embryo sac of *Eremurus himalaicus* showing the development of a helobial endosperm. 6–8, Development of the cellular endosperm in *Villarsia reniformis* in which the first divisions are transverse. 9–11, Cellular endosperms in which the first divisions are longitudinal. 9 and 10, *Adoxa moschatellina*. 11, *Centranthus macrosiphon*. 12, Embryo sac of *Oenothera*. (Adapted from Maheshwari, 1950.)

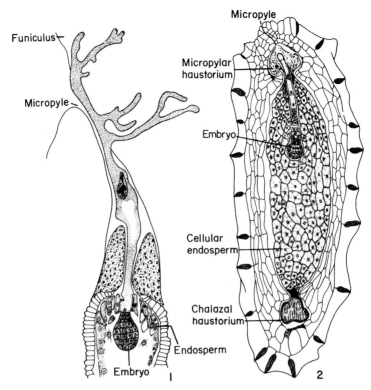

FIG. 275. 1, Upper portion of the ovule of *Impatiens roylei* showing the hypha-like branches of the haustorium which penetrate into the tissue of the funiculus. 2, Longitudinal section of the ovule of *Lobelia amoena* showing the cellular endosperm, and micropylar and chalazal haustoria. (Adapted from Maheshwari, 1950.)

certain parts of the central cell, as in *Capsella bursa-pastoris*. Sometimes a few nuclei may divide at a faster rate than the others and so isolated groups or "nodules" are formed. These "nodules" become surrounded by a distinct cytoplasmic membrane (Fig. 263, no. 4; Fig. 274, no. 1). In this type, as in the following types, there are many variations (Maheshwari, 1950; Di Fulvio, 1983; Vijayaraghavan and Prabhakar, 1984).

Cellular endosperm

In this type the first division of the endosperm nucleus is accompanied by the formation of a wall which is usually horizontal, but which may sometimes be longitudinal or diagonal (Fig. 274, nos. 6–11). The planes of the subsequent divisions may be parallel to that of the first division, but a short while afterwards walls develop in different planes, so that the mature endosperm consists of a tissue the cells of which are orientated in different directions.

In certain plants (e.g. *Thesium, Impatiens, Acanthus, Lobelia,* and *Lobularia*) *haustoria* of peculiar structure develop at one or both poles of the endosperm (Fig. 275, nos. 1, 2). These haustoria may penetrate deep into the neighbouring tissues of the ovule and from these tissues they transfer nutrients to the developing endosperm. In certain plants secondary haustoria develop laterally from endospermal cells close to the micropyle or chalaza. The nucleus of the micropylar haustorium of *Impatiens* was found to become hypertrophied and irregular in outline (Narayana, 1965).

Helobial endosperm

This type of endosperm constitutes an intermediate form between the nuclear and cellular types of endosperm. It occurs in different angiosperm genera, e.g. *Asphodelus, Muscari, Ornitho-*

galum, *Saxifraga*, and *Echium*. Most typically, endosperm of this type develops in the following
manner. The first division of the endosperm nucleus is accompanied by the formation of a horizontal wall which divides the central cell into
two, usually unequal, chambers of which the micropylar one is large and the chalazal small (Fig.
274, nos. 2–5). This is followed by several nuclear
divisions in the micropylar chamber where the
resulting nuclei remain free, while in the chalazal
chamber the nucleus either does not divide or it
undergoes only a few divisions. In the course of
further development the amount of cytoplasm in
the chalazal chamber is reduced and the nuclei
begin to disintegrate. Simultaneously, in many
species, cell walls may appear in the micropylar
chamber.

No doubt intermediate forms exist between the
three above-mentioned endosperm types. It is as
yet not clear whether the course of phylogenetic
development has been from the nuclear to the
cellular type, or vice versa (Maheshwari, 1950).

Development of the embryo

After its formation, the zygote commonly enters
a dormant stage for a certain period. At the same
time the large vacuole which was still present in
the egg cell disappears and the cytoplasm
becomes more homogeneous. Usually the zygote
begins to divide after the division of the endosperm nucleus. There are plants, e.g. *Oryza* and
Crepis, in which the zygote begins to divide a few
hours after fertilization, while in others the first
division takes place only much later. In *Pistacia
vera*, for example, the first division takes place
about 2 months after fertilization (Grundwag and
Fahn, 1969).

The plane of the first division of the zygote is
almost always transverse. As a result of this division two cells are obtained; that closer to the micropyle is termed the *basal cell*, and the other the
terminal cell. During the course of further development the terminal cell may divide transversely
or longitudinally. The basal cell usually divides
transversely. However, in certain genera this cell
does not divide but enlarges to form a large, sac-
like cell. (See also Wardlaw, 1955.)

Dicotyledonous embryo

On the basis of the differences in the manner of
development of the pro-embryo to the four-celled
stage, dicotyledonous embryos have been classified into five main types (Schnarf, 1929; Johansen,
1945; Maheshwari, 1950).

The terminal cell divides longitudinally

(a) The basal cell does not participate in the
formation of the embryo, or it participates
to a small extent only . . . *crucifer type* (Fig.
276, nos. 1–21).
(b) The basal and terminal cells participate in
the formation of the embryo . . . *asterad type*
(Fig. 277, nos. 1–6).

The terminal cell divides transversely

(a) The basal cell does not participate in the
formation of the embryo or it participates
to a small extent only.
  (i) The basal cell develops into a *suspensor*
  which may be two or more cells
  long . . . *solanad type* (Fig. 277, nos.
  7–14).
  (ii) The basal cell does not divide. If a suspensor is present, it develops from the
  terminal cell . . . *caryophyllad type*.
(b) The basal and terminal cells participate in
the formation of the embryo . . . *chenopodiad
type*.

Below a detailed description of the embryo development of *Capsella bursa pastoris* (crucifer
type) is given (Fig. 276, nos. 1–21). The embryo of
this plant was among the first to be studied (Hanstein, 1870; Souèges, 1914, 1919). In this species
the first division of the zygote is transverse, forming a basal cell (*cb*) and a terminal cell (*ca*) (Fig.
276, no. 2). The basal cell divides transversely, and
the terminal longitudinally (Fig. 276, nos. 3–5), to
form a four-celled proembryo. Each of the two
new terminal cells then divides longitudinally so
that the newly formed walls are perpendicular to
the first-formed wall. Thus four cells which are

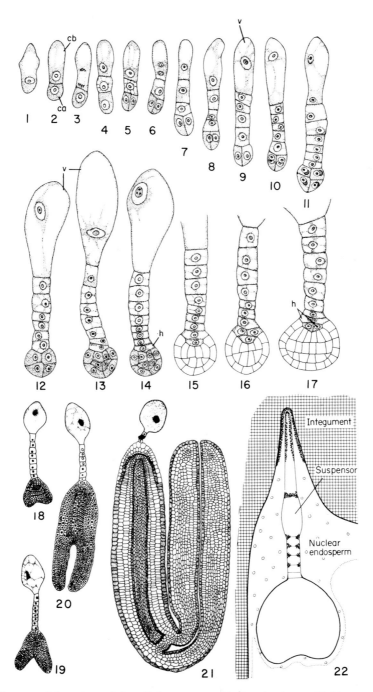

FIG. 276. 1–21, Development of the embryo of *Capsella bursa-pastoris*. Explanation in text. 22, Young embryo of *Diplotaxis erucoides* showing wall protuberances in suspensor cells and in the endosperm wall surrounding the basal cell. (Nos. 1–21 adapted from Souéges, 1914, 1919; Maheshwari, 1950; no. 22 adapted from Simoncioli, 1974.)

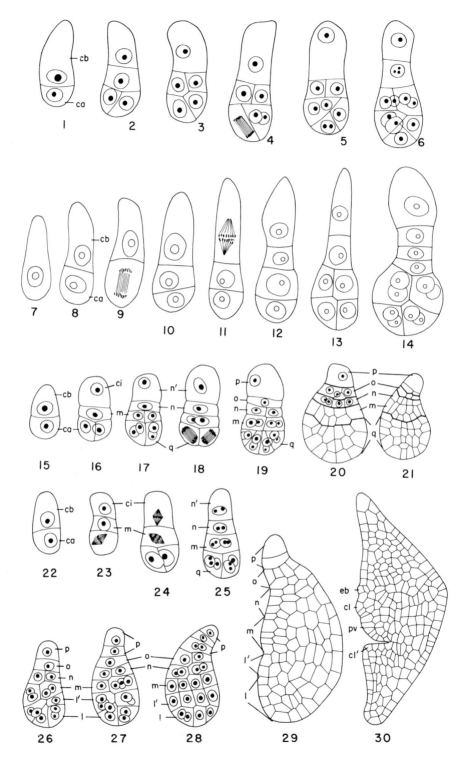

FIG. 277. 1–6, Development of the embryo of *Geum urbanum* (asterad type). 7–14, Development of the embryo of *Nicotiana* (solanad type). 15–21, Development of the embryo of *Muscari comosum*. 22–30, Development of the embryo of *Poa annua*. Explanations in text. (Adapted from Maheshwari, 1950.)

collectively termed the *quadrant* are formed (Fig. ·276, no. 10). The cells of the quadrant then divide transversely to form eight cells—the *octant*—arranged in two tiers (Fig. 276, no. 11; Fig. 263, no. 4). Later each of the octant cells undergoes a periclinal division to form a protodermal and an inner cell (Fig. 276, nos. 13, 14). The subsequent divisions of the protodermal cells are anticlinal. The ground meristem and the procambium of the hypocotyl and the cotyledons develop from the inner cells as a result of further division and differentiation (Fig. 276, nos. 15–21).

Together with the development of the embryo proper, the two cells, derived from the division of the basal cell, divide to form a suspensor which consists of a row of six to ten cells (Fig. 276, nos. 2–19). The suspensor cell closest to the micropyle enlarges and becomes sac-like (*v*), and it apparently functions as a haustorium. The suspensor cell closest to the embryo is termed the *hypophysis* (*h*) and it becomes part of the embryo. This cell undergoes transverse and longitudinal divisions to form two four-celled tiers. The root cap and the neighbouring protoderm develop as a result of further divisions of the cells in the tier closest to the row of the suspensor cells, and the second tier contributes cells to the cortex of the radicle.

Meanwhile, the cells of the embryo proper continue to divide, especially in those two areas where the cotyledons will develop. At this stage the embryo is heart-shaped in longitudinal view (Fig. 276, no. 18). Later the hypocotyl and cotyledons elongate as a result of further cell division (Fig. 276, nos. 19, 20). During still later stages of development the cotyledons bend to conform to the shape of the embryo sac (Fig. 276, no. 21).

The fine structure of developing embryos of *Capsella* was investigated with the aid of the electron microscope. The wall of the micropylar part of the basal cell of the suspensor shows well-developed protuberances (Schulz and Jensen, 1968b, 1969). In *Diplotaxis erucoides* wall protuberances were also observed on the walls of some of the other cells of the suspensor (Fig. 276, no. 22) (Simonciolo, 1974).

In some plants, e.g. *Sedum acre*, the sac-like suspensor cell closest to the micropyle may develop a branched haustorium.

MONOCOTYLEDONOUS EMBRYO

There is no important difference in the first divisions of the mono- and dicotyledonous embryos, but in further developmental stages there are distinct differences. In the mature embryo of nearly all dicotyledons (with a few exceptions, e.g. certain genera of the Umbelliferae and *Ranunculus ficaria*) the shoot apex is found between the bases of the two cotyledons, while in the monocotyledonous embryo the shoot apex is lateral to the single cotyledon.

As examples of the development of monocotyledonous embryos those of *Muscari comosum* (Liliaceae) and *Poa annua* (Gramineae) will be given.

In *Muscari comosum* (Fig. 277, nos. 15–21), according to Souèges (1932) and Maheshwari (1950), the basal cell divides transversely and the terminal cell longitudinally (Fig. 277, no. 16). Following this, the two terminal cells and the cell closest to them (*m*) divide longitudinally, and the cell closest to the micropyle divides transversely into two cells, *n* and *n'* respectively (Fig. 277, no. 17). The cell *n'*, which is now the closest to the micropyle, also divides transversely to form the cells *o* and *p*. After this all the cells, including those of the quadrant (*q*), divide longitudinally or diagonally (Fig. 277, nos. 18, 19). Eventually the cotyledon develops from the tier of cells indicated by *q*, the hypocotyl and shoot tip from *m*, the root initials from *n*, the root cap from *o*, and the suspensor from the cell *p* (Fig. 277, nos. 20, 21).

In *Poa annua* (Fig. 277, nos. 22–30), as described by Souèges (1924) and Maheshwari (1950), the development of the embryo differs from that of *Muscari* principally in the fact that the cells of the quadrant divide transversely to form two tiers *l* and *l'* (Fig. 277, nos. 26, 27), as well as in having differences in the structure of the embryo in the early stages of development (Fig. 277, nos. 29, 30). In *Poa* the *scutellum* and part of the *coleoptile* (*cl'*) develop from the two terminal tiers, *l* and *l'*. The remaining portion of the coleoptile (*cl*) and the shoot and root tips develop from the tier indicated by *m*. The root cap, the *coleorrhiza* and the *epiblast* (*eb*) develop from the tier of cells, *n*. From the remaining cells, *o* and *p*, which are close to the micropyle, the *hypoblast* (termed suspensor in other plants) develops (Fig. 303, no. 2).

The fine structure of the early development of the barley embryo has been described by Norstog (1972).

Explanations and details of the parts of the mature embryo are given in Chapter 21.

In all mature embryos, no matter what the type of development, differentiation into root apex, hypocotyl, cotyledon or cotyledons and shoot apex takes place. However, in certain plants, e.g. the Orchidaceae and the Rafflesiaceae, the embryo remains small and oval, and is undifferentiated (Swamy, 1949a; Rao, 1967).

Variation in structure of the suspensor

In most plants the function of the suspensor is only to push the embryo into the endosperm, but in some plants it may develop into a large haustorium which penetrates between the cells of the endosperm, and, to a certain extent, even between the cells of the tissue surrounding the endosperm. In many genera of the Papilionaceae the suspensor is of the latter type (Guignard, 1881). In *Pisum* and *Orobus*, for example, the suspensor consists of two pairs of multinucleate cells; the cells of the pair closer to the micropyle are large and very elongated, and those of the second pair are almost spherical (Fig. 278, no. 1). In *Cicer* the suspensor consists of two rows of uninucleate cells (Fig. 278, no. 2). The suspensor cells of *Phaseolus* have highly endopolyploid nuclei with giant chromosomes and a well-developed tubular endoplasmic reticulum free of ribosomes. They possess only few mitochondria and dictyosomes as well as a small number of small vacuoles. In the micropylar part, the walls of the suspensor cells have labyrinth-like wall protuberances (Schnepf and Nagl, 1970; Yeung and Clutter, 1979). In the Rubiaceae (Lloyd, 1902; Souèges, 1925) the suspensor at first develops as a multicellular thread, and later the cells closest to the micropyle develop lateral projections. These projections penetrate into the endosperm and their tips swell (Fig. 278, no. 3).

As a result of more recent studies, the suspensor is considered to function as an organ which absorbs nutrients from the surrounding somatic tissues and as a source of supply of important nutrients and growth regulators to the developing embryo (see Natesh and Rau, 1984).

Apomixis

Apomixis is a process of asexual reproduction in which no nuclear fusion takes place and which occurs in place of sexual reproduction. Maheshwari (1950) distinguished four types of apomixis.

1. *Non-recurrent apomixis.* The megaspore mother cell undergoes the regular meiotic division to form the haploid embryo sac. The new embryo develops from the egg cell, i.e. *haploid parthenogenesis*, or from any other cell of the embryo sac, i.e. *haploid apogamy.* As plants thus formed contain only one set of chromosomes they are usually sterile and so the process is not repeated in the next generation.

2. *Recurrent apomixis.* The embryo sac may develop from a sporogenous cell—*generative apospory*—or from other cells of the nucellus—*somatic apospory.* In this case all the cells are diploid and the embryo may develop from the egg cell (*diploid parthenogenesis*) or from any other cell of the embryo sac (*diploid apogamy*).

3. *Adventive embryony.* The embryo does not develop from the cells of the embryo sac but from a cell of the nucellus or the integuments. This type of development is also known as *sporophytic budding.*

4. *Vegetative reproduction.* In this type of development the flowers, or parts of them, are replaced by bulbils or other vegetative propagules, which may germinate while still on the plant.

Polyembryony

The term *polyembryony* refers to the appearance of two or more embryos in a single seed. The processes of apomixis are often accompanied by the formation of a few embryos from the same ovule. Sometimes a normal embryo may develop together with those produced by apomixis. In a single ovule of certain species of *Citrus*, for example, three to twelve apomixial adventitious

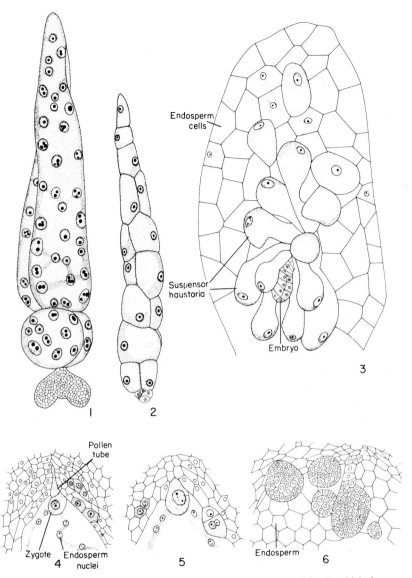

FIG. 278. 1–2, Different types of suspensors in the Papilionaceae. 1, *Orobus angustifolius* in which the suspensor consists of large multinucleate cells. 2, *Cicer arietinum* in which the suspensor consists of two rows of relatively large uninucleate cells. 3, *Asperula* showing young embryo with suspensor haustoria. 4–6, Development of adventitious embryos in *Poncirus trifoliata*. 4, Micropylar portion of the embryo sac showing a fertilized egg cell, pollen tube, endosperm nuclei, and certain nucellar cells (stippled) which are enlarged, rich in cytoplasm and which have large nuclei. 5, As in no. 4, but at a later stage of development. 6, Still later stage, showing numerous embryos in the endosperm. Only the embryo that developed from the zygote has a suspensor. (Adapted from Maheshwari, 1950.)

embryos which originate from cells of the nucellus (Fig. 278, nos. 4–6) may develop alongside the normal embryo (Webber and Batchelor, 1943).

Numerous embryos may develop as a result of the formation of growths on the multicellular body derived from the zygote, which change, during the course of development, into independent embryos. Sometimes additional embryos may be formed by the synergids or the antipodal cells that may be fertilized by sperm cells brought into the embryo sac by additional pollen tubes, or even without fertilization. In *Opuntia elata* the egg aparatus degenerates before pollination, and there are several embryos all of nucellar origin (Naumova, 1978). (For more information on polyembryony see Maheshwari, 1950 and Lakshmanan and Ambegaokar, 1984.)

Pseudo-polyembryony may also result when two or more nucelli, each of which gives rise to a normal embryo, fuse.

In addition to other anatomical characters, the structural and developmental features of embryology have been shown to be of great value in the solution of taxonomic problems (Maheshwari, 1950, 1959, 1963; Palser, 1959; Rau, 1962; Subramanyam, 1962: Johri, 1963; Davis, 1966; Herr, 1984; Tobe and Raven, 1987a, b, and others). Such embryological features include the type of tapetum, the manner of division of the pollen grains, the structure of the ovule, the thickness of the nucellus, the development of the megaspore and embryo sac, the path of entry of the pollen tube into the embryo sac, type and development of the endosperm, and the structure and development of the embryo.

In order to try to understand the phylogeny of the angiosperms attempts were also made to compare the gametophytes of this group of plants with those of the lower groups (Cocucci, 1973; Favre-Duchartre, 1984).

REFERENCES

AGTHE, C. (1951) Über die physiologische Herkunft des Pflanzennektars. *Ber. schweiz. bot. Ges.* **61**: 240–77.

ALON, G. (1986) Pollination mechanism, breeding system, and stigma structure and development in six species of Papilionoideae. *Ph.D. Thesis*, The Hebrew University of Jerusalem, Jerusalem. (In Hebrew with an English summary.)

ARBER, A. (1937) The interpretation of the flower: a study of some aspects of morphological thought. *Biol. Rev.* **12**: 157–84.

ARBER, A. (1950) *The Natural Philosophy of Plant Form.* Cambridge University Press, Cambridge.

BAILEY, I. W. (1956) Nodal anatomy in retrospective. *J. Arnold Arb.* **37**: 269–87.

BAILEY, I. W. and NAST, C. G. (1943a) The comparative morphology of the Winteraceae: I, Pollen and stamens. *J. Arnold Arb.* **24**: 340–6.

BAILEY, I. W. and NAST, C. G. (1943b) The comparative morphology of the Winteraceae: II, Carpels. *J. Arnold Arb.* **24**: 472–81.

BAILEY, I. W., NAST, C. G., and SMITH, A. C. (1943) The family Himantandraceae. *J. Arnold Arb.* **24**: 190–206.

BAILEY, I. W. and SMITH, A. C. (1942) Degeneriaceae, a new family of flowering plants from Fiji. *J. Arnold Arb.* **23**: 356–65.

BAILEY, I. W. and SWAMY, B. G. L. (1949) The morphology and relationships of *Austrobaileya. J. Arnold Arb.* **30**: 211–26.

BAILEY, I. W. and SWAMY, B. G. L. (1951) The conduplicate carpel of dicotyledons and its initial trends of specialization. *Am. J. Bot.* **38**: 373–9.

BAIRD, L. M., JORGENSEN TURANO, M. and WEBSTER, B. D. (1988) Ultrastructural and histochemical characteristics of the stigma of *Cicer arietinum. Am. J. Bot.* **75**: 551–557.

BARNARD, C. (1957) Floral histogenesis in the monocotyledons: I, The Gramineae. *Aust. J. Bot.* **5**: 1–20.

BAUM, H. (1948a) Über die postgenitale Verwachsung in Karpellen. *Öst. Bot. Z.* **95**: 86–94.

BAUM, H. (1948b) Die Verbreitung der postgenitalen Verwachsung im Gynözeum und ihre Bedeutung für die typologische Betrachtung des coenokarpen Gynözeums. *Öst. Bot. Z.* **95**: 124–8.

BAUM, H. (1948c) Postgenitale Verwachsung in und zwischen Karpell- und Staubblattkreisen. *Sber. Akad. Wiss. Wien, Math.-Naturw. Kl.,* Abt. 1, **157**: 17–38.

BAUM, H. (1948d) Ontogenetische Beobachtungen an einkarpelligen Griffeln und Griffelenden. *Öst. Bot. Z.* **95**: 362–72.

BEHRENS, W. J. (1879) Die Nektarien der Blüten. *Flora* **62**: 2–11, 17–27, 49–54, 81–90, 113–23, 145–53, 233–40, 241–7, 305–14, 369–75, 433–57.

BEUTLER, R. (1930) Biologisch-chemische Untersuchungen am Nektar von Immenblumen. *Z. Vergl. Physiol.* **12**: 72–176.

BOKE, N. H. (1949) Development of the stamens and carpels in *Vinca rosea* L. *Am. J. Bot.* **34**: 535–47.

BOKE, N. H. (1966) Ontogeny and structure of the flower and fruit of *Pereskia aculeata. Am. J. Bot.* **53**: 534–42.

BONNIER, G. (1879) *Les Nectaires.* Thèses, Fac. Sci., Paris.

BOUMAN, F. and CALIS, J. I. M. (1977) Integumentary shifting—a third way to unitegmy. *Ber. dt. bot. Ges.* **90**: 15–28.

BOUMAN, F. and SCHIER, S. (1979) Ovule ontogeny and seed coat development in *Gentiana,* with a discussion on the evolutionary origin of the single integument. *Acta bot. neerl.* **28**: 467–78.

BRETT, J. F. and POSLUSHNY, U. (1982) Floral development in *Caulophyllum thalictroides* (Berberidaceae). *Can. J. Bot.* **60**: 2133–2141.

BROWN, R. (1826) In: *Narrative of Travels and Discoveries in Northern and Central Africa,* Major Denham and Captain Clapperton. John Murray, London. App., pp. 208–48.

BROWN, W. H. (1938) The bearing of nectaries on the phylogeny of flowering plants. *Proc. Am. Phil. Soc.* **71**: 549–95.

CANRIGHT, J. E. (1952) The comparative morphology and relationships of the Magnoliaceae. I. Trends of specialization in the stamens. *Am. J. Bot.* **31**: 484–97.

CARLQUIST, S. (1969) Toward acceptable evolutionary interpretations of floral anatomy. *Phytomorphology* **19**: 332–62.

CARR, S. G. M. and CARR, D. J. (1961) The functional significance of syncarpy. *Phytomorphology* **11**: 249–56.

CARR, S. G. M. and CARR, D. J. (1968) Operculum development and the taxonomy of *Eucalyptus. Nature* **219**: 513–15.

CASPARY, R. (1848) *De Nectariis.* A. Marcus, Bonn.

CHEN, Z. K., WANG, F. H. and ZHOU, F. (1988) On the origin, development and ultrastructure of the orbicles and pollenkitt in the tapetum of *Anemarrhena asphodeloides* (Liliaceae). *Grana* **27**: 273–282.

CHEUNG, M. and SATTLER, R. (1967) Early floral development of *Lythrum salicaria*. *Can. J. Bot.* **45**: 1609–18.

CHRISTENSEN, J. E., HORNER, H. T., JR., and LERSTEN, N. R. (1972) Pollen wall and tapetal orbicular wall development in *Sorghum bicolor* (Gramineae). *Am. J. bot.* **59**: 43–58.

CHUTE, H. M. (1930) The morphology and anatomy of the achene. *Am. J. Bot.* **17**: 703–23.

CIAMPOLINI, F., CRESTI, M. and KAPIL, R. N. (1983) Fine structural and cytochemical characteristics of style and stigma in olive. *Caryologia* **36**: 211–230.

CIAMPOLINI, F., CRESTI, M., and PACINI, E. (1978) Caratteristiche ultrastrutturali ed istochimiche del tessuto trasmittente stilare de melo. Seminario sulla "Fertilita delle piante da frutto". *Progetto finalizzato CNR "Biologia della riproduzione"*. Bologna, 15 dicembre, 1978, pp. 544–51.

CIAMPOLINI, F., CRESTI, M., SARFATTI, G. and TIEZZI, A. (1981) Ultrastructure of the styler canal cells of *Citrus limon* (Rutaceae). *Pl. Syst. Evol.* **138**: 263–274.

COCUCCI, A. E. (1973) Some suggestions on the evolution of gametophytes of higher plants. *Phytomorphology* **23**: 109–124.

COCUCCI, A. and JENSEN, W. A. (1969) Orchid embryology: Pollen tetrades of *Epidendrum scutella* in the anther and on the stigma. *Planta* **84**: 215–229.

CORNER, E. J. H. (1946) Centrifugal stamens. *J. Arnold Arb.* **27**: 423–37.

CORNER, E. J. H. (1963) A criticism of the gonophyll theory of flower. *Phytomorphology* **13**: 290–2.

CRESTI, M. and VAN WENT, J. L. (1976) Callose deposition and plug formation in *Petunia* pollen tubes *in situ*. *Planta* **133**: 35–40.

CRESTI, M., CIAMPOLINI, F. and KAPIL, R. N. (1984) Generative cells of some angiosperms with particular emphasis on their microtubules. *J. Submicrosc. Cytol.* **16**: 317–326.

CRESTI, M., CIAMPOLINI, F., VAN WENT, J. L. and WILMS, H. J. (1982). Ultrastructure and histochemistry of *Citrus limon* (L.) stigma. *Planta* **156**: 1–9.

CRESTI, M., KEIJZER, C. J., TIEZZI, A., CIAMPOLINI, F. and FOCARDI, S. (1986) Stigma of *Nicotiana*: Ultrastructural and biochemical studies. *Am. J. Bot.* **73**: 1713–1722.

CRESTI, M., VAN WENT, J. L., PACINI, E. and WILLEMSE, M. T. M. (1976) Ultrastructure of transmitting tissue of *Lycopersicon peruvianum* style: development and histochemistry *Planta* **132**: 305–12.

CRESTI, M., PACINI, E., CIAMPOLINI, F., and SARFATTI, G. (1977) Germination and early tube development *in vitro* of *lycopersicum peruvianum* pollen: ultrastructural features. *Planta* **136**: 239–47.

CRUDEN, R. W. and JENSEN, K. G. (1979) Viscin threads, pollination efficiency and low pollen-ovule ratios. *Am. J. Bot.* **66**: 875–9.

CUSICK, F. (1956) Studies of floral morphogenesis: I, Median bisections of flower primordia in *Primula bulleyana* Forrest. *Trans. R. Soc. Edinb.* **63**: 153–66.

CUSICK, F. (1966) On phylogenetic and ontogenetic fusions. In: *Trends in Plant Morphogenesis* (ed. E. G. Cutter, A. Allsop, F. Cusick, and I. M. Sussex). Longmans, London, pp. 170–83.

DASHEK, W. V., THOMAS, H. R., and ROSEN, W. G. (1971) Secretory cells of lily pistils: II, Electron microscope cytochemistry of canal cells. *Am. J. Bot.* **58**: 909–20.

DAUMANN, E. (1928) Zur Biologie der Blüte von *Nicotiana glauca* Grah. *Biologia Gen.* **4**: 571–88.

DAUMANN, E. (1930a) Das Blütennektarium von *Magnolia* und

die Futterkörper in der Blüte von *Calycanthus*. *Planta* **11**: 108–16.

DAUMANN, E. (1930b) Das Blütennektarium von *Nepenthes*. *Beih. bot. Zbl.* Abt. I, **47**: 1–14.

DAUMANN, E. (1930c) Nectarien und Bienenbesuch bei *Opuntia monacantha* Haw. *Biologia Gen.* **4**: 353–76.

DAUMANN, E. (1931a) Zur Phylogenie der Diskusbildungen. *Beih. bot. Zbl.*, Abt. I, **48**: 183–208.

DAUMANN, E. (1931b) Zur morphologischen Wertigkeit der Blütennektarien von *Laurus*. *Beih. bot. Zbl.*, Abt. I, **48**: 209–13.

DAVIS, G. L. (1966) *Systematic Embryology of the Angiosperms*. Wiley, New York.

DE CANDOLLE, A. P. (1819). *Théorie élémentaire de la botanique*, 2nd edn. Deterville, Paris.

DI FULVIO, T. E. (1983) Los "tipos" de endosperma y de haustorios endospermicos. Su classification. *Kurtziana* **16**: 7–31.

DOUGLAS, G. E. (1944) The inferior ovary. *Bot. Rev.* **10**: 125–86.

DUBUC-LEBREUX, M.-A. and SATTLER, R. (1985) Quantitive distribution of mitotic activity during early corolla development of *Nicotiana tabacum*. *Phytomorphology* **35**: 17–23.

DUMAS, C. and GAUDE, T. (1982) Sérétions et biologie florale. II. Leurs rôle dans l'adhésion et la reconnaissance Pollenstigmate. Dennées récentes, hyphthèses et notion d'immunité végétale. *Bull. Soc. bot. Fr.* **129**: 89–101.

EAMES, A. J. (1931) The vascular anatomy of the flower with refutation of the theory of carpel polymorphism. *Am. J. Bot.* **18**: 147–88.

EAMES, A. J. (1961) *Morphology of the Angiosperms*. McGraw-Hill, New York.

EAMES, A. J. and MacDANIELS, L. H. (1947) *An Introduction to Plant Anatomy*, 2nd edn. McGraw-Hill, New York.

ECHLIN, P. (1968) Pollen. *Scient. Am.* **218** (4): 80–90.

ECHLIN, P. and GODWIN, H. (1969) The ultrastructure and ontogeny of pollen in *Helleborus foetidus* L.: III, The formation of the pollen grain wall. *J. Cell Sci.* **5**: 459–77.

ECKARDT, TH. (1967) Blütenbau und Blütenentwicklung von *Dysphania myriocephala* Benth. *Bot. Jb.* **86**: 20–37.

ENDRESS, P. K. (1980) The reproductive structures and systematic position of the Austrobaileyaceae. *Bot. Jahrb. Syst.* **101**: 393–433.

ENDRESS, P. K. (1983) The early floral development of *Austrobaileya*. *Bot. Jahrb. Syst.* **103**: 481–497.

ENDRESS, P. K. (1987) Floral phyllotaxis and floral evolution. *Bot. Jahrb. Syst.* **108**: 417–438.

ENDRESS, P. K., JENNY, M. and FALLEN, M. E. (1983) Convergent elaboration of apocarpous gynoecia in higher advanced dicotyledons (Sapindaes, Malvales, Gentianales). *Nord. J. Bot.* **3**: 293–300.

ERDTMAN, G. (1952) *Pollen Morphology and Plant Taxonomy: Angiosperms*. Chronica Botanica, Waltham, Mass.

ESAU, K. (1953) *Plant Anatomy*. Wiley, New York.

EYMÉ, J. (1966) Infrastructure des cellules nectarigènes de *Diplotaxis erucoides* D. C., *Helleborus niger* L. et *H. foetidus* L. *CR Séanc. Acad. Sci.* (Paris) **262**: 1629–32.

FAHN, A. (1949) Studies in the ecology of nectar secretion. *Palest. J. Bot., Jerusalem*, ser. 4: 207–24.

FAHN, A. (1952) On the structure of floral nectaries. *Bot. Gaz.* **113**: 464–70.

FAHN, A. (1953a) The origin of the banana inflorescence. *Kew Bull.*, **1953**: 299–306.

FAHN, A. (1953b) The topography of the nectary in the flower and its phylogenetical trend. *Phytomorphology* **3**: 424–6.

FAHN, A. (1979a) *Secretory Tissues in Plants*. Academic Press, London.

FAHN, A. (1979b) Ultrastructure of nectaries in relation to nectar secretion. *Am. J. Bot.* **66**: 977–85.

FAHN, A. (1988) Secretory tissues in vascular plants. *New Phytol.* **108**: 229–257.

FAHN, A. and BENOUAICHE, P. (1979) Ultrastructure, development and secretion in the nectary of banana flowers. *Ann. Bot.* **44**: 85–93.

FAHN, A. and RACHMILEVITZ, T. (1970) Ultrastructure and nectar secretion in *Lonicera japonica*. In: *New Research in Plant Anatomy* (ed. N. K. B. Robson, D. F. Cutler, and M. Gregory). Academic Press, London. *J. Linn. Soc., Bot.* **63**, suppl. 1: 51–56.

FAHN, A., STOLER, S., and FIRST, T. (1963) The histology of the vegetative and reproductive shoot apex of the Dwarf Cavendish banana. *Bot. Gaz.* **124**: 246–50.

FAVRE-DUCHARTRE, M. (1984) Homologies and phylogeny. In: *Embryology of Angiosperms* (ed. B. M. Johri), Springer-Verlag. Berlin, pp. 697–734.

FERNANDEZ, M. C. and RODRIGUEZ-GARCIA, M. I. (1988) Pollen wall development in *Olea europaea* L. *New Phytol.* **108**: 91–99.

FEUER, S., NIEZGODA, C. J. and NEVLING, L. I. (1985) Ultrastructure of *Parkia* polyads (Mimosoideae; Leguminosae). *Am. J. Bot.* **72**: 1871–1890.

FOSTER, A. S. and GIFFORD, E. M., JR. (1959) *Comparative Morphology of Vascular Plants*. W. H. Freeman, San Francisco.

FOSTER, A. S. and GIFFORD, E. M., JR. (1974) *Comparative Morphology of Vascular plants*. 2nd ed. W. H. Freeman, San Francisco.

FRANKEL, R. and GALUN, E. (1977) *Pollination Mechanisms, Reproduction and Plant Breeding*. Springer-Verlag, Berlin.

FREI, E. (1955) Die Innervierung der floralen Nektarien dikotyler Pflanzenfamilien. *Ber. schweiz. bot. Ges.* **65**: 60–114.

FREY-WYSSLING, A. (1933) Über die physiologische Bedeutung der extrafloralen Nektarien von *Hevea brasiliensis* Muell. *Ber. schweiz. bot. Ges.* **42**: 109–22.

FREY-WYSSLING, A. (1955) The phloem supply to the nectaries. *Acta bot. neerl.* **4**: 358–69.

FREY-WYSSLING, A. and HÄUSERMANN, E. (1960) Deutung der gestaltlosen Nektarien. *Ber. schweiz. bot. Ges.* **70**: 151–62.

GALIL, J. and ZERONI, M. (1969) On the organization of the pollinium in *Asclepias curassavica*. *Bot. Gaz.* **130**: 1–4.

GAWLIK, S. R. (1984) An ultrastructural study of transmitting tissue development in the pistil of *Lilium leucanthum*. *Amer. J. Bot.* **71**: 512–521.

GEMMEKE, V. (1982) Entwicklungsgeschichtliche Untersuchungen an Mimosaceen-Blüten. *Bot. Jahrb. Syst.* **103**: 185–210.

GOEBEL, K. (1928–33) *Organographie der Pflanzen*, Vols. 1–3. G. Fischer, Jena.

GRUNDWAG, M. and FAHN, A. (1969) The relation of embryology to the low seed set in *Pistacia vera* (Anacardiaceae). *Phytomorphology* **19**: 225–35.

GUIGNARD, L. (1881) *Recherches anatomiques et physiologiques sur l'embryogénie légumineuses*. G. Masson, Paris.

HANSTEIN, J. (1870) Entwicklungsgeschichte der Keime der Monokotyle und Dikotyle. *Bot. Abhandl., Bonn.* **1**: 1–112.

HAUPT, A. W. (1953) *Plant Morphology*. McGraw-Hill, New York.

HEEL, W. A. VAN (1971) The distally lobed inner integument of *Hernandia peltata* Meissn. in DC (Hernandiaceae). *Blumea* **19**: 147–8.

HELDER, R. J. (1958) The excretion of carbohydrates (nectaries). In: W. Ruhland, *Handbuch der Pflanzenphysiologie* **4**: 978–90.

HENSLOW, G. (1891) On the vascular system of floral organs, and their importance in the interpretation of the morphology of flowers. *J. Linn. Soc., Bot.* **28**: 151–97.

HERR, J. M. Jr. (1984) Embryology and taxonomy. In: *Embryology of Angiosperms* (ed. B. M. Johri), Springer-Verlag, Berlin, pp. 645–696.

HESLOP-HARRISON, J. (1968a) Pollen wall development. *Science* **161**: 230–7.

HESLOP-HARRISON, J. (1968b) Tapetal origin of pollen coat substances in *Lilium*. *New Phytol.* **67**: 779–86.

HESLOP-HARRISON, J. (ed.) (1971) *Pollen: Development and Physiology*. Butterworths, London.

HESLOP-HARRISON, J. (1975) The physiology of the pollen grain surface. *Proc. R. Soc. Lond.* B, **190**: 275–99.

HESLOP-HARRISON, J. (1978) *Cellular Recognition Systems in Plants*. Studies in Biology, no. 100. Edward Arnold, London.

HESLOP-HARRISON, J. and DICKINSON, H. G. (1969) Time relationships of sporopollenin synthesis associated with tapetum and microspores in *Lilium*. *Planta* **84**: 199–214.

HESLOP-HARRISON, J. and HESLOP-HARRISON, Y. (1982a) The specialized cuticles of the receptive surfaces of angiosperm stigmas. In: *The Plant Cuticle* (eds. D. F. Cutler, K. L. Alvin and C. E. Price), Academic Press, London, pp. 99–119.

HESLOP-HARRISON, J. and HESLOP-HARRISON, Y. (1986) Pollen-tube chemotropism: fact or delusion? In: *Biology of Reproduction and Cell Motility in Plants and Animals* (eds. M. Cresti and R. Dallai), University of Siena, Siena, pp. 169–174.

HESLOP-HARRISON, Y. (1981) Stigma characteristics and angiosperm taxonomy. *Nord. J. Bot.* **1**: 401–420.

HESLOP-HARRISON, Y, and HESLOP-HARRISON, J. (1982b) Pollen-Stigma interaction in the Leguminosae: the secretory system of the style in *Trifolium pratense* L. *Ann. Bot.* **50**: 635–645.

HESLOP-HARRISON, Y, and HESLOP-HARRISON, J. (1982c) The microfibrillar component of the pollen intine: some structural features. *Ann. Bot.* **50**: 831–842.

HILLER, G. H. (1884) Untersuchungen über die Epidermis der Blütenblätter. *Jb. wiss. Bot.* **15**: 411–51.

HOROVITZ, A. and COHEN, Y. (1972) Ultraviolet reflectance characteristics in flowers of crucifers. *Am. J. Bot.* **59**: 706–713.

JACKSON, G. (1934) The morphology of the flowers of *Rosa* and certain closely related genera. *Am. J. Bot.* **21**: 453–66.

JENSEN, W. A. (1965) The ultrastructure and histochemistry of the synergids of cotton. *Am. J. Bot.* **52**: 238–56.

JENSEN, W. A. and FISHER, D. B. (1969) Cotton embryogenesis: the tissues of the stigma and style and their relation to the pollen tube. *Planta* **84**: 97–121.

JOHANSEN, D. A. (1945) A critical survey of the present status of plant embryology. *Bot. Rev.* **11**: 87–107.

JOHRI, B. M. (1962) Nutrition of the embryo sac. In: *Proceedings of the Summer School of Botany held June 1960 at Darjeeling* (ed. P. Maheshwari, B. M. Johri, and I. K. Vasil). Ministry of Scientific Research and Cultural Affairs, Govt. of India, New Delhi, pp. 106–18.

JOHRI, B. M. (1963) Embryology and taxonomy. In: *Recent Advances in the Embryology of Angiosperms* (ed. P.

Maheshwari). Univ. Delhi. Int. Soc. of Plant Morphologists, Delhi, pp. 395–444.

JOHRI, B. M. and AGARWAL, S. (1965) Morphological and embryological studies in the family Santalaceae: VIII, *Quinchamalium chilense* Lam. *Phytomorphology* **15**: 360–72.

JOSHI, A. C. (1947) The morphology of the gynoecium. Presidential address. 34th Indian Sci. Congr. Delhi.

KANDASAMY, M. K. and KRISTEN, U. (1987) Developmental aspects of ultrastructure, histochemistry and receptivity of the stigma of *Nicotiana sylvestris*. *Ann. Bot.* **60**: 427–437.

KAPLAN, D. R. (1967) Floral morphology, organogenesis and interpretation of the inferior ovary in *Downingia bacigalupii*. *Am. J. Bot.* **54**: 1274–90.

KAPLAN, D. R. (1968) Structure and development of the perianth in *Downingia bacigalupii*. *Am. J. Bot.* **55**: 406–20.

KAY, Q. O. N. and DAOUD, H. S. (1981) Pigment distribution, light reflection and cell structure in petals. *Bot. J. Linn. Soc.* **83**: 57–84.

KIESSELBACH, T. A. (1949) The structure and reproduction of corn. *Bull. Nebr. Agr. Exp. Stn.* no. 161.

KIRKWOOD, J. E. (1906) The pollen tube in some of the Cucurbitaceae. *Bull. Torrey Bot. Club* **37**: 327–41.

KNOX, R. B. (1984) The pollen grain. In: *Embryology of Angiosperms* (ed. B. M. Johri), Springer-Verlag, Berlin, pp. 196–271.

KNUTH, P. (1898–1905) *Handbuch der Blütenbiologie*. I, II₁, II₂, III₁, III₂. W. Engelmann, Leipzig.

KOCH, M. F. (1930) Studies in the anatomy and morphology of the composite flower: I, The corolla. *Am. J. Bot.* **17**: 938–52.

KONAR, R. N. and LINSKENS, H. F. (1966a) The morphology and anatomy of the stigma of *Petunia hybrida*. *Planta* **71**: 356–71.

KONAR, R. N. and LINSKENS, H. F. (1966b) Physiology and biochemistry of the stigmatic fluid of *Petunia hybrida*. *Planta* **71**: 372–87.

KRESS, W. J., STONE, D. E., and SELLERS, S. C. (1978) Ultrastructure of exine-less pollen: *Heliconia* (Heliconiaceae). *Am. J. Bot.* **65**: 1064–76.

KRISTEN, U., BIEDERMANN, M., LIEBEZEIT, G., DAWSON, R., and BÖHM, L. (1979) The composition of stigmatic exudate and the ultrastructure of the stigma papillae in *Aptenia cordifolia*. *Europ. J. Cell Biol.* **19**: 281–7.

KRONESTEDT, E., WALLES, B, and ALKEMAR, I. (1986) Structural studies of pollen-tube growth in pistil of *Strelitzia reginae*. *Protoplasma* **131**: 224–232.

KUNZE, H. v. (1978) Typologie und Morphogenese des Angiospermen-Staubblattes. *Beitr. Biol. Pfl.* **54**: 239–304.

LAKSHMANAN, K. K. and AMBEGAOKAR, K. B. (1984) Polyembryony. In: *Embryology of Angiosperms* (ed. B. M. Johri), Springer-Verlag, Berlin, pp. 445–474.

LEINFELLNER, W. (1950) Der Bauplan des synkarpen Gynoezeums. *Öst. bot. Z.* **97**: 403–36.

LEROY, Y. F. (1955) Etudes sur les Juglandacee: à la recherche d'une conception morphologique de la fleur femelle et du fruit. *Mém. Muséum Nation. d'Hist. Naturelle*, ser. B, *Bot.* **6**: 1–246.

LERSTEN, N. R. (1971) A review of septate microsporangia in vascular plants. *Iowa St. J. Sci.* **45**: 487–97.

LERSTEN, N. R. and WEMPLE, D. K. (1966) The discontinuity plate, a definitive floral characteristic of the Psoraleae (Leguminosae). *Am. J. Bot.* **53**: 548–55.

LIEBERMAN, S. J., VALDOVINOS, J. G. and JENSEN, T. E. (1982) Ultrastructural, localization of cellulose in abscission cells on tobaco flower pedicels. *Bot. Gaz.* **143**: 32–40.

LINSKENS, H. F. (1975) Incompatibility in *Petunia. Proc. R. Soc. Lond.* B, **188**: 299–311.

LLOYD, F. E. (1902) The comparative morphology of Rubiaceae. *Mem. Torrey Bot. Club* **9**: 1–112.

LORCH, J. (1963) The carpel—case-history of an idea and a term. *Centaurus* **8**: 269–91.

MACDANIELS, L. H. (1940) The morphology of the apple and other pome fruits. *Mem. Cornell Univ. Agr. Exp. Stn.* 230.

MAHESHWARI, P. (1950) *An Introduction to the Embryology of Angiosperms*. McGraw-Hill, New York.

MAHESHWARI, P. (1959) Embryology in relation to taxonomy. In: *Recent Advances in Botany*. Int. Bot. Congr. Montreal, Univ. of Toronto Press, Toronto **1**: 679–82.

MAHESHWARI, P. (1963) Embryology in relation to taxonomy. In: *Vistas in Botany* (ed. W. B. Turrill), Pergamon Press, **4**: 55–97.

MANNING, W. E. (1940). The morphology of the flowers of the Juglandaceae: II, The pistillate flowers and fruits. *Am. J. Bot.* **27**: 838–52.

MARTENS, P. (1934) Recherches sur la cuticule: IV, Le relief cuticulaire et la différentiation épidermique des organes floraux. *Cellule* **43**: 289–320.

MASCARENHAS, J. P. (1975) The biochemistry of angiosperm pollen development. *Bot. Rev.* **41**: 259–314.

MATTSSON, O., KNOX, R. B., HESLOP-HARRISON, J., and HESLOP-HARRISON, Y. (1974) Protein pellicle of stigmatic papillae as a probable recognition site in incompatibility reaction. *Nature* **247**: 298–300.

MCLEAN, R. C. and IVIMEY-COOK, W. R. (1956) *Textbook of Theoretical Botany*, Vol. II. Longmans, London.

MEEUSE, A. D. J. (1966) *Fundamentals to Phytomorphology*. The Ronald Press Company, New York.

MEIRI, L. and DULBERGER, R. (1986) Stamen filament structure in the Asteraceae: The anther collar. *New Phytol.* **104**: 693–701.

MELVILLE, R. (1961) A new theory of the Angiosperm flower. *Nature* **188**: 14–18.

MELVILLE, R. (1962) A new theory of the Angiosperm flower: I, *Kew Bull.* **16**: 1–50.

MEPHAM, R. H. and LANE, G. R. (1970) Observations on the fine structure of developing microspores of *Tradescantia bracteata*. *Protoplasma* **70**: 1–20.

MOGENSEN, H. L. and SUTHAR, H. K. (1979) Ultrastructure of the egg apparatus of *Nicotiana tabacum* (Solanaceae) before and after fertilization. *Bot. Gaz.* **140**: 168–79.

MONTEIRO-SCANAVACCA, W. R. (1974) Vascularizacao do gineceu em Lecythidaceae. *Bol. Botanica. Univ. S. Paulo* **2**: 53–69.

MOSELEY, M. F., JR. (1965) Morphological studies of the Nymphaeaceae: III, The floral anatomy of *Nuphar. Phytomorphology* **15**: 54–84.

MURRAY, M. A. (1945) Carpellary and placental structure in Solanaceae. *Bot. Gaz.* **107**: 243–60.

NARAYANA, L. L. (1965) Contributions to the embryology of Balsaminaceae (2). *Jap. J. Bot.* **40**: 104–16.

NATESH, S. and RAU, M. A. (1984) The embryo. In: *Embryology of Angiosperms* (ed. B. M. Johri), Springer-Verlag, Berlin, pp. 377–443.

NAUMOVA, T. N. (1978) Specifities of development of nucellar tissue and nucellar polyembryony in *Opuntia elata* (Cactaceae). *Bot. Zh. SSSR* **63**: 344–55.

NIEZGODA, C. J., FEUER, S. M. and NEVLING, L. I. (1983) Pollen ultrastructure of the tribe Ingeae (Mimosoideae: Leguminosae). *Am. J. Bot.* **70**: 650–667.

NISHINO, E. (1983) Corolla tube formation in the Primulaceae and Ericales. *Bot. Mag. Tokyo* **96**: 319–342.

NORSTOG, K. (1972) Early development of the barley embryo: fine structure. *Am. J. Bot.* **59**: 123–32.

PACINI, E. and JUNIPER, B. E. (1983) The ultrastructure of the formation and developments of the amoeboid tapetum in *Arum italicum* Miller. *Protoplasma* **117**: 116–129.

PALSER, B. F. (1959) Some aspects of embryology in the Ericales. In: *Recent Advances in Botany.* Int. Bot. Congr. Montreal, Univ. of Toronto Press, Toronto, **1**: 685–98.

PARKIN, J. (1951) The protrusion of the connective beyond the anther and its bearing on the evolution of the stamen. *Phytomorphology* **1**: 1–8.

PATON, J. V. (1921) Pollen and pollen enzymes. *Am. J. Bot.* **8**: 471–501.

PAYER, J. B. (1857) *Traité d'organogénie comparée de la fleur.* V. Masson, Paris.

PERIASAMY, K. and SWAMY, B. G. L. (1956) The conduplicate carpel of *Cananga odorata. J. Arnold Arb.* **37**: 366–72.

PFEIFFER, H. (1928) Die pflanzlichen Trennungsgewebe. In: K. Linsbauer, *Handbuch der Pflanzenanatomie*, Bd. 5, Lief. 22. Gebr. Borntraeger, Berlin.

POPE, M. N. (1946) The course of the pollen tube in cultivated barley. *J. Am. Soc. Agron.* **38**: 432–40.

POPHAM, R. A. and CHAN, A. P. (1952) Origin and development of the receptacle of *Chrysanthemum morifolium. Am. J. Bot.* **39**: 329–39.

PORSCH, O. (1913) Die Abstammung der Monokotylen und die Blütennektarien. *Ber. dt. bot. Ges.* **31**: 580–90.

POSLUSHNY, U., SCOTT, M. G. and SATTLER, R. (1980) Revisions in the technique of epi-illumination on light microscopy for the study of floral and vegetative apices. *Can. J. Bot.* **58**: 2491–2495.

PRIDGEON, A. M. and STERN, W. L. (1983) Ultrastructure of osmophores in *Restrepia* (Orchidaceae). *Am. J. Bot.* **70**: 1233–1243.

PRIDGEON A. M. and STERN, W. L. (1985) Osmophores of *Scaphosepalum* (Orchidaceae). *Bot. Gaz.* **146**: 115–123.

PURI, V. (1951) The role of floral anatomy in the solution of morphological problems. *Bot. Rev.* **13**: 471–557.

PURI, V. (1960) *Carpel Morphology.* Agra Univ. Extension Lectures delivered February 1960, pp. 7–27.

PURI, V. (1962) On the concept of carpellary margins. In: *Proceedings of the Summer School of Botany held June 1960 at Darjeeling* (ed. P. Maheshwari, B. M. Johri, and I. K. Vasil). Ministry of Scientific Research and Cultural Affairs, Govt. of India, New Delhi, pp. 326–33.

RACHMILEVITZ, T. and FAHN, A. (1975) The floral nectary of *Tropaeolum majus* L.—The nature of the secretory cells and the manner of nectar secretion. *Ann. Bot.* **39**: 721–8.

RAMP, E. (1987) Funktionelle Anatomie des Gynoeciums bei *Staphylea. Botanica Helvetica* **97**: 89–98.

RAO, A. N. (1967) Flower and seed development in *Arundina graminifolia. Phytomorphology* **17**: 291–300.

RAU, M. A. (1962) Review of recent work on embryogeny of some families and genera of disputed systematic position. In: *Plant Embryology.* A Symposium, CSIR, New Delhi, pp. 75–80.

RENNER, O. and PREUSS-HERZOG, G. (1943) Der Weg der Pollenschläuche im Fruchtknoten der Oenotheren. *Flora* **36**: 215–22.

RICHARDSON, F. C. (1969) Morphological studies of the Nymphaeaceae: IV, Structure and development of the flower of *Brasenia schreberi* Gmel. *Univ. Calif. Publ. Bot.* **47**: 1–101.

RIESEBERG, L. H. and SCHILLING, E. E. (1985) Floral flavonoids and ultraviolet patterns in *Viquiera* (Compositae). *Am. J. Bot.* **72**: 999–1004.

ROSEN, W. G. (1964) Chemotropism and fine structure of pollen tubes. In: *Pollen Physiology and Fertilization* (ed. H. F. Linskens). North-Holland, Amsterdam, pp. 159–66.

ROSEN, W. G., GAWLIK, S. R., DASHEK, W. V., and SIEGESMUND, K. A. (1964) Fine structure and cytochemistry of *Lilium* pollen tubes. *Am. J. Bot.* **51**: 61–71.

ROSEN, W. G. and THOMAS, H. R. (1970) Secretory cells of lily pistils: I, Fine structure and function. *Am. J. Bot.* **57**: 1108–14.

ROWLEY, J. R. and SOUTHWORTH, D. (1967) Deposition of sporopollenin on lamellae of unit membrane dimensions. *Nature* **213**: 703–4.

SARFATTI, G. and GORI, P. (1969) Embryo sac of *Euphorbia dulcis* L., an ultrastructural study. *G. Bot. Ital.* **103**: 631–2.

SATINA, S. (1944) Periclincal chimeras in *Datura* in relation to development and structure (A) of the style and stigma (B) of calyx and corolla. *Am. J. Bot.* **31**: 493–502.

SATTLER, R. (1978) "Fusion" and "continuity" in floral morphology. *Notes Roy. Bot. Gard. Edinb.* **36**: 397–405.

SATTLER, R. and LACROIX, C. (1988) Development and evolution of basal cauline placentation: *Basella rubra. Am. J. Bot.* **75**: 918–927.

SCHICK, B. (1980) Untersuchungen über die Biotechnik der Apocynaceenblüte. I. Morphologie und Funktion des Nektarkopfes. *Flora* **170**: 394–432.

SCHICK, B. (1982) Untersuchungen über die Bedeutung der Apocynaceenbute. II. Bau and Funktion des Bestaübungsapparates. *Flora* **172**: 347–371.

SCHILL, R. and WOLTER, M. (1986) On the presence of elasto-viscin in all subfamilies of the Orchidaceae and the homology to pollenkitt. *Nord. J. Bot.* **6**: 321–324.

SCHMID, R. (1972) Floral bundle fusion and vascular conservatism *Taxon* **21**: 429–46.

SCHNARF, K. (1927) Embryologie der Angiospermen. In: K. Linsbauer, *Handbuch der Pflanzenanatomie*, Bd. 10, Lief. 21. Gebr. Borntraeger, Berlin.

SCHNARF, K. (1928) Embryologie der Angiospermen. In: K. Linsbauer, *Handbuch der Pflanzenanatomie*, Bd. 10, Lief, 23, Gebr. Borntraeger, Berlin.

SCHNARF, K. (1929) Embryologie der Angiospermen. In: K. Linsbauer, *Handbuch der Pflanzenanatomie*, Bd. 10, Lief. 24. Gebr. Borntraeger, Berlin.

SCHNEPF, E. (1964a) Zur Cytologie und Physiologie pflanzlicher Drüsen. 4. Teil. Licht- und elektronenmikroskopische Untersuchungen an Septalnektarien. *Protoplasma* **58**: 137–71.

SCHNEPF, E. (1964b) Zur Cytologie und Physiologie pflanzlicher Drüsen. 5. Teil. Elektronenmikroskopische Untersuchungen an Cyathialnektarien von *Euphorbia pulcherrima* in verschiedenen Funktioszuständen. *Protoplasma* **58**: 193–219.

SCHNEPF, E. and NAGL, W. (1970) Über einige Strukturbesonderheiten der Suspensorzellen von *Phaseolus vulgaris. Protoplasma* **69**: 133–43.

SCHNEPF, E., WITZIG, F. and SCHILL, R. (1979) Über Bildung und Feinstruktur des Translators der Pollinarien von *Asclepias- curassavica* und *Gomphocarpus fruticosus* (Asclepiadaceae). *Tropische und Subtropische Pflanzenwelt*, **25**: 7–39. Franz Steiner, Wiesbaden.

SCHNIEWIND-THIES, J. (1897) Beiträge zur Kenntnis der Septalnektarien (abstract). *Bot. Zbl.* **69–70**: 216–18.

SCHOCH-BODMER, H. and HUBER, P. (1947) Die Ernährung der Pollenschläuche durch das Leitgewebe. (Untersuchungen an *Lythrum salicaria* L.) *Vjschr. naturf. Ges. Zürich* **92**: 43–48.

SCHOENICHEN, W. (1922) *Mikroskopisches Praktikum der Blütenbiologie*. Quelle & Meyer, Leipzig.

SCHULZ, P. and JENSEN, W. A. (1977) Cotton embryogenesis: the early development of the free nuclear endosperm. *Am. J. Bot.* **64**: 384–94.

SCHULZ, S. R. and JENSEN, W. A. (1968a) *Capsella* embryogenesis: the synergids before and after fertilization. *Am. J. Bot.* **55**: 541–52.

SCHULZ, S. R. and JENSEN, W. A. (1968b) *Capsella* embryogenesis: the early embryo. *J. Ultrastruct. Res.* **22**: 376–92.

SCHULZ, S. R. and JENSEN, W. A. (1969) *Capsella* embryogenesis: the suspensor and the basal cell. *Protoplasma* **67**: 139–63.

SHAW, G. (1971) The chemistry of sporopollenin. In: *Sporopollenin* (ed. J. Brooks, P. R. Grant, M. Muir, P. van Gijzel and G. Shaw). Academic Press, London, pp. 305–50.

SHIVANNA, K. R., HESLOP-HARRISON, Y., and HESLOP-HARRISON, J. (1978) The pollen–stigma interaction: bud pollination in Cruciferae. *Acta bot. neerl.* **27**: 107–19.

SIMONCIOLI, C. (1974) Ultrastructural characteristics of *Diplotaxis erucoides* (L.) DC suspensor. *G. Bot. Ital.* **108**: 175–89.

SKUTCH, A. F. (1932) Anatomy of the axis of banana. *Bot. Gaz.* **93**: 233–53.

SKVARLA, J. J. and LARSON, D. A. (1966) Fine structural studies of *Zea mays* pollen: I, Cell membranes and exine ontogeny. *Am. J. Bot.* **53**: 1112–25.

SKVARLA, J. J. and ROWLEY, J. R. (1970) The pollen wall of *Canna* and its similarity to germinal apertures of other pollen. *Am. J. Bot.* **57**: 519–29.

SMITH, B. N. and MEEUSE, B. J. (1966) Production of volatile amines and skatole at anthesis in some *Arum* Lily species. *Pl. Physiol.* **41**: 343–347.

SMITH, F. H. and SMITH, E. C. (1942) Anatomy of the inferior ovary of *Darbya. Am. J. Bot.* **29**: 464–71.

SOUÈGES, E. C. R. (1914) Nouvelles recherches sur le développement de l'embryon chez les Crucifères. *Annls Sci. nat., Bot.,* ser. 9, **19**: 311–39.

SOUÈGES, E. C. R. (1919) Les premières divisions de l'oeuf et les différentiations du suspenseur chez le *Capsella bursa-pastoris. Annls Sci. nat., Bot.,* ser. 10, **9**: 1–28.

SOUÈGES, E. C. R. (1924) Embryogénie des Graminées. Développement de l'embryon chez le *Poa annua* L. *CR Acad. Sci. (Paris)* **178**: 1307–10.

SOUÈGES, E. C. R. (1925) Développement de l'embryon chez le *Sherardia arvensis* L. *Bull. Soc. bot. Fr.* **72**: 546–65.

SOUÈGES, E. C. R. (1932) Recherches sur l'embryogénie des Liliacées (*Muscari comosum* L.). *Bull. Soc. bot. Fr.* **79**: 11–123.

SPORNE, K. R. (1958) Some aspects of floral vascular systems. *Proc. Linn. Soc. Lond. Bot.* **169**: 75–84.

STANLEY, R. G. and LINSKENS, H. F. (1974) *Pollen: Biology, Biochemistry and Management*. Springer, New York.

STERLING, C. (1966) Comparative morphology of the carpel in the Rosaceae: VIII, Spiraeoideae: Holodisceae, Neillieae, Spiraeeae, Ulmarieae. *Am. J. Bot.* **53**: 521–30.

STRASBURGER, E. (1879) *Die Angiospermen und die Gymnospermen*. Jena.

SUBRAMANYAM, K. (1962) Embryology in relation to systematic botany with particular reference to the Crassulaceae. In:

Plant Embryology. A Symposium. CSIR New Delhi, pp. 94–112.

SWAMY, B. G. L. (1949a) Embryological studies in the Orchidaceae: II, Embryogeny. *Am. Midland Nat.* **41**: 202–32.

SWAMY, B. G. L. (1949b) Further contributions to the morphology of the Degeneriaceae. *J. Arnold Arb.* **30**: 10–38.

TEPFER, S. S. (1953) Floral anatomy and ontogeny in *Aquilegia formosa* var. *truncata* and *Ranunculus repens. Univ. Calif. Publ. Bot.* **25**: 513–648.

TIAGI, Y. D. (1955) Studies in floral morphology: II, Vascular anatomy of the flower of certain species of the Cactaceae. *J. Ind. Bot. Soc.* **34**: 408–28.

TILTON, V. R. and HORNER, H. T., JR. (1980) Stigma, style, and obturator of *Ornithogalum candatum* (Liliaceae) and their function in the reproductive process. *Am. J. Bot.* **67**: 1113–31.

TILTON, V. R., WILCOX, L. W. and PALMER, R. G. (1984a) Postfertilization wandlabrinthe formation and function in the central cell of soybean. *Glycine max* (L.) Merr. (Leguminosae). *Bot. Gaz.* **145**: 334–339.

TILTON, V. R., WILCOX, L. W., PALMER, R. G. and ALBERTSEN, M. C. (1984b) Stigma, style, and obturator of soybean, *Glycine max* (L.) Merr. (Leguminosae) and their function in the reproductive process. *Am. J. Bot.* **71**: 676–686.

TOBE, H. and Raven, P. (1987a) Embryology and systematic position of *Heteropyxis* (Myrtales). *Am. J. Bot.* **74**: 197–208.

TOBE, H. and RAVEN, P. (1987b) The embryology and relationships of *Dactylocladus* (Crypteroniaceae) and a discussion of the family. *Bot. Gaz.* **148**: 103–111.

TROLL, W. (1948) *Allgemeine Botanik*. F. Enke, Stuttgart.

TUCKER, S. C. (1966) The gynoecial vascular supply in *Caltha. Phytomorphology* **16**: 339–42.

TUCKER, S. C. (1972) The role of ontogenetic evidence in floral morphology. In: *Advances in Plant Morphology* (eds. Y. S. Murty *et al.*). Prof. V. Puri Commem. Vol., Sarita Prakashan, Meerut, India, pp. 359–69.

TUCKER, S. C. (1975) Carpellary vasculature and the ovular vascular supply in *Drimys. Am. J. Bot.* **62**: 191–7.

TUCKER, S. C. (1984a) Origin of symetry in flowers. In: *Contemporary Problems in Plant Anatomy* (eds. R. A. White and W. C. Dickison). Academic Press, Orlando.

TUCKER, S. C. (1984b) Unidirectional organ initiation in leguminous flowers. *Am. J. Bot.* **71**: 1139–1148.

TUCKER, S. C. and GIFFORD, E. M., JR. (1966) Carpel development in *Drimys lanceolata. Am J. Bot.* **53**: 671–8.

TUPY, J. (1959) Callose formation in pollen tubes and incompatibility. *Biol. Plant.* **1**: 192–8.

UHL, N. W. and MOORE, H. E., JR. (1977) Centrifugal stamen initiation in phytelephantoid palms. *Am. J. Bot.* **64**: 1152–61.

VAN HEEL (1983) The ascidiform early development of free carpels, A S.E.M.-investigation. *Blumea* **28**: 231–270.

VENTURELLI, M. (1984) Estudos embryologicos em Loranthaceae: *Struthanthus flexicaulis* Mart. *Revta brasil. Bot.* **7**: 107–119.

VIJAYARAGHAVAN, M. R. and PRABHAKAR, K. (1984) The endosperm. In: *Embryology of Angiosperms* (ed. B. M. Johri), Springer Verlag, Berlin, pp. 319–376.

VOGEL, S. (1962) Duftdrüsen in Dienst der Bestäubung. Über Bau und Funktion der Osmophoren. *Akad. Wiss. Lit. Mainz, Abb. Math. naturwiss. KL. Nr.* **10**: 598–763.

WARDLAW, C. W. (1955) *Embryogenesis in Plants*. Methuen, London; Wiley, New York.

WEBBER, H. J. and BATCHELOR, L. D. (1943) *The Citrus Industry*, Vol. I. Univ. of Calif. Press, Berkeley.

WELK, SR. M., MILLINGTON, W. F., and ROSEN, W. G. (1965) Chemotropic activity and pathway of the pollen tube in lily. *Am. J. Bot.* **52**: 774–81.

WENT, J. L., VAN and WILLEMSE, M. T. M. (1984) Fertilization. In: *Embryology of Angiosperms* (ed. B. M. Johri), Springer Verlag, Berlin, pp. 273–317.

WILMS, H. J. (1981) Pollen tube penetration and fertilization in spinach. Acta Bot. Neerl. **30**: 101–122.

WILSON, C. L. (1937) The phylogeny of the stamen. *Am. J. Bot.* **26**: 686–99.

WILSON, C. L. (1942) The telome theory and the origin of the stamen. *Am. J. Bot.* **21**: 759–64.

WODEHOUSE, R. P. (1935) *Pollen Grains*. McGraw-Hill, New York.

WOLTER, M. and SCHILL, R. (1986) Ontogenie von Pollen, Massulae und Pollinien bei den Orchideen. *Trop. Subtrop. Pflanzenwelt* **56**: 7–93. Franz Steiner, Wiesbaden, Stuttgart.

WOLTER, M., SEUFFERT, C. and SCHILL, R. (1988) The ontogeny of pollinia elastoviscin in the anther of *Doritis pulcherrima* (Orchidaceae). *Nord. J. Bot.* **8**: 77–88.

YAMPOLSKY, C. (1934) The cytology of the abscission zone in *Mercurialis annua. Bull. Torrey Bot. Club* **61**: 279–89.

YEUNG, E. C. (1987) Mechanism of pollen aggregation into pollinia in *Epidendrum ibaguense* (Orchidaceae). *Grana* **26**: 47–52.

YEUNG, E. C. and CLUTTER, M. E. (1979) Embryology of *Phaseolus coccineus*: The ultrastructure and development of the suspensor. *Can. J. Bot.* **57**: 120–136.

YEUNG, E. C. and LAW, S. K. (1987) The formation of hyaline caudicle in two vandoid orchids. *Can. J. Bot.* **65**: 1459–1464.

ZIMMERMANN, J. G. (1932) Über die extrafloralen Nektarien der Angiospermen. *Beih. bot. Zbl.* **49**A: 99–196.

ZIMMERMANN, M. (1954) Über die Sekretion saccharosespaltender Transglukosidasen im pflanzlichen Nektar. *Experientia* (*Basel*) **10**: 145–6.

CHAPTER 20

THE FRUIT

THE fruit generally develops from the gynoecium, but in many fruits other organs also participate. Such organs may be the tepals (*Morus*), the receptacle (*Fragaria*), bracts (*Ananas*), the floral tube, which is formed by the floral organs together with the receptacle (*Pyrus malus*), or the enlarged axis of the inflorescence (*Ficus*). In cases where organs other than the gynoecium participate in the formation of the fruit, the fruit is termed as false fruit or *pseudocarp.*

It is generally accepted that the fruit develops after fertilization, but this is not always so. Fruits of many plants, such as certain varieties of *Musa, Citrus,* and *Vitis,* develop without the formation of seeds. This phenomenon is termed *parthenocarpy.* In some plants, e.g. in the peanut (*Arachis hypogaea*), the fruit develops only after the gynophore penetrates into the soil carrying the carpel with the fertilized ovules (Zamski and Ziv, 1976).

There are different methods of fruit classification (Winkler, 1939, 1940; McLean and Ivimey-Cook, 1956; Roth, 1977; and others). The fruits are classified into a few types on the basis of two criteria. The main criterion is the degree of hardness of the *pericarp, i.e.* the fruit wall—whether it is dry and hard or soft and fleshy or juicy. The second criterion is the ability of the fruit to dehisce or not when ripe (Fahn and Werker, 1972).

DRY FRUITS

Dehiscent fruits

(a) Fruits that develop from a single carpel.
  (i) *Follicle*: a pod-like fruit which generally splits down the ventral side (*Delphinium, Brachychiton*).

  (ii) *Legume*: a fruit that splits into two valves along a suture which surrounds the fruit (Leguminosae).

(b) Syncarpous fruits, i.e. those developing from an ovary with two or more carpels.
  (i) *Siliqua*: a pod-like fruit consisting of two carpels which is considered by many to be a special type of capsule (see below). The suture between the carpels' margins forms a thick rib, termed *replum,* around the fruit. From these sutures, which bear the placentae on their inner surfaces, two membranes grow inwards where they fuse to form a false septum which divides the locule into two. When the fruit ripens the two valves separate from the replum to which the seeds remain attached and which itself remains as a frame around the septum. Such fruits occur in most genera of the Cruciferae.
  (ii) *Capsule*: a fruit developing from two or more carpels and dehiscing in different ways that are of taxonomic importance. Usually the split is from apex down and then it may be along the dorsal bundle of each carpel and the dehiscence is said to be *loculicidal* (e.g. *Epilobium, Iris*), or between the carpels and the dehiscence is said to be *septicidal* (e.g. *Hypericum*). If the outer wall of the fruit breaks away from the septa, which remain attached to the axis, the dehiscence is said to be *septifragal.* In some species of *Campanula* and *Papaver,* the dehiscence is *porous,* i.e. by means of small pores which develop in the pericarp. In *Anagallis* and *Hyoscyamus,* the dehiscence is *circumscissile,* i.e. by means of a transverse split which results in the formation

of a lid. When the dehiscence is by means of outwardly flared teeth, it is said to *valvate*. In all the examples given above, the portions into which the fruit splits are termed *valves*.

Indehiscent dry fruits

(a) *Achenium, achene* or *akene*: a single-seeded fruit formed by one carpel (*Ranunculus*).
(b) *Cypsela*: a single-seeded fruit developing from an inferior ovary and thus surrounded by other floral tissues in addition to the ovary wall (Compositae). This is actually a false fruit.
(c) *Nux* or *nut*: a single-seeded fruit that develops from an ovary that originally consists of a few carpels of which all but one, in which a single ovule develops, degenerate (*Valerianella* and *Tilia*). The acorn of *Quercus*, together with its involucre, is a false nut as it develops from an inferior ovary.
(d) *Caryopsis*: a one-seeded fruit in which the seed wall is adnated to the pericarp (Gramineae).
(e) *Samara*: a winged one-seeded fruit (*Ulmus* and *Fraxinus*).

There are also indehiscent dry fruits consisting of several carpels and containing one or more seeds, i.e. *carcerulus* (in a few Cruciferae, as, for example, *Crambe*).

Schizocarpic fruits

Schizocarpic fruits are those that develop from multiloculate ovaries that separate when ripe into akenes, the number of which is equal to the number of carpels. Such a fruit is termed a *schizocarp* and all of the parts into which it separates are termed *mericarps* (e.g. many genera of the Malvaceae such as *Malva* and *Lavatera*).

The *cremocarp* of the Umbelliferae is also a schizocarp, but it is actually a false fruit as it develops from an inferior ovary. In the Labiatae the fruit also separates into four akene-like mericarps (*nutlet* or *coccus*) each of which consists of half a carpel enclosing a single seed.

From a phylogenetic viewpoint the follicle, which develops from an apocarpous gynoecium, is considered the most primitive type of fruit.

FLESHY FRUITS

Berry or *bacca*. A fruit in which the pericarp is usually thick and juicy and in which three strata can be distinguished: the outer stratum which usually contains the pigment of the fruit—*exocarp*; the relatively thick stratum below it—*mesocarp*; and the membranous inner stratum—*endocarp*. This fleshy pericarp may enclose one or many seeds (e.g. grape, tomato). The fruit of *Citrus*, which is also a berry, has been specially termed an *hesperidium*. The fruits of *Coffea, Sambucus, Hedera, Cucumis*, and *Musa* are also berries but, theoretically, they are false fruits as they develop from inferior ovaries; they differ from typical false fruits in that the extra-carpellary parts contribute only a small part in the construction of their pericarp.

Drupe. This fruit differs from the berry in that the endocarp is thick and hard (*Prunus, Mangifera, Pistacia, Juglans*). The "nut" of *Cocos* is a drupe in which the mesocarp consists of fibrous matter.

Aggregate fruits are obtained when the carpels of an apocarpous gynoecium ripen individually but in the course of ripening the individual fruits of a flower aggregate to form a single unit, as, for example, in *Rubus*.

HETEROCARPY AND AMPHICARPY

Some plants produce fruits of more than one kind on the same individual. This phenomenon is called *heterocarpy*. *Aethionema heterocarpum* and *Ae. carneum* (Cruciferae), for instance, produce dehiscent siliquas and one-seeded indehiscent nuts in the same raceme (Fig. 281, nos. 1, 2) (Zohary and Fahn, 1950). Cases in which different fruits are borne above and below the ground are termed *amphicarpy*, e.g. *Trifolium polymorphum*, *Vicia angustifolia* var. *amphicarpa, Amphicarpaea bracteata* (Leguminosae), *Gymnarrhena micranta* (Compositae), and *Emex spinosa* (Polygonaceae). Amphicarpy is in some way related to floral cleisto-gamy (Zohary, 1937; Koller and Roth, 1964; van der Pijl, 1969; Evenari *et al.*, 1977; Schnee and Waller, 1986). In *Cardamine chenopodifolia* (Cruci-ferae) the plants which arise from seeds of the subterranean fruits differ from those which develop from seeds of the aerial fruits (Cheplick, 1983).

HISTOLOGICAL STRUCTURE OF THE PERICARP

There is no distinct tissue differentiation in the pericarp prior to the ripening of the fruit. During the development of the fruit the number of cells increases and the parenchymatous ground tissue may remain as it is, or part of it may become sclerenchymatous. In the mature fruit the above three layers—the exocarp, mesocarp, and endo-carp—can be distinguished but sometimes only the exo- and endocarps are discernible. However, in many fruits the exo- and endocarps may con-sist only of epidermal tissues and the pericarp then consists mainly of mesocarp. The division into these pericarp layers is only of convenience to facilitate the anatomical description of mature fruits, and these layers do not represent separate tissues from the point of view of their origin. In this book this division will also be used in the description of false fruits.

Fruits of many plants have stomata in their epidermis. In *Cucurbita pepo*, for instance, 44 stom-ata were reported to occur per square millimetre (Hayward, 1938). In some, even relatively large fruits, e.g. tomatoes and peppers, no stomata occur in the epidermis. In the capsule of *Eschscholtzia* (Papaveraceae) stomata are present not only on the outer but also on the inner side of the pericarp. The stomata on the inner fruit-wall side as well as those on the seed coat remain constantly open, whereas those on the outer side of the fruit wall were reported to be able to close and open (Jernsteadt and Clark, 1979).

In large fruits vascular bundles, which develop in the ground tissue, are added to the vascular system of the gynoecium (as described in the pre-vious chapter) so that the supply of water and other substances to all parts of the fruit is made possible.

Usually there is a relationship between the manner of fruit and seed dispersal and the histo-logical structure of the pericarp (Guttenberg, 1971; Fahn and Werker, 1972).

Pericarp of dry fruits

Dehiscent Fruits

Follicle. In the follicle of *Delphinium*, for example, the exocarp (outer epidermis and some-times also a hypodermal layer) consists of thick-walled cells, the mesocarp is parenchymatous, and the endocarp (the inner epidermis) consists of thick-walled cells (Fig. 279, no. 1). The vascular bundles have sclerenchymatous sheaths. The peri-carp dries out with the maturation of the fruit and the dehiscence of the follicle, along the line of marginal fusion of the carpel, results from this drying process. The two valves of the follicles of *Banksia* open as a result of stress which develops between the sclerenchymatous cells of the endocarp and those of the mesocarp and exocarp, which differ in the orientation of the cellulose microfibrils in their walls. Of special interest is the adaptation of the follicle opening of *Banksia ornata* to the habitat of this plant. The shrubs *B. ornata* are frequently killed by fires which sweep the heaths in south-eastern Australia and their regeneration is only from seed. The junction between the valves of the follicle of this *Banksia* species, has a layer of interdigitating cells. Inside these cells and between them a resinous material together with phenols is present. When a fire occurs the resin is destroyed, releasing the stress in the valves, as a result of which the follicle opens and releases the seeds (Wardrop, 1983).

Legume. The following is a description of the basic structure of a legume of the Leguminosae. The exocarp usually consists of epidermis only, the mesocarp—of relatively thick parenchyma, the endocarp—of sclerenchyma on the inside of which is usually a thin-walled epidermis or a few parenchyma layers and an epidermis. The vascu-lar bundles are situated in the parenchyma of the mesocarp and they are accompanied by sclerenchyma. Although the general structure of the legume in most species of the Leguminosae is uniform, there may be differences in the relative thickness of the different pericarp tissues, the structure of their cells, the orientation of the different elements, and, sometimes also in the submicroscopic structure (Monsi, 1943; Fahn and Zohary, 1955; Shah *et al.*, 1983). Thus in *Astragalus macrocarpus* the outer epidermal cells are thin-walled (Fig. 279, no. 11), while in *Lupinus hirsutus* (Fig. 279, no. 6) and species of *Vicia* they are thick-walled. The epidermis may be typically uniseriate or a hypodermis may be present, as, for example, in *Lupinus hirsutus*. The tissue of the mesocarp may not always be parenchymatous and in certain legumes it is entirely collenchymatous (*Calycotome villosa*, Fig. 279, no. 4) or it may be partly collenchymatous and partly parenchymatous (*Retama raetam*, Fig. 279, no. 9). In some legumes there are sclereids scattered in the collenchyma (*R. raetam* and *Anagyris foetida*, Fig. 279, nos. 9, 10). The sclerenchyma, which constitutes most of the endocarp, consists either of one zone of fibres which are all arranged in one direction (*Astragalus macrocarpus*, Fig. 279, no. 11; *Acacia raddiana*, Fig. 279, no. 7; *Lupinus hirsutus*, Fig. 279, no. 6) or of two zones in which the longitudinal orientation of their cells differs (*Astragalus hamosus*, Fig. 279, no. 2; *Hymenocarpos circinnatus*). In the legumes of certain species there is no sclerenchymatous region at all (*Glycyrrhiza echinata*; *Trifolium subterraneum*; *Melilotus*, Fig. 280, no. 3). In some species the sclerenchymatous stratum of the endocarp is lined on the inside not by parenchyma but by collenchyma (*Retama raetam*, Fig. 279, no. 9; *Anagyris foetida*, Fig. 279, no. 10). The cell walls of the inner epidermis are usually thin, but in the legumes of certain species they may be slightly thickened (*Trifolium stellatum*) or the cells may be fibre-like (*Trigonella arabica*). Spines, which usually consist of sclerenchyma, develop on certain legumes (*Scorpiurus muricata*, Fig. 280, no. 4; *Hedysarum pallens*, Fig. 280, no. 1).

The two valves of a dried legume usually twist (Fig. 288, no. 2). This is brought about by the anisotropic shrinkage of the thickened walls of the pericarp cells. This feature is a result of the different orientation of the microfibrils of cellulose in the walls. The greatest swelling of the cell walls takes place in the direction at right angles to the longitudinal axis of the microfibrils. Because of this the greatest shrinkage, during the drying out of the valves, is also in this direction. In *Vicia* and in many species of other genera (Fahn and Zohary, 1955) the sclerenchyma cells of the endocarp are orientated at an angle of about 45° to the longitudinal axis of the legume, while the elongated, thick-walled epidermal or epidermal and hypodermal cells are orientated in a similar angle but in the opposite direction. In the valves of these legumes the microfibrillar orientation relative to the cell axis is the same in both the endo- and exocarp, but as the cell axes in these two strata of the pericarp are, themselves, differently orientated, tension develops during the drying out of the valves. This tension results in the twisting of the valves after the forces that keep the cells together in the mature abscission zone are overcome. The legume then dehisces explosively, the valves contort, and the seeds are expelled.

There are many variations in the structure connected with the opening mechanism that has been described above. In the legumes of *Wisteria sinensis* (Monsi, 1943) and *Lupinus angustifolius* (Fahn and Zohary, 1955), for example, all the sclerenchymatous cells of the endocarp have uniform orientation, but two zones, in which the orientation of the cellulose fibrils differ, can be distinguished. These zones, therefore, also differ in the direction of greatest swelling. There are also legumes in which the endocarp sclerenchyma consists of two layers that are distinguished by cell orientation (*Astragalus fruticosus*, Fig. 279, no. 8; *A. hamosus*, Fig. 279, no. 2; *Hedysarum pallens*, Fig. 280, no. 1). In the legumes of other species, e.g. *Ornithopus compressus*, the endocarp sclerenchyma consists of two zones, as mentioned above, but in one zone the orientation of the fibrils is parallel to the longitudinal cell axis and in the second zone at right angles to it. As a result of this the fibrillar orientation, relative to the axis of the legume, is the same in the two zones. In these plants, in addition, the orientation of the fibrils of the epidermal cells is parallel to that of the fibrils in the sclerenchyma and therefore the fruit does not dehisce when dry. Abscission tissue is also not developed in such legumes.

Siliqua. In the pericarp of the siliqua the cells of

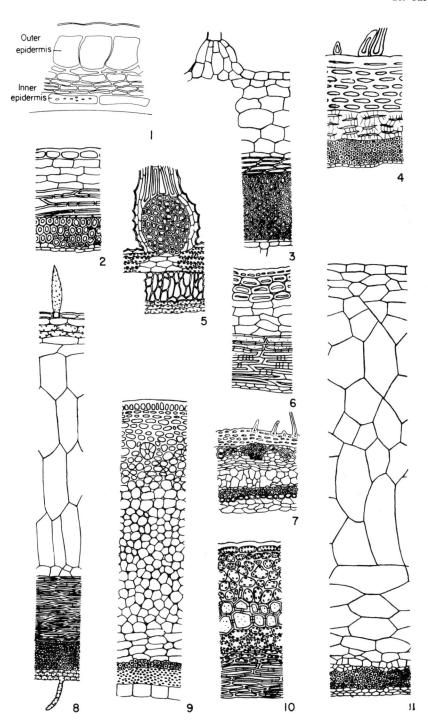

FIG. 279. 1, Portion of a cross-section of the follicle wall of *Delphinium*. 2–11, Portions of sections of the walls of legumes of different species of the Leguminosae. 2, *Astragalus hamosus*, cross-section. 3, *A. amalecitanus*, cross-section. 4, *Calycotome villosa*, oblique section. 5, *Scorpiurus muricata*, cross-section. 6, *Lupinus hirsutus*, oblique section. 7, *Acacia raddiana*, cross-section. 8, *Astragalus fruticosus*, cross-section. 9, *Retama raetam*, oblique section. 10, *Anagyris foetida*, oblique section. 11, *Astragalus macrocarpus*, oblique section. In all the diagrams the outer epidermis is uppermost.

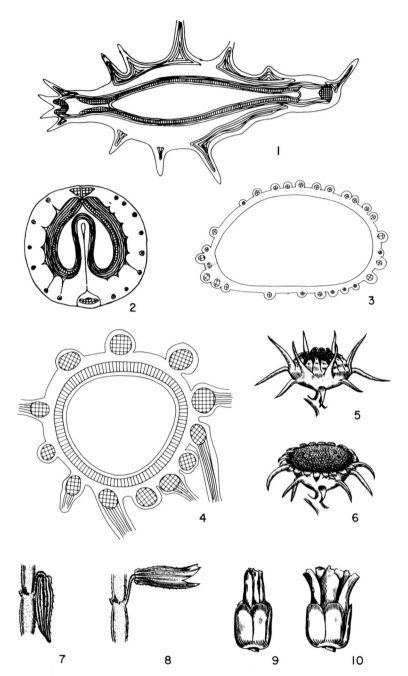

FIG. 280. 1–4, Diagrams of cross-sections of legumes of different species of the Leguminosae. In those places where the sclerenchyma cells are seen in longitudinal section, the sclerenchyma is indicated by parallel lines, and in those places where the cells are cross-sectioned, it is indicated by cross-hatching. 1, *Hedysarum pallens*. 2, *Astragalus fruticosus*. 3, *Melilotus* sp. 4, *Scorpiurus muricata*. 5–10, Hygroscopic movements. 5 and 6, Capitula of *Anvillea garcini*, showing the movement of the involucre bracts. 5, Dry. 6, Moist. 7 and 8, *Salvia horminum*, showing the movement of the pedicel and the mericarp-containing calyx. 7, Dry, showing the pedicel bent downwards, bringing the calyx close to the axis of the inflorescence. 8, Moist. 9 and 10, *Cichorium pumilum*, capitula showing movement of the involucre. 9, Dry. 10, Moist.

the exo- and mesocarp are usually thin-walled and the endocarp tissue is sclerenchymatous. An abscission zone usually develops between the replum and the valves (Fig. 281, no. 2).

Capsule. The epidermal cells of many capsules, which open by teeth or valves, have very thick outer walls while the mesocarp tissue is parenchymatous. Elongated, thick-walled cells may sometimes be present below the epidermis (Fig. 282, no. 1). The inner epidermal cells may also be thick-walled. In the capsules of certain species of the Primulaceae, such as *Lysimachia mauritiana*, the cells of the inner epidermis have particularly thick walls on the side closest to the mesocarp (Guttenberg, 1926). The dehiscence of the capsule is also brought about by the anisotropic swelling of the cell walls. The walls that bring about the dehiscence in this case are mainly the very thick walls of the outer or inner epidermis, and it is they that determine the direction of the bending of the teeth or valves. The thinner walls of the epidermis, or of the sclerenchymatous tissue below the epidermis, or of the sclerenchyma accompanying the vascular bundles, usually constitute the resistance tissues. The swelling and, therefore, the shrinkage of these tissues along the axis of bending are relatively small. These differences in swelling and shrinkage cause the characteristic opening movements. In capsules abscission tissue is developed between the teeth or valves. In capsules that open by means of a lid, e.g. the fruits of *Hyoscyamus, Plantago, Anagallis,* and *Portulaca*, the abscission tissue develops as a ring around the capsule. The rupture between the lid and the rest of the capsule, along this abscission line, is apparently brought about by the fact that the maturing pericarp dries out and shrinks while the volume of seeds, which fill the entire cavity of the fruit, does not change (Rethke, 1946; Subramanyam and Raju, 1953).

In the dispersal of the seeds of dry fruits, sepals and extrafloral organs may sometimes participate, e.g. the bracts of the capitulum (*Cichorium*, Fig. 280, nos. 9, 10; *Anvillea*, Fig. 280, nos. 5, 6), floral pedicels (*Salvia horminum*, Fig. 280, nos. 7, 8), axes of inflorescences (*Plantago cretica*, Fig. 281, nos. 3, 4), and entire branches (*Anastatica hierochuntica*) (Steinbrinck and Schinz, 1908; Zohary and Fahn, 1941; Fahn, 1947). The mechanisms of dis-

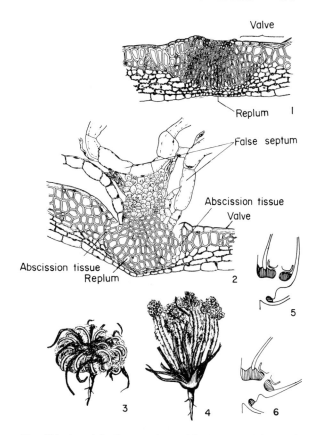

FIG. 281. 1 and 2, Cross-sections of both types of fruits of *Aethionema carneum* showing the region around the replum. 1, Of a young indehiscent siliqua. 2, Of a young dehiscent siliqua. 3–6, *Plantago cretica*. 3, Entire plant in dry condition. 4, As in no. 3, moist. 5 and 6, Diagrams of longitudinal sections through the remnants of a flower and its bracts which remain on the plant even after the ripening of the fruit. Cohesion tissue, represented by hatching. 5, Dry, showing the cohesion tissue, to be contracted. 6, Moist, showing the expansion of the cohesion tissue which causes the sepals and bracts to spread.

persal in these cases are also usually based on the differences in direction of excessive swelling in the different zones of the sclerenchymatous tissues. This *swelling mechanism* also causes the twisting of the "beak" of the mericarp of *Erodium* (Fig. 288, no. 1).

In addition to the swelling mechanism there is the *cohesion mechanism* that is responsible for the movement of organs connected with the dispersal of fruits and seeds. Cohesion tissue occurs, for example, on the outer side of the bracts of many of the Compositae (*Senecio, Tragopogon,* and *Gero-*

pogon, Fig. 282, nos. 2, 3), between the bases of the rays of the umbel, e.g. *Ammi visnaga* (Fig. 282, nos. 4–6), between the tepals, e.g. *Plantago cretica* (Fig. 281, nos. 5, 6), etc. The cells of cohesion tissues are usually elongated and thin-walled; the longitudinal axis of these cells is at a right angles to that of the shrinkage axis of the tissue. As a result of the loss of water the cell walls fold or wrinkle, the cell volume is reduced, and the tissue, in shrinking, draws with it the organ that is attached to it. If water is again absorbed by the cells, they swell and the volume of the tissue increases (Guttenberg, 1926; Zohary and Fahn, 1941).

Indehiscent Fruits

The histological structure of only three dry indehiscent fruits—cypsela, caryopsis, and cremocarp—will be given here.

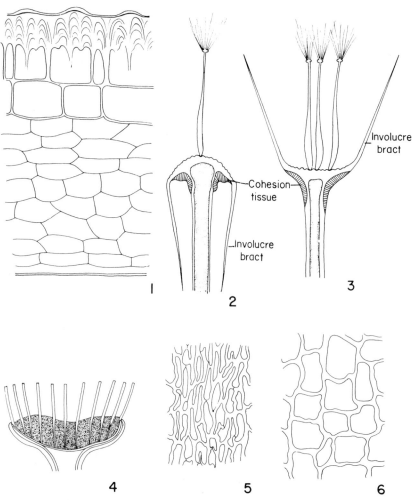

FIG. 282. 1, *Vaccaria pyramidata*, longitudinal section through a tooth of the capsule. The outer epidermis (directed upwards) has very thick walls in which the lamellae are orientated almost perpendicularly to the surface of the teeth. These walls, when wet, are capable of extensive swelling in a longitudinal direction of the tooth, and the elongated cells below this epidermis and the cells of the inner epidermis itself form a tissue that resists this movement. As a result the valves open outwards when they are dry and close when they are moist. 2 and 3, *Geropogon*, showing action of cohesion tissue—hatched areas. 2, Ripe capitulum in dry condition. 3, In moist condition showing that the cohesion tissue, when saturated with water, raises the involucre bracts. 4–6, *Ammi visnaga*. 4, Diagram of a cross-section through the base of the compound umbel, showing the bases of the partial umbels to be embedded in cohesion tissue (stippled). 5 and 6, Tangential sections through the cohesion tissue. 5, Dry. 6, Moist. (Nos. 4–6 adapted from Guttenberg, 1926.)

Cypsela. The anatomical structure and development of the pericarp of *Lactuca sativa* have been described by Borthwick and Robbins (1928). The nucellus has almost completely disappeared prior to pollination and the integument is very thick—from eight to twelve cells in thickness. The pericarp increases in thickness, and its cells are similar to, but smaller than, those of the integument. A distinct outer epidermis is discernible in the pericarp. With the maturation of the flower, lysigenous cavities begin to form in the inner portion of the pericarp (Fig. 283, no. 1). During the process of fruit ripening the inner layers of the pericarp disintegrate. About a week after fertilization, with the growth of the embryo and the development of the endosperm, further disintegration of the pericarp tissues, and also of those of the integument, takes place (Fig. 283, no. 2). About 10 days after fertilization the cells of the inner epidermis of the integument disintegrate and their inner walls may adhere to the endosperm where they form a membrane which, in certain fruits, is suberized. This membrane is considered by Borthwick and Robbins to be semi-permeable. With the ripening of the cypsela, pericarp ribs which consist of sclerenchymatous tissue become obvious (Fig. 283, nos. 2–4). In the mature fruit the rest of the pericarp, apart from the ribs, consists of one or two cell layers as a result of the disintegration of the parenchyma cells. Cypselae differ greatly in

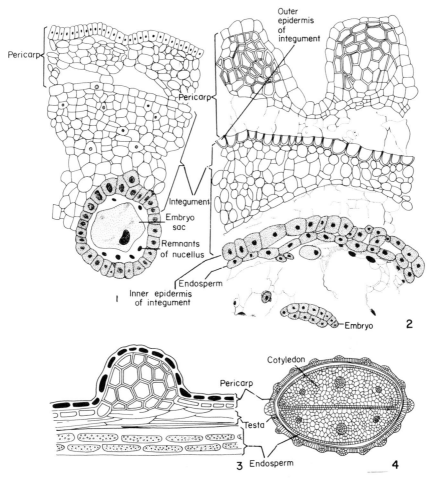

FIG. 283. Development of the pericarp in *Lactuca sativa*. 1, Portion of a cross-section of an ovary 2 hr after fertilization. 2, Portion of a cross-section of the cypsela about a week after fertilization. 3, Mature pericarp. 4, Diagram of a cross-section of an entire cypsela. (Adapted from Borthwick and Robbins, 1928 and Hayward, 1938.)

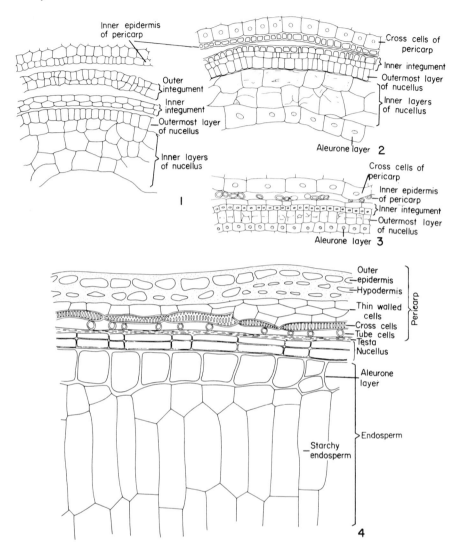

Fig. 284. 1–4, Portions of cross-sections through the pericarp of the caryopsis of *Triticum* at different stages of development. (Adapted from Hayward, 1938.)

colour. These differences are partly due to the presence or absence of pigment in the outer epidermal cells of the pericarp (Fig. 283, no. 3). The integumental cells of the mature seed are compressed and partly obliterated, but the outer epidermis may remain as a layer of thick-walled cells.

Caryopsis. The pericarp and the remains of the integuments of the single seed of the caryopsis are completely fused (Krauss, 1933; Bradbury *et al.*, 1956). In the caryopsis of *Triticum* (Fig. 284, no. 4; Fig. 303), for example, three main parts can be

distinguished: (1) the caryopsis coat which includes the pericarp, the seed coat, and the nucellus; (2) the endosperm; (3) the embryo. Five layers can be distinguished in the pericarp: the outer epidermis, the hypodermis, a zone of thin-walled cells, *cross cells*, and *tube cells*. The outer epidermis and the hypodermis together form the exocarp. The cells of the exocarp are elongated in a direction parallel to the longitudinal axis of the caryopsis, and they become compressed and their walls thicken considerably so that, when the caryopsis is ripe, cell lumina cannot easily be distinguished in them. The cross cells are found below the parenchymatous layer and they have thick walls with pits which are elongated transversely to the cell. The longitudinal axis of these cells is at right angles to that of the exocarp cells. The tube cells constitute the inner epidermis of the pericarp and they occur on the inside of the cross cells. There are large intercellular spaces between the tube cells, the walls of which are pitted and thinner than those of the cross cells. The longitudinal axis of the tube cells is parallel to that of the exocarp cells. The tube cells are only clearly visible on certain parts of the caryopsis (Bradbury *et al.*, 1956). The seed coat, which is

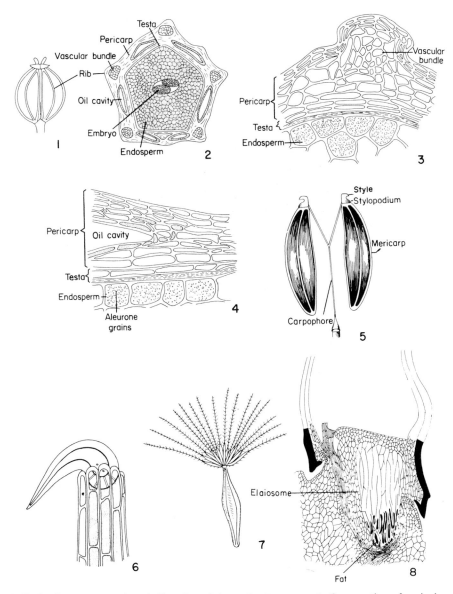

FIG. 285. 1–4, Fruit of *Apium graveolens*. 1, Drawing of the entire cremocarp. 2, Cross-section of a single mericarp. 3, Portion of a cross-section of a mericarp showing the structure of a rib. 4, As in no. 3, in the region of an oil cavity. 5, A mature cremocarp with separated mericarps. 6, Distal portion of spine of the nutlet of *Ranunculus arvensis* showing the pointed, curved cell at its apex by which it becomes attached to the dispersal agents. 7, Cypsela of the Compositae showing a feathery pappus. 8, Longitudinal section of the lower part of a nutlet of *Symphytum*, showing the elaiosome. (Nos. 1–4 adapted from Hayward, 1938; no. 8 adapted from Bresinsky, 1963.)

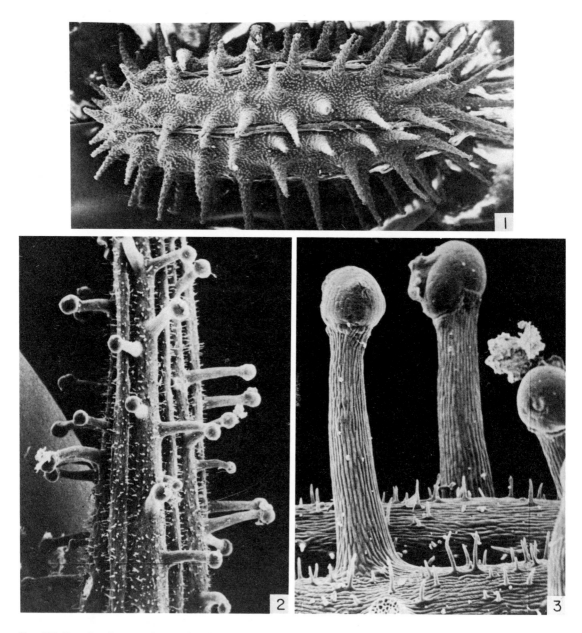

FIG. 286. Scanning electron micrographs of fruit dispersal units. 1, Portion of a mericarp of *Torilis arvensis* showing spines. × 80. 2 and 3, Portion of a calyx of *Plumbago capensis* with trichomes which secrete a sticky substance. 2, × 30; 3, × 120. (Nos. 2 and 3 from Fahn, 1979.)

united with the pericarp, is crushed in the mature caryopsis and therefore it is difficult to discern cells in this zone. When fully mature the seed coat consists only of the inner integument, the outer being completely destroyed. In a section of the caryopsis stained with Sudan IV two cuticles are

discernible: an outer thick one of the single remaining integument, and a thin one which constitutes that of the nucellus together with that of the inner side of the integument. The nucellus tissue may be partially or entirely digested during the development of the caryopsis. In those parts where nucellus is still present it is discernible as one or two layers of thin-walled cells between the aleurone layer and the seed coat. In the mature caryopsis these nucellar cells are usually crushed and they appear as a thin, glassy zone, which is bright and colourless (Fig. 284).

Cremocarp. The fruit of *Apium graveolens* is described here as an example of a cremocarp (Hayward, 1938). Each mericarp has five ribs in which the vascular bundles, accompanied by sclerenchyma, are situated (Fig. 285, nos. 1–3). Between the ribs one to three oil ducts may be present (Fig. 285, nos. 2, 4). The exocarp cells (i.e.

the outer epidermis) are small and isodiametric. In surface view they appear lobed. The mesocarp is parenchymatous and its cells have small papillae and finely striated walls. The oil ducts in the mesocarp are surrounded by polyhedral cells which become brown as the fruit ripens. The slightly elongated cells of the innermost layer of the mesocarp are wider than those of the endocarp. The endocarp (the inner epidermis) consists of narrow cells of which the longitudinal axis is parallel to the transverse axis of the mericarp (Fig. 285, nos. 3, 4), or their arrangement is mosaic. Inside the pericarp is the thin seed coat

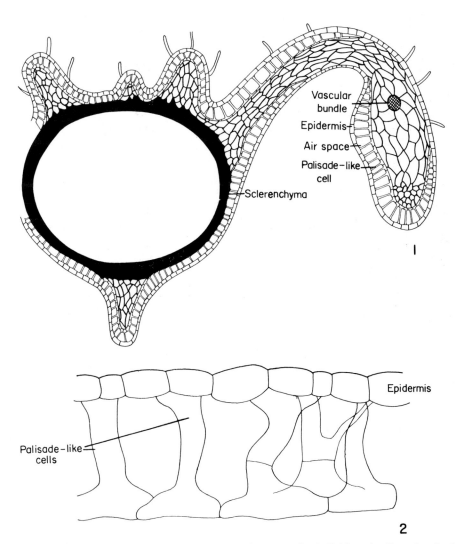

FIG. 287. *Calendula officinalis.* 1, Diagram of a cross-section of the cypsela. 2, Epidermal cells and palisade-like cells, between which there are large air spaces, greatly enlarged. (Adapted from Schoenichen, 1924.)

which surrounds the endosperm in which the embryo is completely enclosed. The maturing mericarps, each of which develops from a single carpel together with a small amount of tissue outside it (inferior ovary), separate one from the other along the area of adnation between them, but they remain attached to the branched *carpophore* (Fig. 285, no. 5). There are different opinions as to the origin of the carpophore—there are investigators who believe that it arises from the floral axis, while others believe that it develops from the carpels. Jackson (1933) claims

FIG. 288. 1, *Erodium*, photograph of a mature mericarp in a dry condition. 2, *Spartium junceum*, photograph of mature legumes which have dehisced as a result of the twisting of the valves. (Courtesy of D. Koller.)

that usually only the basal portion is of receptacular origin while the greater portion of the carpophore is carpellary and contains the ventral carpel bundles. The abscission zone occurs partly between the two mericarps and partly between the mericarps and the carpophore. With the maturation of the cremocarp, an abscission zone also appears within the carpophore, in the lignified tissue between the two ventral veins. This abscission zone splits the carpophore into two.

The dispersal of indehiscent dry fruits and their seeds is brought about in various ways, and here also a correlation can be found between the manner of dispersal and the anatomical structure of the pericarp. Thus, for instance, on mericarps of Umbelliferae ridges with spines (Heywood, 1968) are formed (Fig. 286, no. 1) and on the akenes of *Ranunculus arvensis* strong, hook-like hairs, which consist of extremely thick-walled cells (Fig. 285, no. 6), develop; these hairs enable the fruit to cling to the coats of animals. Cypselae of many of the Compositae develop a characteristic pappus of hairs or bristles which may be branched (Fig. 285, no. 7). The outer cypselae of *Calendula officinalis* are boat-shaped and the margins are widened to form backwardly directed wings. In a cross-section of such a cypsela (Fig. 287, no. 1), it is possible to see that the endocarp consists of a thick sclerenchymatous tissue. The centre of the keel, the wings, and the ribs on the side opposite the keel consist of large parenchyma cells whose pitted walls are slightly thickened. Most of the region between the epidermis and the inner tissues of the cypsela consists of very long palisade-like cells which have large intercellular spaces between them (Fig. 287, no. 2). This tissue together with the thick-walled large parenchyma cells, which are filled with air, aid in the dispersal of these cypselae by wind. The structure of the fruits produced in the female ray flowers described above, is characteristic also of other *Calendula* species (Hilger and Reese, 1983).

An outgrowth, termed *elaiosome*, of large oil-storing cells occurs at the base of the fruit of some plants, e.g. *Hepatica triloba, Ranunculus ficaria, Anemone nemorosa, Adonis vernalis,* and *Fumaria officinalis.* Cypselae of some *Centaurea* species, e.g. *C. behen* and *C. cyanus,* also possess elaiosomes (Dittrich, 1966). In *Ajuga, Borago, Anchusa,* and *Symphytum* the elaiosome develops in the region of the receptacle, apparently from the basal wall of the carpels (Fig. 285, no. 8) (Bresinsky, 1963). Elaiosomes may also occur on seeds (see following chapter). They are thought to be an adaptation to fruit and seed dispersal by means of ants.

There are many other modifications in the histological structure of dry fruits which are adapted to dispersal by air, water, and animals (see Fahn and Werker, 1972).

Pericarp of fleshy fruits

Berry

In a berry all the ground tissue of the ovary wall develops into a fleshy or juicy tissue, and sometimes other organs may also contribute to the formation of this tissue. In *Lycopersicon,* for example, the greater part of the juicy tissue develops from the placenta. In certain berries the locules of the fruit become filled with growths of the pericarp and placenta (*Physalis*) or of the septa (*Bryonia dioica*) (Kraus, 1949). The structure and development of some special fruits which are considered to be berries, in the wide sense of the term, are described below.

Lycopersicon esculentum. This fruit consists of a pericarp and placental tissue on which the seeds are borne. The exocarp consists of an epidermis and three or four layers of collenchyma cells. The epidermal cells are polyhedral and are covered by a thin cuticle. The number of epidermal cells does not increase greatly with the development and growth of the fruit, and so the epidermal cells of the mature fruit are much larger than those in the young fruit. The glandular and other hairs that are usually present on the young fruit are shed as it matures. There are no stomata on the epidermis of the fruit (Rosenbaum and Sando, 1920). The mesocarp consists of a thick layer of large, thin-walled cells which enclose many intercellular spaces. In the early stages of development, soon after pollination, the number of layers in the mesocarp increases rapidly, but the main increase in thickness of the pericarp results from an enormous increase in cell volume in later stages of development. During the

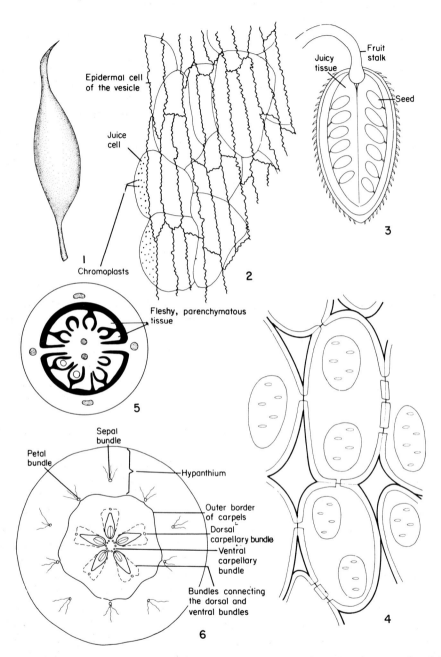

FIG. 289. 1 and 2, *Citrus.* 1, A single juice vesicle. 2, Portion of the vesicle as seen under the microscope. 3 and 4, *Ecballium elaterium.* 3, Diagram of a longitudinally sectioned fruit. 4, A few cells from the inner, white portion of the pericarp. 5, Diagram of a cross-section of the ovary of *Lycopersicon esculentum* after fertilization in which the enlargement of the parenchymatous tissue of the placenta is shown. This tissue forms the fleshy tissue of the fruit. 6, Diagram of a cross-section of the fruit of *Pyrus malus* var. *paradisiaca.* (No. 2 adapted from Schoenichen, 1924; nos. 3 and 4 adapted from Guttenberg, 1926; no. 5 adapted from Hayward, 1938; no. 6 adapted from MacDaniels, 1940.)

process of fruit ripening, some of the cells of the inner and central portion of the carpels may disintegrate. With the development of the ovules, after pollination, the parenchymatous tissue of the placenta grows around the funiculi. This parenchyma continues to grow until it completely encloses the developing seeds (Fig. 289, no. 5). The cells of this tissue are thin-walled and they form a homogenous tissue; they do not fuse with the pericarp but they adhere to it as well as to the seeds. At first this parenchymatous tissue is firm, but as the fruit ripens the cell walls become thinner and the cells are partly destroyed (Hayward, 1938).

Citrus. In this genus the fruit develops from a syncarpous gynoecium with axile placentation. With the development of the fruit the number of cells throughout the ovary increases and, finally, three strata (Fig. 290, nos. 1, 2) can be distinguished (Schoenichen, 1924; Ford, 1942; Scott and Baker, 1947). The exocarp (flavedo) consists of small, dense collenchyma cells which contain chromoplasts. This tissue contains essential oil cavities (Fig. 290, no. 1). The epidermis consists of very small, thick-walled cells, and in surface view it resembles a cobbled surface; the cells contain chromoplasts and oil droplets. A few scattered stomata can be found in the epidermis. The main ultrastructural changes in the exocarp associated with ripening of the orange was found to be the alteration of the chloroplast structure and the subsequent formation of chromoplasts. The mature chromoplasts are characterized by large osmiophilic globules and only few internal membranes (Thomson, 1969). The mesocarp (albedo) consists of loosely connected, colourless cells; this tissue has a spongy nature and is white because of the numerous air spaces in it. The endocarp is relatively thin and consists of very elongated, thick-walled cells which form a compact tissue. The stalked, spindle-shaped juice vesicles, which fill the locules when the fruit ripens, develop from the cells of the inner epidermis and subepidermal layers (Hartl, 1957). Each juice vesicle is covered externally by a layer of elongated epidermal cells which enclose very large, extremely thin-walled juice cells (Fig. 289, nos. 1, 2) (see also Schneider, 1968). The epidermis of the vesicles is covered with a thin cuticle and epicuticular wax with

various structural patterns (Fig. 80, no. 1) (Fahn *et al.,* 1974).

Phoenix dactylifera. In the date the endocarp consists of one layer of small cells, which can be identified easily only in the early stages of fruit development. The mesocarp, which occupies most of the fruit volume, consists of large parenchyma cells and vascular bundles. The mesocarp is divided into an inner and outer part by a 3–10-layered zone of tanniniferous cells. The exocarp consists of the outer epidermis covered by a thick cuticle, one or two layers of tangentially elongated hypodermal cells, below them a 3–5-layered zone of small isodiametric parenchyma cells in which vascular bundles occur, and an innermost zone of radially oriented macroscleroids.

Persea americana. The mesocarp comprises the main part of the pericarp of the avocado fruit. It is composed of large isodiametric, lipid containing parenchyma cells, scattered oil cells (idioblasts), and a few vascular strands. As the fruit ripens, the parenchyma cells loosen, and eventually their walls break down. In the oil cells three distinct layers develop in their walls, the middle one of which is a suberin lamella. It should be mentioned that such a wall structure is characteristic also of oil cells of *Laurus* leaves (Maron and Fahn, 1979). Smooth tubular ER appears to be involved in the oil production (Platt-Aloia and Thomson, 1981; Platt-Aloia *et al.,* 1980, 1983).

Viscin tissue is present in the mistletoe fruit. This tissue, which is part of the pericarp, is closely associated with the seed and enables its adherence to the host branches, during germination. Some other functions, e.g. absorbtion of water at germination and protection against enzymes in the digestive tract of birds, have also been ascribed to the viscin tissue. The ultrastructure and chemistry of the viscin tissue has recently been described (Gedalovich and Kuijt, 1987; Gedalovich *et al.,* 1988).

DRUPE

In the drupe of *Prunus persica* (Addoms *et al.,* 1930) most of the cell divisions, which occur during the development of the pericarp, take place

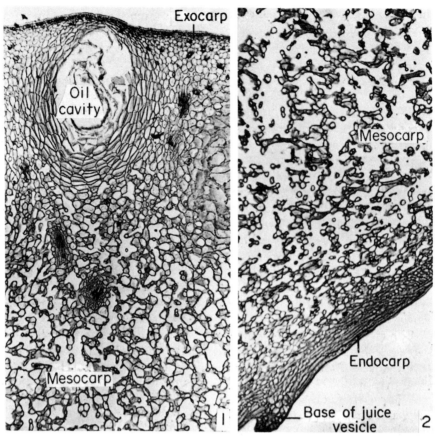

FIG. 290. 1 and 2, Micrographs of portions of the pericarp of *Citrus*. × 50. 1, Outer portion. 2, Inner portion.

before fertilization or immediately after it. The growth of the fruit is mainly accomplished by cell enlargement (Ragland, 1934). At first all the cells enlarge almost equally in all directions, but with further development of the fruit the enlargement is mainly in a radial direction. This feature is particularly noticeable in the inner portion of the mesocarp. The epidermis (the exocarp) of the mature fruit bears a cuticle and many unicellular hairs. The mesocarp consists of a non-compact parenchymatous tissue in which the cell dimensions increase from the periphery inwards. In the same direction the shape of the cells also alters, from oval cells in which the longer axis is parallel to the fruit surface to cylindrical, radially arranged cells. The endocarp consists of sclereids and forms the stone of the fruit. The outer surface of the peach stone is grooved, and in the deep

grooves vascular bundles which branch into the mesocarp are found.

In *Prunus amygdalus* three phases of fruit growth established for other species of *Prunus* (peach and cherry) were distinguished by Sarfatti (1961): (1) growth of pericarp, seed testa, and nucellus—this phase ends after the micropylar portion of the endosperm has become cellular and the seed has reached its final size; (2) growth of endosperm and embryo—the pericarp undergoes a period of greatly reduced growth; (3) resumption of pericarp growth; for a very short period in almond and for a long period in other fleshy fruits of the genus *Prunus*.

In the exocarp *sensu lato* of the drupes of the Anacardiaceae, resin ducts are present (Joel, 1980; Von Teichman, 1987; Von Teichman and Van Wyk, 1988). Joel (1980) suggested that the duct system in the mango fruit provides resistance to the Mediterranean fruit fly.

Fleshy false fruits

The fruit of *Pyrus malus* var. *paradisiaca* may be given as an example of a fleshy fruit developing from an inferior ovary. This fruit has been extensively investigated by numerous workers (MacArthur and Wetmore, 1939, 1941; MacDaniels, 1940; Smith, 1940, 1950) who have reached the conclusion that it mainly develops from the *hypanthium*, i.e. from the basal portions of the perianth and stamens which are fused and adnated to the carpels. The receptacle participates in the formation of only a very small portion of the basal part of the fruit. The outer parenchyma of the fruit develops from the hypanthium and it contains ten vascular bundles—five belonging to the sepals and five belonging to the petals (Fig. 289, no. 6). These vascular bundles are branched and the branches penetrate into the parenchyma where they form a network. The epidermis is covered by a cuticle which increases in thickness as the fruit develops. In early stages of development of the epidermis stomata may be distinguished, but these later cease to function and are replaced by lenticels (Clements, 1935). In young fruits there are unicellular epidermal hairs, which are shed with the maturation of the fruit. In certain varieties, cork develops on part of the surface of the ripening fruit. Cell division in the epidermis continues for a longer time than in the outer parenchyma of the fruit (Skene, 1966). The subepidermal tissue, which develops from the outer portion of the hypanthium, consists of a several-layered, thick-walled collenchyma tissue, the cells of which are tangentially elongated. Intercellular spaces develop in this tissue only shortly before fruit maturation, and they are best developed in the more internal ground parenchyma. In still deeper layers the cells are more or less oval and their longer axis is usually radially orientated. This part of the fruit grows most intensely during fruit development, at the start by division of the cells and the enlargement of the derivatives, and later by the increase in cell volume only (Tukey and Young, 1942). The part of the fruit that de-velops from the ovary is formed by the five-folded, but unfused, carpels. Five dorsal carpel bundles are found on the outer side of the locules and ten ventral bundles in the centre (Fig. 289, no. 6). The dorsal and ventral bundles are interconnected by branches. It is difficult to distinguish distinctly between the ovary and the hypanthium; generally it is possible to state that the border is between the innermost, prominent bundles of the hypanthium and the dorsal bundles of the carpels.

According to MacDaniels (1940) the ovary wall develops into a parenchymatous exocarp and a cartilaginous endocarp which lines the locules. The endocarp consists of elongated sclereids with very thick walls which almost completely obliterate the cell lumen. These sclereids do not develop close to the dorsal bundles. The endocarp is that part of the fruit that first becomes fully matured, and the last-maturing portion is that which develops from the hypanthium.

In the fruits of *Pyrus communis* and *Cydonia oblonga* groups of brachysclereids (see Chapter 6) develop in the parenchyma. As described by Staritsky (1970), in the pear the succulent sugar-containing cells are arranged around such groups of brachysclereids in a radial pattern.

Structural changes occur in the walls of the parenchyma cells of apple and pear fruits during the advanced stages of ripening. The middle lamella disintegrates almost completely and the adjoining outer region of the cell wall also undergoes partial dissolution and its cellulosic fibrils become dispersed. In ripe pears the changes in the wall structure are more pronounced than in the ripe apples (Ben-Arie and Kislev, 1979).

The banana is also a fruit that develops from an inferior ovary, and it is essentially a berry. From the periphery inwards the following parts can be distinguished: the exocarp, which forms the peel of the fruit, consisting of epidermis, hypodermis, and aerenchyma; the mesocarp, consisting of large, radially elongated cells which are rich in starch (Fig. 291, no. 2); and the endocarp, consisting of the inner epidermis only. The mesocarp, which constitutes the edible part of the fruit, develops from the three to five layers of the carpel wall immediately next to the inner epidermis which lines the locules. These cells first elongate extensively in a radial direction and then they

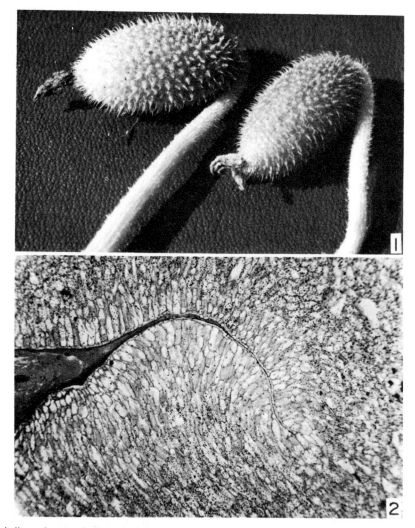

FIG. 291. 1, *Ecballium elaterium* fruits. × 1. 2, Inner portion of mesocarp of a ripening banana fruit lined by a narrow endocarp (inner epidermis). × 100.

mainly undergo periclinal divisions, although anticlinal divisions also occur.

Abscission

The abscission of fruit during its development and ripening is an important physiological process. In peach (*Prunus persica*) the fruit may separate at three different places along the peduncle and fruit receptacle, dependent on activation of the abscission zone (AZ) at these places. The AZ located at base of the peduncle is responsible for the abscission of young fruits; the AZ between peduncle and fruit receptacle and that between the receptacle and fruit are activated in succession during fruit ripening (Rascio *et al.*, 1985).

The structure of the abscission zones and the process of abscission of fruits are similar to those of leaves (Stösser *et al.,* 1969; Gilliland *et al.,* 1976; MacKenzie, 1979; Huberman *et al.*, 1983: see also "Leaf abscission" in Chapter 12).

Structural adaptations to seed dispersal

The seeds of fleshy fruits are usually dispersed by animals that eat the juicy edible portion of the fruit, but there are also fleshy fruits the seeds of which are actively dispersed similarly to those of certain dry, dehiscent fruits. This type of dispersal is called *autochory*. The mechanism of self-dispersal in fleshy fruits is based on turgor pressure. As an example of fruits with this type of mechanism the fruits of *Ecballium elaterium* (Fig. 291, no. 1) and *Impatiens* (Fig. 292) will be described here.

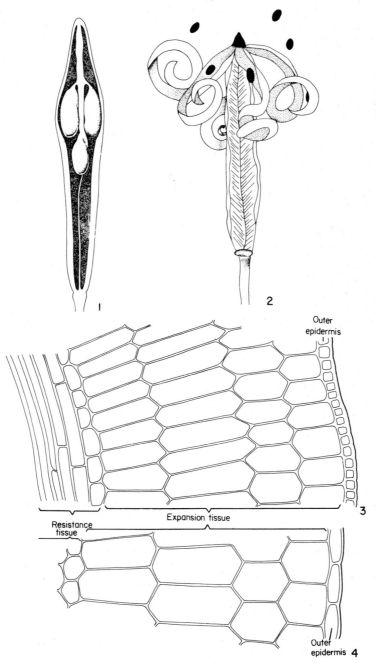

FIG. 292. Fleshy capsule of *Impatiens*. 1, Longitudinally sectioned closed fruit. 2, A fruit the valves of which have curled inwards and in doing so have ejected the seeds. 3, Portion of a longitudinal section of the pericarp showing the tissue active in opening the fruit. 4, As in no. 3, but as seen in cross-section. (Nos. 1, 3, and 4 adapted from Guttenberg, 1926.)

The fruit of *Ecballium* (Fig. 289, nos. 3, 4) is ellipsoidal and is attached to a long stalk, which is bent downwards at an acute angle. The pericarp (which develops from an inferior ovary) is fleshy and its outer portion consists of an epidermis and chloroplast-containing parenchyma, in which vascular bundles are embedded. Further inwards the pericarp is white and consists of elliptical cells which have thick pitted walls that are rich in pectic substances. The longer axis of these cells is at right angles to the longitudinal axis of the fruit and large intercellular spaces are present between them (Fig. 289, no. 4). Still further inwards is the tissue that envelops the seeds, and which consists of large, vesicle-like, extremely thin-walled cells between which there are no intercellular spaces. These cells have a very thin layer of cytoplasm, and their cell sap contains the glucoside, elaterinidin. This substance is present in such large amounts that in the ripe fruit the osmotic pressure of the sap reaches about 27 atmospheres. As a result of the turgor pressure of the elaterinidin-containing cells, the elastic cells of the white portion of the pericarp expand, and this occurs mainly in the direction at right angles to the longitudinal axis of the fruit. Abscission tissue develops, as the fruit matures, around that part of the stalk that is within the pericarp (Fig. 289, no. 3). At the instant when the pressure which develops in the inner juicy tissue surrounding the seeds exceeds that of the force that keeps the cells of the separation layer together, the fruit becomes detached from the stalk. Simultaneously the pericarp, and especially the white portion of it, contracts and the fruit content—the large juicy cells together with the seeds—is ejected with great force through the hole produced by detachment of the "inserted stalk". It was found that the amount of contraction of the pericarp in a transverse direction is 17.3% and in a longitudinal direction 10.8% (Guttenberg, 1926).

The fruit of *Impatiens* is a fleshy capsule in which the septa are extremely delicate. It is cylindrical but somewhat swollen in the upper portion in which the seeds develop (Fig. 292, no. 1). This upper part of the fruit remains inactive, as far as the opening mechanism is concerned, while in the lower portion tension is developed between the outer tissue, which has an expansion poten-

tial, and the inner tissue, which offers resistance. When the fruit is mature, the abscission tissue between the carpels ruptures and each valve abruptly curls inwards and, as a result of this, the seeds are expelled (Fig. 292, no. 2). The expansion tissue is located below the outer epidermis which consists of thick-walled cells. This tissue consists of radially elongated parenchyma cells and lacks intercellular spaces (Fig. 292, nos. 3, 4). The cells have a rich sugar content when the fruit ripens and the osmotic pressure in their cell sap reaches 25–26 atmospheres. This pressure would result in the rounding of the cells were it not for the resistance offered by the inner portion of the pericarp, which consists of two or three layers of collenchyma cells, the longitudinal axes of which are parallel to that of the fruit (Fig. 292, no. 3). These collenchyma cells elongate by 10% as a result of the turgor pressure in the outer tissue, and they contract again to the same extent with the opening of the fruit. The outer tissue elongates parallel to the longitudinal axis of the fruit by 32.25% with the opening of the fruit (Guttenberg, 1926).

Sticky substances serve to adhere fruits and seeds to passing animals, and thus aid in dispersal. In *Plumbago capensis*, for example, these substances are secreted by glandular trichomes present on the persistent calyx which encloses the fruit and is detached with it (Fig. 286, nos. 2, 3).

(For additional knowledge on fruit structures see Roth, 1977.)

REFERENCES

ADDOMS, R. M., NIGHTINGALE, G. T., and BLAKE, M. A. (1930) Development and ripening of peaches as correlated with physical characteristics, chemical composition, and histological structure of the fruit flesh: II, Histology and microchemistry. *Bull. NY Agric. Exp. Stn.*, no. 507.

BEN-ARIE, R. and KISLEV, N. (1979) Ultrastructural changes in the cell wall of ripening apple and pear fruits. *Pl. Physiol.* **64:** 197–202.

BORTHWICK, H. A. and ROBBINS, W. W. (1928) Lettuce seed and its germination. *Hilgardia* **7:** 275–305.

BRADBURY, D., MacMASTERS, M. M. and CULL, J. M. (1956) Structure of mature wheat kernel: II, Microscopic structure of pericarp, seed coat, and other coverings of the endosperm and germ of hard red winter wheat. *Cereal Chem.* **33:** 342–60.

BRESINSKY, A. (1963) Bau, Entwicklungsgeschichte und Inhaltsstoffe der Elaiosomen. *Bible. Bot.*, no. 126.

CHEPLICK, G. P. (1983) Differences between plants arising from aerial and subterranean seeds in the amphicarpic annual *Cardamine chenopodifolia* (Cruciferae). *Bull. Torrey Bot. Club.* **110**: 442–448.

CLEMENTS, H. (1935) Morphology and physiology of the pome lenticels of *Pyrus malus. Bot. Gaz.* **97**: 101–17.

DITTRICH, M. (1966) Karpologische Untersuchungen zur Systematik von *Centaurea* und verwandten Gattungen: I and II. *Bot. Jb.* **88**: 70–122, 123–62.

EVENARI, M., KADOURI, A. and GUTTERMAN, Y. (1977) Eco-physiological investigations on the amphicarpy of *Emex* spinosa (L.). Campd. *Flora* **166**: 223–238.

FAHN, A. (1947) Physico-anatomical investigations in the dispersal apparatus of some fruits. *Palest. J. Bot., Jerusalem ser.* **4**: 36–45.

FAHN, A. (1979) *Secretory Tissues in Plants.* Academic Press, London.

FAHN, A. and WERKER, E. (1972) Anatomical mechanisms of seed dispersal. In: *Seed Biology* (ed. T. T. Kozlowski). Academic Press, New York, pp. 151–221.

FAHN, A. and ZOHARY, M. (1955) On the pericarpial structure of the legumen, its evolution and relation to dehiscence. *Phytomorphology* **5**: 99–111.

FAHN, A., SHOMER, I., and BEN-GERA, I. (1974) Occurrence and structure of epicuticular wax on juice vesicles of *Citrus* fruit. *Ann. Bot.* **38**: 869–72.

FORD, E. S. (1942) Anatomy and histology of the Eureka lemon. *Bot. Gaz.* **104**: 288–305.

GEDALOVICH, E. and KUIJT, J. (1987) An ultrastructural study of the viscin tissue of *Phthirusa pyrifera* (H.B.K.) Eichler (Loranthaceae). *Protoplasma* **137**: 145–155.

GEDALOVICH, E., KUIJT, J. and CARPITA, N. C. (1988) Chemical composition of viscin, an adhesive involved in dispersal of the parasite *Phoradendron californicum* (Viscaceae). *Physiol. Molecular Plant Pathol.* **32**: 61–76.

GILLILAND, M. G., BORNMAN, C. H. and ADDICOTT, F. T. (1976) Ultrastructure and acid phosphatase in pedicel abscission of *Hibiscus. Am. J. Bot.* **63**: 925–935.

GUTTENBERG, H. V. (1926) Die Bewegungsgewebe. In: K. Linsbauer, *Handbuch der Pflanzenanatomie,* Bd. 5, Lief 18. Gebr. Borntraeger, Berlin.

GUTTENBERG, H. V. (1971) Bewegungsgewebe und Perzeptionsorgane. In: K. Linsbauer, *Handbuch der Pflanzenanatomie,* Bd. 5, T. 5. Gebr. Borntraeger, Berlin.

HARTL, D. (1957) Struktur und Herkunft des Endokarps der Rutaceen, *Beitr. Biol. Pfl.* **34**: 35–49.

HAYWARD, H. E. (1938) *The Structure of Economic Plants.* Macmillan, New York.

HEYWOOD, V. H. (1968) Scanning electron microscopy and microcharacters in the fruits of the Umbelliferae–Caucalideae. *Proc. Linn. Soc. London, Bot.* **179**: 287–9.

HILGER, H. H. and REESE, H. (1983) Ontogenie der Strahlbüten und der heterokarpen Achänen von *Calendula arvensis* (Asteraceae). *Beitr. Biol. Pflanzen* **58**: 123–147.

HUBERMAN, M., GOREN, R. and ZAMSKI, E. (1983) Anatomical aspects of hormonal regulation of abscission in citrus—The shoot-peduncle abscission zone in the non-abscising stage. *Physiol. Plant.* **59**: 445–454.

JACKSON, G. (1933) A study of the carpophore of the Umbelliferae. *Am. J. Bot.* **20**: 121–44.

JERNSTEDT, J. A. and CLARK, C. (1979) Stomata on the fruits and seeds of *Eschscholtzia* (Papaveraceae). *Am. J. Bot.* **66**: 586–590.

JOEL, D. M. (1980) Resin ducts in the mango fruit: a defence system. *J. Exp. Bot.* **31**: 1707–1718.

KOLLER, D. and ROTH, N. (1964) Studies of the ecological and physiological significance of amphicarpy in *Gymnarrhena micrantha* (Comp.). *Am. J. Bot.* **51**: 26–35.

KRAUS, G. (1949) Morphologisch-anatomische Untersuchung der entwicklungsbedingten Veränderungen an Achse, Blatt und Fruchtknoten bei einigen Beerenfrüchten. *Öst. Bot. Z.* **96**: 325–60.

KRAUSS, L. (1933) Entwicklungsgeschichte der Früchte von *Hordeum, Triticum, Bromus* und *Poa* mit besonderer Berücksichtigung ihrer Samenschalen. *Jb. wiss. Bot.* **77**: 773–808.

MACARTHUR, M. and WETMORE, R. H. (1939) Developmental studies in the apple fruit in the varieties McIntosh Red and Wagener: I, Vascular anatomy. *J. Pomol. Hort. Sci.* **17**: 218–32.

MACARTHUR, M. and WETMORE, R. H. (1941) Developmental studies of the apple fruit in the varieties McIntosh Red and Wagener: II, An analysis of development. *Can. J. Res., sect. C, Bot. Sci.* **19**: 371–82.

MACDANIELS, L. H. (1940) The morphology of the apple and other pome fruits. *NY Mem. Cornell Univ. Agric. Exp. Stn.,* no. 230.

MACKENZIE, K. A. D. (1979) The structure of the fruit of the red raspberry (*Rubus idaeus* L.) in relation to abscission. *Ann. Bot.* **43**: 355–362.

MARON, R. and FAHN, A (1979) Ultrastructure and development of oil cells in *Laurus nobilis* leaves. *Bot. J. Linn. Soc.* **78**: 31–40.

MCLEAN, R. C. and IVIMEY-COOK, W. R. (1956) *Textbook of Theoretical Botany,* Vol. 2. Longmans, London.

MONSI, M. (1943) Untersuchungen über den Mechanismus der Schleuderbewegung der Sojabohnen-Hülse. *Jap. J. Bot.* **12**: 437–74.

PLATT-ALOIA, K. A. and THOMSON, W. W. (1981) Ultrastructure of the Mesocarp of mature avocado fruit and changes associated with ripening. *Ann. Bot.* **48**: 451–465.

PLATT-ALOIA, K. A., OROSS, J. W. and THOMSON, W. W. (1983) Ultrastructural study of the development of oil cells in the mesocarp of avocado fruit. *Bot. Gaz.* **144**: 49–55.

PLATT-ALOIA, K. A., THOMSON, W. W. and YOUNG, R. E. (1980) Ultrastructural changes in the walls of ripening avocados: transmission, scanning, and freeze fracture microscopy. *Bot. Gaz.* **141**: 366–373.

RAGLAND, C. H. (1934) The development of the peach fruit, with special reference to split-pit and gumming, *Proc. Am. Soc. Hort. Sci.* **31**: 1–21.

RASCIO, N., CASADORO, G., RAMINA, A. and MASIA, A. (1985) Structural and biochemical aspects of peach fruit abscission (*Prunus persica* L. Batsch). *Planta* **164**: 1–11.

RETHKE, R. V. (1946) The anatomy of circumscissile dehiscence. *Am. J. Bot.* **33**: 677–83.

ROSENBAUM, J. and SANDO, C. E. (1920) Correlation between size of the fruit and the resistance of the tomato skin to puncture and its relation to infection with *Macrosporium tomato* Cooke. *Am. J. Bot.* **7**: 78–82.

ROTH, I. (1977) Fruits of Angiosperms. In: K. Linsbauer, *Handbuch der Pflanzenanatomie,* Spez. Teil, Bd. 10, T. 1. Gebr. Borntraeger, Berlin, Stuttgart.

SARFATTI, G. (1961) Accrescimento del pericarpo, seme, endosperma ed embrione in *Prunus amygdalus* Stokes. *Nuovo G. bot. Ital.* **68**: 118–35.

SCHNEE, B. K. and WALLER, D. M. (1986) Reproductive behaviour of *Amphicarpaea bracteata* (Leguminosae), an amphicarpic annual. *Am. J. Bot.* **73**: 376–386.

SCHNEIDER, H. (1968) The anatomy of *Citrus.* In: *The Citrus*

Industry, Vol. II, Rev. edn. (ed. W. Reuther, L. D. Batchelor, and H. J. Webber). Univ. of California, Div. Agric. Sci., Berkeley, pp. 1–85.

SCHOENICHEN, W. (1924) *Biologie der Blütenpflanzen*. T. Fisher, Freiburg i. Br.

SCOTT, F. M. and BAKER, K. C. (1947) Anatomy of Washington navel orange rind in relation to water spot. *Bot. Gaz.* **108:** 459–75.

SHAH, G. L., RANGAYYA, S. and KOTHARI, I. L. (1983) Structure and development of pod of *Calliandra tweedii* Benth. *Flora* **173:** 469–477.

SKENE, D. S. (1966) The distribution of growth and cell division in the fruit of Cox's Orange Pippin. *Ann. Bot.* **30:** 494–512.

SMITH, W. H. (1940) The histological structure of the flesh of the apple in relation to growth and senescence. *J. Pomol. Hort. Sci.* **18:** 249–60.

SMITH, W. H. (1950) Cell-multiplication and cell-enlargement in the development of the flesh of the apple fruit. *Ann. Bot.* **14:** 23–38.

STARITSKY, G. (1970) *The Morphogenesis of the Inflorescence, Flower and Fruit of* Pyrus nivalis Jacquin *var.* orientalis *Terpó*. Veenman & Zonen, Wageningen.

STEINBRINCK, C. and SCHINZ, H. (1908) Über die anatomische Ursache der hygrochastischen Bewegungen der sog. Jerichorosen, usw. *Flora* **98:** 471–500.

STÖSSER, R., RASMUSSEN, H. P., AND BUKOVAC, M. J. (1969) Histochemical changes in the developing abscission layer in fruits of *Prunus cerasus* L. *Planta* **86:** 151–64.

SUBRAMANYAM, K. and RAJU, M. V. S. (1953) Circumscissile dehiscence in some angiosperms. *Am. J. Bot.* **40:** 571–4.

THOMSON, W. W. (1969) Ultrastructural studies on the epicarp of ripening oranges. *Proc. First Int. Citrus Symposium* **3:** 1163–9.

TUKEY, H. B. and YOUNG, J. O. (1942) Gross morphology and histology of developing fruit of the apple. *Bot. Gaz.* **104:** 3–25.

VAN DER PIJL, L. (1969) *Principles of Dispersal in Higher Plants*. Springer-Verlag, Berlin and New York.

VON TEICHMAN, I. (1987) The development and structure of the pericarp of *Lannea discolor* (Sonder) Engl. (Anacardiaceae). *Bot. J. Linn. Soc.* **95:** 125–135.

VON TEICHMAN, I. and VAN WYK, A. E. (1988) The ontogeny and structure of the pericarp and seed-coat of *Harpephyllum caffrum* Bernh. ex Krauss (Anacardiaceae). *Bot. J. Linn. Soc.* **98:** 159–176.

WARDROP, A. B. (1983) The opening mechanism of follicles of some species of *Banksia*. *Aust. J. Bot.* **31:** 485–500.

WINKLER, H. (1939) Versuch eines "natürlichen" Systems der Früchte. *Beitr. Biol. Pfl.* **26:** 201–20.

WINKLER, H. (1940) Zur Einigung und Weiterführung in der Frage des Fruchtsystems. *Beitr. Biol. Pfl.* **27:** 92–130.

ZAMSKI, E. and ZIV, M. (1976) Pod formation and its geotropic orientation in the peanut *Arachis hypogaea* L. in relation to light and mechanical stimulus. *Ann. Bot.* **40:** 631–6.

ZOHARY, M. (1937) Die verbreitungsökologische Verhältnisse der Pflanzen Palästinas. I. Die antitelechorischen Erscheinungen. *Beih. Bot. Zbl.* Abt. A, **56:** 1–155.

ZOHARY, M. and FAHN, A. (1941) Anatomical-carpological observations in some hygrochastic plants of the oriental flora. *Palest. J. Bot., Jerusalem ser.* **2:** 125–35.

ZOHARY, M. and FAHN, A. (1950) On the heterocarpy of *Aethionema*. *Palestine J. Bot.*, Jerusalem ser. **5:** 28–31.

CHAPTER 21

THE SEED

THE seed develops from the ovule. In the mature seed the following parts can be distinguished: the *seed coat*, commonly called *testa*, which develops from one or two integuments; *endosperm*, which may be present in a large or small amount; the *embryo*, which constitutes the partially developed young sporophyte. In some seeds the endosperm is completely absent and such seeds, as well as those that contain a very small amount of endosperm, are termed *exalbuminous seeds* (Fig. 293, nos. 2, 3). In seeds of some plants, e.g. some species of *Citrus*, the embryos contain chloroplasts and are green (Casadoro *et al.*, 1980). In the seeds of certain plants, e.g. species of the Centrospermae, Piperaceae and Nymphaeaceae, the nucellar tissue persists and increases in volume to form the *perisperm*. Seeds with endosperm or perisperm are termed *albuminous seeds*. Several features can be distinguished on the outer surface of the seed. The micropyle may be completely obliterated or it may remain as a distinct pore. In the place where the seed was attached to the funiculus a scar, termed the *hilum* (Fig. 301, no. 4), is present. Water can penetrate with relative ease through the hilum. In anatropous ovules, in which part of the funiculus is fused to the integument, the seed becomes detached together with the fused part of the funiculus which forms a characteristic ridge termed the *raphe*. After the fertilization of the ovule, growths, termed *arils*, develop on the surface of the seeds of certain plants. Arils are very common in tropical and subtropical plants, e.g. in *Intsia bijuga*, *Pithecellobium dulce*, *Durio zibethinus*, *Acacia retivenea*, *Bocconia frutescens*, and *Myristica fragrans*. Arils are organs that are well adapted to seed dispersal by animals (Corner, 1949). Arils that contain oil, such as those of *Chelidonium majus* (Fig. 299, no. 3), *Luzula villosa*, *Gagea lutea*, *Reseda odorata*, *Galan-*

thus nivalis, *Dendromecon* (Fig. 299, nos. 4, 5), and *Ricinus communis*, are *elaiosomes* (see preceding chapter). They are thought to be connected with seed dispersal by ants. For structures adapted for seed dispersal see Fahn and Werker (1972). Some authors have distinguished between true *arils* and *arilloids* (van der Pijl, 1969). True arils are outgrowths of the funiculus near its top, whereas arilloids are outgrowths from other parts of the seed. Arilloids occurring on the raphe are called *strophioles* and those occurring near the micropyle *caruncles*.

Another type of seed appendage is the plug-like structure, formed in the micropylar region, that separates during germination by circumscissile dehiscence. This type of appendage is termed *operculum* (Kapil *et al.*, 1980).

THE SEED COAT

Angiosperm ovules have one or two integuments (see Chapter 19). In order to clarify which parts of the integuments take part in the formation of the seed coat, ontogenetic investigation is necessary. All parts of one or both integuments may take part in the formation of the seed coat, e.g. *Viola tricolor* (Fig. 295, nos. 1, 2). However, in most seeds much of the integumental tissue is destroyed and absorbed by the other developing tissues of the seed, and then the seed coat develops only from the remaining parts of the integuments. The parts that are destroyed are usually the innermost or intermediate middle layers of the integument. The nucellus may also take part in the construction of the seed coat (Fig. 295, no. 4; Fig. 296, no. 2). However, in the development of most seeds the nucellus is apparently completely destroyed (Netolitzky, 1926; Eames and MacDaniels, 1947).

In certain seeds, especially those of indehiscent fruits, only the two or three outermost layers of the integument persist. In the mature seed of the Umbelliferae only the outer unspecialized epidermis of the single integument remains. In the seeds of certain genera of the Compositae, e.g. *Lactuca* (Fig. 283, nos. 3, 4), the integument is represented only by a thin layer of obliterated cells which persists beneath the cypsela coat. The

FIG. 293. 1, A seed of *Pancratium maritimum*. × 4. 2 and 3, Scanning electron micrographs of seeds. 2, *Trigonella arabica*, sectioned seed. × 67. 3, *Portulaca oleracea*. × 95. (No. 1 from Werker and Fahn, 1975; no. 2 from Gutterman, 1978; no. 3 courtesy of Y. Gutterman.)

innermost layers of the latter coat also disintegrate (see Chapter 20). In the seeds of certain monocotyledons, e.g. *Zea*, the integuments are completely destroyed. In seeds developing from ovules with two integuments, the two integuments may be present in the testa or only two or three of the outermost layers of the outer integument may remain. In such seeds the inner integument may constitute the major portion of the seed coat (Fig. 295, no. 1), or the outer one may be the better developed and be specially adapted for protection while the inner is non-specialized (Fig. 296). The former type is characteristic of the Malvaceae, Violaceae, Hypericaceae and Tiliaceae, and the latter of the Cruciferae, Papaveraceae, Berberidaceae and certain genera of the Liliaceae, Iridaceae, and Araceae. In the Onagraceae, Lythraceae, Aristolochiaceae, and others, protective layers develop from both integuments, and the entire nucellus or its outermost layers only also take part in the formation of the seed coat (Fig. 295, no. 4). In most of the Leguminosae, Ranunculaceae and certain genera of the Liliaceae and Amaryllidaceae the inner integument and the nucellus are completely obliterated. In the seeds of only a few plants in which the ovule is surrounded by a single integument does the entire integument form the seed coat. Usually only a few of the outermost cell layers and the inner epidermis persist, e.g. in the Plantaginaceae (Fig. 295, no. 3) and Polemoniaceae.

Histological structure of the seed coat

There are vast differences in histological structure of the seed coats of the different plants. The simplest type of seed coat structure occurs in the seeds of the Orchidaceae where the seed coat consists of a single layer of elongated cells which originate from the outer integument. An air space is present (Fig. 295, no. 6) between this membranous coat and the undifferentiated embryo. This feature enables the dispersal of the seeds, which are about a quarter of a millimetre long, by wind. Minute "dust seeds" without air spaces occur in some other plants, e.g. in *Orobanche* (Joel, 1987). In seeds which are adapted to dispersal by water, e.g. *Pancratium maritimum*, the testa is very thick and consists of large rounded air-filled cells (Fig. 293, no. 1; Fig. 294) (Werker and Fahn, 1975). The structure of seed coats that are very hard or fleshy, and especially of those that are sculptured, is complicated (Fig. 293, nos. 2, 3).

In species of genera of about 30 families stomata occur in the outer epidermis of the testa (outer integument), e.g. in *Ceiba* (Bombacaceae), *Capparis* (Capparidaceae), *Ricinus* (Euphorbiaceae), *Juglans* (Juglandaceae), *Bauhinia* (Leguminosae), *Magnolia* (Magnoliaceae), *Gossypium*

FIG. 294. Portion of a cross-section of the seed coat of *Pancratium maritimum*. The protective layer consists of a few layers of compressed cells impregnated with dark brown material. (From Werker and Fahn, 1975.)

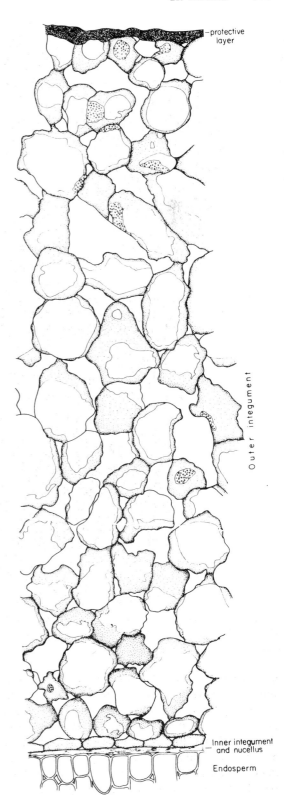

protective layer

Outer integument

Inner integument and nucellus

Endosperm

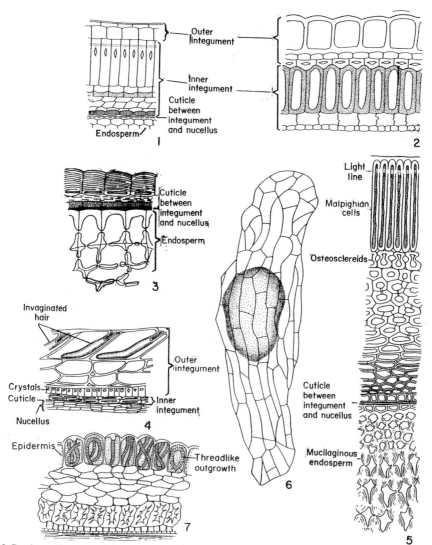

FIG. 295. 1–5, Portions of cross-sections of seed coats. 1, *Malva sylvestris.* 2, *Viola tricolor* 3, *Plantago lanceolata.* 4, *Lythrum salicaria.* 5, *Ceratonia siliqua.* 6, An entire seed of *Orchis* in which the embryo (stippled) can be seen through the transparent seed coat. 7. *Cuphea,* showing epidermis cells with thread-like outgrowths, the walls of which exhibit helically arranged folds. (Nos. 1, 3, and 4 adapted from Netolitzky, 1926; no. 2 adapted from Eames and MacDaniels, 1947; no. 7 adapted from Correns, 1892.)

and *Hibiscus* (Malvaceae), *Eschscholtzia* (Papaveraceae), *Viola* (Violaceae) (Corner, 1976; Jernstedt and Clark, 1979; Rugenstein and Lersten, 1981).

The seed coat of some plants, e.g. certain genera of the Leguminosae, is impermeable to water and oxygen when intact. The structure of the sclerenchymatous cells and the composition of their wall form a barrier to water permeation (Werker, 1980). In *Cercis,* the compressed and reduced inner integument was suggested to be the major barrier to water penetration (Roti-Michelozzi *et al.,* 1987). The presence of substances such as phenols and quinones in a continuous layer of cells may also contribute to this barrier (Spurný, 1973; Werker *et al.,* 1973; Werker and Fahn, 1975; Werker *et al.,* 1979).

In the seed coats of certain plants, a layer of radially elongated cells, which are palisade-like but devoid of intercellular spaces, may be present. These cells have been termed *Malpighian cells* after the investigator who first described them. Because of their shape and the thickness of their walls, these cells are also termed *macrosclereids.* The thickness of the walls is characteristically un-

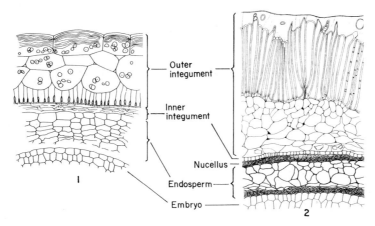

FIG. 296. 1, An almost mature seed coat of *Sinapis alba*. 2, A mature seed coat of *Citrus aurantium*. (No. 1 from Bouman, 1975; no. 2 from Boesewinkel, 1978.)

equal throughout, and the walls may consist of cellulose only or of lignin or cutin as well. Malpighian cells are especially characteristic of the Leguminosae where they constitute the outer epidermis. In seeds of the Leguminosae, the cell lumen of the Malpighian cells is usually widest at the base of the cell. The outer layers of the upper cell walls of the Malpighian cells of some Leguminosae seeds, which are covered by a thin cuticle, form hard caps consisting of pectic substances (Werker *et al.*, 1973, 1979). In *Cercidium floridum*, the walls of such epidermal cells have been observed to be traversed by plasmodesmata (Scott *et al.*, 1962). In a cross section of the seed coat, a thin line which runs across these cells and parallel to the surface of the seed close to the cuticle can be distinguished. This is due to the fact that, along this line, the light refraction differs from that in other parts of the cells. This line is termed the *light line* or *linea lucida* (Fig. 295, no. 5). Some investigations have shown that the orientation of the microfibrils is responsible for the appearance of the light line (Scott *et al.*, 1962). In the seeds of certain species the light line is a result of deposition of wax globules in the cells (Eames and MacDaniels, 1947). Non-epidermal Malpighian cells are mainly found in families other than the Leguminosae, e.g. in the seeds of *Gossypium* (Fig. 298, no. 4) and *Malva* (Fig. 295, no. 1).

In the seeds of many of the Leguminosae one or more layers of cells with unusual shape are found below the Malpighian cells; these cells may be funnel- or bone-shaped for instance, and because they are also thick-walled, they are termed *osteosclereids*. They may, like the epidermal cells, contain pigments or be devoid of them. In certain species of *Phaseolus* these cells also contain crystals of oxalate salts (Netolitzky, 1926). The greater portion of the inner tissue of the integuments disintegrates and the outer cells develop at the expense of the contents of this tissue. The cells that remain after this process of disintegration may sometimes be thick-walled. In the testa of many seeds the inner epidermis may be present and sometimes a cuticle can be distinguished between the seed coat and the remains of the nucellus or endosperm (Fig. 295, nos. 1, 3–5). Relatively well-developed vascular bundles may be found in the testa of certain plants (*Arachis*), while in others (*Pisum* and *Lathyrus*) they can hardly be distinguished. In leguminous seeds the nucellus apparently disappears completely.

Corner (1951, 1976) has drawn attention to the value of the structure of the seed coat in the taxonomy of the Leguminosae. The cells of the seed coats of other families also have different characteristic shapes, wall thickenings, etc.

In the Leguminosae, when the seed breaks from the funicle, an abscission layer is developed so that the hilum is clearly defined. In the bean seed, for instance, the cells of the funicle attached to the palisade layer of the seed coat are also transformed into a layer of palisades similar to those of

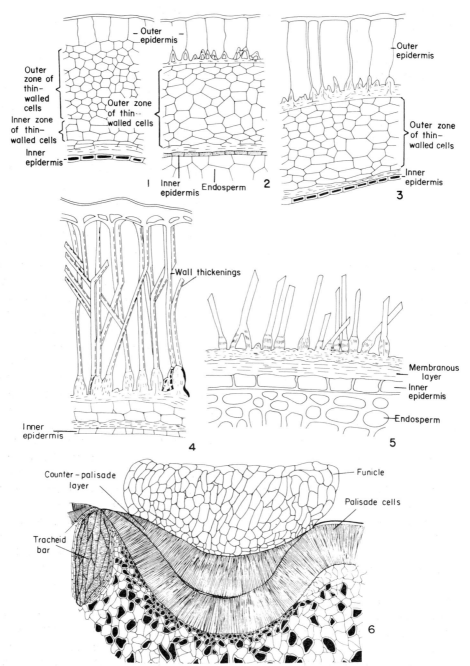

FIG. 297. 1–5, Portions of cross-sections of the seed coat of *Lycopersicon esculentum* showing different stages of development. 6, Median longitudinal section of seed coat through hilum of *Phaseolus aureus*. (Nos. 1–5 adapted from Hayward, 1938; no. 6 adapted from Chowdhury and Buth, 1970.)

the seed coat. This counter-palisade layer is fused with the palisade layer of the hilum. Both these layers are interrupted along the middle line by a very fine groove, which serves as a passage for air and water vapour. This groove leads to a group of tracheids, termed the *tracheid bar*, which is enveloped by a sheath of elongated parenchyma cells (Fig. 297, no. 6). On either side of this structure there is a stellate parenchyma (Corner, 1951; Manning and Van Staden, 1985).

In the seed coat of some plants crystals or silica bodies are present (Boesewinkel and Bouman, 1984). Silica may also occur as incrustation in cell walls (Venturelli and Bouman, 1988).

The testa of *Phoenix* consists of thin-walled cells only. During the course of seed development, the number of cell layers of the integuments, and especially of the inner integument, decreases.

The structure and development of the seed of *Lycopersicon esculentum* has been described by Souèges (1907) and is of great interest. In the thick integument of the young *Lycopersicon* seed the following four parts can be distinguished: an outer epidermis; an intermediate parenchymatous tissue in which inner and outer zones can be distinguished; an inner epidermis which contains pigment (Fig. 297, no. 1). With the development and enlargement of the seed, the cells of the outer parenchymatous zone increase in number, and thickenings develop in the inner tangential walls and at the base of the radial walls of the outer epidermal cells (Fig. 297, nos. 2, 3). When the seeds are partially mature the outer epidermal cells are seen to be very elongated in a radial direction, and longitudinal wall thickenings, which appear more or less in the angles of the cells, are developed. These thickenings increase in length till they reach the outer tangential wall (Fig. 297, no. 4). In the final stage of development of the epidermal cells their whole content is transformed into mucilage. Two to three layers of the placental tissue remain attached to the mucilaginous epidermis of the ripe seed and form a mucilage envelope around the seed (Czaja, 1963). This envelope becomes detached from the seed together with the cell content and the thin wall parts of the epidermal cells of the testa, and only the thickened strands remain. These strands form the hair- or scale-like structures that cover the surface of the mature seed (Fig. 297, no. 5). Together with this process the intermediate parenchyma disintegrates gradually in a centrifugal direction from the inner zone. In the mature seed nothing remains of this parenchyma except for the crushed walls which form a more or less uniform membranous layer. The pigment-containing inner epidermis remains and forms the inner border of the testa.

The surface of the seeds of different species may have special and characteristic features, which are used in classification (Barthlott, 1981). Hairs, ribs, folds, spines, or hooks usually develop from the epidermis of the testa alone, but there are instances in which sub-epidermal cells take part in their formation. The fibres of *Gossypium*, for example, are epidermal cells which elongate to form hairs (Fig. 298, no. 4). Sometimes the seed coat is expanded to form a wing-like structure of thin-walled cells, which become filled with air. Such winged seeds (e.g. of *Fibigia clypeata* and many species of the Bignoniaceae) are easily distributed by wind. The wing may consist of, or include, a layer of tube-like cells with some of the walls thickened or with annular, spiral, or reticulate wall thickenings, e.g. in some members of the Scrophulariaceae, Bignoniaceae, and Rubiaceae. Many seeds, e.g. of most genera of the Asclepiadaceae, Apocynaceae, and Tamaricaceae, bear tufts of hairs which facilitate dispersal by wind. In some such seeds, e.g. those of *Tamarix* (Fig. 299, no. 1) and *Myricaria*, the hairs undergo hygroscopic movements—they spread when dry and converge when wet. This movement is brought about by the specialized structure (Fig. 299, no. 2) of the basal portion of the abaxial wall of each hair (Guttenberg, 1926).

The seed coat of certain plants is juicy (*sarcotestal seeds*). In *Punica*, for example, the juicy edible layer develops entirely from outer epidermal cells which elongate to a very large extent in a radial direction. The sap of these cells develops a turgor pressure which preserves the characteristic external shape of these seeds (Fig. 298, no. 3).

In certain seeds, e.g. those of *Linum usitatissimum*, an outer layer, which becomes mucilaginous when in contact with water, is developed. In the mature seed of *Linum* several zones may easily be distinguished (Hayward, 1938). The outer epidermis is covered by a cuticle. With the development of the seed coat a mucilaginous substance develops on the inner side of the thick outer wall of the epidermal cells so as almost completely to fill the cell lumen (Fig. 298, nos. 1, 2). This substance, which exhibits a stratified texture, swells greatly when it absorbs water. The middle lamellae between the cells are not sufficiently elastic to accommodate this swelling and so they rupture

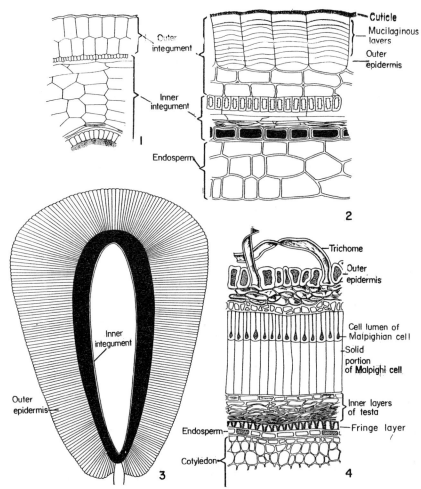

FIG. 298. 1 and 2, Portions of cross-sections of the testa of *Linum usitatissimum*. 1, Developing testa. 2, Mature testa. 3, Diagram of a longitudinal section of the seed of *Punica granatum* showing the radially elongated cells of the outer epidermis of the testa; these cells form the fleshy part of the seed. The inner portion of the outer integument is sclerenchymatous (solid black). 4. Portion of a cross-section of the seed coat of *Gossypium*. According to Ryser *et al.* (1988), the fringe-layer develops from the inner epidermis of the inner integument. The fringe-cells produce protuberances on their lateral walls, resembling those of transfer cells. (Nos. 1 and 2 adapted from Hayward, 1938; nos. 3 and 4 adapted from Eames and MacDaniels, 1947.)

and the outer cutinized wall layers, which are covered with cuticle, become raised and crack. Below the epidermis there are one or two layers of cells whose lumina appear circular in a surface view of the seed. Below these cells is a layer of elongated, pitted sclerenchyma cells (Fig. 298, no. 2). The orientation of these cells is parallel to the longitudinal axis of the seed. Still further inwards there are two zones of parenchyma cells which are elongated at right angles to the sclerenchyma cells. These parenchyma cells contain much starch, which is absorbed by the other tissues dur-

ing the development of the seed, and the cells themselves become crushed and obliterated. The cells of the innermost layer of the testa are cubical or polyhedral with thick pitted walls and they contain in their cell lumina the pigment that determines the colour of the seed.

Corner (1976) in his book *The Seeds of Dicotyledons* introduced an original classification of the seed-coat constituents. The portion of the seed coat produced by the outer integument he termed *testa* and that produced by the inner integument *tegmen*. Seeds with characteristic testa he called

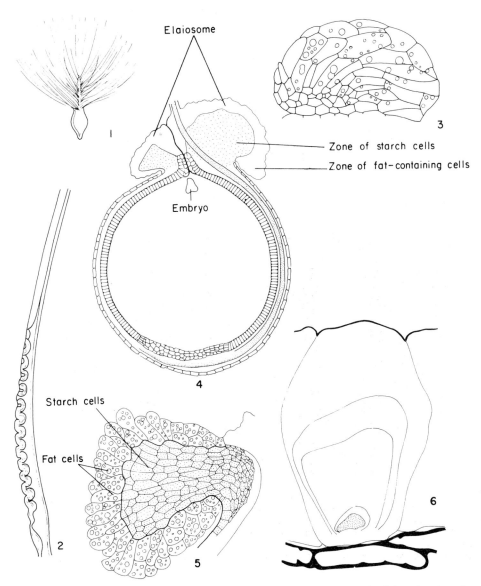

FIG. 299. 1 and 2, *Tamarix*. 1, An entire seed, with tuft of hairs. 2, The base of a single hair of the tuft, enlarged to show the characteristic structure of the abaxial wall. 3, Elaiosome of *Chelidonium majus*. 4, Schematic drawing of a longitudinal section of a seed of *Dendromecon rigida*, showing an elaiosome. 5, Part of an elaiosome of *Dendromecon*. 6, A nearly mature mucilage-containing epidermal cell from an ovule of *Plantago ovata*. (No. 3 adapted from Szemes, 1943; nos. 4 and 5 adapted from Berg, 1966; no. 6 adapted from Hyde, 1970.)

testal and those with characteristic tegmen *tegmic*.

According to which tissue in the seed coat becomes sclerified, Corner suggested the following terms: *exotestal* and *exotegmic* seeds depending on whether the mechanical layer develops from the outer epidermis in the testa or tegmen respect-

ively; *mesotestal* and *mesotegmic* seeds in which the mechanical layer develops in the middle layer of one of the integuments, and *endotestal* and *endotegmic* seeds in which the mechanical layer develops from the inner epidermis of the appropriate integument.

Pulpy and edible testa Corner termed *sarco-testa*.

Mucilage and seed dispersal

Mucilage is common in many seeds, nutlets, and one-seeded fruits. It appears in Cruciferae, Labiatae, Compositae, Plantaginaceae, and other families (Zohary, 1962). The functions of mucilage are many (Fahn and Werker, 1972): when the diaspore is moistened after its release, it may adhere to the soil surface—this feature has the advantage that the adhering diaspore cannot be carried farther away to unfavourable localities by wind or rain; the diaspore may adhere to animals and be thus further dispersed; the mucilage may cause reduction of the specific weight of the diaspore in water; the mucilage may take part in regulation of germination by preventing desiccation of the germinating seed or, in case of excess of water, when the seed becomes entirely covered by mucilage, the mucilage may hinder passage of oxygen, thus preventing germination.

Despite the variability in functions of mucilage, its structure has some characteristics common to all mucilaginous diaspores. Mucilage is the pectinous matrix of cell-wall layers which swells considerably upon wetting. According to Frey-Wyssling (1959) this capacity of the wall matrix to swell so markedly indicates that such cell walls consist of considerable amounts of unesterified galacturonic acid with an especially large capacity for hydration. The mucilage membranes are laid down as such from the start and only rarely develop by metamorphosis from already existing secondary walls (Mühlethaler, 1950).

The mucilage may either contain no cellulose or it may contain cellulose microfibrils. The first type of mucilage was reported in seeds of *Linum usitatissimum* and *Plantago* species. Here, the mucilage constitutes the secondary wall (Fig. 298, no. 2). The second type of mucilage formed upon wetting has more structural strength than pure mucilage. This prevents the mucilage from being washed away. This type was found in the epidermal cells of seeds of Cruciferae, such as *Lepidium sativum*. In this case the cellulose mucilage constitutes the primary wall (Mühlethaler, 1950; Frey-Wyssling, 1959). Upon wetting, the mucilage wall absorbs water and swells. If a layer of cuticle is present on the outer side of the cell it is lifted and/or fissured by the mucilage.

There are cases in which an inner secondary wall is formed on the side walls. The matrix between the cellulosic helical bands of this wall swells forcibly when wetted so that the cells elongate outwards to a considerable extent. The mucilaginous content is then discharged to the outside and the helical strand of the secondary wall is torn and unwound. Such is the case in epidermal cells or epidermal hairs of seeds of *Ruellia* species (Haberlandt, 1918) and of *Cobaea scandens* (Frey, 1928). A similar case is found in the tip cells of the multicellular hair of *Blepharis persica* (Gutterman *et al.*, 1967; Witztum *et al.*, 1969) in which annular or helical secondary wall thickenings are present. The mucilage of the mucilaginous primary wall when examined with the aid of the electron microscope was found to be composed of cellulose microfibrils in a uronic gel. In dry seeds of this plant the hairs are adpressed to the seed surface. Upon wetting the hairs hydrate rapidly, stand erect, and position the seed so that its root end comes in contact with the soil (Fig. 300) (Gutterman and Witztum, 1977).

In nutlets of *Coleus blumei* the thick anticlinal walls of the outer epidermis swell when wetted, causing the outer periclinal walls and underlying tannin-like material to separate (Witztum, 1978).

In species of *Malcolmia* a column is left in the centre of each epidermal cell after the diffusion of the mucilage. The structure of the column may be a diagnostic character (Stork, 1971; Vaughan and Whitehouse, 1971).

In the epidermis of seeds of *Cuphea viscosissima* (Lythraceae) a thread-like outgrowth protrudes into the cell from the outer walls of each cell (Fig. 295, no. 7). This outgrowth consists of a wall with folds forming a dense helix and is filled with a substance that swells when moistened. Upon wetting of the seed a small round section of the outer wall above the thickening opens on one side like a lid, and the inner substance of the threadlike outgrowth swells. Later, the outgrowth is pushed out

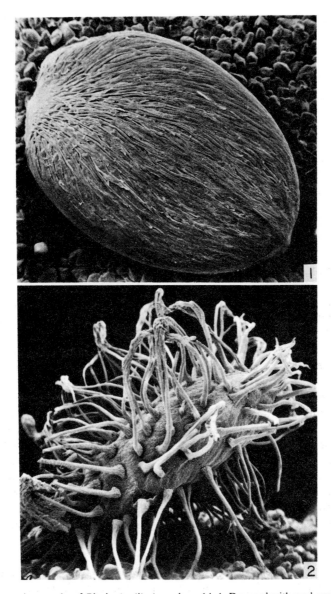

Fig. 300. Scanning electron micrographs of *Blepharis ciliaris* seeds. × 14. 1, Dry seed with seed-coat hairs adpressed to the surface. 2, Wetted seed, the hairs are erect and have positioned the seed so that its root end comes into contact with the soil. (From Gutterman and Witztum, 1977.)

like a finger of a glove and acquires a hair-like structure. The inversion of the thread is a result of the swelling of a substance present in the lumen of the epidermal cell (Correns, 1892). A similar structure occurs in other species of *Cuphea* and the seeds of *Lythrum* (Fig. 295, no. 4) (Netolitzky, 1926; Panigrahi, 1986; Schoenberg and Hofmeister,

1986).

The presence or absence of mucilage in the epidermal cells of seeds may have an inter- or intra-specific taxonomic significance as is the case in the Cruciferae (Vaughan *et al.*, 1963; Vaughan, 1968).

Development of mucilage

In the mucilage-producing epidermal cells of the ovule of *Plantago ovata* the sequence of events during maturation was found to consist of two parts. The first processes which start after pollination are associated with cell enlargement—increase in volume of the nucleus and particularly of the nucleolus, the appearance of rough endoplasmic reticulum and polysomal arrays, synthesis of large starch grains, and an increase of dictyosomes with accompanying vesicles. During these processes the cells increase 30–40 times in volume. The second series of events is associated with further cell expansion and includes the digestion of starch, synthesis and extrusion of mucilage, and degeneration of the protoplast. Deposition of mucilage which takes place inside vacuoles and between the plasmalemma and cell wall is accompanied by a marked increase in the number and size of Golgi vesicles. Fusion occurs between vacuoles and the extra protoplasmic space and the protoplast degenerates almost completely. At maturity there remains only a trace of the protoplast at the base of the cell which is now only a bag of mucilage (Fig. 299, no. 6) (Hyde, 1970).

In the seed epidermis of quince (*Cydonia oblonga*) the mucilage, deposited between the plasmalemma and cell wall contains cellulose microfibrils in addition to amorphous polysaccharide material. At some stage of development, the microfibrils begin to organize and ultimately become helicoidal (Abeysekera and Willison, 1988).

THE ENDOSPERM

The development and types of endosperm have been discussed in Chapter 19; here the mature endosperm will be briefly described. The endosperm tissue in the developing seed may consist of thin-walled cells with large vacuoles which do not contain reserve substances. Such an endosperm is entirely or partially absorbed by the developing embryo, e.g. *Lactuca*. In many other seeds, e.g. those of *Ricinus* and the Gramineae, the endosperm functions as a storage tissue. The surface of this type of endosperm is smooth in most plants. However, in species of some families (e.g. Annon-aceae, Myristicaceae, and Passifloraceae) the mature endosperm shows a very irregular surface. Ingrowths of the seed coat penetrate into the infolds of the surface of this endosperm. The endosperm of such seeds is called *ruminated endosperm* (Fig. 301, no. 8). Seeds with extreme rumination, in which the lobing affects also the embryo, are referred to as *labyrinth seeds* (Boesewinkel and Bouman, 1984).

The storage endosperm may consist of thin-walled cells and then the reserve material is located within the cell, or it may consist of thick-walled cells and then the walls themselves constitute the reserve substance. In the former type mainly oil, starch grains and proteins are the stored substances. There are two main forms in which proteins may be stored in the endosperm—in an amorphous form (glutens), or in the form of aleurone grains. Protein bodies, or as previously called aleuron grains, are composed of an amorphous protein mass. Some enclose crystalloids and some spherical bodies which may contain phytin, others contain both crystalloids and globoids while still others contain none of them. In the caryopses of cereals glutens are found in the starch cells, and protein bodies are restricted to the outermost layer of endosperm cells (the aleurone layer). It has been suggested that rough ER is involved in the synthesis of the protein matrix of the protein bodies in wheat (Morrison *et al.*, 1975). In the aleurone cells are found, in addition to the numerous protein bodies, numerous lipid bodies, also referred to as spherosomes (Fig. 305, no. 2), as well as organelles typical of other plant cells including microbodies and rough ER (Jones, 1969). In *Ricinus*, protein bodies occur throughout the entire endosperm. In endosperm in which there is no starch, oils and fats may be present as reserve substances. (See also Chapter 2.)

Cell walls that constitute reserve material, e.g. in the seeds of *Phoenix* (Fig. 305, no. 3) *Coffea* and *Diospyros*, usually consist of hemicelluloses and other similar carbohydrates. In *Phoenix*, Meier (1958) recorded that these walls apparently contain, in addition, about 6% of cellulose. These cells have a very thick secondary wall.

Some seeds, e.g. those of *Ceratonia*, have a *mucilaginous endosperm* (Fig. 295, no. 5). A stratified thickening of the walls of these endospermal

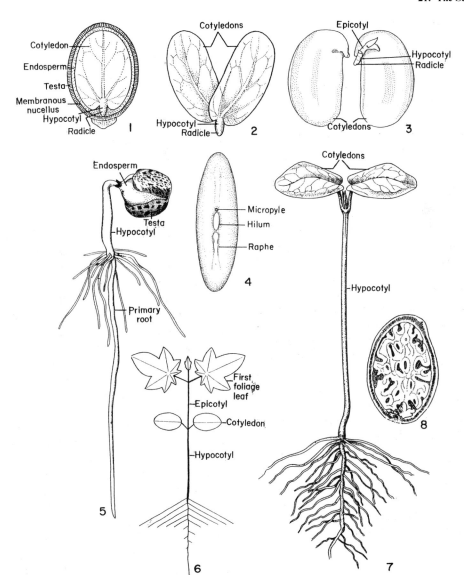

FIG. 301. 1, Longitudinally sectioned seed of *Ricinus communis*. The protuberance at the base of the seed is the caruncle. 2, Embryo of *Ricinus communis* with cotyledons opened out. 3, Embryo of *Phaseolus vulgaris* with cotyledons spread out. 4, A whole bean seed. 5–7, Seedlings of *Ricinus communis* at different stages of development. 5, Emergence of the primary root and hypocotyl. 6, Schematic diagram of a seedling that has already produced foliage leaves. 7, Young seedling with cotyledons expanded. 8, Seed of *Myristica fragrans* with ruminated endosperm. (Nos. 1–3 and 5–7 adapted from Troll, 1948; no. 8 adapted from McLean and Ivimey-Cook, 1956.)

cells can be distinguished. These walls become more or less mucilaginous when in contact with water. In the dry seeds these endospermal cells are hard and, during germination, they function as both a swelling and a nutritional tissue (Netolitzky, 1926).

In the exalbuminous seeds the storage materials

occur in the cotyledons. In *Pharbitis nil* and some other Convolvulaceae, giant cells containing numerous oil drops are present in the cotyledons (Wada *et al.*, 1981).

(For more information on structure and storage materials of seeds see Netolitzky, 1926, Vaughan, 1970; Corner, 1976; Boesewinkel and Bouman,

1984; Mayer and Poljakoff-Mayber, 1989.)

SEEDLINGS

With germination of the seed, the seed coat usually ruptures at the micropylar end, and the radicle emerges. In some plants the embryo bursts from the seed coat by lifting a kind of lid which is called *operculum*. Such lids occur in about 45 families and are more common in monocotyledons (Grootjen, 1983; Bregman and Bouman, 1983; Boesewinkel and Borman, 1984).

Young seedlings of the angiosperms consist of a radicle, plumule (shoot bud), hypocotyl and one or two cotyledons. As mentioned already in Chapter 14, not only seedlings of the Monocotyledonae, but also those of several species of dicotyledoneous families (e.g. Ranunculaceae, Umbelliferae and Gesneriaceae) possess only one cotyledon. It has been suggested that the single cotyledon is phytogenetically derived from an ancestral pair which has fused to provide an apparatus which enables the unexpanded plumule and radicle to be buried deeply in the ground, in the early stage of germination (Haines and Lye, 1979; Clarkson and Clifford, 1987). Generally, the radicle penetrates into the soil, develops root hairs and often lateral roots. After this, further rupturing of the testa takes place.

In many seeds the cotyledons and shoot apex emerge while the hypocotyl elongates as a result of intercalary growth. This type of germination is termed *epigeal germination*; examples of such germination can be seen in *Helianthus, Raphanus, Phaseolus*, and *Ricinus* (Fig. 301, nos. 5–7).

The cotyledons of plants in which the germination is epigeal may vary in form and function. The cotyledons of *Phaseolus vulgaris*, for example, are very thick and function as storage organs (Fig. 301, no. 3). These cotyledons wither early and are shed. The cotyledons of many other dicotyledonous plants in which the germination is epigeal are thinner structures, which, when appearing above ground, more closely resemble foliage leaves (Fig. 301, no. 7).

In many other plants, e.g. *Vicia, Pisum sativum*, and *Quercus*, the thick cotyledons, which contain reserve materials, remain within the testa and the hypocotyl elongates only slightly or not at all. This type of germination is termed *hypogeal germination*. In this type of germination the terminal bud of the embryo is pushed out through the soil by the elongation of the *epicotyl* which is the internode above the cotyledons (Fig. 302, nos. 1, 2).

In addition to the functions of storage and photosynthesis, in the case of epigeal germination, cotyledons may bring about the breakdown, absorption, and transport of nutrient substances from the endosperm to the developing embryo. Cotyledons fulfilling such functions occur in the monocotyledons such as *Allium cepa, Phoenix*, and in the Gramineae. During the germination of the seed of *Allium cepa* (Fig. 302, nos. 3–5), the cotyledon elongates and pushes the radicle tip out of the seed. The emerging root tip is usually directed upward at first, owing to the shape of the seed, and it becomes positively geotropic only at a later stage. The sharp bend thus caused by the change in the direction of growth is known as the "knee" (Fig. 302, no. 5). The other end of the cotyledon remains within the seed where it is in contact with the endosperm and where it functions as an *haustorium* for the absorption of nutrients. The exposed part of the cotyledon becomes green and may carry out photosynthesis. In the final stage of the germination process the remains of the testa and endosperm are shed from the absorbing end of the cotyledon and the cotyledon then straightens.

The germination of the seed of *Phoenix* (Fig. 302, nos. 6–11) is, in general, similar to the above. Here, that end of the cotyledon that remains within the endosperm develops into an haustorium which gradually enlarges with the increased disintegration of the endosperm (Fig. 302, no. 10) (DeMason, 1985; DeMason *et al.*, 1985). In *Phoenix* the terminal bud of the embryo shoot is concealed in the sheathing part of the cotyledon.

In the Gramineae the cotyledon is shield-like and has therefore been termed the *scutellum*. This organ remains within the caryopsis in contact with the endosperm from which it absorbs the nutrient substances (Fig. 303). On the side where the scutellum is in contact with the endosperm it is covered with a well-marked epithelium which is

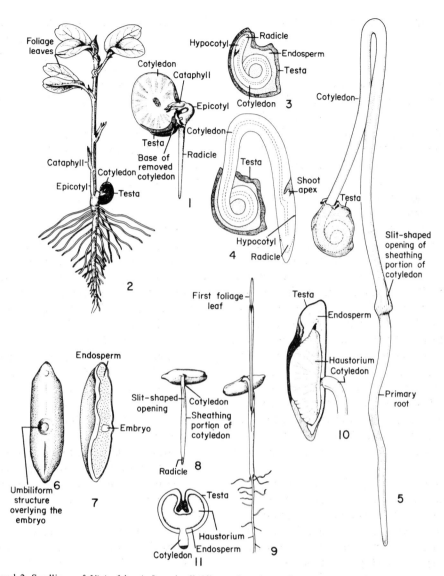

FIG. 302. 1 and 2, Seedlings of *Vicia faba*. 1, Longitudinally sectioned germinating seed. 2, A young plant. 3–5, Seed and seedling of *Allium cepa*. 3, Longitudinally sectioned seed. 4, Germinating seed sectioned longitudinally. 5, Developing seedling. 6–11, Seed and seedling of *Phoenix dactylifera*. 6, An entire seed. 7, Longitudinally sectioned seed. 8, Germinating seed. 9, Mature seedling which has already produced foliage leaves. 10, Longitudinally sectioned seed which has germinated and has already produced a mature seedling. 11, Cross-section of seedling at the same stage as in No. 10. (Adapted from Troll, 1948.)

secretory in function. The surface of the scutellum on the endospermal side is wrinkled and in *Zea mays* deep folds of the epithelium are seen (Fig. 305, no. 1). As described for *Hordeum vulgare,* during the first 3–4 days of germination the epithelial cells detach themselves from one another and grow to twice their former length; aleurone proteins and subsequently the lipids of the lipid bodies are digested. Concomitantly with this process there is a development and temporary increase in activity of the ER, mitochondria, dictyosomes, and leucoplasts. These changes may be related to the production and secretion of enzymes and gibberellins during the first 3 days.

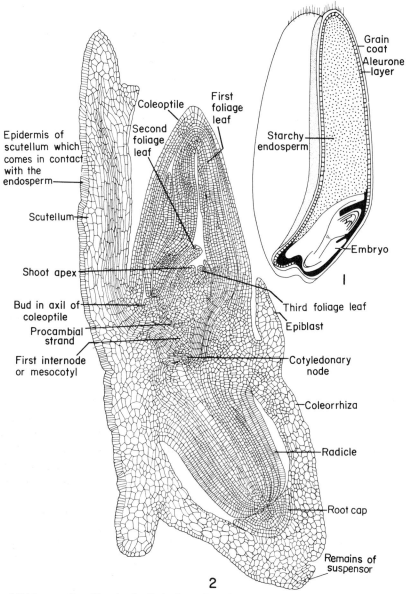

FIG. 303. 1, Grain of *Triticum* sectioned longitudinally in the region of the groove. 2, Longitudinal section of the embryo of *Triticum*. (No. 1 adapted from Esau, 1953; no. 2 adapted from Hayward, 1938.)

Subsequently the function of the epithelium is mainly the uptake and transport of sugars and, probably to a minor extent, amino acids from the endosperm. In a later stage aqueous vacuoles develop and after 3 weeks the cell organelles become smaller and fewer (Nieuwdorp and Buys, 1964).

In *Avena* and some other grasses the scutellum grows, during germination, into and through the endosperm (Fig. 304) (Negbi, 1984; Negbi and Sargent, 1986).

In Gramineae, also, the first organ to emerge from the caryopsis is the radicle. The shoot apex is enclosed in a sheath-like leaf termed the *coleoptile*. At first, the coleoptile grows at the same rate as the bud within it. Only after it reaches a certain length do the internodes of the plumule elongate

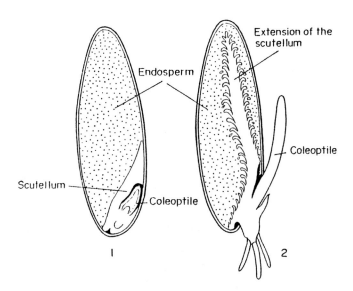

FIG. 304. Schematic drawings of *Avena* type grains at two developmental stages. 1, Dormant. 2, During germination, showing growth of the scutellum into and through the endosperm and development of papillae. (Adapted from Negbi and Sargent, 1986.)

and the foliage leaves enlarge so that they emerge to the exterior through a rupture which forms at the upper end of the coleoptile.

The nature of the coleoptile has been variously interpreted (Eames, 1961). Some authors regard it as the first true leaf, while others believe it to be part of the cotyledon. Various opinions also exist as to the nature of the axial part of the embryo below the coleoptile (Eames, 1961). It is considered by some authors to be a complex structure formed by the fusion of parts of the cotyledon with the hypocotyl, and thus this part of the axis is termed, by them, the *mesocotyl*. Those investigators who consider the coleoptile to be the first leaf regard the above part of the embryonal axis as the first internode. The sheath that surrounds the endogenously developed root is termed the *coleorrhiza*. This sheath is interpreted as being either part of the scutellum or part of the axis. Another characteristic feature of the embryo of certain grass genera is the presence of the *epiblast*. This is a small scale-like organ, opposite the scutellum which is devoid of vascular supply. Different opinions exist as to the homology of the epiblast. According to some investigators it represents a vestigial second cotyledon, while others regard it as being an outgrowth or part of other embryonic organs, such as the coleorrhiza, for example (Roth, 1955; Brown, 1960; Eames, 1961; Foard and Haber, 1962; Negbi and Koller, 1962). A provascular system described in detail by Avery (1930) can be distinguished within the embryo (Fig. 303, no. 2).

The pattern of vascularization of the scutellum was described in wheat by Swift and O'Brien (1970). According to these authors the central bundle of the scutellum and the upper part of some of the lateral strands consist of both xylem and phloem; however, about half the sieve tubes are unaccompanied by xylem and are surrounded solely by a special sheath of parenchyma cells. The sieve tube strands lie in the upper and lateral parts of the scutellum opposite that part of the endosperm which is richest in starch.

(For information on physiology and ecology of germination see Mayer and Poljakoff-Mayber, 1989.)

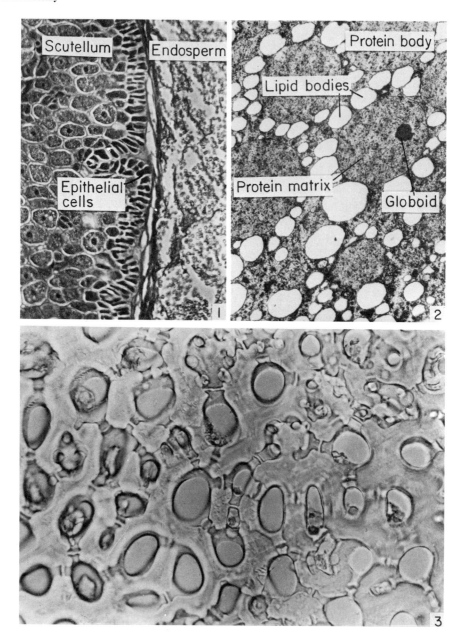

Fig. 305. 1 and 2, Scutellum of *Zea mays*. 1, Portion of scutellum and endosperm of a grain several days after germination. × 115. 2, Electron micrograph showing protein bodies (aleuron grains) surrounded by lipid bodies (spherosomes) in an epithelial cell. × 11,500. 3, Section of the endosperm of *Phoenix dactylifera*. × 600. (Nos. 1, 2 courtesy of E. Zamski.)

REFERENCES

ABEYSEKERA, R. M. and WILLISON, J. H. M. (1988) Development of helical texture in the prerelease mucilage of quince (*Cydonia oblonga*) seed epidermis. *Can. J. Bot.* **66**: 460–467.

AVERY, G. S., JR. (1930) Comparative anatomy and morphology of embryos and seedlings of maize, oats, and wheat. *Bot. Gaz.* **89**: 1–39.

BARTHLOTT, W. (1981) Epidermal and seed surface characters of plants: systematic applicability and some evolutionary aspects. *Nord. J. Bot.* **1**: 345–355.

BERG, R. Y. (1966) Seed dispersal of *Dendromecon*: its ecologic, evolutionary and taxonomic significance. *Am. J. Bot.* **53**: 61–73.

BOESEWINKEL, F. D. (1978) Development of ovule and testa in Rutaceae. III. Some representatives of the Aurantioideae. *Acta Bot. Neerl.* **27**: 341–54.

BOESEWINKEL, F. D. and BOUMAN, F. (1984) The seed: Structure. In: *Embryology of Angiosperms* (ed. B. M. Johri), Springer-Verlag, Berlin, pp. 567–610.

BOUMAN, F. (1975) Integument initiation and testa development in some Cruciferae. *J. Linn. Soc. Bot.* **70**: 213–29.

BREGMAN, R. and BOUMAN, F. (1983) Seed germination in Cactaceae. *Bot. J. Linn. Soc.* **86**: 357–374.

BROWN, W. V. (1960) The morphology of the grass embryo. *Phytomorphology* **10**: 214–34.

CASADORO, G., MARIANI COLOMBO, P., and RASCIO, N. (1980) Plastids in the quiescent embryo and young seedlings of the chloroembryophyte *Citrus nobilis* × *Citrus aurantium amara pumila*. *Ann. Bot.* **45**: 415–8.

CHOWDHURY, K. A. and BUTH, G. M. (1970) Seed coat structure and anatomy of Indian pulses. In: *New Research in Plant Anatomy* (ed. N. K. B. Robson, D. F. Cutler, and M. Gregory). *J. Linn. Soc. Bot.* **63**, Suppl. 1. Academic Press, London, pp. 169–79.

CLARKSON, J. R. and CLIFFORD, H. T. (1987) Germination of *Jedda multicaulis* J. R. Clarkson (Thymelaeaceae). An example of cryptogeal germination in the Australian flora. *Aust. J. Bot.* **35**: 715–720.

CORNER, E. J. H. (1949) The Durian theory or the origin of the modern tree. *Ann. Bot.* **13**: 367–414.

CORNER, E. J. H. (1951) The leguminous seed. *Phytomorphology* **1**: 117–50.

CORNER, E. J. H. (1976) *The Seeds of Dicotyledons*. 2 vols. Cambridge University Press, Cambridge.

CORRENS, C. (1892) Über die Epidermis der Samen von *Cuphea viscosissima*. *Ber. dt. bot. Ges.* **10**: 143–52.

CZAJA, A. TH. (1963) Neue Untersuchungen an der Testa der Tomatensamen. *Planta* **59**: 262–79.

DeMASON, D. A. (1985) Histochemical and ultrastructural changes in the haustorium of date (*Phoenix dactylifera* L.). *Protoplasma* **126**: 168–177.

DeMASON, D. A., SEXTON, R., GORMAN, M. and REID, J. S. G. (1985) Structure and biochemistry of endosperm breakdown in date palm (*Phoenix dactylifera* L.) seeds. *Protoplasma* **126**: 159–167.

EAMES, A. J. (1961) *Morphology of the Angiosperms*. McGraw-Hill, New York.

EAMES, A. J. and MacDANIELS, L. H. (1947) *An Introduction to Plant Anatomy*, 2nd edn. McGraw-Hill, New York.

ESAU, K. (1953) *Plant Anatomy*. Wiley, New York.

FAHN, A. and WERKER, E. (1972) Anatomical mechanisms of seed dispersal. In: *Seed Biology* (ed. T. T. Kozlowski). Academic Press, New York.

FOARD, D. E. and HABER, A. H. (1962) Use of growth characteristics in studies of morphologic relations. I. Similarities between epiblast and coleorhiza. *Am. J. Bot.* **49**: 520–3.

FREY, A. (1928) Das Wesen der Chlorzinkjodreaktion und das Problem des Faserdichroismus. *Jb. wiss. Bot.* **67**: 597.

FREY-WYSSLING, A. (1959) *Die pflanzliche Zellwand*. Springer-Verlag, Berlin.

GROOTJEN, C. J. (1983) Development of ovule and seed in *Cartonema spicatum* R. Br. (Cartonemataceae). *Aust. J. Bot.* **31**: 297–305.

GUTTENBERG, H. V. (1926) Die Bewegungsgewebe. In: K. Linsbauer, *Handbuch der Pflanzenanatomie*, Bd. 5, Lief. 18. Gebr. Borntraeger, Berlin.

GUTTERMAN, Y. (1978) Seed coat permeability as a function of photoperiodical treatments of the mother plants during seed maturation in the desert annual plant: *Trigonella arabica* Del. *J. Arid. Environments* **1**: 141–4.

GUTTERMAN, Y. and WITZTUM, A. (1977) The movement of integumentary hairs in *Blepharis ciliaris* (L.) Burtt. *Bot. Gaz.* **138**: 29–34.

GUTTERMAN, Y., WITZTUM, A., and EVENARI, M. (1967) Seed dispersal and germination in *Blepharis persica* (Burm.) Kunze. *Israel J. Bot.* **16**: 213–34.

HABERLANDT, G. (1918) *Physiologische Pflanzenanatomie*. Engelmann, Leipzig.

HAINES, R. W. and LYE, K. A. (1979) Monocotylar seedlings: a review of evidence supporting an origin by fusion. *Bot. J. Linn. Soc.* **78**: 123–140.

HAYWARD, H. E. (1938) *The Structure of Economic Plants*. Macmillan, New York.

HYDE, B. B. (1970) Mucilage-producing cells in the seed coat of *Plantago ovata*: developmental fine structure. *Am. J. Bot.* **57**: 1197–1206.

JERNSTEDT, J. A. and CLARK, C. (1979) Stomata on the fruits and seeds of *Eschscholzia* (Papaveraceae). *Am. J. Bot.* **66**: 586–90.

JOEL, D. M. (1987) Detection and identification of *Orobanche* seeds using fluorescence microscopy. *Seed Sci. & Technol.* **15**: 119–124.

JONES, R. L. (1969) The fine structure of barley aleurone cells. *Planta* **85**: 359–75.

KAPIL, R. N., BOR, J. and BOUMAN, F. (1980) Seed appendages in angiosperms. I. Introduction. *Bot. Jahrb. Syst.* **101**: 555–73.

MANNING, J. C. and VAN STADEN, J. (1985) The development and ultrastructure of the testa and tracheid bar in *Erythrina lysistemon* Hutch. (Leguminosae: Papilionoideae). *Protoplasma* **129**: 157–167.

MAYER, A. M. and POLJAKOFF-MAYBER, A. (1989) *The Germination of Seeds*, 4th ed. Pergamon Press, Oxford.

MEIER, H. (1958) On the structure of cell walls and cell wall mannans from ivory nuts and from dates. *Biochem. biophys. Acta* **28**: 229–40.

MORRISON, I. N., KUO, J., and O'BRIEN, T. P. (1975) Histochemistry and fine structure of developing wheat aleurone cells. *Planta* **123**: 105–16.

MÜHLETHALER, K. (1950) The structure of plant slimes. *Expl. Cell Res.* **1**: 341–50.

NEGBI, M. (1984) The structure and function of the scutellum of the Gramineae. *Bot. J. Linn. Soc.* **88**: 205–222.

NEGBI, M. and KOLLER, D. (1962) Homologies in the grass embryo—a re-evaluation. *Phytomorphology* **12**: 289–96.

NEGBI, M. and SARGENT, J. A. (1986) The scutellum of *Avena*: a structure to maximize exploitation of endosperm reserves. *Bot. J. Linn. Soc.* **93**: 247–258.

NETOLITZKY, F. (1926) Anatomie der Angiospermen-Samen. In: K. Linsbauer, *Handbuch der Pflanzenanatomie*, Bd. 10, Lief. 14. Gebr. Borntraeger, Berlin.

NIEUWDORP, P. J. and BUYS, M. C. (1964) Electron microscopic structure of the epithelial cells of the scutellum of barley: II, Cytology of the cells during germination. *Acta bot. neerl.* **13**: 559–65.

PANIGRAHI, S. G. (1986) Seed morphology of *Rotala* L., *Ammannia* L., *Nesaea* Kunth and *Hionanthera* Fernandes & Diniz (Lythraceae). *Bot. J. Linn. Soc.* **93**: 389–403.

ROTH, I. (1955) Zur morphologischen Deutung des Grasembryos und verwandter Embryophyten. *Flora* **142**: 564–600.

ROTI-MICHELOZZI, G., SERRATO, G. and RIGGIO BEVILACQUA, L. (1987) Remnants of the inner integument as the major barrier to water entry in *Cercis siliquastrum* L. seed. *Phytomorphology* **37**: 165–171.

RUGENSTEIN, S. R. and LERSTEN, N. R. (1981) Stomata on seeds and fruits of *Bauhinia* (Leguminosae: Caesalpinioideae). *Am. J. Bot.* **68**: 873–876.

RYSER, U., SCHORDERET, M., JAUCH, U. and MEIER, H. (1988) Ultrastructure of the "fringe-layer", the innermost epidermis of cotton seed coats. *Protoplasma* **147**: 81–90.

SCHOENBERG, M. M. and HOFMEISTER, R. M. (1986) A epiderme de semente em *Cuphea calophylla* ssp *mesostemon* (Koehne) Lourt. *Estudos de Biologia*, Publicacao da Universidade Catolica do Parana, No. XVI, pp. 3–10.

SCOTT, F. M., BYSTROM, B. G., and BOWLER, E. (1962) *Cercidium floridum* seed coat, light and electron microscope study. *Am. J. Bot.* **49**: 821–33.

SOUÈGES, E. C. R. (1907) Développement et structure du tégument seminal chez les Solanacées. *Annls Sci. nat., Bot.*, ser. 9, **6**: 1–24.

SPURNÝ, M. (1973) The imbibition process. In: *Seed Ecology*, Proceedings of the 19th Easter School in Agricultural Sciences, University of Nottingham, 1972. Butterworths, London, pp. 367–89.

STORK, A. L. (1971) Seed characters in European taxa of *Malcolmia* R. Br. (Cruciferae). *Svensk bot. Tidskr.* **65**: 283–92.

SWIFT, J. G. and O'BRIEN, T. P. (1970) Vascularization of the scutellum of wheat. *Aust. J. Bot.* **18**: 45–53.

SZEMES, G. (1943) Zur Entwicklung des Elaiosoms von *Chelidonium majus*. *Wiener bot. Z.* **92**: 215–19.

TROLL, W. (1948) *Allgemeine Botanik*. F. Enke, Stuttgart.

VAN DER PIJL, L. (1969) *Principles of Dispersal in Higher Plants*. Springer-Verlag, Berlin and New York.

VAUGHAN, J. G. (1968) Seed anatomy and taxonomy. *Proc. Linn. Soc. Lond.* **179**: 251–5.

VAUGHAN, J. G. (1970) *The Structure and Utilization of Oil Seeds*. Chapman & Hall, London.

VAUGHAN, J. G., HEMINGWAY, J. S., and SCHOFIELD, H. J. (1963) Contributions to a study of variation of *Brassica juncea* Coss. & Czern. *J. Linn. Soc., Bot.* **58**: 435–47.

VAUGHAN, J. G. and WHITEHOUSE, J. M. (1971) Seed structure and the taxonomy of the Cruciferae. *J. Linn. Soc., Bot.* **64**: 383–409.

VENTURELLI, M. and BOUMAN, F. (1988) Development of ovule and seed in Rapateaceae. *Bot. J. Linn. Soc.* **97**: 267–294.

WADA, K., KADOTA, A., TANIHIRA, H. and SUZUKI, Y. (1981) Giant oil cells in the cotyledons of *Pharbitis nil* and other Convolvulacean plants. *Bot. Mag. Tokyo* **94**: 239–247.

WERKER, E. (1980) Review. Seed dormancy as explained by the anatomy of embryo envelopes. *Israel J. Bot.* **29**: 22–44.

WERKER, E. and FAHN, A. (1975) Seed anatomy of *Pancratium* species from three different habitats. *Bot. Gaz.* **136**: 396–403.

WERKER, E., DAFNI, A., and NEGBI, M. (1973) Variability in *Prosopis farcata* in Israel: anatomical features of the seed. *J. Linn. Soc. Bot.* **66**: 223–32.

WERKER, E., MARBACH, I., and Mayer, A. M. (1979) Relation between the anatomy of the testa, water permeability and the presence of phenolics in the genus *Pisum*. *Ann. Bot.* **43**: 765–71.

WITZTUM, A. (1978) Mucilaginous plate cells in the nutlet epidermis of *Coleus blumei* Benth. (Labiatae). *Bot. Gaz.* **139**: 430–5.

WITZTUM A., GUTTERMAN, Y., and EVENARI, M. (1969) Integumentary mucilage as an oxygen barrier during germination of *Blepharis persica* (Burm.) Kuntze. *Bot. Gaz.* **130**: 238–41.

ZOHARY, M. (1962) *Plant Life of Palestine (Israel and Jordan)*. Ronald Press, New York.

GLOSSARY OF TERMS

abaxial, directed outwards from the axis.

abscission layer, a synonym for **separation layer.**

abscission zone, that zone containing the tissues which bring about the abscission of organs such as leaves, fruits and flowers.

accessory or **subsidiary cell,** an epidermal cell which borders on a guard cell of the stoma and which differs from the ordinary epidermal cells.

accessory bud, a bud located above or on either side of the main bud.

acropetal, proceeding towards the apex.

actinocytic, a type of stoma which is enclosed by a circle of radiating subsidiary cells.

actinomorphic, refers to a flower which is radially symmetrical.

actinostele, a protostele in which the xylem, as seen in cross-section, is star-shaped.

adaxial, directed towards the axis.

adaxial meristem, a meristematic tissue on the adaxial side of a leaf primordium which contributes to the increase in thickness of the petiole and midrib (also called *ventral meristem*).

adnation, concrescence of organs or tissues of different nature.

adventitious organ, an organ which developed in an unusual position.

aerenchyma, a parenchymatous tissue characterized by the presence of large intercellular spaces.

aggregate fruit, a fruit developing from an apocarpous ovary in which the development of the carpels is independent, each having its own pericarp, but with the ripening of the fruit a single unit is formed.

albuminous cell, certain cells in the phloem rays or in the phloem parenchyma of Gymnospermae which are connected morphologically and physiologically to the sieve elements. Unlike the companion cells of the Angiospermae, these cells do not usually arise from the same cell as the sieve element. Also called *Strassburger cell.*

albuminous seed, a mature seed which contains endosperm.

alburnum, *see* **sapwood.**

aleurone grain, a characteristic structure of reserve protein present in the seeds of numerous plants (also called *protein body*). In the endosperm of certain plants such grains are found in the outermost cell layer which is then termed the *aleurone layer.*

amphicribral vascular bundle, *see* **vascular bundle.**

amphiphloic siphonostele, *see* **siphonostele.**

amphivasal vascular bundle, *see* **vascular bundle.**

amyloplast, a leucoplast which has become specialized to store starch.

androecium, the collective term for all the stamens of a flower.

androgynophore, a stalk-like elongation of the floral axis between the perianth and stamens, which elevates the androecium and gynoecium.

anisocytic, a type of stomal complex in which the three subsidiary cells are unequal in size.

anisotropic, having different properties along different axes. Optically—having different optical properties, showing birefringence.

anomocytic type (of stomata), a specific pattern of arrangement of the guard cells and other epidermal cells adjacent to them (see p. 167).

anther, the pollen-bearing part of the stamen.

anthesis, the time of maturation of the male and female organs of the flower.

anticlinal, perpendicular to the surface.

antipodal cells, the cells of the female gametophyte present at the chalazal end of the mature embryo sac of the Angiospermae.

aperture (of pollen grain), a depressed area of characteristic shape in the wall; the pollen tube emerges via such an area.

apex, the terminal portion of the shoot or root in which the apical meristem is located.

apical cell, the single initial present in the apical meristem of some roots and shoots; typical of many lower vascular plants.

apocarpy, that condition in a flower and ovary where the carpels are free.

apomixis, process of reproduction in the ovule without fertilization.

apoplast, region of the plant body outside the protoplasts (cell walls and intercellular spaces).

apposition (of cell wall), wall growth as a result of successive addition, layer on layer, of wall material.

areole (in mesophyll), the smallest area of the mesophyll in a leaf delimited by veins. In Cactaceae, small outgrowths on the stem-bearing spines.

aril, a fleshy outgrowth on the surface of certain seeds; *true aril,* an outgrowth of the funiculus near its top; *arilloid,* an outgrowth from other parts of the seed.

astrosclereid, a branched sclereid.

atactostele, a stele which consists of vascular bundles scattered throughout the ground tissue as in the Monocotyledoneae.

autochory, seed dispersal by means of a self-dispersal mechanism.

axial bundle, a main bundle of a shoot; distinguished from leaf traces, cortical and medullary bundles.

axial organ, the root, stem, inflorescence or flower axis without their appendages.

axillary bud, a bud in the axil of a leaf.

back wall, that part of the guard-cell wall which is adjacent to the subsidiary or ordinary neighbouring epidermal cells.

baculum (pl. **bacula**), small rod-shaped structure in the pollen wall.

bark, the layer outside the vascular cambium, consisting of secondary phloem and periderm.

bark, outer, *see* **rhytidome.**

basipetal, proceeding towards the base.

bifacial leaf or **dorsiventral leaf,** a leaf in which palisade parenchyma is present on one side of the blade and spongy parenchyma on the other.

body, primary, that part of the plant which develops from the primary meristems, apical and intercalary.

body, secondary, that part of the plant, comprising the secondary vascular tissues and periderm, which is added to the primary body as a result of the activity of the lateral meristems, i.e. the cambium and phellogen.

brachysclereid *or* **stone cell,** a short, more or less isodiametric sclereid.

bulliform cell, an enlarged epidermal cell common in the leaf of the Gramineae; rows of such cells occur along the leaf.

bundle sheath, a layer or layers of cells surrounding the vascular bundles of leaves; may consist of parenchyma or sclerenchyma.

bundle sheath extension, a strip of ground tissue present along the leaf veins and extending from the bundle sheath to the epidermis; may be present on both sides of the vein or on one side only and may consist of parenchyma or sclerenchyma.

callose, a polysaccharide, β-1,3 glucan, present in sieve areas, walls of pollen tubes, walls of fungal cells, etc.

callus, the tissue formed as a result of wounding or a tissue developing in tissue culture.

calyptrogen, a term arising out of the histogen theory. In the root apex that meristem from which the root cap develops independently of all other initials of the apical meristem.

cambium, a lateral meristem the cells of which divide mostly periclinally.

cambium, vascular, a lateral meristem from which the secondary vascular tissues, i.e. secondary xylem and phloem, develop; *storied,* a cambium in which the fusiform initials, as seen in tangential section, are arranged in horizontal rows; *non-storied,* a cambium in which the fusiform initials, as seen in tangential section, partially overlap one another and are not arranged in horizontal rows.

cambium-like transitional zone, a cyto-histological zone visible in some shoot apices (see pp. 59–61).

carpophore, the split axis of a cremocarp to which the mericarps remain attached after they become separated from each other.

caruncle, an arilloid (type of an outgrowth on a seed) occurring near the micropyle.

caryopsis, a fruit in which the pericarp and testa are fused; characteristic of the Gramineae.

Casparian strip, a band-like structure in the primary wall and middle lamella containing lignin and suberin. Especially characteristic of the endodermal cells of roots where the band is present in the anticlinal walls, both radial and transverse.

cataphyll, a leaf which appears on the basal portions of shoots, e.g. scales on rhizomes, bud scales, etc.

cauline bundle, synonym for **axial bundle.**

cavity, a more or less isodiametric intercellular space, usually secretory.

cell plate, that part of the wall which develops between the two daughter nuclei during telophase.

central mother cells, a cyto-histological zone of the shoot apex (see pp. 59–61).

chalaza, that region in the ovule where the nucellus and integuments connect with the funiculus.

chimera, a combination in a single plant organ of tissues of different genetic composition.

chlorenchyma, a chloroplast-containing parenchyma tissue such as the mesophyll and other green tissues.

chlorophylls, the green pigments present in the chloroplastids.

chloroplast, a plastid in which photosynthesis takes place; contains, among others pigments, the chlorophylls.

chromoplast, a plastid containing pigments other than chlorophyll; usually containing carotenoids.

cisterna, a flattened, sac-like membrane bound cell compartment.

cladode, see *phylloclade.*

cluster, referring to the arrangement of vessels as seen in cross-section of the secondary xylem; several vessels grouped together in both radial and tangential directions.

coenocyte, a group of protoplasmic units; a multinucleate structure. In Spermatophyta often used to refer to multinucleate cells.

cohesion, concrescence of organs or tissues of the same nature.

coleoptile, the sheath surrounding the apical meristem and leaf primordia of the grass embryo.

coleorrhiza, the sheath which surrounds the radicle of the grass embryo.

collenchyma, the supporting tissue of young organs; consisting of more or less elongated cells the walls of which are usually unequally thickened.

colleter, a multicellular hair having a sticky secretion.

columella, the central portion of the root cap in which the cells occur in longitudinal files.

companion cell, a specialized parenchyma cell associated with the sieve tube member in the phloem of Angiospermae; it originates from the same mother cell as the sieve tube member and has a physiological connection with it.

complementary cells, a loose tissue formed towards the periphery by the phellogen of the lenticel; the cells may or may not have suberized walls.

connective, the tissue present between the two lobes of an anther.

cork, see *phellem* in **periderm.**

cork cell, (1) a dead cell which arises from the phellogen and whose walls are impregnated with suberin; has a protective function as the walls are impermeable to water and gases; (2) in an epidermis—a short cell with suberized walls; characteristic of Gramineae.

corpus, according to the tunica-corpus theory, in the shoot apex of Angiospermae that group of cells below the surface layer or layers (i.e. tunica) in which cell divisions take place in various planes; such divisions cause the increase in volume of the shoot apex.

cortex, the tissue between the vascular cylinder and epidermis of the axis.

cotyledon, the first leaf of the embryo.

crassulae (single crassula), in tracheids of Gymnospermae transversely oriented thickenings accompanying the pit-pairs and formed by the intercellular substance and primary wall material. Also called *bars of Sanio.*

cross-field, the rectangle formed by the walls of a ray cell and an axial tracheid as seen in a radial section; mainly used in the description of conifer wood.

cuticle, a layer of cutin, a fatty substance which is almost impermeable to water, on the outer walls of the epidermal cells.

cuticular layer, the outer portions of the epidermal walls which are impregnated with cutin.

cuticularization, process of formation of the cuticle.

cutin, a complex polymer of lipid derivatives, considerably impervious to water; the primary component of the cuticle.

cutinization, the deposition of cutin in cell walls.

cylinder, central or **vascular,** that part of the axis of the plant consisting of vascular tissue and the associated parenchyma. Equivalent to the term *stele,* but without the evolutionary connotation associated with this term.

cystolith, a specific outgrowth of the cell wall on which calcium carbonate is deposited. Characteristic of certain families, e.g. the Moraceae.

cytochimera, a combination, in a single plant organ, of tissues the cells of which are of different chromosome number.

cytokinesis, the process of cell division which results in the formation of two separate cells.

cytosol, the liquid, non-membranous, non-fibrillar part of the cytoplasm, within which all organelles are suspended; also called *hyaloplasm.*

dermatogen *see* **histogen.**

desmotubule, tubule connecting the two endoplasmic reticulum cisternae located at either side of a plasmodesma.

diacytic type, a specific pattern of arrangement of the guard cells and other epidermal cells adjacent to them (see p. 168).

diaphragm, a cross partition in an elongated air cavity.

diarch, that condition in the primary xylem of the root where there are two protoxylem strands.

diaspore, any plant part which is dispersed from the parent plant and functions in reproduction.

dictyosome, a cell organelle composed of stacked cisternae producing vesicles at the periphery; also called *Golgi body.*

dictyostele, a siphonostele in which the leaf-gaps are large and partly overlap one another so as to divide the stele into separate bundles in each of which the phloem surrounds the xylem.

differentiation, the physiological and morphological changes leading to specialization which occur in a cell, tissue, organ or entire plant during the process of development from a meristematic or juvenile state to a mature one.

diffuse-porous wood *see* **wood.**

domain, a portion of a vascular cambium, in which all pseudo transverse divisions have the same orientation.

dorsiventral leaf, *see* **bifacial leaf.**

druse, a compound crystal, more or less spherical in shape and in which the many component crystals protrude from the surface.

duct, an elongated space formed schizogenously, lysigenously or schizo-lysigenously and which may contain secretory substances or air.

duramen, *see* **heartwood.**

eccrine secretion, type of secretion; the secreted material leaves the cytoplasm as individual molecules by a mechanism of active transport through membranes.

ectodesma (pl. ectodesmata), a passage in the walls of the epidermis; mostly in the outer walls; also called *teichode.*

ectomycorrhiza (pl. ectomycorrhizae), a mycorrhiza in which the fungus produces a dense mycelium on the root surface.

ectoplast, *see* **plasmalemma.**

elaioplast, an oil-producing and storing leucoplast.

elaiosome, an outgrowth on a fruit or seed that contains large oil-storing cells.

emergence, a projection of the surface of a plant organ which consists not only of epidermal cells or parts of them, but also of cells derived from underlying tissues.

endarch xylem, in reference to the direction of maturation of the elements in a strand of primary xylem; a strand in which the first-formed elements (the protoxylem) are closest to the centre of the axis, i.e. the maturation is centrifugal.

endocarp, the innermost layer of the pericarp (fruit wall).

endodermis, the layer of cells forming a sheath around the vascular tissue and having Casparian strips on the anticlinal walls; secondary walls may be formed later.

endogenous, developing from internal tissues.

endogenous secretion, type of secretion; the secreted material accumulates in intercellular spaces (ducts or cavities).

endoplasmic reticulum (ER), a system of membranes forming cisternae or tubules that permeate the ground cytoplasm.

endosperm, a nutrient tissue formed within the embryo sac of the Spermatophyta.

endothecium, in the pollen-sac wall; a layer of cells with characteristic wall thickenings situated below the epidermis.

endothelium, the inner epidermis of the integument, next to the nucellus, in those cases where its cells become densely cytoplasmatic and apparently secretory; also called *integumentary tapetum.*

epiblast, a small growth present opposite the scutellum in the embryo of some Gramineae.

epiblem, the outermost cell layer (epidermis) of roots, as termed by some workers.

epicotyl, the stem of an embryo and seedling above the cotyledons.

epidermis, the outermost cell layer of primary tissues of the plant, sometimes comprising more than one layer—*multiseriate epidermis.*

epipetalous stamen, a stamen which is adnated to a petal.

epithelium, a compact layer of cells, often secretory, covering a free surface or lining a cavity or duct.

epithem, the tissues between the vein ending and the secretory pore of a hydathode.

ergastic matter, the non-protoplasmic products of metabolic processes of the protoplasm; starch grains, oil droplets, crystals and certain liquids; found in the cytoplasm, vacuoles, and cell walls.

etioplast, a plastid which has developed in the dark; when exposed to light it may transform to a chloroplast.

euchromatin, the chromatin (DNA with histones) which stains only lightly with histochemical reagents.

eustele, phylogenetically the most advanced type of stele, the vascular tissue of which forms a hollow cylinder built of collateral or bicollateral vascular bundles.

exalbuminous seed, a seed which is devoid of endosperm when mature.

exarch xylem, in reference to the direction of maturation of the elements in a strand of primary xylem; a strand in which the first-formed elements (the protoxylem) are furthest from the centre of the axis, i.e. the maturation is centripetal.

exine, the outer wall of a mature pollen grain.

exocarp, the outermost layer of the pericarp (fruit wall); also termed *epicarp.*

exodermis, in some roots the outermost layer or layers of cells of the cortex, the structure of which is similar to that of the endodermis, i.e. the cell walls are more or less thickened and contain suberin lamellae. A type of hypodermis.

exogenous, developing from external tissues.

exogenous secretion, type of secretion; the secreted substance is discharged to the outside of the plant.

fascicular, being part of or situated in a bundle of vascular tissue.

fibre, an elongated sclerenchymatous cell with tapered ends and with more or less thick secondary walls; the walls may or may not contain lignin and a living protoplast may or may not be retained in the mature fibre.

fibre, gelatinous, a fibre in which the inner layers of the secondary wall have a high capacity to absorb water and to swell.

fibre, libriform, a fibre of the secondary xylem usually with thick walls and simple pits.

fibre, septate, a fibre which becomes subdivided by thin transverse walls after the secondary wall layers have been formed.

fibre-sclereid, a sclerenchyma cell with characteristics intermediate between those of a fibre and a sclereid; occurs in the phloem and develops from parenchyma cells of non-conducting phloem.

fibre-tracheid, in the secondary xylem; a transition form between tracheid and libriform fibre.

filament, the stalk of a stamen.

filiform apparatus, a complex wall outgrowth (a transfer wall) in a synergid cell.

foraminate, having round pores.

fret, the portion of a thylakoid between neighbouring grana in a chloroplast.

funiculus, the stalk by which the ovule is attached to the placenta.

fusiform, of elongated form with tapered ends.

gametophyte, that plant generation which gives rise to the gametes from which, after fertilization, the sporophyte develops.

gap, branch, in a siphonostele, the parenchymatous region in the vascular cylinder above the position where a branch-trace enters a branch.

gap, leaf, in a siphonostele, the parenchymatous region in the vascular cylinder above the position where the leaf-trace enters a leaf.

generative cell, the cell in a pollen grain which on division produces two male gametes.

germination, epigeal, the process of germination in which the cotyledons and epicotyl emerge from the seed as a result of the elongation of the hypocotyl.

germination, hypogeal, the process of germination in which the hypocotyl elongates very little or not at all and the thick cotyledons, which store reserve materials, remain within the testa.

Golgi body, a cell organelle composed of stacked cisternae each producing vesicles at the periphery. Also called *dictyo-some*.

Golgi apparatus, a collective term for all Golgi bodies in a given cell.

gonophyll, according to the gonophyll theory, a sterile leaf together with an ovule-bearing branch from which it is assumed that the carpel has been derived.

granulocrine secretion, type of secretion; the secreted substances accumulate in membrane bound vesicles, which fuse with the plasmalemma and then discharge them from the cytoplasm.

granum (pl. **grana**), green granules in chloroplasts consisting of stacks of thylakoids.

growth, gliding or **sliding,** that type of growth in which the walls of neighbouring cells slide over one another.

growth, intrusive, that type of growth in which the growing cell penetrates between existing cells and in which new areas of contact are formed between the penetrating and neighbouring cells.

growth, mosaic, a theory concerning primary wall growth (see p. 36).

growth, multinet, a theory concerning primary wall growth (see p. 36).

growth, symplastic, the process of uniform growth of neighbouring cells so that the adjacent walls do not alter position relative to each other and no new areas of contact are formed.

growth ring, a clearly distinguishable region in the secondary xylem or phloem which is formed during a single growth season.

guard cells, a pair of specialized epidermal cells which, together with the aperture between them, form the stoma.

gum, a general term for viscous polysaccharide secretion.

gummosis, a condition which is expressed by the formation of gum.

guttation, the secretion of water, in a liquid form, from plants.

gynoecium, all the carpels of a single flower.

gynophore, an elongation of the floral axis between the stamens and the carpels, thus forming a stalk which elevates the gynoecium.

halophyte, a plant that grows in salty soil.

haplostele, a protostele in which the xylem, as seen in cross-section, is more or less circular.

hardwood, a common name for wood of the Dicotyledoneae.

Hartig net, in ectomycorrhizae, hyphae which penetrate between the outermost root cells where they form a mycelium.

haustorium, a specialized organ that draws nutriment from another organ or tissue.

heartwood or **duramen,** the inner layers of wood in the growing tree or shrub, which have lost the ability to conduct and no longer contain living cells. Mostly darker in colour than sapwood.

helobial endosperm, a type of endosperm which develops in a way intermediate between that by which a cellular and a nuclear endosperm develops.

heterocarpy, the production of two or more types of fruit by a plant.

heterochromatin, the chromatin (DNA with histones) which stains more darkly with histochemical reagents than euchromatin.

hilum, (1) that portion of a starch grain around which the starch is laid down in layers; (2) the scar present on a seed resulting from its abscission from the funiculus.

histogen, a term used to refer to those initials in the apical meristem of the root and shoot which are predestined to give rise to a particular and constant tissue system of the organ concerned. Hanstein distinguished three histogens: *dermatogen*, which gives rise to the epidermis; *periblem*, which gives rise to the cortex; *plerome*, which gives rise to the vascular cylinder.

holocrine secretion, type of secretion; the compounds leave the secretory cell as a result of its disintegration.

hydathode, a structure through which water is secreted in liquid form; found mainly on leaves.

hydrophyte, a plant that grows partly or entirely submerged in water.

hygrochastic process, manner of fruit opening or of the movement of other organs, as a result of water uptake; usually connected with the dispersal of seeds or spores.

hypoblast, a term used for the suspensor of the mature grass embryo.

hypodermis, a specific layer or layers of cells beneath the epidermis, which differ structurally from the tissue below them. In the narrow sense of the term, refers only to such layers which arise from a meristem other than the protoderm.

hypophysis, one of the cells of the embryo in its early stages of development (see p. 479).

hypsophyll, an inflorescence bract; a vestigial leaf or any other leaf having a structure different from that of a foliage leaf and occurring near the top of the shoot.

idioblast, a specific cell which is clearly distinguished from the other cells of the tissue in which it appears, either by size, structure or content.

indumentum, a collective term which refers to the whole trichome cover of a plant.

initial, (1) in meristems, a cell that by division gives rise to two cells one of which remains in the meristem and the other is added to the plant body; (2) of an element; a meristematic cell which differentiates into a mature specialized element.

integument, an envelope surrounding the nucellus of the ovule.

integumentary tapetum, see **endothelium.**

intercellular substance, *see* **middle lamella.**

internode, that part of the stem between two nodes.

interxylary, within and surrounded by xylem tissue.

intine, the inner wall of a mature pollen grain.

intraxylary, on the inner side of the xylem relative to the axis of the plant.

intussusception, cell wall growth as a result of the interpolation of new wall material within the wall already formed.

isobilateral or **isolateral leaf,** a leaf having palisade tissue on both sides of the blade.

isotropic, having equal properties along all axes. Optically— having equal optical properties in all directions.

karyokinesis, the process of nuclear division.

karyolymph, the nuclear sap.

knot, a portion of a dead branch, the wood of which has become heartwood and which is embedded in the developing wood of the stem from which the branch arose.

Kranz anatomy, in plants with C_4 pathway of photosynthesis; the term *Kranz* (wreath) refers to a layer of radially-oriented mesophyll cells (sometimes also bundle sheath cells), which surround the vascular bundles.

lacuna, (1) an intercellular space; (2) an interruption in the vascular tissue of the central cylinder.

laticifer, a cell or row of cells containing *latex*, a substance specific to such cells.

lenticel, an isolated area in the periderm consisting of suberized or non-suberized cells with numerous intercellular spaces between them.

leucoplast, a plastid devoid of pigment.

lignin, a mixed polymer containing phenolic derivatives of phenylpropane. Commonly found in secondarily thickened cell walls.

lithocyst, a cell containing a cystolith.

lysigenous, the manner of formation of an intercellular space as a result of the disintegration of cells.

lysosome, an organelle containing acid hydrolitic enzymes; in plants usually represented by vacuoles.

maceration, the artificial separation of the individual cells of a tissue by disintegration of the middle lamella.

macrosclereid, a somewhat elongated sclereid the secondary wall of which is unequally thickened. Common in the seeds of Leguminosae where they represent the epidermis of the testa and where they are termed *Malpighian cells.*

Malpighian cell, *see* **macrosclereid.**

mantle, all those outer cell layers of the shoot apex of the Angiospermae which can be distinguished by their layered arrangement from the cells of the inner portion of the shoot apex.

marginal initials, in marginal meristem of a developing leaf; cells along the margins that contribute cells to the protoderm.

marginal meristem, the meristem along the margin of a developing leaf.

margo, the pit membrane around the torus in bordered pits of conifers.

massula, a large group of adherent pollen grains, which participates in the formation of a pollinium.

matrix, a substance in which another substance is deposited or embedded.

median, situated in the middle.

megaspore, the female spore from which the female gametophyte develops; also called *macrospore.*

medullary bundle, a vascular bundle which is located in the pith.

megagametophyte, female gametophyte; embryo sac within the ovule of angiosperms.

meristele, one of the bundles of a dictyostele; *see also* **vascular bundle.**

meristem, a tissue which produces cells that undergo differentiation to form mature tissues.

meristem, apical, a meristem situated in the apical region of the shoot or root which, as a result of divisions, gives rise to those cells which form the primary tissues of the shoot or root.

meristem, flank or peripheral, one of the cyto-histological regions of the shoot apex (see pp. 57–61).

meristem, ground, a meristematic tissue which originates in the apical meristem and which produces tissues other than epidermis and vascular tissues.

meristem, intercalary, meristematic tissue derived from the apical meristem and which becomes separated from the apex in the course of development of the plant by regions of more or less mature tissues.

meristem, lateral, a meristem which is situated parallel to the circumference of the plant organ in which it occurs.

meristem, peripheral, see **meristem, flank.**

meristem, plate, a parallel layered meristem in which the planes of cell divisions in each layer are perpendicular to the surface of the organ which is usually a flat one.

meristem rib, (1) one of the cyto-histological regions of the shoot apex; (2) a meristem characterized by parallel series of cells in which transverse divisions take place.

meristemoid, a cell or a group of cells constituting an active locus of development surrounded by relatively mature cells.

merocrine secretion, type of secretion; the secreted substance is eliminated from cells which remain intact (as opposed to holocrine secretion).

mesarch xylem, in reference to the direction of maturation of elements in a strand of primary xylem; a strand in which the first-formed elements (the protoxylem) occur in the centre of the strand; i.e. the maturation is both centripetal and centrifugal.

mesocarp, the middle layer of the pericarp (fruit wall).

mesocotyl, often refers to the internode between the scutellar node and the coleoptile in the Gramineae.

mesogenous, see **stoma.**

mesomorphic, having structure characteristic of mesophytes.

mesophyll, the photosynthetic parenchymatous tissue situated between the two epidermal layers of the leaf.

mesophyte, a plant suited to a fairly and continuously moist climate.

mestome sheath, the inner endodermal sheath of a two layered bundle sheath.

metaphloem, that part of the primary phloem which undergoes differentiation after the protophloem.

metaxylem, that part of the primary xylem which undergoes final differentiation after the protoxylem.

micella, the present usage refers to a unit of cellulose in which the molecules are arranged parallel to one another so that the atoms form a crystalline lattice structure.

microfibril, a submicroscopic thread-like constituent of the cell wall; composed in most plants of cellulose molecules.

microfilament, a proteinaceous filament occurring in cells.

microgametophyte, the male gametophyte of a heterosporous plant; mature pollen grain in a seed plant.

micropyle, the opening at the free end of the ovule, in the integuments.

microsporangium, a structure which produces the microspores in a heterosporous plant; pollen sac in angiosperms.

microspore, the male spore from which the male gametophyte develops.

microsporocyte, a cell which differentiates into a microspore.

middle lamella, the lamella present between the walls of two adjacent cells.

mitochondrion (pl. **mitochondria**), a cytoplasmic organelle; contains enzymes involved in respiration.

morphogenesis, the total expression of the morphological phenomena of differentiation and development of tissues and organs.

mother cell, a cell which gives rise to other cells as a result of its division.

multiple (of vessels), referring to the arrangement of vessels as seen in cross-section of the secondary xylem; a group of two or more vessels arranged in radial, oblique, or tangential rows.

mycorrhiza (pl. **mycorrhizae**), the symbiosis between fungi and the roots of higher plants.

myrosin cell, a cell containing the enzyme myrosin; occurs in plants of the Cruciferae and species of some other families.

nacreous wall, a non-lignified wall thickening, often found in sieve elements; term indicating the glistening appearance of the wall.

nectary, a glandular structure which secretes a sugary solution. Found in flowers (*floral nectaries*) or on vegetative plant organs (*extrafloral nectaries*).

nexine, the inner layer of the exine.

node, that portion of the stem where a leaf or leaves are attached; anatomically this region cannot be defined accurately.

nucellus, that tissue within the ovule in which the female gametophyte develops.

obturator, an outgrowth of the funiculus, placenta or stylar canal which brings the transmitting tissue closer to the micropyle.

ontogeny, the process of development of an organism, organ or tissue towards maturation.

orbicules, spherical bodies formed in the tapetum from *pro-orbicules* during maturation of the pollen grains; also called *Ubish bodies*.

osmophore, portion of a flower organ bearing fragrance secreting tissue.

osteosclereid, a spool- or bone-shaped sclereid.

ovule, a structure (in seed plants) consisting of the nucellus which contains the female gametophyte, one or two integuments and the funiculus; differentiates into the seed. A ovule in which the megagametophyte is located deep in the nucellus, is called **crassinucellate ovule;** an ovule in which the mega-metophyte is located close to the upper surface of the nucellus, is called **tenuinucellate ovule.**

P-protein, protein occurring in sieve elements; formerly called slime.

paracytic type (of stomata), a specific pattern of arrangement of the guard cells and other epidermal cells adjacent to them (see p. 168).

parenchyma, a ground tissue composed of living cells which may differ in size, shape, and wall structure.

parenchyma, apotracheal, axial parenchyma of the secondary xylem, typically independent of the vessels although may occasionally be in contact with them. Divided into the following types according to the distribution as seen in cross-section of the secondary xylem: **banded** or **metatracheal**—concentric uni- or multiseriate bands, arcs or entire rings; **diffuse**—single cells distributed irregularly among fibres; **initial**—bands of parenchyma produced at the beginning of a growth ring; **terminal**—bands of parenchyma produced at the end of a growth ring.

parenchyma, axial, parenchyma of the vertical system of secondary xylem, i.e. parenchyma cells derived from fusiform cambial initials.

parenchyma, paratracheal, axial parenchyma of the secondary xylem associated with the vessels or vascular tracheids. Divided into the following types according to the distribution as seen in cross-section: **aliform**—paratracheal parenchyma which is expanded tangentially in the form of wings; **confluent**—groups of aliform parenchyma which become continuous so as to form irregular tangential or diagonal bands; **scanty**—an incomplete sheath or a few parenchyma cells present around the vessels; **vasicentric**—parenchyma forming an entire sheath of variable width around individual vessels or groups of vessels.

parenchyma, wood, *see* **parenchyma, xylem.**

parenchyma, xylem, parenchyma occurring in the secondary xylem, usually in two systems: (1) axial, and (2) radial (ray parenchyma).

parthenocarpy, the production of fruit without fertilization.

pectic compounds (substances), a group of complex carbohydrates, derivatives of polygalacturonic acid; constituting the most important component of the middle lamella.

perforation (in stele), interruption in the vascular tissue of a siphonostele, other than leaf- or branch-gap; developed as a result of secondary reduction.

perforation plate, that portion of the cell wall of a vessel member which is perforated. The following types of perforation plates are distinguished: (1) *foraminate perforation plate*—a plate with numerous, more or less circular perforations; (2) *reticulate perforation plate*—a plate in which the remnants of the wall between the perforations form a net-like structure; (3) *scalariform perforation plate*—a plate with numerous elongated pores which are arranged parallel one near the other; (4) *simple perforation plate*—a plate having one large perforation.

periblem, *see* **histogen.**

pericarp, the fruit wall.

periclinal, parallel to the surface.

pericycle, that portion of the ground tissue of the vascular cylinder between the conducting tissues and the endodermis.

periderm, the secondary protective tissue which replaces the epidermis; consists of *phellem, phellogen,* and *phelloderm.*

perigenous, see **stoma.**

perisperm, a nutrient tissue of the seed, similar to the endosperm, but of nucellar origin.

peroxisome, microbodies which are the sites of oxidation of glycolic acids, a product of carbon dioxide fixation; closely associated with chloroplasts.

phellem (cork), protective tissue composed of nonliving cells with suberized walls and produced centrifugally by the phellogen.

phelloderm, a tissue resembling cortical parenchyma produced centripetally by the phellogen.

phellogen, the cork cambium; a secondary lateral meristem which produces the phellem and phelloderm.

phelloid, a cell within the phellem but lacking suberin in its walls.

phloem, the principal tissue responsible for the transport of assimilates in the vascular plants; consists mainly of sieve elements, parenchyma cells, fibres, and sclereids.

phloem, interxylary, secondary phloem which occurs within the secondary xylem as is the case in certain Dicotyledoneae.

phloem, intraxylary, primary phloem occurring on the inner side of the primary xylem.

phragmoplast, a fibrous structure which appears during mitotic telophase between the two daughter nuclei and within which the cell plate which divides the mother cell in two is formed; consists of microtubules.

phyllocade, a stem which is flattened and resembles a leaf, also called *cladode.*

phyllode, leaf-like photosynthesizing organ which developed from the petiole and rachis.

phyllome, a collective term referring to all types of leaves.

phylogeny, the history of a species or a larger taxonomical group from an evolutionary viewpoint.

piliferous cell, *see* **trichoblast.**

pit, a depression in a cell wall with secondary thickening; in such an area only primary wall and middle lamella are present.

pit, bordered, a pit in which the aperture in the secondary wall is small and conceals below it a dome-shaped chamber which is situated above the pit membrane.

pit cavity, the cavity of a single pit extending from the pit membrane to the aperture bordering the cell lumen.

pit membrane, the middle lamella and primary wall closing the pit cavity on its outer side.

pit-pair, two complementary pits of neighbouring cells.

pit, vestured, a bordered pit having projections, which may be simple or branched, on that part of the secondary wall which forms the border of the pit chamber or the pit aperture; found in certain Dicotyledoneae.

pith, the ground tissue in the centre of the stem and root.

pitting, the type and arrangement of pits in the cell wall.

placenta, the region of attachment of the ovules to the carpel.

placentation, the position of the placenta in the ovary.

plasmalemma, the membrane on the outer surface of the cytoplasm; adjacent to cell wall. Also known as *cell membrane* or *ectoplast.*

plasmodesma (pl. **plasmodesmata**), a thin, cytoplasmic strand which passes through a pore in the cell wall, and which usually connects the protoplasts of two adjacent cells.

plastid, a cytoplasmic organelle fulfilling a specific function. May be concerned with photosynthesis (*chloroplast*) or starch storage (*amyloplast*), or contains orange or yellow pigments (*chromoplast*).

plastochron, that period of time between the commencement of two successive and repetitive phenomena—for example, between the initiation of two successive leaf primordia.

plectostele, a protostele in which the xylem is arranged in longitudinal plates which may be interconnected.

plerome, *see* **histogen.**

plumule, the bud or shoot apex of the embryo.

pneumatode, in a velamen, a group of cells with very dense spiral wall thickenings; enables gas exchange when the root is saturated with moisture.

pneumatophore, an aerial, negatively geotropic root projection serving for gas exchange; produced in swampy habitats.

pollen tube, an elongated projection, covered only by intine, of the vegetative cell of a pollen grain.

pollinium, the entire pollen grain complement of a single pollen sac, when the grains adhere together to form one mass.

polyarch, primary xylem of a root, having many protoxylem poles.

polyderm, a special type of protective tissue composed of alternating bands of endodermis-like cells and non-suberized parenchyma cells.

polyembryony, the presence of more than one embryo in an ovule.

polyribosomes (or **polysome**), aggregation of ribosomes, apparently concerned with protein synthesis as a group.

primary pit-field, a thin portion of the primary wall in which the pores, through which plasmodesmata pass, are concentrated.

primary thickening meristem, a broad region in the apical meristem of the shoot responsible for the primary increase in thickness of the stem; often occurring in monocotyledons.

primexin, the cellulose layer of a pollen cell wall which is deposited within the original callose wall.

primordium, an organ, cell, or organized group of cells in the earliest stage of differentiation.

procambium, a primary meristem which undergoes differentiation to form the primary vascular tissues.

proembryo, the embryo in the earliest stages of development.

prolepsis, the outgrowth of an axillary bud after a period of rest.

promeristem, the initials and their immediate derivatives in the apical meristem.

prophyll, one of the first leaves of a lateral branch.

proplastid, a plastid in the earliest stages of development; a primordial plastid.

protoderm, the meristem of the epidermis.

protophloem, the first-formed elements of the primary phloem.

protostele, the simplest type of stele which comprises a solid core of xylem surrounded by phloem.

protostele, medulated, according to the terminology of some authors, an ectophloic siphonostele without leaf-gaps; occurs in Pteridophyta.

protoxylem, the first-formed elements of the primary xylem.

provascular bundle, a procambium strand.

pseudocarp, a fruit in which organs other than the carpels participate in the formation of its wall.

pulvinus, the swelling at the base of a leaf petiole or of a petiolule of a leaflet.

quiescent centre, a region in the root apical meristem of seed plants in which the mitotic activity is very low or almost nil.

radicle, the embryonic root; the basal continuation of the hypocotyl.

raphe, a ridge along the seed formed by that part of the funiculus which was fused to the ovule.

raphide, a needle-shaped crystal usually occurring in dense bundles.

ray, phloem, that part of the vascular ray which passes through the secondary phloem.

ray, pith, an interfascicular region in a stem.

ray, vascular, a strip of tissue running radially through the secondary xylem and phloem; formed by the cambium.

ray, xylem, that part of the vascular ray which passes through the secondary xylem.

refractive spherule, a small spherical body occurring in the sieve elements of many lower vascular plants.

replum, the ridge surrounding the siliqua of the Cruciferae which remains attached to the false septum, as a frame, on the dehiscence of the fruit.

rhytidome, that part of the bark comprising the periderm and tissues external to it which are cut off by it. Also called *outer bark.*

ribosome, a minute protoplasmic particle composed of protein and RNA; concerned with protein synthesis.

ring-porous wood, *see* wood.

rod cell, *see* **macrosclereid.**

root, contractile, a special root which has the ability to contract and thus bring the developing renewal buds to a definite position in relation to the soil surface.

root cap, a thimble-shaped structure which covers the root apex.

root hair, a type of trichome developing on the epidermis of roots; absorbs solutions from the soil.

ruminated endosperm, a type of endosperm in the mature state of which its surface is very irregular; ingrowths of the seed coat penetrate into the folds of the surface.

sapwood or **alburnum,** that portion of the wood that in the living tree and shrub contains living cells and reserve materials.

scalariform, the parallel arrangement, one near the other, of elongated structures in the cell wall of an element.

schizogenous, the manner of formation of intercellular spaces by the separation of cells along their middle lamellae.

schizo-lysigenous, the manner of formation of intercellular spaces by both cell separation along their middle lamellae and cell disintegration.

sclereid, a sclerenchymatous cell of various shape, but usually not much elongated; has thick, lignified secondary wall which often contains many pits.

sclerenchyma, a supporting tissue composed of fibres and/or sclereids.

sclerification, the process of changing into sclerenchyma by the formation of secondary walls.

scutellum, a part of the embryo of the Gramineae; considered to be homologous to a cotyledon; serves as an organ which transfers nutrients from the endosperm to other parts of the germinating embryo.

separation layer, that layer in the abscission zone the cells of which disintegrate or separate and so cause the abscission of the organ concerned or part of it, e.g. leaf, branch, fruit, fruit valves, etc.

sexine, the outer layer of the exine.

shell zone, a concave zone of cambiform cells at the base of an axillary bud primordium.

shoot, the stem and its appendages.

sieve area, an area on the wall of a sieve element which appears as a depression and which contains pores commonly lined with callose.

sieve cell, a sieve element in which the sieve areas have undergone relatively little differentiation, i.e. they are of more or less uniform structure and have narrow pores and connecting strands.

sieve element, that type of phloem element which takes part in the transport of assimilates. Classified into two types—*sieve cell* and *sieve-tube member.*

sieve plate, that part of the cell wall of a sieve element which contains one or more highly specialized sieve areas. Characteristic of the Angiospermae.

sieve tube, a series of sieve-tube members which are arranged end to end and which are connected through their sieve plates.

sieve-tube member, one of the cells of which a sieve tube is comprised.

silica body, an amorphous, mostly opaline body consisting of silicon dioxide; occurs in cells of certain monocotyledonous taxa.

silica cell, a short epidermal cell filled with silica as occurs in the epidermis of the Gramineae.

siphonostele, a stele in which the vascular tissue comprises a hollow cylinder, i.e. in which the central portion is occupied by pith.

siphonostele, amphiphloic, a siphonostele in which the phloem surrounds the xylem both externally and internally.

siphonostele, ectophloic, a siphonostele in which the phloem surrounds the xylem on the outside only.

slime plug, an accumulation of P-protein on a sieve area; usually with extensions into the pores of the sieve area.

soft wood, a common name for wood of the Gymnospermae, in particular the Coniferae.

solenostele, an amphiphloic siphonostele in which the successive leaf-gaps are considerably distant one from the other.

specialization, the changes during the course of evolution in structure of a cell, tissue, organ, or entire plant, which enable it to carry out a certain function more efficiently.

spongy parenchyma, parenchyma of the leaf mesophyll with conspicuous intercellular spaces.

sporophyte, that plant generation which produces spores from which the gametophyte develops.

starch sheath, referring to the innermost layer of the cortex, when its cells are characterized by presence of a large, more or less constant quantity of starch; homologous with the endodermis.

statolith, starch containing plastid occurring in the root cap cells; suggested to be involved in sensing the gravitational stimulus.

stele, that portion of the plant axis which comprises the vascular system and its associated ground tissue, i.e. pericycle, interfascicular regions, and pith.

stele, polycyclic, a stele consisting of two or more concentric cylinders of vascular tissue.

stereology, a method of quantifying the volumes or surface areas of various components of a certain structure.

stereome, a collective physiological term for all the supporting tissues in the plant, i.e. sclerenchyma and collenchyma.

stigma, the region of a carpel on which the pollen grains germinate.

stoma (pl. **stomata**), an opening in the epidermis bordered by two guard cells and permitting gas exchange; stomata in which the subsidiary cells have a common origin with the guard cells are called **mesogenous stomata**; stomata in which the subsidiary cells develop from cells adjacent to the stomatal mother cell are called **perigenous stomata.**

stomatal complex (or **stomatal apparatus**), a stoma together with the subsidiary cells.

stomium, a fissure or pore in the anther lobe through which the pollen is released.

stone cell, see brachysclereid.

storied cork, a cork found in monocotyledons; the suberized cells occur in radial rows; the cork initials here, unlike those of a typical phellogen, do not form regular uninterrupted layers.

Strassburger cell, *see* **albuminous cell.**

stroma, the ground substance of a plastid.

strophiole, an arilloid (type of an outgrowth on a seed) occurring on the raphe.

style, extension of the upper part of the ovary, terminating at the stigma.

styloid, a long prismatic crystal tapered off at both ends into a blade.

suberization, the deposition of suberin in cell walls.

submarginal initials, in marginal meristem of a developing leaf; cells located interior to the marginal initials: contribute cells to the interior of the leaf.

subsidiary cell, *see* **accessory cell.**

succulent, fleshy, juicy.

supporting tissues, *see* **tissue, mechanical.**

surface layer *or* **surface meristem,** one of the cyto-histological zones of the Gymnospermae shoot apex.

suspensor, that part of the embryo which connects its main part to the basal cell.

syllepsis, the outgrowth of an axillary bud immediately after its formation; without a period of dormancy.

symplast, all the protoplasts of a plant and the plasmodesmata connecting them.

syncarpy, that condition in a flower and ovary where the carpels are fused.

synergids, the cells present alongside the egg cell in a mature embryo sac.

tapetum, the innermost layer of the pollen-sac wall; the contents of its cells are absorbed by the pollen grains during their development.

tapetum, amoeboid, tapetum in which the protoplasts of its cells penetrate between the pollen mother cells and developing pollen grains; **glandular,** tapetum in which the cells remain in their original position until their disintegration.

tectum, the layer of a pollen grain wall formed by the fusing of the heads of the bacula.

tegmen, portion of the seed coat produced by the inner integument.

teichode, see ectodesma.

telome, the ultimate terminal portion of a dichotomously branching axis bearing a sporangium (fertile telome) or lacking a sporangium (sterile telome). According to the *telome theory* the most primitive vascular plants were composed entirely of telome systems.

testa, the seed coat; by some authors refers only to that portion of the seed coat produced by the outer integument.

tetrarch, the primary xylem of root in which the number of protoxylem strands is four.

thylakoids, the membranes inside a chloroplast; they occur in stacks as grana or individually as frets.

tissue, complementary, *see* **complementary cells.**

tissue, conjunctive, (1) a special type of parenchyma associated with included phloem in Dicotyledoneae with anomalous thickening; (2) the parenchyma present between the secondary vascular bundles in Monocotyledoneae with secondary thickening.

tissue, element, each individual cell of a tissue.

tissue, expansion, an intercalary tissue in the outer portion of the inner bark formed mainly by the phloem rays, accommodates the expansion in circumference.

tissue, ground, all the mature plant tissues except the epidermis, periderm, and vascular tissues.

tissue, mature, a tissue which has undergone differentiation.

tissue, mechanical, a tissue comprised of cells, the walls of which are more or less thickened; such tissues give support to the plant body. Also referred to as *supporting tissues.*

tissues, proliferation, a tissue which develops from phloem parenchyma in the outer portion of the inner bark accommodating the expansion in circumference.

tissue, transfusion, in gymnosperm leaves; a tissue consisting of tracheids and parenchyma cells, associated with the vascular bundle. Transfusion tissue which extends laterally into the mesophyll rather than remaining associated with the vascular bundles is called **accessory transfusion tissue.**

tissue, transmitting, a specialized tissue in the style through which the germinating pollen tube penetrates the ovary; connects the stigma and the inside of the ovary.

tonoplast, the cytoplasmic membrane which borders the vacuole.

torus, the thickened central portion of the pit membrane in a bordered pit.

trabecula, a rod-like projection of the cell wall which crosses the cell lumen, usually in a radial direction.

trace, branch, that part of a vascular bundle in the stem extending from the position where it joins the vascular system of the branch to where it joins the vascular system of the main stem.

trace, leaf, that part of a vascular bundle in the stem from the position where it enters the leaf to where it joins the vascular system of the stem.

trachea, *see* **vessel.**

tracheary element, that type of xylem element which takes part in water transport. Classified into two types—*tracheid* and *vessel member.*

tracheid, a tracheary element of the xylem, which, unlike the vessel member, is not perforated.

tracheoid idioblast, a tracheid-like cell occurring in the mesophyll which is supposed to store water.

transfer cell, parenchyma cell with wall ingrowths. Appears to be specialized in short-distance transport of solutes.

triarch, the primary xylem of root in which the number of protoxylem strands is three.

trichoblast, a specialized cell in the root epidermis which gives rise to a root hair.

trichome, an epidermal appendage; may be of various shapes, structures, size, and function; includes hairs, scales, etc.

trichosclereid, a very elongated, somewhat hair-like sclereid; usually branched.

tunica, the outermost layer or layers in the apical shoot meristem of the Angiospermae in which the plane of division is almost entirely anticlinal.

tylosis (pl. **tyloses**), an outgrowth of a ray cell or of an axial parenchyma cell into the lumen of a vessel; such outgrowths partially or completely block the vessels.

tylosoid, a proliferation of an epithelial cell into an intercellular cavity such as resin or gum duct, or of a parenchyma cell into a sieve element.

unifacial leaf, a leaf in which the structure of all sides is alike.

vacuolation, the shape, amount, and size of the vacuome. Also the process of forming vacuoles.

vacuole, a cavity in the cytoplasm bound by a membrane, the tonoplast, and containing an aqueous solution, the *cell sap*.

vacuome, the collective term for all the vacuoles of a single cell.

vascular, an adjective referring to the xylem or phloem or both.

vascular bundle, a strand of conducting tissue. The following types of vascular bundles are recognized: (1) *bicollateral vascular bundle*—vascular bundle in which phloem is present both on the outside and inside of the xylem; (2) *closed vascular bundle*—a bundle lacking vascular cambium; (3) *collateral vascular bundle*—vascular bundle in which phloem is present on one side of the xylem only, commonly external to it; (4) *concentric vascular bundle*—vascular bundle in which the phloem surrounds the xylem (i.e. *amphicribral*) or the xylem surrounds the phloem (i.e. *amphivasal*); (5) *open vascular bundle*—a bundle in which a vascular cambium is present.

vascular tracheid, a short, imperforated cell with annular and helical thickenings.

vein, a strand of vascular tissue in a flat organ, such as a leaf.

velamen, the multiseriate epidermis present on the aerial roots of some tropical epiphytic species of the Orchidaceae and Araceae.

venation, the arrangement of the veins in the leaf blade.

vessel, a series of vessel members joined end to end by their perforated end walls. Also termed *trachea*.

vessel element, a synonym of vessel member.

vessel member, a tracheary element, one of the cells of which a vessel is comprised.

wall layer, terminal, *see* wall layer, tertiary.

wall layer, tertiary, according to some authors, a layer present on the inside of the inner layer of the secondary wall.

wall, special, the first wall formed which separates each of the pollen grains in the tetrad.

wart structure, small granules occurring on the inner surface of the secondary wall of tracheids, fibres, and vessels.

wood, diffuse-porous, secondary xylem in which the vessels are more or less uniform or in which the diameter of vessels alters only slightly across the growth ring.

wood, early, the wood produced in the first part of a growth ring.

wood, late, the wood formed in the later part of a growth ring.

wood, reaction, the secondary xylem with special structure, produced in those parts of trunks and branches which lean or are bent; apparently tends to return these organs to their original positions. *Tension wood* in the Dicotyledoneae; *compression wood* in the Coniferae.

wood, ring-porous, secondary xylem in which the vessels produced at the beginning of a growth season are significantly wider than those produced at the end of the season.

wood, tension, the reaction wood of the Dicotyledoneae formed on the upper side of leaning or bent trunks and branches.

wood compression, the reaction wood of the Coniferae, formed on the lower side of bent or leaning trunks and branches.

woody, an entire plant or a plant organ with well-developed secondary xylem.

xerochastic process, manner of fruit opening or movement of other organs, as a result of loss of water; usually connected with the dispersal of seeds or spores.

xeromorphic, having structure typical of xerophytes.

xerophyte, a plant adapted to arid habitats.

xylem, the tissue mainly responsible for conduction of water in vascular plants; characterized by the presence of tracheary elements. Xylem, especially secondary xylem, may also serve as a supporting tissue.

zygomorphic, irregular flower; bilaterally symmetrical.

zygote, the cell which is produced after an egg cell is fertilized.

AUTHOR INDEX

Page numbers in bold type indicate bibliographic references.

SUBJECT INDEX

Page numbers in **bold** type indicate that the subject is cited in a figure or in a legend to a figure

557